The Heart and Toxins

Meenakshisundaram Ramachandran

Editor
Barking, Havering and Redbridge University Hospitals
NHS Trust, Queens Hospital, Romford, United Kingdom

P. Thirumalaikolundusubramanian

Assistant Editor
Chennai Medical College Hospitals and
Research Centre, Irungalur, Trichy, India

ELSEVIER

AMSTERDAM • BOSTON • HEIDELBERG • LONDON
NEW YORK • OXFORD • PARIS • SAN DIEGO
SAN FRANCISCO • SINGAPORE • SYDNEY • TOKYO
Academic Press is an imprint of Elsevier

Academic Press is an imprint of Elsevier
32 Jamestown Road, London NW1 7BY, UK
525 B Street, Suite 1800, San Diego, CA 92101-4495, USA
225 Wyman Street, Waltham, MA 02451, USA
The Boulevard, Langford Lane, Kidlington, Oxford OX5 1GB, UK

Notices

Knowledge and best practice in this field are constantly changing. As new research and experience broaden our understanding, changes in research methods, professional practices, or medical treatment may become necessary.

Practitioners and researchers must always rely on their own experience and knowledge in evaluating and using any information, methods, compounds, or experiments described herein. In using such information or methods they should be mindful of their own safety and the safety of others, including parties for whom they have a professional responsibility.

With respect to any drug or pharmaceutical products identified, readers are advised to check the most current information provided (i) on procedures featured or (ii) by the manufacturer of each product to be administered, to verify the recommended dose or formula, the method and duration of administration, and contraindications. It is the responsibility of practitioners, relying on their own experience and knowledge of their patients, to make diagnoses, to determine dosages and the best treatment for each individual patient, and to take all appropriate safety precautions.

To the fullest extent of the law, neither the Publisher nor the authors, contributors, or editors, assume any liability for any injury and/or damage to persons or property as a matter of products liability, negligence or otherwise, or from any use or operation of any methods, products, instructions, or ideas contained in the material herein.

British Library Cataloguing-in-Publication Data
A catalogue record for this book is available from the British Library

Library of Congress Cataloging-in-Publication Data
A catalog record for this book is available from the Library of Congress

ISBN: 978-0-12-416595-3

For information on all Academic Press publications
visit our website at http://store.elsevier.com/

Typeset by MPS Limited, Chennai, India
www.adi-mps.com

Contents

2. Cellular and Molecular Perspectives on Cardiac Toxins

Reza Tabrizchi

3. Environmental Toxins and the Heart

Sahand Rahnama-Moghadam, L. David Hillis, and Richard A. Lange

14. The Effects of Active and Passive Smoking and Cardiovascular Disease

Mahmood Ahmad, Emmanuel Selvaraj, and Ramachandran Meenakshisundaram

15. Cardiovascular Toxicity as a Result of Recreational Drugs

Rohin Francis and Azad Ghuran

Foreword

Cardiovascular disease (CVD) is a major cause of morbidity and mortality worldwide. The majority of CVD is caused by risk factors that can be controlled, treated, or modified, such as hypertension, diabetes mellitus, hyperlipidemia, obesity, tobacco use, and lack of physical activity. Although a substantial portion of the population is exposed to toxins that may affect the heart adversely, this is rarely mentioned or considered as a modifiable or treatable risk factor for CVD. Consequently, it is important to understand the myriad ways in which these substances may influence the cardiovascular system.

Despite the global importance of this issue, no comprehensive text has been available previously for the clinician, scientist, or patient who has wanted to learn about the extent, mechanisms, or manifestations of toxin-induced heart disease. This book aims to provide a comprehensive and in-depth review of this subject.

The Heart and Toxins starts off by providing the reader with an appreciation of the epidemiology and molecular perspectives. Subsequent chapters are organized so that the reader can easily locate and learn about the different types of toxin exposure—environmental, envenomation, plants, marine, medicinal—and their effects on the cardiovascular system. Entire chapters are dedicated to specific toxins that are of particular importance because of their widespread use or exposure risk (i.e., chemotherapy, HIV infection and treatment, alcohol, smoking, and recreational drugs), so the reader has ready access to the most recent information concerning them. Detailed information about the effects of toxins on specific patient populations—children and patients undergoing radiologic procedures—and emerging fields (i.e., nanocardiotoxicology) is also provided. Finally, since most medical literature fails to address the ethical and legal aspects of cardiovascular toxicology, this book offers a chapter on the subject.

The book's editor, Dr. Ramachandran Meenakshisundaram, assembled internationally renowned scientists and clinicians from around the world to author the various chapters, and their expertise is evident throughout the book, *The Heart and Toxins*. However, the concept of the book is his; it was he who recognized the need for a practical and all-inclusive book on this topic. The authors are grateful for his participation and assistance from inception to completion and for allowing us the privilege to contribute to this unique project.

Richard A. Lange, MD, MBA

Interventional Cardiologist and President of the Texas, Tech University Health Science Center, and Dean of the Paul L. Foster School of Medicine, El Paso, Texas, USA

Acknowledgments

This book is dedicated to my child, Nemisha Sundar, and my wife, Sweni. The sacrifices they have made by providing me with the time to work on the book and their continuous words of motivation have been truly honorable. Also, I would like to acknowledge my parents, Ramachandran and Amirthavalli, and my brothers, Senthil Kumar and Saravana Kumar, as well as Dharmil Shah for their encouragement and support.

I am most grateful to all the authors for their valuable contributions and the time they have dedicated to this book's chapters. On this occasion, I gratefully acknowledge Shannon Stanton, Editorial Project Manager at Elsevier, and Stacy Masucci, Senior Acquisitions Editor at Elsevier, for their extremely helpful and devoted attention throughout the production process. This project would not have been possible for me without the valuable support of the contributors, Stacy, and Shannon. I would like to acknowledge and extend appreciation to all our reviewers for their valuable time in commenting on the chapters. Finally, I would like to thank Elsevier Inc. for providing me with this wonderful opportunity.

List of Contributors

Subroto Acharjee, Einstein Institute for Heart and Vascular Health, Einstein Medical Center, Philadelphia, Pennsylvania, USA

Mahmood Ahmad, Darent Valley Hospital, Dartford and Gravesham NHS Trust, Dartford, United Kingdom, and Tahir Heart Institute, Rabwah, Pakistan

Hamood Al Kindi, The Royal Victoria Hospital, McGill University Health Centre, Montreal, Quebec, Canada

Kholoud Samy Alsowayigh, Forensic Medicine Specialist, Forensic Medicine Center, Jeddah, Saudi Arabia

Athena Andreosso, Queensland Tropical Health Alliance, School of Public Health and Tropical Medicine, Centre for Biodiscovery and Molecular Development of Therapeutics, and Faculty of Medicine, Health and Molecular Sciences, James Cook University, Cairns, Australia

Omi Bajracharya, Department of Pharmacology, KIST Medical College, Lalitpur, Nepal

Ashish Bhalla, Department of Medicine, Postgraduate Institute of Medical Education and Research, Chandigarh, India

Douglas Ewan Cannie, Department of Cardiology, Epsom and St. Helier University Hospitals NHS Trust, Epsom, United Kingdom

Siow Ann Chong, Research Division, Institute of Mental Health, Singapore

Thomas R. Cochran, University of Kansas School of Medicine, University of Kansas Medical Center, Kansas City, Kansas, USA

Gabriela Cordero-Schmidt, University of California, San Francisco, USA

Carmen Criscitiello, Division of Early Drug Development for Innovative Therapies, Istituto Europeo di Oncologia (European Institute of Oncology), Milan, Italy

Giuseppe Curigliano, Division of Early Drug Development for Innovative Therapies, Istituto Europeo di Oncologia (European Institute of Oncology), Milan, Italy

Harinder Singh Dhindsa, Department of Emergency Medicine, Virginia Commonwealth University, Richmond, Virginia, USA

Angela Esposito, Division of Early Drug Development for Innovative Therapies, Istituto Europeo di Oncologia (European Institute of Oncology), Milan, Italy

Vincent M. Figueredo, Einstein Institute for Heart and Vascular Health, Einstein Medical Center, and Jefferson Medical College, Philadelphia, Pennsylvania, USA

Stacy D. Fisher, Departments of Medicine and Pediatrics, University of Maryland School of Medicine, University of Maryland Comprehensive Heart Center, Baltimore, Maryland, USA

Rohin Francis, University of Cambridge, Cambridge, United Kingdom

Jeffery Fung, Accident and Emergency Department, Tuen Mun Hospital, Hong Kong

Lucia Gelao, Division of Early Drug Development for Innovative Therapies, Istituto Europeo di Oncologia (European Institute of Oncology), Milan, Italy

Sherien Salah Ghaleb, Forensic Medicine and Clinical Toxicology, Cairo University and BeniSuef University, Cairo, Egypt

Azad Ghuran, East and North Hertfordshire NHS Trust, Lister Hospital, Stevenage, United Kingdom

Hanan Hosny Hassan, Forensic Pathology Consultant and Forensic Medicine Authority, Ministry of Justice, Cairo, Egypt

L. David Hillis, Department of Medicine, University of Texas Health Science Center, San Antonio, Texas, USA

Nisha Jha, Department of Pharmacology, KIST Medical College, Lalitpur, Nepal

Magdy A. Kharoshah, Consultant Forensic Medical Examiner, Forensic Medicine Authority, Ministry of Justice, Cairo, Egypt

Richard A. Lange, Department of Medicine, University of Texas Health Science Center, San Antonio, Texas, USA

Steven E. Lipshultz, Carmen and Ann Adams Department of Pediatrics, Wayne State University School of Medicine, University Pediatricians, Children's Hospital of Michigan, Detroit, Michigan, USA

Jennifer A. Lowry, Division of Clinical Pharmacology, Toxicology, and Therapeutic Innovations, Children's Mercy Hospital, and University of Missouri, School of Medicine, Kansas City, Missouri, USA

Mohammed Saleh Madadin, Forensic Medicine Division, Department of Pathology, College of Medicine, University of Dammam, Dammam, Saudi Arabia

Vijayan Manoharan, Center for Nanotechnology and Advanced Biomaterials, School of Chemical and Biotechnology, SASTRA University, Tanjore, India

Catherine M. Mas, Carmen and Ann Adams Department of Pediatrics, Wayne State University School of Medicine, University Pediatricians, Children's Hospital of Michigan, Detroit, Michigan, USA

Ragini Medhi, Department of Genetic Engineering, School of Bioengineering, SRM University, Chennai, India

Ramachandran Meenakshisundaram, Department of Cardiology, Epsom and St. Helier University Hospitals NHS Trust, Epsom, United Kingdom, and Barking, Havering and Redbridge University Hospitals NHS Trust, Queens Hospital, Romford, United Kingdom

Ritesh George Menezes, Forensic Medicine Division, Department of Pathology, College of Medicine, University of Dammam, Dammam, Saudi Arabia

Tracie L. Miller, Department of Pediatrics, University of Miami Miller School of Medicine, Holtz Children's Hospital of the University of Miami, Jackson Memorial Medical Center, and the Sylvester Comprehensive Cancer Center, Miami, Florida, USA

Churchill Lukwiya Onen, Centre for Chronic Diseases, Gaborone, Botswana

Arghya Paul, Harvard-MIT Division of Health Sciences and Technology, Harvard University, Cambridge, Massachusetts, and Department of Chemical and Petroleum Engineering, University of Kansas, Lawrence, Kansas, USA

Satya Prakash, Biomedical Technology and Cell Therapy Research Laboratory, Department of Biomedical Engineering, Faculty of Medicine, McGill University, Montreal, Quebec, Canada

Bhaskar Purushottam, Einstein Institute for Heart and Vascular Health, Einstein Medical Center, Philadelphia, Pennsylvania, USA

Sahand Rahnama-Moghadam, Department of Medicine, University of Texas Health Science Center, San Antonio, Texas, USA

Gwendolyn B. Scott, Department of Pediatrics, University of Miami Miller School of Medicine, Holtz Children's Hospital of the University of Miami, Jackson Memorial Medical Center, and Sylvester Comprehensive Cancer Center, Miami, Florida, USA

P. Ravi Shankar, Department of Pharmacology, Chair, Curriculum Committee and Research Committee, Xavier University School of Medicine, Oranjestad, Aruba, and Honorary Lecturer, Universiti Sains Malaysia, Penang, Malaysia

Emmanuel Selvaraj, London Chest Hospital, Barts Health NHS Trust, London, United Kingdom

Subramanian Senthilkumaran, Sri Gokulam Hospital and Research Centre, Salem, India

Benjamin Seymour, School of Medicine and Dentistry, Faculty of Medicine, Health and Molecular Sciences, James Cook University, Cairns, Australia

Jamie Seymour, Australian Institute of Tropical Health and Medicine, Queensland Tropical Health Alliance, School of Public Health and Tropical Medicine, Centre for Biodiscovery and Molecular Development of Therapeutics, and Faculty of Medicine, Health and Molecular Sciences, James Cook University, Cairns, Australia

Veronica Kaur Sikka, Department of Emergency Medicine, Virginia Commonwealth University, Richmond, Virginia, USA

Surjit Singh, Department of Medicine, Post Graduate Institute of Medical Education and Research, Chandigarh, India

Sari Soghoian, Department of Emergency Medicine, New York University School of Medicine/Bellevue Hospital Center, New York, New York, USA

Gabriel Somarriba, Department of Pediatrics, University of Miami Miller School of Medicine, Holtz Children's Hospital of the University of Miami, Jackson Memorial Medical Center, and Sylvester Comprehensive Cancer Center, Miami, Florida, USA

Mythily Subramaniam, Research Division, Institute of Mental Health, Singapore, and Saw Swee Hock School of Public Health, National University of Singapore, Singapore

Dominique Sum-Tim, The Royal Victoria Hospital, McGill University Health Centre, Montreal, Quebec, Canada

Shah Sweni, Chennai Medical College Hospital and Research Centre, Trichy, India

Reza Tabrizchi, Division of BioMedical Sciences, Faculty of Medicine, Memorial University, St. John's, Newfoundland and Labrador, Canada

Ponniah Thirumalaikolundusubramanian, Department of Medicine, Chennai Medical College Hospital and Research Centre, Trichy, India

Janhavi Ajit Vaingankar, Research Division, Institute of Mental Health, Singapore

James D. Wilkinson, Department of Pediatrics, University of Miami Miller School of Medicine, Holtz Children's Hospital of the University of Miami, Jackson Memorial Medical Center, and the Sylvester Comprehensive Cancer Center, Miami, Florida, USA

Amy See Yuning, Research Division, Institute of Mental Health, Singapore

Heidar Z. Zadeh, Department of Cardiology, Aintree University Hospital, Liverpool, United Kingdom

Chapter 1

Epidemiology of Cardiovascular Toxins

Churchill Lukwiya Onen
Centre for Chronic Diseases, Gaborone, Botswana

1.1 INTRODUCTION

Strictly speaking, the word *toxins* refers to poisonous substances produced during metabolism or growth of certain microorganisms, higher plants, or animals, whereas a poison is any substance that causes injury or illness or death of a living organism, as discussed here, to humans. *Toxicity* refers to the degree to which something is poisonous, and *toxicology* is the study of the adverse effects of chemicals on living organisms. Cardiovascular toxins[1,2] have harmful effects on the circulatory system, resulting in symptoms and signs of injury, and may potentially cause death. A greater understanding of the distribution, determinants, secular trends, and deterrents of cardiovascular toxins may provide the solid epidemiological platform for developing, prioritizing, and evaluating public health programs against morbidity and mortality related to these toxins.

This chapter addresses the epidemiology of major categories of substances with clinically relevant cardiovascular toxic effects ranging from plant toxins, marine toxins, venomous reptiles, trichinellosis, arachnidism, scorpion venoms, air pollution, pesticides, fungicides, household materials, industrial toxins, tobacco use,[3] alcohol, uremic toxins, nonsteroidal antiinflammatory drugs, chemotherapeutic agents and related substances, and finally heavy metals.

1.2 PLANT TOXINS

This section discusses two of the most important plant-derived cardiotoxins, namely the mineralocortioid effects of liquorice and cardiac glycosides. Widespread commercial and medicinal uses of plant sources of these compounds predispose to inadvertant or suicidal exposure.

The Heart and Toxins. DOI: http://dx.doi.org/10.1016/B978-0-12-416595-3.00001-3

1.2.1 Mineralocorticoid Effects of Liquorice

Liquorice comes from the root of *Glycyrrhiza glabra*, a legume related to beans and peas, that is native to the Mediterranean region, southern Europe, and central and southwest Asia. The plant is widely cultivated for commercial and traditional medicinal uses in many parts of the world. The extract of liquorice contains glycyrrhizic acid, a chemical that is 30 to 50 times sweeter than sugar. Traditional medicinal uses include treatment of chronic viral hepatitis in Japan, tuberculosis in China, and peptic ulcers and mouth ulcers in many other parts of the world. Liquorice is a common flavoring in commercial products such as candies, chocolates, tea, spices, tobacco, and liqueurs. In the Netherlands, Finland, and Scandinavian countries, liquorice mixed with ammonium chloride produces the salty taste of the popular "salmiakki." Chinese cuisine commonly uses liquorice as a culinary spice to flavor broths and savory foods.

A linear dose–response relationship exists between amount of liquorice consumed and cardiovascular response, but doses as low as 50 grams consumed for two weeks can cause significant blood pressure elevation. Although any person might be at risk of liquorice mineralocorticoid effects, those who take it for medicinal uses—for example, as a laxative for chronic constipation, or as a habitual indulgence in glycyrrhizin-containing delicacies and flavorings—are at the greatest risk of adverse effects. Large doses of glycyrrhizic acid may lead to hypokalemia and elevated blood pressure due to its mineralocorticoid effects.[4–6] Severe hypokalemia may result in cardiac arrhythmias, cardiac asystole, and risk of sudden death. It is generally recommended that no more than 100 grams of glycyrrhizic acid be consumed per day.

1.2.2 Cardiac Glycoside-Containing Plants

The diverse group of plants that contain cardiac glycosides include *Digitalis purpurea, Digitalis lanata, Nerium oleander, Thevetia peruviana*, and *Strophanthus gratus*. The seeds have the highest concentration of glycoside (4.8%), whereas the leaves, fruit, and milk from the plants contain approximately 0.07%, 0.045%, and 0.036% of glycoside, respectively. Ancient Egyptians and Romans long used cardiac glycoside-containing plants as emetics and for heart ailments. However, their toxicity was only recognized in 1785 after the seminal publication of William Withering.[7] In other countries, oleander has been used as a medicinal plant for the treatment of leprosy, ringworm, malaria, and sexually transmitted diseases, and as abortifacients and appetizers. Toxicity may occur from consuming teas brewed from plant parts or after consuming leaves, flowers, blossoms, sap, berries, or seeds of plants containing cardiac glycosides; or from inappropriate therapeutic self-administration of plant extracts; or during suicide attempts. Toxic manifestations are identical to digoxin overdose and include nausea, vomiting, diarrhea, abnormal cardiac rhythms, sinus nodal dysfunctions, atrioventricular blocks, and premature ventricular contractions.

Ingestion of plants that contain cardiac glycosides is reportedly rare in the United States. Of the 1.33 million exposures to nonpharmaceutical substances reported to the American Association of Poison Control Centers in 2006, only 1405 (0.1%) were due to exposures to cardiac glycoside-containing plants.[8] In Sri Lanka and India, increased suicidal or parasuicidal ingestion of yellow oleander (*Thevetia peruviana*) is associated with case fatality of 5 to 10% in untreated victims.[9] Toxicity occurs with serum digoxin levels of >15 ng/ml. Detection of digoxin poisoning by plant-origin cardiac glycoside is difficult and complicated to interpret, and analyses may not detect all the plant forms of cardiac glycosides.[10] Botanical identification of the suspected plant is helpful. Morbidity related to cardiac glycosides is made worse by advanced age, renal dysfunction, myocardial ischemia, hypothyroidism, hypoxia, and electrolyte imbalances, particularly hypokalemia, hyperkalemia, hypomagnesemia, and hypercalcemia. Plant-specific determinants of morbidity related to cardiac glycoside poisoning include plant species, part of plant ingested, specific type of glycoside contained, and concentration of glycoside in plant parts ingested, but mortality is rare.

Acute digoxin toxicity often occurs in younger patients and is associated with lower mortality risk. Elderly patients have higher mortality risk, particularly those with chronic digoxin toxicity and comorbidities such as cardiac and renal diseases. Prevention of further exposure to plant-origin cardiac glycoside includes removing the plant parts, particularly from patients with suicidal tendencies. Destruction of plant sources, deterrent measures to minimize human access to such plants, and public education regarding the dangers pertaining to injudicious use of cardiac glycoside-containing plants are appropriate public health strategies. Policies that counter potentially harmful botanical dietary supplementation are advocated.[11]

1.3 MARINE TOXINS

Ciguatera poisoning is caused by eating fish that contain toxins produced by the dinoflagellate *Gambierdiscus toxicus*, a one-celled plant-like organism that grows on algae in tropical waters worldwide. These lipid-soluble toxins are transferred through the food chain as carnivorous fish consume contaminated herbivorous reef fish.[12] Toxin concentrations are highest in large, predatory fish such as barracuda, grouper, amberjack, snapper, triggerfish, and shark. Ciguatera is vastly underreported, but estimates of lifetime prevalence range from 7% in Puerto Rico to 70% in the Polynesian Islands. Most cases originate in the tropics and subtropics, between the latitudes 35° north and 35° south. However, many cases of ciguatera also occur in temperate regions because of increasing tourism and fish exportation. Ciguatera outbreaks have been reported in Puerto Rico, the Caribbean, Florida, California, and Guam.

More than 400 different fish species have been associated with ciguatera. Reef-dwelling tropical fish, such as barracuda, moray eel, amberjack, and certain types of grouper, mackerel, parrotfish, and red snapper, are the most common sources of ciguatera toxicity. Rare cases exist of ciguatera occurring after the ingestion of temperate-area fish, including farm-raised salmon. Disruption of the marine environment with resultant survival pressures increases the potential for ciguatoxic biotopes. Also, substantial increases in seafood consumption in recent years, together with globalization of the seafood trade, have increased potential exposure to these agents. In general, however, ciguatera from nontropical fish is extremely rare.

Despite similar acute and long-term symptoms of ciguatoxin poisoning, there are geographical variations in clinical symptomatology.[13] Although gastrointestinal and neurological symptoms dominate clinical presentations, cardiovascular manifestations of ciguatoxin poisoning including bradycardia, heart block, and hypotension occasionally occur. The ciguatoxin case fatality rate is quite low (0.1%) with death usually due to cardiovascular collapse or respiratory failure. It is important to fish in safe harvest waters. Control measures include fish sample bioassay using "cigua-check" test kits.

Although ciguatera poisoning is a global phenomenon, most of it is confined to the warm waters and discrete regions of the Pacific Ocean, western Indian Ocean, and Caribbean Sea. Ciguatoxic fish rarely accumulate sufficient levels of ciguatoxin to be lethal at a single meal, probably because the fish itself succumbs to lethal effects of the toxin.[14] Regional differences in clinical manifestations of ciguatera poisoning reflect differences in ciguatoxin levels in the herbivorous and carnivorous ciguateric fish species. Ciguateric fish from the Indian Ocean are more frequently contaminated by lethal levels of toxin.[15] The extent of human exposure depends on dietary concentration of ciguatoxin and dietary intake rate of ciguatoxic fish. Within the fish, the level of toxin is directly related to the rate of toxin assimilation and inversely related to depuration rates. The biotransformation processes involving oxidation and spiroisomerization have not been easily quantifiable.

Preventive measures include safe storage of fish caught in warm waters, restricting the distribution of potentially ciguatoxic fish, and, at the individual level, restricting fish consumption to <250 g per meal. Other measures applicable to industry and government levels include introduction of fishing bans in high-risk waters and large-scale screening of fish captured in such waters.

A less common fish poisoning caused by scombroid toxicity usually begins within an hour of eating contaminated fish and is characterized by flushing, erythematous rash, palpitations, and tachycardia, resulting from excessive histamine release. Scombroid fish poisoning is a worldwide problem resulting from improper handling of fish, particularly mahi mahi, tuna, sardines, and mackerel; however, it may involve any fish containing high levels of free histidine, which is metabolized into histamine by contaminating bacteria. Control measures include appropriate preservation with proper chilling and temperature controls to limit histidine content to <50 particles per milliliter.

Jellyfish stings are also common in warm and cold coastal waters throughout the world. Most jellyfish envenomation in North America is mild, although some deaths from Portuguese man-of-war stings along the Atlantic coast of the United States have been reported. Irukandji syndrome, resulting from the sting by *Carukia barnesi*, causes a wide range of illness, including cardiac dysfunction and acute pulmonary edema.[16,17] Most stings of another jellyfish, *Chironex fleckeri*, are not life threatening, but a smaller body mass exposed to numerous nematocysts determines morbidity and mortality. However, fatality due to stings from the major Australian box jellyfish *Chironex fleckeri* sometimes occurs because of rapid cardiorespiratory arrest, particularly in children.[18] *Chironex quadrigatus*, a multitentacled box jellyfish, is responsible for fatalities in the Philippines and Japan. Most jellyfish stings occur during "stinger season" when the weather is fine, the waters are warm, and winds are still or mild. The majority of stings occur in shallow waters 1 to 2 m deep, often in the late afternoon or early evening.

The biological basis of the lethal activity of jellyfish venoms remains to be fully elucidated, with specific toxins yet to be characterized.[19,20] Given the harm caused by jellyfish to swimmers, preventive measures to protect swimmers should include warning signs on high-risk beaches during "stinger season," protecting children from swimming in high-risk waters, and erecting offshore barriers at popular beaches to preserve coastal tourism. Skin scrapings and sticky tape samples from sting sites should be properly processed for nematocyst description and identification.

Other marine-based cardiotoxins include ostreolysins from edible oyster mushrooms (*Pleurotus ostreatus*), equinotoxins from sea anemone (*Actinia equine*), and conotoxins derived from marine snails (*Conus geographus*). These toxins cause perturbations of various ion channels; cardiac arrhythmias; and inotropic, chronotropic, and arterial blood pressure changes.[21–23] However, their epidemiological roles are poorly characterized.

1.4 VENOMOUS REPTILES

Venomous reptiles are distributed in select habitats in temperate and tropical areas of the world and have adapted to terrestrial, arboreal, and aquatic environments. Venomous snakes are found in the families *Columbridae*, *Elapidae*, *Hydrophiidae*, and *Viperidae* and the subfamily *Crotalinae*. Sea snakes (*Hydrophiidae*) are a subfamily of elapid snakes that inhabit marine environments. Their common habitats include warm shallow waters, mangroves, muddy estuaries, and open oceans of western Australia and the Indian and western Pacific oceans. Their venoms are predominantly neurotoxic and myotoxic but may also cause cardiotoxicity. Persons at risk of sea snake bites include fishermen, divers, surfers, and swimmers. Most bites occur when sea snakes are disturbed, frightened, harassed, hooked, netted, or entangled. Beached sea snakes may be injured or exhausted but remain potentially venomous. The deadliest sea snakes, *Enhydrina schistosa* and *Enhydrina zweifeli*, are found in lagoons and estuaries of waters stretching from the Arabian Peninsula to Australia, respectively.

Five of the most poisonous land snakes, based on LD_{50} (i.e., a lethal dose of venom that kills 50% of laboratory animals), include the inland taipan (*Oxyuranus microlepidotus*), found only in Australia; the king cobra (*Ophiophagus hannah*), found in India and Southeast Asia; the boomslang (*Dispholidus typus*), a tree dweller found in the wooded grasslands of Sub-Saharan Africa (SSA); the black mamba (*Dendroaspis polylepis*), a large, aggressive, and territorial serpent found in most countries of SSA; and the bushmaster (*Lachesis muta muta*), the largest pit viper with nocturnal habits, that is found in remote, heavily forested tropical jungles of South and Central America. These deadly reptiles produce potent neurotoxic and hemotoxic venoms. Hemotoxic venoms affect the circulatory system by causing hemolysis and venous or arterial thrombosis. Cardiotoxicity is secondary to organ degeneration, generalized tissue damage, and hyperkalemia.

Very few countries possess reliable epidemiological reporting systems capable of providing precise data on snakebites. There is little experimental evidence to convincingly demonstrate that the total chemical or pharmacological effect of a whole venom is equal to the sum of the properties of the individual fractions or functions of the venom identified in toxicologic analyses. An extensive appraisal of the global situation by Chippaux[24] yielded regional profiles on incidence of snakebites, morbidity, and case fatality related to envenomation from various bites. The incidence, morbidity, case fatality, and mortality of snakebites depends on climatic conditions, the habitat of the reptile, interaction between humans and snakes, prevalent species of snakes in an area, type and degree of toxicity of the venom, level of care, and quality of reporting systems. In general, snakebite incidence is higher in warm regions where snakes are abundant and manual agricultural activities are prevalent. Also, rural areas characterized by high peasant farming activities and forests inhabited by hunter-gatherers or rubber tappers tend to have a higher incidence of snakebites.[25]

Hazardous bites refer to snakebites that occur when humans encounter snakes in their natural habitats and are more common in middle- and low-income settings. Illegitimate bites are inflicted by a reptile in captivity or during snake handling and are on the increase in industrialized countries. In SSA, annual incidence of snakebites ranges from 12/100,000 of population in Senegal, 150/100,000 in Kenya, and 450/100,000 in Cameroon to 600/100,000 of population in Nigeria with a case fatality 5.9 to 12.3%. This wide range might reflect, at least in part, variations in estimation and reporting systems in different countries in SSA. The most common culprits include *Causus maculatus* (i.e., spotted night adder), *Naja melanoleuca* (i.e., black cobra), *Dendroaspis* spp. (i.e., green mamba), and *Bitis gabonica* (i.e., Gaboon viper).

Similarly wide variations in snakebite incidence have been reported in Asia. In Japan, for example, the incidence is 340/100,000 of population in the southern region compared to a national annual incidence of 1/100,000. Case fatality ranges from 0.7 to 1.0%. *Trimeresurus* spp. is the predominant culprit in Japan, China, and on the Korean peninsula; Russell's viper accounts for 70% of bites in Myanmar; and in Sri Lanka, *Calloselasma rhodostoma* accounts for bite incidence, ranging from 6 to 18/100,000 of population. In Malaysia, *Agkistrodon blomhoffii* causes a 5% case fatality. In Oceania, Papua New Guinea, and the Pacific Islands, most snakebites are caused by *Pseudonaja* spp., *Notechis* spp., and *Oxyuranus* spp. Annual bite incidents range from 3 to 18/100,000 of population.

The highest mortality rates associated with snakebites occur in the tropical and subtropical regions of the world, especially in India and Oceania (Figure 1.1). Although an association of high annual snakebite mortality and poverty has been proposed,[26,27] this view must be seen in the contextual framework of the warmer tropical and subtropical environments frequently inhabited by these reptiles, their complex interaction with human activities, health-seeking behavior of snakebite victims, and the quality of health care. In Nigeria, 80% of snakebites are first reported to traditional practitioners. In India, Malaysia, Papua New Guinea, and many of the Pacific Islands, for instance, annual snakebite death rate is 1000.9 to 10,000/100,000 of population despite gross disparities in their gross national incomes. Many African

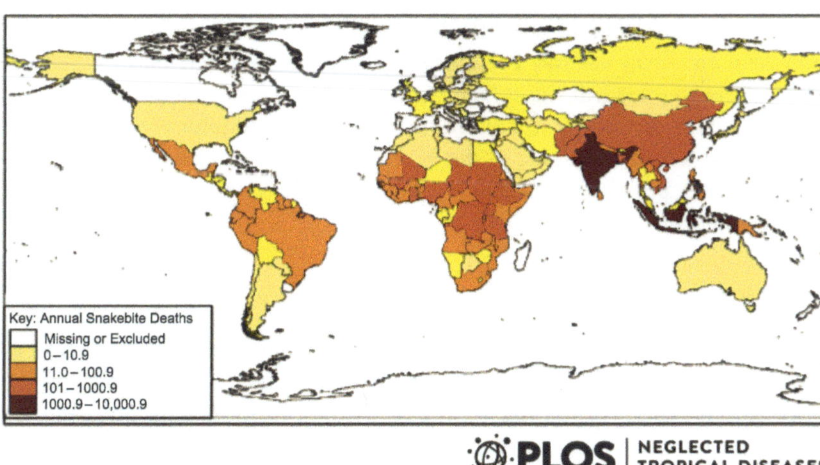

FIGURE 1.1 **Global map of annual snakebite mortality.** **Source:** *Used with permission from Harrison et al., 2009.*[26]

countries with some of the highest human development indices (>0.81) do not necessarily exhibit higher annual mortality rates than some south Asian countries. In high-risk settings, snakebite preventive measures should include provision of protective clothing, gloves, and gumboots for rural farmers; sufficient lighting; and clearing of dumps and bushes around homesteads and other places where humans dwell.

Venomous lizards, belonging to the family *Helodermatidae* of the genus *Heloderma*, are found in the United States and Mexico.[28] Although the Gila monster (*Heloderma suspectum*) and the beaded lizard (*Heloderma horridum*) were believed to be the only two venomous lizards, as many as 100 species of lizards use venom; nine types of lizard toxins are shared with those of venomous snakes. Monitor lizards, found in Africa, Australia, and Asia, though venomous, often tend to avoid confrontation and therefore human bite incidents are rare. The Komodo dragon, found mainly on Komodo Island and smaller islands east of Java in Indonesia, has at least six venom glands on each side of its mandible and multiple ducts between its teeth. Its venom is believed to be similar to the toxin of the inland taipan. However, the effect of a Komodo dragon's toxin is relatively mild in humans due to ineffective delivery methods during occasional human bites. Venoms of various species of monitor lizards found in arid, semiarid, and aquatic environments in the Old World cause hypotension and perturbations in blood clotting. Iguanians constitute nearly 1500 species of iguanas. They live in a wide range of habitats ranging from trees and water edges to arid areas in the Americas, Madagascar, Fiji, and Tonga. Some species of iguanas deliver small amounts of venom when they bite. Persons at risk include intruders into their habitats and pet owners or zoo workers.

Initial postinjection effects of Russell's viper or cobra venom include rapid drop in mean arterial pressure, bradycardia, reduced cardiac output, and increased peripheral vascular resistance. The second phase of envenomation is characterized by a predominant role of vasodilators with increased return of blood to the left side of the heart, resulting in increased cardiac output, decreased peripheral vascular resistance, and return of mean arterial pressure to normal.

1.5 TRICHINELLOSIS

Helminthic infestation by *Trichinella spiralis*, a white intestinal nematode, is widely distributed, mainly in pork-rearing and game-eating regions of the world. Human trichinelosis has been documented in more than 55 countries of both the industrialized and nonindustrialized worlds, including North and South America, Europe, Africa, Asia, and the Pacific region.[29] The prevalence of human trichinellosis (also known as trichiniasis or trichinosis) is the highest in China, Thailand, Mexico, Argentina, Bolivia, the former Soviet Union, Romania, and other parts of central Europe. In 2004, Romania reported the highest incidence of trichinellosis in the world.[30] Seven *Trichinella* species are the ones most implicated in human disease: *T. spiralis*, which is found worldwide[31]; *T. nativa*, found in the arctic regions; *T. nelsoni*, found in SSA; *T. brivoti* found in Europe, western Asia, and western and southern Africa; *T. murelli*, found in the United States and Japan[32]; *T. papuae*, found in Papua New Guinea and Thailand[33]; and *T. pseudospiralis*, found in birds worldwide.[34]

Reservoirs for *Trichinella* include pigs, dogs, cats, rats, horses, and other domestic animals. Adult worms develop rapidly within human intestinal epithelium following larval ingestion in infested pork. Mature female worms produce larvae, which penetrate lymphatics and venules and become widely distributed throughout the body. Larval encapsulation into muscles follows the inability of the parasite to complete its cycle in the human host. Cardiac complications tend to occur 3 to 6 weeks after infestation, and severe cases may lead to myocardial failure and occasionally death.

Those at high risk for trichinellosis include individuals in pig-farming communities and those who ingest game meat, wild pig, or boar; young men (<40 years); and tourists who may consume improperly cooked contaminated meat.[30,35,36]

Cardiovascular complications represent the most important manifestation of moderate to severe trichinellosis.[37] Larvae do not encyst in cardiac muscle but elicit an intense eosinophilic inflammatory response and myocarditis, responsible for the cardiotoxicity of trichinellosis. Common electrocardiographic abnormalities include nonspecific ST-segment and T-wave changes, bundle branch block, and sinus tachycardia. Less frequent electrocardiogram (ECG) changes include sinus bradycardia, right bundle branch block, supraventricular tachycardia, premature ventricular contractions, and low QRS voltages.[38] Mortality in patients with serious trichinellosis is often as a result of heart failure and sometimes, severe arrhythmias.

Measures to mitigate or prevent trichinellosis include basic hygiene such as hand washing with soap by butchers; cleaning meat grinders thoroughly after each use; and avoiding undercooked pork, walrus, horse, bear, and other wild animal meat. Sufficient cooking of all parts of meat to 60 to 71°C or freezing to −15°C also reduces risk of infestation.

1.6 ARACHNIDISM

Envenomations by arachnids cause significant medical illnesses worldwide. Important groups of spiders include the widow spiders (*Latrodectus* spp.), the recluse spiders (*Loxosceles* spp.), the Australian funnel web spider (*Atrax* and *Hadronyche* spp.), and the armed spider (*Phoneutria* spp.) from Brazil.[39] There are approximately 30 species of widow spiders in the world. Widow spiders and their webs are found outdoors in garages, trash piles, potted plants, trashcans, woodpiles, dry storage areas, outhouses, rustic recreational areas, and outdoor furniture. Widow spider bites occur when the spiders or their webs are disturbed by human activity such as picking up trash, potting plants, and moving wood piles; wearing garden shoes; and camping, hiking, or rock climbing.

Most bites in many parts of the world occur outdoors and in young men. In Europe and the Mediterranean regions, widow bites are occupational hazards of wheat farmers who might compress the spider against their bodies as they harvest the crop, and of other outdoor workers, including farm laborers and greenhouse keepers. Approximately 75% of widow bites are on the extremities. Steatoda spiders ("false" black widow spiders) are in the same family (*Theridiidae*) as the black widows and hence share a similar body form, which may cause confusion in identification. The false black widow is more likely to be found inside homes.

The effects of a spider bite depend on the venom's toxicity; the amount injected, which depends on length of mouth parts (*Chelicera*); and the tissue-specific effects of the venom.[40] The majority of venomous spider bites are not associated with toxicity due to insufficient envenoming. Dangerous spiders deliver an adequate dose of potent venom in a single bite. Latrodectism and loxoscelism are the most important clinical syndromes resulting from spider bites. Latrodectism causes local, regional, or generalized pain associated with nonspecific symptoms and autonomic effects such as tachycardia and hypertension. Myocarditis is very rare but has been described.[41] Bites by *Phoneutria* spp. spiders are common in Brazil, although only 0.5 to 1% result in severe envenomation, with most of these occurring in children. Cases of systemic envenomation in adults are very unusual. Life-threatening anaphylactic reactions with shock sometimes occur. Pulmonary edema from spider bites, with potential fatalities, is also rare but has been described.[42,43]

Anxiety reactions caused by the fear of widow spiders may confuse the clinical presentation of some patients. It must be remembered that only 1% of widow spider envenomation has life-threating effects. Expert identification of the biting spider is therefore essential, whenever possible. Prevention involves the use of protective clothing, including gloves, in spider-infested outdoor environments and keeping domestic settings clean. Care must be taken to inspect garden shoes and camping gear before wearing or using them.

1.7 SCORPION ENVENOMATION

Scorpion envenomation is a common medical problem in many parts of the world, particularly in tropical and subtropical countries.[44] Although the most common toxic envenomation offenders arise from the *Buthidae* family of scorpions, cardiovascular effects of various scorpion venoms are remarkably similar. Cardiovascular effects of scorpion venoms have been reported from various parts of the world, including Brazil, Mexico, the southwestern

United States, North Africa, southern parts of Africa, the Middle East, southern Spain, and India.

Effects of envenomation depend on the age of the scorpion, dose of venom delivered, and size of the victim. Reports of cardiovascular manifestations include hypertension (17.5–77%), pulmonary edema (7–32%), and myocardial damage. Hypertension is commonly observed in children and is severe in the majority of them. Electrocardiographic abnormalities include bradytachyarrhythmias, ST–T wave abnormalities (24–39%), QTc prolongation (53%), electrical alternans (13%), complete atrioventricular (AV) blockade (2.5%), shock-like syndrome (7–38%), and sudden cardiac death (7–11%).[45–49] Other electrocardiographic abnormalities include atrial fibrillation, AV dissociation with accelerated junctional rhythm, premature atrial or ventricular contractions, ventricular tachycardia, or fibrillation. Conduction abnormalities such as left bundle branch block and first- and second-degree heart blocks are rare. Echocardiographic and radionuclide studies have shown depressed left ventricular (LV) function and mitral regurgitation after scorpion envenomation. Sudden severe blood pressure elevation may be accompanied by clinical manifestations of hypertensive encephalopathy.

Pathologically, macroscopic examination may reveal no gross cardiac abnormalities, but diffuse microscopic changes with marked mononuclear cellular infiltrates, focal necrosis, and subendocardial hemorrhages may be observed. Pathophysiological mechanisms of scorpion envenomation involve sudden massive release of vasopressor catecholamines,[46] resulting in myocardial inotropic effects, various arrhythmias, enhanced LV contractility, and systemic hypertension. The shock-like phenomenon results from the depressive cholinergic effect of the venom, catecholamine depletion, exaggerated beta-2 vasodilator effect of circulating catecholamines on the peripheral vascular bed, and hypovolemia due to excessive fluid loss.[50] Myocardial stunning may be an important predisposing factor to left ventricular failure. A scorpion sting is the most important arachnid envenomation causing adult morbidity and pediatric mortality.

At the population level, efforts must be made to educate everyone about conditions that foster envenomation such as poor hygiene or neglect of surroundings. Where possible, stones, garbage, and bushy fencing hedges and climbing plants should be removed from domestic areas. Good lighting and use of appropriate protective footwear are recommended. It is a good precaution to check, invert, and shake footwear vigorously before wearing them and to avoid sleeping on the ground at night during the summer in scorpion-prone areas. To decrease morbidity and attendant mortality related to envenomation, local populations can benefit from appropriate first aid measures and effective utilization of nearby health facilities. Care should be standardized at secondary and tertiary care facilities with ready availability of, or access to antivenom, whenever needed. Serious cases of envenomation require hospitalization and appropriate management.

1.8 AIR POLLUTION

Air pollution is a complex heterogeneous mixture of gases, liquids, and particulate matter (PM). Primary air pollutants include sulfur dioxide, oxides of nitrogen, carbon monoxide, volatile organic compounds, and carbonaceous and noncarbonaceous particles. Secondary pollutants are formed by chemical reactions between primary air pollutants and atmospheric oxygen and water, the most familiar of which is ozone. The highest concentrations of classical indicators of air pollution (PM_{10}, $PM_{2.5}$, and sulfur dioxide) are found in Africa, Asia, and Latin America. Determinants of adverse effects of air pollution include fractional deposition of inhaled particles, degree of individual exposure, and variability in cumulative dose. Personal risk factors and deleterious effects of particulate air pollution include low socioeconomic status, a low level of education, extremes of age (the young and the elderly), undernutrition, obesity, diabetes mellitus, preexisting cardiovascular disease, and genetic susceptibility.[51,52]

Several observational studies in the United States and Europe have demonstrated an association between fine particulate air pollution and increased risk for cardiovascular events in men and women.[53–56] These risks are related to the composite effects resulting from both short- and long-term exposure to air pollution and to control measures that aim to reduce PM air pollution.[57–59] The Women's Health Initiative Observational Study[58] examined the database of more than 65,000 postmenopausal women without prior cardiovascular disease to evaluate the relationship between long-term exposure to air pollution and the risk of a cardiovascular event. Potential confounding factors, such as age, body mass index, and traditional cardiovascular risk factors, were corrected for. Each 10 $\mu g/m^3$ increase in pollution was associated with increased risk of any cardiovascular event (hazard ratio 1.24) and death from cardiovascular disease (hazard ratio 1.76).

Jerret et al. analyzed data on 448,850 subjects from the second American Cancer Society Cancer Prevention Study, with 118,777 deaths during the 18-year follow up.[60] Multivariate analysis revealed that concentrations of fine PM that were ≤ 2.5 micron in aerodynamic diameter ($PM_{2.5}$), but not ozone, were significantly associated with 1.2-fold increased relative risk of death from cardiovascular causes. However, Ruidavets et al. in the Toulouse MONICA project[61] showed some marginal increase in risk of acute myocardial infarction (relative risk 1.05; $P = 0.009$) related to 5 $\mu g/m^3$ of ozone concentration. The relative risk was 1.14 in subjects aged 55 to 64 with no prior history of ischemic heart disease.

Short-term exposure to nitrous oxide and sulfur dioxide were not associated with significant risk of acute myocardial infarction. Pope[62] reported a 1.3% increase in cardiovascular mortality for every 5 $\mu g/m^3$ rise in $PM_{2.5}$. A systematic review and meta-analysis done by Mustafic et al. revealed that, with the exception of ozone, all of the main air pollutants were significantly associated with increased risk of acute myocardial infarction.[54] The following are the relative risks associated with exposure: carbon monoxide, 1.048 (95% CI 1.026–1.070); nitrous dioxide, 1.011 (95% CI 1.006–1.016); sulfur dioxide, 1.010 (95% CI 1.003–1.017); PM_{10}, 1.006 (95% CI 1.002–1.0009); and $PM_{2.5}$, 1.025 (95% CI 1.015–1.036). Overall population-attributable risk varied from 0.6 to 4.5% depending on pollutants. All-cause mortality has also been shown to be higher among individuals with greater long-term exposure to $PM_{2.5}$ in hospitalized survivors of acute coronary syndrome in England and Wales.[63]

Peters et al. reported the link between increased PM with specific cardiac events including serious ventricular arrhythmias and myocardial infarction.[64,65] Mechanistic considerations of the effects of fine air pollution have focused on autonomic nervous system alterations, myocardial ischemic responses, ion channel dysfunctions in myocardial cells, inflammatory responses resulting in endothelial dysfunction, release of various cytokines, nitric oxide, interleukins, thromboxane X2, endothelin-1, and eventual atherothrombosis. However, these pathogenic mechanisms are only beginning to be unraveled. Despite these intriguing associations between PM exposure and several parameters of cardiovascular dysfunction, evidence from published studies seems insufficient to conclusively determine their strength of association.[66]

The complexity of conditions that predispose to increased risk of cardiovascular morbidity and mortality makes it difficult to place greater emphasis on the contribution of air pollution relative to other factors. Air pollution is presumably a relatively minor contributing factor embedded within a matrix of inborn factors (e.g., genetic, gender, age); traditional cardiovascular risk factors (e.g., hypertension, diabetes mellitus, obesity, dyslipidemia, psychosocial stress); other environmental factors (e.g., inadequate intake of fruits and vegetables, no or inadequate alcohol intake, tobacco use); and underlying cardiac disease (e.g., subclinical ischemic heart disease, cardiomyopathy, hypertensive heart disease).

Despite the complexity of determinants of susceptibility to air pollution, public information concerning the cardiovascular risk posed by air pollution should be communicated. Health measures to reduce the cardiovascular impact of air pollution should focus on entire communities and populations, particularly targeting the elderly, those with preexisting cardiovascular or respiratory diseases, the poorly educated, and those of low socioeconomic status. Policies that aim to promote cleaner air by reducing levels of pollution, adopting cleaner energy sources, improving air quality, and minimizing human exposure to air pollution must be advocated.[53]

1.9 PESTICIDES

This section deals with cardiotoxicity resulting from the cholinergic effects of substances used industrially or for controlling pests in agricultural or domestic settings.

1.9.1 Organophosphate and Carbamate Poisoning

Although the use of organophosphates as insecticides has largely been replaced by availability of carbamates in the past two decades in developed countries, they are still widely available for industrial, agricultural, and household uses in developing countries because they are less expensive than the newer alternatives. An estimated three million people are exposed to organophosphates and carbamates annually, with some 300,000 fatalities.[67] Organophosphate insecticides include parathion, fenthion, malathion, diazinon, and dursban; methomyl and aldicard are the two commonly encountered carbamates. A number of countries have banned or restricted the use of organophosphorus-containing household roach and ant sprays.

Toxicity to organophosphates typically exhibits a triphasic response. First is the initial acute cholinergic crisis, followed by an intermediate syndrome, both of which are potentially fatal; and then the disabling nonlethal delayed polyneuropathy.[68] Organophosphates and carbamates are potent cholinesterase inhibitors, potentially resulting in cholinergic toxicity following excess exposure through skin, inhalation, or ingestion. The cholinergic phase is due to excessive accumulation of acetylcholine at muscarinic sites, resulting in sinus bradycardia among other muscarinic effects. The effects at nicotinic sites and the central nervous system have no cardiovascular components. The intermediate syndrome, which occurs 1 to 4 days after exposure to toxic levels of organophosphate, is predominantly neuromuscular with proximal muscle weakness, diaphragmatic paralysis, and cranial nerve palsies. Cardiac arrhythmias including heart block, QTc prolongation, and pleomorphic ventricular tachyarrhythmias resulting in torsades de pointes are uncommon but have been reported in some patients experiencing organophosphate poisoning.[69] At-risk groups include farm workers, peasant farmers, handlers, children, and those intending self-harm.[70]

Fungicides are chemical substances that destroy or inhibit the growth of fungi. They have similar uses in agriculture, for lawns, and on golf courses as insecticides. It is important to strictly follow manufacturers' instructions for use and to spray fungicides at specified times and intervals. Accidental contact of the chemical with any part of the body should be promptly washed, and eye splashes should be irrigated with clean water. Whenever possible, fungicides should be sprayed in calm weather conditions to avoid inhalation of aerosols and dust containing the chemicals.

Personal measures against organophosphate exposure involve maximum protective equipment including double layers of clothing and chemically resistant gloves, footwear, head gear, apron, and, where possible, effective masks. Backpack sprayers, working with hands, and knapsacks should be avoided when handling organophosphates and carbamates. The use of human flaggers should be avoided in agricultural fields. Also, the use of organophosphates when growing fruits and vegetables, such as apples, peaches, grapes, green beans, and peas, should be restricted. Such preventive measures are clearly impractical in low-resource settings.

1.10 HOUSEHOLD TOXIC MATERIALS

Many household substances are potentially toxic. They can be grouped into organic solvents (e.g., antifreeze, windshield washer solution, artificial nail polish removers, petroleum, kerosene, lamp oil, paint thinner, furniture polish), disinfectants (e.g., drain cleaner, toilet bowl cleaners, mouthwashes), detergents (e.g., automatic dishwasher detergents, kitchen detergents), deodorants (e.g., perfumes, aftershave lotions), and designated poisons (e.g., antirodents, insecticides). Miscellaneous items include batteries, flaking paints, alcohol, and tobacco products. Some examples of common household materials with potential cardiotoxicity are discussed in this section.

Ethylene glycol (EG), a toxic, colorless, odorless, nonvolatile, and sweet-tasting industrial solvent, is the primary ingredient in antifreeze and hydraulic brake fluid. Its major metabolites are glycolic and oxalic acids, which cause metabolic acidosis and cardiovascular dysfunction. Determination of blood levels of ethylene glycol may not be immediately possible since it requires gas chromatography, and urine calcium oxalate crystals in suspected cases of EG poisoning may be nonspecific. Initial irritant effects of ingested ethylene glycol may cause nausea and vomiting, followed by neurological, cardiopulmonary, and renal toxicity. Cardiotoxicity, manifesting as dysrhythymias, hypotension, myocardial depression with or without heart failure, and sometimes focal myocardial hemorrhages, occurs as a result of metabolic acidosis and hypocalcaemia. Ingestion of 1 g of EG per kg of body weight is considered toxic. Children are often accidentally poisoned by ingestion of 10 to 30 mls because they are enticed by the sweet taste of EG. However, isolated or epidemic human poisoning may occur through suicide or through accidental or malicious adulteration of alcohol with EG. Certainly not all cases of EG poisoning are reported.

In 2002, there were 5816 human exposures to EG in the United States. In Australia, only 17 cases of EG poisoning were reported to the Victorian Poisons Information Centre in 2007 and 30 cases to The Children's Hospital in New South Wales. Inhaled EG is unlikely to result in any significant toxicity. Contact contamination is easily treated by washing the affected area of the skin with mild soap and clean water or irrigating the conjunctiva with clean water. Contact lenses should be removed.

Although a number of U.S. states have advocated adding a bittering agent, such as denatonium benzoate, to minimize accidental and suicidal risks of antifreeze poisoning, studies have cast doubts on the impact that such measures would make.[71,72] Difficulties in ascertaining initial diagnosis of EG poisoning result in life-threatening complications with high case fatality. The mortality rates related to EG poisoning range from

1 to 22%, depending on amount ingested, clinical index of suspicion, and level of care.[73] Presentation with severe metabolic acidosis is easily confused with lactic acidosis, diabetic ketoacidosis, methanol, and salicylic acid poisoning. State legislation also appears to have had little impact. The establishment of poison control centers and Toxic Exposure Surveillance Systems (TESS) as well as helplines and easy access to toxicologists like those in the United States may improve real-time toxicovigilance and reduce morbidity and case fatality related to common poisons.[74] Safe storage of potentially toxic household items should be regularly advocated.

Most disinfectants and antiseptics are locally irritating and potentially corrosive to skin, conjunctivae, and respiratory tract mucosae. External contamination by these chemicals is therefore unlikely to result in any systemic effect. Ingestion of phenolic and carbonic acid compounds accidentally or suicidally may cause cardiac dysrhythmias.[75] Aldehydes are volatile compounds with irritating effects to the eyes. Ingestion of formaldehyde or glutaraldehyde may result in circulatory collapse or death. Phenols and cresols, used as deodorants, have no reported cardiac toxicity.

1.11 PETROL, PARAFFIN, AND LAMP OIL

These products are highly inflammable organic hydrocarbons commonly used as fuel in combustion engines. Unleaded petrol contains benzene, trimethyl benzene, toluene, naphthalene, and methyl tert-butyl ether, which are potentially carcinogenic. Hydrocarbons generally exhibit low toxicity. Although sniffing petroleum products has reached epidemic proportions in low-income communities, especially among indigenous groups in Australia, New Zealand, some of the Pacific Islands, and Canada, there are few reported cases of cardiotoxicity.[76]

1.12 TOXIC MILITARY AND INDUSTRIAL CHEMICALS

Chemicals used by the military include chlorine, ammonia, and hydrogen cyanide. Human exposure may occur during warfare or terrorist attacks; accidental explosions at munitions depots, chemical plants, and/or of transportation vehicles; or at wastewater plants. Industrial metal fumes containing barium have cardiotoxic effects.[77] Other industrial processes, such as metal hardening, arc-welding, ceramic glazing, or enameling, may give rise to barium-containing airborne particulate fumes and aerosols, which have a dose–effect toxic relationship. Other toxic heavy metals (e.g., lead, cadmium, mercury, and chromium) are dealt with in Section 1.19.

1.13 TOBACCO AND CARDIOVASCULAR DISEASES

Cigarette smoking is the leading preventable cause of mortality, responsible for nearly six million deaths worldwide.[78] This toll is projected to rise to more than eight million deaths per year by 2030, with 80% of those deaths occurring in low- and middle-income countries where tobacco use is increasing. Indeed, the World Health Organization projects that tobacco will kill up to one billion people this century if the Framework Convention on Tobacco Control (i.e., WHO FCTC) is not implemented rapidly.[78] As shown in Figure 1.2, the highest smoking rates (>36.5%) are seen in most countries of northern Europe, Chile, Cuba, Greenland, and some of the Pacific Islands. Smoking rates of 28.6 to 36.5% are seen in the United States, Venezuela, Argentina, China, Australia, and New Zealand. Smoking rates are generally low in much of Sub-Saharan Africa except in Namibia, Zimbabwe, and the Republic of South Africa. However, since the introduction of the WHO FCTC: MPOWER strategy in 2008, more than a third of the global population (>2.3 billion people living in 92 countries) is now covered by at least one of the measures at the highest level of achievement. However, two-thirds of countries have not aired antitobacco campaigns to increase awareness on the dangers of tobacco use.

Although nicotine is the most well-known, potently addictive, neuroteratogenic, and toxic cardiovascular substance in all forms of tobacco (smoked or smokeless), tobacco contains myriad other potentially toxic or carcinogenic agents, including heavy metals and metalloids, nitrosoamines, and

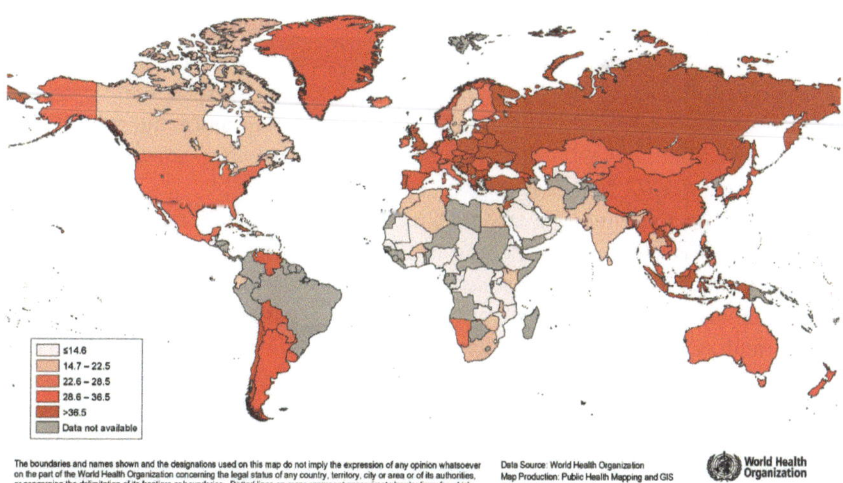

FIGURE 1.2 Global map showing the percentage of tobacco use among adults (age 15 and over), 2005. Source: *From WHO, 2008.*

gaseous complexes. Some of the toxic heavy metals found in tobacco such as lead, cadmium, and arsenic are discussed in Section 1.19.

Nicotine is the principal alkaloid in both commercial and homemade tobacco products, followed by nornicotine, anabasine, anatabine, and many other basic substances that contain a cyclic nitrogenous nucleus. Tobacco types, leaf position on the plant, agricultural practices, fertilizer treatment, degree of ripening, blending recipe, type and amount of additives (e.g., acetaldehyde, ammonia compounds), and product design are among some of the main factors that determine the levels of alkaloids in tobacco. The threshold dose of nicotine necessary to produce dependence has not been firmly established; neither have dose–response relationships between blood nicotine levels and cardiovascular risk. Smoking affects the physiologic, pathologic, hematologic, and metabolic factors leading to the initiation, progression, and sequelae of atherosclerosis. However, several epidemiological studies have yielded inconsistent cardiovascular effects of tobacco. Users of smokeless tobacco ingest levels of nicotine similar to those of smokers and this may be associated with increased risk of myocardial infarction and/or stroke, although the risk is less than from cigarette smoking.

The apparent paradoxic beneficial effects of smoking in patients receiving thrombolytic therapy for acute myocardial infarction in the GUSTO-I trial may be explained by the fact that smokers were significantly younger by a mean of 11 years and had less comorbidity or severe coronary artery disease than nonsmokers.[79] In the INTERHEART study, a case-control examination of acute myocardial infarction involving more than 27,000 participants from 52 countries, current smoking was associated with a 3-fold greater risk of nonfatal acute myocardial infarction compared with never smoking and the risk increased by 5.6% for every additional cigarette smoked.[80] The risk of acute myocardial infarction associated with chewed tobacco and smoking beedies were similar to the risk associated with current smoking, while secondhand smoking showed a dose-related risk ranging from 1.24-fold in those least exposed to 1.62-fold in most-exposed individuals.[80]

Some of the epidemiological studies failed to demonstrate any significantly greater all-cause and disease-specific mortality in users of smokeless tobacco compared with mortality in nonsmokers.[81] However, a 2006 WHO report concluded that all tobacco products were harmful and addictive and that all cause disease and death. Properly implemented, the WHO FCTC and MPOWER strategies are key to the global control of the use of tobacco.[78] In reiterating some of WHO's major messages, education, communication, and training must target the most vulnerable populations in middle- and low-income countries especially children, adolescents, and women; universal implementation of health warnings on all tobacco products should include various languages and graphic images, to cover minorities, the illiterate, and those with sensory disabilities. Cessation of tobacco use should be encouraged and bans on tobacco advertising, promotion, and sponsorships should be enforced across the board.

1.14 CARDIOVASCULAR TOXICITY OF ALCOHOL

Alcoholic beverages are widely consumed worldwide. Consumption of alcohol in moderation has been inversely related to coronary heart disease and possibly all-cause mortality. However, no standardized amounts of ethanol have been established in the commonly stated 1 to 2 drinks per day for women and 2 to 4 drinks per day for men due to differences in beverage brands. A meta-analysis of 34 prospective studies showed a J-shaped relationship between alcohol and total mortality.[82] The maximum protection from 4 g per day of alcohol for women was 18% while 6 g per day gave men a maximum protection of 17%. Higher consumption was associated with increased mortality.

Excessive intake of alcohol, particularly binge drinking or drinking to become intoxicated, has numerous potential harmful effects. Alcohol consumption rates tend to rise during festivities, ceremonies, parties, and holidays with parallel trends in tobacco smoking. The highest levels of ethanol consumption occur in countries within the WHO European Region, the Americas, the Western Pacific, and southern Africa. Countries in North Africa and much of the WHO African, Eastern Mediterranean, and Southeast Asia regions have low alcohol consumption (Figure 1.3). Consumption of homemade or illegally produced alcohol or industrial alcohol (methanol) may be associated with a greater risk of harm because of potentially dangerous impurities and contaminants.

Although there are regional differences, globally the most frequently used alcoholic beverages are spirits (45.7%), beers (36.3%), wine (8.6%), and other

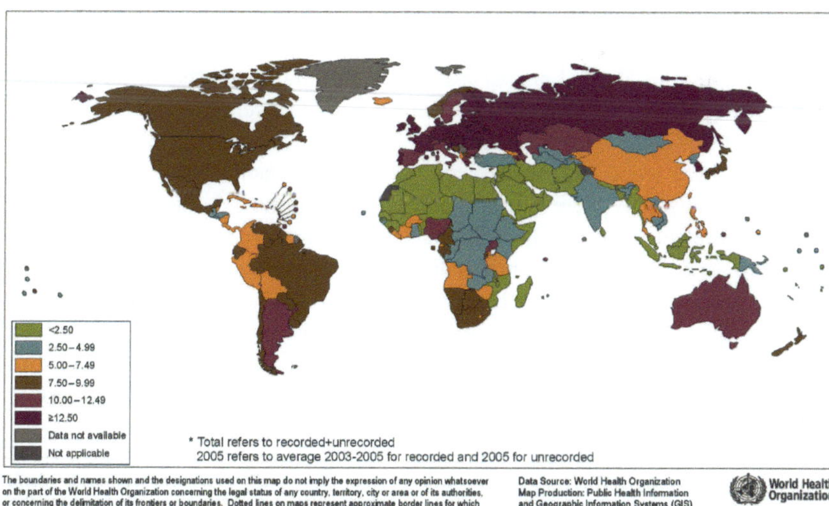

FIGURE 1.3 **Global map showing the total adult (age 15 years and over) per capita consumption of pure alcohol (in liters), 2005. Source:** *From WHO, 2010.*

brews (10.5%). Drinking among young people (13–15 years old) is increasing at alarming rates in most countries. In children, alcohol intake occurs through accidental ingestion and occasionally as a result of child abuse. More than 60 major types of diseases and injuries are causally linked to harmful use of alcohol, resulting in approximately 2.5 million deaths annually (4% of global all-cause mortality). Individuals with a history of hemorrhagic stroke, liver disease, pancreatic disease, alcoholism, gastritis, Barrett's esophagitis, or cancer are at increased risk of the harmful effects of alcohol.

The association between chronic alcohol consumption and alcoholic heart disease in human beings is well recognized. There is a U- or J-shaped relationship between alcohol consumption and survival from cardiac disease. Chronic alcohol consumption is the leading cause of secondary cardiomyopathy, a heart muscle disease associated with long-term alcohol consumption. In the United States, long-term heavy consumption of any type of alcoholic beverage is the leading cause of a nonischemic dilated cardiomyopathy in men and women of all races.

The fact that only a minority of persons with chronic alcoholism have this condition suggests the possibility of genetic vulnerability. Polymorphism of the angiotensin converting enzyme (ACE) gene has been implicated in cardiac dysfunction associated with vulnerability to alcoholic cardiomyopathy.[83] This complication tends to occur in those who have been drinking more than 80 to 90 g of ethanol per day for 5 to 15 years. Despite lower mean lifetime ethanol dose, women appear to be at greater risk of alcoholic cardiomyopathy. This sex differential of risk is partially explained by excessive accumulation of ethanol metabolites in women due to lower gastric metabolism and enhanced hepatic metabolism; both mechanisms contribute to higher acetaldehyde concentration in the blood of women who drink.[84]

Consumption of alcohol is also associated with increased risk of atrial fibrillation (AF) or flutter in men.[85] In the Danish Diet, Cancer, and Health Study of nearly 48,000 middle-aged participants (mean age 56), daily alcohol consumption by men was twice that of women. The greater sensitivity of women to the effects of ethanol may therefore not entirely explain the 1.5-fold increased risk of AF or flutter in men; the disparity in daily alcohol consumption needs to be taken into account. The Framingham Heart study[86] showed significantly increased risk of AF among subjects consuming >36 g per day (>3 drinks/day). The Copenhagen City Heart study of 16,415 women and men also showed no significant association of moderate alcohol consumption and risk of AF. But men who drank 35 g of alcohol or more per week had nearly 1.5-fold increased risk of AF; few women consumed this amount of alcohol.[87]

Furthermore, persons who consume various alcoholic beverages excessively and for a long time also exhibit significantly higher risk of prolonged QTc interval and higher QTc dispersion and other arrhythmias than control groups. Atrial fibrillation occurs in up to 60% of binge drinkers with or without underlying alcoholic cardiomyopathy. There is a positive correlation between alcoholic cardiomyopathy and cirrhosis. Beriberi heart disease, due to nutritional deficiency of thiamine, must be differentiated from alcoholic heart muscle disease. Beriberi heart disease is rapidly reversible after thiamine therapy, manifests predominantly as high-output right heart failure, and exhibits normal myocardial histology.

As with tobacco, control of the harmful effects of alcohol requires determined and concerted efforts of individuals and communities galvanized by institutional, organizational, national, and international policies. Bills, acts, and legislation against drunk driving and the sale of alcohol to minors, and banning of alcohol advertising and marketing all require appropriate power of enforcement with commensurate penalties to reduce alcohol-related harm. High levels of morbidity, mortality, violence, crime, and road traffic accidents, in addition to the negative social impact of alcohol, including family disruption, alcoholism, and a spiral of poverty, are persuasive reasons for public campaigns against excessive drinking.

1.15 CARDIOVASCULAR EFFECTS OF CAFFEINE

Other than water, coffee is arguably the most consumed beverage worldwide. Exposure to caffeine is generally lifelong in most consumers. Despite the widespread belief in the link between coffee consumption and cardiac arrhythmias, there is little conclusive evidence of detrimental cardiovascular effects of coffee, and divergent notions persist regarding its potential harmful versus beneficial effects. An extensive literature search over a decade in the early 1990s by Myers[88] concluded that moderate ingestion of coffee did not increase the frequency or severity of cardiac arrhythmias in normal persons, patients with ischemic heart disease, or those with preexisting serious ventricular ectopy. The Danish Diet, Cancer, and Health Study of nearly 48,000 participants also showed no link between caffeine consumption and the risk of atrial fibrillation or flutter.[89]

Using the lowest quintile of caffeine consumption as the reference, the adjusted hazard ratios for the second, third, fourth, and fifth quintiles (plus 95% confidence interval, CI) were 1.12 (0.87–1.44), 0.85 (0.65–1.12), 0.92 (0.71–1.20), and 0.91 (0.70–1.19), respectively. More recent systematic review and meta-analysis of observational studies also showed no association between caffeine consumption and the risk of atrial fibrillation.[90] In fact, low-dose caffeine might have protective cardiovascular effects. In nonhabitual coffee drinkers, there might be acute blood pressure elevation but this phenomenon is not observed in habitual coffee drinkers.[91]

There has been no conclusive data on the association of caffeine and cause-specific mortality. The Health Professionals follow-up study and the Nurses' Health Study, involving 41,736 men followed up for 18 years and 86,214 women followed up for 24 years respectively, showed no association between regular coffee consumption and all-cause mortality in either men or women.[92] A much larger prospective study by Freedman et al.[93] of more than 400,000 men and women during more than 5 million person-years showed an inverse association between coffee consumption and total and cause-specific mortality. A recent study of a multiethnic urban population in northern Manhattan found a strong inverse association between coffee and vascular-related mortality among Hispanics only.[94]

More recent data from the Aerobics Center Longitudinal Study in South Carolina adds a new twist to the association between coffee consumption and all-cause and cardiovascular mortality.[95] This study enrolled 43,727 participants aged 20 to 87 and followed up for a mean of 17 years (nearly 700,000 person-years). Data were obtained through in-person interviews. During the study period, there were 2512 deaths, one-third of which were due to cardiovascular causes. Multivariate analyses adjusted the data for age, baseline examination, body mass index, hypertension, diabetes mellitus, lipid profiles, use of decaffeinated beverages (coffee, tea), tobacco use, and physical inactivity. Men who drank >28 cups of coffee per week (>4 cups/day) had higher all-cause mortality (hazard ratio (HR) = 1.21, 95% CI 1.0–1.40). In younger age groups (<55 years), excessive intake of coffee (>28 cups/week) was positively associated with higher all-cause and cardiovascular disease mortality in both men and women (HR = 1.56, 95% CI 1.30–1.87 in men; HR = 2.13, 95% CI 1.20–3.59 in women).

Like alcohol, determining the dose of coffee is fraught with lack of standardization of the "cup of coffee," strength, type, method of preparation, and caffeine content. Also, different species and preparations of coffee may have varying chemical compositions. However, this recent observation, plus several case reports suggesting a possible association between intake of unusually large amounts of caffeine and cardiac arrhythmias, makes it prudent to advise susceptible patients to avoid ingesting very large quantities of caffeine, particularly those patients with underlying cardiac disease.

1.16 CARDIOVASCULAR EFFECTS OF ANTIRETROVIRAL DRUGS

The interaction between the human immunodeficiency virus (HIV), acquired immunodeficiency syndrome (AIDS), antiretroviral drugs (ARVs), and cardiovascular disease (CVD) is complex and incompletely understood. Highly active antiretroviral therapy (HAART) is associated with clustering of cardiovascular risk factors resulting from class- and nonclass-specific metabolic effects of ARVs on lipids, glucose, insulin sensitivity, and anthropometric body changes characteristic of lipodystrophy. A paradoxical overall effect on CVD may be observed in untreated or early HIV infection. However, it becomes difficult to disentangle the effects of HIV itself, antiretroviral therapy, and the impact of improved longevity in those on HAART.

Of the drugs considered, only indinavir, ritonavir-boosted lopinavir, didanosine, and abacavir are associated with a significantly increased risk of myocardial infarction.[96] In a recent systematic review, Bavinger et al.[97] found increased relative risk (RR) of myocardial infarction associated with recent exposure (within the last 6 months) to abacavir (RR = 1.92; 95% CI 1.51–2.42) and protease inhibitors (RR = 2.13; 95% CI 1.06–4.28). The risk increased with each additional year of exposure to indinavir (RR = 1.11; 95% CI 1.05–1.17) and lopinavir (RR = 1.22; 95% CI 1.01–1.47). In general, evidence of an association between the use of antiretrovival therapy and cardiovascular disease from most observational studies is mixed. However, it provides sufficient stimulus for further investigation of the relationship.

1.17 UREMIC TOXINS

Chronic kidney disease (CKD) is a worldwide public health problem with increasing incidence and prevalence, poor outcomes, and high cost for care. Atherosclerotic vascular disease risk in end-stage renal disease (ESRD) is 5 to 30 times that in the general population. Many cardiovascular risk factors are more prevalent in ESRD than in the general population, which may explain some but probably not all, of the increased atherosclerotic vascular disease risks in patients with CKD.[98] Also, lower estimated glomerular filtration rate (eGFR) and higher albuminuria are risk factors for all-cause and cardiovascular mortality in high-risk populations, independent of each other and of cardiovascular risk factors.[99] The National Institute of Diabetes' HEMO Study identified ischemic heart disease as a major cause of cardiac hospitalizations and cardiac deaths.[100] In the United States, white individuals who are on dialysis experience much higher overall and cardiovascular mortality rates than black individuals despite a more favorable risk-factor profile, implying an important racial difference.[101]

The European Uremic Toxin Work Group (EUTox) produced guidelines related to identification, characterization, and biological activities of uremic toxins.[102] There are three groups of uremic toxins, divided according to their physicochemical characteristics: small water-soluble compounds (<500 Da), of which urea is a prototype; small protein-bound solutes such as phenols; and middle molecules (>500 Da) such as β-2 microglobulin. Cyanate spontaneously transformed from urea increases as renal function decreases. The potential toxin and active form of cyanate (i.e., *isocyanic acid*). carbamoylated amino acids, proteins, and other molecules may change their structure, charge, and function. These products of carbamoylation can modify the molecular activity of enzymes, cofactors, hormones, low-density lipoproteins, antibodies, receptors, and transport proteins.[103]

Asymmetric dimethylarginine (ADMA), which is significantly increased in ESRD, is the most specific endogenous compound with inhibitory effects on nitric oxide (NO) synthesis.[104] ADMA causes cerebral vasoconstriction and inhibition of acetylcholine-induced vasorelaxation, and has been implicated in the development of hypertension, adverse cardiovascular events, and mortality.[105–107] Symmetrical dimethylarginine (SDMA) is the structural isomer of the endogenous nitric oxide synthase (NOS) inhibitor ADMA. However, SMDA does not directly inhibit NOS but is a competitor of arginine transport and is eliminated through renal excretion.[108] SDMA is also involved in the inflammatory process of CKD, activating NF-κB and resulting in enhanced expression of IL-6 and TNF-α.[109] Hyperphosphatemia promotes the development and progression of secondary hyperparathyroidism and predisposes to metastatic calcification when the product of serum calcium and phosphorus (Ca × PO4) is elevated. Both of these conditions may contribute to the substantial morbidity and mortality seen in patients with ESRD.[110]

In CKD patients on hemodialysis, free p-cresol serum concentrations, mainly the derivative p-cresyl sulfate (PCS), are associated with cardiovascular disease in hemodialysis patients.[111] It is known that PCS causes endothelial dysfunction and possibly plays a role in coronary atherosclerosis.[112] Chronic kidney disease is also associated with significant increase in intracellular S-adenosylhomocysteine (AdoHcy), a potent inhibitor of methyltransferases. AdoHcy accumulation is normally prevented by its enzymatic hydrolysis to homocysteine (Hcy) and adenosine, a reversible reaction catalyzed by AdoHcy hydrolase. Furthermore, there is increased "carbonyl stress" due to nonenzymatic glycation of various serum and tissue proteins, similar to what occurs in diabetes mellitus. Advanced glycation end-products (AGEs), such as pentosidine and carboxymethyllysine, are markedly elevated in both plasma proteins and skin collagen of uremic patients.

Levels of advanced lipoxidation end-products (ALEs), such as malondialdehydelysine, are also elevated in plasma proteins in uremia. This state of carbonyl stress in uremia, propagated by AGEs and ALEs reacting with their specific receptors (RAGE or RALE), may contribute to long-term complications associated with chronic renal failure and dialysis such as dialysis-related amyloidosis and accelerated atherosclerosis.[113] Modified β-2 macroglobulin may play an important role in the formation of dialysis-related amyloid. Thus, the uremic syndrome is a complex mosaic interplay of various retention products of failing kidneys, collectively called uremic toxins. These products probably act together to promote atherosclerosis in patients with CKD when they are on renal replacement therapy such as hemodialysis. There is very little evidence, except for the presence of hyperphosphatemia, to show that uremic toxins are significantly modifiable cardiovascular risk factors, although interventions to reduce their formations might alter mortality.

1.18 EPIDEMIOLOGICAL ASPECTS OF NSAIDS AND CHEMOTHERAPEUTIC AGENTS

There is widespread use of nonsteroidal antiinflammatory drugs (NSAIDs) globally for treatment of acute pain, primary dysmenorrhea, osteoarthrosis, rheumatoid arthritis, and other rheumatological disorders.

1.18.1 Nonsteroidal Antiinflammatory Drugs

The use of most NSAIDs, including nonselective ones and cyclooxygenase-1 and cyclooxygenase-2 (COX-1 and COX-2) inhibitors, is associated with an increased risk of adverse cardiovascular events, including ischemic heart disease, heart failure, elevated blood pressure, and cardiac arrhythmias, particularly atrial fibrillation and atrial flutter. Both nonselective NSAIDs and COX inhibitors cause reduced prostaglandin I_2 (PGI_2) production by vascular endothelium with little or no inhibition of potentially thrombotic platelet thromboxane A_2 (TxA_2) production.

Rofecoxib (Vioxx), a COX-2 inhibitor, was voluntarily withdrawn from the worldwide market by its manufacturer in September 2004 because of adverse cardiovascular outcomes in the APPROVe Study. This was a multicenter, randomized, placebo-controlled double-blind study to determine the effects of three years of treatment with rofecoxib (25 mg/day), compared to a placebo, on the recurrence of neoplastic polyps of the colon in 2600 patients with history of colorectal adenoma. The increased absolute risk of myocardial infarction or stroke was approximately three thrombotic events per 400 patient-years after 18 months of treatment with rofecoxib.[114] Two years earlier, Ray et al. had performed a retrospective analysis of individuals on the Tennessee Medicaid program, of whom 24,132 were on rofecoxib and 151,728 were on nonselective NSAIDs. Users of high-dose rofecoxib (>25 mg/day) were 1.70 times more likely than nonusers to have coronary heart disease.[115]

In a case-control study of 54,475 patients aged 65 years or older, Solomon et al. constructed logistic regression models involving 10,895 cases of acute myocardial infarction (AMI) matched to four controls for patient's age, gender, month of index date of AMI, health-care use, medication use, and cardiovascular risk factors in those on rofecoxib, compared with persons taking no NSAIDs, taking celecoxib, or taking other NSAIDs. The relative risk of AMI was 1.24-fold greater in patients on rofecoxib compared with those on celecoxib and 1.14-fold greater compared to patients not on NSAIDs.[116] Also in 2004, a population-based cohort study revealed that current use of rofecoxib was associated with higher risk of congestive heart failure (CHF) compared with celecoxib or no NSAIDs.[117]

Subsequent studies of celecoxib for the prevention of adenomatous polyps also associated it with increased cardiovascular risk.[118] In a pooled analysis of adjudicated data on 7950 patients in six placebo-controlled trials comparing celecoxib with placebo, with primary end-points including cardiovascular deaths, myocardial infarctions, stroke, heart failure, and thromboembolic events, dose-dependent differential cardiovascular adverse effects of celecoxib emerged whereby 400 mg/day was associated with the lowest hazard ratio (HR 1.1); 200 mg twice daily had an intermediate hazard ratio of 1.8; and 400 mg twice daily had the highest hazard ratio of 3.1.

The hypothesis that increased cardiovascular risks associated with the use of COX-2 inhibitors is a class effect was tested by Andersohn et al. using first- and second-generation COX-2 inhibitors in a nested case-control study involving 486,378 persons in the United Kingdom General Practice database, 2000 through 2004.[119] Current use of etoricoxib was associated with 2.09-fold risk of acute myocardial infarction compared with no NSAIDs during the prior year. Valdecoxib was associated with a 4.6-fold increased relative risk of AMI. Although the relative risk of AMI associated with current use of all COX-2 inhibitors was dose-dependent, there might be subtle class differences in the safety of NSAIDs. For example, a large network analysis of bibliographic databases, conference proceedings, study registers, FDA websites, reference lists, and relevant articles done by Trelle et al. in 2011, involving 116,429 patients in 31 trials, demonstrated relative cardiovascular safety of naproxen among nonselective NSAIDs compared with COX-2 inhibitors.[120] Lumiracoxib was associated with a rate ratio of 2.0 of acute myocardial infarction, while diclofenac and etoricoxib were each associated with a 4-fold increased relative risk of cardiovascular death. The risk of death and cardiovascular morbidity is even greater for patients with preexisting heart failure who are treated with NSAIDs.[121]

The arrhythmogenic potential of NSAIDs was revealed by Schmidt et al. in a population-based case-control study of 32,602 patients with first in-patient or out-patient diagnosis of AF or atrial flutter in Denmark in 2011. Of these, 2925 patients (9%) and 21,871 controls (7%) were current users of either nonselective NSAIDs or COX-2 inhibitors. Compared with nonusers, the unadjusted incidence rate ratio associating current drug use with AF or flutter was 1.33 for nonselective NSAIDs and 1.50 for COX-2 inhibitors. Incidence rate ratios were somewhat lower (1.17 for nonselective NSAIDs; 1.27 for COX-2 inhibitors) when adjusted for age, sex, and preexisting risk factors for AF or flutter.[122]

Thus, the safety of both nonselective NSAIDs and COX inhibitors (includes COX-1 and COX-2 inhibitors) is shrouded in debates regarding the risk of adverse cardiovascular outcome, leading to nonapproval or withdrawal of some newer agents. All COX-2 inhibitors appear to increase the risk of ischemic heart disease in a dose-dependent manner. However, the magnitude of risk may differ between agents and is worsened by preexisting cardiac disease. Therefore, the use of these drugs should be avoided by patients with established heart failure because of increased risk of death or cardiovascular morbidity.

1.18.2 Chemotherapeutic Agents

Anthracyclines (e.g., doxorubicin, daunorubicin, epirubicin, idarubicin) are an important group of chemotherapeutic agents for the treatment of solid tumors, lymphoma, and leukemias. These drugs belong to a class of antineoplastic antibiotics that interfere with cell replication by acting on deoxyribonucleic acid (DNA) at virtually every phase of the cell cycle. Cardiotoxicity of anthracyclines is among the most frequent and serious adverse effect of cancer treatment in both children and adults. There are two broad categories of anthracycline-related cardiotoxicity: cardiomyopathy and arrhythmias.

Risk Factors for Anthracycline-Induced Cardiotoxicity

A cumulative dose of anthracylines is the most significant risk factor for cardiac dysfunction. In one study by Bristow et al.[123] of 3941 patients treated with anthracyclines, symptomatic heart failure occurred in 88 patients, ranging from 0.14% in those who received <400 mg/m^2 to 7% of those who received 550 mg/m^2 and 18% of those who received >700 mg/m^2. At a cumulative dose of 550 mg/m^2, an estimated 20% would experience doxorubicin-related congestive failure. Children and the elderly (>65 years) are the most vulnerable to anthracycline cardiotoxicity at even lower cumulative doses. Among nearly 4000 individuals >65 years with diffuse large cell B-cell lymphoma who were treated with doxorubin, 29% exhibited increased risk of heart failure.[124]

Cardiotoxicity may manifest subclinically or clinically with cardiac arrhythmias, particularly atrial fibrillations (2–10%) or premature ventricular contractions, but ventricular tachycardias or ventricular fibrillations are rare. Three clinical patterns of anthracycline cardiotoxicity have been described: acute toxicity, subacute (early-onset progressive) cardiomyopathy, and chronic (late-onset progressive) cardiomyopathy. Long-term cardiac dysfunctions have been reported in about 5% of doxorubicin-treated children receiving cumulative doses >300 mg/m^2 as late as 15 years after treatment completion.[125,126] Doxorubicin-related CHF occurred with greater frequency in older patients (>65) than in those <65 years after cumulative dosing of 400 mg/m^2, as noted in a retrospective analysis of three trials done by Swain et al.[127]

Other risk factors for anthracycline-related cardiotoxicity include underlying cardiac disease, hypertension, and diabetes mellitus. Although continuous doxorubicin infusion appears to be less toxic than bolus doses in adults, the mode of parental administration of doxorubicin did not affect cardiac outcome in children with leukemia.[128,129] Further, females are more vulnerable than males at comparative doses of anthracyclines. Concomitant irradiation and administration of other potentially cardiotoxic chemotherapeutic agents may result in synergistic cardiotoxicity, particularly in women with left breast cancer. Structural analogs of doxorubicin (i.e., epirubicin and mitoxantrone) as well as liposome-encapsulated doxorubin are believed to be less cardiotoxic and are generally preferred for the treatment of solid tumors. However, a systematic review and pooled meta-analysis of 55 randomized controlled trials failed to show sufficiently robust evidence to support clear recommendations regarding various anthracycline treatment regimens or routine use of cardioprotective agents or liposomal formulations.[130]

1.18.3 Trastubumab-Associated Cardiotoxicity

Moja et al. assessed evidence of the efficacy and the safety of concurrent or sequential therapy with trastuzumab in 11,991 women with human epidermal growth factor receptor-2 positive (HERS2+) early breast cancer. Although trastuzumab-containing regimens improved overall survival and disease-free survival, trastuzumab significantly increased the risk of congestive heart failure 5-fold, and there was decline in left ventricular ejection fraction of nearly 2-fold.[131] A separate population-based, retrospective study involved 12,500 women, mean age 60 years (range 22–99) with incident, invasive breast cancer. One-third were treated with either anthracycline alone (29.6%), trastuzumab alone (0.9%), or the combination of anthracycline and trastuzumab (3.5%); 19.5% were being treated with other chemotherapeutic agents and 46% received no chemotherapy. The risk of heart failure or cardiomyopathy (HF/CM) increased by 40% in the group on anthracycline alone compared to women on no chemotherapy, while the use of trastuzumab alone increased the risk of HF/CM 4-fold and the combination of the two drugs resulted in a 7-fold increased risk in this group of relatively younger women with fewer comorbidities.[132]

The study by Chen et al. involved 45,537 relatively older women, mean age 76.2 years (range 67—94) with early-stage breast cancer.[133] Adjusted three-year HF/CM incidence rates were significantly higher for patients receiving trastuzumab (32.1/100) and the combination of anthracycline and trastuzumab (41.9/100) compared to no adjuvant therapy (18/100). Pooled data analysis of 3698 patients from 44 clinical studies showed low levels of cardiotoxicity for lapatinib—that is, another oral HERS2-inhibitor/tyrosine kinase epidermal growth factor (ErbB1) inhibitor. Most adverse effects were asymptomatic or reversible reductions in left ventricular ejection fractions.[134]

1.18.4 Cardiovascular Toxicity of Molecularly Targeted Antiangiogenic Agents

Vascular endothelial growth factor receptor (VEGFR) tyrosine kinase inhibitors (TKIs) have become cornerstones in the treatment of many solid cancers. More than 16 TKIs have been approved for use in oncology since the end of September 2012. The cardiovascular safety of these drugs pose major public health concerns. In a systematic review and meta-analysis including 16 randomized controlled trials, Ranpura et al. reported fatal adverse effects in 2.5% of 10,217 patients with various advanced solid tumors by the addition of bevacizumab compared to those treated with chemotherapy alone.[135] Four specific cardiovascular aspects of TKIs are addressed in the following subsection: drug-related systemic arterial hypertension; arterial and venous thromboembolism; left ventricular dysfunction; and induction of cardiac arrhythmias, particularly prolongation of QTc interval.

Hypertension in Cancer Patients Treated with Antiangiogenic Agents

Analyses of various randomized controlled trials and systematic reviews of the literature published in English indicate an overall incidence of induced hypertension of 8.0 to 36.0%. In China, a meta-analysis by An et al.[136] of nearly 13,000 cancer patients treated with bevacizumab (2.5–5.0 mg/week) showed hypertension incidence of 8.0 (95% CI 6.0–10) with a relative risk of 5.38 (95% CI 3.63–7.97). An et al. showed that the relative risks of bevacizumab-related hypertension in patients with renal cell carcinoma (RCC) was 13.77 (95% CI 2.28–83.15) while women with breast cancer had a relative risk of 18.83 (95% CI 1.123–292). The risk of developing significantly raised blood pressure was nearly twice as great in patients treated with 5 mg/kg per week of bevacizumab compared to those receiving 2.5 mg/kg per week.

Earlier analyses of nearly 10,000 U.S. cancer patients with RCC or other solid tumors showed an overall hypertension incidence of 21.6% and 23.4% associated with the use of sunitinib and sorafenib, respectively.[137,138] High-grade hypertension occurred in 6.8% and 5.7% in the two studies. The highest incidence of drug-related hypertension was more recently reported by Qi et al. in a meta-analysis of 1651 cancer patients with RCC and various non-RCC solid tumors.[139] Those treated with 800 mg per day of pazopanib had an overall hypertension incidence of 35.9%, while high-grade hypertension developed in 6.5%, giving relative risks of approximately 5- and 3-fold for all grades of hypertension and high-grade hypertension, respectively. The use of sorafanib and sunitinib was not associated with any significantly increased risk of hypertension. Axitinib is another second-line treatment option for advanced renal cell carcinoma.

Arterial and Venous Thromboembolic Complications of Treatment with Antiangiogenic Agents

There are various reports of drug-related hemostatic complications. In 2007, Scappaticci et al. reported a 2-fold increased risk of arterial thromboembolism (ATE) in a pooled analysis of 1745 patients with metastatic cancers treated with bevacizumab in the United States. No significant risk of venous thromboembolic events (VTEs) was observed in this study.[140] A year later, Nalluri et al. reported the incidence of all grades of VTEs in 11.9% of about 8000 New York cancer patients with a variety of advanced solid tumors treated with bevacizumab compared to controls.[141] Later studies appear to confirm the risk of ATE and myocardial ischemia but there is controversy regarding drug exposure–adjusted risk of venous thromboembolism.

Schutz et al. showed a nearly 50% increase relative of arterial thromboembolism in a wide literature search and meta-analysis, including 20 randomized controlled studies involving 13,026 cancer patients treated with bevacizumab.[142] In another analysis of 10,255 patients with RCC and non-RCC who were treated with sunitinib or sorafenib, Choueiri et al. showed a 1.4% incidence of ATE with a 3-fold increased relative risk of systemic hemostatic complications compared to controls.[143] Three recent reports suggest no statistically significant increases in unadjusted or exposure-adjusted incidences of all grades of venous thromboembolism for TKIs compared to controls.[144–146] These authors argue that the risk of VTEs is driven predominantly by tumor types, host factors, and concomitant use of anticancer agents.

Thalidomide and its analogs, lenalidomide and pomalidomide, are other agents with antiangiogenic properties. Given in combination with multiagent chemotherapy, particularly anthracycline-containing regimens and dexamethasone, thalidomide and lenalidomide are associated with a significantly increased risk of deep vein thrombosis (DVT) in patients with newly diagnosed, relapsed, or refractory multiple myeloma (MM).[147] Pooled venous thromboembolism incidence was 2.7 (95% CI 1.1–4.3) for thalidomide whereas the combination of lenalidomide and dexamethasone was associated with a risk of VTE of 0.8 (95% CI 0.07–2.0) per 100 patient-cycles and 0.7 (95% CI 0.4–0.9) per 100 patient-cycles in those with newly diagnosed and previously treated multiple myeloma, respectively.

The individual risk factors for thrombosis associated with thalidomide- or lenalidomide-based therapy include age, history of VTEs, central venous catheter, infections, diabetes, cardiac disease, immobilization, surgery, and inherited thrombophilia. Myeloma-specific related risk factors include hyperviscosity. Thalidomide may be procoagulant by altering the expression of the protease activated receptor-1 (PAR-1) on injured endothelium, resulting in endothelial dysfunction. Also, thalidomide protects endothelial cells from doxorubicin-induced apoptosis but alters cell morphology. Patients with newly diagnosed MM treated with thalidomide in combination with dexamethasone have a VTE risk of 4.1 (95% CI 2.8–5.9) per 100 patient-cycles. Deep vein thrombosis and pulmonary embolism occurred in only 0.53% of 7764 patients during the first two years of commercial use of lenalidomide in the United States.[148]

The immunomodulatory drugs lenalidomide and pomalidomide increase risk for venous thromboembolism by mechanisms that are unknown. The risk of DVT was highest in patients with acquired protein C (APC) resistance to thalidomide.[149] There is a paucity of data regarding the risk of VTE among patients with multiple myeloma who are treated with the thalidomide-analog pomalidomide. Venous thromboembolism rates were $\leq 5\%$ in phase studies.[150]

Left Ventricular Dysfunction in Patients Treated with Antiangiogenic Agents

Declines in left ventricular function can be seen in patients treated with VEGFR tyrosine kinase inhibitors, particularly those treated with sunitinib and pazopanib. Myocardial ischemia has been reported in patients treated with sorafenib and vandetanib. Chu et al. noted their findings from a small retrospective study of 75 patients with imatinib-resistant metastatic gastrointestinal stromal tumors (GIST), who were enrolled in a phase I/II efficacy trial of sunitinib. Ten of 36 patients had absolute reductions in left ventricular ejection fractions (LVEF) of at least 10% and 7 had a 15% reduction in LVEF.[151] The authors speculated that left ventricular dysfunction may be caused by direct cardiomyocyte toxicity, exacerbated by hypertension. Others later proposed mitochondrial injury and cardiomyocyte apoptosis.[152] In a meta-analysis of nearly 7000 patients with RCC or GIST done by Richards et al. in Boston, the overall incidence of congestive heart failure was 4.1% and high-grade CHF incidence was 1.5% in sunitinib-treated patients compared to those given placebo.[153] The overall relative risk of CHF and high-grade CHF in sunitinib-treated patients compared with placebo-treated patients increased nearly 2- and 3-fold, respectively.

Prolongation of QTc Interval in Cancer Patients Treated with Antiangiogenic Agents

The propensity of tyrosine kinase inhibitors to induce QT-interval prolongation has been persuasively demonstrated. A systematic review and meta-analysis involving nine trials by Zang et al. of 2188 patients with nonthyroid cancer treated with vandetanib (300 mg/day) showed an all-grade prolongation of corrected QT (QTc) interval incidence of 16.4% and a high-grade QTc prolongation incidence of 3.7%. In patients with thyroid cancers, all-grade and high-grade incidences of QTc prolongation were 18% and 12%, respectively.[154] Treatment durations and different cancer categories may affect the risk of developing prolonged QTc interval. Careful selection of patients for therapy with angiogenesis inhibitors based on performance status, blood pressure level, absence of serious cardiovascular disease in the past year, close monitoring, and prompt intervention may alleviate the risk of toxicity posed by these agents. Engagement of a multidisciplinary specialized team including oncologists, cardiologists, and physicians may effectively mitigate potential cardiovascular toxicities by VEGFR/TKIs.

1.19 HEAVY METAL POISONING AND CARDIOVASCULAR DISEASE

Due to the fact that traditional risk factors cannot entirely account for all the morbidity and mortality associated with cardiovascular diseases, contributory roles of environmental, dietary, and lifestyle factors must be considered. The potential association between CVD and chronic heavy metal exposure, such as to arsenic, lead, cadmium, and mercury, remains ill-defined. The mechanism through which heavy metals act to increase cardiovascular risk factors remain largely unclear, although impaired antioxidant metabolism and oxidative stress may play a role. Altered gene expression by environmental influence, particularly dietary components over gene regulation, may also be responsible for heavy metal toxicity. The effects of confounding variables as potential mediators, moderators, direct causes, or other factors in the complex mechanistic pathways must also be considered. These pathways include individual characteristics such as age, sex, ethnicity, and race; behavioral risk factors including dietary habits; biological factors, particularly genetics; social factors; timing and dose (i.e., acute and cumulative) of heavy metals; comorbid health conditions (e.g., diabetes, heart disease, hypertension, renal dysfunction); and cognitive decline and other biological markers predictive of disease.

Heavy metals, commonly defined as those having a specific density of more than 5 g/cm^3 are widely distributed in the Earth's crust, but they are present at very low concentrations in the body. Their presence, even in trace amounts in the atmosphere, soil, and water, can cause serious problems to all organisms. Their impact on human health is principally through occupational exposure, environmental contamination, and accumulation in food, particularly in vegetables grown on contaminated soil. Environmental dissemination results from extensive use of heavy metals and their compounds in industry and consumer products. Because metals are not biodegradable, they can persist in the environment and produce a variety of adverse effects. Although arsenic, cadmium, mercury, and lead are the most common causes of heavy metal-related disease, quite often populations are concomitantly exposed to mixtures of heavy metals. Heavy metals generate many of their adverse health effects through the formation of free radicals, resulting in DNA damage, lipid peroxidation, and depletion of protein sulfhydryl groups such as glutathione.[155]

Insufficient epidemiological data, variations in population intake of dietary antioxidants, and genetic differences complicate the determination of threshold values for heavy metal toxicity. While the harmful effects of high levels of exposure to heavy metals are understandable, there are mounting concerns about the possibility that subtle exposures such as occur in the cosmetic industry (e.g., in lipsticks, eye makeup, talcum powder, skin-lightening creams) may also have chronic adverse effects.[156,157] Other hidden sources may include ethnic folk remedies, toys, candies, and spices, especially for children.[158] Tobacco plants have a special ability to absorb cadmium from soil and to accumulate it in the leaf.[159] Heavy metals have no known metabolic functions; they are poorly absorbed into the body and are slowly and poorly excreted, thus resulting in a harmful accumulation that causes organ damage. Metallothioneins, small proteins rich in cysteine residues, play an important role in binding, distribution, and storage of heavy metals in the body. The cardiovascular effects of heavy metal exposure remain poorly defined despite significant ecological and epidemiological associations.

1.19.1 Arsenic

Elemental arsenic is ubiquitous in nature and human exposure occurs through environmental, medicinal, occupational, and marine-fish sources. Chronic arsenic intoxication poses serious public health problems in Argentina, Bangladesh, Chile, China, Hungary, India, and Japan.[160] Arsenic has been documented as the major risk factor of peripheral vascular disease in endemic areas of arseniasis, such as Taiwan, and in several other countries. Other arsenic-induced cardiovascular effects include hypertension, diabetes mellitus, atherosclerosis, coronary heart disease, and stroke in a dose-dependent manner.[161–163] Long-term high-level arsenic exposures in drinking water have also been associated with hypertension in Bangladesh,[164] type 2 diabetes mellitus in Taiwan,[165] and increased age-adjusted mortality from ischemic heart disease in Chile.[166] However, the strength of association between arsenic exposure and cardiovascular disease has been inadequate to firmly establish a causal inference due to methodological limitations of ecological studies and inconclusive occupational population studies. A few U.S. epidemiologic studies have reported the association of arsenic exposure with cardiovascular endpoints at low to moderate chronic levels in drinking water.[167,168] Suggested mechanisms by which arsenic exerts its toxic effect include impairment of cellular respiration by inhibition of several enzymes involved with carbohydrate metabolism, uncoupling of oxidative phosphorylation, a range of effects related to oxidative stress, and vascular inflammation.[169,170]

1.19.2 Lead

Lead exposures occur in various parts of the world through several sources including foods contaminated by soil lead, agricultural technology, food-can soldering; drinking water from lead pipes; lead paints; fuel; ceramic glazing; industrial pollution; and traditional remedies such as *litargirio* in the Dominican Republic, *ba-baw-san* in China, *ghasard* in India, and *daw tway* in Thailand.[171] Health risks associated with lead exposure prevail across a wide range of doses, with no clearly defined safe level. Although public health concerns have related the risks of cardiovascular events, such as elevated blood pressure, coronary heart disease, stroke, and peripheral arterial disease, to the effects of environmental lead exposure, the exact role of lead in CVD is still incompletely understood.[172] A meta-analysis by Nawrot et al. of nearly 60,000 individuals in 19 surveys from general populations and 12 studies in occupationally lead-exposed groups, adjusted for confounding factors (e.g., age, body mass index, use of alcohol or medications), revealed only weak association between blood pressure (BP) and blood lead levels in men and women. A 2-fold increase in blood lead was associated with a 1.0 and a 0.6 mmHg rise in systolic and diastolic blood pressures, respectively.[173]

Lead appears to act in a dose-graded manner at multiple sites within the cardiovascular system, resulting in oxidative stress, reduced availability of nitric oxide (NO), inflammation, and deranged lipid metabolism, leading to atherosclerosis and death.[174] However, epidemiological studies of association between lead exposure and cardiovascular disease suffer from similar limitations due to toxicokinetic factors, confounding effects of coexposure to other heavy metals, comorbidity, and differences among study populations, as mentioned earlier for arsenic poisoning. Nutrition is an important modifier of susceptibility to lead toxicity. Essential elements, such as calcium, zinc, iron, selenium, and antioxidant vitamins, have been shown to counteract the toxic effects of lead.[175]

1.19.3 Cadmium

Human exposure to cadmium results from widespread natural environmental contamination or industrial use of the heavy metal in countries such as Belgium, China, Japan, Sweden, and the United Kingdom. Routes of human cadmium exposure include foods (e.g., leafy vegetables, grains, organ meats, crustaceans), drinking water, inhalation of polluted air or tobacco smoke, and occupational exposures in cadmium-processing industries. Although the European Commission has set a threshold value for safe dietary cadmium exposure level at below 2.5 g/kg body weight per week, cadmium may exert effects on the cardiovascular system at extremely low exposure levels. Chronic cadmium exposure is associated with hypertension, diabetes, and kidney disease.[176,177]

Epidemiological studies in the United States have shown an association between cadmium exposure and increased mortality in men from all causes, cancers, and cardiovascular diseases but not in women.[178] In the Third National Health and Nutrition Examination Survey (NHANES III), high levels of urinary cadmium have also been associated with peripheral arterial disease.[179] The exact mechanisms by which cadmium exerts its adverse cardiovascular effects are poorly understood. Cadmium toxicity might result from partial agonism of calcium channels, direct vasoconstriction, or inhibition of vasodilators such as nitric oxide, resulting in cardiac dysfunction, atherosclerosis, and myocardial infarction.[180,181] Other plausible mechanisms of cadmium toxicity include oxidative stress, depletion of glutathione, and alteration of sulfhydryl homeostasis.[155] Cadmium affects cell cycle progression, proliferation, differentiation, DNA replication and repair, as well as apoptotic pathways.

1.19.4 Mercury

Environmental pollution with mercury, especially in water systems (e.g., lakes, rivers, oceans) and aquatic food chains such as fatty fish, are quite common. Mercury may exist in metallic (elemental), organic, or inorganic forms. Elemental mercury is used in thermometers, sphygmomanometers, and dental amalgam fillings. Methylmercury is the organic mercury that is found mainly in fish, while ethylmercury may be used in some vaccines. Evaporation of mercury vapor is responsible for the global cycle of volatile mercury, which repollutes the ambient air and environment. Epidemiological studies have confirmed the association of high mercury concentration with increased atherosclerosis and risk of cardiovascular events, particularly in men.[182–184] A small nested case-control study in northern Sweden involving 78 index cases of myocardial infarction and 156 controls showed an inverse relationship between risk of MI and the biomarkers of fish intake.[185] The limitations of this study included small sample size, the element of high fish consumption, and the use of erythrocyte-mercury levels rather than human hair or nail clippings as markers of long-term mercury exposure, given the rapid red cell turnover. However, incriminating evidence for the exact adverse effects of mercury is complicated to derive and highly dependent on other metabolic conditions. Mercury can lead to oxidative stress, deletion of sulfhydryl groups, altered mitochondrial function, and apoptosis.[186]

1.20 CONCLUSION

Worldwide, increasing use of tobacco and alcohol with environmental exposure has led to an increase in the incidence of cardiovascular toxicity from such toxins. Figures 1.2 and 1.3 depict the use of tobacco and alcohol among adults worldwide and are adapted from World Health Organization publications.[187,188] Moreover, the toxicities are acquired from drugs and toxins from biological and nonbiological materials. The clinical manifestations of each toxin are discussed in detail in the following chapters.

REFERENCES

1. Loewenson R. Globalization and occupational health: a perspective from southern Africa. *Bull World Health Organ* 2001;**79**(9):863–8.
2. Broughton E. The Bhopal disaster and its aftermath: a review. *Environ Health* 2005;**4**(1):6.
3. Kaleta D, Koziel A, Miskiewicz P. *MPOWER-strategia na rzecz walki ze swiatowa epidema uzywania tytoniu* [MPOWER—strategy for fighting the global tobacco epidemic.] *Med Pr* 2009;**60**(2):145–9.
4. van Uum SH. Liquorice and hypertension. *Neth J Med* 2005;**63**(4):119–20.
5. Armanini D, Calo L, Semplicini A. Pseudohyperaldosteronism: pathogenetic mechanisms. *Crit Rev Clin Lab Sci* 2003;**40**(3):295–335.

6. Sigurjonsdottir HA, Franzson L, Manhem K, et al. Liquorice-induced rise in blood pressure: a linear dose–response relationship. *J Hum Hypertens* 2001;**15**(8):549–52.
7. Bessen HA. Therapeutic and toxic effects of digitalis: William Withering, 1785. *J Emerg Med* 1986;**4**(3):243–8.
8. Bronstein AC, Spyker DA, Cantilena Jr LR, et al. 2006 Annual report of the American association of poison control centers' National Poison Data System (NPDS). *Clin Toxicol* 2007;**45**(8):815–917.
9. Eddleston M, Ariaratnam CA, Sjostrom L, et al. Acute yellow oleander (*Thevetia peruviana*) poisoning: cardiac arrhythmias, electrolyte disturbances, and serum cardiac glycoside concentrations on presentation to hospital. *Heart* 2000;**83**(3):301–6.
10. Cheung K, Hinds JA, Duffy P. Detection of poisoning by plant-origin cardiac glycoside with the Abbott TDx analyzer. *Clin Chem* 1989;**35**(2):295–7.
11. Cardellina JH, 2nd. Challenges and opportunities confronting the botanical dietary supplement industry. *J Nat Prod* 2002;**65**(7):1073–84.
12. Whittle K, Gallacher S. Marine toxins. *Br Med Bull* 2000;**56**(1):236–53.
13. Holt RJ, Miro G, Del Valle A. An analysis of poison control center reports of ciguatera toxicity in Puerto Rico for one year. *J Toxicol Clin Toxicol* 1984;**22**(2):177–85.
14. Lewis RJ. Ciguatoxins are potent ichthyotoxins. *Toxicon* 1992;**30**(2):207–11.
15. Habermehl GG, Krebs HC, Rasoanaivo P, et al. Severe ciguatera poisoning in Madagascar: a case report. *Toxicon* 1994;**32**(12):1539–42.
16. Huynh TT, Seymour J, Pereira P, et al. Severity of Irukandji syndrome and nematocyst identification from skin scrapings. *Med J Aust* 2003;**178**(1):38–41.
17. Macrokanis CJ, Hall NL, Mein JK. Irukandji syndrome in northern Western Australia: an emerging health problem. *Med J Aust* 2004;**181**(11–12):699–702.
18. Currie BJ, Jacups SP. Prospective study of *Chironex fleckeri* and other box jellyfish stings in the "Top End" of Australia's Northern Territory. *Med J Aust* 2005;**183**(11–12):631–6.
19. Currie BJ. Marine antivenoms. *J Toxicol Clin Toxicol* 2003;**41**(3):301–8.
20. Bailey PM, Little M, Jelinek GA, et al. Jellyfish envenoming syndromes: unknown toxic mechanisms and unproven therapies. *Med J Aust* 2003;**178**(1):34–7.
21. Berne S, Krizaj I, Pohleven F, et al. Pleurotus and Agrocybe hemolysins, new proteins hypothetically involved in fungal fruiting. *Biochim Biophys Acta* 2002;**1570**(3):153–9.
22. Macek P, Lebez D. Isolation and characterization of three lethal and hemolytic toxins from the sea anemone *Actinia equina* L. *Toxicon* 1988;**26**(5):441–51.
23. Nielsen KJ, Schroeder T, Lewis R. Structure-activity relationships of omega-conotoxins at N-type voltage-sensitive calcium channels. *J Mol Recognit* 2000;**13**(2):55–70.
24. Chippaux JP. Snake-bites: appraisal of the global situation. *Bull World Health Org.* 1998;**76**(5):515–24.
25. Pierini SV, Warrell DA, de Paulo A, et al. High incidence of bites and stings by snakes and other animals among rubber tappers and Amazonian Indians of the Jurua Valley, Acre State, Brazil. *Toxicon* 1996;**34**(2):225–36.
26. Harrison RA, Hargreaves A, Wagstaff SC, et al. Snake envenoming: a disease of poverty. *PLoS Negl Trop Dis* 2009;**3**(12):e569.
27. Kasturiratne A, Wickremasinghe AR, de Silva N, et al. The global burden of snakebite: a literature analysis and modelling based on regional estimates of envenoming and deaths. *PLoS Med* 2008;**5**(11):e218.
28. Kunkel DB, Curry SC, Vance MV, et al. Reptile envenomations. *J Toxicol Clin Toxicol* 1983;**21**(4–5):503–26.

29. Pozio E. World distribution of *Trichinella* spp. infections in animals and humans. *Vet Parasitol* 2007;**149**(1–2):3–21.
30. Marincu I, Neghina AM, Calma CL, et al. New foci of trichinellosis in western Romania, 2011. *Acta Trop* 2012;**121**(1):47–9.
31. Murrell KD, Pozio E. Trichinellosis: the zoonosis that won't go quietly. *Int J Parasitol* 2000;**30**(12–13):1339–49.
32. Pozio E, La Rosa G. *Trichinella murrelli n.* sp: etiological agent of sylvatic trichinellosis in temperate areas of North America. *J Parasitol* 2000;**86**(1):134–9.
33. Pozio E, Owen IL, La Rosa G, et al. *Trichinella papuae n.* sp. (Nematoda), a new non-encapsulated species from domestic and sylvatic swine of Papua, New Guinea. *Int J Parasitol* 1999;**29**(11):1825–39.
34. Jongwutiwes S, Chantachum N, Kraivichian P, et al. First outbreak of human trichinellosis caused by *Trichinella pseudospiralis*. *Clin Infect Dis* 1998;**26**(1):111–5.
35. Barrett-Connor E, Davis CF, Hamburger RN, et al. An epidemic of trichinosis after ingestion of wild pig in Hawaii. *J Infect Dis* 1976;**133**(4):473–7.
36. Bura MW, Willett WC. An outbreak of trichinosis in Tanzania. *East Afr Med J* 1977;**54**(4):185–93.
37. Neghina R, Neghina AM, Marincu I. Reviews on trichinellosis (III): cardiovascular involvement. *Foodborne Pathog Dis* 2011;**8**(8):853–60.
38. Puljiz I, Beus A, Kuzman I, et al. Electrocardiographic changes and myocarditis in trichinellosis: a retrospective study of 154 patients. *Ann Trop Med Parasitol* 2005;**99**(4):403–11.
39. Isbister GK, Graudins A, White J, et al. Antivenom treatment in arachnidism. *J Toxicol Clin Toxicol* 2003;**41**(3):291–300.
40. Diaz JH. The global epidemiology, syndromic classification, management, and prevention of spider bites. *Am J Trop Med Hyg* 2004;**71**(2):239–50.
41. Levine M, Canning J, Chase R, et al. Cardiomyopathy following latrodectus envenomation. *West J Emerg Med* 2010;**11**(5):521–3.
42. La Grange MA. Pulmonary oedema from a widow spider bite. *S Afr Med J* 1990;**77**(2):110.
43. Bucaretchi F, Mello SM, Vieira RJ, et al. Systemic envenomation caused by the wandering spider *Phoneutria nigriventer*, with quantification of circulating venom. *Clin Toxicol* 2008;**46**(9):885–9.
44. Gueron M, Ilia R, Sofer S. The cardiovascular system after scorpion envenomation. A review. *J Toxicol Clin Toxicol* 1992;**30**(2):245–58.
45. Bawaskar HS, Bawaskar PH. Stings by red scorpions (*Buthotus tamulus*) in Maharashtra State, India: a clinical study. *Trans R Soc Trop Med Hyg* 1989;**83**(6):858–60.
46. Gueron M, Yaron R. Cardiovascular manifestations of severe scorpion sting. Clinicopathologic correlations. *Chest* 1970;**57**(2):156–62.
47. Sofer S, Gueron M. Vasodilators and hypertensive encephalopathy following scorpion envenomation in children. *Chest* 1990;**97**(1):118–20.
48. Poon-King T. Myocarditis from scorpion stings. *Br Med J* 1963;**1**(5327):374–7.
49. Curry SC, Vance MV, Ryan PJ, et al. Envenomation by the scorpion *Centruroides sculpturatus*. *J Toxicol Clin Toxicol* 1983;**21**(4–5):417–49.
50. Gueron M, Adolph RJ, Grupp IL, et al. Hemodynamic and myocardial consequences of scorpion venom. *Am J Cardiol* 1980;**45**(5):979–86.
51. O'Neill MS, Jerrett M, Kawachi I, et al. Health, wealth, and air pollution: advancing theory and methods. *Environ Health Perspect* 2003;**111**(16):1861–70.

52. Gwynn RC, Thurston GD. The burden of air pollution: impacts among racial minorities. *Environ Health Perspect* 2001;**109**(Suppl. 4):501–6.
53. Brook RD, Franklin B, Cascio W, et al. Air pollution and cardiovascular disease: a statement for healthcare professionals from the expert panel on population and prevention science of the American Heart Association. *Circulation* 2004;**109**(21):2655–71.
54. Mustafic H, Jabre P, Caussin C, et al. Main air pollutants and myocardial infarction: a systematic review and meta-analysis. *JAMA* 2012;**307**(7):713–21.
55. Samet JM, Dominici F, Curriero FC, et al. Fine particulate air pollution and mortality in 20 U.S. cities, 1987–1994. *N Engl J Med* 2000;**343**(24):1742–9.
56. Forastiere F, Stafoggia M, Picciotto S, et al. A case-crossover analysis of out-of-hospital coronary deaths and air pollution in Rome, Italy. *Am J Respir Crit Care Med* 2005;**172**(12):1549–55.
57. Pope 3rd CA, Muhlestein JB, May HT, et al. Ischemic heart disease events triggered by short-term exposure to fine particulate air pollution. *Circulation* 2006;**114**(23):2443–8.
58. Miller KA, Siscovick DS, Sheppard L, et al. Long-term exposure to air pollution and incidence of cardiovascular events in women. *N Engl J Med* 2007;**356**(5):447–58.
59. Clancy L, Goodman P, Sinclair H, et al. Effect of air-pollution control on death rates in Dublin, Ireland: an intervention study. *Lancet* 2002;**360**(9341):1210–4.
60. Jerrett M, Burnett RT, Pope 3rd CA, et al. Long-term ozone exposure and mortality. *N Engl J Med* 2009;**360**(11):1085–95.
61. Ruidavets JB, Cournot M, Cassadou S, et al. Ozone air pollution is associated with acute myocardial infarction. *Circulation* 2005;**111**(5):563–9.
62. Pope 3rd CA. Epidemiology of fine particulate air pollution and human health: biologic mechanisms and who's at risk? *Environ Health Perspect* 2000;**108**(Suppl 4):713–23.
63. Tonne C, Wilkinson P. Long-term exposure to air pollution is associated with survival following acute coronary syndrome. *Eur Heart J* 2013;**34**(17):1306–11.
64. Peters A, Liu E, Verrier RL, et al. Air pollution and incidence of cardiac arrhythmia. *Epidemiol* 2000;**11**(1):11–7.
65. Peters A, Dockery DW, Muller JE, et al. Increased particulate air pollution and the triggering of myocardial infarction. *Circulation* 2001;**103**(23):2810–5.
66. Utell MJ, Frampton MW, Zareba W, et al. Cardiovascular effects associated with air pollution: potential mechanisms and methods of testing. *Inhal Toxicol* 2002;**14**(12):1231–47.
67. Eyer P. The role of oximes in the management of organophosphorus pesticide poisoning. *Toxicol Rev* 2003;**22**(3):165–90.
68. Kamanyire R, Karalliedde L. Organophosphate toxicity and occupational exposure. *Occup Med (Chic Ill)* 2004;**54**(2):69–75.
69. Wang MH, Tseng CD, Bair SY. Q-T interval prolongation and pleomorphic ventricular tachyarrhythmia ("Torsade de pointes") in organophosphate poisoning: report of a case. *Hum Exp Toxicol* 1998;**17**(10):587–90.
70. London L, Flisher AJ, Wesseling C, et al. Suicide and exposure to organophosphate insecticides: cause or effect? *Am J Ind Med* 2005;**47**(4):308–21.
71. White NC, Litovitz T, White MK, et al. The impact of bittering agents on suicidal ingestions of antifreeze. *Clin Toxicol* 2008;**46**(6):507–14.
72. White NC, Litovitz T, Benson BE, et al. The impact of bittering agents on pediatric ingestions of antifreeze. *Clin Pediatr (Phila)* 2009;**48**(9):913–21.
73. Karlson-Stiber C, Persson H. Ethylene glycol poisoning: experiences from an epidemic in Sweden. *J Toxicol Clin Toxicol* 1992;**30**(4):565–74.

74. Watson WA, Litovitz TL, Belson MG, et al. The Toxic Exposure Surveillance System (TESS): risk assessment and real-time toxicovigilance across United States poison centers. *Toxicol Appl Pharmacol* 2005;**207**(2 Supp. 1):604–10.
75. Warner MA, Harper JV. Cardiac dysrhythmias associated with chemical peeling with phenol. *Anesthesiol* 1985;**62**(3):366–7.
76. Brady Jr. WJ, Stremski E, Eljaiek L, et al. Freon inhalational abuse presenting with ventricular fibrillation. *Am J Emerg Med* 1994;**12**(5):533–6.
77. Hicks R, Caldas LQ, Dare PR, et al. Cardiotoxic and bronchoconstrictor effects of industrial metal fumes containing barium. *Arch Toxicol Suppl* 1986;**9**:416–20.
78. World Health Organization. Report on the Global Tobacco Epidemic, 2011: Warning about the dangers of tobacco. Available from: <www.who.int/tobacco/global_report/2011/en/>.
79. Barbash GI, Reiner J, White HD, et al. Evaluation of paradoxic beneficial effects of smoking in patients receiving thrombolytic therapy for acute myocardial infarction: mechanism of the "smoker's paradox" from the GUSTO-I trial, with angiographic insights. Global utilization of streptokinase and tissue-plasminogen activator for occluded coronary arteries. *J Am Coll Cardiol* 1995;**26**(5):1222–9.
80. Teo KK, Ounpuu S, Hawken S, et al. Tobacco use and risk of myocardial infarction in 52 countries in the INTERHEART study: a case-control study. *Lancet* 2006;**368**(9536):647–58.
81. Accortt NA, Waterbor JW, Beall C, et al. Chronic disease mortality in a cohort of smokeless tobacco users. *Am J Epidemiol* 2002;**156**(8):730–7.
82. Di Castelnuovo A, Costanzo S, Bagnardi V, et al. Alcohol dosing and total mortality in men and women: an updated meta-analysis of 34 prospective studies. *Arch Intern Med* 2006;**166**(22):2437–45.
83. Fernandez-Sola J, Nicolas JM, Oriola J, et al. Angiotensin-converting enzyme gene polymorphism is associated with vulnerability to alcoholic cardiomyopathy. *Ann Intern Med* 2002;**137**(5 Part 1):321–6.
84. Urbano-Marquez A, Estruch R, Fernandez-Sola J, et al. The greater risk of alcoholic cardiomyopathy and myopathy in women compared with men. *JAMA* 1995;**274**(2):149–54.
85. Frost L, Vestergaard P. Alcohol and risk of atrial fibrillation or flutter: a cohort study. *Arch Intern Med* 2004;**164**(18):1993–8.
86. Djousse L, Levy D, Benjamin EJ, et al. Long-term alcohol consumption and the risk of atrial fibrillation in the Framingham Study. *Am J Cardiol* 2004;**93**(6):710–3.
87. Mukamal KJ, Tolstrup JS, Friberg J, et al. Alcohol consumption and risk of atrial fibrillation in men and women: the Copenhagen City Heart Study. *Circulation* 2005;**112**(12):1736–42.
88. Myers MG. Caffeine and cardiac arrhythmias. *Ann Intern Med* 1991;**114**(2):147–50.
89. Frost L, Vestergaard P. Caffeine and risk of atrial fibrillation or flutter: the Danish Diet, Cancer, and Health study. *Am J Clin Nutr* 2005;**81**(3):578–82.
90. Caldeira D, Martins C, Alves LB, et al. Caffeine does not increase the risk of atrial fibrillation: a systematic review and meta-analysis of observational studies. *Heart* 2013;**99**(19):1383–9.
91. Corti R, Binggeli C, Sudano I, et al. Coffee acutely increases sympathetic nerve activity and blood pressure independently of caffeine content: role of habitual versus nonhabitual drinking. *Circulation* 2002;**106**(23):2935–40.
92. Lopez-Garcia E, van Dam RM, Li TY, et al. The relationship of coffee consumption with mortality. *Ann Intern Med* 2008;**148**(12):904–14.
93. Freedman ND, Park Y, Abnet CC, et al. Association of coffee drinking with total and cause-specific mortality. *N Engl J Med* 2012;**366**(20):1891–904.

94. Gardener H, Rundek T, Wright CB, et al. Coffee and tea consumption are inversely associated with mortality in a multiethnic urban population. *J Nutr* 2013;**143**(8):1299–308.
95. Liu J, Sui X, Lavie CJ, et al. Association of coffee consumption with all-cause and cardiovascular disease mortality. *Mayo Clin Proc* 2013;**88**(10):1066–74.
96. Worm SW, Sabin C, Weber R, et al. Risk of myocardial infarction in patients with HIV infection exposed to specific individual antiretroviral drugs from the 3 major drug classes: the data collection on adverse events of anti-HIV drugs (D:A:D) study. *Scand J Infect Dis* 2010;**201**(3):318–30.
97. Bavinger C, Bendavid E, Niehaus K, et al. Risk of cardiovascular disease from antiretroviral therapy for HIV: a systematic review. *PloS One* 2013;**8**(3):e59551.
98. Longenecker JC, Coresh J, Powe NR, et al. Traditional cardiovascular disease risk factors in dialysis patients compared with the general population: the CHOICE Study. *J Am Soc Nephrol* 2002;**13**(7):1918–27.
99. van der Velde M, Matsushita K, Coresh J, et al. Lower estimated glomerular filtration rate and higher albuminuria are associated with all-cause and cardiovascular mortality. A collaborative meta-analysis of high-risk population cohorts. *Kidney Int* 2011;**79**(12):1341–52.
100. Cheung AK, Sarnak MJ, Yan G, et al. Cardiac diseases in maintenance hemodialysis patients: results of the HEMO Study. *Kidney Int* 2004;**65**(6):2380–9.
101. Parekh RS, Zhang L, Fivush BA, et al. Incidence of atherosclerosis by race in the dialysis morbidity and mortality study: a sample of the US ESRD population. *J Am Soc Nephrol* 2005;**16**(5):1420–6.
102. Vanholder R, Glorieux G, De Smet R, et al. European Uremic Toxin Work Group. New insights in uremic toxins. *Kidney Int Suppl* 2003;**84**:S6–10.
103. Kraus LM, Kraus Jr. AP. Carbamoylation of amino acids and proteins in uremia. *Kidney Int Suppl* 2001;**78**:S102–7.
104. MacAllister RJ, Rambausek MH, Vallance P, et al. Concentration of dimethyl-L-arginine in the plasma of patients with end-stage renal failure. *Nephrol Dial Transplant* 1996;**11**(12):2449–52.
105. Faraci FM, Brian Jr JE, Heistad DD. Response of cerebral blood vessels to an endogenous inhibitor of nitric oxide synthase. *Am J Physiol* 1995;**269**(5 Pt 2):H1522–7.
106. Kielstein JT, Zoccali C. Asymmetric dimethylarginine: a cardiovascular risk factor and a uremic toxin coming of age? *Am J Kidney Dis* 2005;**46**(2):186–202.
107. Zoccali C, Bode-Boger S, Mallamaci F, et al. Plasma concentration of asymmetrical dimethylarginine and mortality in patients with end-stage renal disease: a prospective study. *Lancet* 2001;**358**(9299):2113–7.
108. Bode-Boger SM, Scalera F, Kielstein JT, et al. Symmetrical dimethylarginine: a new combined parameter for renal function and extent of coronary artery disease. *J Am Soc Nephrol* 2006;**17**(4):1128–34.
109. Schepers E, Barreto DV, Liabeuf S, et al. Symmetric dimethylarginine as a proinflammatory agent in chronic kidney disease. *Clin J Am Soc Nephrol* 2011;**6**(10):2374–83.
110. Block GA, Hulbert-Shearon TE, Levin NW, et al. Association of serum phosphorus and calcium x phosphate product with mortality risk in chronic hemodialysis patients: a national study. *Am J Kidney Dis* 1998;**31**(4):607–17.
111. Meijers BK, Claes K, Bammens B, et al. p-Cresol and cardiovascular risk in mild-to-moderate kidney disease. *Clin J Am Soc Nephrol* 2010;**5**(7):1182–9.
112. Wang CP, Lu LF, Yu TH, et al. Serum levels of total p-cresylsulphate are associated with angiographic coronary atherosclerosis severity in stable angina patients with early stage of renal failure. *Atherosclerosis* 2010;**211**(2):579–83.

113. Miyata T, van Ypersele de Strihou C, Kurokawa K, et al. Alterations in nonenzymatic biochemistry in uremia: origin and significance of "carbonyl stress" in long-term uremic complications. *Kidney Int* 1999;**55**(2):389–99.
114. Bresalier RS, Sandler RS, Quan H, et al. Cardiovascular events associated with rofecoxib in a colorectal adenoma chemoprevention trial. *N Engl J Med* 2005;**352**(11):1092–102.
115. Ray WA, Stein CM, Daugherty JR, et al. COX-2 selective non-steroidal anti-inflammatory drugs and risk of serious coronary heart disease. *Lancet* 2002;**360**(9339):1071–3.
116. Solomon DH, Schneeweiss S, Glynn RJ, et al. Relationship between selective cyclooxygenase-2 inhibitors and acute myocardial infarction in older adults. *Circulation* 2004;**109**(17):2068–73.
117. Mamdani M, Juurlink DN, Lee DS, et al. Cyclo-oxygenase-2 inhibitors versus non-selective non-steroidal anti-inflammatory drugs and congestive heart failure outcomes in elderly patients: a population-based cohort study. *Lancet* 2004;**363**(9423):1751–6.
118. Solomon SD, Pfeffer MA, McMurray JJ, et al. Effect of celecoxib on cardiovascular events and blood pressure in two trials for the prevention of colorectal adenomas. *Circulation* 2006;**114**(10):1028–35.
119. Andersohn F, Suissa S, Garbe E. Use of first- and second-generation cyclooxygenase-2-selective nonsteroidal antiinflammatory drugs and risk of acute myocardial infarction. *Circulation* 2006;**113**(16):1950–7.
120. Trelle S, Reichenbach S, Wandel S, et al. Cardiovascular safety of non-steroidal anti-inflammatory drugs: network meta-analysis. *BMJ* 2011;**342**:c7086.
121. Gislason GH, Rasmussen JN, Abildstrom SZ, et al. Increased mortality and cardiovascular morbidity associated with use of nonsteroidal anti-inflammatory drugs in chronic heart failure. *Arch Intern Med* 2009;**169**(2):141–9.
122. Schmidt M, Christiansen CF, Mehnert F, et al. Non-steroidal anti-inflammatory drug use and risk of atrial fibrillation or flutter: population based case-control study. *BMJ* 2011;**343**:d3450.
123. Bristow MR, Mason JW, Billingham ME, et al. Dose-effect and structure-function relationships in doxorubicin cardiomyopathy. *Am Heart J* 1981;**102**(4):709–18.
124. Hershman DL, McBride RB, Eisenberger A, et al. Doxorubicin, cardiac risk factors, and cardiac toxicity in elderly patients with diffuse B-cell non-Hodgkin's lymphoma. *J Clin Oncol* 2008;**26**(19):3159–65.
125. Lipshultz SE, Colan SD, Gelber RD, et al. Late cardiac effects of doxorubicin therapy for acute lymphoblastic leukemia in childhood. *N Engl J Med* 1991;**324**(12):808–15.
126. Kremer LC, Caron HN. Anthracycline cardiotoxicity in children. *N Engl J Med* 2004;**351**(2):120–1.
127. Swain SM, Whaley FS, Ewer MS. Congestive heart failure in patients treated with doxorubicin: a retrospective analysis of three trials. *Cancer* 2003;**97**(11):2869–79.
128. Legha SS, Benjamin RS, Mackay B, et al. Reduction of doxorubicin cardiotoxicity by prolonged continuous intravenous infusion. *Ann Intern Med* 1982;**96**(2):133–9.
129. Lipshultz SE, Giantris AL, Lipsitz SR, et al. Doxorubicin administration by continuous infusion is not cardioprotective: the Dana-Farber 91-01 Acute Lymphoblastic Leukemia Protocol. *J Clin Oncol* 2002;**20**(6):1677–82.
130. Smith LA, Cornelius VR, Plummer CJ, et al. Cardiotoxicity of anthracycline agents for the treatment of cancer: systematic review and meta-analysis of randomised controlled trials. *BMC Cancer* 2010;**10**:337.
131. Moja L, Tagliabue L, Balduzzi S, et al. Trastuzumab containing regimens for early breast cancer. *Cochrane Database Syst Rev* 2012;4:CD006243.

132. Bowles EJ, Wellman R, Feigelson HS, et al. Risk of heart failure in breast cancer patients after anthracycline and trastuzumab treatment: a retrospective cohort study. *J Natl Cancer Inst* 2012;**104**(17):1293–305.
133. Chen J, Long JB, Hurria A, et al. Incidence of heart failure or cardiomyopathy after adjuvant trastuzumab therapy for breast cancer. *J Am Coll Cardiol* 2012;**60**(24):2504–12.
134. Perez EA, Koehler M, Byrne J, et al. Cardiac safety of lapatinib: pooled analysis of 3689 patients enrolled in clinical trials. *Mayo Clin Proc* 2008;**83**(6):679–86.
135. Ranpura V, Hapani S, Wu S. Treatment-related mortality with bevacizumab in cancer patients: a meta-analysis. *JAMA* 2011;**305**(5):487–94.
136. An MM, Zou Z, Shen H, et al. Incidence and risk of significantly raised blood pressure in cancer patients treated with bevacizumab: an updated meta-analysis. *Eur J Clin Pharmacol* 2010;**66**(8):813–21.
137. Zhu X, Stergiopoulos K, Wu S. Risk of hypertension and renal dysfunction with an angiogenesis inhibitor sunitinib: systematic review and meta-analysis. *Acta Oncol (Madr)* 2009;**48**(1):9–17.
138. Wu S, Chen JJ, Kudelka A, et al. Incidence and risk of hypertension with sorafenib in patients with cancer: a systematic review and meta-analysis. *Lancet Oncol* 2008;**9**(2):117–23.
139. Qi WX, Lin F, Sun YJ, et al. Incidence and risk of hypertension with pazopanib in patients with cancer: a meta-analysis. *Cancer Chemother Pharmacol* 2013;**71**(2):431–9.
140. Scappaticci FA, Skillings JR, Holden SN, et al. Arterial thromboembolic events in patients with metastatic carcinoma treated with chemotherapy and bevacizumab. *J Natl Cancer Inst* 2007;**99**(16):1232–9.
141. Nalluri SR, Chu D, Keresztes R, et al. Risk of venous thromboembolism with the angiogenesis inhibitor bevacizumab in cancer patients: a meta-analysis. *JAMA* 2008;**300**(19):2277–85.
142. Schutz FA, Je Y, Azzi GR, et al. Bevacizumab increases the risk of arterial ischemia: a large study in cancer patients with a focus on different subgroup outcomes. *Ann Oncol* 2011;**22**(6):1404–12.
143. Choueiri TK, Schutz FA, Je Y, et al. Risk of arterial thromboembolic events with sunitinib and sorafenib: a systematic review and meta-analysis of clinical trials. *J Clin Oncol* 2010;**28**(13):2280–5.
144. Hurwitz HI, Saltz LB, Van Cutsem E, et al. Venous thromboembolic events with chemotherapy plus bevacizumab: a pooled analysis of patients in randomized phase II and III studies. *J Clin Oncol* 2011;**29**(13):1757–64.
145. Sonpavde G, Je Y, Schutz F, et al. Venous thromboembolic events with vascular endothelial growth factor receptor tyrosine kinase inhibitors: a systematic review and meta-analysis of randomized clinical trials. *Crit Rev Oncol Hematol* 2013;**87**(1):80–9.
146. Qi WX, Min DL, Shen Z, et al. Risk of venous thromboembolic events associated with VEGFR-TKIs: a systematic review and meta-analysis. *Int J Cancer* 2013;**132**(12):2967–74.
147. Prince HM, Schenkel B, Mileshkin L. An analysis of clinical trials assessing the efficacy and safety of single-agent thalidomide in patients with relapsed or refractory multiple myeloma. *Leuk Lymphoma* 2007;**48**(1):46–55.
148. Yang X, Brandenburg NA, Freeman J, et al. Venous thromboembolism in myelodysplastic syndrome patients receiving lenalidomide: results from postmarketing surveillance and data mining techniques. *Clin Drug Investig* 2009;**29**(3):161–71.

149. Carrier M, Le Gal G, Tay J, et al. Rates of venous thromboembolism in multiple myeloma patients undergoing immunomodulatory therapy with thalidomide or lenalidomide: a systematic review and meta-analysis. *J Thromb Haemost* 2011;**9**(4):653–63.
150. Richardson PG, Siegel D, Baz R, et al. Phase 1 study of pomalidomide MTD, safety, and efficacy in patients with refractory multiple myeloma who have received lenalidomide and bortezomib. *Blood* 2013;**121**(11):1961–7.
151. Chu TF, Rupnick MA, Kerkela R, et al. Cardiotoxicity associated with tyrosine kinase inhibitor sunitinib. *Lancet* 2007;**370**(9604):2011–9.
152. Khakoo AY, Kassiotis CM, Tannir N, et al. Heart failure associated with sunitinib malate: a multitargeted receptor tyrosine kinase inhibitor. *Cancer* 2008;**112**(11):2500–8.
153. Richards CJ, Je Y, Schutz FA, et al. Incidence and risk of congestive heart failure in patients with renal and nonrenal cell carcinoma treated with sunitinib. *J Clin Oncol* 2011; **29**(25):3450–6.
154. Zang J, Wu S, Tang L, et al. Incidence and risk of QTc interval prolongation among cancer patients treated with vandetanib: a systematic review and meta-analysis. *PloS One* 2012;**7**(2):e30353.
155. Valko M, Morris H, Cronin MT. Metals, toxicity and oxidative stress. *Curr Med Chem* 2005;**12**(10):1161–208.
156. Sainio EL, Jolanki R, Hakala E, et al. Metals and arsenic in eye shadows. *Contact Derm* 2000;**42**(1):5–10.
157. Al-Ashban RM, Aslam M, Shah AH. Kohl (surma): a toxic traditional eye cosmetic study in Saudi Arabia. *Public Health* 2004;**118**(4):292–8.
158. Meyer PA, Brown MJ, Falk H. Global approach to reducing lead exposure and poisoning. *Mutat Res* 2008;**659**(1-2):166–75.
159. Chiba M, Masironi R. Toxic and trace elements in tobacco and tobacco smoke. *Bull World Health Org* 1992;**70**(2):269–75.
160. Smith AH, Smith MM. Arsenic drinking water regulations in developing countries with extensive exposure. *Toxicol* 2004;**198**(1-3):39–44.
161. Tseng CH, Chong CK, Tseng CP, et al. Long-term arsenic exposure and ischemic heart disease in arseniasis-hyperendemic villages in Taiwan. *Toxicol Lett* 2003;**137**(1-2): 15–21.
162. Axelson O, Dahlgren E, Jansson CD, et al. Arsenic exposure and mortality: a case-referent study from a Swedish copper smelter. *Br J Ind Med* 1978;**35**(1):8–15.
163. Chiou HY, Huang WI, Su CL, et al. Dose-response relationship between prevalence of cerebrovascular disease and ingested inorganic arsenic. *Stroke* 1997;**28**(9).1717–23.
164. Rahman M, Tondel M, Ahmad SA, et al. Hypertension and arsenic exposure in Bangladesh. *Hypertension* 1999;**33**(1):74–8.
165. Tseng CH, Tai TY, Chong CK, et al. Long-term arsenic exposure and incidence of non-insulin-dependent diabetes mellitus: a cohort study in arseniasis-hyperendemic villages in Taiwan. *Environ Health Perspect* 2000;**108**(9):847–51.
166. Yuan Y, Marshall G, Ferreccio C, et al. Acute myocardial infarction mortality in comparison with lung and bladder cancer mortality in arsenic-exposed region II of Chile from 1950 to 2000. *Am J Epidemiol* 2007;**166**(12):1381–91.
167. Zierold KM, Knobeloch L, Anderson H. Prevalence of chronic diseases in adults exposed to arsenic-contaminated drinking water. *Am J Public Health* 2004;**94**(11):1936–7.
168. Meliker JR, Wahl RL, Cameron LL, et al. Arsenic in drinking water and cerebrovascular disease, diabetes mellitus, and kidney disease in Michigan: a standardized mortality ratio analysis. *Environ Health* 2007;**6**:4.

169. Lee TC, Ho IC. Modulation of cellular antioxidant defense activities by sodium arsenite in human fibroblasts. *Arch Toxicol* 1995;**69**(7):498–504.
170. Lee PC, Ho IC, Lee TC. Oxidative stress mediates sodium arsenite-induced expression of heme oxygenase-1, monocyte chemoattractant protein-1, and interleukin-6 in vascular smooth muscle cells. *Toxicol Sci* 2005;**85**(1):541–50.
171. Markowitz M. Lead poisoning. *Pediatr Rev* 2000;**21**(10):327–35.
172. Navas-Acien A, Guallar E, Silbergeld EK, et al. Lead exposure and cardiovascular disease: a systematic review. *Environ Health Perspect* 2007;**115**(3):472–82.
173. Nawrot TS, Thijs L, Den Hond EM, et al. An epidemiological re-appraisal of the association between blood pressure and blood lead: a meta-analysis. *J Hum Hypertens* 2002;**16**(2):123–31.
174. Schober SE, Mirel LB, Graubard BI, et al. Blood lead levels and death from all causes, cardiovascular disease, and cancer: results from the NHANES III mortality study. *Environ Health Perspect* 2006;**114**(10):1538–41.
175. Ahamed M, Singh S, Behari JR, et al. Interaction of lead with some essential trace metals in the blood of anemic children from Lucknow, India. *Clin Chim Acta* 2007;**377**(1-2):92–7.
176. Edwards JR, Prozialeck WC. Cadmium, diabetes and chronic kidney disease. *Toxicol Appl Pharmacol* 2009;**238**(3):289–93.
177. Gallagher CM, Meliker JR. Blood and urine cadmium, blood pressure, and hypertension: a systematic review and meta-analysis. *Environ Health Perspect* 2010;**118**(12):1676–84.
178. Menke A, Muntner P, Silbergeld EK, et al. Cadmium levels in urine and mortality among U.S. adults. *Environ Health Perspect* 2009;**117**(2):190–6.
179. Navas-Acien A, Silbergeld EK, Sharrett R, et al. Metals in urine and peripheral arterial disease. *Environ Health Perspect* 2005;**113**(2):164–9.
180. Messner B, Knoflach M, Seubert A, et al. Cadmium is a novel and independent risk factor for early atherosclerosis mechanisms and *in vivo* relevance. *Arterioscler Thromb Vasc Biol* 2009;**29**(9):1392–8.
181. Everett CJ, Frithsen IL. Association of urinary cadmium and myocardial infarction. *Environ Res* 2008;**106**(2):284–6.
182. Salonen JT, Seppanen K, Nyyssonen K, et al. Intake of mercury from fish, lipid peroxidation, and the risk of myocardial infarction and coronary, cardiovascular, and any death in eastern Finnish men. *Circulation* 1995;**91**(3):645–55.
183. Salonen JT, Seppanen K, Lakka TA, et al. Mercury accumulation and accelerated progression of carotid atherosclerosis: a population-based prospective 4-year follow-up study in men in eastern Finland. *Atherosclerosis* 2000;**148**(2):265–73.
184. Yoshizawa K, Rimm EB, Morris JS, et al. Mercury and the risk of coronary heart disease in men. *N Engl J Med* 2002;**347**(22):1755–60.
185. Hallgren CG, Hallmans G, Jansson JH, et al. Markers of high fish intake are associated with decreased risk of a first myocardial infarction. *Br J Nutr* 2001;**86**(3):397–404.
186. Clarkson TW. The pharmacology of mercury compounds. *Annu Rev Pharmacol* 1972;**12**:375–406.
187. World Health Organization. *Map Production: Public Health Mapping and Geographic Information Systems (GIS)*. Geneva, Switzerland: WHO; 2008.
188. World Health Organization. *Map Production: Public Health Information and Geographic Information Systems (GIS)*. Geneva, Switzerland: WHO; 2010.

Chapter 2

Cellular and Molecular Perspectives on Cardiac Toxins

Reza Tabrizchi
Division of BioMedical Sciences, Faculty of Medicine, Memorial University, St. John's, Newfoundland and Labrador, Canada

2.1 INTRODUCTION

The manifestation of toxic effects by drugs in myocardium can occur from overdose due to accidental and/or inappropriate and incorrect use. In addition, polypharmacy can at times also lend itself to the occurrence of adverse reactions in the heart. It has become evident that the myocardium is a site of toxicity for a large number of medicinal and recreational drugs and chemicals. This can occur by the interaction of chemicals at a number of sites, including ion channels, membrane-bound receptors, ionic pumps, and intracellular organelles such as microsomes, mitochondria, and sarcoplasmic reticulum (SR) (Figure 2.1). Generally, substances that produce toxic effects in the myocardium do so by influencing rhythm, rate, and contractile function. In broad terms, unwanted toxic effects in the heart can be categorized as acute or chronic in nature.

There are many substances that influence myocardial function, be it via overt positive or negative inotropic, chronotropic, or dromotropic effects. This chapter provides an overview of a range of chemicals and different classes of drugs that have been noted to produce undesirable effects in mammalian hearts. In general, groups of chemicals that seem to have a high incidence of cardiac toxicity associated with them are the tricyclic antidepressants, cardiac glycosides, Ca channel antagonists, β-blockers, anthracyclines, recreational drugs (e.g., sympathomimetics), and certain toxins. In addition, a host of other chemicals seem to inhibit the rapid delayed rectifier K^+ current (i.e., hERG potassium channel), prolonging repolarization, which can then lead to ventricular tachycardia of a polymorphic nature that sometimes can be fatal.

Unaccounted for properties of chemicals not specifically associated with normal pharmacological actions of a substance can also play a role in toxic

The Heart and Toxins. DOI: http://dx.doi.org/10.1016/B978-0-12-416595-3.00002-5

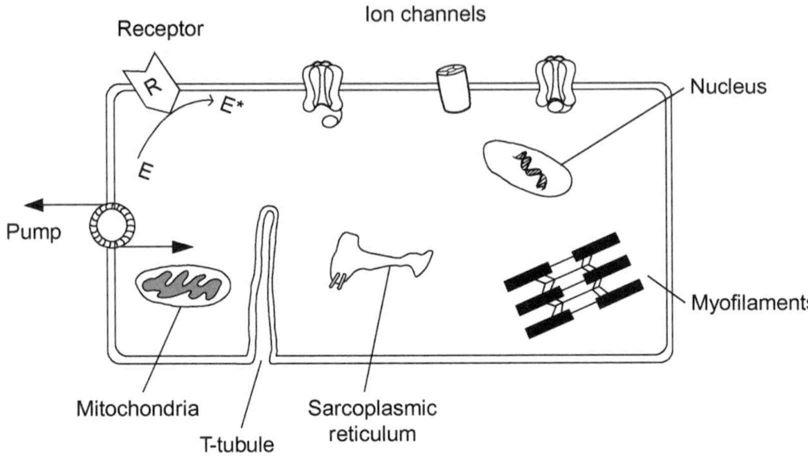

FIGURE 2.1 Schematic of cardiomyocytes showing sites at which drugs can act to produce toxic effects.

actions of drugs in the myocardium. An interesting example is the outcome reported with the selective α_1-adrenoceptor antagonist doxazosin. It has been noted that doxazosin in comparison to chlorthalidone in the antihypertensive and lipid-lowering treatment to prevent heart attack trail (ALLHAT) was associated with significantly more cardiac events.[1] Thus, the doxazosin arm of the trial was abandoned and it was suggested that treatment of high-risk hypertensive patients with a selective α_1-adrenoceptor blocker is associated with greater occurrence of heart failure.[1]

Since then, evidence has been presented to suggest that prazosin, doxazosin, and terazosin are capable of inhibiting the human Ether-à-go-go related gene (hERG) K channels, and this may be linked to events leading to myocardial death.[2] Furthermore, a doxazosin-mediated blockade of hERG channels has been suggested to trigger apoptosis of myocardial cells, supporting the view that this α_1-adrenergic receptor antagonist may cause heart failure independent of its action as an α-adrenoceptor blocker.[3] Moreover, a novel functional role has been implicated for hERG K channels and K conductance in some types of cancerous cells.[4] Such data have led to the idea that hERG potassium channel inhibitors may be useful in the treatment of certain forms of cancer. This, of course, raises the question of cardiac safety issues in patients that who may be treated with known hERG blockers.

2.2 RECREATIONAL DRUGS

The cause of cardiac toxicity by recreational drugs is multifactorial but ultimately the result is a disturbance in the process of excitation–contraction coupling. For instance, one of the most common side effects noted with the use of crack cocaine is respiratory complication, which is a common side effect of opioids. But, respiratory complication is not the major adverse effect associated with the use of cocaine. It seems that the cause of 25% of chest pains in all young adults presented to emergency departments in the United States is the use of cocaine.[5] Moreover, evidence from a prospective analysis of acute coronary syndrome in patients (30 years and younger) admitted to a coronary care unit in Spain revealed that 25% appear to be cocaine users.[6] Certainly, aside from other complications, use of sympathomimetics, such as cocaine, can lead to severe cardiac arrhythmias that can be fatal.

The basis for the observed toxic effects of cocaine leading to inhibition, generation, and/or progression of electrical events in the heart is likely a combination of overstimulation of the cardiomyocytes, membrane stabilization (i.e., local anesthetic) effects, and overt ischemia, secondary to reduced coronary blood flow. Experimentally, it has been found that administration of toxic amounts of cocaine to anesthetized dogs and guinea pigs led to atrioventricular (AV) dissociation, ectopic pacemaker activity, irregular heart rate, and gradual increase in PR intervals in parallel with overt hypotension.[7,8] The effect of cocaine on cardiac dynamics is a significant reduction in left ventricular contractility and increased left ventricular end-diastolic pressure accompanied by left ventricular dilation and its associated coronary artery vasoconstriction and thus reduced myocardial blood flow.[9] Intravenous administration of cocaine to conscious dogs has also been noted to produce depression of left ventricular function, which is likely due to increased oxygen consumption not matched by increased coronary blood flow.[10] Moreover, administration of cocaine intravenously (IV) to anesthetized dogs was reported to cause an increase in QRS duration and QTc interval, AV conduction time, and atrial effective refractory period.[11] Not surprisingly, such occurrences lead to disturbances in cardiac rhythm and contractility.

Intracoronary infusion of cocaine in humans has been found to result in left ventricular systolic and diastolic dysfunction.[12] Cocaine is capable of indirectly causing overstimulation of α-adrenoceptors in coronary blood vessels.[13] In addition, cocaine has been implicated in the promotion of platelet aggregation and possibly thrombosis.[14,15] Patients intoxicated with cocaine have also been noted to have significantly increased plasma and urine levels of endothelin-1.[16] Certainly, it is possible that the overt coronary artery vasoconstriction noted in cocaine users is facilitated by the release of endothelin-1, as it has been recognized that endothelin is a very potent constrictor of the coronary arteries. Clearly, these events form the preamble for an extensive reduction in myocardial perfusion, leading to ischemia, impairment in electrical conduction and coupling, and subsequent alteration in the process of excitation–contraction coupling. However, aside from its sympathomimetic and Na channel–blocking properties, cocaine has been found to be an hERG K channel blocker.[17,18]

Interestingly, two distinct types of cardiac toxicity have been ascribed to toxic levels of cocaine: (1) monomorphic ventricular tachycardia or idioventricular rhythm occurs due to extensive Na channel blockade, and (2) polymorphic tachycardia and torsades de pointes (TdP) can develop as sequelae to prolong QT intervals as a result of hERG K channel blockade, especially in individuals with inherited long QT syndrome.[19]

The primary cardiovascular toxic effects of amphetamines are a result of an increased concentration of catecholamines and direct stimulation of peripheral β- and α-adrenoceptors, leading to a significant rise in pulse rate and blood pressure. It is not surprising that a comparable pattern of cardiac toxic effects of cocaine can manifest with the use of amphetamine-like drugs.[20,21]

2.3 TRICYCLIC ANTIDEPRESSANTS

In connection with myocardial tissue, a variety of toxicities have been ascribed to tricyclic antidepressants in mammals. The ability of these compounds to cause cardiac toxicity is multifaceted. The nature of the toxicity of this group of drugs varies depending on the substance, but in general these compounds have been noted for inhibition of noradrenaline reuptake and anticholinergic activity causing tachycardia.[22] However, suggested modes of action for tricyclic antidepressant related to toxicity in the heart include inhibition of hERG K channels, voltage-gated Ca channels, and Na channels, and impairment of calcium release and/or inhibition of Ca^{2+} uptake by sarcoplasmic reticulum (Figure 2.2). Early indications suggested that impramine

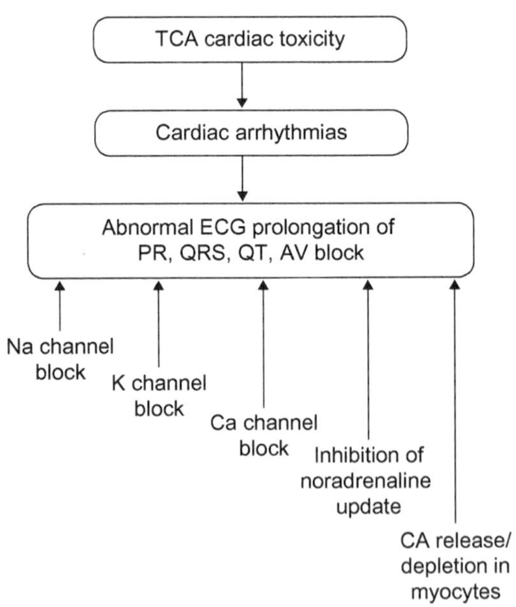

FIGURE 2.2 **Graphic of tricyclic antidepressants that act to produce cardiac toxic effects. These are possible sites of actions.**

produced "quinidine-like" actions in the atrial muscle of rabbits.[23] Changes in sinoatrial (SA) node and atrial action potential similar to the changes caused by local anesthetics were reported for imipramine in hearts from rabbits.[23]

Further investigations in Purkinje fibers from sheep revealed a decrease in action potential duration and a reduction in V_{max} and conduction velocity with imipramine in line with depression of membrane responsiveness.[24] In addition, a decrease in the amplitude of action potential, V_{max}, a decrease in conduction velocity and prolongation of effective refractory period, a reduction in slope of phase 4 diastolic depolarization, and a slow spontaneous rate of depolarization were reported in Purkinje fibers from bovine heart with imipramine.[25] Isenberg and Tamargo,[26] using bovine ventricular myocytes, showed that imipramine reduced Ca^{2+} conductance without affecting the inactivation time constant, voltage dependence of peak Ca^{2+} current, or reversal potential while inward rectifying K^+ currents were reduced.

Accordingly, imipramine was reported to suppress Na^+ currents in guinea pig ventricular myocytes, with the possibility of two molecules binding to the channel and causing perturbation of action potential.[27] In the same tissue, imipramine was found to show slow onset and offset of actions in inducing Na channel block.[28] In rabbit isolated atrial cells, imipramine was found to block transient outward K^+ currents without any effect on the resting membrane potential.[29] Amitriptyline was reported to block human cardiac Na channels (hH1).[30] In this study, use-dependent block was shown to occur with greater inhibition at rest than in an inactivated state. Substitution of phenylalanine with lysine at F1760 K of the hH1 channel led to reduction in the affinity of amitriptyline. Moreover, amitriptyline seemed to bind to a similar site as local anesthetics.[30] Collectively, the data seem to indicate that imipramine and amitriptyline are capable of inhibiting the generation of an action potential by the attenuation of Na^+ currents in cardiomyocytes.

In rabbit atria myocytes, while amitriptyline was noted to block Na^+ current in a use-dependent manner, an increase in extracellular pH from 7.3 to 8.0 enhanced time of recovery from the block, but the same magnitude increase in the intracellular pH did not influence the block.[31] Imipramine was found to inhibit Na^+ current in isolated human atrial myocytes, and binding occurred at both activated and inactivated states. Both intracellular and extracellular alkalization (pH 7.2–7.6) led to antagonism of the actions of imipramine on the closed state, but only an intracellular increase in pH reversed the binding to the open state.[32] Watts et al.,[33] using rat ventricular myocytes, reported that imipramine was able to inhibit the increase in intracellular Ca^{2+} levels in an electrically paced cell and this effect could be prevented by alkalization. To this end, while both imipramine and amitriptyline are able to block the Na channel, this inhibition can be reversed by alkalization of the cell milieu.

Tricyclic antidepressants, such as amitriptyline, have also been reported to cause depletion of Ca from intracellular stores, notably from sarcoplasmic reticulum. Amitriptyline has been reported to cause release of Ca^{2+} and/or inhibition of sarcoplasmic-endoplasmic reticulum Ca^{2+}-ATPase (SERCA) mediated Ca^{2+} uptake in cardiac ventricular myocytes from cats and rabbits.[34] An initial increase in Ca^{2+} spark was followed by a decrease in frequency due to the action of amitriptyline. Moreover, this action seems to be mediated by an increase in the open probability of cardiac ryanodine receptors in the ventricular myocytes.[34] Chopra et al.[35] also reported that in mice and sheep cardiomyocytes, amitriptyline binds to ryanodine RyR2 channels causing it to open and release Ca^{2+} from sarcoplasmic reticulum. Tricyclic antidepressants have also been shown to block hERG K channels and inhibit trafficking. For example, amitriptyline was reported to block hERG K channels, showing voltage-dependent inhibition,[36] while doxepin-induced block did not seem to show voltage dependence.[37]

In summary, it is evident that multiple sites of action can account for the occurrence of cardiac toxicity resulting from tricyclic antidepressant use. These seem to include direct action on membrane-bound receptors (i.e., cholinoceptors) and ion channels (i.e., Na, K, and Ca) at intracellular levels on the Ca^{2+}-release process and/or inhibition of hERG forward trafficking from endoplasmic reticulum, as well as hERG endocytosis and degradation. However, one notable action associated with the antidepressive action of this class of compounds (i.e., inhibition of catecholamine neuronal reuptake) can also contribute to their toxicity in the myocardium. For example, desipramine has been found to cause inhibition of noradrenaline neuronal reuptake in human and rat atria.[38] This imposing effect in the myocardium in addition to other actions of the tricyclic antidepressants on receptors, ion channels, and intracellular organelles can lead to complex electrical and mechanical disturbances in the heart.

2.4 CALCIUM CHANNEL ANTAGONISTS

Organic calcium channel antagonists that influence the cardiovascular system can be divided into two general subgroups: (1) vascular-selective compounds (e.g., nifedipine, amlodipine, and nicardipine) and (2) "cardio-selective" compounds, namely D600, diltiazem, and verapamil. The mechanical, electrical, and rhythmic activity of the myocardium is dependent on Ca^{2+}. As such, pacemaker activity, conduction of electrical signals, and excitation–contraction coupling in the heart in part rely on the availability of Ca^{2+} and optimal operation of a variety of Ca channels and exchangers. While vascular-selective Ca channel antagonists have limited direct action on cardiomyocytes, cardio-selective blockers can have profound effects on cardiac contractility and electrical events. It is recognized that the slow Ca^{2+} current is mediated in part via the voltage-gated L-type Ca channels (Figure 2.3), and it can be influenced by Ca channel antagonists such as verapamil, D600, and diltiazem. Cardiac toxicity attributed to the actions of

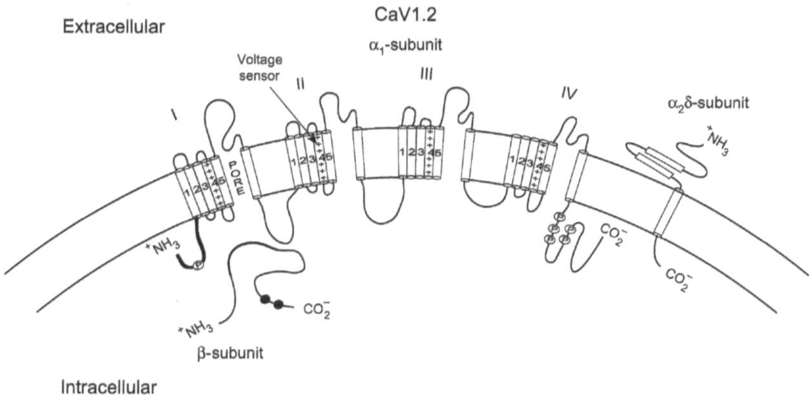

FIGURE 2.3 Schematic representation of α_1-, β-, and $\alpha_2\delta$-subunits of cardiac calcium channel. The channel has six transmembrane-spanning domains where S4 is the *voltage sensor* and the *pore* is located between the S5 and S6 regions.

Ca channel blockers can result in the disruption of rhythm and rate, as well as contraction and relaxation of the heart.

It has been found that the Ca channel antagonist verapamil can cause the disappearance of the P wave, and impairment of AV conduction in chloralose-anesthetized dogs.[39] Cranefield et al.[40] also reported that verapamil could suppress spontaneous activity in canine isolated cardiac Purkinje fibers and abolish the sustained rhythmic activity evoked by externally applied depolarizing pulses, thus significantly diminishing the rate at which quiescent fibers can be driven. Furthermore, the intra-arterial administration of verapamil and D600 in sodium pentobarbital-anesthetized dogs was noted to be capable of depressing the frequency of sinus node discharge, lengthening AV nodal conduction time and prolonging AV nodal refractoriness.[41] The study noted that the actions were in parallel to those of inorganic Ca channel antagonists, $MnCl_2$ and $LaCl_3$.[41] Of special concern is the ability of these compounds to impair AV nodal conduction, leading to a 2-to-1 or Wenckebach supra-His AV block. Zipes and Fischer[41] also reported that in anesthetized dogs the actions of organic as well as inorganic Ca channel antagonists can be reversed by the administration of the β-adrenoceptor agonists (i.e., isoprenaline or adrenaline).

Verapamil has also been reported to significantly reduce SA nodal action potentials with a modest but significant decline in the maximum diastolic potential in rabbit isolated cardiac tissue.[42] Also, verapamil was found to significantly depress AV nodal action potentials, prolonging AV nodal conduction time without any discernible effects on the action potentials generated in the bundle of His. The negative dromotropic effects of verapamil in rabbit isolated cardiac tissue were reversed by adrenaline.[42] In contrast, the presence of a β-adrenergic antagonist (i.e., propranolol) enhanced the pharmacological actions of the Ca channel antagonists on the heart, resulting in a substantial depression of the electrical activity in the pacemaker cells and in the conduction fibers. Not surprisingly, the extensive depression of the contractile function was ultimately found to be lethal.[43] There is evidence in the literature that indicates that these synergistic actions of the Ca channel antagonists and β-adrenergic blockers can be in part attenuated by the administration of Ca^{2+}.[43,44]

Differences have been revealed in the inhibitory actions of the Ca channel blocker verapamil on electrical activity versus contractile effects in cardiac tissue. Essentially, verapamil (≥2 mg/l) was shown to preferentially impair the electrical events in SA node and AV node, reducing conduction velocity while very modestly affecting contractility in the isolated rabbit atrial muscle.[42,45]

Toxicity that is associated with Ca channel blockers (e.g., verapamil) can manifest as impairment in the conduction ability of conducting fibers in the myocardium. For example, high doses of verapamil have been reported to produce an AV conduction block in a preparation such as isolated and cross-circulated canine AV nodal cells.[46] However, verapamil can also impair ventricular contractility, as reported in sodium pentobarbital anesthetized dogs.[47] Zaritsky et al.[48] reported that organic Ca channel blockers can impair contractility in rat isolated perfused Langendorff preparation, but that this effect can be reversed by simultaneous perfusion with glucagon. In sodium pentobarbital anesthetized dogs, the type III phosphodiesterase inhibitor amrinone was reported to be capable of reversing cardiac toxicity caused by verapamil.[49] Of interest was the ability of amrinone to reverse the mechanical depressive action of verapamil by increasing cardiac index without having a significant effect on the bradycardia induced by verapamil.[49] The toxic cardiac depressive actions of Ca channel blockers, such as verapamil and diltiazem, can be substantially accentuated by the copresence of β-blockers, leading to profound bradycardia and depression of myocardial functional response and, if not addressed with appropriate intervention, death.[50]

Depressive actions of Ca channel blockers on myocardial contractility may be partially alleviated by the infusion of calcium glucagon, among other supportive measures, that could be employed in an individual intoxicated with such agents.[50,51] More recently, the use of fat emulsion treatment has been found helpful to combat intoxication resulting from ingestion of Ca channel antagonists such as verapamil.[52,53] However, the use of fat emulsion has been noted to result in some adverse effects such as lipemia, acute lung injury, and renal failure.[54] The use of fat emulsion as an antidote against Ca channel blocker-related toxicity requires further investigation.

To summarize, cardiac toxicity produced by Ca channel antagonists is linked to impairment of Ca^{2+} entry through the slow Ca channels that can interfere with pacemaker cells in the sinus node. In addition, Ca influx associated with AV node conductivity can be inhibited, leading to an AV block. Finally, inhibition of the phase 2 plateau of the action potential in the myocardium can precipitate impediment of the ability of the myocyte to contract.

2.5 β-ADRENOCEPTOR ANTAGONISTS

The toxic effects of β-adrenoceptor antagonists (blockers) can be linked to their ability to block β-adrenergic receptors in the myocardium (Table 2.1). Pharmacological properties of β blockers are described again in Chapter 8. However, this only partly accounts for the manifestion of toxicity following exposure to high levels of β-adrenoceptor antagonists. Nevertheless, this class of drugs shows a low level of toxicity in the myocardium, but accidental

TABLE 2.1 Pharmacological Properties of Some β-Adrenoceptor Blockers

Blocker	Cardio-Selectivity	Intrinsic Sympathomimetic Activity	Local Anesthetic-like Activity
Alprenolol	−	+	+
Atenolol	+	−	−
Acebutolol	+	+	−
Betaxolol	+	−	−
Bisoprolol	+	−	−
Bucindolol	−	−	−
Carvedilol[a]	−	−	−
Celiprolol	+	+	−
Esmolol[b]	+	−	−
Labetalol[a]	−	−	−
Landiolol[b]	+	−	−
Mepindolol	−	+	−
Metoprolol	+	−	−
Nadolol	−	−	−
Nebivolol[c]	+	−	−
Oxprenolol	−	+	+
Pindolol	−	+	+
Propranolol	−	−	+
Sotalol	−	−	−
Timolol	−	−	−

+ effect; − no effect.
[a]*Alpha-adrenceptor blocking ability.*
[b]*Ultrashort acting.*
[c]*Nitric oxide potentiating ability.*

overdose and attempted suicide account for a large number of β-adrenoceptor antagonist toxicities that sometimes lead to fatal outcomes. Aside from their ability to block the β-adrenergic receptors, some of these compounds have membrane-stabilizing effects and are able to act as a local anesthetic and show "quinidine-like" actions at higher concentrations in the myocardium. The high plasma concentration of β-blockers can lead to a manifestation of negative inotropic effects and associated electrocardiogram (ECG) changes, such as first-degree AV block and disappearance of P waves, and a possibility of prolongation of the QRS interval.[55]

An earlier study to determine the additional effects of β-blockers on the myocardium revealed that, for example, propranolol may block myocardial conduction velocity as well as nerve conduction via local anesthetic effects.[56] However, the ability to act as a local anesthetic is not equally shared among known β-blockers, and compounds (e.g., atenolol, metoprolol, nadolol, practolol, sotalol, and timolol) seem to be devoid of this activity.[55] Clearly, propranolol is noted for strong membrane-stabilizing effects and this feature accounts for reduction in myocardial contraction at much lower concentrations when compared to timolol, which has no membrane-stabilizing effects. Propranolol but not timolol was found to be capable of increasing coronary flow in an isolated heart, an effect also most likely due to the membrane-stabilizing action of the former.

The action of propranolol reduces atrial rate and myocardial contractility in either spontaneously beating or electrically stimulated tissue, which may be explained by its local anesthetic effects; those of timolol, however, remain an anomaly.[57] The concept of a reduction in heart rate affecting inotropy is interesting and may be responsible for the negative inotropic effects of β-blockers. De Pauw et al.[58] provided some evidence to indicate that the heart function is regulated in part by a force–frequency relationship. Of note was the observation that during sympathetic nerve stimulation, chronotropic response plays a major role in the development of inotropic response. This phenomenon adds to toxicity with β-blocker in overdose, where both rate and force of contraction are affected in an interrelated manner.

Evidence has been presented that a high concentration of propranolol affects Ca^{2+} transport by subcellular organelles such as microsomes and mitochondria. This has been suggested as a possible mechanism for the negative inotropic effects of propranolol in the myocardium.[59] Another possible noted site for the cardiac inhibitory actions of propranolol is the sarcoplasmic reticulum.[60]

The view that a β-blocker such as propranolol may show differential actions on atria and ventricular myocytes in comparison to sinus and atrioventricular nodes has been proposed.[61] However, at toxic concentrations any differential action would most likely not be present. Symptomatic electrocardiographic changes associated with β-blocker toxicity have been prolonged at PR, QRS, and QTc intervals.[62]

Sotalol is a unique β-blocker classified as a class III antiarrhythmic agent and is noted for prolongation of QT intervals.[63] Sotalol is capable of prolongation of action potential duration and effective refractory period.[64] It has been recognized that sotalol is capable of inhibiting hERG K channels and has proarrhythmic actions, and that it can cause a polymorphic ventricular tachycardia known as TdP,[65,66] which can lead to short episodes of loss of consciousness and/or sudden death.

Certain β-blockers at high concentrations have been found to inhibit Ca and Na channels in the myocardium. Fallouh et al.[67] reported that the short-acting β-blocker esmolol was capable of inhibiting L-type Ca and fast Na currents with an IC_{50} of 450 ± 5.0 and 170 ± 2.5 μM, respectively. In electrically driven cardiomyocytes, esmolol was reported to decrease contraction without affecting myofilament Ca^{2+} sensitivity.[67] A subsequent investigation also revealed that esmolol was able to reduce action potential duration in guinea pig heart, an effect in part linked to inhibition of the inward rectifier K^+ and L-type Ca^{2+} currents, and that it was possibly responsible for the negative inotropic effects of this compound.[68]

In summary, while β-adrenoceptor blockers have a very good safety margin in the human with respect to their unwanted actions on the myocardium, toxicity associated with this class of drugs is not unusual due to widespread clinical use. At toxic concentrations, this class of drugs can have significant negative chronotropic, inotropic, and chronotropic effects on mammalian hearts. The cellular mechanism(s) responsible for the toxic effects of beta blockers extends well beyond their ability to block the actions of catecholamines on β-adrenoceptors in the myocardium due to the inhibition of ion movement across the cell membrane that could affect pacemaker cells, conducting fibers, and/or contracting cardiomyocytes.

2.6 SODIUM CHANNEL ACTIVATOR TOXINS

The relationship between transmembrane Na^+ movement in the myocardium is tightly coupled with the process of excitation–contraction coupling. The toxins that increase Na^+ permeability in myocytes have profound effects on cardiac function. For example, a group of lipid-soluble, plant-derived toxins (e.g., grayanotoxin, veratridine, and aconitine) have been found to be capable of increasing Na^+ permeability in mammalian cardiomyocytes and to bind to site 2 on the Na channel (Figure 2.4). Batrachotoxin, which is found in *Phyllabates* frog skin, is another lipid-soluble Na channel activator that also binds to site 2 of this channel.[69] These compounds can have pronounced effects on cardiac function.

For instance, the grayanotoxin has been noted to produce positive inotropic effects and cardiac arrhythmias that can be blocked by tetrodotoxin.[70] Garyanotoxin is noted to produce afterpotentials in feline cardiac Purkinje fibers and this could account for its arrhythmic activities.[71] In SA node cells

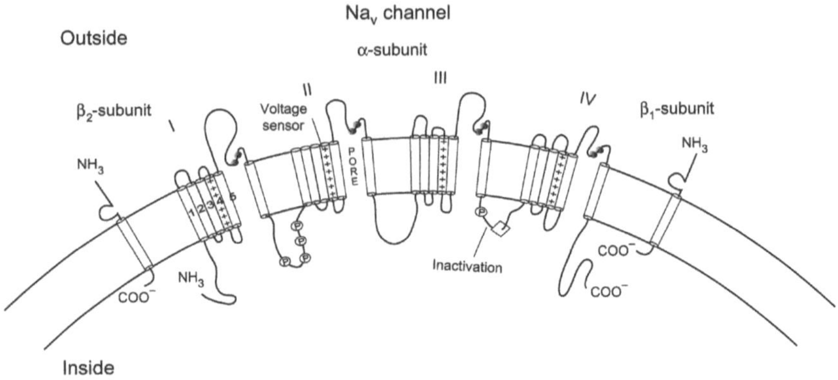

FIGURE 2.4 **Schematic representation of α- and β-subunits of the cardiac sodium channel.** The channel has six transmembrane-spanning domains where S4 is the *voltage sensor* and the *pore* is located between the S5 and S6 regions.

from rabbits, garyanotoxin was shown to modify Na^+ but not K^+ currents. It was able to increase Na^+ permeability effectively by inhibiting the inactivation process for this cation.[72] In addition, investigation of the actions of grayanotoxin in canine Purkinje fibers and guinea pig ventricular muscle revealed that this toxin produced depolarization of the myocytes that was antagonized by the presence of tetrodotoxin.[73]

Furthermore, studies using ^{23}Na nuclear magnetic resonance spectroscopy in isolated perfused guinea pig heart using the Langendorff method have indicated an increase in intracellular Na^+ by grayanotoxin that paralleled the positive inotropic effect following the addition of this toxin.[74] The binding sites for grayanotoxin have been explored using human embryonic kidney cells transfected with cardiac Na channel alpha-subunits using patch-clamp techniques. It has been noted that this toxin binds to Ile-433 in the D1S6 transmembrane segments and to Ser251 of the D1S4–S5 regions on the cardiac Na channel.[75] In addition, mutation of Leu224 in D1S4–S5 and D4S4–S5 of the cardiac Na channel alpha-subunit has revealed an important binding site for grayanotoxin.[76] Collectively, the evidence is in line with the view that sites involved in the activation of this channel show good affinity toward grayanotoxin and that this toxin binds to the open state while an unbinding reaction only seems to occur in the closed state of the Na channel.[77] Thus, assuming the mechanism for activation of voltage-gated Na channels involves S4–S5 domains, it is not unreasonable to believe that the binding site for grayanotoxin to the voltage sensor may increase the propensity of the Na channel to open, thus allowing the passage of Na^+ into the cell.

Yet another toxin that has been noted to increase intracellular concentration of Na^+ in cardiomyocytes is veratridine. Veratridine has been found to increase contractility and prolong the action potential duration in guinea pig papillary muscle, and these actions were inhibited by tetrodotoxin.[78] In sheep

Purkinje fibers, it has been demonstrated that veratridine increases intracellular concentration of Na^+ and action potential. These actions of veratridine are also inhibited by tetrodotoxin.[79] Evidence presented using rat ventricular myocytes seems to suggest that veratridine activates Na channels by interacting with a site at the intracellular level.[80]

Another toxin, ciguatoxin, can cause an increase in the contractile force of atria, and it has also been found to prolong cardiac action potential and facilitate Na^+ current in cardiomyocytes.[81] Similar actions have been noted for aconitine toxin, which has been reported to attenuate hyperpolarization and increase inward Na^+ current in cardiomyocytes.[82] Aconitine has positive inotropic effects on the myocardium[83] and is reported to produce delay after depolarization in cardiac tissue.[84] Cardiac effects of aconitine intoxication are ventricular ectopics, ventricular tachycardia, and ventricular fibrillation.[85]

The pharmacology of batrachotoxin can be attributed to its ability to depolarize cardiac tissue, as noted in studies carried out on Purkinje fibers where cells depolarized after the application of the toxin.[86] In feline papillary muscle, batrachotoxin has been shown to produce spontaneous contractions accompanied by an oscillation in membrane potential.[87] The positive inotropic effects of batrachotoxin were associated with an increase in force development of myocardial tissue. Tetrodotoxin was found to inhibit the stimulatory actions of batrachotoxin in the papillary muscle.[87] In parallel, batrachotoxin was reported to cause positive inotropic action on guinea pig papillary muscle and to increase force of contraction while prolonging relaxation time.[88] At low concentrations (<60 nM), the stimulatory actions of batrachotoxin on the action potential seem to appear after several responses are elicited.[88] The ability of batrachotoxin to bind to Na channels seems to be regulated by cyclic adenosine monophosphate (cAMP). Reportedly, an increase in the concentration of cAMP reduces the number of batrachotoxin binding sites in rat cardiac ventricular myocytes.[89]

Overall, site 2 toxins seem to shift Na^+ current activation toward a more hyperpolarized potential. This in turn seems to lead to an increase in Na^+ permeability in cardiomyocytes, eventually destabilizing the resting membrane potential, resulting in the disruption in rhythm of the heart causing arrhythmias. Some degree of hyperpolarization shift in voltage dependence of channel activation occurs, manifesting in repetitive action potentials in the cardiac tissue. This may produce premature contractions and less than optimal cardiac function.

2.7 CARDIOGLYCOSIDES

One of the major sites of action for cardioglycosides in the myocardium is the Na^+/K^+-ATPase ionic pump.[90] However, these compounds have been noted to affect system and input of the heart by causing undesirable effects on function. Cardioglycoside influence in the heart can be direct, through the autonomic nervous system, as well as indirect.[91] For example, digitalis has been noted to produce bradycardia by acting on a number sites, namely on nodose ganglia and efferent vagal pathways as well as on carotid sinus receptors and the central vagal nucleus.[92] The negative chronotropic effects of cardioglycosides were also found to occur via their direct actions on the SA node.[93] The pharmacological basis for the chronotropic actions of cardioglycosides is mainly attributed to the slowing of the sinus node impulse, hyperpolarization of cells in the atria, enhanced atrial conduction, slow atrioventricular nodal effective period, and prolongation of atrioventricular nodal effective refectory period.[94]

Accordingly, the Na^+/K^+-ATPase is a critical site of action for toxicity produced by cardioglycosides. Due to the narrow therapeutic index of this class in compounds, overinhibition of the pump can cause electrical disturbances, leading to severe diastolic and systolic dysfunction. This largely occurs as a result of inappropriate levels of inhibition of the Na^+/K^+-ATPase pump by this class of drugs; this leads to the disruption of intracellular K^+ and Ca^{2+} homeostasis due to the lowering of intracellular K^+ and the raising of intracellular levels of free Ca^{2+}, thus causing electrical and mechanical abnormalities in the myocardium.[95,96] As reported, progressive lowering of intracellular K^+ leads to an increase in automaticity and subsequently causes inexcitability in myocytes.[90] Cardioglycosides have also been found to cause a shortening of the action potential; thus, a combination of block and shortened action potential can set the stage for reentry, which in turn can manifest as fibrillation.[90,95,96] An extension of these events is a major disturbance of electrical events due to occurrence of early and delayed after-depolarization that in turn can precipitate life-threatening cardiac arrhythmias.[97–99] Cardioglycoside-induced bradycardia further compounds development of the undesirable effect via direct and indirect vagomimetic actions, which then further accentuate mechanical dysfunction in the heart.[91,94]

The resultant interaction of cardioglycosides with Na^+/K^+-ATPase in cardiomyocytes seems to increase intracellular Ca^{2+} by increasing the influx of this cation near the sarcolemmal surface and the T-tubules.[90,100] The Na^+/Ca^{2+} exchangers play a pivotal role in the increase in intracellular concentration of Ca^{2+} brought on by inhibition of the Na^+/K^+-ATPase pump induced by cardioglycosides.[101] In addition, activation of protein kinases seems to occur.[102] It also seems that activation of cAMP dependency leads to inhibition of Na^+/K^+-ATPase via the activation of NADPH oxidase.[103]

Na^+/K^+-ATPase is a ubiquitous plasma membrane-bound enzyme that belongs to the P-type ATPase family of isozymes. It is composed of two major polypeptides with two catalytic units, α and β.[104] These subunits each consist of multiple isoforms (i.e., $\alpha1$, $\alpha2$, $\alpha3$, $\alpha4$, $\beta1$, and $\beta2$).[105,106] The isoforms of α-subunits that have been found to be expressed in cardiomyocytes are $\alpha1$, $\alpha2$, and $\alpha3$.[107–109] They are present in different proportions in cardiac tissue depending on species; however, the $\alpha1$-subunit is predominant in cardiac tissue and seems to be uniformly distributed.[110] The α3-subunit has been suggested to be localized between cardiomyocytes and the heart conduction system.[111] Suggestions have been made that the α2-subunit is preferentially localized in the cardiomyocyte T-tubules,[110] is functionally coupled to Na^+/Ca^{2+} exchanger,[112] and is able to preferentially modulate Ca^{2+} transients and sarcoplasmic Ca^{2+} release.[113]

Cardiac function is also influenced by the presence of β1-subunits, and its absence causes a decrease in myocardial contractile function.[114] There are functional and structural similarities between the Na^+/K^+-ATPase and non-heavy metal (P2) subtypes of P-type ATPase, which include H^+/K^+-ATPase, SERCA, and plasma membrane Ca^{2+}-ATPase.[104,115] The toxic effects of cardioglycosides may result from the impairment of physiological function in cardiomyocytes. Data in the literature point to the binding of cardioglycosides and the Na^+/K^+-ATPase to the activation of c-Src.[104,115] In addition, the activation of Na^+/K^+-ATPase and c-Src has been linked to the transactivation of epithelial growth factor receptor, leading to production of intracellular messengers (e.g., diacylglycerol, inositol 1,4,5-trisphosphate, and the reactive oxygen species) and disturbance in Ca^{2+} homeostasis in cardiomyocytes.[104,115] Taken together, the involvement of many of these processes and/or factors could account for the electrical abnormalities associated with actions of toxic levels of cardioglycosides in cardiomyocytes.

Although the exact nature of the adverse effects of cardioglycosides remains to be fully elucidated, toxic actions of these compounds may extend beyond an increase in the intracellular concentration of Na^+ via inhibition of the Na^+/K^+-ATPase pump, leading to an abnormal increase in the intracellular concentration of Ca^{2+}. The associated toxic effects may involve the activation of second messenger cascades, disturbances in Ca^{2+} homeostasis, and activation of gene expression, leading to the disruption of normal physiological cell function.

2.8 ANTHRACYCLINES

Anthracycline compounds, such as daunorubicin and doxorubicin, employed in the treatment of cancer as chemotherapeutic agents can produce cardiac toxicity. However, the mechanisms of the action for this class of drugs appear to be fairly complex and are yet to be fully delineated.[116] In broad terms, myriad cellular processes have been proposed to account for the toxic actions of such compounds—that is, the generation of reactive oxygen species,[117] lipid peroxidation,[118,119] impairment of mitochondrial function,[120,121] and modifications of gene transcription[122] (Figure 2.5). However, a discord is evident in the correlation of the doses of anthracyclines used experimentally and those employed clinically that lead to toxic cardiac events.

A point of discussion as to the basis of this discrepancy of experimental versus clinical observations is the view that certain anthracyclines can selectively accumulate in cardiac tissue via active processes.[123,124] Thus, this specific

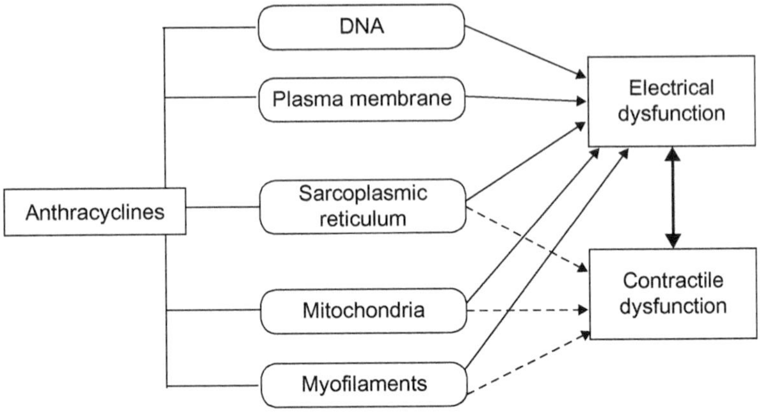

FIGURE 2.5 **Possible sites in myocardial cells at which anthracyclines can act to produce cardiac toxic effects.**

circumstance may result in the accumulation of higher concentrations of the drug in cardiomyocytes than would be normally expected based on the administered dose and could thus precipitate toxic events associated with this class of drugs. Retrospective examination of cardiac tissue from patients treated with anthracyclines found myocardial necrosis, fibrosis, and myocytolysis.[125]

Bristow et al.[126,127] cataloged and described the impact of anthracyclines as causing congestive heart failure dose-dependently. More than 20% of patients treated with anthracyclines experienced substantial degrees of heart failure.[126,128] Moreover, the toxic effects of anthracyclines appeared to be partially a result of the formation of toxic metabolites.[129]

A report by Nakamae et al.[130] provided evidence of correlation between QTc interval increase, left ventricular dilation, and systolic dysfunction in patients treated with anthracycline. Reduction in repolarization reserve due to exposure to anthracycline may account for prolongation of the QTc interval. It has been reported that short periods of treatment of patients with doxorubicin could result in a significant increase in ventricular recovery time dispersion indexes.[131] Such occurrences may manifest in sustained ventricular tachycardia. In addition, experimental data presented by Milberg et al.[132] seem to support the view that treatment of rabbits with doxorubicin reduces repolarization reserve and may facilitate TdP by drugs that block hERG. Prolongation of QTc interval has been reported six months post-anthracycline therapy, and changes in ECG have been linked to left ventricular dysfunction.[133] Collectively, evidence seems to indicate that treatment with anthracycline is a risk factor in the development of TdP. Low levels of K in plasma (serum) increase the risk of TdP if a patient is taking a drug that can prolong QT interval. Females have a higher risk of developing TdP than males when using drugs that are known to prolong QT interval.

2.8.1 Sarcoplasmic Reticulum

The acute effects of anthracyclines on cardiac tissue appear to be biphasic in vitro. An initial early effect of increased contractile force is followed by a delayed decrease in cardiac function.[135] Chemicals that have toxic effects on intracellular organelles in the myocardium can produce profound effects on the process of excitation–contraction coupling. Evidence has been provided to indicate that tension of permeabilized cardiac trabeculae was significantly reduced in animals chronically treated with doxorubicin when compared to saline for four weeks.[136] In addition, cardiac tissue from doxorubicin-treated animals had a lower Ca^{2+} sensitivity in permeabilized tissue, but the amplitude of contractions induced by caffeine were similar in doxorubicin compared to saline-treated animals.[136]

Taken together, the findings can be interpreted as suggesting that treatment with doxorubicin may impair actin–myosin interaction but does not significantly attenuate SR functionality. Data have also been presented to indicate that ventricular myocytes exposed to doxorubicin show concentration-dependent increases in myofilament disarray.[137] Accordingly, it would appear that the detrimental actions of doxorubicin, at least in part, can be attributed to a disturbance in the contractile elements in the myocardial tissue.

Doxorubicin has been found to produce contractions in isolated permeabilized and intact rabbit cardiac tissue similar to the effects of caffeine by releasing Ca^{2+} from SR without directly affecting myofilament function.[138] Daunorubicin has also been found to affect SR Ca^{2+} handling, but in a free radical-independent manner in rabbit cardiac tissue.[139] Alterations in Ca^{2+} release from the SR attributed to anthracyclines may play a role in modifying the inotropic and lusitropic action of the myocardium, thus causing cardiac dysfunction.

2.8.2 Mitochondria

A disruption in mitochondrial function seems to be at the center of the toxic effects associated with anthracyclines. There have been reports that chronic treatment of rats with doxorubicin produces Ca^{2+} cycling and disturbances in mitochondrial Ca^{2+} homeostasis.[140] Damage to cardiac but not liver mitochondria has been reported in mice treated with doxorubicin; specifically, swelling and disruption of the mitochondrial cristae were found to occur.[141] Tokarska-Schlattner et al.[142] have reported that anthracyclines impair mitochondrial creatine kinase activity in a dose- and time-dependent way. Doxorubicin has been found to cause matrix swelling and cristae disorganization of the mitochondria based on electron microscopy evidence.[143] Childs et al.[144] noted that doxorubicin induces oxidative stress and mitochondria-mediated apoptosis. In addition, data in the current literature seem to indicate that rats injected with doxorubicin have damaged cardiac mitochondrial DNA.[145] Both reactive oxygen and nitrogen products have been found to be responsible for acute doxorubicin-induced cardiac mitochondrial injury,[146] but it seems that oxidative damage precedes nitrative damage in this tissue.[147] Collectively, information in the literature implies that the cardiac mitochondria is a specific site of toxic action by anthracyclines.[148] Recent findings in rabbits with left ventricular dilation and congestive heart failure induced by daunorubicin seem to indicate that global oxidative stress is not the critical ingredient for mitochondrial dysfunction. However, severe downregulation of mitochondrial genes has been reported both at nuclear and mitochondrial genome levels.[149]

2.8.3 Deoxyribonucleic Acid/Messenger Ribonucleic Acid

In addition to the effects of anthracyclines on various organelles in cardiomyocytes, this class of compounds may also have a negative impact on DNA/mRNA. For example, doxorubicin, daunorubicin, and epirubicin have been found to decrease α-actin mRNA.[150] Also, an examination of the ultrastructure of cardiac tissue has been found to show loss of myofilaments.[150] Production of superoxide and associated peroxide and hydroxyl radicals may provide the means by which anthracyclines cause DNA damage in cardiac tissue.[139] However, damage caused by anthracyclines to DNA and the mitochondria may not only occur in series but also seem to occur in parallel. Moreover, it has been suggested that the damage caused by doxorubicin to DNA is proximal to damage caused to mitochondria in cardiac tissue.[151]

More recent data have also linked sarcomere disruption to interference by anthracycline, doxorubicin, and mRNA. Cardiac ankyrin repeat protein (CARP) is associated with sarcomere organization. Doxorubicin was found to inhibit CARP mRNA transcription, leading to sarcomere disruption and disorganization in cardiomyocytes.[152] In hearts of animals treated with doxorubicin, 5′-adenosine monophosphate–activated protein kinase (AMPK) was found to be inhibited.[153] AMPK is believed to be a sensor and regulator in cellular energy-signaling systems, and its inhibition by anthracycline was reported to be largely trigged by damage to DNA.[153] In light of the cellular actions of anthracycline, substantial collateral damage to the cardiac DNA moiety will ultimately compromise myocardial cell function. The broad basis of the toxic effects of anthracycline in the myocardium can be mechanical and/or electrical. It can be categorized as biophysical in nature. These chemicals seem to affect many facets of cellular function particularly relating to energy production and utilization. Cell dysfunction and death seem to be imminent once toxic levels of these compounds have made contact with the myocyte.

Among the adjuvants, one drug (i.e., trastuzumab) is a recombinant antibody directed against the human epidermal growth receptor tyrosine kinase (HER2/ErbB2) and has been used together with anthracyclines in the treatment of breast cancer. The use of trastuzumab seems to increase the likelihood of cardiac dysfunction, especially for high-risk patients and if used concomitantly with anthracyclines.[154–156] The mechanisms of action by which trastuzumab produces cardiac dysfunction remain to be fully resolved. However, evidence in the current literature points to multiple cellular actions that underpin the occurrence of cardiac toxicity. For example, suggestions have been made that trastuzumab may activate the mitochondrial apoptosis pathway via the proapoptotic Bcl-1 family protein Bcl-xS, which causes an impairment of mitochondrial function and integrity, leading to disruption of cellular energies.[157]

Nonetheless, as noted by electron microscopy, cardiac toxicity associated with clinical use of trastuzumab does not seem to lead to cellular death and abnormal ultrastructure of myocytes;[158] this is in contrast to anthracyclines. In general, the cardiac toxic effects of trastuzumab are related to the inhibition of HER2 signaling pathways, which reduces cell-protective and growth-promoting processes. Crone and colleagues[159] demonstrated the development of cardiomyopathy and dysfunction with contraction and relaxation processes, as well as reduced response to β-adrenoceptor agonists and decreased stress tolerance in ErbB2 knockout mice. Thus, an impairment in ErbB2-related cellular function by treatment with trastuzumab could lead to cardiac dysfunction. Trastuzumab has also been noted to cause myofilament degradation comparable to that produced by inhibitors of the mitogen-activated protein (MAP) kinase signaling system; thus, one potential mechanism of action may be an alteration in MAP kinase and/or extracellular signal-regulated kinases (ERKs) half pathways.[160]

2.9 MISCELLANEOUS hERG K-CHANNEL INHIBITORS

Many drugs that seem to be capable of causing fatal cardiac arrhythmias have been associated with prolongation of QT interval (Figure 2.6), and a large number of drugs have been noted to inhibit hERG K channels (Figure 2.7). These include antimicrobial, antifungal, psychiatric, gastrointestinal, antimalarial, and antihistamine drugs. The link between inhibition of hERG K channel, prolongation of QT interval, and fatal cardiac arrhythmias is not simple. The rapidly activating delayed-rectifier current (I_{Kr}) in the myocardium is considered to be generated by hERG K channels.[161,162] Inhibition of this channel, causing prolongation of QT intervals, has been associated with the development of polymorphic tachycardia of the cardiac arrhythmia's TdP. The occurrence of TdP can cause death. Examples of drugs that have been found to produce TdP are listed in Table 2.2.

Apart from the direct block produced by compounds that have been found to interfere with the hERG K^+ current, inhibition of trafficking has been identified as a mechanism that can also impede the development and generation of this current. For example, arsenic trioxide can directly block hERG, a K channel, but has also been noted for attenuating channel actions by reducing its expression in human embryonic kidney (HEK) cell lines. Arsenic trioxide was reported to interfere with channel trafficking by

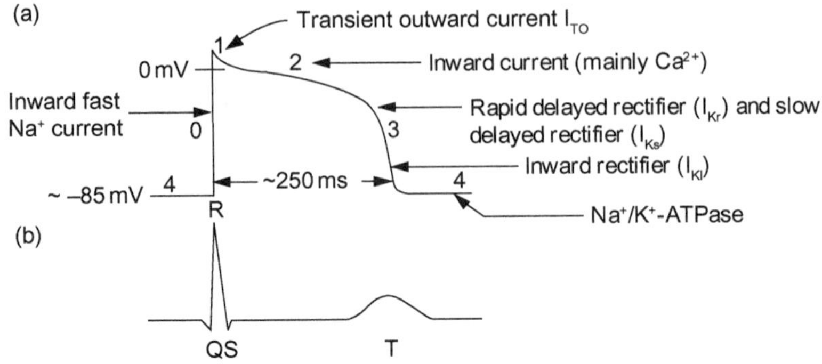

FIGURE 2.6 Representation of (a) typical cardiac (ventricular) action potential and (b) the corresponding electrocardiogram.

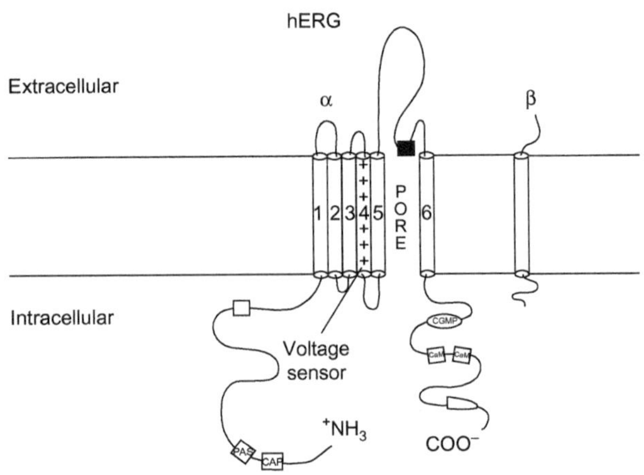

FIGURE 2.7 Schematic representation of α- and β-subunits of the cardiac hERG potassium channel. The channel has six transmembrane-spanning domains where S4 is the *voltage sensor* and the *pore* is located between the S5 and S6 regions.

inhibition of the hERG chaperone complex, leading to a decrease in current density in guinea pig cardiomyocytes.[163] Geldanamycin, an inhibitor of Hsp90, was found to prevent maturation of hERG K channels, and thus Hsp90 has been identified as a possible site of interference with maturation of hERG K channels.[164] Among drugs that have been identified to interfere with development of hERG K channels are pentamidine,[165,166] some cardioglycosides,[167,168] ketoconazole,[169] amoxapine,[170] desipramine,[171] and fluconazole.[172] It is evident that many drugs are able to attenuate I_{Kr} and this can cause significant toxicity in the myocardium, leading to cardiac arrest.

TABLE 2.2 Examples of Drugs That Have Been Noted to Cause TdP

Class	Drug
Psychiatric	Chlorpromazine Clozapine Haloperidol Iloperidone Pimozide
Antiarrhythmic	Disopyramide Dofetilide Ibutilide Procainamide Quinidine Sotalol
Antimicrobial	Ciprofloxacin Clarithromycin Erythromycin Ketoconazole Olfoxacin Pentamidine
Antihistamine	Astemizole Diphenhydramine Loratadine Terfenadine
Antidepressant	Amitriptyline (TCA) Citalopram (SSRI) Fluoxetine (SSRI) Imipramine (TCA) Nortriptyline (TCA)
Others	Arsenic trioxide (anticancer) Bepridil (antianginal) Chloroquine (antimalarial) Cisapride (gastrointestinal stimulant) Terodiline (antimuscarinic)

SSRI, selective serotonin reuptake inhibitors; TCA, tricyclic antidepressants

2.10 CONCLUSION

Many chemicals, drugs, and toxins can cause toxic effects in the myocardium. It is also recognized that the nature of the toxic effects produced by compounds in the heart can be multifaceted and fairly complex. There is usually a positive correlation between the dose/concentration of drugs and their ability to cause toxic effects in cardiomyocytes. Drugs that affect ion channel (i.e., Na, K, and Ca) functions seem to be the most likely candidates to produce toxic effects in the heart. It also appears that a vast number of chemicals with diverse chemical structures can inhibit hERG K channels and cause adverse cardiac events. However, in many cases, toxic effects of chemicals on the myocardium may be avoided and/or minimized by suitable use of therapeutic agents and careful monitoring of the cardiovascular system thereafter.

Key Learning Points

- Chemicals produce toxicity in the heart by affecting many cellular events.
- The major classes of drugs that produce cardiac toxicity are β-blockers, tricyclic antidepressants, anthracyclines, and cardioglycosides.
- The other major chemicals that produce toxicity are channel blockers, including a host of drugs with diverse chemical structures that block I_{kr} and prolong QT interval.

REFERENCES

1. ALLHAT Collaborative Research Group. Major cardiovascular events in hypertensive patients randomized to doxazosin vs. chlorthalidone: the antihypertensive and lipid-lowering treatment to prevent heart attack trial (ALLHAT). *JAMA* 2000;**283**(15):1967–75.
2. Thomas D, Wimmer AB, Wu K, et al. Inhibition of human ether-à-go-go-related gene potassium channels by alpha 1-adrenoceptor antagonists prazosin, doxazosin, and terazosin. *Naunyn Schmiedebergs Arch Pharmacol* 2004;**369**(5):462–72.
3. Thomas D, Bloehs R, Koschny R, et al. Doxazosin induces apoptosis of cells expressing hERG K^+ channels. *Eur J Pharmacol* 2008;**579**(1–3):98–103.
4. Jehle J, Schweizer PA, Katus HA, et al. Novel roles for hERG K^+ channels in cell proliferation and apoptosis. *Cell Death Dis* 2011;**2**:e193.
5. Qureshi AI, Suri MF, Guterman LR, et al. Cocaine use and the likelihood of nonfatal myocardial infarction and stroke: data from the Third National Health and Nutrition Examination Survey. *Circulation* 2001;**103**(4):502–6.
6. Carrillo X, Curós A, Muga R, et al. Acute coronary syndrome and cocaine use: 8-year prevalence and inhospital outcomes. *Eur Heart J* 2011;**32**(10):1244–50.
7. Seifen E, Plunkett LM, Kennedy RH. Cardiovascular and lethal effects of cocaine in anesthetized dogs and guinea pigs. *Arch Int Pharmacodyn Ther* 1989;**300**:241–53.
8. Plunkett LM, Seifen E, Kennedy RH. Effects of morphine pretreatment on cocaine cardiotoxicity in anesthetized guinea pigs. *Arch Int Pharmacodyn Ther* 1989;**297**:60–7.
9. Hale SL, Alker KJ, Rezkalla S, et al. Adverse effects of cocaine on cardiovascular dynamics, myocardial blood flow, and coronary artery diameter in an experimental model. *Am Heart J* 1989;**118**(5 Pt 1):927–33.
10. Fraker Jr TD, Temesy-Armos PN, Brewster PS, et al. Mechanism of cocaine-induced myocardial depression in dogs. *Circulation* 1990;**81**(3):1012–6.
11. Temesy-Armos PN, Fraker Jr TD, Brewster PS, et al. The effects of cocaine on cardiac electrophysiology in conscious, unsedated dogs. *J Cardiovasc Pharmacol* 1992;**19**(6):883–91.
12. Pitts WR, Vongpatanasin W, Cigarroa JE, et al. Effects of the intracoronary infusion of cocaine on left ventricular systolic and diastolic function in humans. *Circulation* 1998;**97**(13):1270–3.

13. Lange RA, Cigarroa RG, Yancy Jr CW, et al. Cocaine-induced coronary-artery vasoconstriction. *N Engl J Med* 1989;**321**(23):1557–62.
14. Moliterno DJ, Lange RA, Gerard RD, et al. Influence of intranasal cocaine on plasma constituents associated with endogenous thrombosis and thrombolysis. *Am J Med* 1994;**96**(6): 492–6.
15. Rezkalla SH, Mazza JJ, Kloner RA, et al. Effects of cocaine on human platelets in healthy subjects. *Am J Cardiol* 1993;**72**(2):243–6.
16. Wilbert-Lampen U, Seliger C, Zilker T, et al. Cocaine increases the endothelial release of immunoreactive endothelin and its concentrations in human plasma and urine: reversal by coincubation with sigma-receptor antagonists. *Circulation* 1998;**98**(5):385–90.
17. Ferreira S, Crumb Jr WJ, Carlton CG, et al. Effects of cocaine and its major metabolites on the HERG-encoded potassium channel. *J Pharmacol Exp Ther* 2001;**299**(1):220–6.
18. Zhang S, Rajamani S, Chen Y, et al. Cocaine blocks HERG, but not KvLQT1 + minK, potassium channels. *Mol Pharmacol* 2001;**59**(5):1069–76.
19. Bauman JL, DiDomenico RJ. Cocaine-induced channelopathies: emerging evidence on the multiple mechanisms of sudden death. *J Cardiovasc Pharmacol Ther* 2002;**7**(3):195–202.
20. O'Connor AD, Rusyniak DE, Bruno A. Cerebrovascular and cardiovascular complications of alcohol and sympathomimetic drug abuse. *Med Clin North Am* 2005;**89**(6):1343–58.
21. Devlin RJ, Henry JA. Clinical review: major consequences of illicit drug consumption. *Crit Care* 2008;**12**(1):202–8.
22. Thanacoody HK, Thomas SH. Tricyclic antidepressant poisoning: cardiovascular toxicity. *Toxicol Rev* 2005;**24**(3):205–14.
23. Matsuo S. Comparative effects of imipramine and propranolol on the transmembrane potentials of the isolated rabbit atria. *Jpn J Pharmacol* 1967;**17**(2):279–86.
24. Rawling DA, Fozzard HA. Effects of imipramine on cellular electrophysiological properties of cardiac Purkinje fibers. *J Pharmacol Exp Ther* 1979;**209**(3):371–5.
25. Rodriguez S, Tamargo J. Electrophysiological effects of imipramine on bovine ventricular muscle and Purkinje fibres. *Br J Pharmacol* 1980;**70**(1):15–23.
26. Isenberg G, Tamargo J. Effect of imipramine on calcium and potassium currents in isolated bovine ventricular myocytes. *Eur J Pharmacol* 1985;**108**(2):121–31.
27. Ogata N, Narahashi T. Block of sodium channels by psychotropic drugs in single guinea-pig cardiomyocytes. *Br J Pharmacol* 1989;**97**(3):905–13.
28. Habuchi Y, Furukawa T, Tanaka H, et al. Block of Na^+ channels by imipramine in guinea-pig cardiac ventricular cells. *J Pharmacol Exp Ther* 1991;**256**(3):1072–81.
29. Delpón E, Tamargo J, Sánchez-Chapula J. Effects of imipramine on the transient outward current in rabbit atrial single cells. *Br J Pharmacol* 1992;**106**(2):464–9.
30. Nau C, Seaver M, Wang SY, et al. Block of human heart hH1 sodium channels by amitriptyline. *J Pharmacol Exp Ther* 2000;**292**(3):1015–23.
31. Barber MJ, Starmer CF, Grant AO. Blockade of cardiac sodium channels by amitriptyline and diphenylhydantoin. Evidence for two use-dependent binding sites. *Circ Res* 1991;**69**(3): 677–96.
32. Bou-Abboud E, Nattel S. Molecular mechanisms of the reversal of imipramine-induced sodium channel blockade by alkalinization in human cardiomyocytes. *Cardiovasc Res* 1998; **38**(2):395–404.
33. Watts JA, Yates KM, Badar SK, et al. Mechanisms of Ca^{2+} antagonism in imipramine-induced toxicity of isolated adult rat cardiomyocytes. *Toxicol Appl Pharmacol* 1998;**153** (1):95–101.

34. Zima AV, Qin J, Fill M, et al. Tricyclic antidepressant amitriptyline alters sarcoplasmic reticulum calcium handling in ventricular myocytes. *Am J Physiol Heart Circ Physiol* 2008; **295**(5):H2008–16.
35. Chopra N, Laver D, Davies SS, et al. Amitriptyline activates cardiac ryanodine channels and causes spontaneous sarcoplasmic reticulum calcium release. *Mol Pharmacol* 2009;**75**(1): 183–95.
36. Jo SH, Youm JB, Lee CO, et al. Blockade of the HERG human cardiac K^+ channel by the antidepressant drug amitriptyline. *Br J Pharmacol* 2000;**129**(7):1474–80.
37. Duncan RS, McPate MJ, Ridley JM, et al. Inhibition of the HERG potassium channel by the tricyclic antidepressant doxepin. *Biochem Pharmacol* 2007;**74**(3):425–37.
38. Kurz T, Richardt G, Hagl S, et al. Two different mechanisms of noradrenaline release during normoxia and simulated ischemia in human cardiac tissue. *J Mol Cell Cardiol* 1995;**27** (5):1161–72.
39. Garvey HL. The mechanism of action of verapamil on the sinus and AV nodes. *Eur J Pharmacol* 1969;**8**(2):159–66.
40. Cranefield PF, Aronson RS, Wit AL. Effect of verapamil on the normal action potential and on a calcium-dependent slow response of canine cardiac Purkinje fibers. *Circ Res* 1974;**34**(2):204–13.
41. Zipes DP, Fischer JC. Effects of agents which inhibit the slow channel on sinus node automaticity and atrioventricular conduction in the dog. *Circ Res* 1974;**34**(2):184–92.
42. Wit AL, Cranefield PF. Effect of verapamil on the sinoatrial and atrioventricular nodes of the rabbit and the mechanism by which it arrests reentrant atrioventricular nodal tachycardia. *Circ Res* 1974;**35**(3):413–25.
43. Vick JA, Kandil A, Herman EH, et al. Reversal of propranolol and verapamil toxicity by calcium. *Vet Hum Toxicol* 1983;**25**(1):8–10.
44. Henry M, Kay MM, Viccellio P. Cardiogenic shock associated with calcium-channel and β blockers: reversal with intravenous calcium chloride. *Am J Emerg Med* 1985;**3**(4):334–6.
45. Okada T, Konishi T. Effects of verapamil on SA and AV nodal action potentials in the isolated rabbit heart. *Jpn Circ J* 1975;**39**(8):913–7.
46. Iijima T, Taira N. Modification by manganese ions and verapamil of the responses of the atrioventricular node to norepinephrine. *Eur J Pharmacol* 1976;**37**(1):55–62.
47. Smith HJ, Goldstein RA, Griffith JM, et al. Regional contractility. Selective depression of ischemic myocardium by verapamil. *Circulation* 1976;**54**(4):629–35.
48. Zaritsky AL, Horowitz M, Chernow B. Glucagon antagonism of calcium channel blocker-induced myocardial dysfunction. *Crit Care Med* 1988;**16**(3):246–51.
49. Koury SI, Stone CK, Thomas SH. Amrinone as an antidote in experimental verapamil overdose. *Acad Emerg Med* 1996;**3**(8):762–7.
50. DeWitt CR, Waksman JC. Pharmacology, pathophysiology and management of calcium channel blocker and β-blocker toxicity. *Toxicol Rev* 2004;**23**(4):223–38.
51. Ashraf M, Chaudhary K, Nelson J, et al. Massive overdose of sustained-release verapamil: a case report and review of literature. *Am J Med Sci* 1995;**310**(6):258–63.
52. Cave G, Harvey M. Intravenous lipid emulsion as antidote beyond local anesthetic toxicity: a systematic review. *Acad Emerg Med* 2009;**16**(9):815–24.
53. Christian MR, Pallasch EM, Wahl M, et al. Lipid Rescue 911: are poison centers recommending intravenous fat emulsion therapy for severe poisoning? *J Med Toxicol* 2013;**9**(3):231–4.
54. Geib AJ, Liebelt E, Manini AF. Toxicology Investigators' Consortium (ToxIC). Clinical experience with intravenous lipid emulsion for drug-induced cardiovascular collapse. *J Med Toxicol* 2012;**8**(1):10–4.

55. Heath A. Beta-adrenoceptor blocker toxicity: clinical features and therapy. *Am J Emerg Med* 1984;**2**(6):518–25.
56. Hellenbrecht D, Lemmer B, Wiethold G, et al. Measurement of hydrophobicity, surface activity, local anaesthesia, and myocardial conduction velocity as quantitative parameters of the non-specific membrane affinity of nine adrenergic blocking agents. *Naunyn Schmiedebergs Arch Pharmacol* 1973;**277**(2):211–26.
57. de Wildt D, Sangster B, Langemeijer J, et al. Different toxicological profiles for various beta-blocking agents on cardiac function in isolated rat hearts. *J Toxicol Clin Toxicol* 1984; **22**(2):115–32.
58. De Pauw M, Vilaine JP, Heyndrickx GR. Role of force–frequency relation during AV-block, sinus node block and beta-adrenoceptor block in conscious animals. *Basic Res Cardiol* 2004;**99**(5):360–71.
59. Dhalla NS, Lee SL. Comparison of the actions of acebutolol, practolol and propranolol on calcium transport by heart microsomes and mitochondria. *Br J Pharmacol* 1976;**57**(2):215–21.
60. Noack E, Kurzmack M, Verjovski-Almeida S, et al. The effect of propranolol and its analogs on Ca^{++} transport by sarcoplasmic reticulum vesicles. *J Pharmacol Exp Ther* 1978; **206**(2):281–8.
61. Chang MS, Zipes DP. Differential sensitivity of sinus node, atrioventricular node, atrium, and ventricle to propranolol. *Am Heart J* 1988;**116**(2 Pt 1):371–8.
62. Love JN, Enlow B, Howell JM, et al. Electrocardiographic changes associated with beta-blocker toxicity. *Ann Emerg Med* 2002;**40**(6):603–10.
63. Singh BN, Vaughan Williams EM. A third class of anti-arrhythmic action. Effects on atrial and ventricular intracellular potentials, and other pharmacological actions on cardiac muscle of MJ 1999 and AH 3474. *Br J Pharmacol* 1970;**39**(4):675–87.
64. Strauss HC, Bigger Jr JT, Hoffman BF. Electrophysiologial and beta-receptor blocking effects of MJ 1999 on dog and rabbit cardiac tissue. *Circ Res* 1970;**26**(6):661–78.
65. Haverkamp W, Breithardt G, Camm AJ, et al. The potential for QT prolongation and proarrhythmia by non-antiarrhythmic drugs: clinical and regulatory implications. Report on a policy conference of the European Society of Cardiology. *Eur Heart J* 2000;**21**(15):1216–31.
66. Hondeghem LM, Lu HR, van Rossem K, et al. Detection of proarrhythmia in the female rabbit heart: blinded validation. *J Cardiovasc Electrophysiol* 2003;**14**(3):287–94.
67. Fallouh HB, Bardswell SC, McLatchie LM, et al. Esmolol cardioplegia: the cellular mechanism of diastolic arrest. *Cardiovasc Res* 2010;**87**(3):552–60.
68. Shibata S, Okamoto Y, Endo S, et al. Direct effects of esmolol and landiolol on cardiac function, coronary vasoactivity, and ventricular electrophysiology in guinea-pig hearts. *J Pharmacol Sci* 2012;**118**(2):255–65.
69. Wang SY, Wang GK. Voltage-gated sodium channels as primary targets of diverse lipid-soluble neurotoxins. *Cell Signal* 2003;**15**(2):151–9.
70. Akera T, Ku DD, Frank M, et al. Effects of grayanotoxin I on cardiac Na^+K^+-adenosine triphosphatase activity, transmembrane potential and myocardial contractile force. *J Pharmacol Exp Ther* 1976;**199**(1):247–54.
71. Brown BS, Akera T, Brody TM. Mechanism of grayanotoxin III-induced afterpotentials in feline cardiac Purkinje fibers. *Eur J Pharmacol* 1981;**75**(4):271–81.
72. Nakao M, Seyama I. Effect of alpha-dihydro-grayanotoxin-II on the electrical activity of the rabbit sino-atrial node. *J Physiol* 1984;**357**:79–91.
73. Ito K, Saruwatari N, Mitani K, et al. Characterization of depolarization induced by palytoxin and grayanotoxin-I in isolated cardiac tissues from dogs and guinea pigs. *Naunyn Schmiedebergs Arch Pharmacol* 1985;**330**(1):67–73.

74. Hotta Y, Ando H, Takeya K, et al. Direct measurement of increased myocardial cellular 23Na NMR signals in perfused guinea-pig heart induced by dihydroouabain and grayanotoxin-I. *Mol Cell Biochem* 1994;**139**(1):59–70.
75. Kimura T, Yamaoka K, Kinoshita E, et al. Novel site on sodium channel alpha-subunit responsible for the differential sensitivity of grayanotoxin in skeletal and cardiac muscle. *Mol Pharmacol* 2001;**60**(4):865–72.
76. Maejima H, Kinoshita E, Yuki T, et al. Structural determinants for the action of grayanotoxin in D1 S4-S5 and D4 S4-S5 intracellular linkers of sodium channel alpha-subunits. *Biochem Biophys Res Commun* 2002;**295**(2):452–7.
77. Yuki T, Yamaoka K, Yakehiro M, et al. State-dependent action of grayanotoxin I on Na^+ channels in frog ventricular myocytes. *J Physiol* 2001;**534**(Pt 3):777–90.
78. Honerjäger P, Reiter M. The relation between the effects of veratridine on action potential and contraction in mammalian ventricular myocardium. *Naunyn Schmiedebergs Arch Pharmacol* 1975;**289**(1):1–28.
79. Brill DM, Wasserstrom JA. Intracellular sodium and the positive inotropic effect of veratridine and cardiac glycoside in sheep Purkinje fibers. *Circ Res* 1986;**58**(1):109–19.
80. Honerjäger P, Dugas M, Zong XG. Mutually exclusive action of cationic veratridine and cevadine at an intracellular site of the cardiac sodium channel. *J Gen Physiol* 1992;**99**(5): 699–720.
81. Seino A, Kobayashi M, Momose K, et al. The mode of inotropic action of ciguatoxin on guinea-pig cardiac muscle. *Br J Pharmacol* 1988;**95**(3):876–82.
82. Nilius B, Boldt W, Benndorf K. Properties of aconitine-modified sodium channels in single cells of mouse ventricular myocardium. *Gen Physiol Biophys* 1986;**5**(5):473–84.
83. Honerjäger P, Meissner A. The positive inotropic effect of aconitine. *Naunyn Schmiedebergs Arch Pharmacol* 1983;**322**(1):49–58.
84. Sawanobori T, Hirano Y, Hiraoka M. Aconitine-induced delayed afterdepolarization in frog atrium and guinea pig papillary muscles in the presence of low concentrations of Ca^{2+}. *Jpn J Physiol* 1987;**37**(1):59–79.
85. Chan TY. Aconite poisoning. *Clin Toxicol* 2009;**47**(4):279–85.
86. Hogan PM, Albuquerque EX. The pharmacology of batrachotoxin. 3. Effect on the heart Purkinje fibers. *J Pharmacol Exp Ther* 1971;**176**(3):529–37.
87. Shotzberger GS, Albuquerque EX, Daly JW. The effects of batrachotoxin on cat papillary muscle. *J Pharmacol Exp Ther* 1976;**196**(2):433–44.
88. Honerjäger P, Reiter M. Sarcolemmal sodium permeability and contractile force of guinea pig papillary muscle: effects of germitrine. *Circ Res* 1977;**40**(1):90–8.
89. Taouis M, Sheldon RS, Hill RJ, et al. Cyclic AMP-dependent regulation of the number of [^{3}H]batrachotoxinin benzoate binding sites on rat cardiomyocytes. *J Biol Chem* 1991;**266**(16):10300–4.
90. Langer GA. Effects of digitalis on myocardial ionic exchange. *Circulation* 1972;**46**(1): 180–7.
91. Gillis RA, Quest JA. The role of the nervous system in the cardiovascular effects of digitalis. *Pharmacol Rev* 1979;**31**(1):19–97.
92. Chai CY, Wang HH, Hoffman BF, et al. Mechanisms of bradycardia induced by digitalis substances. *Am J Physiol* 1967;**212**(1):26–34.
93. Ten Eick RE, Hoffman BF. Chronotropic effect of cardiac glycosides in cats, dogs, and rabbits. *Circ Res* 1969;**25**(4):365–78.
94. Rosen MR, Wit AL, Hoffman BF. Electrophysiology and pharmacology of cardiac arrhythmias. IV. Cardiac antiarrhythmic and toxic effects of digitalis. *Am Heart J* 1975;**89**(3):391–9.

95. Saunders JH, Ferrier GR, Moe GK. Conduction block associated with transient depolarizations induced by acetylstrophanthidin in isolated canine Purkinje fibers. *Circ Res* 1973;**32**(5):610–7.
96. Rosen MR, Gelband H, Merker C, et al. Mechanisms of digitalis toxicity: effects of ouabain on phase four of canine Purkinje fiber transmembrane potentials. *Circulation* 1973;**47**(4):681–9.
97. Ferrier GR, Saunders JH, Mendez C. A cellular mechanism for the generation of ventricular arrhythmias by acetylstrophanthidin. *Circ Res* 1973;**32**(5):600–9.
98. Marban E, Robinson SW, Wier WG. Mechanisms of arrhythmogenic delayed and early afterdepolarizations in ferret ventricular muscle. *J Clin Invest* 1986;**78**(5):1185–92.
99. Sicouri S, Antzelevitch C. Drug-induced afterdepolarizations and triggered activity occur in a discrete subpopulation of ventricular muscle cells (M cells) in the canine heart: quinidine and digitalis. *J Cardiovasc Electrophysiol* 1993;**4**(1):48–58.
100. Barry WH, Hasin Y, Smith TW. Sodium pump inhibition, enhanced calcium influx via sodium-calcium exchange, and positive inotropic response in cultured heart cells. *Circ Res* 1985;**56**(2):231–41.
101. Dostanic I, Schultz JJ, Lorenz JN, et al. The $\alpha 1$ isoform of Na/K-ATPase regulates cardiac contractility and functionally interacts and co-localizes with the Na/Ca exchanger in heart. *J Biol Chem* 2004;**279**(52):54053–61.
102. Tian J, Cai T, Yuan Z, et al. Binding of Src to Na^+/K^+-ATPase forms a functional signaling complex. *Mol Biol Cell* 2006;**17**(1):317–26.
103. White CN, Liu CC, Garcia A, et al. Activation of cAMP-dependent signaling induces oxidative modification of the cardiac Na^+-K^+ pump and inhibits its activity. *J Biol Chem* 2010;**285**(18):13712–20.
104. Pierre SV, Xie Z. The Na^+/K^+-ATPase receptor complex: its organization and membership. *Cell Biochem Biophys* 2006;**46**(3):303–16.
105. McDonough AA, Velotta JB, Schwinger RH, et al. The cardiac sodium pump: structure and function. *Basic Res Cardiol* 2002;**97**(Suppl. 1):I19–24.
106. Lingrel J, Moseley A, Dostanic I, et al. Functional roles of the alpha isoforms of the Na/K-ATPase. *Ann NY Acad Sci* 2003;**986**:354–9.
107. McDonough AA, Zhang Y, Shin V, et al. Subcellular distribution of sodium pump isoform subunits in mammalian cardiomyocytes. *Am J Physiol* 1996;**270**(4 Pt 1):C1221–7.
108. Wang J, Schwinger RH, Frank K, et al. Regional expression of sodium pump subunit isoforms and Na^+-Ca^{++} exchanger in the human heart. *J Clin Invest* 1996;**98**(7):1650–8.
109. Harada K, Lin H, Endo Y, et al. Subunit composition and role of Na^+/K^+-ATPase in ventricular myocytes. *J Physiol Sci* 2006;**56**(1):113–21.
110. Despa S, Bers DM. Functional analysis of Na^+/K^+-ATPase isoform distribution in rat ventricular myocytes. *Am J Physiol Cell Physiol* 2007;**293**(1):C321–7.
111. Zahler R, Sun W, Ardito T, et al. The $\alpha 3$ isoform protein of the Na^+/K^+-ATPase is associated with the sites of cardiac and neuromuscular impulse transmission. *Circ Res* 1996;**78**(5):870–9.
112. Swift F, Tovsrud N, Enger UH, et al. The Na^+/K^+-ATPase alpha2-isoform regulates cardiac contractility in rat cardiomyocytes. *Cardiovasc Res* 2007;**75**(1):109–17.
113. Despa S, Lingrel JB, Bers DM. Na^+/K^+-ATPase α_2-isoform preferentially modulates Ca^{2+} transients and sarcoplasmic reticulum Ca^{2+} release in cardiomyocytes. *Cardiovasc Res* 2012;**95**(4):480–6.
114. Barwe SP, Jordan MC, Skay A, et al. Dysfunction of ouabain-induced cardiac contractility in mice with heart-specific ablation of Na/K-ATPase β_1-subunit. *J Mol Cell Cardiol* 2009;**47**(4):552–60.

115. Xie Z, Cai T. Na^+/K^+-ATPase-mediated signal transduction: from protein interaction to cellular function. *Mol Interv* 2003;**3**(3):157–68.
116. Olson RD, Mushlin PS. Doxorubicin cardiotoxicity: analysis of prevailing hypotheses. *FASEB J* 1990;**4**(13):3076–86.
117. Bachur NR, Gordon SL, Gee MV, et al. NADPH cytochrome P-450 reductase activation of quinone anticancer agents to free radicals. *Proc Natl Acad Sci USA* 1979;**76**(2): 954–7.
118. Goodman J, Hochstein P. Generation of free radicals and lipid peroxidation by redox cycling of adriamycin and daunomycin. *Biochem Biophys Res Commun* 1977;**77**(2): 797–803.
119. Myers CE, McGuire WP, Liss RH, et al. Adriamycin: the role of lipid peroxidation in cardiac toxicity and tumor response. *Science* 1977;**197**(4299):165–7.
120. Ferrero ME, Ferrero E, Gaja G, et al. Adriamycin: energy metabolism and mitochondrial oxidations in the heart of treated rabbits. *Biochem Pharmacol* 1976;**25**(2):125–30.
121. Seraydarian MW, Artaza L, Goodman MF. Adriamycin: effect on mammalian cardiac cells in culture. I. Cell population and energy metabolism. *J Mol Cell Cardiol* 1977;**9**(5):375–82.
122. Kim Y, Ma AG, Kitta K, et al. Anthracycline-induced suppression of GATA-4 transcription factor: implication in the regulation of cardiomyocyte apoptosis. *Mol Pharmacol* 2003;**63** (2):368–77.
123. Kang W, Weiss M. Kinetic analysis of saturable myocardial uptake of idarubicin in rat heart: effect of doxorubicin and hypothermia. *Pharm Res* 2003;**20**(1):58–63.
124. Kang W, Weiss M. Caffeine enhances myocardial uptake of idarubicin but reverses its negative inotropic effect. *Naunyn Schmiedebergs Arch Pharmacol* 2003;**367**(2):151–5.
125. Cascales A, Pastor-Quirante F, Sánchez-Vega B, et al. Association of anthracycline-related cardiac histological lesions with NADPH oxidase functional polymorphisms. *Oncologist* 2013;**18**(4):446–53.
126. Bristow MR, Billingham ME, Mason JW, et al. Clinical spectrum of anthracycline antibiotic cardiotoxicity. *Cancer Treat Rep* 1978;**62**(6):873–9.
127. Bristow MR, Mason JW, Billingham ME, et al. Dose-effect and structure-function relationships in doxorubicin cardiomyopathy. *Am Heart J* 1981;**102**(4):709–18.
128. Singal PK, Iliskovic N. Doxorubicin-induced cardiomyopathy. *N Engl J Med* 1998;**339** (13):900–5.
129. Zucchi R, Danesi R. Cardiac toxicity of antineoplastic anthracyclines. *Curr Med Chem Anticancer Agents* 2003;**3**(2):151–71.
130. Nakamae H, Tsumura K, Akahori M, et al. QT dispersion correlates with systolic rather than diastolic parameters in patients receiving anthracycline treatment. *Intern Med* 2004; **43**(5):379–87.
131. Sarubbi B, Orditura M, Ducceschi V, et al. Ventricular repolarization time indexes following anthracycline treatment. *Heart Vessels* 1997;**12**(6):262–6.
132. Milberg P, Fleischer D, Stypmann J, et al. Reduced repolarization reserve due to anthracycline therapy facilitates torsade de pointes induced by I_{Kr} blockers. *Basic Res Cardiol* 2007;**102**(1):42–51.
133. Horacek JM, Jakl M, Horackova J, et al. Assessment of anthracycline-induced cardiotoxicity with electrocardiography. *Exp Oncol* 2009;**31**(2):115–7.
134. Arbel Y, Swartzon M, Justo DQT. Prolongation and Torsades de Pointes in patients previously treated with anthracyclines. *Anticancer Drugs* 2007;**18**(4):493–8.
135. Hagane K, Akera T, Berlin JR. Doxorubicin: mechanism of cardiodepressant actions in guinea pigs. *J Pharmacol Exp Ther* 1988;**246**(2):655–61.

136. Bottone AE, Voest EE, de Beer EL. Impairment of the actin-myosin interaction in permeabilized cardiac trabeculae after chronic doxorubicin treatment. *Clin Cancer Res* 1998;**4**(4):1031–7.
137. Sawyer DB, Zuppinger C, Miller TA, et al. Modulation of anthracycline-induced myofibrillar disarray in rat ventricular myocytes by neuregulin-1β and anti-erbB2: potential mechanism for trastuzumab-induced cardiotoxicity. *Circulation* 2002;**105**(13):1551–4.
138. Boucek Jr RJ, Buck SH, Scott F, et al. Anthracycline-induced tension in permeabilized cardiac fibers: evidence for the activation of the calcium release channel of sarcoplasmic reticulum. *J Mol Cell Cardiol* 1993;**25**(3):249–59.
139. Shadle SE, Bammel BP, Cusack BJ, et al. Daunorubicin cardiotoxicity: evidence for the importance of the quinone moiety in a free-radical-independent mechanism. *Biochem Pharmacol* 2000;**60**(10):1435–44.
140. Solem LE, Henry TR, Wallace KB. Disruption of mitochondrial calcium homeostasis following chronic doxorubicin administration. *Toxicol Appl Pharmacol* 1994;**129**(2):214–22.
141. Papadopoulou LC, Theophilidis G, Thomopoulos GN, et al. Structural and functional impairment of mitochondria in adriamycin-induced cardiomyopathy in mice: suppression of cytochrome c oxidase II gene expression. *Biochem Pharmacol* 1999;**57**(5):481–9.
142. Tokarska-Schlattner M, Wallimann T, Schlattner U. Multiple interference of anthracyclines with mitochondrial creatine kinases: preferential damage of the cardiac isoenzyme and its implications for drug cardiotoxicity. *Mol Pharmacol* 2002;**61**(3):516–23.
143. Ogihara M, Tanno M, Izumiyama N, et al. Increase in DNA polymerase gamma in the hearts of adriamycin-administered rats. *Exp Mol Pathol* 2002;**73**(3):234–41.
144. Childs AC, Phaneuf SL, Dirks AJ, et al. Doxorubicin treatment in vivo causes cytochrome C release and cardiomyocyte apoptosis, as well as increased mitochondrial efficiency, superoxide dismutase activity, and Bcl-2:Bax ratio. *Cancer Res* 2002;**62**(16):4592–8.
145. Serrano J, Palmeira CM, Kuehl DW, et al. Cardioselective and cumulative oxidation of mitochondrial DNA following subchronic doxorubicin administration. *Biochim Biophys Acta* 1999;**1411**(1):201–5.
146. Chaiswing L, Cole MP, Ittarat W, et al. Manganese superoxide dismutase and inducible nitric oxide synthase modify early oxidative events in acute adriamycin-induced mitochondrial toxicity. *Mol Cancer Ther* 2005;**4**(7):1056–64.
147. Chaiswing L, St Cole MP, Clair DK, et al. Oxidative damage precedes nitrative damage in adriamycin-induced cardiac mitochondrial injury. *Toxicol Pathol* 2004;**32**(5):536–47.
148. Lebrecht D, Walker UA. Role of mtDNA lesions in anthracycline cardiotoxicity. *Cardiovasc Toxicol* 2007;**7**(2):108–13.
149. Jirkovsky E, Popelová O, Kriváková-Stanková P, et al. Chronic anthracycline cardiotoxicity: molecular and functional analysis with focus on nuclear factor erythroid 2-related factor 2 and mitochondrial biogenesis pathways. *J Pharmacol Exp Ther* 2012;**343**(2):468–78.
150. Papoian T, Lewis W. Anthracyclines selectively decrease alpha cardiac actin mRNA abundance in the rat heart. *Am J Pathol* 1992;**141**(5):1187–95.
151. L'Ecuyer T, Sanjeev S, Thomas R, et al. DNA damage is an early event in doxorubicin-induced cardiomyocyte death. *Am J Physiol Heart Circ Physiol* 2006;**291**(3):H1273–80.
152. Chen B, Zhong L, Roush SF, et al. Disruption of a GATA4/Ankrd1 signaling axis in cardiomyocytes leads to sarcomere disarray: implications for anthracycline cardiomyopathy. *PLoS One* 2012;**7**(4):e35743.
153. Gratia S, Kay L, Potenza L, et al. Inhibition of AMPK signalling by doxorubicin: at the crossroads of the cardiac responses to energetic, oxidative, and genotoxic stress. *Cardiovasc Res* 2012;**95**(3):290–9.

154. Klein PM, Dybdal N. Trastuzumab and cardiac dysfunction: update on preclinical studies. *Semin Oncol* 2003;**30**(5 Suppl. 16):49–53.
155. Ewer SM, Ewer MS. Cardiotoxicity profile of trastuzumab. *Drug Saf* 2008;**31**(6):459–67.
156. Russell SD, Blackwell KL, Lawrence J, et al. Independent adjudication of symptomatic heart failure with the use of doxorubicin and cyclophosphamide followed by trastuzumab adjuvant therapy: a combined review of cardiac data from the National Surgical Adjuvant breast and Bowel Project B-31 and the North Central Cancer Treatment Group N9831 clinical trials. *J Clin Oncol* 2010;**28**(21):3416–21.
157. Grazette LP, Boecker W, Matsui T, et al. Inhibition of ErbB2 causes mitochondrial dysfunction in cardiomyocytes: implications for herceptin-induced cardiomyopathy. *J Am Coll Cardiol* 2004;**44**(11):2231–8.
158. Ewer MS, Vooletich MT, Durand JB, et al. Reversibility of trastuzumab-related cardiotoxicity: new insights based on clinical course and response to medical treatment. *J Clin Oncol* 2005;**23**(31):7820–6.
159. Crone SA, Zhao YY, Fan L, et al. ErbB2 is essential in the prevention of dilated cardiomyopathy. *Nat Med* 2002;**8**(5):459–65.
160. Pentassuglia L, Timolati F, Seifriz F, et al. Inhibition of ErbB2/neuregulin signaling augments paclitaxel-induced cardiotoxicity in adult ventricular myocytes. *Exp Cell Res* 2007;**313**(8):1588–601.
161. Hancox JC, Witchel HJ, Varghese A. Alteration of HERG current profile during the cardiac ventricular action potential, following a pore mutation. *Biochem Biophys Res Commun* 1998;**253**(3):719–24.
162. Witchel HJ, Hancox JC. Familial and acquired long QT syndrome and the cardiac rapid delayed rectifier potassium current. *Clin Exp Pharmacol Physiol* 2000;**27**(10):753–66.
163. Ficker E, Kuryshev YA, Dennis AT, et al. Mechanisms of arsenic-induced prolongation of cardiac repolarization. *Mol Pharmacol* 2004;**66**(1):33–44.
164. Ficker E, Dennis AT, Wang L, et al. Role of the cytosolic chaperones Hsp70 and Hsp90 in maturation of the cardiac potassium channel hERG. *Circ Res* 2003;**92**(12):e87–100.
165. Kuryshev YA, Ficker E, Wang L, et al. Pentamidine-induced long QT syndrome and block of hERG trafficking. *J Pharmacol Exp Ther* 2005;**312**(1):316–23.
166. Dennis AT, Wang L, Wan H, et al. Molecular determinants of pentamidine-induced hERG trafficking inhibition. *Mol Pharmacol* 2012;**81**(2):198–209.
167. Wang L, Wible BA, Wan X, et al. Cardiac glycosides as novel inhibitors of human ether-à-go-go-related gene channel trafficking. *J Pharmacol Exp Ther* 2007;**320**(2):525–34.
168. Wang L, Dennis AT, Trieu P, et al. Intracellular potassium stabilizes human ether-à-go-go-related gene channels for export from endoplasmic reticulum. *Mol Pharmacol* 2009;**75**(4):927–37.
169. Takemasa H, Nagatomo T, Abe H, et al. Coexistence of hERG current block and disruption of protein trafficking in ketoconazole-induced long QT syndrome. *Br J Pharmacol* 2008;**153**(3):439–47.
170. Obers S, Staudacher I, Ficker E, et al. Multiple mechanisms of hERG liability: K^+ current inhibition, disruption of protein trafficking, and apoptosis induced by amoxapine. *Naunyn Schmiedebergs Arch Pharmacol* 2010;**381**(5):385–400.
171. Dennis AT, Nassal D, Deschenes I, et al. Antidepressant-induced ubiquitination and degradation of the cardiac potassium channel hERG. *J Biol Chem* 2011;**286**(39):34413–25.
172. Han S, Zhang Y, Chen Q, et al. Fluconazole inhibits hERG K^+ channel by direct block and disruption of protein trafficking. *Eur J Pharmacol* 2011;**650**(1):138–44.

Chapter 3

Environmental Toxins and the Heart

Sahand Rahnama-Moghadam, L. David Hillis, and Richard A. Lange
Department of Medicine, University of Texas Health Science Center, San Antonio

3.1 AIR POLLUTION

Exposure to particulate matter in the air (i.e., air pollution) can lead to increased cardiovascular morbidity and mortality from atherosclerosis, stroke, acute coronary syndromes, arrhythmias, and death.[1–10] Most research has focused on pollutants that result from the combustion of fossil fuels. Although the specific fossil fuel-derived pollutants that are responsible for the cardiotoxic effects of particulate matter have not yet been identified, the known particles released during combustion are a complex mixture of elemental carbon, organic carbon compounds, and reactive components such as transition metals, metal oxides, acid condensates, sulfates, and nitrates. Fine particulate pollutants (defined as being $<2.5\ \mu m$ in size) have been shown to have the most consistent relationship with disease across epidemiologic studies; this is the particle size most likely to reach the alveoli.[5,11,12] The health risks of air pollution appear to be linearly related to exposure with no safe lower limit of exposure to particulate pollution.[5]

Inhalation of particulate air pollution creates systemic inflammation and oxidative stress, provokes vascular injury and atherosclerosis, and induces autonomic dysfunction.[11,13] Animal experiments show that long-term exposure to aerosolized fine particulate matter induces the development and progression of atherosclerosis.[14,15] In humans, air pollution results in increases in serum fibrinogen levels, platelet activation, blood viscosity, and other meditators of aggregation.[11,16] Additionally, exposure to air particulate matter (PM) is associated with increases in inflammation and atherogenesis.[17,18] Numerous studies have demonstrated an increased risk of myocardial infarction with exposure to particulate matter, both in the short term (within hours of exposure to traffic or to an increase in local pollutant levels) and over a longer period of time (months to years).[3–5,13,19–23]

The Heart and Toxins. DOI: http://dx.doi.org/10.1016/B978-0-12-416595-3.00003-7

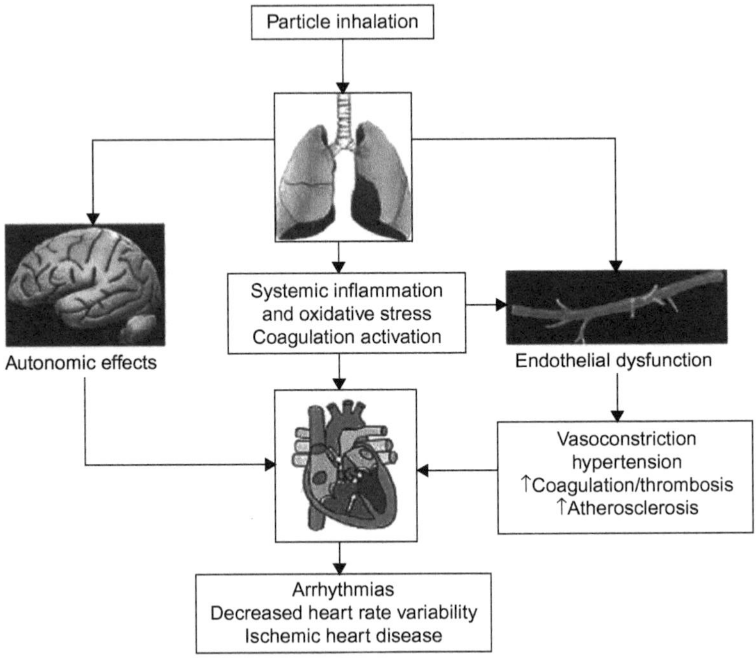

FIGURE 3.1 **Effects of particulate matter on the cardiovascular system. Source:** *Reproduced with permission from Franchini and Mannucci, 2009.*[35]

Congestive heart failure is also exacerbated by exposure to particulate matter.[24–26] The number of hospital admissions for congestive heart failure (CHF) increase when levels of airborne particulate matter are elevated,[26] perhaps because particulate air matter impairs the ability of the lung to clear edema fluid.[27,28] Clinical decompensation may also be precipitated by arrhythmias, as exposure to pollution has been associated with ventricular arrhythmias,[29,30] atrial fibrillation,[31] and implantable defibrillator discharges.[1] Exposure to particulate matter has also been shown to reduce heart rate variability, a purported marker for cardiac autonomic function.[11,32–34] Heart rate variability reduction has been linked to poor cardiovascular outcomes[5] thought to be related to decreased parasympathetic input to the heart (Figure 3.1).[36,37]

3.2 ANABOLIC DRUGS

Performance-enhancing drugs are used by professional, amateur, and recreational athletes. Anabolic androgens are among the most commonly used "doping agents," and they have been associated with cardiovascular morbidity and mortality in young adults with no other known cardiac risk factors.[37–40] The cardiovascular complications associated with the use of anabolic steroids include unfavorable metabolic changes in serum low-density lipoprotein (LDL) and high-density lipoprotein (HDL) cholesterol levels, hypertension, concentric left ventricular hypertrophy, cardiomyopathy, acute coronary syndromes, arrhythmias, and sudden cardiac death.[37,39–46] There is evidence that the associated left ventricular hypertrophy improves when anabolic steroid use ceases.[47] The cardiovascular risks of other doping agents, such as cortisol, human growth hormone, prolactin, and platelet-derived products, have not been well studied.[48]

Among professional athletes and celebrities, clenbuterol has been gaining popularity as a performance-enhancing and/or weight-loss drug. It is a long-acting sympathomimetic agonist that is primarily used in veterinary medicine as a bronchodilator. In animals, the β-2 agonist activity stimulates skeletal muscle anabolism, and stimulation of β-3 adrenoreceptors promotes lipolysis.[38,49,50] Cardiac toxicities linked to short-term use of clenbuterol include myocardial infarction and supraventricular tachycardias, some of which have been lethal.[38,39,51,52] Long-term use of clenbuterol has resulted in left ventricular hypertrophy and the consequent increased risk of arrhythmias.[38] This drug is on the list of substances that are banned by the International Olympic Committee and other sports regulatory agencies. Detection of clenbuterol and anabolic androgen steroids is accomplished with immunoaffinity, mass spectrometry, and chromatography testing.[53,54]

3.3 ACONITE

Aconite is derived from the ubiquitous plant *Aconitum napellus* (Figure 3.2), which is also known by these common names: monkshade, wolfsbane, and "the devil's helmet."[55–58] Aconite is a toxic alkaloid that has been used worldwide in traditional herbal medicines for its analgesic and antiinflammatory effects.[56] Toxicity may occur from ingestion following improper processing of the plant for use in complementary medicines, mistaking it for an edible species, and intentional suicide and homicide attempts.[56,59–61]

FIGURE 3.2 Photo of an *Aconitum napellus* plant. The toxic alkaloid aconite is derived from this plant.

On a molecular level, aconite causes persistent activation of the voltage-gated sodium channels, leading to sustained depolarization and resistance to excitation. It is toxic to the following:

- The neurologic system, causing descending paresthesias, numbness, and mild weakness
- The gastrointestinal tract, inducing nausea, vomiting, and diarrhea
- The cardiovascular system

The first symptom, perioral numbness, begins within minutes to hours of ingestion.[55,56,62–66] Thereafter, sustained depolarization of sodium channels has been shown to lead to hypotension and arrhythmias. In fact, aconite has been used to induce arrhythmias experimentally to assess the efficacy of antiarrhythmic agents.[56,65] A diagnosis of aconite poisoning is usually established clinically, but it can be confirmed with chromatographic and mass spectrometric analysis of serum and urine alkaloids and metabolites.[56–59,65]

The treatment of aconite poisoning is supportive, since no proven antidote is available. Because of the persistent activation of voltage-gated sodium channels, class I antiarrhythmic agents (i.e., sodium channel-blocking agents) have been recommended, but have not been shown to be consistently efficacious. When arrhythmias are resistant to pharmacologic therapy or direct cardioversion, mechanical support with cardiopulmonary extracorporeal bypass or a left ventricular assist device has been suggested as a salvage measure.[55,56,64,66,67]

3.4 ANTIDOTES

Medications administered as antidotes for therapeutic purposes may occasionally have adverse cardiotoxic effects.

3.4.1 Adenosine

Adenosine is an endogenous nucleoside with a short plasma half-life that has been shown to cause transient reduction in sinus nodal discharge and atrioventricular nodal block when administered intravenously. While pharmacologic doses of adenosine are generally considered safe, serious cardiovascular complications have been reported following its administration. Prolonged atrioventricular block that results in asystole lasting more than 4 seconds may occur in up to 7% of patients who receive therapeutic doses of adenosine.[68]

Atrial fibrillation,[68,69] supraventricular tachycardia,[70] and ventricular fibrillation have all been reported following adenosine therapy.[68,69,71] In individuals with structurally normal hearts and in children with congenital long QT syndromes, adenosine usage has been associated with polymorphic ventricular tachycardia (torsades de pointes).[69,71] The mechanisms for these adenosine-induced complications may be a transient increase in sympathetic tone or stimulation of spatial and temporal inhomogeneity of ventricular and atrial refractoriness.[69]

3.4.2 Diphenhydramine

Diphenhydramine is a relatively nontoxic antagonist of the histamine H1 receptor. However, sedative and anticholinergic effects become prominent in individuals who overdose on this drug. This autonomic disturbance has been linked to hypertension, hypotension, tachycardia, ventricular arrhythmia, and cardiac arrest.[72] Death may occur within 2 hours of overdose, with a fatal dose for children estimated to be 500 mg and for adults 20 to 40 mg/kg.[73] Because of diphenhydramine's cholinergic effects, cautious use of physostigmine as an antidote has been suggested.[73] Diphenhydramine rapidly distributes from the plasma to the tissue so that forced diuresis and hemodialysis are unlikely to be efficacious in facilitating its clearance.[72,73]

3.4.3 Protamine

Heparin is used during extracorporeal circulation for cardiopulmonary bypass; at termination of bypass, protamine is used to reverse the anticoagulation. Protamine is a natural product of fish and in an occasional individual its administration is associated with anaphylactic shock.[74] Moreover, protamine use is associated with hypotension, bradycardia, and fatal ventricular arrhythmias.[74–76]

3.5 ANTIMONY

The metal antimony is found in trivalent, pentavalent, and gaseous forms. The gaseous form (e.g., stibine gas, SbH_3) is the most toxic, causing massive hemolysis. The next most toxic form is antimony potassium tartrate, a trivalent form of the metal that is available in many countries as tartar emetic (used for treatment of alcohol abuse).[77] The least toxic forms are pentavalent antimonials, which are used widely for the treatment of subjects with leishmaniasis and schistosomiasis; antimony inhibits phosphofructokinase, the rate-limiting step in the parasites' glycolytic pathway. In the workplace, elemental antimony is used in the manufacture of semiconductors, infrared detectors, diodes, fire retardants, and plastics.[78] The occurrence of antimony toxicity is related to (1) the form ingested, (2) the amount, and (3) the duration of use.[79]

Antimony toxicity is associated with electrocardiographic abnormalities, with QT interval prolongation, and with T-wave flattening or inversion appearing first. With higher exposure, chest pain, bradycardia, hypotension, ventricular arrhythmias, and sudden death have been reported.[80,81] Left ventricular systolic dysfunction after intravenous administration of trivalent antimony has been reported.[82] Although the mechanism of antimony cardiotoxicity has not been completely elucidated, it is most likely the result of altered cardiac calcium channel excitability.[83]

3.6 ARSENIC

Historically, arsenic toxicity has been associated with pesticide poisoning. Arsenic poisoning has been linked to electrocardiographic abnormalities (e.g., QT interval prolongation with T-wave inversions), myocarditis, and pericardial effusion. More recently, attention has focused on population-level environmental exposure to arsenic (Figure 3.3) and the resultant cardiovascular sequelae.[84–86] Specifically, environmental arsenic exposure is associated with an increased incidence of ischemic heart disease, cerebrovascular disease, and hypertension.[84,86–89]

A large cohort study in Bangladesh has demonstrated a direct correlation between the magnitude of exposure to arsenic in water and ischemic heart disease mortality.[87] In addition, an interaction between smoking and arsenic exposure has been noted, with the combination increasing the risk of ischemic heart disease beyond the additive risk of each individual component. In experimental animal models, a relationship between arsenic exposure and vascular inflammation, atherosclerosis,[90,91] hypertension, and left ventricular hypertrophy[92,93] has been observed.

In addition to its direct cardiac toxicity, arsenic causes peripheral arterial disease,[88,89] manifested most dramatically by so-called "blackfoot disease," which is endemic in Taiwanese villagers who consume artesian well water

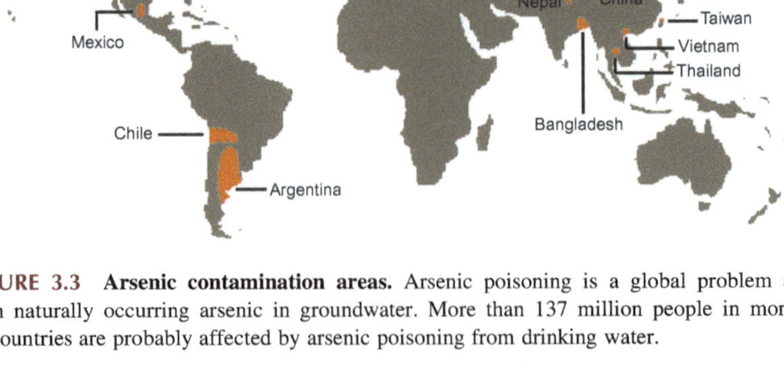

FIGURE 3.3 Arsenic contamination areas. Arsenic poisoning is a global problem arising from naturally occurring arsenic in groundwater. More than 137 million people in more than 70 countries are probably affected by arsenic poisoning from drinking water.

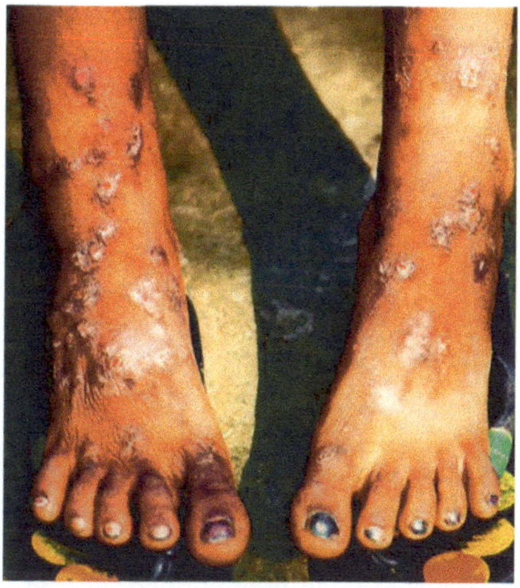

FIGURE 3.4 **Blackfoot disease.** Peripheral arterial disease (the so-called "blackfoot disease") is endemic in villagers who consume artesian well water that has a high arsenic content. **Source:** *Courtesy of Tseng, 1977.*[94]

with a high arsenic content. These subjects have systemic atherosclerosis, thromboangiitis obliterans, and dry gangrene, at times leading to spontaneous amputations of affected extremities (Figure 3.4).[89,95] In experimental animals, arsenic lowers endothelial nitrous oxide production and increases oxidative stress.[96] Furthermore, studies with rats have shown that blood vessels exposed to arsenite have a blunted vasodilator response to acetylcholine infusion, thereby possibly contributing to the hypertension associated with arsenic exposure.[91]

Clinically, arsenic trioxide is administered as a salvage therapy to those patients who have acute promyelocytic leukemia. In these individuals, its use has resulted in electrocardiographic QT interval prolongation,[97–99] pericardial effusion,[99] and conduction block.[100] In experiments with animals, arsenic exposure induces myocardial fibrosis and myocarditis.[101,102]

3.7 ARECA NUT

The areca (betel) nut is the seed of the areca palm (*Areca catechu*), which grows in much of the tropical Pacific, Asia, and parts of East Africa. It is commonly referred to as betel nut, as it is often chewed wrapped in betel leaves (Figure 3.5). Betel nut is the fourth (behind alcohol, nicotine, and caffeine) most widely used addictive substance in the world, with its chewers making up $\geq 10\%$ of the world's population (mostly in Asia and South Asia).[103–105] Betel nut is ingested due to its ability to increase stimulation and improve the feeling of well-being, with users describing the stimulating symptoms of betel as similar to tobacco or cocaine.[106] It has also been used as a sexual stimulant, laxative, and diuretic.[105,107] The active ingredients—arecoline, arecaidine, guvacine, and guvacoline—are alkaloids that mimic

FIGURE 3.5 **Areca nut.** Here three are wrapped in betel leaves to be chewed.

the action of acetylcholine centrally on peripheral nicotinic and muscarinic receptors and inhibit the gamma-aminobutyric acid receptor.[104,105,107]

Betel nut has been associated with oral/esophageal inflammatory disease and malignancy, obesity, type 2 diabetes mellitus, and hyperlipidemia.[104,108–110] With regard to cardiovascular toxicities, betel nut use can induce hypertension, sinus and supraventricular tachycardia, and acute myocardial infarction.[105,110] Compared to nonusers, individuals who chew betel nuts have a higher incidence of cardiovascular mortality,[104] which is thought to be a result of the increased incidence of comorbidities (e.g., diabetes mellitus, hypertension, and obesity), oxidative stress, contamination by trace heavy metals (e.g., arsenic and manganese), induction of the sympathetic nervous system leading to secretion of cathecholamines, and periodontal disease, a known risk factor for cardiovascular disease.[104,111–115]

3.8 BISMUTH

Bismuth compounds are present in various cosmetics, pigments, and pharmaceuticals. The most commonly prescribed bismuth-containing pharmaceutical agents are Pepto-Bismol, Kaopectate, and Devrom, which are oral preparations used to treat individuals with gastrointestinal (GI) disorders (e.g., peptic ulcer disease, diarrhea, flatulence) (Figure 3.6). Historically, bismuth compounds were administered parenterally to those with treponomal infections such as syphilis and yaws.[117]

Although isolated reports of cardiovascular toxicity with bismuth administered parenterally have appeared, no reports of such have been noted with its oral administration. In 1948, Goodman described three subjects with sudden cardiovascular collapse following intravenous sodium bismuth tartrate as a treatment for yaws.[118] In this report, "slowing of the heart" and "a direct depressant action upon the heart with irregularities of which heart-block is the commonest" are mentioned. Another case report suggested that parenterally administered sodium bismuth tartrate precipitated heart failure.[119] Other than these reports from seven decades previously, no evidence links bismuth to cardiovascular toxicity.

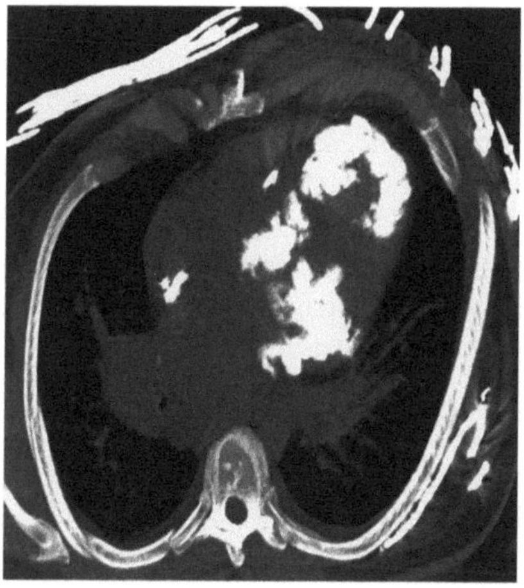

FIGURE 3.6 **Bony heart.** Image of a computed tomographic scan that shows prominent calcification of the left ventricle as a result of chronic hypercalcemia. **Source:** *Used with permission from Shackley et al., 2011.*[116]

3.9 CADMIUM

Cadmium exposure most often results from inhalation of cigarette smoke; the gases that are produced in metal-smelting factories; and the gaseous products of facilities where batteries, paints, and pigments are manufactured.[120–123] Epidemiologic studies suggest that cadmium exposure is associated with atherosclerosis, vascular injury, and increased cardiovascular mortality.[120,122–125] At relatively low, sublethal concentrations, cadmium damages myocardial and vascular endothelial cells via its effects on cell adhesion molecules, metal ion transporters, protein kinase signaling pathways, and oxidative stress. Through these molecular mechanisms, as well as its direct deposition into cardiovascular tissue, it induces myocyte and endothelial damage in mice.[121,123,126–130] Various antioxidants exert a protective effect against cadmium-induced oxidative stress in such animal models.[126–129]

3.10 CAFFEINE

Caffeine is a natural alkaloid methylxanthine and adenosine receptor antagonist that increases heart rate and blood pressure.[131] Caffeine is found in beverages such as coffee, tea, cocoa, and soft drinks, as well as commercial stimulants, analgesics, and cold remedies. Adverse effects of caffeine are typically seen after ingestion of doses larger than 200 mg,[132] with fatal ingestions usually involving a dose of more than 5 g.[131,133]

Caffeine binds to the adenosine class of G protein-coupled receptors on the surface of myocytes and activates a second messenger system that leads to an increase in cyclic adenosine monophosphate within the cells, thereby mimicking the effects of epinephrine. Caffeine enhances glycolysis and it increases the amount of adenosine triphosphate available for muscle contraction and relaxation. This results in positive inotropy and chronotropy.[132,134] Caffeine immediately increases blood pressure and peripheral vascular resistance, in part because of sympathetic stimulation and also by stimulating renin release.[131,135,136] With sympathetic stimulation there is the propensity for sinus tachycardia, supraventricular tachycardia, ventricular tachycardia, and ventricular fibrillation.[131,136–139]

Treatment of caffeine overdose with nonselective beta blockers (e.g., propranolol, esmolol, or metoprolol), vasopressin, procainamide, lidocaine, and hemodialysis have been described in various case reports.[134,136,138–140]

3.11 CALCIUM

The cardiac consequences of hypercalcemia have been known to occur in the clinical setting as hyperparathyroidism, vitamin D intoxication, chronic renal failure, and malignancy. The electrocardiographic manifestations of hypercalcemia often include QT interval shortening, T-wave flattening, atrioventricular block (including complete heart block), atrial premature beats, ventricular tachycardia, and J waves (the so-called "Osborne" waves) in the normothermia setting.[141–147]

Hypercalcemia may cause structural heart disease due to calcium deposition. Calcification in the mitral valve annulus is most common,[143] but calcium deposition may also occur in the tricuspid, pulmonic, and aortic valves, causing stenosis when it is severe.[143,148] Extensive involvement may lead to myocardial calcification, progression of atherosclerosis, and myocardial infarction.[116,143,149–151] In such cases, cardiac computed tomography may demonstrate a "white" heart, described as "bony" or "petrified" (see Figure 3.6).[149,152] Myocardial calcification has been observed in newborns and infants with hypercalcemia as well as in adults, which indicates that calcification is due to a calcium excess rather than the result of a chronic myocardial injury or senescence.[116,153]

3.12 CARBON MONOXIDE

Carbon monoxide (CO) poisoning, which occurs after sufficient exposure to combustion gases from furnaces, engine exhaust, and burning charcoal, results in 50,000 emergency department visits annually in the United States.[154] The harmful effects of CO are mediated by tissue hypoxia as well as direct toxic effects. Carbon monoxide binds to hemoglobin with high affinity (i.e., 200–250 times more rapidly than oxygen), thereby interfering with oxygen delivery to tissue.[154–156] In addition, it interferes with various stages of cellular respiration, including inhibition of the cytochrome c oxidase enzyme.[154–157]

Inhalation of CO is associated with a hypercoagulable state that may lead to myocardial ischemia and/or infarction (even in the absence of atherosclerotic coronary artery disease) and stent thrombosis.[157–160] The prothrombotic effects of carbon monoxide poisoning are caused by CO-induced endothelial dysfunction, increased platelet aggregation, and alterations of the fibrinolytic pathway.[158] Myocardial ischemia also may occur because of tissue hypoxia, increased oxidative stress, systemic arterial hypotension (with resultant hypoperfusion),[161] and lipid peroxidation.[155,162] Left ventricular systolic dysfunction and myocarditis also have been reported following CO exposure.[161,163–167] Finally, CO exposure may lead to various arrhythmias, including sinus tachycardia, premature atrial or ventricular beats, supraventricular tachycardia, atrial fibrillation, atrioventricular block, and unmasking of a familial predisposition to arrhythmias.[157,161,168,169]

The treatment of CO poisoning is the administration of 100% supplemental oxygen. If toxicity and sequelae are severe (i.e., transient or prolonged unconsciousness, abnormal neurologic findings, cardiovascular dysfunction, severe acidosis, age >36 years, CO exposure for 24 hours or more, or a serum carboxyhemoglobin level $\geq 25\%$), treatment with hyperbaric oxygen is recommended.[154,162] Among survivors of carbon monoxide poisoning, those with evidence of myocardial injury (e.g., elevated serum troponin or creatine kinase-MB concentrations) have a 2-fold increased risk of dying over the ensuing months and years when compared to those without evidence of myocardial injury.[170]

3.13 CARDENOLIDES

Cardenolides are naturally occurring cardiac glycosides found in plant species throughout the world and in some of the butterflies that feed on the plants.[171–173] Ingestion of cardenolides may lead to serious dysrhythmias, including second- or third-degree heart block and cardiac arrest, with the paragon being toxicity from digitalis cardenolides (digoxin and digitoxin). Cardenolides isolated from common plants have been used in insecticides and rodenticides for centuries.

In South Asia, cardenolide poisoning from yellow, pink, or white oleander and fruits from the *Cerbera manghas* family (e.g., sea mango, pink-eyed cerbera, odollam tree) is a leading cause of self-harm, with thousands of cases a year and a 5 to 10% case fatality ratio.[173–176] All parts of the plant, no matter how prepared (fresh, dried, or boiled) are toxic. Fatality in humans occurs after ingestion of one leaf by children[177] and occurs in adults after eating 8 to 10 seeds, 15 to 20 g of the root, or 5 to 15 leaves.[178,179]

The sites of toxicity are the cardiovascular and autonomic nervous systems. The common molecular mechanism is inhibition of the Na^+/K^+-ATPase channel.[174] While initial symptoms may show up within minutes after ingestion, prolonged hospitalization and observation are recommended after cardenolide ingestion, since the occurrence of dangerous dysrhythmias may be delayed up to 72 hours after ingestion.[173]

A meta-analysis of the possible treatments concluded that multiple doses of activated charcoal within 24 hours of toxin ingestion and administration of antidigoxin Fab antitoxin reduces the risk of cardiac dysrhythmias.[173] Hemodialysis or hemoperfusion are not effective against toxicity because of the large volume of distribution of the toxin.[180]

3.14 CATECHOLAMINES AND BETA-RECEPTOR AGONISTS

The direct cardiotoxic effects of catecholamines (i.e., epinephrine and norepinephrine) are mediated by increased free radical production, altered autonomic tone, enhanced lipid mobility, calcium overload, increased sarcolemmal permeability, and mitochondrial toxicity.[181,182] In addition, myocardial damage may be due to a myocardial oxygen supply–demand imbalance that is caused by catecholamine-induced coronary arterial vasoconstriction and/or platelet aggregation.

Catecholamine excess may manifest clinically as acute myocarditis (with microscopic evidence of contraction band necrosis), supraventricular and ventricular tachyarrhythmias, and cardiomyopathy. Such clinical scenarios have been observed following excessive endogenous catecholamine production by neuroendocrine tumors (i.e., pheochromocytoma or paraganglioma) and following the administration of exogenous catecholamines (e.g., intravenous dobutamine or epinephrine)[183,184] as well as beta-adrenergic agonist inhalants and methylxanthines[185] in subjects with severe pulmonary disease. Catecholamines may precipitate lethal arrhythmias in individuals who have a hereditary predisposition to sudden cardiac death.[186]

Excessive endogenous catecholamine production that occurs during stress[187–190] or cerebral subarachnoid hemorrhage,[191–193] as well as the exogenous administration of intravenous beta-receptor agonists,[194,195] have been associated with the development of *takotsubo*, or *stress cardiomyopathy*, which is characterized by electrocardiographic T-wave inversions in the anterior leads and transient left ventricular apical dyskinesis.

3.15 CHROMIUM

Chromium is required for glucose, lipid, and protein metabolism. Although chromium overexposure is associated with carcinogenesis, cardiovascular toxicities have not been reported.[196] Deficiency of chromium has been associated with impaired glucose tolerance, hyperglycemia, and abnormal lipid concentrations.[197,198] Subjects with coronary artery disease have a lower serum concentration of chromium than those without coronary artery disease,[199] and low serum or tissue concentrations of chromium are associated with an increased risk of cardiovascular disease.[197,199–201] In rats, the administration of chromium (III) chloride causes regression of atheromatous plaques in the aorta and coronary arteries.[202,203]

3.16 COBALT

In the mid-1960s, an acute and fulminant form of dilated cardiomyopathy with a 40% fatality rate was described in heavy beer drinkers in Quebec City, Canada, and in several American cities.[204,205] Epidemiologic investigations and animal studies implicated as the etiology cobalt chloride salt that was added to the beer to stabilize the foam. Autopsy demonstrated high levels of cobalt in endomyocardial tissue along with myocardial tissue degeneration, increased vacuolation, and interstitial edema[205,206]; all chambers of the heart were affected, with an atrial predilection.[205,207] Concomitant polycythemia, pericardial effusion, hypothyroidism, lactic acidosis, gastrointestinal ulcerations, and elevated liver enzymes may distinguish cobalt-induced cardiomyopathy from other etiologies of dilated cardiomyopathy.[205–207]

Although the addition of cobalt to beer was discontinued in the late 1960s, cobalt-induced dilated cardiomyopathy has been described following occupational exposures to this element[207,208] and in patients who have had a hip replacement with metal-on-metal acetabular surface joints.[204]

3.17 COPPER AND ZINC

Copper is an essential micronutrient and acts as a cofactor in several oxidation reactions, including those catalyzed by cytochrome c oxidase, copper-zinc superoxide dismutase, and tyrosinase.[209] Exposure to copper occurs from dietary ingestion, including dietary supplements; environmental exposures (e.g., copper released into the air from volcano eruptions and forest fires); medical treatments (e.g., copper tubing often used for hemodialysis and prolonged intravenous total parenteral nutrition)[209]; and industrial sources. Industrial exposure comes from copper smelters, iron and steel production, municipal incinerators, and pesticide production. Mortality related to serum copper levels has a "U-shaped relationship" with excessively low and high levels of copper being deleterious.[210]

Increased serum copper levels are associated with the production of free radicals, which is believed to be the mechanism of its association with increased serum LDL level and decreased HDL concentrations and resultant atherogenesis.[210–216] A common laboratory method of initiating LDL oxidation in vitro involves incubation of LDL with copper.

Elevated serum copper concentrations are associated with diabetic complications, including diabetic cardiomyopathy, microvascular disease, and hypertension.[217,218] Hyperglycemia is purported to hinder the ability of albumin and ceruloplasmin to bind copper, thereby leading to enhanced free radical production, increased oxidative stress, and myocardial fibrosis.[217]

Both increased serum copper concentrations and decreased zinc concentrations have been reported in individuals with reduced ventricular systolic function as well as those with idiopathic dilated cardiomyopathy.[212] In mouse models, cocksackie infection induces the accumulation of trace elements in the heart and may explain the observation of increased levels of trace metals and the observation of increased tissue concentrations of trace metals in patients with a dilated cardiomyopathy believed to be viral in origin.[212] High serum copper concentrations are present in patients with dilated cardiomyopathy compared to normal, with the severity of symptoms correlating with the concentration.[219] Furthermore, elevated serum copper concentrations are linked to increased mortality in these individuals.[219–221]

A clear causal relationship between serum copper concentrations and cardiovascular disease has not been proved; they have only been associated in observational studies. Elevated serum concentrations of copper and ceruloplasmin (the main carrier of serum copper) are associated with an increased risk of cardiovascular disease, myocardial infarction, and cerebral vascular accidents.[215,216,222,223] Additionally, acute myocardial infarction patients with high serum copper concentrations experience more postinfarction complications than those whose serum copper concentrations are low.[224] Subjects with copper overload are treated with high-dose zinc, tetrathiomolybdate, ascorbic acid (which decreases the intestinal absorption of copper), and, if necessary, d-penicillamine.[209]

Low serum copper concentrations have been linked to atherosclerosis; copper is required for antioxidant function (through the copper-zinc–dependent enzyme superoxide dismutase).[210,214,225–227] In addition to dyslipidemia, low serum copper concentrations are associated with increased systemic arterial pressure (due to impaired endothelium-dependent relaxation[218]), serum uric acid concentrations, serum glucose concentrations, and cardiovascular death.[210,227–229] It is postulated that decreased capillary density in chronic ischemia may be mediated by low tissue copper levels in the heart.[230]

Copper deficiency leads to cardiac mitochondrial structural damages, to myofibrillar enlargement, and to disturbances in oxidative phosphorylation.[212–214,230,231] These changes have been linked to cardiac enlargement through concentric hypertrophy.[209,213,214,229,232] In experiments with animals, the enlargement can affect all chambers and also lead to ventricular aneurysms[213]; the disturbances can be reversed by copper supplementation.[233] Human subjects fed a diet low in copper may experience severe tachycardia, heart block, and myocardial infarction.[229] With acute myocardial infarction, serum copper concentrations increase (and serum zinc levels concomitantly fall), suggesting that these trace elements may be linked to the magnitude of myocardial damage.[218]

Zinc, an essential cofactor in several enzymatic processes, competes with copper; high serum concentrations of zinc simulate copper deficiency and are associated with atherosclerosis and hypertension. Conversely, zinc deficiency—similar to copper excess—has been linked to atherosclerosis, insulin resistance, hypertension, and congestive heart failure.[218,226,231,234,235]

3.18 ENERGY DRINKS

Since their introduction in the United States during the 1990s, energy drinks have grown in popularity, with hundreds of different brands now available. Moreover, the consumption of energy beverages has been associated with high-risk behaviors such as cigarette smoking, alcohol abuse, illicit drug use, sexual risk taking, fighting, failure to use seat belts, and taking risks on a dare.[132,236,237] Most energy beverages are proprietary blends of caffeine, taurine, glucuronolactone, guaranine, and B vitamins.[132,238] By far, caffeine is the most extensively studied of these compounds; its cardiovascular toxicities are reviewed elsewhere in this book. The caffeine content of energy drinks ranges from 50 to 505 mg per container (in comparison, an 8-ounce cup of drip-made coffee contains 110–150 mg of caffeine).[132]

One of the common elements in energy beverages is taurine, a naturally occurring sulfonic acid that is associated with positive cardiac inotropy.[239] These effects are additive to those of caffeine.[238,240] Another compound of interest in energy drinks is guaranine, which is obtained from the guarana rainforest vine that has been domesticated as a source of caffeine.[132,241,242] Guarana seeds (Figure 3.7) contain more caffeine—about four times as much as in coffee—than any other plant worldwide.[241] The guarana plant also is a source of theobromine and theophylline, which are chronotropes and inotropes, respectively.[241,242] Most energy drinks contain guarana products, and while the amount of these substances is below the amount typically considered to have any physiologic effect, hospitalization due to consumption of beverages with guarana has been reported.[132,241]

Various adverse cardiovascular effects have been observed in individuals who ingest energy beverages: for example, hypertension,[132,239,242] enhanced platelet aggregation and endothelial dysfunction,[243] reduced heart rate variability (a sign of autonomic dysfunction and a risk factor for sudden cardiac death),[244] increased heart rate, sinus and supraventricular tachycardia,[245–247] reversible postural tachycardia with syncope,[248] subarachnoid hemorrhage with concomitant cerebral vasculopathy,[249] transient dilated cardiomyopathy,[247] and cardiac arrest.[250]

3.19 FLUORIDE

Hydrofluoric acid is used in glass etching as well as the electronic and chemical industries, whereas fluoride salts may be found in insecticides, as a catalyst to produce higher-octane fuel during oil refining, and as a component in many household and commercial rust removers.[251] Almost all fluoride toxicity results from accidental exposure.[251,252] Hydrofluoric acid toxicity can occur after inhalation of vapors or absorption through the skin.[251] In addition, fluoride salts have been used as rat poison, and toxicity has occurred when they were mistaken for table salt or baking soda and ingested.[253]

FIGURE 3.7 **Photo of Guarana seeds.** They contain more caffeine—about four times as much as in coffee—than any other plant worldwide. **Source:** *Used with permission from http://mr-ginseng.com/en/guarana/.*

Systemic toxicity is directly related to the amount of fluoride that is absorbed systemically. Death has been reported following the oral ingestion of as little as one teaspoon (15 mL) of a 9% hydrofluoride solution and exposure of only 2.5% of body surface area from a hydrofluoride chemical burn.[251,252] Systemic toxicity can occur with exposure of more than 1% of the body surface area to concentrated hydrofluoride (i.e., approximately the size of the palm of a hand).[254] Systemic toxicity results in hypocalcemia, hypomagnesemia, hyperkalemia, and direct myocardial toxicity.[251,255,256] Cardiovascular toxicity most often causes QT interval prolongation, tachy- and bradyarrhythmias, and refractory ventricular fibrillation.[251,252] Direct myocardial toxicity has been demonstrated in experimental animals[252,255] and occurs via adenylate cyclase activation, leading to increased cyclic adenosine monophosphate formation, with resultant myocardial irritability.[254,257] The fluoride toxicity therapy has included milk ingestion (to dilute the acid and to bind the fluoride in the GI tract), gastric lavage with fluoride binders, parenteral calcium and magnesium administration, and hemodialysis.[251–254,257,258]

3.20 FUMIGANTS AND PESTICIDES

Pesticides and fumigants—chemicals employed as a pesticide or disinfectant in a gaseous state—are used to sterilize soil before planting, treat infested crops, and treat harvested products that have been infested. As such, they may be used in greenhouses, storage facilities, and on open fields. Fumigants are also sometimes used for sterilization at medical facilities and for equipment decontamination.

3.20.1 Aluminum Phosphide

Aluminum phosphide (sold as pellets and tablets) is an inexpensive and widely available rodenticide used to protect grains (i.e., so-called "rice tablets"). When exposed to water or acid, they release highly toxic phosphine gas. Aluminum phosphide may be absorbed via the skin or gastrointestinal tract, and the phosphine gas can be absorbed via inhalation.[259,260] Phosphine's gaseous form and extreme toxicity make it a potential agent for chemical terrorism. In addition to accidental industrial exposures, intentional ingestion of aluminum phosphide for homicide and suicide is common, especially in the Indian subcontinent[261,262]; mortality with poisoning approaches 50%.[260,261,263]

The clinical features of phosphine poisoning include severe vomiting, epigastric pain, pulmonary edema, refractory hypotension, and metabolic acidosis.[259,261–267] A smell of garlic or decaying fish may be noted and is a result of the impurities in the technical grade of the aluminum phosphide; pure phosphine gas is odorless.

Cardiac manifestations include myocarditis, myocardial infarction, left ventricular systolic dysfunction, and congestive heart failure.[268] Over time, ventricular function may recover.[263,268] Hypotension has been estimated to occur in 75 to 100% of individuals exposed to phosphine gas and is multifactorial in origin, due to the combined effects of depressed systolic ventricular function, peripheral vasodilation, and intravascular fluid loss.[268] Myocardial necrosis may lead to

- T-wave inversion and ST segment elevation or depression on the 12-lead electrocardiogram
- Elevated serum cardiac enzymes
- Dilated cardiomyopathy[261–263,265,267,269,270]

Hypokalemia may occur with phosphine gas exposure and predisposes one to arrhythmias. In addition, phosphine gas exposure may cause supraventricular tachycardia, ventricular tachycardia, and heart block.[259,261,262,264,266]

Phosphine gas can be measured in the serum using gas chromatography or mass spectrometry.[268,271] A bedside diagnosis can be made by applying gastric aspirate or exhaled breath to silver nitrate-impregnated paper, looking for it to turn a black color, which indicates the formation of silver phosphide.

Phosphine toxicity disrupts cytochrome c oxidase, denatures hemoglobin, impairs mitochondrial function, and leads to the production of free radicals, thereby interfering with cellular respiration.[260–262,268,269,271] Treatment consists of

- Early gastric lavage with activated charcoal or vegetable oil
- Neutralization of gastric contents with dilute potassium permanganate (0.01%) to oxidize phosphine to nontoxic phosphate
- Intravenous magnesium sulfate to prevent oxidative stress (and subsequent accelerated lipid peroxidation during the first six hours of aluminum phosphide poisoning)
- Supportive care[259,261–263,268,271,272]

N-acetylcysteine reduces myocardial oxidative damage in animals that have been exposed to aluminum phosphide; its efficacy in humans is unknown.[259] Despite these measures, survival is unlikely if more than 1.5 g of aluminum phosphide has been ingested,[268] with death occurring within $<$72 hours.[273]

3.20.2 Endosulfan

Endosulfan is a chlorinated hydrocarbon (an organochlorine) of the cyclodiene group used as an insecticide. Toxicity is mainly through oral ingestion, but it is also described following dermal exposure and inhalation.[274] The mechanism of endosulfan toxicity is likely from inhibition of Ca^{++}-ATPase and Na^{+}-ATPase and antagonism of chloride ion transport in gamma-aminobutyric acid receptors.[274]

The clinical manifestations of endosulfan toxicity include rhabdomyolysis, anion gap metabolic acidosis, acute kidney injury, altered mental status, hypotension, cerebral edema, and seizures.[274,275] The cardiac manifestations include myocardial infarction, ventricular systolic dysfunction (that may show recovery during convalescence), ventricular fibrillation, and death.[274,275] Electrocardiographic manifestations include premature beats, QTc interval prolongation, ST segment abnormalities, sinus tachycardia, atrial fibrillation, atrioventricular block, and ventricular fibrillation.[274,275]

Chromatographic analysis of serum, urine, and tissue specimens may confirm the ingestion of endosulfan.[274] No specific antidote to its toxicity exists; gastric lavage with activated charcoal and supportive care are advised.[274]

3.20.3 Organophosphates

Organophosphates and carbamates are used throughout the world as pesticides. Toxicity has occurred following accidental exposure and with chemical warfare. These compounds are lipid soluble, and intoxication may occur via inhalation, absorption from skin contact, or orally, as occurs with ingestion of food recently sprayed with these compounds. Organophosphates and carbamates include more than 50,000 compounds.[276] Organophosphates irreversibly inhibit cholinesterase, whereas carbamates reversibly bind to cholinesterase. Both lead to a massive parasympathetic surge.[277–279] Fat-soluble organophosphates, such as fenthion and chlorfenthion, may lead to cholinergic overactivity for days to weeks due to prolonged systemic release from subcutaneous adipose tissue; this can also manifest as a relapse of toxic symptoms after successful recovery.[279]

Classically, three clinical stages of poisoning occur. First, a brief period of sympathetic activity—attributed to an agonist effect on nicotinic receptors—is manifested as hypertension and sinus tachycardia. Second, a period of extreme cholinergic activity ensues, which is characterized by bradycardia, hypotension, and electrocardiographic ST- and T-wave changes, possibly with life-threatening arrhythmias. Finally, prolongation of the QTc interval with an attendant increased risk of sudden death may occur.[277,278,280–282] The severity of systemic poisoning may be estimated by measuring plasma or urine organophosphate concentration, cholinesterase activity, and serum ß-glucuronidase concentration.[279,283]

Various life-threatening arrhythmias may occur, including bradycardia, atrioventricular and intraventricular block, atrial fibrillation, and polymorphic ventricular tachycardia (so-called torsades de pointes).[277,280,282,284–286] ST-segment abnormalities may occur acutely and persist for weeks after drug exposure,[280] and life-threatening arrhythmias may occur as late as 20 days after exposure.[279–281] Late-onset arrhythmias may represent healing myocardium or increased free fatty acid levels in the myocardium.[276,280,287] Poor prognostic factors include a combination of ST segment and T-wave changes with concomitant low-voltage complexes.[276]

It is difficult to predict which patients are likely to develop the cardiac manifestations of organophosphate poisoning and when after exposure these toxicities may manifest.[279,280,287] Rarely, myocardial infarction may occur as a result of catecholamine release, coronary vasospasm, leukocytosis, hypoxemia, electrolyte disturbances, or possibly a direct toxic effect of the organophosphates.[276–278,280,284,287] Postmortem examination of the heart reveals focal areas of micronecrosis, pericarditis, and separate areas of myocarditis.[276,287] Usually, ventricular function is not affected by organophosphate poisoning[287]; however, takotsubo cardiomyopathy has been reported.[288]

Treatment consists of administering (1) atropine and pralidoxime to antagonize the parasympathetic effects and (2) benzodiazepines to treat seizures induced by the organophosphates.[279,287] Beta blockers, lidocaine, and cardiac pacing have not been found to be effective in the treatment of organophosphate poisoning.[280] Since carbamate reversibly binds to cholinesterase, poisonings with it usually resolve sooner and are associated with less morbidity and mortality than organophosphate poisoning.[289]

The evidence of cardiac risk following chronic, low-level exposure to organophosphates is not strong.[290–292] In fact, a recent prospective study of a large number of pesticide applicators chronically exposed to organophosphates and carbamates showed no increased risk of myocardial infarction.[293] Exposure to imidacloprid, a newer organophosphate-related insecticide comprised of neonicotinoid compounds that stimulate the nicotinic acetylcholine receptor, has been linked to reports of fatal ventricular fibrillation.[294]

3.20.4 Amitraz

Amitraz is a formamidine derivative insecticide that dissolves in the organic solvents acetone, toluene, and xylene.[295,296] Intoxication with it has been reported via inhalation, oral ingestion, dermal application, and intravenous injection.[295–297] The main clinical effects of amitraz are derived from alpha-2 adrenergic agonist activity (similar to clonidine poisoning), although poisoning also leads to secondary monoamine oxidase inhibition and inhibition of prostaglandin synthesis.[295–300] Signs of poisoning usually occur within minutes to an hour of exposure, with survivors recovering in a few hours to a few days.[295,296,298,300] Subjects exposed to amitraz may require ICU monitoring and mechanical ventilation. The toxic effects of amitraz include central nervous system (CNS) depression, respiratory depression, hypotension, bradycardia, hyperglycemia, elevation of liver function enzymes, mydriasis, and hypothermia. Electrocardiographic changes include bradycardia, ST segment changes, ventricular arrhythmias, and torsades de pointes.[295–297,301]

In animal studies, the alpha-2-adrenergenic antagonists (e.g., yohimbine, atipamezole, and phentolamine) have been helpful in treating amtiraz-induced toxicity. In humans, treatment is mainly supportive, since recovery without sequelae almost always occurs.[296–302] Atropine should be avoided for those who have ingested amitraz, as it may precipitate ventricular arrhythmias.[295,296]

3.20.5 Pyrethroid Insecticides

Pyrethroids are insecticides that are routinely used against household pests (e.g., mosquitoes, houseflies, and cockroaches). They are considered axonal excitoxins, with their effects mediated by inhibiting closure of the voltage-dependent sodium channels, voltage-gated chloride channels, and gamma-aminobutyric-gated chloride channels.[303] Inhalation or oral ingestion of pyrethrin insecticides may lead to paresthesias, headache, nausea, vomiting, diarrhea, melena, epigastric pain, dyspnea, bronchospasm, chest pain, neural excitability, and seizures.[303,304]

Cardiovascular toxicities reported with pyrethroid exposure include hypotension, sinus tachycardia, sinus arrest (with junctional escape rhythm),[304] aortic dissection, depressed left ventricular systolic function,[305] and takotsubo cardiomyopathy.[288] Pyrethroid-induced toxicity most often resolves spontaneously within 48 hours after exposure with supportive care alone.[305]

3.20.6 Sulfuryl Fluoride

Sulfuryl fluoride (i.e., trade name Vikane) is a colorless, odorless fumigant used against wood-boring insects. Accidental and intentional poisonings have been reported, with inhalation the most common route of exposure. The effects of the poisoning are equivalent to that of being exposed to systemic fluoride poisoning, namely electrolyte disturbances (i.e., hypokalemia, hypocalcemia, and hypomagnesemia), refractory dysrhythmias, and death.[306,307] Poisoning with sulfuryl fluoride is confounded by its coadministration with chloropicrin, a lacrimating irritant pesticide that is used to provide warning that sulfuryl fluoride is in the air.[307,308] Chloropicrin is associated with skin and corneal irritation, noncardiogenic pulmonary edema, and death.[308,309] Treatment for sulfuryl fluoride exposure is similar to other means of intoxication with systemic fluoride: (1) oral administration of dilute calcium hydroxide or calcium chloride to prevent further absorption and (2) injection of calcium gluconate to increase the blood calcium concentration.[307]

3.20.7 Hydrogen Sulfide

Hydrogen sulfide is a water-soluble, colorless gas with the distinct odor of rotten eggs. People exposed to hydrogen sulfide at concentrations above 100 to 150 ppm may lose the ability to smell hydrogen sulfide after 2 to 15 min of continuous exposure due to olfactory fatigue.[310,311] Hydrogen sulfide is produced from sewage sludge, liquid manure, sulfur hot springs, and natural gas. It is also is a byproduct of various industrial processes such as petroleum refining, wood pulp processing, rayon manufacturing, manure processing, sugar beet processing, fish processing, and hot asphalt paving.[310] The mechanism of its toxicity is via interruption of mitochondrial cellular respiration through inactivation of cytochrome oxidase.[310,311]

Patients with hydrogen sulfide poisoning may present with mucosal damage, hemoptysis, tachycardia, cyanosis, lactic acidosis, and loss of consciousness. Myocardial infarction, myocarditis, and dilated cardiomyopathy have all been described following hydrogen sulfide exposure.[310] Detection of hydrogen sulfide in patients suspected of exposure can be accomplished with measurement of thiosulfate levels via chromatography of the blood or tissue.[311,312]

Victims of hydrogen sulfide poisoning should be removed from the site of exposure, and oxygen should be administered. Hyperbaric oxygen may be of use, although the evidence of its efficacy is anecdotal.[310,313] Inhaled amyl nitrite and intravenous sodium nitrite have been used as antidotes and are effective if administered within minutes of hydrogen sulfide exposure. Nitrites induce methemoglobin, which competitively binds the sulfide ion, thereby liberating the cytochrome oxidase. The resultant compound, sulfmethemoglobin, can be metabolized and excreted.[312] However, caution must be used because nitrites can induce hypotension, and methemoglobin may reduce oxygen delivery.

3.21 GOLD

Gold-coated stents gained popularity because of their radiographic visibility and animal studies demonstrating reduced thrombogenicity, neointimal tissue growth, and stent endothelialization with their use compared to noncoated bare metal stents.[314–316] However, it quickly became clear that patients who received gold-coated stents were more likely to develop a contact allergy to them, and they experienced higher rates of in-stent restenosis compared to those who received stainless steel stents.[316–320] It is believed that leakage of gold into the blood leads to sensitization, an allergic response, and local vascular inflammation.[316,320–322] Similarly, gold dental implants are associated with detectable serum gold concentrations and an increased incidence of contact allergy to gold.[321,322] Interestingly, the release of gold into the blood from dental restorations is enhanced by cigarette smoking.[323]

In a randomized trial of gold-coated versus stainless steel stents, no difference in outcomes was noted at 30 days, but at a year those who received gold-coated stents had higher rates of restenosis and repeat revascularization, as well as lower rates of event-free survival.[318] Angiographic studies conducted six months after stent placement showed more neointimal proliferation and a smaller minimal luminal diameter in the subjects receiving gold-covered stents.[314,315] It is important to note that the use of a gold-coated stent as opposed to an uncoated stent was associated with higher rates of death, myocardial infarction, or target vessel revascularization at five years.[316] As a result, gold-covered stents are no longer used.[323]

Apart from stent-related complications, gold therapy for rheumatoid arthritis, or the ingestion of gold found in health supplements, has been associated with myocardial infarction, dilated cardiomyopathy, and ventricular tachycardia.[324–326]

3.22 HOUSEHOLD CHEMICALS

Household chemicals are routinely used for cleaning, disinfection, and general hygiene purposes. The following subsections describe some of them.

3.22.1 Camphor

Camphor is a pleasant-smelling terpene used in skin lotions and in many ayurvedic medicines intended for oral use as an analgesic, abortifacent (i.e., contraceptive), aphrodisiac, antiseptic, and antipruritic.[327] Its strong aroma and mild anesthesia may be mistaken for effective medicine, especially when applied topically. Ingestion of 2 g of camphor is sufficient to produce toxic effects in adults.[328] Initial symptoms may occur within 5 to 15 min after ingestion and include nausea and vomiting, oral and epigastric burning, a feeling of warmth, and a headache. These symptoms may progress to altered mental status, convulsions, coma, and death. Death usually is attributed to respiratory failure or neurologic complications. Clinical toxicity typically resolves within 24 hours in survivors.[328,329]

Cardiovascular toxicities include tachycardia, prolonged QTc and QRS intervals, atrioventricular conduction block, and ST segment changes.[328] Myocarditis and depressed ventricular systolic function may occur following camphor ingestion.[328] Treatment is largely supportive; hemodialysis has not been shown to be of benefit.[328,329]

3.22.2 Detergents

Suicidal intoxication with detergents has been associated with hypotension, refractory ventricular fibrillation, depressed ventricular systolic function, and pulmonary edema.[330] Treatment is supportive.

3.22.3 Disinfectants and Cleansers

Chlorhexidine, used in disinfectants, has been implicated in coronary artery vasospasm in angiographically normal arteries, leading to myocardial infarction.[331] Use of household sprays and scented items has been linked to reduced heart rate variability, a marker of autonomic dysfunction.[332] In a study of hair salon workers, higher concentrations of serum C-reactive protein (CRP) and lower scores on measures of heart rate variability were observed on the days the subjects were at work (i.e., exposed to different airborne pollutants and chemicals) compared to days they were not at work.[333]

Alcohol-based hand sanitizers usually contain ethanol and isopropyl alcohol. They have been associated with similar toxicities to ethanol intoxication. Cardiovascular complications reported include ST segment and T-wave changes, ventricular fibrillation, and cardiac arrest.[334,335]

3.22.4 Dettol

Dettol is a liquid household disinfectant containing 4.8% chloroxylenol, pine oil, and isopropyl alcohol. Most toxicity is related to its CNS suppressant effects, local corrosion of the GI tract, as well as acute respiratory distress syndrome. Hypotension, tachyarrythmias, and bradyarrhythmias have been described as a result of Dettol poisoning.[336–338]

3.23 INHALANTS

Inhalants include organic solvents, organic nitrites (e.g., amyl nitrite or amyl butyl), and nitrous oxide. The organic solvents include toluene (found in airplane glue, rubber cement, and paint thinner), Freon, kerosene, gasoline, carbon tetrachloride, acrylic paint sprays, shoe polish, degreasers, nail polish remover, correction fluid, adhesives, permanent markers, room fresheners, deodorants, dry-cleaning agents, and lighter fluids. These solvents most often are inhaled by children or young adolescents (so-called "huffing," "sniffing," or "dusting"), or they may be absorbed through the skin. On occasion, they have been used intentionally for homicide and suicide attempts, most prominently with trichloromethane (chloroform).[339–344]

Symptoms and signs of their use include sneezing, salivation, skin flushing, cough, nausea, vomiting, photophobia, tinnitus, diplopia, headache, ataxia, slurred speech, depressed reflexes, nystagmus, and altered consciousness. Inhalant abuse has become a global concern for intoxication of adolescents and teenagers, since inhalants are inexpensive and widely available.[345]

Acute or chronic inhalant use occasionally has been reported to induce cardiac abnormalities, the most common of which is dysrhythmias; rarely, it has been associated with myocarditis, myocardial infarction, and sudden death.[339,344–349] The inhalation of halogenated chemicals can sensitize the myocardium to catecholamines and depress innate conduction automaticity, which explains why fatal arrhythmias have been reported to occur when the user is startled during inhalation or experiences a catecholamine surge from any source.[344,345,347,350–352]

Each exposure to inhalants puts one at risk of sudden cardiac death no matter whether the person is a first-time or chronic user.[351] The arrhythmias produced by inhalants include sinus bradycardia or tachycardia, atrioventricular block, and premature ventricular beats; an electrocardiogram (ECG) often shows a prolonged QTc interval, which places the individual at risk of sudden cardiac death.[339,344,349,352–354] Freon may decompose into phosgene and hydrochloric acid, the inhalation of which may lead to inflammation, pulmonary edema, and myocardial infarction.[355]

Halocarbons (e.g., Freon) can activate hyperpolarizing potassium channels, reduce gap junction conductance between cells, alter voltage-gated calcium channel activity, increase calcium release from the sarcoplasmic reticulum, and depress the sodium current.[339,344,350,352] The simultaneous hypoxia induced by inhalants enhances their proarrhythmic effects and renders the resulting arrhythmia more refractory to treatment.[351]

In addition, the technique of spraying Freon from a compressed container directly onto the palate may be arrhythmogenic. When released from a pressured container, the rapid expansion results in a gas that may be as cold as −20°C.[351] The application of the cold gas to the palate, larynx, and/or pharynx can induce a vasovagal discharge, leading to profound bradycardia and possibly even asystole.[345,351]

Chronic cardiac damage from inhalant abuse can lead to a dilated cardiomyopathy, with histologic changes of myocarditis, myofibril rupture, edema, and fibrosis.[344,346,348] Refrigeratation workers chronically exposed to fluorocarbons are at a higher risk for premature beats.[349,353] Detection of the inhalants is accomplished with gas chromatography.[356] Treatment of inhalant abuse is supportive, but avoidance of sympathomimetic agents is advisable, and beta blockers may be administered.[356]

3.24 IRON

Iron plays a catalytic role in the generation of highly reactive oxygen species via conversion of superoxide and hydrogen peroxide into a toxic hydroxyl radical, the so-called Fenton and Haber-Weiss reaction.[357–360] Iron promotes lipid peroxidation in vitro and amplifies the prooxidant capacity of vascular cells. Thus, it is hypothesized that the toxic effects of iron may be most prominent in patients with elevated serum LDL concentrations.[358,361,362] In experimental models, iron administration exacerbates ischemic myocardial injury, and iron chelation can ameliorate this effect.[362,363] Some epidemiologic studies have demonstrated an association between body iron stores and cardiovascular events,[359,362–365] whereas others have failed to show a relationship between them.[359,366–370]

Several surrogate markers of cardiovascular disease have been associated with measures of increased body iron stores. Increased serum ferritin levels are associated with subclinical coronary atherosclerosis determined by coronary artery calcium computed tomography (CT) score, independent of traditional cardiovascular risk factors.[371] They have also been associated with increased vessel stiffness—as measured by pulse wave velocity[372]—and hypertension.[373,374] Serum ferritin and transferrin levels have been modestly associated with peripheral arterial disease[358] and carotid artery atherosclerosis (i.e., both its presence and progression).[361] Iron excess may play a role in the pathogenesis of peripheral insulin resistance[371,375–377] and diabetes mellitus.[371,376–379]

Iron concentrations in human pathologic specimens of atherosclerosis are higher than those found in healthy arterial tissue.[380] Although iron chelation therapy has been shown to improve endothelial dysfunction in the forearm arteries of patients with coronary artery disease,[381] reduction of body iron stores in patients with symptomatic peripheral arterial disease does not decrease mortality, nonfatal myocardial infarction, or stroke in patients with symptomatic peripheral arterial disease.[357]

Many studies have examined the role of iron in the prognosis for patients with stroke. Increased iron stores are linked to poorer outcomes with thrombolytic therapy of ischemic stroke.[382] Increased serum iron levels are experimentally linked to cerebral arterial endothelial dysfunction,[381] with a potential role in worsened brain edema after intracerebral hemorrhage[383] and larger cerebral infarct volumes.[384] Although studies show that increased iron body stores and dietary iron intake are associated with poorer outcomes after a stroke,[382,385] iron chelation therapy has not been shown definitively to improve outcomes in acute stroke patients.[360]

Iron overload states occur in hereditary hemochromatosis, cirrhosis, sickle-cell anemia, myelodysplastic syndrome, and severe thalassemia. These conditions have been associated with iron deposition in the myocardium, which may lead to restrictive or dilated cardiomyopathy.[386–389] Iron overload cardiomyopathy may present initially as diastolic dysfunction and later as dilated cardiomyopathy with systolic dysfunction.[390,391] Conduction system disturbances may also be present and range from minor arrhythmias to sudden cardiac death.[392]

Serum ferritin levels and liver iron concentrations are good indicators of total body iron stores but are insufficient to allow an estimate of myocardial iron accumulation. Cardiac magnetic resonance imaging with T2 weighting is a noninvasive technique that provides rapid and direct assessment of myocardial iron content.[389,392,393] Early detection of myocardial iron overload and chelation treatment may improve and reverse the pathologic state, although by then the prognosis is often poor.[386–389,392,394,395]

Iron deficiency is a risk factor for stroke and carotid thrombosis, which may be the result of an associated hypercoagulable state, thrombocytosis, or anemic hypoxia; this effect has been studied more in children but appears to be present in adults as well.[396–400] The relationship between iron body stores and cerebral vascular accident levels may be U-shaped because both deficiency and excess have been associated with increased risk.[374,400] Iron deficiency anemia may lead to a hypercoagulable state directly related to iron deficiency and/or anemia; thrombocytosis secondary to iron deficiency anemia; and anemic hypoxia, whereby a mismatch between oxygen supply and end-organ demand leads to ischemia and infarction[396,397,399,401]

3.25 LEAD

Lead exposure from leaded gasoline, lead-based paints, polluted water sources, and industrial emissions has been associated with developmental delays, neurologic deficits, and renal disease. Several epidemiologic studies have linked lead exposure to hypertension,[402–407] with no evidence of a threshold level of toxicity; any exposure appears to be associated with an increase in systemic arterial blood pressure.[407,408] Lead may contribute to cardiovascular disease by increasing oxidative stress and causing endothelial dysfunction, inflammation, downregulation of NO production, and renal damage.[409,410]

Apart from hypertension, lead exposure appears to be associated with higher rates of left ventricular hypertrophy;[406,411] decreased ventricular systolic function;[411] atrioventricular conduction block; a prolonged QTc interval;[412–414] atherosclerotic disease of the peripheral arteries,[407,409] the coronary arteries,[402,406,407,410,415–418] and the cerebral arteries;[407,410] and cardiac mortality.[407,408,410,416]

3.26 MAD HONEY

Grayanotoxin ("mad honey") is a natural compound found in the honey of nectar that is derived from various species of rhododendron, including *Rhododendron luteum*, *R. ponticum*, and *R. simsii*.[419–423] The rhododendrons associated with grayanotoxin-containing mad honey are found in the Black Sea region of Eastern Turkey (where mad honey intoxication has been described the most often) as well as in North America, Europe, and eastern Asia.[420–422,424] Some cases have also been reported in Germany, Austria, Switzerland, and Korea.[420,424] Mad honey has been used as a herbal medicine for the treatment of sexual dysfunction,[419,422,423,425] hypertension, heart disease, diabetes mellitus, and gastrointestinal disorders;[422,423,426] such usage has contributed to episodes of accidental poisoning.

Grayanotoxin's toxic effects are mediated through binding of voltage-dependent sodium channels in their open state and resultant prevention of the channels' inactivation; thus, the channels remain in a state of depolarization. This effect on peripheral and cardiac branches of the vagus nerve leads to an increase of parasympathetic tone, causing bradycardia, hypotension, and various degrees of atrioventricular block. Atrial fibrillation, asystole, and myocardial infarction have rarely been observed following mad honey ingestion.[420,423,426–433]

The cardiac and general cholinergic symptoms of grayanotoxin poisoning (e.g., nausea, vomiting, syncope, diplopia) generally occur within minutes to a few hours of mad honey ingestion,[420,422,423,430,432–434] after which they last less than 24 hours. If treatment is necessary, atropine sulfate, intravenous fluids, vasopressors, and temporary pacing have all been used successfully until the toxic effects of grayanotoxin dissipate[419,422–424,432–435]

3.27 MAGNESIUM

Magnesium toxicity occurs most commonly in the setting of renal failure (since excretion occurs mainly via the kidney) or iatrogenic administration. Magnesium acts as a physiologic calcium blocker, which results in conduction abnormalities. In experimental studies with animals, high serum magnesium concentrations cause transient tachycardia, followed by bradycardia and prolongation of the PR, QRS, and QT intervals.[436] In humans, excess magnesium has been associated with hypotension,[437–440] heart block,[436] decreased cardiac contractility,[438,440] and asystole.[439]

The clinical consequences of magnesium toxicity are related to the serum concentration (Table 3.1), with a serum concentration >4.0 mEq/L causing hyporeflexia, >5.0 mEq/L leading to prolonged atrioventricular conduction, >10.0 mEq/L causing complete heart block, and >13.0 mEq/L resulting in cardiac arrest.

3.28 MANGANESE

Manganese (Mn) is a trace element that has been linked to various essential enzymes, including superoxide dismutase.[441] Exposure to manganese occurs

TABLE 3.1 Clinical Consequences of Magnesium Toxicity and Their Relationship to Serum Concentration

Serum Magnesium	Clinical Consequences
>4.0 mEq/L	Hyporeflexia
>5.0 mEq/L	Prolonged atrioventricular conduction
>10.0 mEq/L	Complete heart block
>13.0 mEq/L	Cardiac arrest

during the welding process; production of manganese alloys with other metals, cell batteries, and various organometallic chemicals; and long-term parenteral nutrition.[442,443] Less often, toxicity to manganese can occur through administration of ionic contrast medium for magnetic resonance imaging, contact with a manganese-containing fungicide (e.g., manganese ethylenebis-dithiocarbamate, so-called MANEB) and ingestion of the street drug "bazooka," a cocaine-based drug contaminated with manganese-carbonate during the free-base preparation method.[444]

Manganese exposure has been shown to be responsible for neuropsychiatric syndromes; extrapyramidal dysfunction; and a Parkinsonism-like syndrome,[442] with toxicity thought to be related to oxidative stress, damage to mitochondria, and inhibition of oxidative phosphorylation.[441]

Although data supporting the cardiovascular effects of manganese have been obtained from experimenting with animals, little evidence exists linking manganese exposure to toxicity in humans. In animal studies, manganese toxicity can cause depression of systolic function (e.g., negative inotropy), vasodilation, increased heart rate and systolic arterial pressure (due to release of catecholamines), dysrhythmias, myonecrosis, and histologic evidence of damage to the mitochondria.[445] Some of these effects occur because manganese is a calcium antagonist.[444,446,447]

In humans, the infusion of manganese-containing contrast material causes a slight increase in systolic arterial pressure and heart rate after 3 to 4 min, but these variables return to normal when the infusion is discontinued.[446] In factory workers, the concentration of manganese oxide in the ambient environment is associated with sinus arrhythmia, ST segment and T-wave abnormalities, and reduced blood pressure.[442] Subjects with manganese toxicity may have symptoms of orthostasis along with Parkinsonism; heart rate variability is also reduced.[448]

Treatment of manganese toxicity consists of removal from exposure. Chelation therapy is successful in increasing manganese excretion in urine, but it has not been shown to be associated with clinical improvement.[444]

3.29 MERCURY

Mercury is a heavy metal released from coal-burning electric power plants, chlorine production, dental amalgams, thermometers, and batteries. Once released into the environment, mercury is transfixed by microorganisms into methylmercury (MeHg), which bioaccumulates into larger, long-lived predators higher in the food chain.[449] Compared with elemental or inorganic mercury that is poorly absorbed from the intestinal tract, MeHg is readily absorbed from the GI system and actively transported to tissues by a widely distributed amino acid carrier protein[449]; as a result, it is toxic at relatively low levels of exposure.

Although chronic low-level mercury exposure from the consumption of fish (the most common source of MeHg) has raised concerns about public health,[450] evidence that MeHg exposure from eating fish adversely affects health is lacking.[449–453] Studies with experimental animals have shown that exposure to mercury can lead to negative cardiovascular effects such as mitochondrial dysfunction,[454,455] enhanced production of free radicals,[455,456] lipid peroxidation,[457] impairment of the clotting cascade and platelet function,[458] decreased cardiac myocyte contractile force,[459] and hypertension.[455,460]

In humans, mercury exposure can lead to hypertension,[461,462] reduced heart rate variability and autonomic dysfunction,[462–465] left ventricular diastolic dysfunction,[466] and accelerated progression of carotid atherosclerosis.[455,467,468] The data regarding the effects of mercury exposure on primary cardiovascular endpoints have been mixed. Several prospective studies observed no link between serum mercury levels and risk of myocardial infarction,[449,452,469–471] whereas another one observed an association.[472] Similarly, prospective studies have failed to demonstrate a link between blood mercury levels and stroke[452,473] or mercury exposure and the risk of coronary heart disease.[451]

With acute intoxication, mercury binds to and inactivates the sulfhydryl donor S-adenosyl methionine, a necessary cofactor for catecholamine-0-methyl transferase (COMT). Since COMT is required for the metabolism of norepinephrine, epinephrine, and dopamine, its inhibition results in a clinical syndrome that resembles a pheochromocytoma crisis with malignant hypertension.[454,474] Serum and urine catecholamine levels and urinary mercury levels are increased in the patient with acute mercury intoxication.[474]

3.30 METHYL BROMIDE

There are no published studies of the impact of methyl bromide on the cardiovascular system in humans.

3.31 MOLYBDENUM

The primary cardiovascular effect of concern with molybdenum exposure is an inflammatory vascular reaction that may lead to “allergic angina” and acute coronary syndrome—the so-called Kounis syndrome—whereby unstable angina or acute myocardial infarction occurs as a result of mast cell activation and the release of inflammatory mediators following an allergic, hypersensitivity, anaphylactic, or anaphylactoid (resembling anaphylactic) reaction.[475–477] (For a complete explanation, see the next section on nickel.)

Concern has been raised that stents made from 316L stainless steel containing molybdenum ($\sim 2\%$) may lead to an inflammatory, hypersensitivity reaction that predisposes patients to in-stent restenosis.[478,479] Although some studies suggest that individuals with contact dermatitis to molybdenum have an increased risk of in-stent restenosis with stainless steel stents,[478,480] others have found no correlation between the two.[481,482]

3.32 NICKEL

Placement of medical devices composed of metal alloys containing nickel have been implicated in the development of allergic angina and acute coronary syndrome, so-called Kounis syndrome, whereby unstable angina or acute myocardial infarction occurs as a result of mast cell activation and release of inflammatory mediators following an allergic, hypersensitivity, anaphylactic, or anaphylactoid (resembling anaphylactic) reaction.[475,477]

Stent thrombosis due to nickel allergy is supported by isolated reports of peripheral eosinophilia, markedly elevated serum IgE concentrations, intracoronary mast cell infiltration, and eosinophilic infiltration of coronary thrombi in patients who died of acute stent thrombosis.[475] Accordingly, the package inserts of the new-generation stents recommend avoidance of such implantations in patients with known hypersensitivity to any of the stent components. In North America, nickel allergy (as determined by patch testing) has a prevalence of 16.9%,[483] whereas in Europe the prevalence is around 25%.[484]

In addition to reports of Kounis syndrome with stents, some case studies have described

- Nickel allergy, chest pain, dyspnea, bronchospasm, and pericarditis in patients with atrial septal defect and patent foramen ovale who underwent implantation of an occluder device made of nickel[485,486]
- Failure and incompetence of nickel alloy mitral and aortic valve prostheses in patients with a known nickel allergy[487,488]
- Restenosis of nickel-containing stents in patients with contact dermatitis to nickel (Figure 3.8)[478]

Nickel salts are known to (1) cause local inflammation, (2) induce endothelial damage, and (3) upregulate endothelium adhesion molecules such as intercellular adhesion molecule-1 (ICAM-1), vascular cell adhesion molecule-1 (VCAM-1), and endothelial leukocyte adhesion molecule-1 (ELAM-1).[490]

Stent restenosis and the possible link to nickel allergy has received considerable attention due to the high number of stents placed worldwide. Most early-generation intracoronary stents were made from 316L stainless steel that contains metals associated with contact dermatitis, including nickel (~12%), chromium (as chromate, 17%), and molybdenum (~2%)[478]; the new-generation stents avoid nickel and use chromium and cobalt, which are less allergenic and associated with fewer complications. However, data concerning the contribution of nickel hypersensitivity to coronary restenosis are inconclusive.[475,476,478,491] Some retrospective studies demonstrate a link between nickel allergy (as determined by patch testing) and 316L stainless steel stent restenosis,[478,480] whereas others[481,482,492,493] show no increased risk of restenosis with stents containing nickel.

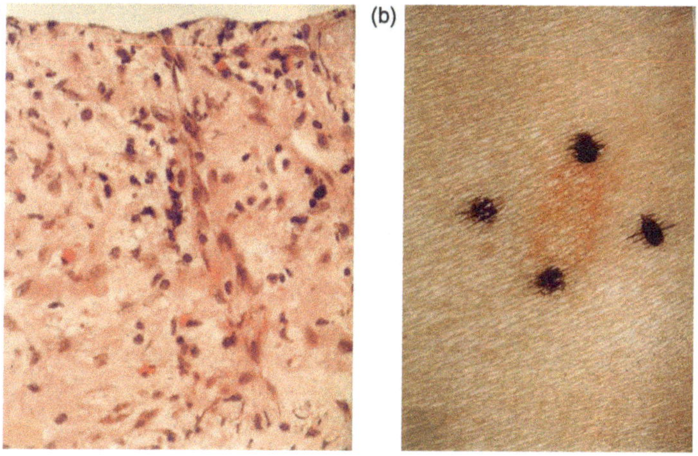

FIGURE 3.8 **Allergic in-stent restenosis.** (a) Granulation tissue with eosinophil infiltration in the restenotic lesion after placement of a stainless steel coronary stent. (b) Positive skin patch test to 316L stainless steel (i.e., the stent components). **Source:** *Used with permission from Kawano et al., 2004.*[489]

3.33 PHOSPHOROUS

Dietary phosphorous ingestion has been on the rise as a result of increased consumption of processed "fast foods" that contain phosphorous additives[494] such as processed meat, ham, sausages, canned fish, baked goods, and soft drinks. High serum phosphate concentrations are associated with

- Coronary artery calcification[495]
- Incident cardiovascular disease (e.g., angina, myocardial infarction, stroke, peripheral vascular disease, or congestive heart failure), as noted in the Framingham Offspring Study[496]
- Incident heart failure, myocardial infarction, and cardiovascular death in subjects with coronary artery disease[497,498]
- Myocardial infarction and death of individuals with chronic kidney disease[494,499]

These effects may be due to arterial calcification and endothelial dysfunction associated with high serum phosphate concentrations.[496,500] In mice, genetic mutations that lead to hyperphosphatemia (e.g., fibroblast growth factor-23 and klotho gene knockouts) are associated with premature aging and vascular calcification.[494,496,500,501]

3.34 POTASSIUM

Hyperkalemia is one of the more common life-threatening electrolyte disturbances that has been shown to have a significant impact on the cardiac

TABLE 3.2 Electrocardiographic Changes Associated with Hyperkalemia

Serum Potassium Level	Expected ECG Finding
5.5–6.5 mEq/L	Peaked T waves (most prominent in precordial leads)
6.5–8.0 mEq/L	Peaked T waves Prolonged PR interval Widened QRS interval Decreased P-wave amplitude
>8.0 mEq/L	Absence of P wave Atrioventricular, intraventricular, fascicular, bundle branch blocks QRS axis shift Widening of QRS complex leading to "sine-wave" (sinoventricular) pattern Ventricular fibrillation Asystole

conduction system. Hyperkalemia is often due to renal insufficiency but can also be seen in crush injuries, metabolic acidosis, and certain medications (e.g., potassium-sparing diuretics, angiotensin-converting enzyme inhibitors, angiotensin receptor blockers, potassium supplements). The classic progression of electrocardiographic changes is shown in Table 3.2; however, manifestations may vary between patients. Various degrees of atrioventricular and intraventricular block have been described as well.

3.35 SELENIUM AND SODIUM

Selenium toxicity has not been shown to be associated with cardiac complications to date. Isolated hypernatremia due to sodium has not been associated with cardiac complications in humans to date.

3.36 SULFUR DIOXIDE

There are no published studies describing the impact of sulfur dioxide on the cardiovascular system in humans.

3.37 THALLIUM

Thallium was initially used as an insecticide, rodenticide, and antiinfective for the treatment of syphilis, gonorrhea, tuberculosis, and ringworm.[502–504] Its use as a household rodenticide was banned in the United States in 1965 following multiple episodes of unintentional poisoning.[503,505] Unfortunately, unintentional poisonings still occur in other countries where thallium continues to be used as a rodenticide and ant killer.

At present, thallium is used in the manufacture of electronic components, optical lenses, semiconductor materials, alloys, gamma radiation-detection equipment, imitation jewelry, artists' paints, low-temperature thermometers, and green fireworks. Thallium exposure may occur at smelters in the maintenance and cleaning of ducts and flues and through contamination of cocaine, heroin, and herbal products. Criminal and unintentional thallium poisonings are still reported, some leading to death.[503,506–509]

Thallium is toxic with a dose of as little as 0.2 to 1 g when inhaled, ingested, or absorbed through the skin[504]; generally, a dose of >10 to 15 mg/kg is lethal.[503,510,511] It is handled by the body in a manner similar to that of potassium; as a result, it accumulates in tissues with a high potassium concentration. Following absorption, thallium is distributed to soft tissues, with the highest concentrations found in scalp hair, kidney, and heart. Because of its enterohepatic circulation, it has a slow elimination half-life of 3 to 30 days. Thus, it is considered a cumulative poison (i.e., small doses over time build up to toxic and, eventually, fatal levels).[502,510,511]

Thallium's toxic effects are due to multiple mechanisms, including

- Impaired glucose metabolism
- Sodium-potassium ATPase inhibition
- Mitochondrial damage
- Accumulation of lipid peroxides
- Impairment of electron transport
- Interruption of ribosomal function and protein synthesis

Cardiovascular toxicities include these: myocarditis and cardiac arrhythmias; sudden cardiac death related to arrhythmias has been noted up to two months after acute intoxication.[503] Early therapy with Prussian blue, forced diuresis, and hemodialysis may improve the prognosis.[503,504,510,511]

3.38 VITAMINS

Low levels of certain vitamins have been shown to be associated with cardiovascular disease and increased mortality. Unfortunately, vitamin supplementation has not proven useful in reducing these risks.

3.38.1 Vitamin A

Low-serum vitamin A concentrations have been shown to be associated with coronary artery disease.[512–515] Nevertheless, administration of vitamin A supplements (1) concomitantly with vitamin C and E does not reduce coronary artery calcification[516] and (2) concomitantly with beta-carotene is associated with a probable increase in cardiovascular mortality.[517]

The carboxylic acid form of vitamin A—all trans-retinoic acid (ATRA)—is used to treat acute promyelocytic leukemia, and sometimes leads to ATRA differentiation syndrome characterized by respiratory distress, acute pulmonary edema, hypotension, and heart failure.[518] It is thought that the rapid increase in leukocytes and cytokines that occurs as a result of differentiation of the promyelocytes induces a systemic inflammatory response, capillary permeability, and endothelial disruption. Other complications of ATRA treatment include pericardial effusion,[518,519] myocardial stunning,[520,521] myocarditis,[518,519] and valvular dysfunction.[518] These conditions may improve when ATRA is discontinued.[518,521]

Vitamin A is teratogenic in animal models and retinol metabolites, such as all-trans and 13-cis retinoic acid, are teratogenic in humans.[522] Some epidemiologic studies demonstrate an increased risk of fetal cardiac malformations with high maternal intake of vitamin A[523,524] while others do not.[525,526] Maternal ingestion of isotretinoin or retinol (>10,000 IU daily), both vitamin A derivatives, is associated with transposition of the great arteries and tetralogy of Fallot in the newborn.[522]

3.38.2 Vitamin C

Vitamin C toxicity has not been shown to be associated with cardiac complications in humans to date.

3.38.3 Vitamin D

Although myonecrosis has been noted in animal studies of vitamin D overdose, no published reports have described this association in humans. In a case report about vitamin D intoxication, an elderly individual presented with symptoms of hypercalcemia (see Section 3.11) and heart block that did not resolve when the hypercalcemia was corrected.[147]

Vitamin D deficiency has been associated with hypertension, stroke, congestive heart failure, and myocardial infarction, as well as a variety of other cardiovascular-related diseases such as diabetes mellitus, peripheral vascular disease, atherosclerosis, and endothelial dysfunction.[527]

3.38.4 Vitamin E

Vitamin E is a fat-soluble vitamin with antioxidant properties that is found in nuts, seeds, margarine, mayonnaise, salad dressing, and breakfast cereals. The most naturally abundant and biologically active form of it is alpha-tocophorel. Vitamin E prevents LDL oxidation in vitro, and in experimental animals it prevents atherosclerotic plaque formation.[528] Whereas some studies in humans suggested a modest cardiovascular benefit for vitamin E supplements,[529–531] the preponderance of reports have shown it to be associated with increased cardiovascular events, including hemorrhagic stroke,[532] heart failure,[533] and mortality.[534] Accordingly, vitamin E supplementation is not recommended.[528,535]

3.38.5 Vitamin K

Adverse reactions to intravenous vitamin K use include symptoms of allergic reactions; among them are flushing, bronchoconstriction, and wheezing. Symptoms that can be mistaken for myocardial infarction may also occur, including chest pain, dyspnea, tachycardia, hypotension, and death.[536,537]

REFERENCES

1. Anderson HR, et al. Air pollution and activation of implantable cardioverter defibrillators in London. *Epidemiol* 2010;**21**(3):405–13.
2. Berger A, et al. Runs of ventricular and supraventricular tachycardia triggered by air pollution in patients with coronary heart disease. *J Occup Environ Med* 2006;**48**(11): 1149–58.
3. Bhaskaran K, et al. The effects of hourly differences in air pollution on the risk of myocardial infarction: case crossover analysis of the MINAP database. *BMJ* 2011;**343**:d5531.
4. Bhaskaran K, et al. Effects of air pollution on the incidence of myocardial infarction. *Heart* 2009;**95**(21):1746–59.
5. Brook RD. Cardiovascular effects of air pollution. *Clin Sci (Lond)* 2008;**115**(6):175–87.
6. Brook RD, Rajagopalan S. Air pollution and cardiovascular events. *N Engl J Med* 2007; **356**(20):2105–6.
7. Dockery DW, et al. Association of air pollution with increased incidence of ventricular tachyarrhythmias recorded by implanted cardioverter defibrillators. *Environ Health Perspect* 2005;**113**(6):670–4.
8. Huang W, et al. Air pollution and autonomic and vascular dysfunction in patients with cardiovascular disease: interactions of systemic inflammation, overweight, and gender. *Am J Epidemiol* 2012;**176**(2):117–26.
9. Kunzli N, et al. Ambient air pollution and atherosclerosis in Los Angeles. *Environ Health Perspect* 2005;**113**(2):201–6.
10. Liao D, et al. Association of higher levels of ambient criteria pollutants with impaired cardiac autonomic control: a population-based study. *Am J Epidemiol* 2004;**159**(8):768–77.
11. Dockery DW, Stone PH. Cardiovascular risks from fine particulate air pollution. *N Engl J Med* 2007;**356**(5):511–3.
12. Osornio-Vargas AR, et al. Proinflammatory and cytotoxic effects of Mexico City air pollution particulate matter in vitro are dependent on particle size and composition. *Environ Health Perspect* 2003;**111**(10):1289–93.
13. Routledge HC, Ayres JG, Townend JN. Why cardiologists should be interested in air pollution. *Heart* 2003;**89**(12):1383–8.
14. Chen LC, Nadziejko C. Effects of subchronic exposures to concentrated ambient particles (CAPs) in mice. V. CAPs exacerbate aortic plaque development in hyperlipidemic mice. *Inhal Toxicol* 2005;**17**(4–5):217–24.
15. Sun Q, et al. Long-term air pollution exposure and acceleration of atherosclerosis and vascular inflammation in an animal model. *JAMA* 2005;**294**(23):3003–10.

16. Pekkanen J, et al. Daily concentrations of air pollution and plasma fibrinogen in London. *Occup Environ Med* 2000;**57**(12):818–22.
17. Diez Roux AV, et al. Long-term exposure to ambient particulate matter and prevalence of subclinical atherosclerosis in the multi-ethnic study of atherosclerosis. *Am J Epidemiol* 2008;**167**(6):667–75.
18. Diez Roux AV, et al. Recent exposure to particulate matter and C-reactive protein concentration in the multi-ethnic study of atherosclerosis. *Am J Epidemiol* 2006;**164**(5):437–48.
19. Brook RD, et al. Air pollution and cardiovascular disease: a statement for healthcare professionals from the Expert Panel on Population and Prevention Science of the American Heart Association. *Circulation* 2004;**109**(21):2655–71.
20. Miller KA, et al. Long-term exposure to air pollution and incidence of cardiovascular events in women. *N Engl J Med* 2007;**356**(5):447–58.
21. Peters A, et al. Exposure to traffic and the onset of myocardial infarction. *N Engl J Med* 2004;**351**(17):1721–30.
22. Rosenlund M, et al. Long-term exposure to urban air pollution and myocardial infarction. *Epidemiol* 2006;**17**(4):383–90.
23. Tonne C, et al. A case-control analysis of exposure to traffic and acute myocardial infarction. *Environ Health Perspect* 2007;**115**(1):53–7.
24. Dominici F, et al. Fine particulate air pollution and hospital admission for cardiovascular and respiratory diseases. *JAMA* 2006;**295**(10):1127–34.
25. Wellenius GA, et al. Particulate air pollution and the rate of hospitalization for congestive heart failure among medicare beneficiaries in Pittsburgh, Pennsylvania. *Am J Epidemiol* 2005;**161**(11):1030–6.
26. Wellenius GA, Schwartz J, Mittleman MA. Particulate air pollution and hospital admissions for congestive heart failure in seven United States cities. *Am J Cardiol* 2006;**97**(3):404–8.
27. Mutlu GM, Bellmeyer A, Budinger GR. Air pollution impairs lung's ability to clear edema fluid. *Am J Cardiol* 2006;**98**(3):423–4.
28. Mutlu GM, et al. Airborne particulate matter inhibits alveolar fluid reabsorption in mice via oxidant generation. *Am J Respir Cell Mol Biol* 2006;**34**(6):670–6.
29. Van Hee VC, et al. Association of long-term air pollution with ventricular conduction and repolarization abnormalities. *Epidemiol* 2011;**22**(6):773–80.
30. Ljungman PL, et al. Rapid effects of air pollution on ventricular arrhythmias. *Eur Heart J* 2008;**29**(23):2894–901.
31. Liao D, et al. Fine particulate air pollution is associated with higher vulnerability to atrial fibrillation: the APACR study. *J Toxicol Environ Health A* 2011;**74**(11):693–705.
32. Holguin F, et al. Air pollution and heart rate variability among the elderly in Mexico City. *Epidemiol* 2003;**14**(5):521–7.
33. Shields KN, et al. Traffic-related air pollution exposures and changes in heart rate variability in Mexico City: a panel study. *Environ Health* 2013;**12**:7.
34. Vallejo M, et al. Ambient fine particles modify heart rate variability in young healthy adults. *J Expo Sci Environ Epidemiol* 2006;**16**(2):125–30.
35. Franchini M, Mannucci PM. Particulate air pollution and cardiovascular risk: short-term and long-term effects. *Semin Thromb Hemost* 2009;**35**(7):665–70.
36. Link MS, Dockery DW. Air pollution and the triggering of cardiac arrhythmias. *Curr Opin Cardiol* 2010;**25**(1):16–22.
37. Appleby M, Fisher M, Martin M. Myocardial infarction, hyperkalaemia and ventricular tachycardia in a young male body-builder. *Int J Cardiol* 1994;**44**(2):171–4.

38. Daubert GP, et al. Acute clenbuterol overdose resulting in supraventricular tachycardia and atrial fibrillation. *J Med Toxicol* 2007;**3**(2):56–60.
39. Kierzkowska B, Stanczyk J, Kasprzak JD. Myocardial infarction in a 17-year-old body builder using clenbuterol. *Circ J* 2005;**69**(9):1144–6.
40. Lunghetti S, et al. Cardiogenic shock complicating myocardial infarction in a doped athlete. *Acute Card Care* 2009;**11**(4):250–1.
41. Achar S, Rostamian A, Narayan SM. Cardiac and metabolic effects of anabolic-androgenic steroid abuse on lipids, blood pressure, left ventricular dimensions, and rhythm. *Am J Cardiol* 2010;**106**(6):893–901.
42. Ahlgrim C, Guglin M. Anabolics and cardiomyopathy in a bodybuilder: case report and literature review. *J Card Fail* 2009;**15**(6):496–500.
43. Dickerman RD, McConathy WJ, Zachariah NY. Testosterone, sex hormone-binding globulin, lipoproteins, and vascular disease risk. *J Cardiovasc Risk* 1997;**4**(5-6):363–6.
44. Kennedy C. Myocardial infarction in association with misuse of anabolic steroids. *Ulster Med J* 1993;**62**(2):174–6.
45. Rossi GP, et al. Drug-related hypertension and resistance to antihypertensive treatment: a call for action. *J Hypertens* 2011;**29**(12):2295–309.
46. Vanberg AD. Androgenic anabolic steroid abuse and the cardiovascular system. *Handb Exp Pharmacol* 2010;**195**:411–57.
47. Urhausen A, Albers T, Kindermann W. Are the cardiac effects of anabolic steroid abuse in strength athletes reversible? *Heart* 2004;**90**(5):496–501.
48. Lippi G, Banfi G. Doping and thrombosis in sports. *Semin Thromb Hemost* 2011;**37**(8): 918–28.
49. Prather ID, et al. Clenbuterol: a substitute for anabolic steroids? *Med Sci Sports Exerc* 1995;**27**(8):1118–21.
50. Spann C, Winter ME. Effect of clenbuterol on athletic performance. *Ann Pharmacother* 1995;**29**(1):75–7.
51. Goldstein DR, et al. Clenbuterol and anabolic steroids: a previously unreported cause of myocardial infarction with normal coronary arteriograms. *South Med J* 1998;**91**(8): 780–4.
52. Hoffman RJ, et al. Clenbuterol ingestion causing prolonged tachycardia, hypokalemia, and hypophosphatemia with confirmation by quantitative levels. *J Toxicol Clin Toxicol* 2001; **39**(4):339–44.
53. Parr MK, Opfermann G, Schanzer W. Analytical methods for the detection of clenbuterol. *Bioanalysis* 2009;**1**(2):437–50.
54. Kintz P. Testing for anabolic steroids in hair: a review. *Leg Med (Tokyo)* 2003;**5**(Suppl 1): S29–33.
55. Gottignies P, et al. Successful treatment of monkshood (aconite napel) poisoning with magnesium sulfate. *Am J Emerg Med* 2009;**27**(6):755 e1–4.
56. Chan TY. Aconite poisoning. *Clin Toxicol (Phila)* 2009;**47**(4):279–85.
57. Fujita Y, et al. Five cases of aconite poisoning: toxicokinetics of aconitines. *J Anal Toxicol* 2007;**31**(3):132–7.
58. Lowe L, Matteucci MJ, Schneir AB. Herbal aconite tea and refractory ventricular tachycardia. *N Engl J Med* 2005;**353**(14):1532.
59. Liu Q, et al. Seven cases of fatal aconite poisoning: forensic experience in China. *Forensic Sci Int* 2011;**212**(1–3):e5–9.
60. Van Landeghem AA, et al. Aconitine involvement in an unusual homicide case. *Int J Legal Med* 2007;**121**(3):214–9.

61. Kim JY, et al. Aconitine poisoning from leaves of aconitum pseudo-leave var. erectum, not the root, in humans. *Clin Toxicol (Phila)* 2009;**47**(8):836.
62. Yim KM, Tse ML, Lau FL. Reversible intraventricular conduction defect in aconitine poisoning. *Singapore Med J* 2009;**50**(8):e302–5.
63. Fatovich DM. Aconite: a lethal Chinese herb. *Ann Emerg Med* 1992;**21**(3):309–11.
64. Lin CC, Chan TY, Deng JF. Clinical features and management of herb-induced aconitine poisoning. *Ann Emerg Med* 2004;**43**(5):574–9.
65. Pullela R, et al. A case of fatal aconitine poisoning by Monkshood ingestion. *J Forensic Sci* 2008;**53**(2):491–4.
66. Chan TY. Aconite poisoning presenting as hypotension and bradycardia. *Hum Exp Toxicol* 2009;**28**(12):795–7.
67. Strzelecki A, et al. Acute toxic herbal intake in a suicide attempt and fatal refractory ventricular arrhythmia. *Basic Clin Pharmacol Toxicol* 2010;**107**(2):698–9.
68. Camaiti A, et al. Prospective evaluation of adenosine-induced proarrhythmia in the emergency room. *Eur J Emerg Med* 2001;**8**(2):99–105.
69. Smith JR, Goldberger JJ, Kadish AH. Adenosine induced polymorphic ventricular tachycardia in adults without structural heart disease. *Pacing Clin Electrophysiol* 1997;**20** (3 Pt 1):743–5.
70. Mallet ML. Proarrhythmic effects of adenosine: a review of the literature. *Emerg Med J* 2004;**21**(4):408–10.
71. Celiker A, et al. Adenosine induced torsades de pointes in a child with congenital long QT syndrome. *Pacing Clin Electrophysiol* 1994;**17**(11 Pt 1):1814–7.
72. Kamijo Y, et al. Fatal diphenhydramine poisoning with increased vascular permeability including late pulmonary congestion refractory to percutaneous cardiovascular support. *Clin Toxicol (Phila)* 2008;**46**(9):864–8.
73. Krenzelok EP, Anderson GM, Mirick M. Massive diphenhydramine overdose resulting in death. *Ann Emerg Med* 1982;**11**(4):212–3.
74. Weiler JM, et al. Serious adverse reactions to protamine sulfate: are alternatives needed? *J Allergy Clin Immunol* 1985;**75**(2):297–303.
75. Fadali MA, et al. Mechanism responsible for the cardiovascular depressent: effect protamine sulfate. *Ann Surg* 1974;**180**(2):232–5.
76. Collins C, O'Donnell A. Does an allergy to fish pre-empt an adverse protamine reaction? A case report and a literature review. *Perfusion* 2008;**23**(6):369–72.
77. Macias Konstantopoulos W, Burns Ewald M, Pratt DS. Case records of the Massachusetts General Hospital. Case 22-2012. A 34-year-old man with intractable vomiting after ingestion of an unknown substance. *N Engl J Med* 2012;**367**(3):259–68.
78. Sundar S, Chakravarty J. Antimony toxicity. *Int J Environ Res Public Health* 2010;**7** (12):4267–77.
79. Alvarez M, et al. Antimony-induced cardiomyopathy in guinea pig and protection by L-carnitine. *Br J Pharmacol* 2005;**144**(1):17–27.
80. Rijal S, et al. Sodium stibogluconate cardiotoxicity and safety of generics. *Trans R Soc Trop Med Hyg* 2003;**97**(5):597–8.
81. T'Ao SC. Cardiac manifestations of the toxic action of potassium antimony tartrate in schistosomiasis patients: paroxysmal ventricular tachycardia and fibrillation. *Chin Med J* 1957;**75**(5):365–78.
82. Badr MH, Zayadi GM, Harras FK. Left ventricular dysfunction after intravenous tartar emetic. *Egypt J Bilharz* 1978;**5**(1–2):9–18.

83. Kuryshev YA, et al. Antimony-based antileishmanial compounds prolong the cardiac action potential by an increase in cardiac calcium currents. *Mol Pharmacol* 2006;**69**(4):1216–25.
84. Gong G, O'Bryant SE. Low-level arsenic exposure, AS3MT gene polymorphism and cardiovascular diseases in rural Texas counties. *Environ Res* 2012;**113**:52–7.
85. Chen Y, et al. Arsenic exposure from drinking water and mortality from cardiovascular disease in Bangladesh: prospective cohort study. *BMJ* 2011;**342**:d2431.
86. Kurppa K, et al. Chemical exposures at work and cardiovascular morbidity. Atherosclerosis, ischemic heart disease, hypertension, cardiomyopathy and arrhythmias. *Scand J Work Environ Health* 1984;**10**(6 Spec No):381–8.
87. Bhatnagar A. Environmental cardiology: studying mechanistic links between pollution and heart disease. *Circ Res* 2006;**99**(7):692–705.
88. Chen CJ, et al. Increased prevalence of hypertension and long-term arsenic exposure. *Hypertension* 1995;**25**(1):53–60.
89. Tseng CH. Blackfoot disease and arsenic: a never-ending story. *J Environ Sci Health C Environ Carcinog Ecotoxicol Rev* 2005;**23**(1):55–74.
90. Srivastava S, et al. Arsenic exacerbates atherosclerotic lesion formation and inflammation in ApoE$^{-/-}$ mice. *Toxicol Appl Pharmacol* 2009;**241**(1):90–100.
91. Lee MY, et al. Arsenic-induced dysfunction in relaxation of blood vessels. *Environ Health Perspect* 2003;**111**(4):513–7.
92. Sanchez-Soria P, et al. Chronic low-level arsenite exposure through drinking water increases blood pressure and promotes concentric left ventricular hypertrophy in female mice. *Toxicol Pathol* 2012;**40**(3):504–12.
93. Chakraborty S, Ray M, Ray S. Effects and dose–response relationships of skin cancer and blackfoot disease with arsenic. Arsenic toxicity: a heart-breaking saga of a freshwater mollusc. *Tissue Cell* 2012;**44**(3):151–5.
94. Tseng W. Effects and dose–response relationships of skin cancer and blackfoot disease with arsenic. *Environ Health Perspect* 1977;**19**:109–19.
95. Tseng CH, et al. Dose-response relationship between peripheral vascular disease and ingested inorganic arsenic among residents in blackfoot disease endemic villages in Taiwan. *Atherosclerosis* 1996;**120**(1–2):125–33.
96. Kumagai Y, Pi J. Molecular basis for arsenic-induced alteration in nitric oxide production and oxidative stress: implication of endothelial dysfunction. *Toxicol Appl Pharmacol* 2004;**198**(3):450–7.
97. Unnikrishnan D, et al. Cardiac monitoring of patients receiving arsenic trioxide therapy. *Br J Haematol* 2004;**124**(5):610–7.
98. Raghu KG, et al. Evaluation of adverse cardiac effects induced by arsenic trioxide, a potent anti-APL drug. *J Environ Pathol Toxicol Oncol* 2009;**28**(3):241–52.
99. Vizzardi E, et al. QT prolongation: a case of arsenical pericardial and pleural effusion. *Cardiovasc Toxicol* 2008;**8**(1):41–4.
100. Kathirgamanathan K, et al. Cardiac conduction block at multiple levels caused by arsenic trioxide therapy. *Can J Cardiol* 2013;**29**(1): 130 e5–6
101. Das AK, et al. Arsenic-induced myocardial injury: protective role of *Corchorus olitorius* leaves. *Food Chem Toxicol* 2010;**48**(5):1210–7.
102. Chu W, et al. Arsenic-induced interstitial myocardial fibrosis reveals a new insight into drug-induced long QT syndrome. *Cardiovasc Res* 2012;**96**(1):90–8.
103. Boucher BJ, Mannan N. Metabolic effects of the consumption of *Areca catechu*. *Addict Biol* 2002;**7**(1):103–10.

104. Lin WY, et al. Betel nut chewing is associated with increased risk of cardiovascular disease and all-cause mortality in Taiwanese men. *Am J Clin Nutr* 2008;**87**(5):1204–11.
105. Deng JF, et al. Acute toxicities of betel nut: rare but probably overlooked events. *J Toxicol Clin Toxicol* 2001;**39**(4):355–60.
106. Nelson BS, Heischober B. Betel nut: a common drug used by naturalized citizens from India, Far East Asia, and the South Pacific Islands. *Ann Emerg Med* 1999;**34**(2):238–43.
107. Norton SA. Betel: consumption and consequences. *J Am Acad Dermatol* 1998;**38**(1):81–8.
108. Trivedy C, Warnakulasuriya S, Peters TJ. Areca nuts can have deleterious effects. *BMJ* 1999;**318**(7193):1287.
109. Chang WC, et al. Betel nut chewing and other risk factors associated with obesity among Taiwanese male adults. *Int J Obes (Lond)* 2006;**30**(2):359–63.
110. Dwivedi G, Dwivedi S. Betel quid seller syndrome. *Occup Environ Med* 2010;**67**(2):144.
111. Heck JE, et al. Betel quid chewing in rural Bangladesh: prevalence, predictors and relationship to blood pressure. *Int J Epidemiol* 2012;**41**(2):462–71.
112. Yen AM, et al. A population-based study of the association between betel-quid chewing and the metabolic syndrome in men. *Am J Clin Nutr* 2006;**83**(5):1153–60.
113. Lin WY, et al. Betel nut chewing is strongly associated with general and central obesity in Chinese male middle-aged adults. *Obesity (Silver Spring)* 2009;**17**(6):1247–54.
114. Chu NS. Cardiovascular responses to betel chewing. *J Formos Med Assoc* 1993;**92**(9):835–7.
115. Hung DZ, Deng JF. Acute myocardial infarction temporally related to betel nut chewing. *Vet Hum Toxicol* 1998;**40**(1):25–8.
116. Shackley BS, et al. Idiopathic massive myocardial calcification: a case report and review of the literature. *Cardiovasc Pathol* 2011;**20**(2):e79–83.
117. Salvador JA, et al. Bismuth compounds in medicinal chemistry. *Future Med Chem* 2012;**4**(11):1495–523.
118. Goodman L. Sudden death after intravenous sodium bismuth tartrate. *Br Med J* 1948(1):978–9.
119. Krayer O, Farah A. Action of cysteine and of dimercaptopropanol in heart failure caused by sodium bismuth tartrate. *Fed Proc* 1948;**7**(1 Pt 1):235.
120. Messner B, Bernhard D. Cadmium and cardiovascular diseases: cell biology, pathophysiology, and epidemiological relevance. *Biometals* 2010;**23**(5):811–22.
121. Ferramola ML, et al. Myocardial oxidative stress following sub-chronic and chronic oral cadmium exposure in rats. *Environ Toxicol Pharmacol* 2011;**32**(1):17–26.
122. Menke A, et al. Cadmium levels in urine and mortality among U.S. adults. *Environ Health Perspect* 2009;**117**(2):190–6.
123. Tellez-Plaza M, et al. Cadmium exposure and incident cardiovascular disease. *Epidemiol* 2013;**24**(3):421–9.
124. Tellez-Plaza M, et al. Cadmium exposure and hypertension in the 1999–2004 National Health and Nutrition Examination Survey (NHANES). *Environ Health Perspect* 2008;**116**(1):51–6.
125. Tellez-Plaza M, et al. Cadmium exposure and all-cause and cardiovascular mortality in the U.S. general population. *Environ Health Perspect* 2012;**120**(7):1017–22.
126. Donpunha W, et al. Protective effect of ascorbic acid on cadmium-induced hypertension and vascular dysfunction in mice. *Biometals* 2011;**24**(1):105–15.
127. Ferramola ML, et al. Cadmium-induced oxidative stress and histological damage in the myocardium. Effects of a soy-based diet. *Toxicol Appl Pharmacol* 2012;**265**(3):380–9.
128. Milton PS, Muthumani M, Shagirtha K. Quercetin potentially attenuates cadmium induced oxidative stress mediated cardiotoxicity and dyslipidemia in rats. *Eur Rev Med Pharmacol Sci* 2013;**17**(5):582–95.

129. Sompamit K, et al. Reversal of cadmium-induced vascular dysfunction and oxidative stress by meso-2,3-dimercaptosuccinic acid in mice. *Toxicol Lett* 2010;**198**(1):77–82.
130. Prozialeck WC, Edwards JR, Woods JM. The vascular endothelium as a target of cadmium toxicity. *Life Sci* 2006;**79**(16):1493–506.
131. Holmgren P, Norden-Pettersson L, Ahlner J. Caffeine fatalities: four case reports. *Forensic Sci Int* 2004;**139**(1):71–3.
132. Higgins JP, Tuttle TD, Higgins CL. Energy beverages: content and safety. *Mayo Clin Proc* 2010;**85**(11):1033–41.
133. Kerrigan S, Lindsey T. Fatal caffeine overdose: two case reports. *Forensic Sci Int* 2005; **153**(1):67–9.
134. Schmidt A, Karlson-Stiber C. Caffeine poisoning and lactate rise: an overlooked toxic effect? *Acta Anaesthesiol Scand* 2008;**52**(7):1012–4.
135. Robertson D, et al. Effects of caffeine on plasma renin activity, catecholamines and blood pressure. *N Engl J Med* 1978;**298**(4):181–6.
136. Emohare O, Ratnam V. Multiple cardiac arrests following an overdose of caffeine complicated by penetrating trauma. *Anaesthesia* 2006;**61**(1):54–6.
137. Chopra A, Morrison L. Resolution of caffeine-induced complex dysrhythmia with procainamide therapy. *J Emerg Med* 1995;**13**(1):113–7.
138. Kapur R, Smith MD. Treatment of cardiovascular collapse from caffeine overdose with lidocaine, phenylephrine, and hemodialysis. *Am J Emerg Med* 2009;**27**(2): 253, e3–6.
139. Price KR, Fligner DJ. Treatment of caffeine toxicity with esmolol. *Ann Emerg Med* 1990;**19**(1):44–6.
140. Holstege CP, et al. Massive caffeine overdose requiring vasopressin infusion and hemodialysis. *J Toxicol Clin Toxicol* 2003;**41**(7):1003–7.
141. Douglas PS, Carmichael KA, Palevsky PM. Extreme hypercalcemia and electrocardiographic changes. *Am J Cardiol* 1984;**54**(6):674–5.
142. Otero J, Lenihan DJ. The "normothermic" Osborn wave induced by severe hypercalcemia. *Tex Heart Inst J* 2000;**27**(3):316–7.
143. Roberts WC, Waller BF. Effect of chronic hypercalcemia on the heart. An analysis of 18 necropsy patients. *Am J Med* 1981;**71**(3):371–84.
144. Schwartz KV. Letter: Heart-block in renal failure and hypercalcemia. *JAMA* 1976;**235**(15):1550.
145. Sataline L, Donaghue G. Hypercalcemia, heart-block, and hyperthyroidism. *JAMA* 1970; **213**(8):1342.
146. Ginsberg H, Schwartz KV. Letter: Hypercalcemia and complete heart block. *Ann Intern Med* 1973;**79**(6):903.
147. Garg G, et al. Vitamin D toxicity presenting as hypercalcemia and complete heart block: An interesting case report. *Indian J Endocrinol Metab* 2012;**16**(Suppl. 2):S423–5.
148. McFarland KF, et al. Hypercalcemia and idiopathic hypertrophic subaortic stenosis. *Ann Intern Med* 1978;**88**(1):57–8.
149. Kloeppel R, et al. Acute hypercalcemia of the heart ("bony heart"). *J Comput Assist Tomogr* 2001;**25**(3):407–11.
150. Rostand SG, et al. Myocardial calcification and cardiac dysfunction in chronic renal failure. *Am J Med* 1988;**85**(5):651–7.
151. McClure J, et al. Myocardial fibre calcification. *J Clin Pathol* 1981;**34**(10):1167–74.
152. Mana M, et al. Petrified myocardium: the age of stone? *Circulation* 2012;**126**(9):1139–42.
153. Dudink J, Walther FJ, Beekman RP. Subcutaneous fat necrosis of the newborn: hypercalcaemia with hepatic and atrial myocardial calcification. *Arch Dis Child Fetal Neonatal Ed* 2003;**88**(4):F343–5.

154. Weaver LK. Clinical practice. Carbon monoxide poisoning. *N Engl J Med* 2009;**360**(12): 1217–25.
155. Ernst A, Zibrak JD. Carbon monoxide poisoning. *N Engl J Med* 1998;**339**(22):1603–8.
156. Hajsadeghi S, et al. Electrocardiographic findings and serum troponin I in carbon monoxide poisoned patients. *Acta Med Iran* 2012;**50**(3):185–91.
157. Marius-Nunez AL. Myocardial infarction with normal coronary arteries after acute exposure to carbon monoxide. *Chest* 1990;**97**(2):491–4.
158. Dileo PA, et al. Late stent thrombosis secondary to carbon monoxide poisoning. *Cardiovasc Revasc Med* 2011;**12**(1):56–8.
159. Isik T, et al. ST-elevation myocardial infarction after acute carbon monoxide poisoning. *Anadolu Kardiyol Derg* 2012;**12**(3):278–9.
160. Satran D, et al. Cardiovascular manifestations of moderate to severe carbon monoxide poisoning. *J Am Coll Cardiol* 2005;**45**(9):1513–6.
161. Gandini C, et al. Carbon monoxide cardiotoxicity. *J Toxicol Clin Toxicol* 2001;**39**(1): 35–44.
162. Tibbles PM, Edelsberg JS. Hyperbaric-oxygen therapy. *N Engl J Med* 1996;**334**(25):1642–8.
163. Leroy A, et al. An unusual "winter case" of myocarditis. *Int J Cardiol* 2011;**148**(3): e68–9.
164. Serrao MG, et al. Transient cardiomyopathy secondary to carbon monoxide poisoning. *Rev Port Cardiol* 2008;**27**(6):833–8.
165. Kalay N, et al. Cardiovascular effects of carbon monoxide poisoning. *Am J Cardiol* 2007;**99**(3):322–4.
166. Lee SJ, et al. A case report of carbon monoxide poisoning induced cardiomyopathy complicated with left ventricular thrombus. *J Cardiovasc Ultrasound* 2011;**19**(2):83–6.
167. Morris RD, Naumova EN. Carbon monoxide and hospital admissions for congestive heart failure: evidence of an increased effect at low temperatures. *Environ Health Perspect* 1998;**106**(10):649–53.
168. Cetin M, et al. A case of carbon monoxide poisoning presenting with supraventricular tachycardia. *Intern Med* 2011;**50**(21):2607–9.
169. Canpolat U, Aytemir K, Oto A. Familial Brugada syndrome unmasked by carbon monoxide intoxication. *Int J Cardiol* 2012;**156**(1):e18–9.
170. Henry CR, et al. Myocardial injury and long-term mortality following moderate to severe carbon monoxide poisoning. *JAMA* 2006;**295**(4):398–402.
171. Mebs D, Reuss E, Schneider M. Studies on the cardenolide sequestration in African milkweed butterflies (*Danaidae*). *Toxicon* 2005;**45**(5):581–4.
172. Rasmann S, et al. Cardenolides, induced responses, and interactions between above- and belowground herbivores of milkweed (*Asclepias* spp.). *Ecol* 2009;**90**(9):2393–404.
173. Roberts DM, Buckley NA. Antidotes for acute cardenolide (cardiac glycoside) poisoning. *Cochrane Database Syst Rev* 2006;(4):CD005490.
174. Eddleston M, Haggalla S. Fatal injury in eastern Sri Lanka, with special reference to cardenolide self-poisoning with *Cerbera manghas* fruits. *Clin Toxicol (Phila)* 2008;**46** (8):745–8.
175. Eddleston M, Persson H. Acute plant poisoning and antitoxin antibodies. *J Toxicol Clin Toxicol* 2003;**41**(3):309–15.
176. Eddleston M, Sheriff MH, Hawton K. Deliberate self harm in Sri Lanka: an overlooked tragedy in the developing world. *BMJ* 1998;**317**(7151):133–5.
177. Al B, et al. A case of non-fatal oleander poisoning. *BMJ Case Rep* 2010;**2010**. Available from: http://casereports.bmj.com/content/2010/bcr.02.2009.1573.

178. Pietsch J, et al. A non-fatal oleander poisoning. *Int J Legal Med* 2005;**119**(4):236–40.
179. Shriyan A, et al. Fatal flower. *Indian J Pediatr* 2011;**78**(3):364–5.
180. Rajapakse S. Management of yellow oleander poisoning. *Clin Toxicol (Phila)* 2009;**47**(3):206–12.
181. Mukherjee D, et al. Melatonin protects against isoproterenol-induced alterations in cardiac mitochondrial energy-metabolizing enzymes, apoptotic proteins, and assists in complete recovery from myocardial injury in rats. *J Pineal Res* 2012;**53**(2):166–79.
182. Singh A, et al. Exploring takotsubo cardiomyopathy in an elderly patient with acute anxiety attack. *WV Med J* 2012;**108**(5):14–7.
183. von Knobelsdorff-Brenkenhoff F, Abdel-Aty H, Schulz-Menger J. Takotsubo cardiomyopathy after nasal application of epinephrine: a magnetic resonance study. *Int J Cardiol* 2010;**145**(2):308–9.
184. Patankar GR, Donsky MS, Schussler JM. Delayed takotsubo cardiomyopathy caused by excessive exogenous epinephrine administration after the treatment of angioedema. *Proc Baylor Univ Med Cent* 2012;**25**(3):229–30.
185. Bybee KA, Prasad A. Stress-related cardiomyopathy syndromes. *Circulation* 2008;**118**(4):397–409.
186. Philips B, et al. High prevalence of catecholamine-facilitated focal ventricular tachycardia in patients with arrhythmogenic right ventricular dysplasia/cardiomyopathy. *Circ Arrhythm Electrophysiol* 2013;**6**(1):160–6.
187. Akashi YJ, et al. Reversible left ventricular dysfunction "takotsubo" cardiomyopathy related to catecholamine cardiotoxicity. *J Electrocardiol* 2002;**35**(4):351–6.
188. Wittstein IS, et al. Neurohumoral features of myocardial stunning due to sudden emotional stress. *N Engl J Med* 2005;**352**(6):539–48.
189. Apostolakis E, Parissis H, Dougenis D. Brain death and donor heart dysfunction: implications in cardiac transplantation. *J Card Surg* 2010;**25**(1):98–106.
190. Berman M, et al. Is stress cardiomyopathy the underlying cause of ventricular dysfunction associated with brain death? *J Heart Lung Transplant* 2010;**29**(9):957–65.
191. Caretta G, et al. The link between intracranial haemorrhage and cardiogenic shock: a case of takotsubo cardiomyopathy. *Acta Cardiol* 2012;**67**(3):363–5.
192. Bagga S, Sharma YP, Jain M. Cardiac dysfuntion after acute subarachnoid hemorrhage: neurogenic stress cardiomyopathy or takotsubo cardiomyopathy. *Neurol India* 2011;**59**(2):304–6.
193. Konrad FM, Unertl KE, Schroeder TH. Takotsubo cardiomyopathy after cerebral aneurysm rupture. *J Neurosurg Anesthesiol* 2010;**22**(2):181–2.
194. Joao I, et al. Cardiac rupture during exercise stress echocardiography: a case report. *J Am Soc Echocardiogr* 2000;**13**(8):785–7.
195. Nixon JL, et al. Impact of high-dose inotropic donor support on early myocardial necrosis and outcomes in cardiac transplantation. *Clin Transplant* 2012;**26**(2):322–7.
196. Zhitkovich A. Chromium in drinking water: sources, metabolism, and cancer risks. *Chem Res Toxicol* 2011;**24**(10):1617–29.
197. Guallar E, et al. Low toenail chromium concentration and increased risk of nonfatal myocardial infarction. *Am J Epidemiol* 2005;**162**(2):157–64.
198. Roeback Jr JR, et al. Effects of chromium supplementation on serum high-density lipoprotein cholesterol levels in men taking beta-blockers. A randomized, controlled trial. *Ann Intern Med* 1991;**115**(12):917–24.
199. Newman HA, et al. Serum chromium and angiographically determined coronary artery disease. *Clin Chem* 1978;**24**(4):541–4.

200. Hummel M, Standl E, Schnell O. Chromium in metabolic and cardiovascular disease. *Horm Metab Res* 2007;**39**(10):743–51.
201. Simonoff M. Chromium deficiency and cardiovascular risk. *Cardiovasc Res* 1984;**18**(10): 591–6.
202. Price Evans DA, et al. Chromium chloride administration causes a substantial reduction of coronary lipid deposits, aortic lipid deposits, and serum cholesterol concentration in rabbits. *Biol Trace Elem Res* 2009;**130**(3):262–72.
203. Schroeder HA, Balassa JJ. Influence of chromium, cadmium, and lead on rat aortic lipids and circulating cholesterol. *Am J Physiol* 1965;**209**:433–7.
204. Machado C, Appelbe A, Wood R. Arthroprosthetic cobaltism and cardiomyopathy. *Heart Lung Circ* 2012;**21**(11):759–60.
205. Morin Y, Tetu A, Mercier G. Cobalt cardiomyopathy: clinical aspects. *Br Heart J* 1971; **33**(Suppl):175–8.
206. Bonenfant JL, Miller G, Roy PE. Quebec beer-drinkers' cardiomyopathy: pathological studies. *Can Med Assoc J* 1967;**97**(15):910–6.
207. Jarvis JQ, et al. Cobalt cardiomyopathy: a report of two cases from mineral assay laboratories and a review of the literature. *J Occup Med* 1992;**34**(6):620–6.
208. Kennedy A, Dornan JD, King R. Fatal myocardial disease associated with industrial exposure to cobalt. *Lancet* 1981;**1**(8217):412–4.
209. Gaetke LM, Chow CK. Copper toxicity, oxidative stress, and antioxidant nutrients. *Toxicol* 2003;**189**(1–2):147–63.
210. Kok FJ, et al. Serum copper and zinc and the risk of death from cancer and cardiovascular disease. *Am J Epidemiol* 1988;**128**(2):352–9.
211. Adam B, et al. The interaction between copper and coronary risk indicators. *Jpn Heart J* 2001;**42**(3):281–6.
212. Topuzoglu G, et al. Concentrations of copper, zinc, and magnesium in sera from patients with idiopathic dilated cardiomyopathy. *Biol Trace Elem Res* 2003;**95**(1):11–7.
213. Saari JT, Schuschke DA. Cardiovascular effects of dietary copper deficiency. *Biofactors* 1999;**10**(4):359–75.
214. Nath R. Copper deficiency and heart disease: molecular basis, recent advances and current concepts. *Int J Biochem Cell Biol* 1997;**29**(11):1245–54.
215. Tang WH, et al. Clinical and genetic association of serum ceruloplasmin with cardiovascular risk. *Arterioscler Thromb Vasc Biol* 2012;**32**(2):516–22.
216. Ford ES. Serum copper concentration and coronary heart disease among US adults. *Am J Epidemiol* 2000;**151**(12):1182–8.
217. Aneja A, et al. Diabetic cardiomyopathy: insights into pathogenesis, diagnostic challenges, and therapeutic options. *Am J Med* 2008;**121**(9):748–57.
218. Ghayour-Mobarhan M, et al. The relationship between established coronary risk factors and serum copper and zinc concentrations in a large Persian cohort. *J Trace Elem Med Biol* 2009;**23**(3):167–75.
219. Kosar F, et al. Trace element status (Se, Zn, Cu) in heart failure. *Anadolu Kardiyol Derg* 2006;**6**(3):216–20.
220. Shokrzadeh M, et al. Serum zinc and copper levels in ischemic cardiomyopathy. *Biol Trace Elem Res* 2009;**127**(2):116–23.
221. Malek F, et al. Difference of baseline serum copper levels between groups of patients with different one year mortality and morbidity and chronic heart failure. *Cent Eur J Public Health* 2003;**11**(4):198–201.

222. Reunanen A, Knekt P, Aaran RK. Serum ceruloplasmin level and the risk of myocardial infarction and stroke. *Am J Epidemiol* 1992;**136**(9):1082–90.
223. Salonen JT, et al. Serum copper and the risk of acute myocardial infarction: a prospective population study in men in eastern Finland. *Am J Epidemiol* 1991;**134**(3):268–76.
224. Singh MM, et al. Serum copper in myocardial infarction: diagnostic and prognostic significance. *Angiol* 1985;**36**(8):504–10.
225. Wang XL, et al. Plasma extracellular superoxide dismutase levels in an Australian population with coronary artery disease. *Arterioscler Thromb Vasc Biol* 1998;**18**(12): 1915–21.
226. Islamoglu Y, et al. The relationship between serum levels of Zn and Cu and severity of coronary atherosclerosis. *Biol Trace Elem Res* 2011;**144**(1–3):436–44.
227. Aliabadi H. A deleterious interaction between copper deficiency and sugar ingestion may be the missing link in heart disease. *Med Hypotheses* 2008;**70**(6):1163–6.
228. Klevay LM. Dietary copper and risk of coronary heart disease. *Am J Clin Nutr* 2000;**71** (5):1213–4.
229. Reiser S, et al. Indices of copper status in humans consuming a typical American diet containing either fructose or starch. *Am J Clin Nutr* 1985;**42**(2):242–51.
230. He W, James Kang Y. Ischemia-induced copper loss and suppression of angiogenesis in the pathogenesis of myocardial infarction. *Cardiovasc Toxicol* 2013;**13**(1):1–8.
231. Kazemi-Bajestani SM, et al. Serum copper and zinc concentrations are lower in Iranian patients with angiographically defined coronary artery disease than in subjects with a normal angiogram. *J Trace Elem Med Biol* 2007;**21**(1):22–8.
232. da Cunha S, et al. Thiamin, selenium, and copper levels in patients with idiopathic dilated cardiomyopathy taking diuretics. *Arq Bras Cardiol* 2002;**79**(5):454–65.
233. Klevay LM. Heart failure improvement from a supplement containing copper. *Eur Heart J* 2006;**27**(1):117–8.
234. Weber KT, Weglicki WB, Simpson RU. Macro- and micronutrient dyshomeostasis in the adverse structural remodelling of myocardium. *Cardiovasc Res* 2009;**81**(3):500–8.
235. Ghaemian A, et al. Zinc and copper levels in severe heart failure and the effects of atrial fibrillation on the zinc and copper status. *Biol Trace Elem Res* 2011;**143**(3):1239–46.
236. Thombs DL, et al. Event-level analyses of energy drink consumption and alcohol intoxication in bar patrons. *Addict Behav* 2010;**35**(4):325–30.
237. Miller KE. Energy drinks, race, and problem behaviors among college students. *J Adolesc Health* 2008;**43**(5):490–7.
238. Seifert SM, et al. Health effects of energy drinks on children, adolescents, and young adults. *Pediatrics* 2011;**127**(3):511–28.
239. Franks AM, et al. Comparison of the effects of energy drink versus caffeine supplementation on indices of 24-hour ambulatory blood pressure. *Ann Pharmacother* 2012;**46** (2):192–9.
240. Baum M, Weiss M. The influence of a taurine containing drink on cardiac parameters before and after exercise measured by echocardiography. *Amino Acids* 2001;**20**(1):75–82.
241. Smith N, Atroch AL. Guarana's journey from regional tonic to aphrodisiac and global energy drink. *Evid Based Complement Alternat Med* 2010;**7**(3):279–82.
242. Usman A, Jawaid A. Hypertension in a young boy: an energy drink effect. *BMC Res Notes* 2012;**5**:591.
243. Worthley MI, et al. Detrimental effects of energy drink consumption on platelet and endothelial function. *Am J Med* 2010;**123**(2):184–7.

244. Wiklund U, et al. Influence of energy drinks and alcohol on post-exercise heart rate recovery and heart rate variability. *Clin Physiol Funct Imaging* 2009;**29**(1):74–80.
245. Steinke L, et al. Effect of "energy drink" consumption on hemodynamic and electrocardiographic parameters in healthy young adults. *Ann Pharmacother* 2009;**43**(4):596–602.
246. Nagajothi N, et al. Energy drink-related supraventricular tachycardia. *Am J Med* 2008;**121**(4):e3–4.
247. Peake ST, Mehta PA, Dubrey SW. Atrial fibrillation-related cardiomyopathy: a case report. *J Med Case Rep* 2007;**1**:111.
248. Terlizzi R, et al. Reversible postural tachycardia syndrome due to inadvertent overuse of Red Bull. *Clin Auton Res* 2008;**18**(4):221–3.
249. Worrall BB, Phillips CD, Henderson KK. Herbal energy drinks, phenylpropanoid compounds, and cerebral vasculopathy. *Neurology* 2005;**65**(7):1137–8.
250. Berger AJ, Alford K. Cardiac arrest in a young man following excess consumption of caffeinated "energy drinks". *Med J Aust* 2009;**190**(1):41–3.
251. Bertolini JC. Hydrofluoric acid: a review of toxicity. *J Emerg Med* 1992;**10**(2):163–8.
252. Caravati EM. Acute hydrofluoric acid exposure. *Am J Emerg Med* 1988;**6**(2):143–50.
253. Yolken R, Konecny P, McCarthy P. Acute fluoride poisoning. *Pediatrics* 1976;**58**(1):90–3.
254. Bjornhagen V, et al. Hydrofluoric acid-induced burns and life-threatening systemic poisoning—favorable outcome after hemodialysis. *J Toxicol Clin Toxicol* 2003;**41**(6):855–60.
255. McIvor ME, et al. Sudden cardiac death from acute fluoride intoxication: the role of potassium. *Ann Emerg Med* 1987;**16**(7):777–81.
256. Gradinger R, et al. Toxic myocarditis due to oral ingestion of hydrofluoric acid. *Heart Lung Circ* 2008;**17**(3):248–50.
257. Abukurah AR, et al. Acute sodium fluoride poisoning. *JAMA* 1972;**222**(7):816–7.
258. Berman L, et al. Inorganic fluoride poisoning: treatment by hemodialysis. *N Engl J Med* 1973;**289**(17):922.
259. Bogle RG, et al. Aluminium phosphide poisoning. *Emerg Med J* 2006;**23**(1):e3.
260. Khosla SN, Nand N, Kumar P. Cardiovascular complications of aluminum phosphide poisoning. *Angiology* 1988;**39**(4):355–9.
261. Jadhav AP, et al. Unresponsive ventricular tachycardia associated with aluminum phosphide poisoning. *Am J Emerg Med* 2012;**30**(4): 633, e3–5.
262. Proudfoot AT. Aluminium and zinc phosphide poisoning. *Clin Toxicol (Phila)* 2009;**47**(2):89–100.
263. Akkaoui M, et al. Reversible myocardial injury associated with aluminum phosphide poisoning. *Clin Toxicol (Phila)* 2007;**45**(6):728–31.
264. Kaushik RM, Kaushik R, Mahajan SK. Subendocardial infarction in a young survivor of aluminium phosphide poisoning. *Hum Exp Toxicol* 2007;**26**(5):457–60.
265. Nayyar S, Nair M. Brugada pattern in toxic myocarditis due to severe aluminum phosphide poisoning. *Pacing Clin Electrophysiol* 2009;**32**(11):e16–7.
266. Shah V, Baxi S, Vyas T. Severe myocardial depression in a patient with aluminium phosphide poisoning: a clinical, electrocardiographical and histopathological correlation. *Indian J Crit Care Med* 2009;**13**(1):41–3.
267. Siddaiah L, et al. Intra-aortic balloon pump in toxic myocarditis due to aluminum phosphide poisoning. *J Med Toxicol* 2009;**5**(2):80–3.
268. Gurjar M, et al. Managing aluminum phosphide poisonings. *J Emerg Trauma Shock* 2011;**4**(3):378–84.
269. Wilson R, et al. Acute phosphine poisoning aboard a grain freighter: epidemiologic, clinical, and pathological findings. *JAMA* 1980;**244**(2):148–50.

270. Gupta MS, Malik A, Sharma VK. Cardiovascular manifestations in aluminium phosphide poisoning with special reference to echocardiographic changes. *J Assoc Physicians India* 1995;**43**(11):773–4, 779–80.
271. Anger F, et al. Fatal aluminum phosphide poisoning. *J Anal Toxicol* 2000;**24**(2):90–2.
272. Singh UK, Chakraborty B, Prasad R. Aluminium phosphide poisoning: a growing concern in pediatric population. *Indian Pediatr* 1997;**34**(7):650–1.
273. Soltaninejad K, et al. Fatal aluminum phosphide poisoning in Tehran, Iran, from 2007 to 2010. *Indian J Med Sci* 2012;**66**(3-4):66–70.
274. Moon JM, Chun BJ. Acute endosulfan poisoning: a retrospective study. *Hum Exp Toxicol* 2009;**28**(5):309–16.
275. Eyer F, et al. Acute endosulfan poisoning with cerebral edema and cardiac failure. *J Toxicol Clin Toxicol* 2004;**42**(6):927–32.
276. Dalvi CP, Abraham PP, Iyer SS. Correlation of electrocardiographic changes with prognosis in organophosphorus poisoning. *J Postgrad Med* 1986;**32**(3):115–9.
277. Saadeh AM, Farsakh NA, al-Ali MK. Cardiac manifestations of acute carbamate and organophosphate poisoning. *Heart* 1997;**77**(5):461–4.
278. Karasu-Minareci E, et al. What may happen after an organophosphate exposure: acute myocardial infarction? *J Forensic Leg Med* 2012;**19**(2):94–6.
279. Bardin PG, et al. Organophosphate and carbamate poisoning. *Arch Intern Med* 1994;**154**(13):1433–41.
280. Roth A, et al. Organophosphates and the heart. *Chest* 1993;**103**(2):576–82.
281. Chacko J, Elangovan A. Late onset, prolonged asystole following organophosphate poisoning: a case report. *J Med Toxicol* 2010;**6**(3):311–4.
282. Yurumez Y, et al. Electrocardiographic findings of acute organophosphate poisoning. *J Emerg Med* 2009;**36**(1):39–42.
283. Shadnia S, et al. Prognostic value of long QT interval in acute and severe organophosphate poisoning. *J Med Toxicol* 2009;**5**(4):196–9.
284. Karki P, et al. Cardiac and electrocardiographical manifestations of acute organophosphate poisoning. *Singapore Med J* 2004;**45**(8):385–9.
285. Gul EE, Can I, Kusumoto FM. Case report: an unusual heart rhythm associated with organophosphate poisoning. *Cardiovasc Toxicol* 2012;**12**(3):263–5.
286. Topacoglu H, et al. An unusual cause of atrial fibrillation: exposure to insecticides. *Am J Ind Med* 2007;**50**(1):48–9.
287. Anand S, et al. Cardiac abnormalities in acute organophosphate poisoning. *Clin Toxicol (Phila)* 2009;**47**(3):230–5.
288. Lin CC, et al. Takotsubo cardiomyopathy related to carbamate and pyrethroid intoxication. *Resuscitation* 2010;**81**(8):1051–2.
289. Siegal D, et al. Complete heart block following intentional carbamate ingestion. *Can J Cardiol* 2009;**25**(8):e288–90.
290. Dayton SB, et al. Pesticide use and myocardial infarction incidence among farm women in the agricultural health study. *J Occup Environ Med* 2010;**52**(7):693–7.
291. Fleming LE, et al. National Health Interview Survey mortality among U.S. farmers and pesticide applicators. *Am J Ind Med* 2003;**43**(2):227–33.
292. Lee E, et al. Proportionate mortality of crop and livestock farmers in the United States, 1984–1993. *Am J Ind Med* 2002;**42**(5):410–20.
293. Mills KT, et al. Pesticides and myocardial infarction incidence and mortality among male pesticide applicators in the Agricultural Health Study. *Am J Epidemiol* 2009;**170**(7):892–900.

294. Huang NC, et al. Fatal ventricular fibrillation in a patient with acute imidacloprid poisoning. *Am J Emerg Med* 2006;**24**(7):883–5.
295. Demirel Y, et al. Acute amitraz intoxication: retrospective analysis of 45 cases. *Hum Exp Toxicol* 2006;**25**(10):613–7.
296. Hu SY, et al. Torsades de pointes in amitraz poisoning. *Resuscitation* 2010;**81**(3):366–7.
297. Ulukaya S, Demirag K, Moral AR. Acute amitraz intoxication in humans. *Intensive Care Med* 2001;**27**(5):930–3.
298. Gursoy S, et al. Intravenous amitraz poisoning. *Clin Toxicol (Phila)* 2005;**43**(2):113–6.
299. Chakraborty J, et al. An uncommon but lethal poisoning: Amitraz. *Aust Med J* 2011;**4**(8):439–41.
300. Yilmaz HL, Yildizdas DR. Amitraz poisoning, an emerging problem: epidemiology, clinical features, management, and preventive strategies. *Arch Dis Child* 2003;**88**(2):130–4.
301. Avsarogullari L, et al. Acute amitraz poisoning in adults: clinical features, laboratory findings, and management. *Clin Toxicol (Phila)* 2006;**44**(1):19–23.
302. Elinav E, et al. Near-fatal amitraz intoxication: the overlooked pesticide. *Basic Clin Pharmacol Toxicol* 2005;**97**(3):185–7.
303. Ray DE, Forshaw PJ. Pyrethroid insecticides: poisoning syndromes, synergies, and therapy. *J Toxicol Clin Toxicol* 2000;**38**(2):95–101.
304. Bhaskar EM, et al. Cardiac conduction disturbance due to prallethrin (pyrethroid) poisoning. *J Med Toxicol* 2010;**6**(1):27–30.
305. Pagel PS, et al. A 2-day history of shortness of breath and intermittent dull substernal chest pain in a 42-year-old man: a case of acute insecticide toxicity or something more ominous? *J Cardiothorac Vasc Anesth* 2009;**23**(5):732–4.
306. Scheuerman EH. Suicide by exposure to sulfuryl fluoride. *J Forensic Sci* 1986;**31**(3):1154–8.
307. Schneir A, et al. Systemic fluoride poisoning and death from inhalational exposure to sulfuryl fluoride. *Clin Toxicol (Phila)* 2008;**46**(9):850–4.
308. Taxay EP. Vikane inhalation. *J Occup Med* 1966;**8**(8):425–6.
309. Gonmori K, et al. A case of homicidal intoxication by chloropicrin. *Am J Forensic Med Pathol* 1987;**8**(2):135–8.
310. Lee EC, et al. Hydrogen sulfide intoxication with dilated cardiomyopathy. *J Occup Health* 2009;**51**(6):522–5.
311. Ballerino-Regan D, Longmire AW. Hydrogen sulfide exposure as a cause of sudden occupational death. *Arch Pathol Lab Med* 2010;**134**(8):1105.
312. Knight LD, Presnell SE. Death by sewer gas: case report of a double fatality and review of the literature. *Am J Forensic Med Pathol* 2005;**26**(2):181–5.
313. Yalamanchili C, Smith MD. Acute hydrogen sulfide toxicity due to sewer gas exposure. *Am J Emerg Med* 2008;**26**(4):518, e5–7.
314. vom Dahl J, et al. Effects of gold coating of coronary stents on neointimal proliferation following stent implantation. *Am J Cardiol* 2002;**89**(7):801–5.
315. Danzi GB, et al. Patterns of in-stent restenosis after placement of NIR gold-coated stents in unselected patients. *Catheter Cardiovasc Interv* 2002;**55**(2):157–62.
316. Pache J, et al. Sustained increased risk of adverse cardiac events over 5 years after implantation of gold-coated coronary stents. *Catheter Cardiovasc Interv* 2006;**68**(5):690–5.
317. Svedman C, et al. A correlation found between contact allergy to stent material and restenosis of the coronary arteries. *Contact Dermatitis* 2009;**60**(3):158–64.

318. Kastrati A, et al. Increased risk of restenosis after placement of gold-coated stents: results of a randomized trial comparing gold-coated with uncoated steel stents in patients with coronary artery disease. *Circulation* 2000;**101**(21):2478–83.
319. Svedman C, et al. Coronary restenosis and contact allergy to stent material. *J Invas Cardiol* 2011;**23**(4):e94.
320. Svedman C, et al. Contact allergy to gold in patients with gold-plated intracoronary stents. *Contact Dermatitis* 2005;**52**(4):192–6.
321. Ekqvist S, et al. Does gold concentration in the blood influence the result of patch testing to gold? *Br J Dermatol* 2009;**160**(5):1016–21.
322. Ekqvist S, et al. High frequency of contact allergy to gold in patients with endovascular coronary stents. *Br J Dermatol* 2007;**157**(4):730–8.
323. Ekqvist S, et al. A correlation found between gold concentration in blood and patch test reactions in patients with coronary stents. *Contact Dermatitis* 2008;**59**(3):137–42.
324. Archer SL. Dilated cardiomyopathy and left bundle branch block associated with ingestion of colloidal gold and silver is reversed by British antiLewisite and vitamin E: the potential toxicity of metals used as health supplements. *Can J Cardiol* 2008;**24**(5):397–9.
325. Gottlieb NL, Brown Jr. HE. Acute myocardial infarction following gold sodium thiomalate induced vasomotor (nitritoid) reaction. *Arthritis Rheum* 1977;**20**(4):1026–30.
326. Scharf J, Nahir M, Nirenberg L. Ventricular tachycardia caused by gold overdose. *Arthritis Rheum* 1976;**19**(6):1373–4.
327. Gibson DE, Moore GP, Pfaff JA. Camphor ingestion. *Am J Emerg Med* 1989;**7**(1):41–3.
328. Bhaya M, Beniwal R. Camphor induced myocarditis: a case report. *Cardiovasc Toxicol* 2007;**7**(3):212–4.
329. Ragucci KR, et al. Camphor ingestion in a 10-year-old male. *South Med J* 2007;**100**(2):204–7.
330. Okumura T, et al. Intravenous detergent poisoning. *J Toxicol Clin Toxicol* 2000;**38**(3):347–50.
331. Conraads VM, et al. Coronary artery spasm complicating anaphylaxis secondary to skin disinfectant. *Chest* 1998;**113**(5):1417–9.
332. Mehta AJ, et al. Heart rate variability in association with frequent use of household sprays and scented products in SAPALDIA. *Environ Health Perspect* 2012;**120**(7):958–64.
333. Ma CM, et al. Volatile organic compounds exposure and cardiovascular effects in hair salons. *Occup Med (Lond)* 2010;**60**(8):624–30.
334. Shetty RK, et al. An unusual cause of cardiac arrest in a hospitalized patient. *J Pharmacol Pharmacother* 2013;**4**(1):72–3.
335. Thanarajasingam G, Diedrich DA, Mueller PS. Intentional ingestion of ethanol-based hand sanitizer by a hospitalized patient with alcoholism. *Mayo Clin Proc* 2007;**82**(10):1288–9.
336. Chan TY, Lau MS, Critchley JA. Serious complications associated with Dettol poisoning. *Q J Med* 1993;**86**(11):735–8.
337. Joubert P, Hundt H, Du Toit P. Severe Dettol (chloroxylenol and terpineol) poisoning. *Br Med J* 1978;**1**(6117):890.
338. Meek D, Gabriel R, Piercy DM. Fatal self-poisoning with Dettol. *Postgrad Med J* 1977;**53**(618):229–31.
339. Zhou Y, et al. Ionic mechanisms underlying cardiac toxicity of the organochloride solvent trichloromethane. *Toxicol* 2011;**290**(2–3):295–304.
340. Gaillard Y, et al. A case of drug-facilitated sexual assault leading to death by chloroform poisoning. *Int J Legal Med* 2006;**120**(4):241–5.

341. Giusti GV, M. Chiarotti. Double "suicide" by chloroform in a pair of twins. *Med Sci Law* 1981;**21**(1):2–3.
342. Nashelsky MB, Dix JD, Adelstein EH. Homicide facilitated by inhalation of chloroform. *J Forensic Sci* 1995;**40**(1):134–8.
343. Schroeder HG. Acute and delayed chloroform poisoning. A case report. *Br J Anaesth* 1965;**37**(12):972–5.
344. Turkoglu C, et al. Slow heart-slow brain: consequence of short-term occupational exposure to toluene in a young woman: what is the real mechanism? *Clin Cardiol* 2010;**33**(2):E68–71.
345. Meadows R, Verghese A. Medical complications of glue sniffing. *South Med J* 1996;**89**(5):455–62.
346. Banathy LJ, Chan LT. Fatality caused by inhalation of "Liquid Paper" correction fluid. *Med J Aust* 1983;**2**(12):606.
347. Bass M. Sudden sniffing death. *JAMA* 1970;**212**(12):2075–9.
348. Wiseman MN, Banim S. "Glue sniffer's" heart? *Br Med J (Clin Res Edition)* 1987;**294**(6574):739.
349. Sabik LM, et al. Cardiotoxicity of Freon among refrigeration services workers: comparative cross-sectional study. *Environ Health* 2009;**8**:31.
350. Jiao Z, et al. A possible mechanism of halocarbon-induced cardiac sensitization arrhythmias. *J Mol Cell Cardiol* 2006;**41**(4):698–705.
351. Steffee CH, Davis GJ, Nicol KK. A whiff of death: fatal volatile solvent inhalation abuse. *South Med J* 1996;**89**(9):879–84.
352. Cruz SL, et al. Inhibition of cardiac sodium currents by toluene exposure. *Br J Pharmacol* 2003;**140**(4):653–60.
353. Edling C, et al. Cardiac arrhythmia in refrigerator repairmen exposed to fluorocarbons. *Br J Ind Med* 1990;**47**(3):207–12.
354. Alper AT, et al. Glue (toluene) abuse: increased QT dispersion and relation with unexplained syncope. *Inhal Toxicol* 2008;**20**(1):37–41.
355. Sjogren B, Gunnare S, Sandler H. Inhalation of decomposed chlorodifluoromethane (freon-22) and myocardial infarction. *Scand J Work Environ Health* 2002;**28**(3):205–7.
356. Anderson CE, Loomis GA. Recognition and prevention of inhalant abuse. *Am Fam Physician* 2003;**68**(5):869–74.
357. Zacharski LR, et al. Reduction of iron stores and cardiovascular outcomes in patients with peripheral arterial disease: a randomized controlled trial. *JAMA* 2007;**297**(6):603–10.
358. Menke A, et al. The association of biomarkers of iron status with peripheral arterial disease in U.S. adults. *BMC Cardiovasc Disord* 2009;**9**:34.
359. Kim KS, et al. Associations of serum ferritin and transferrin % saturation with all-cause, cancer, and cardiovascular disease mortality: Third National Health and Nutrition Examination Survey follow-up study. *J Prev Med Public Health* 2012;**45**(3):196–203.
360. Ma J, You C, Hao L. Iron chelators for acute stroke. *Cochrane Database Syst Rev* 2012;**9**:CD009280.
361. Kiechl S, et al. Body iron stores and the risk of carotid atherosclerosis: prospective results from the Bruneck study. *Circulation* 1997;**96**(10):3300–7.
362. Salonen JT, et al. High stored iron levels are associated with excess risk of myocardial infarction in eastern Finnish men. *Circulation* 1992;**86**(3):803–11.
363. Tuomainen TP, et al. Association between body iron stores and the risk of acute myocardial infarction in men. *Circulation* 1998;**97**(15):1461–6.
364. Zhang W, et al. Associations of dietary iron intake with mortality from cardiovascular disease: the JACC study. *J Epidemiol* 2012;**22**(6):484–93.

365. Salonen JT, Nyyssonen K, Salonen R. Body iron stores and the risk of coronary heart disease. *N Engl J Med* 1994;**331**(17):1159–60.
366. Sempos CT, et al. Serum ferritin and death from all causes and cardiovascular disease: the NHANES II Mortality Study. National Health and Nutrition Examination Study. *Ann Epidemiol* 2000;**10**(7):441–8.
367. Danesh J, Appleby P. Coronary heart disease and iron status: meta-analyses of prospective studies. *Circulation* 1999;**99**(7):852–4.
368. Gupta R, et al. Dietary and serum iron, body iron stores and coronary heart disease. *J Assoc Physicians India* 2000;**48**(5):489–92.
369. Sun Q, et al. Excessive body iron stores are not associated with risk of coronary heart disease in women. *J Nutr* 2008;**138**(12):2436–41.
370. Baer DM, Tekawa IS, Hurley LB. Iron stores are not associated with acute myocardial infarction. *Circulation* 1994;**89**(6):2915–8.
371. Sung KC, et al. Ferritin is independently associated with the presence of coronary artery calcium in 12,033 men. *Arterioscler Thromb Vasc Biol* 2012;**32**(10):2525–30.
372. Lee KR, et al. Serum ferritin is linked with aortic stiffness in apparently healthy Korean women. *Crit Pathw Cardiol* 2010;**9**(3):160–3.
373. Oshaug A, et al. Associations between serum ferritin and cardiovascular risk factors in healthy young men. A cross-sectional study. *Eur J Clin Nutr* 1995;**49**(6):430–8.
374. Kim MK, et al. Increased serum ferritin predicts the development of hypertension among middle-aged men. *Am J Hypertens* 2012;**25**(4):492–7.
375. Hramiak IM, Finegood DT, Adams PC. Factors affecting glucose tolerance in hereditary hemochromatosis. *Clin Invest Med* 1997;**20**(2):110–8.
376. Fernandez-Real JM, Ricart W. Insulin resistance and chronic cardiovascular inflammatory syndrome. *Endocr Rev* 2003;**24**(3):278–301.
377. Salonen JT, et al. Relation between iron stores and non-insulin dependent diabetes in men: case-control study. *BMJ* 1998;**317**(7160):727.
378. Jehn M, Clark JM, Guallar E. Serum ferritin and risk of the metabolic syndrome in U.S. adults. *Diabetes Care* 2004;**27**(10):2422–8.
379. Lee BK, Kim Y, Kim YI. Association of serum ferritin with metabolic syndrome and diabetes mellitus in the South Korean general population according to the Korean National Health and Nutrition Examination Survey, 2008. *Metabolism* 2011;**60**(10):1416–24.
380. Sullivan JL. Iron in arterial plaque: modifiable risk factor for atherosclerosis. *Biochim Biophys Acta* 2009;**1790**(7):718–23.
381. Duffy SJ, et al. Iron chelation improves endothelial function in patients with coronary artery disease. *Circulation* 2001;**103**(23):2799–804.
382. Millan M, et al. Increased body iron stores are associated with poor outcome after thrombolytic treatment in acute stroke. *Stroke* 2007;**38**(1):90–5.
383. Huang FP, et al. Brain edema after experimental intracerebral hemorrhage: role of hemoglobin degradation products. *J Neurosurg* 2002;**96**(2):287–93.
384. Castellanos M, et al. Iron intake increases infarct volume after permanent middle cerebral artery occlusion in rats. *Brain Res* 2002;**952**(1):1–6.
385. Kaluza J, Wolk A, Larsson SC. Heme iron intake and risk of stroke: a prospective study of men. *Stroke* 2013;**44**(2):334–9.
386. Rusovici A, et al. Extensive myocardial iron deposition in a patient with hepatitis C. *Tex Heart Inst J* 2012;**39**(2):281–3.
387. Cagirci G, Tufekcioglu O, Yetim M. Iron chelating in thalassaemia with left ventricular dysfunction. *Heart* 2006;**92**(10):1440.

388. Fabio G, et al. Reversal of cardiac complications by deferiprone and deferoxamine combination therapy in a patient affected by a severe type of juvenile hemochromatosis (JH). *Blood* 2007;**109**(1):362–4.
389. Chow CH, et al. Reversal of severe biventricular dysfunction from cardiac hemochromatosis with iron removal. *Circ Heart Fail* 2013;**6**(1):e14–5.
390. Liu P, Olivieri N. Iron overload cardiomyopathies: new insights into an old disease. *Cardiovasc Drugs Ther* 1994;**8**(1):101–10.
391. Voskaridou E, Christoulas D, Terpos E. Sickle-cell disease and the heart: review of the current literature. *Br J Haematol* 2012;**157**(6):664–73.
392. Giesbrandt KJ, et al. Diffuse diseases of the myocardium: MRI-pathologic review of cardiomyopathies with dilatation. *AJR Am J Roentgenol* 2013;**200**(3):W274–82.
393. Cassinerio E, et al. Cardiac iron removal and functional cardiac improvement by different iron chelation regimens in thalassemia major patients. *Ann Hematol* 2012;**91**(9):1443–9.
394. Fernandes JL. Iron chelation therapy in the management of transfusion-related cardiac iron overload. *Transfusion* 2012;**52**(10):2256–68.
395. McAllen PM, Coghill NF, Lubran M. The treatment of haemochromatosis; with particular reference to the removal of iron from the body by repeated venesection. *Q J Med* 1957;**26**(102):251–76.
396. Maguire JL, deVeber G, Parkin PC. Association between iron-deficiency anemia and stroke in young children. *Pediatrics* 2007;**120**(5):1053–7.
397. Baptist EC, Castillo SF. Cow's milk-induced iron deficiency anemia as a cause of childhood stroke. *Clin Pediatr (Phila)* 2002;**41**(7):533–5.
398. Mehta PJ, et al. Acute ischemic stroke secondary to iron deficiency anemia: a case report. *Case Rep Neurol Med* 2012;**2012**:487080.
399. Saxena K, et al. Fatal stroke in a child with severe iron deficiency anemia and multiple hereditary risk factors for thrombosis. *Clin Pediatr (Phila)* 2005;**44**(2):175–80.
400. Dubyk MD, et al. Iron deficiency anemia prevalence at first stroke or transient ischemic attack. *Can J Neurol Sci* 2012;**39**(2):189–95.
401. Alexander MB. Iron deficiency anemia, thrombocytosis, and cerebrovascular accident. *South Med J* 1983;**76**(5):662–3.
402. Vaziri ND, Gonick HC. Cardiovascular effects of lead exposure. *Indian J Med Res* 2008;**128**(4):426–35.
403. Pirkle JL, et al. The relationship between blood lead levels and blood pressure and its cardiovascular risk implications. *Am J Epidemiol* 1985;**121**(2):246–58.
404. Harlan WR, et al. Blood lead and blood pressure: relationship in the adolescent and adult U.S. population. *JAMA* 1985;**253**(4):530–4.
405. Pocock SJ, et al. The relationship between blood lead, blood pressure, stroke, and heart attacks in middle-aged British men. *Environ Health Perspect* 1988;**78**:23–30.
406. Schwartz J. Lead, blood pressure, and cardiovascular disease in men and women. *Environ Health Perspect* 1991;**91**:71–5.
407. Navas-Acien A, et al. Lead exposure and cardiovascular disease: a systematic review. *Environ Health Perspect* 2007;**115**(3):472–82.
408. Schober SE, et al. Blood lead levels and death from all causes, cardiovascular disease, and cancer: results from the NHANES III mortality study. *Environ Health Perspect* 2006;**114**(10):1538–41.
409. Navas-Acien A, et al. Lead, cadmium, smoking, and increased risk of peripheral arterial disease. *Circulation* 2004;**109**(25):3196–201.

410. Menke A, et al. Blood lead below 0.48 micromol/L (10 microg/dL) and mortality among U.S. adults. *Circulation* 2006;**114**(13):1388–94.
411. Kasperczyk S, et al. Function of heart muscle in people chronically exposed to lead. *Ann Agric Environ Med* 2005;**12**(2):207–10.
412. Cheng Y, et al. Electrocardiographic conduction disturbances in association with low-level lead exposure (the Normative Aging Study). *Am J Cardiol* 1998;**82**(5):594–9.
413. Eum KD, et al. Prospective cohort study of lead exposure and electrocardiographic conduction disturbances in the Department of Veterans Affairs Normative Aging Study. *Environ Health Perspect* 2011;**119**(7):940–4.
414. Silver W, Rodriguez-Torres R. Electrocardiographic studies in children with lead poisoning. *Pediatrics* 1968;**41**(6):1124–7.
415. Revis NW, Zinsmeister AR, Bull R. Atherosclerosis and hypertension induction by lead and cadmium ions: an effect prevented by calcium ion. *Proc Natl Acad Sci USA* 1981;**78**(10):6494–8.
416. Lustberg M, Silbergeld E. Blood lead levels and mortality. *Arch Intern Med* 2002;**162**(21):2443–9.
417. Jain NB, et al. Lead levels and ischemic heart disease in a prospective study of middle-aged and elderly men: the VA Normative Aging Study. *Environ Health Perspect* 2007;**115**(6):871–5.
418. Weisskopf MG, et al. A prospective study of bone lead concentration and death from all causes, cardiovascular diseases, and cancer in the Department of Veterans Affairs Normative Aging Study. *Circulation* 2009;**120**(12):1056–64.
419. Dubey L, Maskey A, Regmi S. Bradycardia and severe hypotension caused by wild honey poisoning. *Hellenic J Cardiol* 2009;**50**(5):426–8.
420. Saritas A, et al. Paroxysmal atrial fibrillation and intermittent left bundle branch block: an unusual electrocardiographic presentation of mad honey poisoning. *Clinics (São Paulo)* 2011;**66**(9):1651–3.
421. Weiss TW, et al. The honey man: second degree heart block after honey intoxication. *Int J Cardiol* 2010;**142**(1):e6–7.
422. Gunduz A, et al. Mad honey poisoning. *Am J Emerg Med* 2006;**24**(5):595–8.
423. Kalkan A, Gokce M, Memetoglu ME. An unusual clinical state: atrial fibrillation due to mad-honey intoxication. *Anadolu Kardiyol Derg* 2012;**12**(4):365–6.
424. Lennerz C, et al. Sinus arrest from mad honey disease. *Ann Intern Med* 2012;**157**(10):755–6.
425. Yarlioglues M, et al. Mad-honey sexual activity and acute inferior myocardial infarctions in a married couple. *Tex Heart Inst J* 2011;**38**(5):577–80.
426. Jansen SA, et al. Grayanotoxin poisoning: "mad honey disease" and beyond. *Cardiovasc Toxicol* 2012;**12**(3):208–15.
427. Akinci S, et al. An unusual presentation of mad honey poisoning: acute myocardial infarction. *Int J Cardiol* 2008;**129**(2):e56–8.
428. Sayin MR, et al. Transient ST segment elevation and left bundle branch block caused by mad-honey poisoning. *Wien Klin Wochenschr* 2012;**124**(7-8):278–81.
429. Osken A, et al. Slow ventricular response atrial fibrillation related to mad honey poisoning. *J Cardiovasc Dis Res* 2012;**3**(3):245–7.
430. Bayram NA, et al. A rare cause of atrial fibrillation: mad honey intoxication. *J Emerg Med* 2012;**43**(6):e389–91.
431. Yildirim N, et al. Clinical presentation of non-ST-segment elevation myocardial infarction in the course of intoxication with mad honey. *Am J Emerg Med* 2008;**26**(1):108 e1–2.
432. Gunduz A, et al. Mad honey poisoning-related asystole. *Emerg Med J* 2007;**24**(8):592–3.

433. Gunduz A, et al. Does mad honey poisoning require hospital admission? *Am J Emerg Med* 2009;**27**(4):424–7.
434. Okuyan E, Uslu A, Ozan Levent M. Cardiac effects of "mad honey": a case series. *Clin Toxicol (Phila)* 2010;**48**(6):528–32.
435. Dursunoglu D, Gur S, Semiz E. A case with complete atrioventricular block related to mad honey intoxication. *Ann Emerg Med* 2007;**50**(4):484–5.
436. Mordes JP, Wacker WE. Excess magnesium. *Pharmacol Rev* 1977;**29**(4):273–300.
437. Douban S, et al. Significance of magnesium in congestive heart failure. *Am Heart J* 1996;**132**(3):664–71.
438. Kurtzman JL, et al. Do nifedipine and verapamil potentiate the cardiac toxicity of magnesium sulfate? *Am J Perinatol* 1993;**10**(6):450–2.
439. Richards A, Stather-Dunn L, Moodley J. Cardiopulmonary arrest after the administration of magnesium sulphate. A case report. *S Afr Med J* 1985;**67**(4):145.
440. Thorp Jr. JM, et al. Nifedipine enhances the cardiac toxicity of magnesium sulfate in the isolated perfused Sprague-Dawley rat heart. *Am J Obstet Gynecol* 1990;**163**(2):655–6.
441. Miller KB, Caton JS, Finley JW. Manganese depresses rat heart muscle respiration. *Biofactors* 2006;**28**(1):33–46.
442. Jiang Y, Zheng W. Cardiovascular toxicities upon manganese exposure. *Cardiovasc Toxicol* 2005;**5**(4):345–54.
443. Fell JM, et al. Manganese toxicity in children receiving long-term parenteral nutrition. *Lancet* 1996;**347**(9010):1218–21.
444. Crossgrove J, Zheng W. Manganese toxicity upon overexposure. *NMR Biomed* 2004;**17**(8):544–53.
445. Shao JJ, et al. The disruption of mitochondrial metabolism and ion homeostasis in chicken hearts exposed to manganese. *Toxicol Lett* 2012;**214**(2):99–108.
446. Fernandes JL, et al. Preliminary assessment of cardiac short term safety and efficacy of manganese chloride for cardiovascular magnetic resonance in humans. *J Cardiovasc Magn Reson* 2011;**13**:6.
447. Jynge P, et al. Cardiovascular safety of MnDPDP and MnCl2. *Acta Radiol* 1997;**38**(4 Pt 2):740–9.
448. Barrington WW, et al. Autonomic function in manganese alloy workers. *Environ Res* 1998;**78**(1):50–8.
449. Mozaffarian D. Fish, mercury, selenium and cardiovascular risk: current evidence and unanswered questions. *Int J Environ Res Public Health* 2009;**6**(6):1894–916.
450. Wennberg M, et al. Myocardial infarction in relation to mercury and fatty acids from fish: a risk-benefit analysis based on pooled Finnish and Swedish data in men. *Am J Clin Nutr* 2012;**96**(4):706–13.
451. Mozaffarian D, Rimm EB. Fish intake, contaminants, and human health: evaluating the risks and the benefits. *JAMA* 2006;**296**(15):1885–99.
452. Mozaffarian D, et al. Mercury exposure and risk of cardiovascular disease in two U.S. cohorts. *N Engl J Med* 2011;**364**(12):1116–25.
453. Park K, Mozaffarian D. Omega-3 fatty acids, mercury, and selenium in fish and the risk of cardiovascular diseases. *Curr Atheroscler Rep* 2010;**12**(6):414–22.
454. Houston MC. The role of mercury and cadmium heavy metals in vascular disease, hypertension, coronary heart disease, and myocardial infarction. *Altern Ther Health Med* 2007;**13**(2):S128–33.
455. Houston MC. Role of mercury toxicity in hypertension, cardiovascular disease, and stroke. *J Clin Hypertens (Greenwich)* 2011;**13**(8):621–7.

456. Jansson G, Harms-Ringdahl M. Stimulating effects of mercuric- and silver ions on the superoxide anion production in human polymorphonuclear leukocytes. *Free Radic Res Commun* 1993;**18**(2):87–98.
457. Lin TH, Huang YL, Huang SF. Lipid peroxidation in liver of rats administrated with methyl mercuric chloride. *Biol Trace Elem Res* 1996;**54**(1):33–41.
458. Kostka B, et al. Blood coagulation changes in rats poisoned with methylmercuric chloride (MeHg). *Pol J Pharmacol Pharm* 1989;**41**(2):183–9.
459. Oliveira EM, et al. Mercury effects on the contractile activity of isolated heart muscle. *Toxicol Appl Pharmacol* 1994;**128**(1):86–91.
460. Wakita Y. Hypertension induced by methyl mercury in rats. *Toxicol Appl Pharmacol* 1987;**89**(1):144–7.
461. Fillion M, et al. A preliminary study of mercury exposure and blood pressure in the Brazilian Amazon. *Environ Health* 2006;**5**:29.
462. Valera B, Dewailly E, Poirier P. Cardiac autonomic activity and blood pressure among Nunavik Inuit adults exposed to environmental mercury: a cross-sectional study. *Environ Health* 2008;**7**:29.
463. Valera B, Dewailly E, Poirier P. Impact of mercury exposure on blood pressure and cardiac autonomic activity among Cree adults (James Bay, Quebec, Canada). *Environ Res* 2011;**111**(8):1265–70.
464. Grandjean P, et al. Cardiac autonomic activity in methylmercury neurotoxicity: 14-year follow-up of a Faroese birth cohort. *J Pediatr* 2004;**144**(2):169–76.
465. Yaginuma-Sakurai K, et al. Intervention study on cardiac autonomic nervous effects of methylmercury from seafood. *Neurotoxicol Teratol* 2010;**32**(2):240–5.
466. Poreba R, et al. Left ventricular diastolic function in workers occupationally exposed to mercury vapour without clinical presentation of cardiac involvement. *Toxicol Appl Pharmacol* 2012;**263**(3):368–73.
467. Salonen JT, et al. Mercury accumulation and accelerated progression of carotid atherosclerosis: a population-based prospective 4-year follow-up study in men in eastern Finland. *Atherosclerosis* 2000;**148**(2):265–73.
468. Choi AL, et al. Methylmercury exposure and adverse cardiovascular effects in Faroese whaling men. *Environ Health Perspect* 2009;**117**(3):367–72.
469. Ahlqwist M, et al. Serum mercury concentration in relation to survival, symptoms, and diseases: results from the prospective population study of women in Gothenburg, Sweden. *Acta Odontol Scand* 1999;**57**(3):168–74.
470. Hallgren CG, et al. Markers of high fish intake are associated with decreased risk of a first myocardial infarction. *Br J Nutr* 2001;**86**(3):397–404.
471. Yoshizawa K, et al. Mercury and the risk of coronary heart disease in men. *N Engl J Med* 2002;**347**(22):1755–60.
472. Virtanen JK, et al. Mercury, fish oils, and risk of acute coronary events and cardiovascular disease, coronary heart disease, and all-cause mortality in men in eastern Finland. *Arterioscler Thromb Vasc Biol* 2005;**25**(1):228–33.
473. Wennberg M, et al. Fish intake, mercury, long-chain n-3 polyunsaturated fatty acids and risk of stroke in northern Sweden. *Br J Nutr* 2007;**98**(5):1038–45.
474. Torres AD, Rai AN, Hardiek ML. Mercury intoxication and arterial hypertension: report of two patients and review of the literature. *Pediatrics* 2000;**105**(3): e34.
475. Kounis NG, et al. Kounis syndrome is the likely culprit in devastating stent thrombosis. *J Invasive Cardiol* 2011;**23**(3):88.

476. Kounis NG, Hahalis G, Theoharides TC. Coronary stents, hypersensitivity reactions, and the Kounis syndrome. *J Interv Cardiol* 2007;**20**(5):314–23.
477. Tan W, Cheng KL, Chen QX. Hypersensitivity to drug-eluting stent and stent thrombosis: Kounis or not Kounis syndrome? *Chin Med J (Engl)* 2009;**122**(19):2390–3.
478. Koster R, et al. Nickel and molybdenum contact allergies in patients with coronary in-stent restenosis. *Lancet* 2000;**356**(9245):1895–7.
479. Kounis NG, et al. Eosinophilic responses to stent implantation and the risk of Kounis hypersensitivity associated coronary syndrome. *Int J Cardiol* 2012;**156**(2):125–32.
480. Saito T, et al. Metal allergic reaction in chronic refractory in-stent restenosis. *Cardiovasc Revasc Med* 2009;**10**(1):17–22.
481. Hillen U, et al. Evaluation of metal allergies in patients with coronary stents. *Contact Dermatitis* 2002;**47**(6):353–6.
482. Norgaz T, et al. Is there a link between nickel allergy and coronary stent restenosis? *Tohoku J Exp Med* 2005;**206**(3):243–6.
483. Nguyen SH, et al. Prevalence of patch test results from 1970 to 2002 in a multi-centre population in North America (NACDG). *Contact Dermatitis* 2008;**58**(2):101–6.
484. Group EW. The European Surveillance System of Contact Allergies (ESSCA): results of patch testing the standard series, 2004. *J Eur Acad Dermatol Venereol* 2008;**22**(2): 174–81.
485. Lai DW, et al. Pericarditis associated with nickel hypersensitivity to the Amplatzer occluder device: a case report. *Catheter Cardiovasc Interv* 2005;**66**(3):424–6.
486. Kounis NG, et al. Metal allergy, atrial septal occluder devices and the risk of Kounis syndrome. *Ann Thorac Surg* 2010;**90**(6):2087–8.
487. Lyell A, Bain WH, Thomson RM. Repeated failure of nickel-containing prosthetic heart valves in a patient allergic to nickel. *Lancet* 1978;**2**(8091):657–9.
488. Lusini M, et al. Aortic valve replacement in a patient with severe nickel allergy. *J Card Surg* 2011;**26**(6):618–20.
489. Kawano H, et al. Granulation tissue with eosinophil infiltration in the restenotic lesion after coronary stent implantation. *Circ J* 2004;**68**(7):722–3.
490. Goebeler M, et al. Nickel chloride and cobalt chloride, two common contact sensitizers, directly induce expression of intercellular adhesion molecule-1 (ICAM-1), vascular cell adhesion molecule-1 (VCAM-1), and endothelial leukocyte adhesion molecule (ELAM-1) by endothelial cells. *J Invest Dermatol* 1993;**100**(6):759–65.
491. Kounis GN, et al. Kounis syndrome and simultaneous multivessel acute coronary syndromes after successful drug-eluting stent implantation. *Int J Cardiol* 2008;**127**(1):146–8.
492. El-Mawardy R, et al. Does nickel allergy play a role in the development of in-stent restenosis? *Eur Rev Med Pharmacol Sci* 2011;**15**(11):1235–40.
493. Thyssen JP, et al. No association between metal allergy and cardiac in-stent restenosis in patients with dermatitis-results from a linkage study. *Contact Dermatitis* 2011;**64**(3):138–41.
494. Calvo MS, Uribarri J. Public health impact of dietary phosphorus excess on bone and cardiovascular health in the general population. *Am J Clin Nutr* 2013;**98**(1):6–15.
495. Foley RN, et al. Serum phosphorus levels associated with coronary atherosclerosis in young adults. *J Am Soc Nephrol* 2009;**20**(2):397–404.
496. Dhingra R, et al. Relations of serum phosphorus and calcium levels to the incidence of cardiovascular disease in the community. *Arch Intern Med* 2007;**167**(9):879–85.
497. Tonelli M, et al. Relation between serum phosphate level and cardiovascular event rate in people with coronary disease. *Circulation* 2005;**112**(17):2627–33.

498. Larsson TE, et al. Conjoint effects of serum calcium and phosphate on risk of total, cardiovascular, and noncardiovascular mortality in the community. *Arterioscler Thromb Vasc Biol* 2010;**30**(2):333–9.
499. Kestenbaum B, et al. Serum phosphate levels and mortality risk among people with chronic kidney disease. *J Am Soc Nephrol* 2005;**16**(2):520–8.
500. Ritz E, et al. Phosphate additives in food: a health risk. *Dtsch Arztebl Int* 2012;**109**(4):49–55.
501. Razzaque MS. Phosphate toxicity: new insights into an old problem. *Clin Sci (Lond)* 2011;**120**(3):91–7.
502. Cvjetko P, Cvjetko I, Pavlica M. Thallium toxicity in humans. *Arh Hig Rada Toksikol* 2010;**61**(1):111–9.
503. Moore D, House I, Dixon A. Thallium poisoning. Diagnosis may be elusive but alopecia is the clue. *BMJ* 1993;**306**(6891):1527–9.
504. Roby DS, et al. Cardiopulmonary effects of acute thallium poisoning. *Chest* 1984;**85**(2): 236–40.
505. Galvan-Arzate S, Santamaria A. Thallium toxicity. *Toxicol Lett* 1998;**99**(1):1–13.
506. Atsmon J, et al. Thallium poisoning in Israel. *Am J Med Sci* 2000;**320**(5):327–30.
507. Riyaz R, et al. A fatal case of thallium toxicity: challenges in management. *J Med Toxicol* 2013;**9**(1):75–8.
508. Rusyniak DE, Furbee RB, Kirk MA. Thallium and arsenic poisoning in a small midwestern town. *Ann Emerg Med* 2002;**39**(3):307–11.
509. Saddique A, Peterson CD. Thallium poisoning: a review. *Vet Hum Toxicol* 1983;**25**(1): 16–22.
510. Pau PW. Management of thallium poisoning. *Hong Kong Med J* 2000;**6**(3):316–8.
511. Pelclova D, et al. Two-year follow-up of two patients after severe thallium intoxication. *Hum Exp Toxicol* 2009;**28**(5):263–72.
512. Cebi A, et al. Trace elements, heavy metals and vitamin levels in patients with coronary artery disease. *Int J Med Sci* 2011;**8**(6):456–60.
513. Yildiran H, et al. Serum antioxidant vitamin levels in patients with coronary heart disease. *Int J Vitam Nutr Res* 2011;**81**(4):211–7.
514. Liu S, et al. Intake of vegetables rich in carotenoids and risk of coronary heart disease in men: The Physicians' Health Study. *Int J Epidemiol* 2001;**30**(1):130–5.
515. Gey KF, et al. Low plasma retinol predicts coronary events in healthy middle-aged men: the PRIME Study. *Atherosclerosis* 2010;**208**(1):270–4.
516. Hatzigeorgiou C, et al. Antioxidant vitamin intake and subclinical coronary atherosclerosis. *Prev Cardiol* 2006;**9**(2):75–81.
517. Omenn GS, et al. Effects of a combination of beta carotene and vitamin A on lung cancer and cardiovascular disease. *N Engl J Med* 1996;**334**(18):1150–5.
518. Isik P, et al. All-transretinoic acid (ATRA) treatment-related pancarditis and severe pulmonary edema in a child with acute promyelocytic leukemia. *J Pediatr Hematol Oncol* 2010; **32**(8):e346–8.
519. van Rijssel RH, et al. A case of ATRA-induced isolated myocarditis in the absence of circulating malignant cells: demonstration of the t(15;17) translocation in the inflammatory infiltrate by in situ hybridisation. *Leuk Res* 2010;**34**(7):e142–4.
520. de Santis GC, et al. Cardiac stunning as a manifestation of ATRA differentiation syndrome in acute promyelocytic leukemia. *Med Oncol* 2012;**29**(1):248–50.
521. Manna A, et al. Reversible cardiac dysfunction without myocytolysis related to all-trans retinoic acid administration during induction therapy of acute promyelocytic leukemia. *Ann Hematol* 2009;**88**(1):91–2.

522. Botto LD, et al. Vitamin A and cardiac outflow tract defects. *Epidemiol* 2001;**12**(5):491–6.
523. Rothman KJ, et al. Teratogenicity of high vitamin A intake. *N Engl J Med* 1995;**333**(21):1369–73.
524. Werler MM, et al. Maternal vitamin A supplementation in relation to selected birth defects. *Teratol* 1990;**42**(5):497–503.
525. Mills JL, et al. Vitamin A and birth defects. *Am J Obstet Gynecol* 1997;**177**(1):31–6.
526. Mastroiacovo P, et al. High vitamin A intake in early pregnancy and major malformations: a multicenter prospective controlled study. *Teratol* 1999;**59**(1):7–11.
527. McGreevy C, Williams D. New insights about vitamin D and cardiovascular disease: a narrative review. *Ann Intern Med* 2011;**155**(12):820–6.
528. Saremi A, Arora R. Vitamin E and cardiovascular disease. *Am J Ther* 2010;**17**(3):e56–65.
529. Stephens NG, et al. Randomised controlled trial of vitamin E in patients with coronary disease: Cambridge Heart Antioxidant Study (CHAOS). *Lancet* 1996;**347**(9004):781–6.
530. Boaz M, et al. Secondary prevention with antioxidants of cardiovascular disease in endstage renal disease (SPACE): randomised placebo-controlled trial. *Lancet* 2000;**356**(9237):1213–8.
531. Pocobelli G, et al. Use of supplements of multivitamins, vitamin C, and vitamin E in relation to mortality. *Am J Epidemiol* 2009;**170**(4):472–83.
532. The Alpha-Tocopherol, Beta Carotene Cancer Prevention Study Group. The effect of vitamin E and beta carotene on the incidence of lung cancer and other cancers in male smokers. *N Engl J Med* 1994;**330**(15):1029–35.
533. Lonn E, et al. Effects of long-term vitamin E supplementation on cardiovascular events and cancer: a randomized controlled trial. *JAMA* 2005;**293**(11):1338–47.
534. Miller III ER, et al. Meta-analysis: high-dosage vitamin E supplementation may increase all-cause mortality. *Ann Intern Med* 2005;**142**(1):37–46.
535. American Heart Association Nutrition Committee, et al. Diet and lifestyle recommendations revision 2006: a scientific statement from the American Heart Association Nutrition Committee. *Circulation* 2006;**114**(1):82–96.
536. Songy Jr. KA, Layon AJ. Vitamin K-induced cardiovascular collapse. *J Clin Anesth* 1997;**9**(6):514–9.
537. Bosse GM, Mallory MN, Malone GJ. The safety of intravenously administered vitamin K. *Vet Hum Toxicol* 2002;**44**(3):174–6.

Chapter 4

Problems and Paradoxes of Animal Toxins and the Heart

Subramanian Senthilkumaran,[1] Ramachandran Meenakshisundaram,[2] and Ponniah Thirumalaikolundusubramanian[3]

[1]*Sri Gokulam Hospitals and Research Centre, Salem, India*

[2]*Epsom and St. Helier University Hospitals NHS Trust, Epsom, United Kingdom*

[3]*Chennai Medical College Hospital and Research Centre, Trichy, India*

4.1 INTRODUCTION

Venoms are complex concentrates of biologically highly active molecules and exist mainly as peptides and proteins that interfere with one or more fundamental metabolic function of the prey. The venom affecting cardiac physiology is very effective in the sense of defense and especially in capturing prey. The venom can disturb the electrical activity of the heart which in turn affects impluse conduction of the heart. Also, some of the venoms have a direct cardiotoxic effect, which acts on cardiac myocytes. Overall, these can affect the pumping capacity of the heart and thus cause myocardial ischemia. Snakes, scorpions, bees, wasps, centipedes, and frogs are just a few of the animals that use venoms. In addition, such venoms and toxins are multitude and varied from species to species. The aim of this chapter is to describe the mechanisms by which venom affects cardiac function, its targets in the heart tissue, and the cardiac effects of envenomations. Thus, we focus only on bites and stings. Considering the scope of chapter, we have not dealt with cardiac toxicities from infections or toxins from infectious organisms.

4.2 SNAKES

Snakebite is a neglected tropical disease of global importance. India has a unique snake fauna including some with highly potent venoms. The major Indian snakes are the Indian cobra (*Naja naja*), common krait (*Bungarus caeruleus*), Russell's viper (*Daboia Russelii*), saw-scaled viper (*Echis carinatus*), and hump-nosed pit viper (*Hypnale hypnale*). Cardiac involvement is not a prominent feature of a snakebite and the clinical picture is usually dominated by the neurological,

The Heart and Toxins. DOI: http://dx.doi.org/10.1016/B978-0-12-416595-3.00004-9

hematological, and vascular damages. Snake venom cannot be regarded as a single toxin. The chemical makeup of the venom varies from species to species, which explains the clinical diversity of snake envenomation. There is also considerable variation in the relative proportion of different venom constituents within a single species within a geographical region at different seasons.[1] Snake venom contains a variety of enzymes, such as phospholipases A2, serine proteinases, metalloproteinases, acetylcholinesterases, L-amino acid oxidases, nucleotidases, hyaluronidases, and nonenzymatic components, with different biologic activities.[2] Therefore, it is not difficult to appreciate the complexity of cardiovascular changes produced by different snake venoms.[3]

In most cases of snakebite, circulatory disturbance is one of the most frequently encountered events. The site of action of these toxins includes cardiac muscle, vascular smooth muscle, and the capillary vascular bed. Most viperidae venoms cause a fall in blood pressure following envenomation. It has been claimed that this effect is due to depression of the central autonomic vasoregulatory mechanism.[4] However, some snake venoms (e.g., *Bothrops jararaca*) contain peptides that inhibit the angiotensin-converting enzyme and potentiate the biological actions of bradykinin. These bradykinin-potentiating peptides in the venom inhibit the breakdown of the endogenous vasodilator bradykinin and also inhibit synthesis of the endogenous vasoconstrictor angiotensin II, thus leading to a fall in blood pressure.[5]

Many snake venoms contain the structural and functional equivalents of mammalian natriuretic peptides, which elicit natriuretic, diuretic, and vasorelaxant effects.[6] Snakes, such as the mamba (*Dendroaspis*), have also evolved toxins that block L-type Ca^{2+} currents (e.g., calciseptine, FS2 toxins, C10S2C2, and S4C8).[7] The potency of the hypotensive activity induced by mamba toxins has a prolonged effect, which is much higher when compared with nifedipine. Another calcium channel blocker, calcicludine, also has been identified in the venom of the green mamba. This toxin blocks different Ca-channels (e.g., L-, N-, and P-type) and its three-dimensional structure is similar to that of the K^+ channel blockers.[8] Snake venom proteins have also been shown to increase vascular permeability and it is remarkable to see the emergence of a new set of selected molecules. One such protein, called increasing capillary permeability protein (ICPP), has been isolated from the venom of *Vipera lebetina*.[9] ICPP is an extremely potent permeability factor with a structure highly homologous to the vascular endothelial growth factor (VEGF) that results in pooling of blood and fluids in the microcirculation and may contribute to hypotension.[10]

Although it is very rare, hypertension has been reported following elapid envenomation[11,12] and also noted in a number of studies.[13–15] Hypertension was probably ascribed to dysautonomia due to snake venom. The pathogenesis for autonomic dysfunction in snakebite remains unclear. However, it may be attributable to the presynaptic α-2 adrenoreceptor inhibition by elapid neurotoxin, thereby blocking inhibition of the neutrally mediated release of norepinephrine. Thus, it gives rise to sympathetic overactivity and decreased parasympathetic stimulation.[13–15] Malina et al. described hypertension in a patient who was bitten by *Vipera berus bosniensis* in Hungary and the victim responded to angiotensin-converting enzyme inhibitor.[14] The venom of a certain *Vipera* population may contain cardiotoxins, which have the potential to act at or via autonomic synapses; moreover, snake venom releases catecholamines that contribute to elevated blood pressure.[11,12] Indeed, hypertension without neurotoxic symptoms also has been observed in patients with western Russell's viper envenomation.[13–15] Further attributable causes for hypertension are stress, pain, distress, or hypoxia. However, severe hypertension in such cases is extremely rare. Additionally, the normalization of blood pressure with improvements in autonomic function increases the possibility of a relationship between venom and hypertension.[11,12]

Snake venoms comprise a natural library of valuable bioactive substances for hemostasis and thrombosis. Viperine bites accounts for nearly half of snakebites in India. Viper venom is predominantly hemotoxic and contains components such as factor V and factor X activators, which create high procoagulant potential within the bite victim, thus giving rise to venom-induced consumption coagulopathy with thrombotic occlusion in previously normal mid-sized coronary arteries or elsewhere such as cerebral and pulmonary vasculature.[16] The nonenzymatic fraction hemorrhagins, the complement-mediated toxic components of Viperidae snake venom, may result in severe vascular spasm, endothelial damage, and increased vascular permeability, all of which contribute to vascular occlusion.[17]

Indeed, sarafotoxins are a unique group of 21-residue short peptides that are found in the venoms of snakes from *Atractaspis* spp. that display potent vasoconstriction properties. These may induce coronary vasoconstriction and act either directly on the phosphoinositide phosphodiesterase system or on a novel receptor. Neither their binding nor their activity is affected by blockers or activators of known receptors and ion channels.[18] Their amino acid sequence shows a high degree of homology with that of endothelin. In addition, the direct action of venom causes vascular endothelial injury, with the release of VEGF and von Willebrand's factor that produce toxic vasculitis, promoting thrombus formation and subsequent infarcts.

In addition, the toxic components of venom may result in severe vascular spasm, which contributes to infarction.[19] Procoagulants in viper venom, such as hydrolase and arginine esterase, activate the procoagulant pathway and cause aggregation of platelets, thus leading to intraluminal thrombus formation.[20] Hyperviscosity caused by hypovolemia, hypoperfusion, and coagulopathy may also contribute to vessel occlusion.[21] Few cases of acute myocardial infarction following snakebite have been reported.[16,17]

Similarly, ST segment elevation myocardial infarctions with normal coronaries have been observed in viper bite victims who developed coronary artery spasm following Kounis syndrome, which is a concurrence of acute coronary syndromes associated with mast cell activation including allergic and anaphylactic insults.[22] Clinicians should keep in mind the possibility of coronary vasospasm with allergic reactions. Three variants were described: type 1 was characterized by vasospasm with normal coronary arteries; type 2 the development of atheromatous plaques and coronary thrombosis; and type 3, a variant, includes patients with coronary artery stent thrombosis. Recently, Satish et al. reported for the first time stent thrombosis following snakebite and related it to the consequences of anaphylaxis.[23] Interestingly, Murase and Takagi alert clinicians managing snakebite to consider not only the cardioinhibitory effect of snake venom but also takotsubo cardiomyopathy as a differential diagnosis if the victim develops cardiac dysfunction following snakebite.[24]

Venom contains hemorrhagin, a metalloproteinase and kinase insert domain containing receptors (KDR), which binds to VEGF. These elements dissect the endothelial lining of blood vessels from their muscular coat, thus producing ischemic damage by occlusion or rupture and result in hemorrhage into the pericardium.[25] We have reported a case of cardiac tamponade following pericardial hemorrhage secondary to coagulopathy caused by snakebite (Russell's viper).[26] Similarly, pleuropericardial hemorrhage following envenomation by viper bite was documented earlier.[27] The direct cardiotoxic effect of snake venom can result in myocarditis and extensive myocardial necrosis that have been reported in two horses injected with *Vipera palaestinae* venom for commercial production of antibodies.[28]

Singh et al. documented nonbacterial thrombotic endocarditis of the aortic valve with multiple embolic infarcts of the brain, spleen, and kidneys, acute tubular necrosis, and disseminated intravascular coagulopathy (DIC) in a young male following viperine bite and attributed it to phospholipase A2.[29] Cardiotoxicity is an unusual manifestation of severe neurotoxic snake envenomation. Agarwal et al. reported cardiogenic pulmonary edema after neurotoxic snake envenoming by *Bungarus caeruleus*.[30] Cardiac involvement in neurotoxic snake envenomation remains unclear but is likely to be due to one of the myriad toxins seen in snake venom that can cause morphological changes, enzyme alterations, ultrastructural disturbances, and genetic alterations of the myocardial tissue.

It is believed that snake venom modifies the electrophysiological properties of the cardiac cell membrane and can have profound effects on impulse generation and conduction throughout the heart. Phospholipase A2 in snake venom has a significant and rapid myotoxic effect. It has been reported that there is an uneven distribution of histamine and mast cells in various parts of the heart, these being more in the areas surrounding the sinoatrial node and the atrioventricular node.[31] Furthermore, the venom reduces resting membrane potential, amplitude, duration of action potential, and contractility of the rat heart. The mechanism is independent of phospholipid hydrolysis or hydrolytic products such as arachidonic acid.[32] The neurotoxins and cardiotoxins of elapid and hydrophid venom are of small molecular size and carry a strong positive charge, which is reflected in their low antigenicity and rapid absorption.

The lethal potency of cobra venom cardiotoxin is 1/20 of its neurotoxin. The cardiotoxins (CTX) are made up of 60 to 62 amino acid polypeptides and are members of the "three-finger fold " toxin group.[33] These toxins exhibit cytolytic activity and can cause cell death. They have often been reported to produce myonecrosis and are thus also called "direct lytic factor." CTX induces the contracture of isolated cardiac muscle cells, leading to systolic arrest and an irreversible decrease of the resting potential and results in loss of the fast phase of the action potential. CTX has been postulated to increase membrane permeability by causing pore formation, thus allowing Ca^{2+} influx and causing irreversible membrane depolarization and asystolic cardiac arrest.[34]

Venom-induced electrophysiological changes, as noted in electrocardiogram (ECG) following *Naja* spp. envenomation, are prolonged QTc, varying degrees of atrioventricular (AV) block (e.g., sinus arrest, junctional escape rhythm, second-degree block, and bradycardia-dependent left bundle branch block).[35] Also, tachyarrhythmias, such as atrial fibrillation, ventricular bigeminy, and ventricular tachycardia following snake envenomation, have been reported.[36–38] In a prospective study of 69 patients who were bitten by taipan (*Oxyuranus scutellatus*) snakes, ECG changes were observed in 36 (52%) and 8.3% of patients had markedly elevated troponin as a marker of myocardial damage.[39] Common ECG changes observed among such patients were septal T-wave inversion, bradycardia, and variable degrees of AV block.[39] The taicatoxin, a calcium channel blocker in taipan snakes, may be responsible for such changes.[39] From the experimental models, it is known that venom induces the alteration of genes expressed in the heart, which is responsible for the mediators of immune response, apoptosis, ion transport, signal transduction, hypotension, energy metabolism, and electron transport. Thus, the occurrence of cardiotoxicity depends on the venom-induced alterations of genes expressed in the heart of the susceptible individual.[40]

There are a vast array of snake toxins with potent cardiovascular activity. Snake venoms represent an extraordinary source of these proteins, and peptides have proved to be highly selective tools in the study of basic biological processes. Others have been used as probes of potential therapeutic targets or as lead compounds in the development of therapeutic agents. It is advisable to refer to the treatment guidelines for the management of snakebites and scorpion stings in the WHO publication[41] and the Tamil Nadu State Health publication.[42]

4.3 SCORPIONS

Scorpions are the oldest arachnids for which their fossil remains have been discovered. They are reputed to be one of the first land animals on earth. Human beings have known of the scorpion species for centuries, as evident from mythological data. There are about 1500 scorpion species worldwide, of which 50 are known as dangerous to humans. Scorpion stings are a major public health problem in many underdeveloped tropical and subtropical countries, especially India, Africa, the Middle East, Mexico, and South America. The estimated annual number of scorpion stings is 1.2 million leading to 3250 deaths (27%). For every person killed by a poisonous snake, 10 are killed by a poisonous scorpion.[43] Almost all potentially lethal scorpions belong to the *Buthide* family. Three species, *Mesobuthus tamulus* (Indian red scorpion), *Palamneus swammerdami* (black scorpion), and *Heterometrus bengalensis*, are of medical importance.[44]

Scorpion venom[45] is a water-soluble, antigenic, heterogeneous mixture; it is composed of varying concentrations of neurotoxin, cardiotoxin, nephrotoxin, hemolytic toxin, phosphodiesterases, phospholipases, hyaluronidases, glycosaminoglycans, histamine, serotonin, tryptophan, cytokine releasers, and so on. Scorpion stings cause a wide range of conditions, from severe local skin reactions to neurologic, respiratory, and cardiovascular collapse.[45] Scorpion venom stimulates the autonomic nervous system, predominantly the sympathetic, which leads to sudden liberation of endogenous catecholamine resulting in an early transient rise in blood pressure, bradycardia, and increased vascular resistance. Clinical manifestations depend on time lapse between sting and hospitalization and/or treatment received at a periphery location. According to clinical manifestations, they are divided into three grades: II, III, and IV. Cardiovascular manifestations have been reported irrespective of the various species of scorpion. All cases had initial signs and symptoms suggestive of autonomic storm.[46]

- Grade II: hypertension or transient hypotension, tachycardia, bradycardia, and cold extremities
- Grade III: hypertension, hypotension, tachycardia and pulmonary edema, and respiratory failure
- Grade IV: tachycardia, hypotension and pulmonary edema with warm extremities, called warm shock

Morbidity and mortality due to a scorpion sting is related to acute pulmonary edema, cardiogenic shock, and multiorgan failure.[46] Hemodynamic consequences of scorpion envenomation occur in two phases.[47] The early one is a hyperdynamic phase with elevated blood pressure and enhanced left ventricular (LV) contractility related to catecholamine oversecretion. This phase is frequently followed by a hypokinetic phase characterized by hypotension and impaired LV systolic function. The most accepted hypothesis, based mainly on animal studies, ascribes cardiac dysfunction to an increase in systemic arterial impedance and a decrease in the left ventricular ejection fraction consecutive to catecholamine discharge.[47] Karnad has documented predominant left ventricular dysfunction with normal systemic vascular resistance causing pulmonary edema.[47] Abroug et al. have reported cardiac dysfunction and pulmonary edema due to a reduction in left ventricular compliance and increased impendence to left ventricular emptying.[48] Increased systemic pressure without rise in systemic vascular resistance is attributed to a rise in cardiac output.

The primary targets of scorpion venom are voltage-dependent ion channels, of which sodium channels are the best studied.[49] Venom alters these channels and the side chains of scorpion venom that are positively charged. This is important in their ability to bind to specific membrane channels. Alpha (of *Buthus* spp.) and beta (of *Centruroides* spp.) venom act on the sodium channel. Scyllatoxin, a charybdotoxin of the *Leiurus* species, and *Tityus* toxin act primarily on potassium channels. The venom acts by opening the sodium channel at presynaptic nerve terminals and inhibits calcium-dependent potassium channels. Autonomic storm is thus initiated by releasing adrenaline and noradrenaline from neurons, ganglia, and adrenals, thereby increasing myocardial oxygen demand by direct inotropic and chronotropic effect on already-compromised myocardial blood supply; this results in "scorpionic myocarditis" characterized by nonspecific ultrastructural changes, including intracellular edema, destruction of I bands, and lipid deposition.[50] Many end-organ effects are secondary to this excessive excitation.[50]

Cardiac sarcolemmal defects and depletion of the glycogen content of the heart, liver, and skeletal muscles have been observed in experimental animals with acute myocarditis that is produced by Indian red scorpion venom. Sundararaman et al. have observed scorpion envenomation as a risk factor for the subsequent development of idiopathic dilated cardiomyopathy.[51] Few cases of infective endocarditis in adults and children following scorpion stings have been reported.[52,53] However, a child developed a septic pulmonary embolism following scorpion sting.[53] Unusual organisms are responsible for infective endocarditis in such cases.[52]

Alpha receptor stimulation by the toxin plays a major role, resulting in hypertension, tachycardia, myocardial dysfunction, pulmonary edema, and cool extremities through peripheral vasodilation.[54] Raised angiotensin I levels have also been documented, which further facilitate the sympathetic outflow through conversion to angiotensin II. Excess catecholamines cause accumulation of endothelins and vasoconstriction. These unopposed effects of α-receptor stimulation lead to suppression of insulin secretion, hyperglycemia, hyperkalemia, free fatty acids, and free radical accumulation, all of which are injurious to myocardium.[55]

Electrocardiographic abnormalities are frequently observed. In most victims, sinus tachycardia was initially recorded.[56] However, a few patients also presented with sinus bradycardia. Tall-tented T waves mimicking an acute myocardial infarction pattern have been recorded without any electrolyte disturbances and persisted for more than five days (Figure 4.1). Other changes, such as bizarre, broad-notched biphasic T-wave changes with additional ST changes, and electrical alternans of QRS, have been documented.[56]

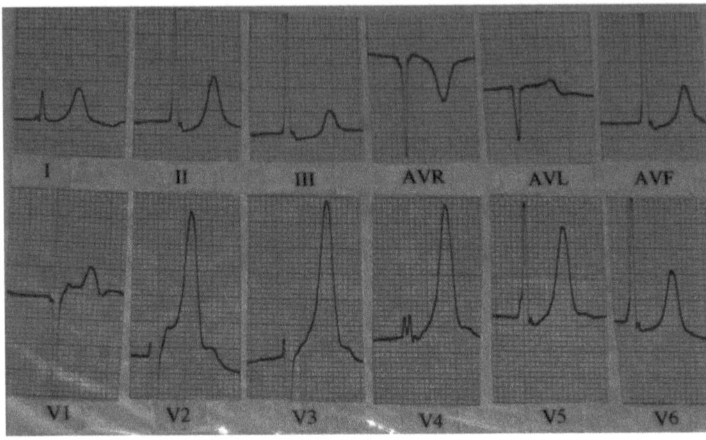

FIGURE 4.1 **Tall-tented T waves following scorpion sting. Source:** *Courtesy of Dr. H. S. Bawaskar, Bawaskar Hospital and Research Center, Mahad District, Raigad, Maharashtra, India.*

Electrophysiological changes induced by scorpion included variable degrees of AV block, paroxysmal supraventricular tachycardia, atrial flutter, atrial fibrillation, and premature atrial/ventricular beats; however, these are less common.[56,57] Echocardiogram reveals poor global myocardial contractility for 12 to 15 hours after sting, with low-ejection fraction, decreased left ventricular performance, trivial mitral regurgitation, and abnormal diastolic filling for variable periods of time from 5 days to 4 weeks.[56,58] However, in a child bitten by a scorpion, who presented with myocarditis and pulmonary edema, 4 months follow-up with 14 serial echocardiograms showed severe changes despite the clinical improvement seen within the first 24 hours of sting.[59] Thus, clinical course and outcome have been found to be variable.

Diminished or hypokinetic left ventricular with decrease in systolic function was noted in scintigraphic study.[60] There is a good correlation between the clinical improvement and return of LV function.[60] Abroug et al. reported gradual normalization of wall motion abnormality in the left ventricle and septum and also complete restoration of the systolic function in cases with severe scorpion envenomations.[48] There have been few cases of severe myocarditis induced by scorpion stings.[61] In some cases, right ventricular dysfunction was noted along with LV impairment, which may be from possible myocarditis or myopathic effect.[50,61,62]

In addition, scorpion venom inhibits an angiotensin-converting enzyme, resulting in the accumulation of bradykinin, which is implicated in the development of pulmonary edema.[63] Unilateral occurrence of pulmonary edema, though rare (Figure 4.2), can be attributed to a simultaneous and localized

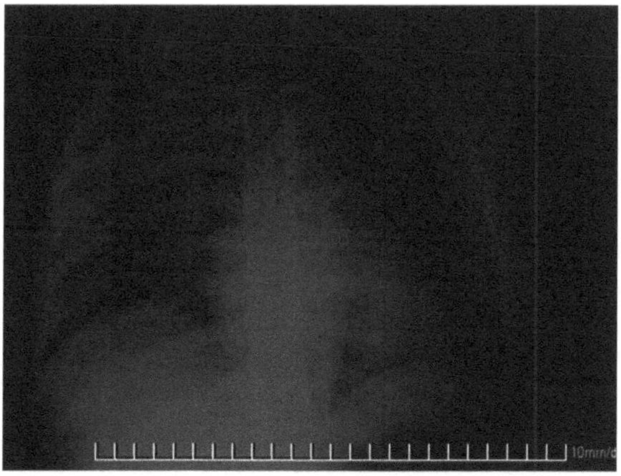

FIGURE 4.2 **Unilateral pulmonary edema following scorpion sting. Source:** *Courtesy of Dr. S. Senthilkumaran, Sri Gokulam Hospitals, Salem, Tamil Nadu, India.*

increase in pulmonary vascular permeability induced by scorpion venom.[64] The release of vasoactive, inflammatory, and thrombogenic peptides and amine constituents (e.g., histamine, serotonin, bradykinin, leukotrienes, thromboxane), which act on the coronary vasculature and induce coronary artery vasospasm, platelet aggregation, and thrombosis, leads to myocardial ischemia or infarction.[65] Release of allergenic proteins causes anaphylactic shock leading to hypotension with vasodilation and decreased intravascular volume with reduced myocardial perfusion.

The mechanism of the cardiogenic shock is not clear. It has been hypothesized that the depletion of myocardial norepinephrine stores due to the increased sympathetic stimulation of the heart and downregulation of the β receptors makes the myocardium less sensitive to catecholamines. In addition to the depressive cholenergic effect of the venom, hypovolemia and an exaggerated β-2 vasodilator effect of the circulating pressure amines on the peripheral blood vessels are thought to be involved.[66] The existence of myocardial depressant substances has been postulated to explain the reduction in cardiac function during hemorrhagic and endotoxic shock in animals. Recent studies suggest that these circulating myocardial-depressant substances have biochemical and pathogenic characteristics similar to those of a number of cytokines and particularly tumor necrosis factor (TNF). TNF produces a concentration-dependent depression of myocardial cell shortening and may be involved in the cardiovascular effects caused by scorpion venom.[67] Moreover, lack of elevation of systemic vascular resistance in response to a decrease in cardiac index was observed by Nouira et al., who attributed it to the release of vasodilating mediators such as TNF.[50]

4.4 HYMENOPTERA STING ENVENOMATIONS

Wasps, bees, and ants are some of the stinging insects that belong to the order Hymenoptera, and envenomations are quite common. Hymenoptera stings are responsible for more deaths than those from all other poisonous creatures. Apart from local cutaneous manifestations or generalized anaphylaxis, stings can cause a variety of systemic complications. The most frequent clinical events are hypotension, dyspnea, anaphylactic shock, and angioedema. There have been few reports dealing with cardiac toxicities such as rhythm abnormalities and myocardial infarction after hymenoptera stings.

More recently a new neurotoxin, α-pompilidotoxin (α-PMTX), has been isolated from the venom of a solitary wasp and shown to slow or block the inactivation of the voltage-sensitive Na^+ channels resulting in rhythm abnormalities such as supraventricular arrhythmias (atrial flutter/atrial fibrillation), premature ventricular complexes, junctional rhythm, and right-bundle branch block during initial stages after the sting.[68,69] Although the exact mechanism is unknown, it may involve vagal-mediated shortening of atrial refractoriness, the venom itself, and/or the catecholamine surge to bee stings. Also, the toxin contains many other pharmacological constituents such as histamine, dopamine, noradrenaline, serotonin, apamin, phospholipase, and mellitin.[68–70] Many of these constituents are vasoactive substances and have proarrhythmic effects.[69] In addition, a case report of ventricular fibrillation following bee sting has been reported.[71] This may be due to proarrhythmic potential or ischemic effect or direct toxicity of venom on cardiac myocytes.

Pharmacologically active constituents in the toxins isolated from hymenoptera venom can cause inflammatory reactions, induce coronary artery vasospasm, and facilitate platelet aggregation and thrombosis.[72] Bee venom induces an endogenous amine release similar to the mediators released from mast cells during allergic reactions, which normally produce vasodilatation. However, paradoxical coronary vasoconstriction is a possibility, especially in the presence of endothelial damage that may result in acute coronary syndrome.[73,74] Multiple case reports have demonstrated the occurrence of acute coronary syndrome and myocardial injury following bee stings.[75,76] An acute stent thrombosis in a drug-eluting stent caused by wasp and bee stings leading to myocardial infarction has been reported.[77,78]

In another report, a patient developed acute myocardial infarction following a wasp sting with evidence of thrombotic lesion in the coronary artery, although without any evidence for hypersensitivity reaction.[79] This patient required revascularization therapy with a stent, thus indicating the thrombogenic potential of wasp sting venom.[79] Recently, Kounis syndrome presented as very late stent thrombosis in a drug (e.g., everolimus) eluting stent following wasp stings.[80] In a recent report, a patient presented with transient inferior ST elevation myocardial infarction following a wasp sting, and coronary angiography revealed no obstructive lesions in the coronaries.[81] This may be due to coronary vasospasm and coronary vasodilators, such as calcium channel blockers, may be effective in treating such events.[81] Possible mechanisms include inflammatory, allergic or anaphylactic reactions, and endothelial dysfunction, which precede ischemia, vasospasm, and thrombosis.[80] The mechanism may be secondary to severe yet temporary vasospasm or in situ thrombus secondary to active constituents of hymenoptera venom.

4.5 CENTIPEDE BITES

Centipedes are slim, multisegmented arthropods. Their venoms are complex mixtures containing 5-hydroxytryptamine, histamine, lipids, polysaccharides, and various enzymes such as proteinases and esterases. It is thought that the venom is a lipid-toxin complex similar to that of scorpion venom, which facilitates local cellular penetration and absorption. Gomes et al. used animal models to demonstrate significant cardiovascular effects of *Scolopendra* venom that were mediated by histamine and a cardiodepressant factor designated as Toxin-S, which is a high-molecular acidic and heat-labile cardiotoxic protein.[82]

In the English medical literature, only three cases of acute coronary syndrome after centipede envenomation have been reported.[83–85] In some of them, coronary angiogram was entirely normal, which reveals the fact of coronary vasospasm.[85] Even though the exact pathophysiology in both cases was unidentified, the most probable mechanisms may be acute release of inflammatory mediators, inducing increased capillary permeability, inflammatory changes, hypotension, and coronary artery spasm; this may have led to acute myocardial infarction with raised cardiac biomarkers.[85] In these patients, bradycardia was noted despite hypotension.[83,85] Possible hypothesis for this clinical manifestation is a direct toxic effect on the conduction pathway or increased vagal response from the venom.[85]

4.6 BLACK WIDOW SPIDERS

Spiders are natural enemies of insects; however, in humans many spiders can induce local damage of variable severity and a few can inflict dangerous bites with generalized symptoms including death. The cardiovascular manifestations rarely have been observed.[86] Hypertension within the few minutes after the bite has been reported in a few patients. Electrocardiographic abnormalities, such as slurring of the QRS with ST- and T-segment depression and prolonged QT interval, were documented in some patients.[87] The exact mechanism for the electrocardiographic changes has not been elucidated. Experimental evidence suggests that the α-latrotoxin, the most active component of the venom, causes the release and depletion of acetylcholine, norepinephrine, and other neurotransmitters from synaptic terminals and inhibits their reuptake.[88] Uncommon cardiac manifestations following a spider bite include ischemia, myocarditis, myocardial necrosis, pulmonary edema, hypertension, and varying degrees of tachybrady arrhythmias.[87,89–91]

4.7 TOAD POISONING

Toads (*Bufo melanostictus*) have a long history of use in alternative medicine and are considered as a magic treatment for strengthening the heart, eliminating toxins, reducing edema, aphrodisiac, and relieving pain.[92] Toads have a wide geographic distribution across Southeast Asia. The species thrives in anthropogenically modified environments, including cities and agricultural lands, and is frequently encountered by people. Toad mouthing, entire toad ingestion, toad licking, toad soup, and toad egg eating have been reported to cause severe and frequently fatal toxicity. Venom is present in the entire body of the toad, with its highest concentration in the skin, followed by the parotid gland, viscera, and eggs, in descending order. The bufadienolides have a similar structure and biochemical activity to digitalis and a diversity of alkaloid toxins, amines, bufogenins, proteins, mucins, and peptides.

Once bufadienolides are ingested, around 50% are passively absorbed in the intestine and then extensively bound to the subunit of the Na^+/K^+-ATPase in myocardium, skeletal muscle, and kidney, as well as red blood cells. Myocardium has the highest concentration of Na^+/K^+-ATPase (505–760 pmol/g), thus accounting for the conspicuous cardiotoxicity of bufadienolides.[93] Clinical features of toad poisoning in humans resemble digoxin toxicity and include profuse salivation, perioral numbness, nausea, vomiting, dysrhythmias (e.g., ventricular fibrillation, bradycardia, and varying degrees of heart block), hypotension, hypertension, hyperkalemia, and seizures.[94] Elevation of the serum digoxin level is universally presented in patients with toad venom poisoning because of structural cross-reactivity on assays.

4.8 CONCLUSION

Acute envenomation causes functional changes in cardiac anatomy, physiology, and electrophysiology with or without myocardial ischemia. The venom targets multiple pathways including the ionic channels in cardiac cells and produces an alteration in the action potential (e.g., depolarization, repolarization, alterations in its duration) or causes significant changes in the coronary vasculature. In addition to the conventional treatment for cardiac complications with such envenomation, primary treatment should focus on eliminating the particular toxin (refer to the standard guidelines for the management of particular toxins).

Key Learning Points

- Snake and arthropod envenomations are neglected diseases of global importance. Such envenomation is prevalent worldwide and is more common in the developing world.
- Envenomation from snakes and arthropods causes cardiac toxicities such as hypotension, hypertension, conduction disturbances, varying degrees of tachybrady arrhythmias, myocardial ischemia/infarction, myocardial injury, myocarditis, and pericardial effusion.
- During the management of such envenomation, the primary focus should be on eliminating the particular toxin in addition to conventional treatment for managing cardiac complications.
- Refer to standard treatment guidelines for managing a particular toxin.

Acknowledgment We would like to thank Dr. K. Arthanari for his logistic support and Dr. H. S. Bawaskar and Dr. S. Senthilkumaran for giving us permission to include their figures.

REFERENCES

1. David S, Matathia S, Christopher S. Mortality predictors of snake bite envenomation in southern India: a ten-year retrospective audit of 533 patients. *J Med Toxicol* 2012;**8**(2):118–23.
2. Meenakshisundaram R, Sweni S, Thirumalaikolundusubramanian P. Hypothesis of snake and insect venoms against human immunodeficiency virus: a review. *AIDS Res Ther* 2009;**6**:25.
3. Nayak KC, Jain AK, Sharda DP, et al. Profile of cardiac complications of snake bite. *Indian Heart J* 1990;**42**(3):185–8.
4. Karaye KM, Mijinyawa MS, Yakasai AM, et al. Cardiac and hemodynamic features following snakebite in Nigeria. *Int J Cardiol* 2012;**156**(3):326–8.
5. Camargo AC, Ianzer D, Guerreiro JR, et al. Bradykinin-potentiating peptides: beyond captopril. *Toxicon* 2012;**59**(4):516–23.
6. Bdolah A. Hypertensive and hypotensive snake venom components. In: Kini RM, Clemetson KJ, Markland FS, et al, editors. *Toxins and hemostasis: from bench to bedside*. New York: Springer; 2010. p. 655–72.
7. de Weille JR, Schweitz H, Maes P, et al. Calciseptine, a peptide isolated from black mamba venom, is a specific blocker of the L-type calcium channel. *Proc Natl Acad Sci USA* 1991;**88**(6):2437–40.
8. Gilquin B, Lecoq A, Desne F, et al. Conformational and functional variability supported by the BPTI fold: solution structure of the Ca^{2+} channel blocker calcicludine. *Proteins* 1999;**34**(4):520–32.
9. Gasmi A, Bourcier C, Aloui Z, et al. Complete structure of an increasing capillary permeability protein (ICPP) purified from *Vipera lebetina* venom. ICPP is angiogenic via vascular endothelial growth factor receptor signalling. *J Biol Chem* 2002;**277**(33):29992–8.

10. Gasmi A, Abidi F, Srairi N, et al. Purification and characterization of a growth factor-like which increases capillary permeability from *Vipera lebetina* venom. *Biochem Biophys Res Commun* 2000;**268**(1):69–72.
11. Meenakshisundaram R, Senthilkumaran S, Grootveld M, et al. Severe hypertension in elapid envenomation. *J Cardiovasc Dis Res* 2013;**1**:65–7.
12. Senthilkumaran S, Khamis A, Manikam R, et al. Snakebite and severe hypertension: Looking for the Holy Grail. *Indian J Crit Care Med* 2014;**18**(3):186.
13. Agarwal R, Aggarwal AN, Gupta D. Elapid snakebite as a cause of severe hypertension. *J Emerg Med* 2006;**30**(3):319–20.
14. Malina T, Krecsak L, Warrell DA. Neurotoxicity and hypertension following European adder (*Vipera berus berus*) bites in Hungary: case report and review. *QJM* 2008;**101**(10): 801–6.
15. Hung DZ, Wu ML, Deng JF, et al. Multiple thrombotic occlusions of vessels after Russell's viper envenoming. *Pharmacol Toxicol* 2002;**91**(3):106–10.
16. Silva A, Pilapitiya S, Siribaddana S. Acute myocardial infarction following a possible direct intravenous bite of Russell's viper (*Daboia russelli*). *BMC Res Notes* 2012;**5**:500.
17. Maheshwari M, Mittal SR. Acute myocardial infarction complicating snake bite. *J Assoc Physicians India* 2004;**52**:63–4.
18. Sokolovsky M. Structure-function relationships of endothelins, sarafotoxins, and their receptor subtypes. *J Neurochem* 1992;**59**(3):809–21.
19. Sajevic T, Leonardi A, Krizaj I. Haemostatically active proteins in snake venoms. *Toxicon* 2011;**57**(5):627–45.
20. Clemetson KJ, Lu Q, Clemetson JM. Snake venom proteins affecting platelets and their applications to anti-thrombotic research. *Curr Pharm Des* 2007;**13**(28):2887–92.
21. Boviatsis EJ, Kouyialis AT, Papatheodorou G, et al. Multiple hemorrhagic brain infarcts after viper envenomation. *Am J Trop Med Hyg* 2003;**68**(2):253–7.
22. Frangides C, Kouni S, Niarchos C, et al. Hypersersensitivity and Kounis syndrome due to a viper bite. *Eur J Intern Med* 2006;**17**(3):215–6.
23. Satish R, Kanchan R, Yashawant R, et al. Acute MI in a stented patient following snake bite-possibility of stent thrombosis—a case report. *Indian Heart J* 2013;**65**(3):327–30.
24. Murase K, Takagi K. Takotsubo cardiomyopathy in a snake bite victim: a case report. *Pan Afr Med J* 2012;**13**:51.
25. Yamazaki Y, Nakano Y, Imamura T, et al. Augmentation of vascular permeability of VEGF is enhanced by KDR-binding proteins. *Biochem Biophys Res Commun* 2007;**355**(3): 693–9.
26. Senthilkumaran S, Meenakshisundaram R, Thirumalaikolundusubramanian P. Cardiac tamponade in Russell viper (*Daboia russelii*) bite: case report with brief review. *J Emerg Med* 2012;**42**(3):288–90.
27. Lakhotia M, Kothari D, Choudhary DR, et al. A case of saw scale viper snakebite presenting as pleuro-pericardial haemorrhage. *JIACM* 2002;**3**(4):392–4.
28. Hoffman A, Levi O, Orgad U, et al. Myocarditis following envenoming with *Vipera palaestinae* in two horses. *Toxicon* 1993;**31**(12):1623–8.
29. Singh S, Dass A, Jain S, et al. Fatal non-bacterial thrombotic endocarditis following viperine bite. *Intern Med* 1998;**37**(3):342–4.
30. Agarwal R, Singh AP, Aggarwal AN. Pulmonary oedema complicating snake bite due to Bungarus caeruleus. *Singapore Med J* 2007;**48**(8):e227–30.
31. Wolff AA, Levi R. Histamine and cardiac arrhythmias. *Circ Res* 1986;**58**(1):1–16.

32. Gueron M, Ilia R, Margulis G. Arthropod poisons and the cardiovascular system. *Am J Emerg Med* 2000;**18**(6):708–14.
33. Chang LS, Huang HB, Lin SR. The multiplicity of cardiotoxins from *Naja naja atra* (Taiwan cobra) venom. *Toxicon* 2000;**38**(8):1065–76.
34. Slotta KH, Vick JA. Identification of the direct lytic factor from cobra venom as cardiotoxin. *Toxicon* 1969;**6**(3):167–73.
35. Pahlajani DB, Iya V, Tahiliani R, et al. Sinus node dysfunction following cobra bite. *Indian Heart J* 1987;**39**(1):48–9.
36. Ismail AK, Weinstein SA, Auliya M, et al. Ventricular bigeminy following a cobra envenomation. *Clin Toxicol* 2012;**50**(6):518–21.
37. Francis J. Ventricular tachycardia following snakebite envenoming. *EMJ* 2009;**26**(2):151.
38. Gupta OP, Mewar SH, Kalantri SP, et al. Reversible atrial fibrillation following snake bite. *J Assoc Physicians India* 1987;**35**(7):535–6.
39. Lalloo DG, Trevett AJ, Nwokolo N, et al. Electrocardiographic abnormalities in patients bitten by taipans (*Oxyuranus scutellatus canni*) and other elapid snakes in Papua New Guinea. *Trans R Soc Trop Med Hyg* 1997;**91**(1):53–6.
40. Senthilkumaran S, Meenakshisundaram R, Thirumalaikolundusubramanian P, et al. Cardiac toxicity following cobra envenomation. *Clin Toxicol* 2012;**50**(9):862.
41. World Health Organization. *Guidelines for the treatment of snake bites*. Geneva, Switzerland: WHO; 2009. pp. 1–162.
42. Tamilnadu State Govt I. *Handbook on treatement guidelines for snake bite and scorpion stings*. Chennai, India: Tamil Nadu Health System Project, Health and Family Welfare Department; 2008. pp. 1–129.
43. Rajarajeswari G, Sivaprakasam S, Viswanathan J. Morbidity and mortality pattern in scorpion stings (a review of 68 cases). *J Indian Med Assoc* 1979;**73**(7-8):123–6.
44. Mahadevan S. Scorpion sting. *Indian Pediatr* 2000;**37**(5):504–14.
45. Bawaskar HS, Bawaskar PH. Indian red scorpion envenoming. *Indian J Pediatr* 1998;**65**(3):383–91.
46. Bawaskar HS, Bawaskar PH. Scorpion envenoming and the cardiovascular system. *Trop Doct* 1997;**27**(1):6–9.
47. Karnad DR. Haemodynamic patterns in patients with scorpion envenomation. *Heart* 1998;**79**(5):485–9.
48. Abroug F, Boujdaria R, Belghith M, et al. Cardiac dysfunction and pulmonary edema following scorpion envenomation. *Chest* 1991;**100**(4):1057–9.
49. Meves H, Simard JM, Watt DD. Interactions of scorpion toxins with the sodium channel. *Ann NY Acad Sci* 1986;**479**:113–32.
50. Nouira S, Abroug F, Haguiga H, et al. Right ventricular dysfunction following severe scorpion envenomation. *Chest* 1995;**108**(3):682–7.
51. Sundararaman T, Olithselvan M, Sethuraman KR, et al. Scorpion envenomation as a risk factor for development of dilated cardiomyopathy. *J Assoc Physicians India* 1999;**47**(11):1047–50.
52. Wheatley 3rd GH, Wait MA, Jessen ME. Infective endocarditis associated with a scorpion sting. *Ann Thorac Surg* 2005;**80**(4):1489–90.
53. Prakash V, Krishnamurthy S, Mahadevan S, et al. Infective endocarditis and septic pulmonary embolism following scorpion sting envenoming in an 11-year-old child. *Paediatr Int Child Health* 2014;**34**(2):145–7.
54. Bawaskar HS, Bawaskar PH. Management of scorpion sting. *Heart* 1999;**82**(2):253–4.

55. Murthy KR, Billimoria FR, Khopkar M, et al. Acute hyperglycaemia and hyperkalaemia in acute myocarditis produced by scorpion (*Buthus tamulus*) venom injection in dogs. *Indian Heart J* 1986;**38**(1):71–4.
56. Alan S, Ulgen MS, Soker M, et al. Electrocardiologic and echocardiographic features of patients exposed to scorpion bite. *Angiology* 2004;**55**(1):79–84.
57. Deman AL, Lerecouvreux M, Miandrisoa MR, et al. *Envenimation scorpionique avec atteinte cardiaque: a propos d'une observation de flutter atrial*. [Cardiac damage due to scorpion envenomation: case involving atrial flutter.] *Med Trop (Mars)* 2009;**69**(3): 309–10.
58. Sofer S, Zucker N, Bilenko N, et al. The importance of early bedside echocardiography in children with scorpion envenomation. *Toxicon* 2013;**68**:1–8.
59. Brand A, Keren A, Kerem E, et al. Myocardial damage after a scorpion sting: long-term echocardiographic follow-up. *Pediatr Cardiol* 1988;**9**(1):59–61.
60. Maheshwari M, Tanwar CP. Scorpion bite induced myocardial damage and pulmonary edema. *Heart Views* 2012;**13**(1):16–8.
61. Sagarad SV, Kerure SB, Thakur B, et al. Echocardiography guided therapy for myocarditis after scorpion sting envenomation. *JCDR* 2013;**7**(12):2836–8.
62. Lin GM, Han CL. Letter regarding rapidly improving, severe acute myocarditis after a scorpion bite. *Am J Emerg Med* 2011;**29**(3):346–7.
63. Bahloul M, Chaari A, Dammak H, et al. Pulmonary edema induced by scorpion venom: evidence of cardiogenic nature. *Int J Cardiol* 2012;**158**(2):292–3.
64. Goncalves E, Maia BT, Martelli-Junior H. Scorpion sting-induced unilateral pulmonary edema. *Rev Soc Bras Med Trop* 2012;**45**(3):419.
65. Patra S, Satish K, Singla V, et al. Acute myocardial infarction following scorpion sting in a case with obstructive coronary artery disease. *BMJ Case Rep* 2013;pii: bcr2013009865. Available from: http://dx.doi.org/10.1136/bcr-2013-009865.
66. Dias LS, Vivek G, Manthappa M, et al. Delayed presentation of scorpion sting with cardiogenic shock. *BMJ Case Rep* 2012;pii: bcr2012006827. Available from: http://dx.doi.org/10.1136/bcr-2012-006827.
67. Abdel-Haleem AH, Meki AR, Noaman HA, et al. Serum levels of IL-6 and its soluble receptor, TNF-alpha and chemokine RANTES in scorpion envenomed children: their relation to scorpion envenomation outcome. *Toxicon* 2006;**47**(4):437–44.
68. Fisher BA, Antonios TF. Atrial flutter following a wasp sting. *J Postgrad Med* 2003;**49**(3): 254–5.
69. Law DA, Beto RJ, Dulaney J, et al. Atrial flutter and fibrillation following bee stings. *Am J Cardiol* 1997;**80**(9):1255.
70. Habermann E. Bee and wasp venoms. *Science* 1972;**177**(4046):314–22.
71. Wojtowicz M, Biernat C. *Przypadek migotania komor w wyniku uzadlenia przez pszezole*. [Case of ventricular fibrillation following a bee sting.] *Pol Tyg Lek (Wars)* 1976;**31**(52): 2229–30.
72. Cerne K, Kristan KC, Budihna MV, et al. Mechanisms of changes in coronary arterial tone induced by bee venom toxins. *Toxicon* 2010;**56**(3):305–12.
73. Taggar JS, Watson T, Musarrat K, et al. Kounis syndrome presenting as ST-segment elevation myocardial infarction following a hymenoptera (bee) sting. *Int J Cardiol* 2009;**136**(2): e29–30.
74. Senthilkumaran S, David SS, Menezes RG, Thirumalaikolundusubramanian P. Acute myocardial infarction triggered by bee sting: an alternative view. *EMA* 2013;**25**(6):615.

75. Puttegowda B, Chikkabasavaiah N, Basavappa R, et al. Acute myocardial infarction following honeybee sting. *BMJ Case Rep* 2014;pii: bcr2014203832. Available from: http://dx.doi.org/10.1136/bcr-2014-203832.
76. Nisahan B, Selvaratnam G, Kumanan T. Myocardial injury following multiple bee stings. *Trop Doct* 2014; available from: http://dx.doi.org/10.1177/0049475514525606.
77. Greif M, Pohl T, Oversohl N, et al. Acute stent thrombosis in a sirolimus eluting stent after wasp sting causing acute myocardial infarction: a case report. *Cases J* 2009;**2**:7800.
78. Kahraman F, Ozaydin M, Varol E, et al. Acute stent thrombosis and/or atrial fibrillation occurring after bee sting. *Blood Coagul Fibrinolysis* 2014;**25**(4):387–8.
79. Bongo AS, Fornaro G, Sansa M, et al. *Infarto miocardico acuto da puntura di vespa in assenza di reazione anafilattica.* [Acute myocardial infarction after wasp sting without anaphylactic reaction.] *Ital Heart J Suppl* 2005;**6**(3):178–82.
80. Min JH, Kang MH. Kounis syndrome presenting as very late stent thrombosis in an everolimus-eluting stent following wasp stings. *Korean Circ J* 2013;**43**(8):561–4.
81. Rekik S, Andrieu S, Aboukhoudir F, et al. ST elevation myocardial infarction with no structural lesions after a wasp sting. *J Emerg Med* 2012;**42**(4):e73–5.
82. Gomes A, Datta A, Sarangi B, et al. Pharmacodynamics of venom of the centipede *Scolopendra subspinipes dehaani Brandt. Indian J Exp Biol* 1982;**20**(8):615–8.
83. Ozsarac M, Karcioglu O, Ayrik C, et al. Acute coronary ischemia following centipede envenomation: case report and review of the literature. *Wilderness Environ Med* 2004;**15**(2):109–12.
84. Yildiz A, Biceroglu S, Yakut N, et al. Acute myocardial infarction in a young man caused by centipede sting. *EMJ* 2006;**23**(4):e30.
85. Senthilkumaran S, Meenakshisundaram R, Michaels AD, et al. Acute ST-segment elevation myocardial infarction from a centipede bite. *J Cardiovasc Dis Res* 2011;**2**(4):244–6.
86. Weitzman S, Margulis G, Lehmann E. Uncommon cardiovascular manifestations and high catecholamine levels die to "black widow" bite. *Am Heart J* 1977;**93**(1):89–90.
87. Erdur B, Turkcuer I, Bukiran A, et al. Uncommon cardiovascular manifestations after a *Latrodectus* bite. *Am J Emerg Med* 2007;**25**(2):232–5.
88. Sari I, Zengin S, Davutoglu V, et al. Myocarditis after black widow spider envenomation. *Am J Emerg Med* 2008;**26**(5):630.e1–3.
89. Isbister GK, Fan HW. Spider bite. *Lancet* 2011;**378**(9808):2039–47.
90. Torda TA, Loong E, Greaves I. Severe lung oedema and fatal consumption coagulopathy after funnel-web bite. *Med J Aust* 1980;**2**(8):442–4.
91. Malajati H, Bensaad H, Abidi K, et al. The contribution of MRI in the diagnosis of acute myocarditis following a spider bite: a case report. *Diagn Interv Imaging* 2012;**93**(11):887–9.
92. Kuo HY, Hsu CW, Chen JH, et al. Life-threatening episode after ingestion of toad eggs: a case report with literature review. *EMJ* 2007;**24**(3):215–6.
93. Cheng CJ, Lin CS, Chang LW, et al. Perplexing hyperkalaemia. *Nephrol Dial Transplant* 2006;**21**(11):3320–3.
94. Keomany S, Mayxay M, Souvannasing P, et al. Toad poisoning in Laos. *Am J Trop Med Hyg* 2007;**77**(5):850–3.

Chapter 5

Plant Toxins and the Heart

Subramanian Senthilkumaran,[1] Ramachandran Meenakshisundaram,[2] and Ponniah Thirumalaikolundusubramanian[3]
[1]*Sri Gokulam Hospital and Research Centre, Salem, India*
[2]*Epsom and St. Helier University Hospitals NHS Trust, Epsom, United Kingdom*
[3]*Chennai Medical College Hospital and Research Centre, Trichy, India*

5.1 INTRODUCTION

Nature is beautiful; however, hazardous and risky materials are lurking in the green foliage, including a variety of toxic plants. Plants and their byproducts have been used by humans to combat diseases since the existence of human civilization. They are also used by people living in urban, rural, and remote areas for a varity of purposes including religious ceremonies, black magic, health care, food, clothing, shelter, agriculture, agrochemicals, pharmaceuticals, narcotics, and more. The art of healing with plant medicine is still widely practiced across the globe. Much of this lore is indigenous and has been passed by word of mouth from one generation to the next by priests, witchdoctors, and medicine men; yet there are strong indications of Indian, Hebrew, and Egyptian, as well as Greek and other Arabic, influences.[1]

Sales and transactions of plant medicine run into billions of U.S. dollars.[2] In India, 960 plant species are used by the country's herbal industry, of which 178 are of high volume exceeding 100 metric tons per year.[3] In contrast, in parts of the developing world, plant poisoning is favored by many—mainly by women—as a method of deliberate self-harm.[4] Easy availability in their backyard, free access to highly toxic plants, and limited access to medical facilities contribute to high mortality and result in a major clinical problem. Plant toxicology began in view of significant poisoning of livestock. In fact, animal studies have helped researchers understand the effects of plant toxins.

Establishing a diagnosis of poisoning by a plant in patients can be a very demanding task. It relies mainly on evidence of ingestion of cardiotoxic plant materials, access, and/or suspicion generated by cardiac dysrhythmias.[5] Getting a history may be difficult in clinical situations because the patient's mental capacity is usually impaired. Indeed, laboratory analyses for cardiac glycosides are available; however, various digoxin immunoassays developed

The Heart and Toxins. DOI: http://dx.doi.org/10.1016/B978-0-12-416595-3.00005-0

for the detection of digoxin potentially cross-react with other cardiac glycosides, such as oleandrin, which trigger false-positive results and pose a therapeutic dilemma for the treating physician.

This chapter describes and discusses some of the plants that are not routinely used for medical purposes and have the potential to be toxic to the cardiovascular system.

5.2 CARDIOACTIVE STEROIDS

Cardioactive steroids (CAS) are a class of compounds with a steroid nucleus and have a specific inotropic, chronotropic, and dromotropic effects that are found in plants, animals, and insects.[6] Cardioactive steroids are classified as a cardenolide (with a 5-membered lactone ring) or a bufadienolide (with a 6-membered lactone ring). The term *cardiac glycoside* refers to a subgroup of CAS that also contains sugar residues. Most cardenolides are derived from plants[7] and are found in 11 plant families: Apocynaceae, Asclepiadaceae, Celastraceae, Cruciferae, Liliaceae, Leguminosae, Moraceae, Ranunculaceae, Scrophulariaceae, Sterculiaceae, and Tiliaceae. Cardenolides from these families include the following, with photos of them:

- Foxglove (*Digitalis purpurea*; Figure 5.1)
- Oleander (*Nerium oleander*; Figure 5.2)
- Odollam tree (*Cerbera odollam*; Figure 5.3)
- Yew (*Taxus baccata*; Figure 5.4)

FIGURE 5.1 **Photo of foxglove (*Digitalis purpurea*). Source:** *Courtesy of Ms. Elizabeth Dauncey, Royal Botanic Gardens, Kew, United Kingdom.*

FIGURE 5.2 **Photo of pink oleander (*Nerium odorum*). Source:** *Courtesy of Dr. K. Arthanari, Sri Gokulam Hospitals and Research Centre, Salem, India.*

FIGURE 5.3 Photo of odollam tree (*Cerbera odollam*). Source: *Courtesy of Mr. A. Subramanian, Salem, India.*

FIGURE 5.4 Photo of yew (*Taxus baccata*). Source: *Courtesy of Ms. Elizabeth Dauncey, Royal Botanic Gardens, Kew, United Kingdom.*

The adverse effects of cardiac glycosides have been known for centuries; however, Liu et al.[8] described the adverse effects of ouabain on mitochondrial energetics and CGP-37157, which is an inhibitor of the mitochondrial Na^{+}/Ca^{2+} exchanger, abrogates the negative effect on energetics and attenuates ouabain-induced arrhythmias. Human endogenous digoxin-like immune reactive substance (EDLIS) is likely a combination of the cardenolide ouabain and bufadienolides such as marinobufagenin.[9] Many centuries before William Withering studied the pharmacologic properties of purple foxglove, *Digitalis purpurea* was well known to medieval herbalists who used it as the principal ingredient in a herbal preparation to treat congestive heart failure.[10]

The pharmacokinetics and pharmacodynamics of regulated products, such as digoxin and digitoxin, are well known, and the clinical features of overdose are well described (discussed in Chapters 6 and 8). However, plants collected by foraging or used in herbal preparations present a true challenge for clinicians who must identify and treat CAS poisoning caused by nonpharmaceutical sources. These practices present the possibility of "mistaken identity," as when *Digitalis lanata* (containing digoxin) was mistaken for a plantain plant and incorporated into a herbal product.[11] It is important for emergency care physicians to recognize that these CAS may be incorporated into several readily available legal and illegally prepared products. Severe toxicity from CAS has been documented in products sold for increased sexual performance and male enhancement, herbal colon cleansing, and as rodenticides.[12] In Southeast Asia, especially in India and Sri Lanka, CAS poisoning caused by ingestion of yellow oleander (*Thevetia peruviana*) and their seeds are commonly used for suicide.[13]

Digitalis (a prototypical cardenolide) and oleander-derived cardenolides share similar physiological and structural characteristics.[14] These plant cardenolides have carbohydrate or sugar moiety. The sugar moieties of cardenolides are specific, and this portion of the compound defines the name of the cardenolide (e.g., digitoxose, digitoxin, oleandrose, oleandrin, neriantin, rosagenin).[15] The entire oleander plant is toxic either fresh or dried form; however, the concentration of oleandrin varies widely between various strains, species, and season. During the flowering stage for red oleander, the concentrations of total cardiac glycosides are high. The concentration profile of cardenolide in oleander plant in descending order is seed, stems, roots, fruit, and leaves.

Indeed, Oji et al. identified many cardenolides in the bark, leaves, kernel of the seed, or sap of yellow oleander that are highly toxic cardiac glycosides.[16] Their poisonous effect and toxicity following oral, transcutaneous,[17] and inhalation[18] routes have been described in the literature. Multiple lay print media have reported that poisoning may occur by consuming food cooked either due to accidental stirring of the food with the oleander stem or smoke from oleander extracts settling on cooked foods.

CAS inhibits the sodium–potassium adenosine triphosphatase enzyme on the cardiac myocyte membrane, which in turn increases intracellular sodium and decreases potassium, leading to an accumulation of intracellular calcium via the sodium–calcium exchange system. This raises the intracellular calcium within the myocyte and, in response to this, the sarcoplasmic reticulum releases additional calcium intracellularly, resulting in depolarization of the cell.[19] Positive inotropy is achieved by increased available calcium to bind troponin, actin, and myosin. CAS can also affect the parasympathetic nervous system through acetylcholine release from the vagus nerve, which reduces sinoatrial firing rate and reduces conduction velocity of electrical impulses through the atrioventricular (AV) node.[20]

The toxicokinetics depend on numerous components such as dyselectrolytemia, drug interactions, renal/cardiac dysfunction, other comorbid illnesses, and disruption of the enteric bacterium Eggerthella lenta (*Eubacterium lentum*). Hypokalemia, in particular, may result in increased cardiac automaticity because it inhibits skeletal Na^+/K^+-ATPase and reduces its functional capability. Hypomagnesemia and hypercalcemia may also aggravate the toxicity.[21] CAS toxicity can cause cardiac, neurological, gastrointestinal, and visual manifestations. Nausea, vomiting, abdominal pain, and confusion are the most common extracardiac manifestations.

Electrocardiography (ECG) findings vary widely and may result in almost any type of cardiac dysrhythmia associated with increased automaticity and decreased AV conduction, which includes any form of tachyarrhythmia and bradyarrhythmia such as sinus bradycardia, atrial tachycardia/fibrillation/flutter with slow ventricular response, variable degrees of AV block, junctional tachycardia, and ventricular tachycardia or fibrillation.[22] Biventricular tachycardia is considered to be pathognomonic, but it is rarely seen.[23] The most dreadful complication of oleander poisoning is complete heart block. Cardiogenic shock, ventricular fibrillation, and death may occur following severe oleander toxicity. Conduction delays may persist up to 3 to 6 days and these cardiac effects resemble typical digitalis toxicity.[24]

The presence of a positive digoxin immunoassay is in accord with the ingestion of oleander, because oleandrin cross-reacts due to structural similarities.[25] However, the identity of these cardiac glycosides and the extent of these cross-reactions are not well defined. The estimated concentrations vary depending on the assay. Acute CAS poisoning may present with elevated serum potassium levels due to Na^+/K^+-ATPase pump inhibition, and hyperkalemia has been shown as a poor prognostic marker.[26] If therapy for hyperkalemia is required, sodium bicarbonate, insulin, and glucose can be used safely. Concurrent calcium therapy for the treatment of hyperkalemia is to be avoided when CAS is suspected or confirmed, as intracellular calcium is already elevated, and administration of calcium salts is believed to cause tetanic myocardial contractions or "stone heart," reported to be fatal.[27]

After they are absorbed into the systemic circulation, cardiac glycosides are secreted into the gut lumen by the action of P-glycoprotein.[28] The reason for starting multidose-activated charcoal in these patients is to prevent enterohepatic recycling.[29] In the gut, activated charcoal binds the secreted glycoside and thereby enhances glycoside excretion and reduces the half-life of the cardenolides. Symptomatic bradycardia should be treated initially with atropine. Patients who do not respond to supportive care should receive digoxin-specific Fab fragments, which has revolutionized the treatment of CAS toxicity.[30] Interestingly, fructose-1, 6-diphosphate (FDP) is a phosphorylated sugar that has been proven to be effective in treating life-threatening arrhythmias and reversal of hyperkalemia due to oleander poisoning. It has also been shown to stimulate Na^+/K^+-ATPase activity and inhibit potassium efflux from myocardial cells.[31]

Before the advent of antidigoxin Fab antitoxin, temporary transvenous pacing was the mainstay of treatment for significant digitalis-induced dysrhythmias. French retrospective studies compared the outcomes between antitoxin and pacing.[32] They observed increased complications with temporary pacing than with immunotherapy. In the case of CAS toxicity, transvenous pacing should be avoided, as the myocardium is hyperexcitable, and an alternative transcutaneous pacing may be attempted with low energy.[33] Transthoracic defibrillation is relatively contraindicated in CAS toxicity as it has been associated with the induction of potentially lethal dysrhythmias similar to digoxin-toxic rhythms. This effect seems to be related to the severity of toxicity and the amount of current used. However, direct-current (DC) cardioversion (with minimal joules) is indicated for those with hemodynamic instability, failed chemical cardioversion, and near-fatal ventricular dysrhythmias.[34]

5.3 *CLEISTANTHUS COLLINUS*

Cleistanthus collinus is a toxic deciduous shrub belonging to the family Euphorbiaceae (Figure 5.5). It is one of the suicide poisons used by young women in rural parts of southern India. It is popularly known as oduvanthalai (Tamil), nilapala (Malayalam), vadisaku (Telugu), kadagargari (Kannada), and garari (Hindi). All parts of the plant are potentially toxic, but it is the leaves that are commonly used for deliberate self-harm.[35] Patients consume it by chewing fresh leaves, decoction prepared by boiling the leaves in water, or as a freshly ground leaf paste.[36] Boiled extract with jaggery consumption seems to be associated with increased mortality.[37] The active toxic principles isolated from the various parts of the plant are arylnaphthalene lignan lactones, ellagic acid, diphyllin, and glycosides such as cleistanthin A and B and collinusin. However, the toxicity of the plant has been attributed primarily to cleistanthin A and B.

The main clinical features of poisoning are hypokalemia, acidosis, cardiac rhythm abnormalities, hypotension, and respiratory failure. Both cleistanthin A and cleistanthin B showed a dose-dependent hypotensive effect; cleistanthin A is more potent that B. Hypotension caused by this poison is ascribed to the $\alpha 1$ adrenergic receptor antagonist effect on the peripheral vascular system and can further progress to vasodilatory shock.[38] The involvement of the cardiogenic component in shock is ambiguous. Cardiac rhythm abnormalities are seen more commonly because the principal toxin is glycosides. The most common ECG abnormalities are sinus tachycardia, sinus bradycardia, flat P waves, prolonged corrected QT intervals, ST segment depression, and inverted T waves.[37]

From clinical practice it has been noted that tachyarrhythmias and premature ventricular ectopics are the most common dysrhythmias seen initially in patients.[37] Appearances of ST- and T-wave changes indicate a poor prognosis. Malignant cardiac rhythms, such as ventricular fibrillation and asystole, have also been reported.[39] Elevated cardiac enzymes have been observed, suggesting a possibility of direct cardiac toxicity; however, the extent to which the dominant metabolic derangements affect cardiac function is still imprecise. Enzyme-linked immunosorbent assays for cleistanthin A and B have been developed and the kits can detect even low levels of cleistanthin A and B.[40] Death occurs even with patients who have prophylactic temporary transvenous pacemakers.

FIGURE 5.5 **Photo of *Cleistanthus collinus* shrub.** **Source:** *Courtesy of Mr. A. Subramanian, Salem, India.*

The accurate role of cardiac pacing in oduvan poisoning is a matter of assumption because the clinical evidence for preventing deaths through prophylactic pacing is scarce. The aggressive preemptive correction of hypokalemia, hypomagnesemia, and acidosis along with multidose-activated charcoal reduces mortality.[41] N-acetylcysteine has been suggested as a possible antidote for *C. collinus* toxicity.[42] The terminal events in these patients are acute respiratory distress syndrome (ARDS), fatal ventricular dysrhythmia, asystole, and subsequent cardiac arrest.[43]

5.4 ACONITE

Aconite is a crude extract of dried leaves and roots from various species of *Aconitum* (Figure 5.6). The genus *Aconitum* contains more than 350 species of flowering plants belonging to the family Ranunculaceae. It can be found throughout the world, and it has long been known as a poison.[44] Several species of *Aconitum* have been used as arrow poisons, for suicide and homicide because of its high toxicity. Common causes of the toxicity of aconite are contamination, adulteration, and inadequately processed aconite roots; large doses; tincture preparations; and when it is mistakenly consumed instead of edible wild plants. The roots and seeds are freely sold in native medicine stores for treating musculoskeletal pain; however, the margin of safety is low, so intoxication is not rare when aconite is used in native medicine (see Chapter 6).

Active components are aconite and related alkaloids such as benzoylaconines, aconines, and the aconitines (e.g., aconitine, napelline, hypaconitine, jesaconitine, and mesaconitine).[45] Alkaloids account for up to 1.5% of the dry weight of the plant. Aconitine is hydrolyzed to picraconitine, which hydrolyzes to aconine; several minor alkaloids have been isolated from the various species of it. These minor alkaloids include sinomontanitines, lappaconitine, and ranaconitine. In addition to the cardioactive alkaloids, *Aconitum* roots contain physiologically active catecholamine analogs.[46]

Aconitine and mesaconitine bind with high affinity to the open state of the voltage-sensitive sodium channels at site 2 of the α subunit, thereby causing a persistent activation of the sodium channels, which become refractory with excitation.[47] The electrophysiological mechanism of arrhythmia induction is

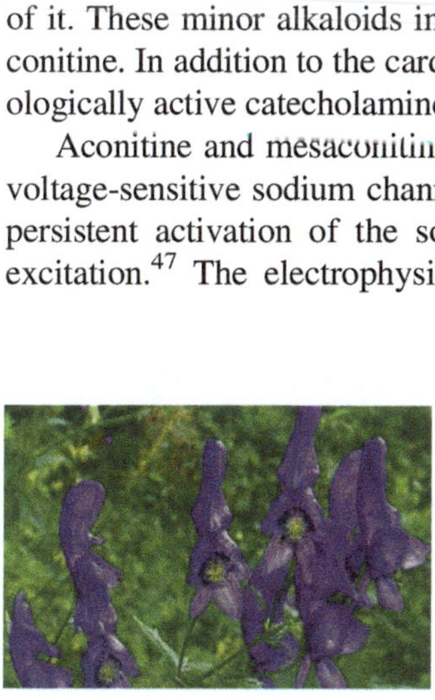

FIGURE 5.6 Photo of monkshood (aconite of the genus *Aconitum*). Source: *Courtesy of Mr. John David Peter, Cleveland, Ohio.*

triggered activity that is a result of delayed after-depolarization and early after-repolarization. The arrhythmogenic properties of aconitine are in part due to its cholinolytic (anticholinergic) effects that are mediated by the vagus nerve. Aconitine has a positive inotropic effect by prolonging sodium influx during the action potential.[48] It has hypotensive and bradycardic actions due to activation of the ventromedial nucleus of the hypothalamus.[49] The toxin acts on the voltage-sensitive sodium channels in the axons and blocks neuromuscular transmission by decreasing the evoked quantal release of acetylcholine.[50]

Cardiovascular toxicity includes hypotension, bradycardia, varying degrees of AV block, and tachyarrhythmias such as atrial fibrillation and ventricular tachycardia.[51] Other electrophysiological abnormalities include right bundle branch block and prolonged QT interval.[52] Biventricular tachycardia is rare and is typically associated with digitalis toxicity, but also has been observed in aconite poisoning.[53] Aconite poisoning sometimes results in death due to cardiac arrest caused by fatal arrhythmia such as ventricular fibrillation; however, no gross and histological cardiac abnormalities have been observed in the few autopsied cases.[54]

Management of aconite poisoning should be supportive. Inotropic therapy is required if hypotension persists and atropine should be used to treat bradycardia. Aconite-induced ventricular arrhythmias are often refractory to DC cardioversion and antiarrhythmic drugs.[55] Available clinical evidence suggests that amiodarone and flecainide are reasonable first-line treatments. Magnesium is sometimes effective for polymorphic ventricular tachycardia.[56] The early use of cardiopulmonary bypass is recommended in refractory ventricular arrhythmias and cardiogenic shock.[57] Since aconite is a life-threatening plant used in all cultures, every clinician should be aware of this poison.

5.5 KHAT

Khat (*Catha edulis*) is a beautiful, small deciduous flowering plant belonging to the Celastraceae family. Khat leaves have been chewed for centuries by people living in the eastern part of Africa and the Arabian Peninsula, where it is grown abundantly; nowadays it is also grown by expatriate communities living in Europe and North America.[58] The khat plant is known by a variety of names such as kat, qat, and chat, depending on the language. Fresh leaves from khat trees are chewed daily by more than 20 million people in Yemen, Kenya, Somalia, Ethiopia, Djibouti, Eritrea, Somalia, and Uganda. In these countries, khat is traded openly and used socially across a range of age and class groups, particularly in regions/societies where alcohol is not permitted.

The traditional use of khat may be considered largely functional: to assist with religious studies, arduous work demands, food shortages, and social cohesion; and to self-medicate for a range of ailments including depression.[59]

More than 40 compounds have been identified in khat extract, among which are the controlled substances cathinone (Schedule I) and cathine (Schedule IV), both sympathomimetic amines with an identical structure to amphetamine and adrenaline.[60] These release serotonin and dopamine in the central nervous system and noradrenaline from peripheral sympathetic neurons.[61]

Regular khat use is associated with elevation in arterial blood pressure and heart rate with corresponding proportionate levels of cathinone in the plasma, which peak 1.5 to 3.5 hours after chewing.[62] The increase in blood pressure after chewing khat is not blocked by the $\alpha 1$-adrenoceptor antagonist indoramine, confirming the lack of involvement of $\alpha 1$-adrenoceptors in the pressor response. There is probably a significant cardiac component to the increase in blood pressure after chewing khat through tachycardia, and this effect is counteracted by beta blockers. It is likely, therefore, that the residual pressor response is due to stimulation of trace amine-associated receptors by cathinone.[63]

Cardiovascular complications from khat chewing may therefore be similar to those of amphetamine.[64] Acute myocardial infarction (AMI) could be precipitated by the increased myocardial oxygen demand from cardiac stimulation followed by catecholamine-mediated platelet aggregation and peripheral and coronary vasoconstriction by cathinone.[65] Saha and colleagues noticed increasing numbers of patients presenting with AMI in the evening either during or after a khat-chewing session, in contrast to presentation in the early hours of the morning by nonchewers, suggesting a causal link between khat and heart attacks.[66]

A case-control study in Yemen comparing 100 patients with acute myocardial infarction to age- and sex-matched controls showed a 39-fold increased myocardial infarction risk in heavy khat chewers.[67] A significant reduction of bleeding time was observed in khat chewers who were taking aspirin (100 mg daily) on a long-term basis compared with nonkhat chewers taking the same dose of aspirin. This finding suggested that constituents of khat attenuate the antiplatelet aggregating properties of aspirin, and thereby neutralize the beneficial actions of aspirin.[68]

Cathinone shares similar properties of amphetamine and may have the same effects on the myocardium and also produces dilated cardiomyopathy.[69] Interestingly, an East African man who had chewed khat constantly for 2 to 3 days presented to the emergency department with AMI and during the follow-up echocardiography revealed severely impaired biventricular function consistent with ischemic cardiomyopathy.[70] Khat-related dilated cardiomyopathy warrants further studies with imaging modalities to confirm the extent of myopathy. No standard guidelines or consensus exist for the optimal management of khat-induced coronary ischemia, however; it all depends on the clinical presentation along with risk factors to decide the need for angiography along with conventional medical treatment.

5.6 YEW

The ancient Britons planted yews near their temples; it was considered a sacred tree and symbol of immortality. Yews (*Taxus* spp., Taxaceae) are evergreen ornamental shrubs often planted on sacred sites and are common throughout the Europe. The *Taxus* species includes *T. baccata* (English yew), *T. brevifolia* (Western yew), *T. cuspidata* (Japanese yew), *T. wallichiana* (Himalayan yew), and *T. sumatrana* (Sumatra yew). All parts of *Taxus* plants are toxic except the edible pulp or scarlet aril. More than 350 naturally occurring taxoids (taxane diterpenoids) were identified in the *Taxus* species, of which the toxic alkaloid (taxine A and B, isotaxine B, paclitaxel), taxane-derived substances (taxol A and B), and glycosides (taxicatine) appear to account for the toxicity.[71,72] It also contains nitriles, ephedrine, and irritant oils. Canadian yew (*Taxus canadensis)* differs from other yews because this species contains a variety of uncommon taxanes.[72] The taxine content of yew plants varies with season and species, with the highest taxine concentrations occurring during winter.

Minor ingestions typically result only in gastrointestinal effects, while larger ingestions may cause severe cardiac toxicity. Taxine B, believed to be the primary cardiotoxic alkaloid, directly antagonizes both calcium and sodium transport across cell membranes, causing a cardiotoxic increase in cytosolic calcium concentrations.[73] This toxin reduces the maximum rate of depolarization of the action potential.[74] Cardiovascular toxic manifestations include decreased myocardial contractility, hypotension, bradycardia, widened QRS complexes, variable degrees of AV conduction block, ventricular tachycardia, and ventricular fibrillation even in the presence of a normal serum electrolyte level.[75] Interestingly, a Brugada pattern has been noted on the ECG during toxicity and it resolved with recovery.[76]

There is no known antidote for this toxin; only supportive measures and monitoring are advised. Decontamination with activated charcoal may also be effective. On one occasion, digoxin-specific Fab antibody fragments were used because the taxine molecule has a structural similarity to digitalis. However, there is no consensus report on their usage, dose, and duration.[77] Miller et al. were able to successfully reverse severe cardiotoxicity and a wide-complex QRS interval with intravenous administration of hypertonic sodium bicarbonate.[78] but a similar effect was not reproduced in the swine model.[79] The ventricular tachyarrhythmias were resistant to DC cardioversion, but responded to amiodarone infusion.[80]

From a recent case report,[81] a patient who presented with ventricular tachycardia (VT) following yew berry toxicity failed DC cardioversion and amiodarone infusion; however, VT improved after sodium bicarbonate infusion. The authors concluded that the possible effects are taxine clearance following sodium bicarbonate infusion and/or the synergistic effect of amiodarone and sodium bicarbonate in controlling arrhythmias. Thus, sodium bicarbonate can be considered in such circumstances if there are no contraindications for their use.[81] The bradycardia was resistant to transvenous cardiac pacing; however, it can be managed by temporary external transthoracic pacing in the initial hours.[77]

5.7 *VERATRUM* ALKALOIDS

In the early 1950s, *Veratrum* alkaloids were tested as antihypertensive agents but they fell into disfavor because of their adverse effects.[82] The genus *Veratrum* belongs to the Melanthiaceae family and it contains up to 45 species of flowering plants. The most common species are *V. album*, *V. californicum*, *V. nigrum*, *V. fimbriatum*, and *V. viride*. It is generally known as hellebore, green false hellebore, white hellebore, corn cabbage, and indian poke. It is most commonly mistaken for *Gentiana lutea* (yellow gentian) and results in toxicity. The entire plant of the *Veratrum* species contains a mixture of more than 200 lipid-soluble steroidal alkaloids. The concentration varies significantly between species and plant parts; the roots and rhizomes are more toxic than the leaves. Also the genera *Schoenocaulon* and *Zygadenus* are sources of *Veratrum* alkaloids. The principal toxins are steroid alkaloids; some have a modified steroid template, whereas others differ in their esterified acid moieties.

The alkaloids occur as glycophenol aglycones or in the form of esters with various acids divided into seven groups. The first two most important groups are (1) alkaloids of the jervanine and veratranine type with a steroidian skeleton (e.g., veratridine, veratramine, veratrosine, jervine, pseudojervine, and isorubrijervine) and (2) alkaloids with the cevanine skeleton (e.g., cevadine, cevine, zygadenine, veracerine, germine, germerine, germidine, protoverine, and their esters).

These alkaloids act by increasing the permeability of the sodium channels of excitable cells, causing them to fire prematurely and then leaving them refracted, which results in hyperexcitability and depolarization.[83] The afferent vagal fibers of the coronary sinus and posterior wall of left ventricle appear to be the most sensitive to the depolarizing effects of *Veratrum* alkaloids. Cardiac parasympathetic stimulation results in a decrease in the peripheral α-adrenergic tone that causes a triad of responses (apnea, bradycardia, and hypotension) known as Bezold–Jarisch reflex.[84] The *Veratrum* alkaloids activate all known receptors in the carotid-aortic and cardiopulmonary areas.

The hypotensive effect of *Veratrum* alkaloids is independent of heart rate, and the use of atropine only partially reverses the hypotension associated with the use of *Veratrum* alkaloids.[85] The onset of symptoms occurs between 30 min and 4 hours after the ingestion of the *Veratrum* plant. The most common clinical findings are nausea, vomiting, abdominal pain, paresthesia, numbness, headache, visual disturbance, diaphoresis, bradycardia, hypotension, seizures, and coma.[86] The electrophysiological changes observed are bradycardia, bundle branch block, atrioventricular block, QT prolongation, junctional, and idioventricular escape rhythms.[87] Patients must be monitored closely for change in airway status. It is rationale to administer activated charcoal if the patient ingests large amounts of *Veratrum* alkaloids, but use of gastric lavage is dubious.[88] The vomiting produced by the plant appears to limit absorption of the toxic alkaloids. Atropine reverses the bradycardia; fluid replacement and vasopressors may be needed to treat the hypotension.[89]

5.8 BROOMS

The broom was used as a heraldic badge by Plantagenet kings of England as a royal emblem. The "broomscod" was the personal emblem of Charles VI of France. It is a perennial leguminous shrub native to western and central Europe and belongs to the family Fabaceae. The four most common species are Scotch broom (*Cytisus scoparius*), French broom (*Genista monspessulana*), Spanish broom (*Spartium junceum*), and Portuguese broom (*Cytisus striatus*). All parts of the plant contain quinolizidine alkaloids (sparteine and cytisine), which are the toxic principle. Both alkaloids bind to the nicotinic cholinergic receptors on the sympathetic and parasympathetic ganglion.

Initial transitory stimulation of nicotinic receptors is followed by a more persistent depression and paralysis as a result of desensitization.[90] Both the stimulatory and the inhibitory action affect several systems. Moreover, sparteine exerts quinidine-like effects on the heart.[91] It reduces the sensitivity and conductivity of the cardiac muscle. The clinical presentation of toxicity varies depending on time, dose, and route of exposure and also depends on individual health status.[92] Symptomatic and supportive measures remain the only possible options as there is no antidote available.

5.9 MISTLETOE

Mistletoe is a hemiparasite normally found on the branches of a variety of trees. It can live either as a parasite or as a perennial plant capable of photosynthesis, and it has white or translucent berries that can be quite sticky. There are two plants with the common name "mistletoe," and they are the American (*Phoradendron serotinum* or *Phoradendron flavescens*) and the European (*Viscum album*). These plants are common adornments and holiday symbols most commonly sold around Christmas time. All sections of the plants are potentially harmful.

The major constituents of mistletoe are gastric-irritating alkaloids, cardiac toxins (viscotoxins, phoratoxins), and lectins. Viscotoxins are toxic peptides that are physiologically similar to cobra venom, but they differ in the sequences of amino acid and acetylcholine agonist/antagonist properties.[93] Although *Phoradendron* and *Viscum* spp. both contain the common name mistletoe, the *Phoradendron* spp. has a relatively lower toxicity compared to *Viscum* spp.[94] It causes significant gastrointestinal irritation following ingestion. Vasoconstriction, bradycardia, and negative inotropic effects have been observed in an animal study.[95] The probable cardiac effects hypothetically exist based on the presence of cardiotoxins in mistletoe, but there is limited data about human responses to indicate that these cardiotoxins may cause clinically significant cardiac abnormalities following causal ingestions.[96–98] Emergency department (ED) management should be directed toward providing supportive care. Patients with dysrhythmias or an electrolyte imbalance should be hospitalized for observation and managed symptomatically.

5.10 MISCELLANEOUS

Acokanthera oppositifolia (bushman poison bush, boesmansgif) sap contains cardenolides and has been used by the San people for applying to the tips of their hunting arrows.[97] Rhododendron poisoning is much more common. Toxic manifestation related to plants belonging to the *Rhododendron* (family Ericaceae) are seen often in animals and occasionally in humans.

A cardiac syndrome of domestic ruminants found in South Africa, called gousiekte (or quick disease), is caused by ingestion of plants containing the polyamine pavettamine belonging to Rubiaceae or the coffee family.[99,100] Pavettamine may have an influence on gene transcription and chromatin remodeling. It alters methodical activity, produces changes in lysosomes, and degrades cytoskeleton and myofibril proteins, all of which lead to ultrastructural changes in myocardium and loss of myofilaments. As a result, myofibers become disintegrated and are accompanied by replacement fibrosis. Thus, pavettamine causes degradation of a number of cardiac proteins that are involved in cardiac contractility and cell signaling.[101] Not only are epidemiological and clinical data available but also molecular, experimental, pharmacological, and other information from animal studies. These have helped facilitate a better understanding of plant toxins and encouraged further collaborative and/or translational research.

There are other plants that are known to have cardiotoxic effects. These are the Bangkok calachuche (Apocynaceae), cape tulips (Iridaceae), *Camellia sinensis* (Theaceae), the *Chimonanthus* species (Calycanthaceae), the *Cinchona* species (Rubiaceae), *Chenopodium ambrosioides* (Amaranthaceae), the Dogbane species (Apocynaceae), the figwort family (Scrophulariaceae), *Harpephyllum caffrum* (Anacardiaceae), the lily species (Liliaceae), the milkweed species (Apocynaceae), the mother of millions (Crassuleaceae), *Persea americana* Mill. (Lauraceae), the plumeria species (Apocynaceae), *Pulsatilla koreana* (Ranunculaceae), the *Ornithogalum* species (Asparagaceae), and others. Plants contain various xenobiotics such as alkaloids; glycosides; phenols and phenyl propanoids; proteins, peptides, and lectins; and many other identified and unidentified chemicals. These act independently or in concert. In addition, there are secondary metabolites may add to plants' toxic effects. Dissimilar molecules from diverse pharmacognosy classes share many of the clinical effects. Although many other plants have cardiotoxic effects, a few are summarized in Table 5.1.

5.11 COMMUNITY EDUCATION

Although botanical and/or plant products are considered generally safe, they may be toxic at times and may also interact with prescribed drugs[111] or other ingested herbs or associated comorbid illnesses. This is extremely important when treating cardiac patients.[112] Many factors play a role in the toxic manifestations, as listed in Table 5.2.

5.11.1 Plant Toxins Present in Breast Milk

Plant toxins consumed are excreted through the milk secreted from the mammary glands.[113] Many organic compounds that have lipid partitioning get into mammary gland secretions. Thus, babies may be exposed to toxins during breastfeeding.

5.11.2 Identification of Plants

Whenever plant toxicity is suspected, the caregiver or person accompanying the patient may be asked to bring the entire plant. The treating physician can communicate with an expert botanist, medical toxicologist, or a poison center to get appropriate information about the plant, if needed. It is suggested that a digital image of it be transmitted to the poison center or sent by fax.[114,115] Sometimes a suspected plant can be identified with a field guide of flora. The Poisonous Plant database, which is managed by the U.S. Food and Drug Administration (FDA), can be used for identification and comparison purposes.[116]

The following are some of the websites with information about poisonous plants:

- The American Association of Poison Control Centers[118] (National Poison Data System), *www.aapcc.org/data-system/*
- Canadian Poisonous Plants Information System,[117] *www.cbif.gc.ca/pls/pp/poison?p_x=px*
- U.S. Food and Drug Administration,[116] *www.accessdata.fda.gov/scripts/plantox/*

5.12 CHALLENGES AND REGULATIONS

There is international diversity in terms of definitions and categorization of plants and herbs. Thus, there is a need to regulate products from them and the use of various plants or herbs, including their quality, safety, and scientific aspects.[3,119] Legislation on herbal medicine is in place in Europe,[120]

TABLE 5.1 Cardiotoxicity of Plants

Botanical Name	Common Name	Active Compounds	Uses	Cardiac Effects	Reference
Acokanthera oppositifolia, Acokanthera schimperi	Bushman	Acoremoside	Snakebite, headache, common cold	Cardiotoxicity; similar to digitalis	102
Asclepias curassavica	Red cotton bush	Resinoid, galitoxin	Snakebite remedy, contraceptive, emesis, laxative,	Cardio toxicity; similar to digitalis	103
Citrus aurantium	Bitter orange	Synephrine	Dietary supplement for weight loss	Similar to ephedrine	104
Convallaria majalis	Lily of the valley	Convallarin (concallamarin), a cardiac glycoside similar to foxglove plants	Arrhythmia, cardiac insufficiency	Arrhythmia, cardiac shock	105
Erythroxylum coca	Epadu Ypadu	Cocaine, cinnamoylcocaine	Stimulant	Tachycardia, hypertension	106
Hyoscyamus niger	Henbane	Hyoscyamine (tropane alkaloids)	Stomach complaints, toothaches, ulcers, tumors	Irregular heartbeat	107
Mitragyna speciosa	Kratum	Alkaloids, mitrgynine	Antidiarrhoeal	Hypotension	108
Tussilago farfara	Coltsfoot Kuan-Dong hua	Senkirkine, a pyrrozolidine alakaloid, diterpine	Antitussive agent	Increase in peripheral vascular resistance	109
Urginea maritima	Squill	Cardiac glycoside	For cardiac insufficiency, arrhythmia	Arrhythmia, AV block	110

TABLE 5.2 Influencing Factors for Plant Toxins

Plant	Environment	Individual Organisms/ Animals
Species diversity	Temperature	Latency period
Parts of the plant	Humidity	Quantity
Stage of harvesting	Harvesting and storage	Quality
Season	Drying	Route of entry and absorption
Solubility	Extraction and processing	Enterohepatic circulation and reabsorption
Active compounds	Methods of preparation	Metabolism of toxins
Multiple toxins	Soil status and geography Residual fertilizers and/or pesticides Contamination (e.g., bacteria, fungus, heavy metals, toxic elements, etc.) Adulteration Substitution	Health status Co-morbidities Consumption of other herbs, plants, and/or chemical agents

Note: The influencing factors for the plant toxins are related to plants, environment, individual organisms, and/or animals.

and there has been significant progress in research on complementary and alternative medicine (CAM) in Europe and the United States. See also Chapter 6, Section 6.8, "Safety and Regulations."

5.12.1 Prevention of Plant Poisoning

Plant poisoning contributes to 3 to 10% of all poisoning. Since plant toxins affect various parts of the body including risk for life, the best strategy is to avoid plant poisoning altogether. Some suggestions for prevention include the following:

- Children and the public have to be educated and informed "never put any plant parts/materials (e.g., twigs, berries, flowers, leaves) in the mouth" and make them aware that it is not safe to do so.
- People should be informed not to suck nectar from flowers or make tea or decoction from unknown plant materials.
- People should be aware of the names of household plants and should learn about new plants if brought home.

- Keep plants and plant materials out of the reach of the children.
- Educate the community not to be lured by folklores and proverbs on plants or plant materials.
- Avoid exposure to smoke from burning plant materials as the toxins may enter the system through inhalation.
- Inform the community that heating or cooking toxic plants or plant materials does not destroy the toxins or toxic substances always.
- When someone develops symptoms and signs of toxicity following exposure to plants, bring all parts of the plant sample when the patient is brought for treatment.
- Educate the community not to apply plant medicine over raw surfaces of the skin and mucosa, as the toxins may get absorbed transcutaneously.

5.12.2 Proposed Action Plan

Plant medicine use is prevalent all over the globe. Thus, there is a need to assess the pattern of assistance required to ensure patient safety and efficacy with regard to the use of plant medicine. These are considered under different phases; the action plan is shown in Table 5.3.

5.12.3 Future Directions and Limitations

There is a need to develop legislation, policies, and programs to regulate and promote plant medicines. Research on them have to be strengthened so as to develop new drugs and new modes of delivery along with safety and efficacy. Similarly, health systems have to develop training programs and modules to teach and train traditional practitioners/healers and bring them under modern health systems. Biodiversity acts may be enforced to protect the natural wealth of plants from extinction. Animals pasturing on toxic plants may pass on the toxic compounds through milk, posing a hazard for the general public. So, laboratory methods have to be developed to find out and monitor the natural toxicants of plant origin in milk or milk products and eliminate exposure to plant toxins. Educational modules can be developed to examine plant medicine, incorporating the social, cultural, religious, ecological, educational, economical, scientific, clinical, legal, and ethical aspects. (See also Chapter 6, Section 6.9.)

The list of cardiotoxic plants is exhaustive and can never be completed because newer plants are identified each year as a result of toxicological studies. Toxicokinetics, toxicodynamics, toxicogenetics, and toxicogenomics related to poisonous plants have not been discussed. The threshold required to produce adverse effects from poisonous plants are not considered because, for each plant, it is influenced by many factors.

TABLE 5.3 Phases and Proposed Action for Plant Medicine

Phase	Course of Action
Preparedness (collection)	Sociodemographic and cultural aspects related to use of plant medicine Awareness, attitude, and behavior of the prescribers and users, and probable reasons for the gap
Analysis	Influencing factors for the use of plant medicine Usefulness of the plant medicine Data on basic research related to plant medicine Physiological, pharmacological, and toxicological aspects including contaminants and adulterants or substitution
Planning	Estimation of economical aspects Education and training on plant medicine for patients, public, prescribers, professionals, and policymakers Capacity building to use plant medicine Explore the phytochemical studies on plant medicine and toxicity
Networking	Community, industry, and professional networking to identify useful ones from toxic ones Intra- and international country networking to document and share data on the usefulness, limitations, and adverse effects of plant medicines
Monitoring and evaluation	Monitoring adverse effects Qualify control and drug delivery system Patient safety and efficacy of medicines

5.13 CONCLUSION

Plants and herbs are grown all over the world and used by billions of people. Issues related to plant medicines are quality assessment (QA), quality control (QC), good manufacturing practices (GMP), knowledge of mechanism of actions, bioavailability, chemical constituents, safety issues, herb–herb or herb–drug or herb–disease interactions, and so on. The commercial aspects of the ethnobotanical market is estimated to reach more than U.S. $60 billion,[121] and so there is an urgent need for international regulations and monitoring.

Key Learning Points

- Plants are used for medicinal and nonmedicinal purposes but their safety and efficacy have to be ascertained.
- Cardiotoxicity due to plants has to be remembered and warrants extensive evaluation including a thorough history, especially among patients presenting with one or more cardiac manifestations.

- Community education is essential for prevention of plant toxicity.
- Research on plant toxins is a potential area to explore newer investigative methods and therapeutic targets.

Acknowledgments We would like to thank Ms. Elizabeth Dauncey, Dr. K. Arthanari, Mr. A. Subramanian, and Mr. John David Peter for providing the figures to include in this chapter.

REFERENCES

1. Oehme FW, Keyler DE. Plant and animal toxins. In: Hayes AW, editor. *Principles and methods of toxicology*. 5th ed. Boca Raton B, FL: Informa Healthcare, CRC Press; 2008. p. 983–1052.
2. Eisenberg DM, Davis RB, Ettner SL, et al. Trends in alternative medicine use in the United States, 1990–1997: results of a follow-up national survey. *JAMA* 1998;**280**(18): 1569–75.
3. Sahoo N, Manchikanti P, Dey S. Herbal drugs: standards and regulation. *Fitoterapia* 2010; **81**(6):462–71.
4. Keyvanara M, Haghshenas A. The sociocultural contexts of attempting suicide among women in Iran. *Health Care Women Int* 2010;**31**(9):771–83.
5. Lockie T, Willcocks M, Harrington D. An unusual case of atrioventricular heart block in a young vegan. *Int J Cardiol* 2008;**125**(1):e12–3.
6. Schoner W. Endogenous cardiac glycosides, a new class of steroid hormones. *Eur J Biochem* 2002;**269**(10):2440–8.
7. Hollman A. Plants and cardiac glycosides. *Br Heart J* 1985;**54**(3):258–61.
8. Liu T, Brown DA, O'Rourke B. Role of mitochondrial dysfunction in cardiac glycoside toxicity. *J Mol Cell Cardiol* 2010;**49**(5):728–36.
9. Bagrov AY, Shapiro JI. Endogenous digitalis: pathophysiologic roles and therapeutic applications. *Nat Clin Pract Nephrol* 2008;**4**(7):378–92.
10. Wade OL. Digoxin 1785–1985. I. Two hundred years of digitalis. *J Clin Hosp Pharm* 1986; **11**(1):3–9.
11. Slifman NR, Obermeyer WR, Aloi BK, et al. Contamination of botanical dietary supplements by *Digitalis lanata*. *N Engl J Med* 1998;**339**(12):806–11.
12. Barrueto Jr. F, Jortani SA, Valdes Jr. R, et al. Cardioactive steroid poisoning from an herbal cleansing preparation. *Ann Emerg Med* 2003;**41**(3):396–9.
13. Eddleston M, Ariaratnam CA, Meyer WP, et al. Epidemic of self-poisoning with seeds of the yellow oleander tree (*Thevetia peruviana*) in northern Sri Lanka. *Trop Med Int Health* 1999;**4**(4):266–73.
14. Maringhini G, Notaro L, Barberi O, et al. Cardiovascular glycoside-like intoxication following ingestion of Thevetia nereifolia/peruviana seeds: a case report. *Ital Heart J* 2002;**3**(2):137–40.
15. Radford DJ, Gillies AD, Hinds JA, et al. Naturally occurring cardiac glycosides. *Med J Aust* 1986;**144**(10):540–4.
16. Oji O, Okafor QE. Toxicological studies on stem bark, leaf and seed kernel of yellow oleander (*Thevetia peruviana*). *Phytother Res* 2000;**14**(2):133–5.

17. Senthilkumaran S, Saravanakumar S, Thirumalaikolundusubramanian P. Cutaneous absorption of oleander: fact or fiction. *J Emerg Trauma Shock* 2009;**2**(1):43–5.
18. Senthilkumaran S, Meenakshisundaram R, Michaels AD, et al. Electrocardiographic changes during inhalational oleander toxicity. *J Electrocardiol* 2011;**44**(4):470–2.
19. Levi AJ, Dalton GR, Hancox JC, et al. Role of intracellular sodium overload in the genesis of cardiac arrhythmias. *J Cardiovasc Electrophysiol* 1997;**8**(6):700–21.
20. Chen KK, Henderson FG. Pharmacology of sixty-four cardiac glycosides and aglycones. *J Pharmacol Exp Ther* 1954;**111**(3):365–83.
21. Saravanapavananthan N, Ganeshamoorthy J. Yellow oleander poisoning: a study of 170 cases. *Forensic Sci Int* 1988;**36**(3–4):247–50.
22. Eddleston M, Ariaratnam CA, Sjostrom L, et al. Acute yellow oleander (*Thevetia peruviana*) poisoning: cardiac arrhythmias, electrolyte disturbances, and serum cardiac glycoside concentrations on presentation to hospital. *Heart* 2000;**83**(3):301–6.
23. Kummer JL, Nair R, Krishnan SC. Images in cardiovascular medicine. Bidirectional ventricular tachycardia caused by digitalis toxicity. *Circulation* 2006;**113**(7):e156–7.
24. Bose TK, Basu RK, Biswas B, et al. Cardiovascular effects of yellow oleander ingestion. *J Indian Med Assoc* 1999;**97**(10):407–10.
25. Roberts DM, Southcott E, Potter JM, et al. Pharmacokinetics of digoxin cross-reacting substances in patients with acute yellow oleander (*Thevetia peruviana*) poisoning, including the effect of activated charcoal. *Ther Drug Monit* 2006;**28**(6):784–92.
26. Bismuth C, Gaultier M, Conso F, et al. Hyperkalemia in acute digitalis poisoning: prognostic significance and therapeutic implications. *Clin Toxicol* 1973;**6**(2):153–62.
27. Rajapakse S. Management of yellow oleander poisoning. *Clin Toxicol* 2009;**47**(3):206–12.
28. Caldwell JH, Caldwell PB, Murphy JW, et al. Intestinal secretion of digoxin in the rat. Augmentation by feeding activated charcoal. *Naunyn-Schmiedeberg's Arch Pharmacol* 1980;**312**(3):271–5.
29. de Silva HA, Fonseka MM, Pathmeswaran A, et al. Multiple-dose activated charcoal for treatment of yellow oleander poisoning: a single-blind, randomised, placebo-controlled trial. *Lancet* 2003;**361**(9373):1935–8.
30. Eddleston M, Rajapakse S, Rajakanthan, et al. Anti-digoxin Fab fragments in cardiotoxicity induced by ingestion of yellow oleander: a randomised controlled trial. *Lancet* 2000;**355**(9208):967–72.
31. Gawarammana I, Mohamed F, Bowe SJ, et al. Fructose-1, 6-diphosphate (FDP) as a novel antidote for yellow oleander-induced cardiac toxicity: a randomized controlled double blind study. *BMC Emerg Med* 2010;**10**:15.
32. Taboulet P, Baud FJ, Bismuth C, et al. Acute digitalis intoxication: is pacing still appropriate? *J Toxicol Clin Toxicol* 1993;**31**(2):261–73.
33. Senthilkumaran S, Balamurugan N, Suresh P, Thirumalaikolundusubramanian P. Safety and efficacy of transcutaneous temporary cardiac pacing in oleander seed poisoning. In: *Proceedings of the Thirteenth International Conference of Emergency Medicine*, Kolkota, West Bengal, India; 2011.
34. Albertson TE, Dawson A, de Latorre F, et al. TOX-ACLS: toxicologic-oriented advanced cardiac life support. *Ann Emerg Med* 2001;**37**(4 Suppl.):S78–90.
35. Thomas K, Dayal AK, Gijsbers A, et al. Oduvanthalai leaf poisoning. *J Assoc Physicians India* 1987;**35**(11):769–71.
36. Nagaraj S. Cardiac toxicity of oduvanthalai (*Cleistanthus collinus*) common leaves poisoning in Tamil Nadu (report of 25 cases). *Antiseptic* 1987;**84**:33–5.

37. Benjamin SP, Fernando ME, Jayanth JJ, et al. *Cleistanthus collinus* poisoning. *J Assoc Physicians India* 2006;**54**:742–4.
38. Parasuraman S, Raveendran R, Selvaraj RJ. Effects of cleistanthins A and B on blood pressure and electrocardiogram in Wistar rats. *Z Naturforschung C* 2011;**66**(11–12):581–7.
39. Thomas K, Dayal AK, Narasimhan, et al. Metabolic and cardiac effects of *Clistanthus collinus* poisoning. *J Assoc Physicians India* 1991;**39**(4):312–4.
40. Ragupathi G, Prabhasankar P, Sekharan PC, et al. Enzyme-linked immunosorbent assay (ELISA) for the determination of the toxic glycoside cleistanthin B. *Forensic Sci Int* 1992; **56**(2):127–36.
41. Thirumavalavan R. Aggressive potassium correction may halt death in *Cleistanthus collinus* poisonining. *J Toxicol Clin Toxicol* 2004;**42**(5):801.
42. Anapoorani KS, Damodharan C, Chandrasekharan P. A promising antidote to *Cleistanthus collinus* poisoning. *J Sci Soc Ind* 1984;**2**:36.
43. Chrispal A. *Cleistanthus collinus* poisoning. *J Emerg Trauma Shock* 2012;**5**(2):160–6.
44. Chan TY. Aconitine poisoning: a global perspective. *Vet Hum Toxicol* 1994;**36**(4):326–8.
45. Konno C, Shirasaka M, Hikino H. Cardioactive principle of aconitum carmichaeli roots. *Planta Med* 1979;**35**(2):150–5.
46. Wang FP, Peng CS, Jian XX, et al. Five new norditerpenoid alkaloids from *Aconitum sinomontanum*. *J Asian Nat Prod Res* 2001;**3**(1):15–22.
47. Friese J, Gleitz J, Gutser UT, et al. *Aconitum* sp. alkaloids: the modulation of voltage-dependent Na^+ channels, toxicity and antinociceptive properties. *Eur J Pharmacol* 1997; **337**(2–3):165–74.
48. Kapoor SC, Sen AK. Cardiovascular aspects of aconite poisoning in human beings. *Indian Heart J* 1969;**21**(3):329–38.
49. Chan TY. Aconite poisoning presenting as hypotension and bradycardia. *Hum Exp Toxicol* 2009;**28**(12):795–7.
50. Chan TY. Aconite poisoning. *Clin Toxicol* 2009;**47**(4):279–85.
51. Weijters BJ, Verbunt RJ, Hoogsteen J, et al. Salade malade: malignant ventricular arrhythmias due to an accidental intoxication with *Aconitum napellus*. *Neth Heart J* 2008;**16**(3): 96–9.
52. Imazio M, Belli R, Pomari F, et al. Malignant ventricular arrhythmias due to *Aconitum napellus* seeds. *Circulation* 2000;**102**(23):2907–8.
53. Smith SW, Shah RR, Hunt JL, et al. Bidirectional ventricular tachycardia resulting from herbal aconite poisoning. *Ann Emerg Med* 2005;**45**(1):100–1.
54. Dickens P, Tai YT, But PP, et al. Fatal accidental aconitine poisoning following ingestion of Chinese herbal medicine: a report of two cases. *Forensic Sci Int* 1994;**67**(1):55–8.
55. Kolev ST, Leman P, Kite GC, et al. Toxicity following accidental ingestion of *Aconitum* containing Chinese remedy. *Hum Exp Toxicol* 1996;**15**(10):839–42.
56. Lin CC, Chan TY, Deng JF. Clinical features and management of herb-induced aconitine poisoning. *Ann Emerg Med* 2004;**43**(5):574–9.
57. Niinuma H, Aoki H, Suzuki T, Shibata M, et al. Two survival cases of severe aconite poisoning by percutaneous cardiopulmonary support system and cardiopulmonary bypass for fatal arrhythmia: a case report. *Internet J Pharmacol* 2002;**2**(1).
58. Griffiths P. Qat use in London: a study of Qat use among a sample of Somalis living in London. London: NCJRS; 1998. Available from: <www.ncjrs.gov/App/Publications/abstract.aspx?ID = 180284> [accessed 2014].
59. Al-Motarreb A, Baker K, Broadley KJ. Khat: pharmacological and medical aspects and its social use in Yemen. *Phytother Res* 2002;**16**(5):403–13.

60. Kalix P, Braenden O. Pharmacological aspects of the chewing of khat leaves. *Pharmacol Rev* 1985;**37**(2):149–64.
61. Brenneisen R, Fisch HU, Koelbing U, et al. Amphetamine-like effects in humans of the khat alkaloid cathinone. *Br J Clin Pharmacol* 1990;**30**(6):825–8.
62. Halket JM, Karasu Z, Murray-Lyon IM. Plasma cathinone levels following chewing khat leaves (*Catha edulis Forsk.*). *J Ethnopharmacol* 1995;**49**(2):111–3.
63. Hassan NA, Gunaid AA, El-Khally FM, et al. Khat chewing and arterial blood pressure. A randomized controlled clinical trial of alpha-1 and selective beta-1 adrenoceptor blockade. *Saudi Med J* 2005;**26**(4):537–41.
64. Ali WM, Al Habib KF, Al-Motarreb A, et al. Acute coronary syndrome and khat herbal amphetamine use: an observational report. *Circulation* 2011;**124**(24):2681–9.
65. Al-Motarreb AL, Broadley KJ. Coronary and aortic vasoconstriction by cathinone, the active constituent of khat. *Auton Autacoid Pharmacol* 2003;**23**(5–6):319–26.
66. Saha S, Dollery C. Severe ischaemic cardiomyopathy associated with khat chewing. *J R Soc Med* 2006;**99**(6):316–8.
67. Al-Motarreb A, Briancon S, Al-Jaber N, et al. Khat chewing is a risk factor for acute myocardial infarction: a case-control study. *Br J Clin Pharmacol* 2005;**59**(5):574–81.
68. Alkadi HO, Al-Kamarany M, Al-Kadi H, et al. Khat–aspirin interaction. *Yemen J Pharm Biol Sci* 2008;**2**:32–9.
69. Kalix P. The pharmacology of khat. *Gen Pharmacol* 1984;**15**(3):179–87.
70. Saha S, Dollery C. Severe ischemic cardiomyopathy associated with khat chewing. *J R Soc Med* 2006;**99**(6):316–8.
71. Shi QW, Sauriol F, Mamer O, et al. New taxanes from the needles of *Taxus canadensis*. *J Nat Prod* 2003;**66**(4):4706.
72. Zhang J, Sauriol F, Mamer O, et al. Taxoids from the needles of the Canadian yew. *Phytochemistry* 2000;**54**(2):221–30.
73. Wilson CR, Sauer J, Hooser SB. Taxines: a review of the mechanism and toxicity of yew (*Taxus* spp.) alkaloids. *Toxicon* 2001;**39**(2–3):175–85.
74. Pietsch J, Schulz K, Schmidt U, et al. A comparative study of five fatal cases of *Taxus* poisoning. *Int J Legal Med* 2007;**121**(5):417–22.
75. Jones R, Jones J, Causer J, et al. Yew tree poisoning: a near-fatal lesson from history. *Clin Med* 2011;**11**(2):173–5.
76. Veltmann C, Borggrefe M, Schimpf R, et al. Images in cardiovascular medicine. Yew causes Brugada ECG. *Circulation* 2009;**119**(13):1836–7.
77. Cummins RO, Haulman J, Quan L, et al. Near-fatal yew berry intoxication treated with external cardiac pacing and digoxin-specific Fab antibody fragments. *Ann Emerg Med* 1990;**19**(1):38–43.
78. Miller MB EJ, Curry SC. Sodium bicarbonate for *Taxus*-induced dysrhythmia. *J Toxicol Clin Toxicol* 2000;**38**(5):572.
79. Ruha AM, Tanen DA, Graeme KA, et al. Hypertonic sodium bicarbonate for *Taxus* media-induced cardiac toxicity in swine. *Acad Emerg Med* 2002;**9**(3):179–85.
80. Yersin B, Frey JG, Schaller MD, et al. Fatal cardiac arrhythmias and shock following yew leaves ingestion. *Ann Emerg Med* 1987;**16**(12):1396–7.
81. Pierog J, Kane B, Kane K, et al. Management of isolated yew berry toxicity with sodium bicarbonate: a case report in treatment efficacy. *J Med Toxicol* 2009;**5**(2):84–9.
82. Kupchan SM. Hypotensive *Veratrum* ester alkaloids. *J Pharmaceut Sci* 1961;**50**:273–87.
83. Wang SY, Wang GK. Voltage-gated sodium channels as primary targets of diverse lipid-soluble neurotoxins. *Cell Signal* 2003;**15**(2):151–9.

84. Aviado DM, Guevara Aviado D. The Bezold–Jarisch reflex. A historical perspective of cardiopulmonary reflexes. *Ann NY Acad Sci* 2001;**940**:48–58.
85. Kimura I, Islam A, Honda R, et al. Blood-pressure lowering, positive chronotropy and inotropy by the *Veratrum* alkaloids germidine and germerine but negative chronotropy by veratridine in mice. *J Asian Nat Prod Res* 2000;**2**(2):133–44.
86. Gilotta I, Brvar M. Accidental poisoning with *Veratrum album* mistaken for wild garlic (*Allium ursinum*). *Clin Toxicol* 2010;**48**(9):949–52.
87. Marinov A, Koev P, Mirchev N. Electrocardiographic studies of patients with acute hellebore (*Veratrum album*) poisoning. *Vutr Boles* 1987;**26**(6):36–9. [*Elektrokardiografski prouchvaniia pri bolni s ostro otraviane s chemerika* (Veratrum album).]
88. Quatrehomme G, Bertrand F, Chauvet C, et al. Intoxication from *Veratrum album. Hum Exp Toxicol* 1993;**12**(2):111–5.
89. Jaffe AM, Gephardt D, Courtemanche L. Poisoning due to ingestion of *Veratrum viride* (false hellebore). *J Emerg Med* 1990;**8**(2):161–7.
90. Boido CC, Tasso B, Boido V, et al. Cytisine derivatives as ligands for neuronal nicotine receptors and with various pharmacological activities. *Farmaco* 2003;**58**(3):265–77.
91. Raschack M. Actions of sparteine and sparteine derivatives on the heart and circulation. *Arzneimittel-Forschung* 1974;**24**(5):753–9. [*Wirkungen von Soartein und Sparteinderivaten auf Herz und Kreislauf.*]
92. Riccardi A, Frumento F, Ghinatti M, et al. Spanish broom flower ingestion: a very unusual poisoning. *Eur J Emerg Med* 2006;**13**(5):317–8.
93. Schaller G, Urech K, Grazi G, et al. Viscotoxin composition of the three European subspecies of *Viscum album. Planta Med* 1998;**64**(7):677–8.
94. Mellstrand ST, Samuelsson G. Phoratoxin, a toxic protein from the mistletoe *Phoradendron tomentosum* subsp. macrophyllum (Loranthaceae). The amino acid sequence. *Acta Pharm Suec* 1974;**11**(4):347–60.
95. Andersson KE, Johannsson M. Effects of viscotoxin on rabbit heart and aorta, and on frog skeletal muscle. *Eur J Pharmacol* 1973;**23**(3):223–31.
96. Hall AH, Spoerke DG, Rumack BH. Assessing mistletoe toxicity. *Ann Emerg Med* 1986;**15**(11):1320–3.
97. van der Bijl Jr P, van der Bijl Sr P. Cardiotoxicity of plants in South Africa. *Cardiovasc J Afr* 2012;**23**(9):476–7.
98. Krenzelok EP, Jacobsen TD, Aronis J. American mistletoe exposures. *Am J Emerg Med* 1997;**15**(5):516–20.
99. Van Elst D, Nuyens S, van Wyk B, et al. Distribution of the cardiotoxin pavettamine in the coffee family (Rubiaceae) and its significance for gousiekte, a fatal poisoning of ruminants. *Plant Physiol Biochem* 2013;**67C**:15–9.
100. Fourie N, Erasmus GL, Schultz RA, et al. Isolation of the toxin responsible for gousiekte, a plant-induced cardiomyopathy of ruminants in southern Africa. *Onderstepoort J Vet Res* 1995;**62**(2):77–87.
101. Ellis CE, Naicker D, Basson KM, et al. Damage to some contractile and cytoskeleton proteins of the sarcomere in rat neonatal cardiomyocytes after exposure to pavettamine. *Toxicon* 2010;**55**(6):1071–9.
102. Gottlieb OR. Ethnopharmacology versus chemosystematics in the search for biologically active principles in plants. *J Ethnopharmacol* 1982;**6**(2):227–38.
103. Patnaik GK, Kohler E. Pharmacological investigation on asclepin: a new cardenolide from *Asclepias curassavica.* Part II. Comparative studies on the inotropic and toxic effects of asclepin, g-strophantin, digoxin and digitoxin. *Arzneimittel-Forschung* 1978;**28**(8):1368–72.

104. Hansen DK, George NI, White GE, et al. Cardiovascular toxicity of *Citrus aurantium* in exercised rats. *Cardiovasc Toxicol* 2013;**13**(3):208–19.
105. Lehmann HD. Effect of plant glycosides on resistance and capacitance vessels. *Arzneimittel-Forschung* 1984;**34**(4):423–9. [*Zur Wirkung pflanzlicher Glykoside auf Widerstandsgefasse und Kapazitatsgefasse.*]
106. Stolberg VB. The use of coca: prehistory, history, and ethnography. *J Ethn Subst Abuse* 2011;**10**(2):126–46.
107. Khan AU, Gilani AH. Cardiovascular inhibitory effects of *Hyoscyamus niger*. *Methods Find Exp Clin Pharmacol* 2008;**30**(4):295–300.
108. Prozialeck WC, Jivan JK, Andurkar SV. Pharmacology of kratom: an emerging botanical agent with stimulant, analgesic and opioid-like effects. *J Am Osteopath Assoc* 2012;**112**(12):792–9.
109. Li YP, Wang YM. Evaluation of tussilagone: a cardiovascular-respiratory stimulant isolated from Chinese herbal medicine. *Gen Pharmacol* 1988;**19**(2):261–3.
110. Tuncok Y, Kozan O, Cavdar C, et al. *Urginea maritima* (squill) toxicity. *J Toxicol Clin Toxicol* 1995;**33**(1):83–6.
111. Lin MC, Nahin R, Gershwin ME, et al. State of complementary and alternative medicine in cardiovascular, lung, and blood research: executive summary of a workshop. *Circulation* 2001;**103**(16):2038–41.
112. Amabile CM, Spencer AP. Keeping your patient with heart failure safe: a review of potentially dangerous medications. *Arch Intern Med* 2004;**164**(7):709–20.
113. Panter KE, James LF. Natural plant toxicants in milk: a review. *J Anim Sci* 1990;**68**(3):892–904.
114. Kuhlmann S, Ahlers-Schmidt CR, Steinberger E. TXT@WORK: Pediatric hospitalists and text messaging. *Telemed J E Health* 2014;**20**(7):647–52.
115. McKinney PE, Gomez HF, Phillips S, et al. The fax machine: a new method of plant identification. *J Toxicol Clin Toxicol* 1993;**31**(4):663–5.
116. FDA. Poisonous Plant Database, 2007. Available from: <www.accessdata.fda.gov/scripts/plantox/>.
117. CBIF. CPPI System. Ottawa, ON, Canada. Available from: <www.cbif.gc.ca/pls/pp/poison?p_x=px>.
118. AAoPC Centers. National Poison Data System. Available from: <www.aapcc.org/data-system/>.
119. Sahoo N, Choudhury K, Manchikanti P. Manufacturing of biodrugs: need for harmonization in regulatory standards CBIF. *Bio Drugs* 2009;**23**(4):217–29.
120. Calapai G. European legislation on herbal medicines: a look into the future. *Drug Saf* 2008;**31**(5):428–31.
121. Tilburt JC, Kaptchuk TJ. Herbal medicine research and global health: an ethical analysis. *Bull World Health Organ* 2008;**86**(8):594–9.

Chapter 6

Native Medicines and Cardiovascular Toxicity

Ashish Bhalla,[1] Ponniah Thirumalaikolundusubramanian,[2] Jeffery Fung,[3] Gabriela Cordero-Schmidt,[4] Sari Soghoian,[5] Veronica Kaur Sikka,[6] Harinder Singh Dhindsa,[6] and Surjit Singh[1]

[1]*Department of Medicine, Postgraduate Institute of Medical Education and Research, Chandigarh, India*

[2]*Department of Medicine, Chennai Medical College Hospital and Research Centre, Trichy, India*

[3]*Accident and Emergency Department, Tuen Mun Hospital, Hong Kong*

[4]*University of California, San Francisco*

[5]*Department of Emergency Medicine, New York University School of Medicine/Bellevue Hospital Center, New York*

[6]*Department of Emergency Medicine, Virginia Commonwealth University, Richmond, Virginia*

6.1 INTRODUCTION

Native medicines are traditional medicines that belong to a particular place or vicinity and originate from naturally occurring biologic materials with or without other nonbiologic materials. The term *native medicine* is used interchangeably as home medicine, remedy, or recipe and/or botanical, herbal, phytomedicine, traditional medicine, or complementary and alternative medicine (CAM). Native medicines paved the way for the development of pharmacotherapeutic agents such as quinine, artemisinin compounds, digoxin, and aspirin.[1–3] Herbal medicines include herbs, herbal materials, herbal preparations, or finished herbal products. Contrary to what many people believe, native medicines or herbal medicines may have harmful effects too.[4]

Traditional medicine refers to health practices, approaches, knowledge, and beliefs incorporating plant, animal, and mineral-based medicines; spiritual therapies; manual techniques; and exercises applied singularly or in combination to treat, diagnose, and prevent illnesses or maintain well-being.[5] Traditional herbal medicines are used for various illnesses by the majority of the population in India, South America, North America, Sub-Saharan Africa, China, and many other countries and use is more common in rural areas.[6,7] In these places, it is

The Heart and Toxins. DOI: http://dx.doi.org/10.1016/B978-0-12-416595-3.00006-2

culturally acceptable as a choice of health care. The use of native medicines is related to sociodemographic factors, chronicity of illness, beliefs in native medicine, economical aspects of medicines, the assurance given by traditional practitioners or healers, influence of family members, availability, quality, personal characteristics of users, and cost and adverse effects of modern medicine. The most common illnesses for which traditional herbal medicines are used are diabetes[8,9] and hypertension.[10,11] In Sub-Saharan Africa, the World Health Organization (WHO) estimates that 80% of the population consult traditional healers regularly.[5] In Mahi (West Africa) there used to be one traditional healer for every 500 inhabitants, however, only 0.04 physicians of modern medicine for the same population.[12]

The National Center for Complementary and Alternative Medicine (NCCAM) defines CAM as a "group of diverse medical and health care systems, practices and products that are not generally considered a part of conventional medicine." WHO estimates that 80% of the world's population relies on CAM.[5] The terms *complementary medicine* or *alternative medicine* are used interchangeably with *traditional medicine* in many countries. The term *alternative medicine* means any form of medicine that is outside the mainstream of Western, modern, or conventional medicine or allopathy.[13] CAM includes more than 100 systems of alternative medicines with no less than 135 therapies. Every country, region, or area has its own traditional system of health practice such as acupuncture in China, herbalism in English-speaking countries, Ayurveda and siddha in India, shiatsu in Japan, and so on. Herbal medicines are used for treatment of a variety of cardiovascular diseases.[14]

Integrative, integrated, and collaborative medicine are emerging models that include a combination of conventional medicine and individual complementary therapy. Collectively, these are known as *complementary and integrative medicine* (CIM); it is not discussed in this chapter.

Native American healers treat cardiovascular diseases and these need clinical scrutiny to understand the benefits.[15] This chapter focuses on the cardiotoxic effects of a few of the commonly used native, herbal, traditional, complementary, and alternative medicines, and some of the aspects of their mechanisms with potential side effects and drug interactions.

6.2 MATERIALS USED

Materials used in native medicines are plant- and/or animal-based, and other biological and/or herbomineral materials. Preparations include home recipes such as mixed with water, sugar, milk, buttermilk or curd, ghee, oil, salt, spices, selected food materials, and so on. Native/traditional medicines are available in the form of powder, liquid, drops, paste, and enema, which are administered through different routes. Although many hundreds of native medicines are prepared and dispensed, few of them are cardiotoxic. In this chapter, native medicines are broadly categorized based on their preparations: plant-based, animal-based, biological, and herbomineral-based and are described in the subsections that follow.

6.2.1 Plant-Based Preparations

Plants are the gift of nature and are believed to have medicinal values and, thus, are used for illnesses. In the ensuing paragraphs, plant-based preparations are highlighted. Among the plant-based materials, toxic effects from cardiac glycosides, steroid alkaloids, and grayanotoxins are frequently encountered.

Cardiac Glycosides, Steroid Alkaloids, Aconites, and Grayanotoxins

Cardiotoxic plants and herbs are grown all over the world with different names in local languages. These plant preparations are prescribed to treat heart failure and/or cardiac ailments. Toxicity is attributed mostly to cardiac glycoside, which produces cardiac arrhythmias. The cardiac glycosides are present in 11 plant families (see Chapter 5).[16] Toxic compounds are present in almost all parts of at least 34 genera of the plant families; these materials, when consumed, produce either acute effects or chronic cumulative toxicity. The best known cardiac glycosides are digoxin and digitoxin, derived from the parent glycoside digitalis found in foxgloves. Digitalis is now produced from the commercially grown foxglove (*Digitalis lanata*).

Toxins produced by plants mainly affect the heart through modification in the functioning of ion channels, which leads to electrical (i.e., producing arrhythmias) and mechanical (i.e., producing pump failure) instability of the cardiac fibers/myocytes. The cardiac glycosides inhibit the cellular membrane-bound enzyme system, Na^+/K^+-ATPase, increasing intracellular Na^+ and Ca^{2+} levels. This leads to increased automaticity. The increased intracellular Ca^{2+} interacts with the troponin C of the actin–myosin complex, resulting in increased contractions. The increased contractions appear as premature ventricular contractions both clinically and on electrocardiogram (EKG or ECG). Inhibition of Na^+/K^+-ATPase in baroreceptor and skeletal muscle cells results in increased vagal tone. This clinically manifests as bradycardia and hyperkalemia.[17]

Aconite preparations are prescribed by practitioners of native medicine, Chinese medicine, Ayurveda, and homeopathy. The preparations are available in India, China, Taiwan, Japan, and many other parts of the world. Alkaloids of the *Aconitum* species (monkshood) act on sodium channels by opening them. Steroid alkaloids are commonly found in these species. Please refer to Figure 5.6 in Chapter 5 for an illustration of *Aconitum napellus*. Alkaloids activate voltage-sensitive sodium channels and affect excitable membranes of neural, cardiac, and muscle tissues.[18] This results in prolonged depolarization in cardiac muscles, which induces delayed and early after-depolarizations. Delayed after-depolarizations result in increased automaticity such as premature ventricular beats.[19] Early after-depolarizations produce lengthening of the QT interval. Hypotension and bradycardia are induced by activation of the hypothalamus.[20] All these contribute to various forms of tachy-brady arrhythmias.[20–23]

The most serious complication in aconite poisoning is ventricular arrhythmias, and fatality is usually rare as a result of such poisoning.[20,24,25] The possible mechanisms through which aconite-induced cardiovascular toxicity have been demonstrated in experimental models include positive inotropic effect (through the Na^+/Ca^{2+} exchanger), increase in intracellular calcium, stimulation of alpha and beta receptor agonists, decrease in nitric oxide, histamine release, increased dopamine (increases synthesis and also aconite includes methyldopamine), and induction of signal transduction.[26]

Grayanotoxins are found in rhododendrons and azaleas, and they produce toxicity similar to that seen with steroid alkaloids and aconite. The vast genus Rhododendron includes species that have been used in traditional medicine for the treatment of pain, inflammatory conditions, gastrointestinal disorders, the common cold, asthma, skin diseases, and so on. These plants grow in China and contain grayanotoxin. Similar grayanotoxin is present in mad honey.

Grayanotoxins are also known as andromedotoxin, acetylandromedol, and rhodotoxin. There are 18 forms of grayanotoxins. These have a unique class of toxic diterpenoids, which are polyhydroxylated cyclic hydrocarbons, that can be detected from a suspected commodity by paper electrophoresis, thin-layer chromatography, gas chromatography, or mass spectrometry.[27] These toxins block sodium channels in cell membranes and prevent inactivation; thus the cells are in a stable depolarization, during which the entry of calcium into the cells may be facilitated. Some toxic plants, such as the ones belonging to Solanaceae family and *Brugmansia* genus, have a very potent combination of anticholinergic substances in them. Intoxication/poisoning with these compounds results in anticholinergic syndrome.

Clinical Presentation

The onset of symptoms and duration of action vary with the dose ingested. The clinical manifestations are well known for certain glycoside preparations, such as digoxin, digitoxin, and ouabain, but may vary depending on toxic plant ingestions. Generally, cardiac glycoside toxicity produces nausea, vomiting, diarrhea, abdominal discomfort, visual changes (i.e., appearance of yellow and green colors, "halos," photophobia), and changes in mental status (e.g., disorientation, psychosis, lethargy, stupor, dysarthria, weakness, dizziness, restlessness, seizures). Cardiac manifestations are common, usually in the form of arrhythmias. Arrhythmias from this kind of toxicity, such as premature ventricular beats, paroxysmal atrial tachycardia, junctional tachycardia, various degrees of AV block, bidirectional ventricular tachycardia, and ventricular arrhythmias, have been described (refer the Chapters 5 and 8).

Myocardial depression can present with cardiogenic shock. Hyperkalemia is generally seen and may be lethal if not detected or treated promptly. The ECG may reveal nonspecific ST-segment and T-wave changes, similar to digoxin-induced changes. Serum digoxin levels may be elevated after exposure to plants containing cardiac glycosides, and patients with impaired renal function are more prone to such toxicity. Nonspecific antibody-based digoxin assays show significant cross-reactivity with many cardiac glycosides. Hyperkalemia often correlates with severity of cardiac glycoside toxicity.

The clinical presentation of alkaloids/aconite toxicity begins 10 to 90 min after the ingestion of *Aconitum*; generally the toxic effects manifest after administration of a second or third dose[25] and may remain in the system for several days. Mild aconite poisoning lasts for 1.5 to 2.5 days, whereas those with cardiovascular complications need about 7 to 9 days in order to recover. Occurrences of neurotoxicity and cardiotoxicity together distinguish it from glycoside toxicity. Cardiac effects are clinically similar to cardiac glycoside toxicity with enhanced vagal tone.[22,28] Hypokalemia; metabolic or respiratory acidosis; respiratory alkalosis; and renal, hepatic, and central nervous system impairment may be noted.

Symptoms of grayanotoxin toxicity include diaphoresis, hypersalivation, nausea, vomiting, bradycardia, AV block, junctional rhythms, chest pain (probably due to coronary vasospasm), hypotension, shock, dizziness, syncope, circumoral and extremity paresthesia, incoordination, and muscular weakness.

Treatment

Treatment of all these poisonings is symptomatic along with antidote for a particular toxin, if available or known. Management includes immediate discontinuation of further exposure to the toxic medicine/material. Toxicity from plant ingestion may be mitigated by prevention of absorption and elimination from the gastrointestinal tract. Persistent vomiting is a common feature that helps to get rid of the toxin from the upper gastrointestinal tract. Gastric decontamination has to be attempted with gastric lavage after securing the airway. A single dose or multiple repeated doses of activated charcoal was found to be effective in preventing toxin absorption. Supportive management in the form of maintenance of fluid and electrolyte supplementation is an important component of therapy.

Cardiac Glycosides Dyselectrolytemia, primarily potassium imbalance, is common with glycoside plant ingestion; thus, electrolyte levels should be monitored closely. Correction of hyperkalemia with insulin and dextrose is recommended. Calcium gluconate and/or chloride should be avoided. All such patients need to have continuous cardiac monitoring for at least 24 hours because of the potential risk for arrhythmias. Digoxin levels may be elevated with cardiac glycoside ingestion but levels do not correlate with severity or clinical symptoms. The definitive therapy for cardiac glycoside poisoning is digoxin-specific Fab fragments (e.g., Digibind and DigiFab). The Fab antibody fragments bind to the circulating cardiac glycoside and limit its toxicity. Administration of the digoxin-specific Fab antibody has been shown to be successful in several animal and human studies, especially after yellow oleander poisoning. (The complications and management of digoxin toxicity are discussed in Chapters 3 and 8.)

Bradyarrhythmias have to be managed with atropine and β-adrenergic agents. However, there is an increased risk for inducing tachyarrhythmias with these agents in such patients, so it should be used with caution. Ventricular tachycardia is often resistant to cardioversion, and electrical cardioversion may precipitate ventricular fibrillation or asystole. Therefore, electrical cardioversion should be reserved for resistant cases. Lidocaine is the drug of choice for tachyarrhythmias. The role of magnesium is controversial and it should be avoided. Amiodarone, quinidine, and calcium-channel blockers are contraindicated because these may increase the concentration of glycosides. Beta blockers are contraindicated because they may worsen heart block.

Steroid Alkaloids (Aconite) and Grayanotoxins Treatment of poisoning with these is mainly supportive. There is no definite antidote. Correction of fluid excess and electrolyte imbalance is paramount. Atropine for bradycardia and vasopressors for fluid-resistant hypotension are often required. Mechanical ventilation is required in severe cases when there is respiratory muscle paralysis. Ventricular dysrhythmias may be resistant to cardioversion and antiarrhythmic agents. Magnesium has been shown to suppress polymorphic ventricular tachycardia. Even though lidocaine, procainamide, flecainide, and amiodarone are useful to treat ventricular arrhythmias, amiodarone is the preferred agent.[25] Experimental evidence has revealed that tetrodotoxin, a membrane sodium channel blocker, antagonized the effects of cardiotoxicity of aconite; however, tetrodoxin itself has cardiovascular toxicity.[29] The treatment of toxicity due to grayanotoxins is discussed in mad honey poisoning (Section 6.3.1).

Abrus precatorius It is also known as jequirity, Crab's eye, or precatory pea or bean. In native medicine, the paste prepared from the seeds are used for baldness, vitiligo, pain, pruritus, and as an abortifacient.[30] Oral administration of agglutinins isolated from the seeds has been tried for the treatment of hepatitis and acquired immunodeficiency syndrome (AIDS). Seeds contain abrin (a toxalbumin), indole derivatives, anthocyanins, sterols, and terpenes. Abrin causes agglutination of erythrocytes, hemolysis, and enlargement of lymph glands. The roots and the leaves also contain toxins but are less in concentration. Cardiovascular toxic manifestations of such poisoning by this plant include venous thrombosis (due to hemolysis, dehydration, and endothelial cell damage), pulmonary edema (due to endothelial damage and increased capillary permeability), hypertension, and arrhythmias (due to electrolyte imbalance and through one or other associated mechanisms as noted in other pathologies).[31–33] Treatment is mainly symptomatic and supportive. Some investigators mentioned that antiabrin used to be available as an antidote.[32]

Anamirta cocculus It is commonly found in the Khasi Hills, Orissa, and the Indian peninsular regions and some other parts of India. It is also known as fish berry, levant berry, poison berry, or the crow killer. The stem and roots contain quaternary alkaloids such as berberine, palmatine, magnoflorine, and columbamine. The seeds deliver picrotoxin, a sesquiterpene, while the seed shells contain the tertiary alkaloids menispermine and paramenispermine.[34] In Thai medicine, it has a role for fever, as a central nervous system stimulant, for antitumor activity, immune stimulation, and hemostasis.[35] It is used as parasiticide to kill lice and other parasites. Picrotoxin is very toxic: a dose of 20 mg (2–3 berries) is lethal. Toxicity manifests as hypertension (due to vasoconstriction, stimulation of autonomic nervous system) and arrhythmias (due to sympathetic stimulation, cholinergic excitation) in experimental models. Neurological manifestations dominate in such toxicity.[36,37] Treatment is symptomatic and supportive.

Antiaris toxicaria This substance is prevalent in Australia, China, India, Indonesia, Malaysia, the Philippines, and Vietnam. It is a deciduous tree and when cut the twigs yield thin creamy latex that becomes darker on exposure to air. The fruit is edible. In traditional medicine, the leaves and root are used to treat mental illnesses. In various parts of Asia, seeds, leaves, and bark are used as astringents and the seeds to treat dysentery. The active components are cardenolides and alkaloids, which include cardiac glycoside, coumarins, and flavanones.[38] The most important toxin is a cardiac glycoside named antiarin. Clinical manifestations are similar to digitalis overdose and the management is supportive.[39]

Areca catechu It (i.e., the areca nut) is found along the coasts of the Indian states of Karnataka, Kerala, Tamil Nadu, West Bengal, Assam, and Maharashtra. Traditionally, it has been used as an aphrodisiac, appetite suppressant, digestive aid, and diuretic; and as a treatment for asthma, cough, dermatitis (used on the skin), fainting, glaucoma, impotence, intestinal worms, leprosy, and toothache. It has also been indicated for use with leucorrhoea (vaginal discharge) and vaginal laxity. Betel nut chewing is common in the developing world and the main ingredient is the areca nut. The nut contains several alkaloids belonging to the pyridine group, the most important being arecoline. The others are arecaidine, guvacine, and isoguvacine. Arecaidine has no parasympathomimetic effects, only stimulating properties. It is a sedative when taken in higher doses. Isoguvacine produces hypotension. Its use can precipitate bronchial asthma due to bronchoconstrictive effects of the alkaloid arecoline. Areca nut is used as a major constituent of Indian medicine for gynecological disorders. Poisoning by it causes cardiovascular toxicity, including hypotension, hypertension, acute myocardial infarction, variation of RR interval, and any forms of tachy-brady arrhythmias.[40]

In a case control study among Taiwanese man, areca nut chewing was an independent risk factor for obstructive coronary artery disease, hypertension, and dyslipidemia.[41] The possible explanations for such pathologies include inflammatory reaction, COX upregulation, oxidative stress, promotion of oxidation of low-density cholesterol, endothelial dysfunction, stimulation of sympathetic and parasympathetic nervous system, and increased secretion of epinephrine/norepinephrine and histamine, which have been demonstrated in experimental models and human studies.[41,42] The treatment due to areca nut toxicity is supportive.

Cannabis sativa It is an annual herbaceous plant in the *Cannabis* genus, a species of the Cannabaceae family. The active ingredient in the plant is tetrahydrocannabinol. Another name for it is ijuana. Medical ijuana is used for neuropathic pain, muscle spasm, fibromyalgia, and chemotherapy-induced nausea/vomiting.[43] However, the U.S. Food and Drug Administration (FDA) does not allow the use of cannabis for medical purposes except under a few conditions where synthetic cannabinoids are approved for therapeutic use.[44] Recently, several states in the United States have approved *Cannabis sativa* use for a number of medical problems. See Section 15.5 for toxicity and management of ijuana toxicity under the section "Cannabis."

Cerbera manghas It is found in the geographic area between the Seychelles Islands in the Indian Ocean eastward to French Polynesia, and also in Sri Lanka, southern India, and South China. The leaves and fruits of *Cerbera manghas* contain the potent cardiac glycoside cerberin that is extremely poisonous if ingested. In traditional medicine, it is used as a cardiac stimulant, analgesic, antiinflammatory, and anticarcinogenic. The toxic content of cerberin is similar to cardiac glycoside, and management and complications are similar to digitalis toxicity.[45–47]

Chan Su Extract It also known as Venenum bufonis, is an extract traditionally used in Chinese medicine and applied topically for anesthetic effect and as an antiarrhythmic drug.[48] The active ingredient is bufadienolides, such as bufalin, cinobufagin, and resibufogenin. Bufalin (structurally similar to digoxin) is known to increase vasoconstriction, vascular resistance, and blood pressure, probably by inhibiting Na/K-ATPase activity and/or interfering with calcium ion concentration.[48] Chan Su in high doses was found to cause cardiac arrhythmias in women followed by death after the consumption of Chinese herbal tea containing Chan Su.[49]

In a study among healthy volunteers, serum digoxin level was estimated for those who took a Chan Su extract pill, and a case of death was reported following ingestion of it with a level of 4.9 ng/ml.[50] Chan Su can also interfere in serum digoxin measurement by immunoassays. Bick et al. concluded that Chan Su acts as a highly competitive inhibitor of Na/K-ATPase and it can be extremely cardiotoxic, even in small doses, and acts rapidly to alter intracellular calcium stores from cardiomyocytes. The treatment with Digibind, and the possible use of beta blockers and calcium release channel blockers, may "rescue" myocytes before irreversible damage occurs.[48]

Dandelion It (i.e., *Taraxacum officinale*) is used in home remedies and Mexican and North American folklore medicine. The use of the plant has been associated with prolonged QT interval with frequent episodes of sustained polymorphic ventricular tachycardia as a result of electrolyte disturbances and its direct toxic effect on the nodal pathway.[51] Treatment is supportive.

Datura innoxia It also is known as thorn-apple, downy thorn-apple, Indian-apple, lovache, and moonflower and is found in the Western Himalayas and hilly regions of western parts of the Indian peninsula, abundantly in Maharashtra. The plant is the source of alkaloid scopolamine, which is used as a preanesthetic during surgery and childbirth, in ophthalmology, and for the prevention of motion sickness. Atropine, hyoscyamine, hyoscine, and meteloidine are found in the leaves, flowers, pericarp, and seeds of the plant. The various parts of the plant are used in treatment of headache, hemiplegia, epilepsy, delirium, convulsions, cramps, rigid thigh muscles, and rheumatism. Seeds are used for treatment of asthma and diarrhea, and if accidentally consumed cause toxicity, manifesting as tachycardia, mydriasis, and dry skin. Hallucinations, delirium, or coma can be life threatening. There might be blurred vision, hyperthermia, or urinary retention.

Cardiovascular toxic manifestations are similar to anticholinergic toxicity and, predominantly, antimuscarinic effects such as acute myocardial infarction (primarily vasospasm), tachyarrhythmias, and pulmonary edema (due to decreased contractility through channelopathy).[52,53] Diagnosis of toxicity is clinical; in addition, sometimes toxicological analysis of ingredients (i.e., atropine/hyoscyamine) in the serum or urine samples and/or in the ingested content (if known/available) is possible.[53,54] Neostigmine or physostigmine can rapidly reverse the side effects. Respiratory failure and arrhythmias are the serious complications, and patients may need ventilatory support. Benzodiazepines may be needed for seizure control.

Edible Mushroom *Flammulina velutipes* is an edible mushroom that is widely eaten in the orient. Preparations from it are canned for local consumption. Flammutoxin, which is a cardiotoxin, has been isolated from this edible mushroom. This toxin is known to interfere with voltage-gated ion channels.[55] It causes depression of the ST segment and inversion of T waves.[56] Similarly, volvatoxin A2, a novel pore-forming cardiotoxic protein was isolated from another edible mushroom, *Volvariella volvacea*.[57] Volvatoxin A causes cardiac arrest by activation of Ca^{2+}-dependent ATPase activity. It also causes inhibition of Ca^{2+} accumulation in microsomal fraction of the sarcoplasmic reticulum of guinea pig ventricular muscle.[57,58] Ostreolysin, an acidic protein from the edible oyster mushroom (*Pleurotus ostreatus*), is a toxic, pore-forming cytolysin.[59]

In experimental models, ostreolysin has been demonstrated to cause a transient increase in arterial blood pressure followed by a progressive fall to mid-circulatory pressure accompanied by bradycardia, myocardial ischemia, ST elevation or depression in ECG, endothelial dysfunction, sustained contraction of the coronary arteries, ventricular extrasystoles, arrhythmia (due to electrolyte disturbances in addition to channelopathy), and varying degrees of AV block.[59,60] Calcium sensitizers isolated from the edible pine mushroom, *Tricholoma matsutake* (S. Ito and Imai) Singer, act in the ventricular cells of rat and can significantly increase the calcium ion concentration (Ca^{2+}) in myocytes, which is dependent on the influx of extracellular Ca^{2+}.[61] The treatment of toxicity is supportive.

Ephedra sinica It is used in traditional Chinese medicine and Ayurvedic medicine for cough, cold, asthma, obesity, and edema. Because it enhances performance through brain stimulation, athletes misuse these agents.[62] The plants contain many alkaloids and nonalkaloidal compounds. The effective substance in the plants are ephedrine and pseudoephedrine.[63] Cardiovascular toxic manifestations of *Ephedra* include arrhythmias (e.g., ventricular tachycardia/fibrillation), acute myocardial infarction, vasoconstriction, hypersensitivity myocarditis, coronary artery aneurysm, thrombosis, and vasculitis.[64–67] Pathogenesis include endothelial dysfunction, stimulation of alpha and beta receptor agonist with increased secretion of catecholamines, and promoting inflammatory process. In experimental models, *Ephedra* activates the potassium ion channel and shortens the QT interval.[68] There is no specific antidote for overdose and treatment is supportive.

Hypothetically, since *Ephedra sinica* stimulates alpha and beta agonist and the use of a beta blocker in such situations may cause rebound phenomenon through stimulating alpha receptors, combined alpha and beta blockers are preferred.[69]

Ergot Alkaloids These have various medicinal uses that include dopamine agonists (e.g., bromocriptine, lisuride, pergolide, and cabergoline), migraine therapy (e.g., methysergide, ergotamine, and dihydroergotamine), drugs for cognitive and neurosensory deficits (e.g., nicergoline, dihydroergocryptine), and so on.[70] Ergot alkaloids are associated with valvular abnormalities of both regurgitant and stenotic type. Mechanisms for the adverse effect of ergot alkaloids are attributable to norepinephrine and serotonin contents present in it. These substances activate protein kinases and in turn promote transforming growth factor beta, which results in fibroblast proliferation and collagen synthesis and thus contributes to valvular pathology. Also, ergot alkaloids are known to precipitate peripartum cardiomyopathy, coronary vasospasm-induced acute myocardial infarction, and sinica conduction disturbances.[71–75] Treatment is withdrawal of the offending drugs and managing particular complications.

Home Remedies

Turmeric, ginger, capsicum, asafoetida, beetroot, and castor oil are some commonly plant materials used at home. Here we focus on the cardiotoxic effects of such ingredients. The treatment is invariably symptomatic and supportive.

Turmeric

Turmeric is a plant material often used as the main spice in at home preparation of curry. The medicinal values of turmeric are known and used by medical practitioners of Siddha, Ayurveda, and Indian systems of medicine. It is also used as a home remedy for many gastrointestinal, musculoskeletal, and skin disorders. Different forms of preparations, such as stick, powder, paste, liquid, and so on, are available. Adverse reactions to turmeric are known, though generally people believe it as harmless.

Turmeric is also called curcumin, *Curcuma*, *Curcuma aromatica*, and many other names. Curcumin (diferuloylmethane) is a polyphenol responsible for the yellow color of turmeric. It has antioxidant, antiinflammatory, anticarcinogenic, antithrombotic, and cardiovascular protective effects. The possible effect of curcumin is through stabilizing calcium ion homeostasis, which may play a role in the prevention of ventricular arrhythmias through the conduction pathway.[76] Also, curcumin inhibits hERG potassium channels (causing prolongation of cardiac repolarization) in vitro, which in turn is believed to have antiarrhythmic or proarrhythmic effects (through QT prolongation).[77] Interestingly, transient complete heart block was reported in a male after intake of curcumin-containing pills for one month and no further conduction disturbances were noticed on cessation of the pills.[78]

Ginger

Ginger (*Zingiber officinale*) has been used since time immemorial as a home remedy for many ailments. It has an effect on diuresis, heart rate (through channelopathy), and blood pressure. Ginger is believed to exert negative inotropic and chronotropic effects along with vasodilator mechanism through voltage-dependent calcium channel blockade.[79] Gul et al. reported bradycardia and hypotension in a 59-year-old male who consumed ginger at a dose of 450 mg per day for flu for 5 days; he was treated conservatively.[80]

Capsicum annum

The plant *Capsicum annum*, belonging to the family Solanaceae, is found in the kitchen gardens and cultivated areas. It is available in the form of bell peppers, cayenne peppers, jalapeños, and other hot peppers with capsicum being the active ingredient. It contains capsaicin and dihydrocapsaicin. It is used as a home remedy and in Chinese medicine for colds, bronchitis, improving blood circulation, indigestion, and as an appetizer. The available preparations include pills, powder, paste, and inhalers. There are reports available for capsicum-induced acute myocardial infarction and supraventricular tachycardia (AF) and among them, capsicum was administered as oral pills or topical paste or through inhalation.[81–83] Capsaicin activates the sympathetic system and also influences the cardiac action potential in atrial and ventricular myocytes through potassium- and calcium-gated channels.[84]

Asafoetida

Asafoetida is indigenous to Iran and western Afghanistan. It is a form of a resin obtained from the roots and rhizomes of the plant *Ferula asafoetida*. It has different chemical compounds and is frequently prescribed for abdominal colic by native medicine practitioners. In experimental models, asafoetida causes hypotension and thinning of blood through vasodilatation and increased fibrinolytic activity. As a result, asafoetida enhances the anticoagulation effect of warfarin.[85–87] In rabbit, asafoetida induced a lethargic effect on heart.[88]

Beetroot Extract

Practitioners of native medicine and naturopathy prescribe raw beetroot juice (500 ml/day for six days) to reduce blood pressure, and also a few studies conducted among healthy volunteers showed reduction in blood pressure.[89–91] Possible mechanisms include diuresis and vasodilation. Nitric oxide formed after the ingestion of raw beetroot juice (obtained through a reduction process) is known to exert vascular smooth muscle relaxation and vasodilatation. In addition, beetroot is known to exert endothelium independent vasodilatation.[92]

Castor Oil

Castor oil is extracted from castor seeds that are obtained from the plant *Ricinus communis*. In native and home medicine, it is given for constipation. Castor oil contains ricin (a toxic alkaloid) and toxalbumin ricin. Cardiotoxic effects are rare. Inversion of the T wave in leads II, III, and aVF, and bradycardia were reported.[93] The mechanism is generally due to fluid and electrolyte imbalance with direct cytotoxicity. In experimental models, ricin decreases systolic and diastolic left ventricular function, influences calcium hemostasis in the heart, and increases vascular permeability and hypotension.[94,95]

Kalanchoe laciniata

This plant is grown in a few regions of India (e.g., Maharashtra and Bengal) and traditional healers use it for fever, dyspepsia, skin allergy, diabetes, joint pain, bilious diarrhea/dysentery, wound healing, asthma, and bronchitis. Leaves have astringent and antiseptic properties and can be applied to insect bites. A variety of cardiotoxic bufadienolides are present in all parts of the *Kalanchoe laciniata*, also known as the Christmastree plant, more so in the flowers. The cardiotoxins, including bryotoxins, bryophyllins, bersalgenins, flavonoids, and glycosides, which have a similar action to the cardiac glycosides found in foxglove and oleander; cardiac toxic manifestations are similar to cardiac glycoside poisoning.[96] Their primary effect is inhibition of Na^+/K^+-adenosine triphosphatase (ATPase) thereby decreasing the transportation of sodium and potassium across cell membranes, which decreases cardiac function. Although there is considerable variation in the toxicity of the *Kalanchoe* species, all plants should be considered toxic until proven otherwise.[62] The treatment is supportive.

Licorice

Licorice is the extract of *Glycyrrhiza glabra* roots and is often used in ancient Siddha medicine and is approved by the German commission. Licorice is used for gastritis, cough, bronchitis, ulcers, inflammation, and epilepsy.[62] The association between the use of licorice and hypertension is well established. It is attributed to sodium retention, edema, and hypokalemia. Licorice ingestion/toxicity causes tachy-brady arrhythmias, fatal arrhythmias (possibly due to hypokalemia), and cardiomyopathy.[97–99] The hypertensive effect of licorice antagonizes the beneficial effects of angiotensin-converting-enzyme inhibitors (ACEIs) and beta blockers used to treat heart failure. It is suggested that patients with heart failure avoid licorice preparations due to its potential harmful effects.[100] Treatment is supportive including correction of electrolytes.

Neem

The neem tree (*Azadirachita indica*), though growing in tropical and subtropical regions, is a native tree of India. Various parts of the tree are well known for their medicinal properties and are prescribed by Ayurvedic, Siddha, and herbal medicine practitioners in India. In general, it is believed to be safe. Neem leaf has various properties such as antibacterial, antiviral, antiparasitic, antiinflammatory, anticarcinogenic, antioxidant, immune upregulation, and anti-snake venom activities; it is used for skin diseases and digestive system disorders.[101] In experimental studies, neem extracts influenced inotropic and chronotropic mechanisms, which in turn induced variable degrees of AV disturbances, arrhythmias, and hypotension.[102] Also, a case report revealed ventricular fibrillation and cardiac arrest following neem leaf poisoning.[103] The toxic effects are attributable to various terpenoids present in the plant's materials.

Oleander

Oleander (*Thevetia peruviana* or *Thevetia neriifolia*) is a potent cardiotoxic plant. Preparations from the plant are used in native medicines for cardiac diseases, myalgia, parasitic infestations, wound healing, abortifacient, and so on. Oleander paste applied to raw wounds and inhalation of smoke from burning twigs of the plants have resulted in cardiotoxicity, thereby indicating that the cardiotoxins of oleander are likely to get absorbed transcutaneously and via inhalation routes as well.[104,105] All parts of the oleander plant in either wet or dried form contain highly toxic cardenolides that are structurally and functionally similar to those of cardiac (digitalis) glycosides. Thus, toxic manifestations are similar to cardiac glycosides, as discussed earlier in this chapter. In experimental models, garlic extract reduced case fatality as a therapeutic and prophylactic agent. Therefore, it is likely that garlic extract can be considered as a potential and affordable antidote in oleander poisoning, in addition to the conventional measures for management of cardiac complications. Further studies are warranted, however, to establish the usefulness of garlic for oleander toxicity.[106]

Rauwolfia serpentina

Rauwolfia serpentina (*Ophioxylon serpentinum*) is prescribed by traditional medicine and Ayurvedic practitioners for a wide variety of clinical disorders including hypertension.[107] It inhibits the sympathetic system and reduces blood pressure. It also has parasympathomimetic actions. Cardiotoxic manifestations are bradycardia and ventricular premature contractions.[108]

Stephania tetrandra and Aristolochia fangchi

Stephania tetrandra is used in Chinese medicine for a variety of purposes. Tetrandrine, the active part of *Stephania tetrandra* radix extracts, has broad pharmacological activity, including effects on the cardiovascular system. In guinea pigs, tetradrine increases left ventricular preload. In a small study among patients who took *Stephania tetrandra* preparations along with *Aristolachia fangchi* for medicinal purposes, 30% were found to experience aortic regurgitation, which raised the suspicion for their association.[109,110]

Strophanthus

Strophanthus, which is of the Apocynaceae family, is a flowering plant that grows in tropical Africa, South Africa, southern India, the Philippines, Laos, Vietnam, and South China. *Strophanthus* has been used medicinally as a cardiac stimulant and in the treatment of peptic ulcer and snakebites. The extract possesses antimicrobial, wound-healing, antioxidant, analgesic, and anticarcinogenic properties.[111] The plant contains toxic alkaloids and cardiac glycosides (e.g., g-strophanthin (ouabain), k-strophanthin, and e-strophanthin). The clinical manifestations and management are similar to those for digitalis overdose.[112,113]

Tupistra chinensis

Tupistra chinensis is a flowering plant found in South Asia, from South China to Sumatra. It is traditionally used for the treatment of inflammatory diseases in southwestern China. The plant contains toxic saponins that are structurally similar to cardiac glycosides and this toxin is present in fruits, roots, rhizomes, leaves, bark, and stems of the plant. The clinical presentation and management of toxicity are similar to the other cardiac glycosides.[114]

6.2.2 Other Plant-Based Preparations

A vast number of plants and plant preparations are used as native medicines by South Americans. A few of them have serious cardiovascular toxicity, including ayahuasca, coca, and *Cecropia pachystachya*. These plants may increase diastolic blood pressure and heart rate or cause hypotension.[115–117] Botanical remedies were the mainstay of treatment (i.e., more than 200 different species) used by Native Americans.[7] They applied the homeopathic principle of "like cures like" with red flowers for bleeding disorders, yellow flowers for jaundice, the plant snakeroot for snakebites, and so on. Cathartics and emetics were used extensively by Native Americans to rid the body of "evil spirits." Some of these plants have potential risk for cardiovascular toxicity (refer Chapter 5).[118]

6.2.3 Animal-Based Preparations

Animal-based preparations are used and/or prescribed in native medicine for physical disabilities, weakness, and sexual disorders.

Coconut Crab Poisoning

Indo-Pacific area native islanders traditionally eat coconut crab (*Birgus latro L.*) and poisoning due to it has been described in Japan and French Polynesia. The claws, legs, fleshy parts, and oily abdominal sac are considered the edible parts of the crustaceans. The crustaceans contains nerifolin and patients who consume them manifest acute digitalis intoxication and may develop acute renal failure. Those affected respond well with digoxin-specific Fab antibody fragments.[119,120]

Seafood Material

Rarely, seafood materials are given by traditional healers or as home remedies for general well-being or sexual disorders. Among the sea materials used, the most potent toxins are maitotoxin and ciguatoxins. The toxicological manifestations of such toxins are discussed in Chapter 7.

Toad Toxin

Toad soup is prescribed for an aphrodisiac effect by native medicine practitioners in many places, more so in Laos and in some parts of India. Toad secretions from skin and parotid contain several toxic substances, similar to cardiac glycosides. The complications and management of toad poisoning are similar to toad toxin (see Section 4.7 on toad poisoning). Education of traditional practitioners and the community on the seriousness of toad toxins is a useful public health measure to avoid poisoning.[49]

6.3 BIOLOGICAL MATERIALS AND HERBOMINERAL-BASED PREPARATIONS

As evolution advanced, humans started separating minerals from natural resources and found its values. Later, along with plant and/or herbal preparations, practitioners started mixing in various forms of mineral salts, either to enhance or to improve medicines' potency. Thus, herbomineral preparations entered the portal of native medicine.

6.3.1 Mad Honey Poisoning

Mad honey, also called bitter honey, was known even before the birth of Christ. It causes a sharp burning sensation in the throat. Mad honey is used as a household remedy and alternative medicine in the Black Sea region for the treatment of stomach pain, bowel disturbances, diabetes, hypertension, and sexual disorders. Reports about mad honey poisoning have come from Turkey and a few other parts of the world. Bees collect honey from the toxic *Rhododendron* plant species; the nectar contains grayanotoxins. The sources for mad honey are from a certain group of plant species[27] such as *Agauria*, *Andromeda*, and *Kalmia*; and *Rhododendron ponticum*, *Rhododendron flavum*, *Azalea pontica*, and *Kalmia latifolia*, all of which belong to the family Ericaceae. One another species, *Paullinia australis*, belonging to the family Sapindaceae, is also included for mad honey. These toxins block sodium channels in cell membranes and prevent inactivation, thus the excitable cells are in a state of depolarization, during which time the entry of calcium into cells may be facilitated.

The honey's toxic effects are dose-dependent, which may vary from 5 to 30 g. Severe toxicity may lead to life-threatening cardiac complications. However, clinical manifestations and cardiac complications may differ from case to case. Since these toxins are metabolized and excreted rapidly, patients may feel better within hours, blood pressure and heart rate may return to normal within 2 to 9 hours, and regain consciousness. Rarely do these toxins cause seizures and death. Clinical details about grayanotoxins were discussed earlier in this chapter. Grayanotoxins affect blood glucose and lipid levels. The decrease in blood sugar is attributable to the effects of mad honey, which cause islets of Langerhans to secrete insulin by stimulating the parasympathetic nervous system or M2 muscarinic receptors.

Mad honey causes sinus bradycardia, nodal rhythm, and various degrees of AV block including asystole, prolonged QT, junctional rhythm with escape, atrial fibrillation, transient coronary vasospasm (manifested as transient ST elevation and left bundle branch block), inferior ST-segment elevation myocardial infarction (STEMI) and non-STEMI with ST depression.[121–126] Most of these respond well with atropine injection, IV fluids, and vasopressors, if needed. Practitioners and emergency physicians should remember mad honey toxicity if they see a case of unexplained bradycardia, arrhythmia, or hypotension, especially in areas of Middle Eastern countries, and if the patient has consumed mad honey. (Also refer to Chapter 3 for other information about mad honey poisoning.)

6.3.2 Mercury

Mercury salts are used along with herbal materials in traditional Tibetan medicine[127] and Siddha medicine, an ancient system practiced by the Tamil-speaking population found in the southern peninsula of India. It is also used in Ayurvedic medicine.[128] Mercury affects the heart in various ways, including cardiac muscle contractility, electrical conductivity, and regulation of cardiac activity. Refer to Chapter 3 for information about mercury and poisoning by other minerals.

6.4 DRUG INTERACTIONS AS A RESULT OF HERBS

Herbal medicines often interact with prescribed drugs and medicines. As many patients with cardiovascular disease use CAM, some of the herbal remedies may enhance or reduce the pharmacological effects of anticoagulant and antiplatelet drugs with systemic or local consequences.[129,130] The phytoestrogen content of herbal medicine may also induce unintended adverse effects[131] and interact with prescribed medicines. Some home recipes and herbs are known to interact with cardiovascular drugs. In a study conducted among 250 patients with atrial fibrillation on warfarin, the international normalized ratio (INR) was found to be subtherapeutic for the patients who were frequent users of garlic, ginger, fenugreek, papaya, green tea, and psyllium seed.[132] From a recent review on herb and recipe intraction with warfarin based on the available clinical evidence, cannabis, chamomile, cranberry, garlic, ginkgo, grapefruit, lycium, red clover, and St. John's wort were evaluated and shown to have major severe interactions with warfarin.[133]

Bello et al.[134] highlighted the potential interaction between herbal medicines, such as garlic, and prescribed drugs aggravating the hypotensive effect of lisinopril, increasing the antihypertensive activity of ACE inhibitors and beta blockers, and enhancing the inhibitory chronotropism of dilitiazem. Similarly, linseed increases the risk of hypotension of ramipril; clove oil enhances the risk of bleeding while the patient is on antiplatelets or dicoumarin or NASIDs; the flavanoids of herbal tea may increase the oral absorption of digoxin and quinine; and peppermint may cause additive effects. Fennel used with loop diuretics may cause photosensitivity and increased risk of sunburn due to its interaction with enalapril.[134] Health care professionals, therefore, should encourage their patients to disclose the nature of native medicine or CAM used by them and understand its effects on the medicines prescribed among them.

Patients may consume one or more plant derivatives along with their modern medicines with the fond hope of early recovery or to reduce the dose of prescribed drugs. As a result, the constituents of plant medicines may affect the absorption, distribution, metabolism, and excretion of the drug and exert pharmacodynamic interactions. Since plant medicines act mostly through cytochrome P450 (CYP), which has a genetic polymorphism, the potency or adverse effects vary from person to person and it is difficult to assess or predict these effects clinically. Please refer to Section 8.21 for drug interactions.

6.5 SAFETY, COSTS, AND REGULATIONS

Native medicines' use is practiced all over the world and involves numerous expenditures for the public and health care providers. However, their safety and their regulation have not been updated or monitored; thus, there is an urgent need to evaluate the facts with regard to the relevant issues.

6.5.1 Safety of Native Medicine for Children

The Canadian Pediatric Surveillance Program (CPSP) states that use of health products and spinal manipulations are popular therapies for children.[135] The survey revealed the occurrence of adverse events after the use of natural health products. Serious adverse events are likely to be under-recognized and underreported. Moreover, limited data are available on the interaction between CAM and other modes of therapy for children. Therefore pediatricians, in a nonjudgmental manner, should be motivated to inquire about the use of CAM as a part of their routine history-taking from parents or caregivers who accompany children. Disclosure of native medicine/CAM use and early recognition not only prevent adverse reactions and herb–drug interactions, but also avoid untoward events. Moreover, many of the native/herbal medicines have been adulterated with heavy metals or undisclosed prescribed drugs and thereby constitute serious health risks.[136]

6.5.2 Economic Aspects

Worldwide, CAM is used by a vast majority of patients. In the United States, the expenditure for complementary and alternative medicine was $14 billion in 1990, $27 billion in 1997, and $34 billion in 2007.[137,138] The ethnobotanical market is estimated to reach $60 billion (please refer Chapter 5). Cost-effectiveness, cost–utility, and cost–benefit analyses of CAM revealed that few of the medications are cost-effective for some of the clinical situations. After analyzing 338 economical evaluations of CIM, Herman et al. evaluated the implications to policymakers, clinicians, and future researchers.[138]

6.5.3 Legal Status and Regulations

WHO has made an attempt to collect the legal status of traditional and complementary/alternative medicine from 191 member states.[139] A total of 123 countries responded that they practiced one or more forms of traditional medicine. Health authorities have made some efforts to ensure necessary regulations and legal status to promote and maintain good practice and ensure authenticity, safety, and efficacy. In India for centuries, Ayurveda, Siddha, and Unani systems of medicine have been practiced. Siddha has close similarities to Ayurveda in practice, the main difference between them is linguistic: Tamil versus Sanskrit.[139]

Most native medicine/CAM preparations are not licensed as medicines and therefore evidence of their quality, efficacy, and safety have not been assessed by various countries' governing bodies.[140] Traditional Chinese medicine has an established history in Australia; it now accounts for 3.2% of the total use of complementary medicine.[141] The quality of herbal medicine refers to the intrinsic properties of the herb, which accounts for the active constituents present in it. The influencing factors for the quality of herbal medicines are species of the plant, place of cultivation, season, agricultural system adopted, time of harvest, processing method, good manufacturing practices, and others (refer to Table 5.2). The status of safety and quality assessment of herbal medicines have been described in general and also in particular disorders in two studies.[142,143]

The issues associated with the quality and the safety of herbal medicines include toxic herbs, contamination or adulteration with heavy metals (e.g., mercury and arsenic), and interactions with prescription or nonprescription drugs (e.g., paracetamol, indomethacin, antidiabetics, antihypertensives, antihistamines, steroids). Chinese herbal products are regulated by the therapeutic goods administration (TGA) in Australia, which assesses their quality and safety. Australian practitioners prescribe herbal medicines even though the country's regulations for them do not specify their prescribing and dispensing functions.[144] In contrast, in India doctors of modern medicine registered under the Indian Medical Council Act are permitted to prescribe only allopathic medicine (i.e., orthodox, scientific, Western, modern) as per a Supreme Court of India judgment. If they happen to prescribe nonallopathic medicines, doctors are liable to be prosecuted under civil and/or criminal laws, resulting in cancelation of registration and/or heavy fines and/or imprisonment.

6.6 FUTURE DIRECTIONS

Clinicians and practitioners should analyze the risks of native/herbal medicines and present the results to patients, the public, professionals, and policymakers.

6.6.1 Clinical Aspects

Identification of host genetic factors that influence absorption, metabolism, and the action of native medicines at the receptor level may help to prevent adverse effects. Currently available pharmacovigilance programs should include documentation of unwanted effects of native medicines/CAM. Cardiologists and emergency physicians should be encouraged to elicit a clinical history of coadministration of native medicines with prescribed cardiovascular drug, as native/herbal medicines are likely to potentiate or reduce the effect of the ones prescribed, which may lead to complications.

6.6.2 Education and Training

Teaching and training programs for medical students should include the potential interaction between native medicine/CAM and conventional drugs as well as underlying diseases. Incorporation of complementary and alternative medicine as a subject in undergraduate curricula may benefit health care practitioners regarding its usefulness, drug–herb interactions, adverse effects, and limitations. Adequate interactive educational programs should be designed to teach and train practitioners and the public about the applications and importance, as well as risks, of native medicines/CAM.

6.6.3 Research

Health science research should highlight the usefulness and limitations of native medicines. Various forms of economical evaluation, including out-of-pocket expenses for native medicine, should be undertaken, and its implications in terms of clinical, professional, and policy matters have to be identified at inter- and intracountry levels. Basic research studies on mad honey and methods to distinguish it from nontoxic honey is the need of the hour. The embroyotoxic, fetotoxic, and abortifacient risks of herbal/plant medicines have to be explored. Efforts should be undertaken to elucidate the risk of native/plant medicines and to find newer therapeutic strategies to treat overdose and/or poisoning. Various forms of bioavailability techniques may be applied to improve the bioavailabilities of native/herbal medicines.

6.6.4 Regulatory Aspects and Limitations

A surveillance should be done on current legal regulations, processing, and prescription status of native medicines. There is a need to explore and assess the knowledge, attitude, behavior, and practice of the public, patients, health care professionals, and policymakers about CAM; in addition, it is necessary to elicit inter- or intracountry variations, so as to introduce regulations and standard guidelines for complementary and alternative medicines' use at the global level. Marketing of native medicines through the Internet should be analyzed and regulated. Standards need to focus on safety, effectiveness, availability, and accessibility of native medicine/CAM to the public.

Present reports are based mostly on published materials made available via PubMed and well-known databases that describe the cardiotoxic effects of native medicines. Poor documentation; lack of scientific rigor on the reported cases; sociocultural influences; unawareness among patients, the public, and professionals; inadequacies in the standardization of the products; and interobserver or individual biases may have influenced the presentations in the published reports.

6.7 CONCLUSION

Native medicines are used all over the world due to their continued popularity and inherent and associated myths and facts. These are marketed frequently across the borders and sold on the Internet. Native medicine, however, is not always harmless; it produces life-threatening complications at times. Some of the medicines contain not only botanical but also mineral or animal-derived ingredients. Before evaluating and identifying native or herbal medicines to use, consider the ingredients in them. Cardiotoxic manifestations are attributable to the quantity and quality of what is in native medicine, their processing, and other health factors related to individuals. Thus, in a given clinical situation, history related to consumption of a native medicine and its probable influence on physiological, pharmacological, and possible toxic effects, as well as interaction with prescribed drugs and/or an associated illness have to be considered and elucidated as much as possible, along with proper documentation. Future directives should focus on clinical, educational, economical, regulatory, and basic research aspects to ensure safety for patients and medications' effectiveness.

Key Learning Points

- Native medicines are used globally and some of them have cardiotoxicity.
- The cardiotoxic manifestations are attributable to the quantity and quality of the contents of native medicines, their potency, pathogenesis, and other health factors related to an individual.
- Acute myocardial infarction with normal coronary arteries in an appropriate setting should raise the suspicion of poisoning.

- The need of the hour is to create awareness about adverse events and interactions between native medicines and prescribed drugs or illnesses among patients, public, professionals, and policymakers; guidelines should be introduced for prescribers and users.
- Regulations and legal systems have to be strengthened to ensure patient safety and drug effectiveness.
- Marketing of native medicines through the Internet needs to be monitored.

REFERENCES

1. Raghavendra HL, Yogesh HS, Gopalakrishna B, et al. An overview of herbal medicine. *Int J Ph Sci* 2009;**1**(1):1–20.
2. Rates SM. Plants as source of drugs. *Toxicon* 2001;**39**(5):603–13.
3. Raskin I, Ribnicky DM, Komarnytsky S, et al. Plants and human health in the twenty-first century. *Trends Biotechnol* 2002;**20**(12):522–31.
4. Ernst E. Harmless herbs? A review of the recent literature. *Am J Med* 1998;**104**(2):170–8.
5. WHO. *Traditional medicine*. Geneva, Switzerland: WHO; 2003 [accessed 04-04-14]. Available from: *www.who.int/mediacentre/factsheets/2003/fs134/en/*.
6. Bussmann RW, Sharon D. Traditional medicinal plant use in northern Peru: tracking two thousand years of healing culture. *J Ethnobiol Ethnomed* 2006;**2**:47.
7. Borchers AT, Keen CL, Stern JS, et al. Inflammation and Native American medicine: the role of botanicals. *Am J Clin Nutr* 2000;**72**(2):339–47.
8. Bailey CJ, Day C. Traditional plant medicines as treatments for diabetes. *Diabetes Care* 1989;**12**(8):553–64.
9. DiNardo MM, Gibson JM, Siminerio L, et al. Complementary and alternative medicine in diabetes care. *Curr Diab Rep* 2012;**12**(6):749–61.
10. Osamor PE, Owumi BE. Complementary and alternative medicine in the management of hypertension in an urban Nigerian community. *BMC Complement Altern Med* 2010;**10**:36.
11. Hughes GD, Aboyade OM, Clark BL, et al. The prevalence of traditional herbal medicine use among hypertensives living in South African communities. *BMC Complement Altern Med* 2013;**13**:38.
12. Gronhaug TE, Glaeserud S, Skogsrud M, et al. Ethnopharmacological survey of six medicinal plants from Mali, West-Africa. *J Ethnobiol Ethnomed* 2008;**4**:26.
13. NIH. *Complementary, Alternative, or Integrative Health: What's in a Name?* Bethesda, MD: NIH; 2008 [accessed 5-4-14]. Available from: *http://nccam.nih.gov/health/whatiscam*.
14. Mashour NH, Lin GI, Frishman WH. Herbal medicine for the treatment of cardiovascular disease: clinical considerations. *Arch Intern Med* 1998;**158**(20):2225–34.
15. Nauman E. Native American medicine and cardiovascular disease. *Cardiol Rev* 2007;**15** (1):35–41.
16. Hollman A. Plants and cardiac glycosides. *Br Heart J* 1985;**54**(3):258–61.
17. Page E. The actions of cardiac glycosides on heart muscle cells. *Circulation* 1964;**30**: 237–51.
18. Amran MS, Hashimoto K, Homma N. Effects of sodium-calcium exchange inhibitors, KB-R7943 and SEA0400, on aconitine-induced arrhythmias in guinea pigs in vivo, in vitro, and in computer simulation studies. *J Pharmacol Exp Ther* 2004;**310**(1):83–9.

19. Yamanaka H, Doi A, Ishibashi H, et al. Aconitine facilitates spontaneous transmitter release at rat ventromedial hypothalamic neurons. *Br J Pharmacol* 2002;**135**(3):816–22.
20. Tai YT, But PP, Young K, et al. Cardiotoxicity after accidental herb-induced aconite poisoning. *Lancet* 1992;**340**(8830):1254–6.
21. Dhesi P, Ng R, Shehata MM, et al. Ventricular tachycardia after ingestion of Ayurveda herbal antidiarrheal medication containing *Aconitum*. *Arch Intern Med* 2010;**170**(3):303–5.
22. Smith SW, Shah RR, Hunt JL, et al. Bidirectional ventricular tachycardia resulting from herbal aconite poisoning. *Ann Emerg Med* 2005;**45**(1):100–1.
23. Chan TY. Aconite poisoning presenting as hypotension and bradycardia. *Hum Exp Toxicol* 2009;**28**(12):795–7.
24. Chan TY, Tomlinson B, Tse LK, et al. Aconitine poisoning due to Chinese herbal medicines: a review. *Vet Hum Toxicol* 1994;**36**(5):452–5.
25. Lin CC, Chan TY, Deng JF. Clinical features and management of herb-induced aconitine poisoning. *Ann Emerg Med* 2004;**43**(5):574–9.
26. Zhao D, Wang J, Cui Y, et al. Pharmacological effects of Chinese herb aconite (fu zi) on cardiovascular system. *J Tradit Chin Med* 2012;**32**(3):308–13.
27. Koca I, Koca AF. Poisoning by mad honey: a brief review. *Food Chem Toxicol* 2007;**45** (8):1315–8.
28. Guha S, Dawn B, Dutta G, et al. Bradycardia, reversible panconduction defect and syncope following self-medication with a homeopathic medicine. *Cardiol* 1999;**91**(4):268–71.
29. Ono T, Hayashida M, Tezuka A, et al. Antagonistic effects of tetrodotoxin on aconitine-induced cardiac toxicity. *J Nippon Med Sch* 2013;**80**(5):350–61.
30. Vaidya AD, Raut AA, Vaidya RA. *Abrus precatorius*, Gaertin: an Ayurvedic potent phytomedicine. *J Assoc Physicians India* 2005;**53**:739–40.
31. Vinod KV, Thabah MM, Venkatesh T, et al. A rare cause of dural venous sinus thrombosis. *Neurol India* 2013;**61**(6):669–70.
32. Pillay VV, Bhagyanathan PV, Krishnaprasad R, et al. Poisoning due to white seed variety of *Abrus precatorius*. *J Assoc Physicians India* 2005;**53**:317–9.
33. Dickers KJ, Bradberry SM, Rice P, et al. Abrin poisoning. *Toxicol Rev* 2003;**22**(3): 137–42.
34. Verpoorte R, Siwon J, Tieken MEM, Svendsen AB. Studies on Indonesian medicinal plants, V. The alkaloids of *Anamirta cocculus*. *J Nat Prod* 1981;**44**(2):221–4.
35. Palasuwan A, Sooganun S, Lertlum T, et al. Inhibition of Heinz body induction in and in vitro model and total antioxidant activity of medicinal Thai plants. *Asian Pacific J Cancer Prev* 2005;**6**:458–63.
36. El Idrissi A, Okeke E, Yan X, et al. Taurine regulation of blood pressure and vasoactivity. *Adv Exp Med Biol* 2013;**775**:407–25.
37. Wang J, Wang X, Irnaten M, et al. Endogenous acetylcholine and nicotine activation enhances GABAergic and glycinergic inputs to cardiac vagal neurons. *J Neurophysiol* 2003; **89**(5):2473–81.
38. Shi LS, Kuo SC, Sun HD, et al. Cytotoxic cardiac glycosides and coumarins from *Antiaris toxicaria*. *Bioorg Med Chem* 2014;**22**(6):1889–98.
39. Shi LS, Liao YR, Su MJ, et al. Cardiac glycosides from *Antiaris toxicaria* with potent cardiotonic activity. *J Nat Prod* 2010;**73**(7):1214–22.
40. Deng JF, Ger J, Tsai WJ, et al. Acute toxicities of betel nut: rare but probably overlooked events. *J Toxicol Clin Toxicol* 2001;**39**(4):355–60.
41. Tsai WC, Wu MT, Wang GJ, et al. Chewing areca nut increases the risk of coronary artery disease in Taiwanese men: a case-control study. *BMC Public Health* 2012;**12**:162.

42. Chu NS. Effect of betel chewing on RR interval variation. *J Formos Med Assoc* 1995;**94**(3):106–10.
43. Kashyap S, Kashyap K. Medical ijuana: a panacea or scourge. *Lung India* 2014;**31**(2): 145–8.
44. FDA. *Drug Enforcement Administration*. Washington, DC: U.S. FDA; 2013 [accessed 2014]. Available from: *www.accessdata.fda.gov/scripts/cdrh/cfdocs/cfcfr/CFRSearch.cfm?fr = 1308.11.*
45. Eddleston M, Haggalla S. Fatal injury in eastern Sri Lanka, with special reference to cardenolide self-poisoning with *Cerbera manghas* fruits. *Clin Toxicol* 2008;**46**(8):745–8.
46. Cheenpracha S, Karalai C, Rat APY, et al. New cytotoxic cardenolide glycoside from the seeds of *Cerbera manghas. Chem Pharm Bull (Tokyo)* 2004;**52**(8):1023–5.
47. Eddleston M, Persson H. Acute plant poisoning and antitoxin antibodies. *J Toxicol Clin Toxicol* 2003;**41**(3):309–15.
48. Bick RJ, Poindexter BJ, Sweney RR, et al. Effects of Chan Su, a traditional Chinese medicine, on the calcium transients of isolated cardiomyocytes: cardiotoxicity due to more than Na, K-ATPase blocking. *Life Sci* 2002;**72**(6):699–709.
49. Ko RJ, Greenwald MS, Loscutoff SM, et al. Lethal ingestion of Chinese herbal tea containing ch'an su. *West J Med Surg* 1996;**164**(1):71–5.
50. Panesar NS. Bufalin and unidentified substance(s) in traditional Chinese medicine cross-react in commercial digoxin assay. *Clin Chem* 1992;**38**(10):2155–6.
51. Agarwal SC, Crook JR, Pepper CB. Herbal remedies: how safe are they? A case report of polymorphic ventricular tachycardia/ventricular fibrillation induced by herbal medication used for obesity. *Int J Cardiol* 2006;**106**(2):260–1.
52. Krenzelok EP. Aspects of *Datura* poisoning and treatment. *Clin Toxicol* 2010;**48**(2): 104–10.
53. Papoutsis I, Nikolaou P, Athanaselis S, et al. Mass intoxication with *Datura innoxia*: case series and confirmation by analytical toxicology. *Clin Toxicol* 2010;**48**(2):143–5.
54. Konishi T, Akaki K, Hatano K. Determination of hyoscyamine and scopolamine in serum and urine of humans by liquid chromatography with tandem mass spectrometry. *[Shokuhin eiseigaku zasshi.] J Food Hyg Soc Japan* 2008;**49**(4):266–71.
55. Tadjibaeva G, Sabirov R, Tomita T. Flammutoxin, a cytolysin from the edible mushroom *Flammulina velutipes*, forms two different types of voltage-gated channels in lipid bilayer membranes. *Biochim Biophys Acta* 2000;**1467**(2):431–43.
56. Lin JY, Lin YJ, Chen CC, et al. Cardiotoxic protein from edible mushrooms. *Nature* 1974; **252**(5480):235–7.
57. Weng YP, Lin YP, Hsu CI, et al. Functional domains of a pore-forming cardiotoxic protein, volvatoxin A2. *J Biol Chem* 2004;**279**(8):6805–14.
58. Fassold E, Slade AM, Lin JY, et al. An effect of the cardiotoxic protein volvatoxin A on the function and structure of heart muscle cells. *J Mol Cell Cardiol* 1976;**8**(7):501–19.
59. Zuzek MC, Macek P, Sepcic K, et al. Toxic and lethal effects of ostreolysin, a cytolytic protein from edible oyster mushroom (*Pleurotus ostreatus*), in rodents. *Toxicon* 2006;**48**(3): 264–71.
60. Juntes P, Rebolj K, Sepcic K, et al. Ostreolysin induces sustained contraction of porcine coronary arteries and endothelial dysfunction in middle- and large-sized vessels. *Toxicon* 2009;**54**(6):784–92.
61. Hou Y, Sun S, Wu L, et al. Calcium sensitizers isolated from the edible pine mushroom, *Tricholoma matsutake* (S. Ito & Imai) Singer. *Z. Naturforsch C* 2013;**68**(3-4):113–7.
62. Brown AC. Potentially life threatening herbs: reported cases in MEDLINE of liver toxicity, renal toxicity, cardiotoxicity, cancer, and death. *Exp Biol* 2002;(April):20–4.

63. Barrueto Jr. F, Kirrane BM, Cotter BW, et al. Cardioactive steroid poisoning: a comparison of plant- and animal-derived compounds. *J Med Toxicol* 2006;**2**(4):152–5.
64. Samenuk D, Link MS, Homoud MK, et al. Adverse cardiovascular events temporally associated with ma huang, an herbal source of ephedrine. *Clinic Proc* 2002;**77**(1):12–6.
65. Cockings JG, Brown M. Ephedrine abuse causing acute myocardial infarction. *Med J Aust* 1997;**167**(4):199–200.
66. Zaacks SM, Klein L, Tan CD, et al. Hypersensitivity myocarditis associated with *Ephedra* use. *J Toxicol Clin Toxicol* 1999;**37**(4):485–9.
67. Flanagan CM, Kaesberg JL, Mitchell ES, et al. Coronary artery aneurysm and thrombosis following chronic *Ephedra* use. *Int J Cardiol* 2010;**139**(1):e11–3.
68. Jing H, Luo L, Li H, et al. Ephedrine controls heart rhythms by activating cardiac I(ks) currents. *J Cardiovasc Pharmacol* 2010;**55**(2):145–52.
69. Thomas JE, Munir JA, McIntyre PZ, et al. STEMI in a 24-year-old man after use of a synephrine-containing dietary supplement: a case report and review of the literature. *Texas Heart Inst J* 2009;**36**(6):586–90.
70. Drug-related cardiac valve disease. *Prescrire Int* 2013;22(144):292–95. Available from: www.ncbi.nlm.nih.gov/pubmed/24600729#.
71. Y-Hassan S, Jernberg T. Bromocriptine-induced coronary spasm caused acute coronary syndrome, which triggered its own clinical twin-takotsubo syndrome. *Cardiol* 2011;**119**(1): 1–6.
72. Andrejak M, Tribouilloy C. Drug-induced valvular heart disease: an update. *Arch Cardiovasc Dis* 2013;**106**(5):333–9.
73. Aursulesei V. Peripartum cardiomyopathy: a challenge for cardiologist. *Rev Med Chir Soc Med Nat Iasi* 2013;**117**(2):358–67.
74. Patel SM, Ohori NP, Badhwar V, et al. From headache to heartache: ergotamine-induced aortic and mitral valvulopathy. *J Am Coll Cardiol* 2013;**62**(22):2144.
75. Gowri V, Al Hinai A. Postpartum second degree heart block induced by Methergine. *Int J Gynaecol Obstet* 2003;**81**(2):227–9.
76. Wongcharoen W, Phrommintikul A. The protective role of curcumin in cardiovascular diseases. *Int J Cardiol* 2009;**133**(2):145–51.
77. Hu CW, Sheng Y, Zhang Q, et al. Curcumin inhibits hERG potassium channels in vitro. *Toxicol Lett* 2012;**208**(2):192–6.
78. Lee SW, Nah SS, Byon JS, et al. Transient complete atrioventricular block associated with curcumin intake. *Int J Cardiol* 2011;**150**(2):e50–2.
79. Ghayur MN, Gilani AH. Ginger lowers blood pressure through blockade of voltage-dependent calcium channels. *J Cardiovasc Pharmacol* 2005;**45**(1):74–80.
80. Gul EE, Erdogan HI, Erer M, et al. Herbal syncope: ginger-provoked bradycardia. *Am J Emerg Med* 2012;**30**(2):386.e5–7.
81. Sogut O, Kaya H, Gokdemir MT, et al. Acute myocardial infarction and coronary vasospasm associated with the ingestion of cayenne pepper pills in a 25-year-old male. *Int J Emerg Med* 2012;**5**:5.
82. Akcay AB, Ozcan T, Seyis S, et al. Coronary vasospasm and acute myocardial infarction induced by a topical capsaicin patch. *Turk Kardiyoloji Dernegi arsivi: Turk Kardiyoloji Derneginin yayin organidir* 2009;**37**(7):497–500.
83. Cil H, Atilgan ZA, Islamoglu Y, et al. Is the pepper spray a triggering factor in myocardial infarction? A case report. *Eur Rev Med Pharmacol Sci* 2012;**16**(Suppl 1):73–4.
84. D'Alonzo AJ, Grover GJ, Darbenzio RB, et al. In vitro effects of capsaicin: antiarrhythmic and antiischemic activity. *Eur J Pharmacol* 1995;**272**(2-3):269–78.

85. Heck AM, DeWitt BA, Lukes AL. Potential interactions between alternative therapies and warfarin. *AJHP* 2000;**57**(13):1221–7 quiz 8-30
86. Fatehi M, Farifteh F, Fatehi-Hassanabad Z. Antispasmodic and hypotensive effects of *Ferula asafoetida* gum extract. *J Ethnopharmacol* 2004;**91**(2-3):321–4.
87. Mahendra P, Bisht S. *Ferula asafoetida*: traditional uses and pharmacological activity. *Pharmacognosy Rev* 2012;**6**(12):141–6.
88. Soudamini KK, Unnikrishnan MC, Sukumaran K, et al. Mutagenicity and anti-mutagenicity of selected spices. *Indian J Physiol Pharmacol* 1995;**39**(4):347–53.
89. Lansley KE, Winyard PG, Fulford J, et al. Dietary nitrate supplementation reduces the O_2 cost of walking and running: a placebo-controlled study. *J Appl Physiol* 2011;**110**(3):591–600.
90. Ferreira LF, Behnke BJ. A toast to health and performance! Beetroot juice lowers blood pressure and the O_2 cost of exercise. *J Appl Physiol* 2011;**110**(3):585–6.
91. Coles LT, Clifton PM. Effect of beetroot juice on lowering blood pressure in free-living, disease-free adults: a randomized, placebo-controlled trial. *Nutr J* 2012;**11**:106.
92. Hobbs DA, Goulding MG, Nguyen A, et al. Acute ingestion of beetroot bread increases endothelium-independent vasodilation and lowers diastolic blood pressure in healthy men: a randomized controlled trial. *J Nutr* 2013;**143**(9):1399–405.
93. Al-Tamimi FA, Hegazi AE. A case of castor bean poisoning. *Sultan Qaboos Univ Med J* 2008;**8**(1):83–7.
94. Christiansen VJ, Hsu CH, Dormer KJ, et al. The cardiovascular effects of ricin in rabbits. *Pharmacol Toxicol* 1994;**74**(3):148–52.
95. Ma L, Hsu CH, Patterson E, et al. Ricin depresses cardiac function in the rabbit heart. *Toxicol Appl Pharmacol* 1996;**138**(1):72–6.
96. Jhuma Deb GKD. *Kalanchoe laciniata* (L.) DC: a lesser known indian medicinal plant. *IJSIT* 2013;**2**(2):158–62.
97. Eriksson JW, Carlberg B, Hillorn V. Life-threatening ventricular tachycardia due to liquorice-induced hypokalaemia. *J Intern Med* 1999;**245**(3):307–10.
98. Hasegawa J, Suyama Y, Kinugawa T, et al. Echocardiographic findings of the heart resembling dilated cardiomyopathy during hypokalemic myopathy due to licorice-induced pseudoaldosteronism. *Cardiovasc Drugs Ther* 1998;**12**(6):599–600.
99. Crean AM, Abdel-Rahman SE, Greenwood JP. A sweet tooth as the root cause of cardiac arrest. *Can J Cardiol* 2009;**25**(10):e357–8.
100. Amabile CM, Spencer AP. Keeping your patient with heart failure safe: a review of potentially dangerous medications. *Arch Intern Med* 2004;**164**(7):709–20.
101. Kumar VS, Navaratnam V. Neem (*Azadirachta indica*): prehistory to contemporary medicinal uses to humankind. *Asian Pacific J Trop Biomed* 2013;**3**(7):505–14.
102. Koley KM, Lal J. Pharmacological effects of *Azadirachta indica* (neem) leaf extract on the ECG and blood pressure of rat. *Indian J Physiol Pharmacol* 1994;**38**(3):223–5.
103. Sivashanmugham R, Bhaskar N, Banumathi N. Ventricular fibrillation and cardiac arrest due to neem leaf poisoning. *J Assoc Physicians India* 1984;**32**(7):610–1.
104. Senthilkumaran S, Saravanakumar S, Thirumalaikolundusubramanian P. Cutaneous absorption of oleander: fact or fiction. *J Emerg, Trau, Shock* 2009;**2**(1):43–5.
105. Senthilkumaran S, Meenakshisundaram R, Michaels AD, et al. Electrocardiographic changes during inhalational oleander toxicity. *J Electrocardiol* 2011;**44**(4):470–2.
106. Fattahi M, Dalir-Naghadeh B, Maham M. Prophylactic and therapeutic effects of garlic extract on *Nerium oleander*-induced arrhythmia: a new approach to antiarrhythmic therapy in an ovine model. *Clin Toxicol* 2013;**51**(8):737–47.

107. Pathania S, Randhawa V, Bagler G. Prospecting for novel plant-derived molecules of *Rauvolfia serpentina* as inhibitors of *Aldose Reductase*, a potent drug target for diabetes and its complications. *PLoS One* 2013;**8**(4):e61327.
108. Wilson BN, Wimberley Jr NA. Production of premature ventricular contractions by *Rauwolfia*. *J Am Med Assoc* 1955;**159**(14):1363–4.
109. Vanherweghem JL. Association of valvular heart disease with Chinese-herb nephropathy. *Lancet* 1997;**350**(9094):1858.
110. Wang XX, Chen JZ, Cheng LX. Nerium oleander. Effect of tetradrine on electrophysilogic changes caused by rising of left ventricular preload in guinea pigs. [*Zhongguo Zhong yao za zhi = Zhongguo zhongyao zazhi.*] *China J Chinese Mater Med* 2003;**28**(11):1054–6.
111. Agyare C, Dwobeng AS, Agyepong N, et al. Antimicrobial, Antioxidant, and Wound Healing Properties of *Kigelia africana (Lam.) Beneth.* and *Strophanthus hispidus* DC. *Adv Pharmacol Sci* 2013;**2013**:692613.
112. Davies MK, Hollman A. Digitalis and strophanthus: cardiac glycosides. *Heart* 1998;**80**(1):4.
113. Vikaramaditya, Sarkar M, Rashmi R, et al. Some distinguishing features of a few *Strophanthus* species. *Ancient Sci Life* 1995;**15**(2):145–9.
114. Pan ZH, Li Y, Liu JL, et al. A cytotoxic cardenolide and a saponin from the rhizomes of *Tupistra chinensis*. *Fitoterapia* 2012;**83**(8):1489–93.
115. Riba J, Valle M, Urbano G, et al. Human pharmacology of ayahuasca: subjective and cardiovascular effects, monoamine metabolite excretion, and pharmacokinetics. *J Pharmacol Exp Ther* 2003;**306**(1):73–83.
116. Wilbert J. Does pharmacology corroborate the nicotine therapy and practices of South American shamanism? *J Ethnopharmacol* 1991;**32**(1-3):179–86.
117. Consolini AE, Ragone MI, Migliori GN, et al. Cardiotonic and sedative effects of *Cecropia pachystachya t. (ambay)* on isolated rat hearts and conscious mice. *J Ethnopharmacol* 2006;**106**(1):90–6.
118. Zitkevicius V, Smalinskiene A, Lesauskaite V, et al. Influence of *Echinacea purpurea* (L.) *Moench* extract on the toxicity of cadmium. *Ann NY Acad Sci* 2007;**1095**:585–92.
119. Maillaud C, Lefebvre S, Sebat C, et al. Double lethal coconut crab (*Birgus latro* L.) poisoning. *Toxicon* 2010;**55**(1):81–6.
120. Maillaud C, Barguil Y, Mikulski M, et al. First successful curative use of digoxin-specific Fab antibody fragments in a life-threatening coconut crab (*Birgus latro* L.) poisoning. *Toxicon* 2012;**60**(6):1013–7.
121. Yarlioglues M, Akpek M, Ardic I, et al. Mad-honey sexual activity and acute inferior myocardial infarctions in a married couple. *Tex Heart Inst J* 2011;**38**(5):577–80.
122. Popescu R, Kopp B. The genus Rhododendron: an ethnopharmacological and toxicological review. *J Ethnopharmacol* 2013;**147**(1):42–62.
123. Ozhan H, Akdemir R, Yazici M, et al. Cardiac emergencies caused by honey ingestion: a single centre experience. *EMJ* 2004;**21**(6):742–4.
124. Weiss TW, Smetana P, Nurnberg M, et al. The honey man: second degree heart block after honey intoxication. *Int J Cardiol* 2010;**142**(1):e6–7.
125. Yorgun H, Ulgen A, Aytemir K. A rare cause of junctional rhythm causing syncope; mad honey intoxication. *J Emerg Med* 2010;**39**(5):656–8.
126. Sayin MR, Karabag T, Dogan SM, et al. Transient ST segment elevation and left bundle branch block caused by mad-honey poisoning. *Wien Klin Wochenschr* 2012;**124**(7-8):278–81.
127. Sallon S, Namdul T, Dolma S, et al. Mercury in traditional Tibetan medicine: panacea or problem? *Hum Exp Toxicol* 2006;**25**(7):405–12.

128. Saper RB, Phillips RS, Sehgal A, et al. Lead, mercury, and arsenic in US- and Indian-manufactured Ayurvedic medicines sold via the Internet. *JAMA* 2008;**300**(8):915–23.
129. Tsai HH, Lin HW, Lu YH, et al. A review of potential harmful interactions between anticoagulant/antiplatelet agents and Chinese herbal medicines. *PloS One* 2013;**8**(5):e64255.
130. Posadzki P, Watson L, Ernst E. Herb-drug interactions: an overview of systematic reviews. *Br J Clin Pharmacol* 2013;**75**(3):603–18.
131. Sheehan DM. Herbal medicines, phytoestrogens and toxicity: risk:benefit considerations. *Proceedings of the Society for Experimental Biology and Medicine* 1998;**217**(3):379–85.
132. Chan HT, So LT, Li SW, et al. Effect of herbal consumption on time in therapeutic range of warfarin therapy in patients with atrial fibrillation. *J Cardiovasc Pharmacol* 2011;**58**(1):87–90.
133. Ge B, Zhang Z, Zuo Z. Updates on the clinical evidence of herb–warfarin interactions. *eCAM* 2014;**2014**:957362.
134. Bello N, Winit-Watjana W, Baqir W, et al. Disclosure and adverse effects of complementary and alternative medicine used by hospitalized patients in the North East of England. *Pharmacy Pract* 2012;**10**(3):125–35.
135. Canadian Paediatric Surveillance Program. CPSP highlights. *Paediatr and child health* 2004;9(10):718–748.
136. Ernst E. Toxic heavy metals and undeclared drugs in Asian herbal medicines. *Trends Pharmacol Sci* 2002;**23**(3):136–9.
137. Eisenberg DM, Davis RB, Ettner SL, et al. Trends in alternative medicine use in the United States, 1990-1997: results of a follow-up national survey. *JAMA* 1998;**280**(18):1569–75.
138. Herman PM, Poindexter BL, Witt CM, et al. Are complementary therapies and integrative care cost-effective? A systematic review of economic evaluations. *BMJ opin* 2012;**2**:5.
139. WHO. *Legal Status of Traditional Medicine and Complementary/Alternative Medicine: A Worldwide Review*. Geneva, Switzerland: WHO; 2001 [accessed 03.23.14]. Available from: *http://apps.who.int/medicinedocs/pdf/h2943e/h2943e.pdf*.
140. Barnes J. Quality, efficacy and safety of complementary medicines: fashions, facts and the future: Part I. Regulation and quality. *Br J Clin Pharmacol* 2003;**55**(3):226–33.
141. MacLennan AH, Wilson DH, Taylor AW. The escalating cost and prevalence of alternative medicine. *Prev Med* 2002;**35**(2):166–73.
142. Speijers G, Bottex B, Dusemund B, et al. Safety assessment of botanicals and botanical preparations used as ingredients in food supplements: testing an European Food Safety Authority-tiered approach. *Mol Nutr Food Res* 2010;**54**(2):175–85.
143. Hou YZ, Wang S, Zhao ZQ, et al. Clinical assessment of complementary treatment with *Qishen Yiqi* dripping pills on ischemic heart failure: study protocol for a randomized, double-blind, multicenter, placebo-controlled trial (CACT-IHF). *Trials* 2013;**14**:138.
144. George Q, Li CCD, Basil D. Roufogalis. The quality and safety of traditional traditional Chinese medicines. *Aust Prescr* 2003;**26**(6):128–30.

Chapter 7

Cardiovascular Toxicity from Marine Envenomation

Benjamin Seymour,[1] Athena Andreosso,[2] and Jamie Seymour[3]
[1]*School of Medicine and Dentistry, Faculty of Medicine, Health and Molecular Sciences, James Cook University, Cairns, Australia*
[2]*Queensland Tropical Health Alliance, School of Public Health and Tropical Medicine, Centre for Biodiscovery and Molecular Development of Therapeutics, and Faculty of Medicine, Health and Molecular Sciences, James Cook University, Cairns, Australia*
[3]*Australian Institute of Tropical Health and Medicine, Queensland Tropical Health Alliance, School of Public Health and Tropical Medicine, Centre for Biodiscovery and Molecular Development of Therapeutics, Faculty of Medicine, Health and Molecular Sciences, James Cook University, Cairns, Australia*

7.1 MARINE CARDIOTOXINS

Within the marine realm there exist numerous toxins that are employed by both fauna and flora for protection, prey acquisition, or settlement of battles, including dissuading epiphytes from living on organisms through to acquiring and/or holding substrates for growth. Like all toxins in organisms, they can be divided into poisons and venoms. For such toxins to be defined as poisons, they need to be ingested or absorbed across the epidermis. For them to be venoms, the toxins need to be delivered into organisms through specific structures such as spines (e.g., stonefish), teeth (e.g., snakes), or stinging organelles (e.g., jellyfish). Apart from the delivery differences, there are also differences in the toxin profiles of poisons and venoms, with many poisonous animals having single compounds in their toxins; while the venom profiles of venomous animals are often cocktails of many complex proteins.

Within toxins are compounds that have direct cardiotoxic effects, usually causing death to prey or predators (e.g., box jellyfish) or those that have indirect cardiotoxic effects such as creation of catecholamine storms that result in hypertensive crisis (Irukandji jellyfish). Still others have no effect at all on the cardiac system but, by causing voluntary muscle paralysis, failure of the respiratory system occurs and results in cardiac complications. An examples is tetrodotoxin (TTX) from blue-ringed octopus and fugufish. Those that

The Heart and Toxins. DOI: http://dx.doi.org/10.1016/B978-0-12-416595-3.00007-4

have no effect on the cardiac system will not be included in this chapter; only the toxins that cause direct and indirect cardiotoxic effects will be discussed. As such, the chapter is subdivided into poisons and venoms.

7.2 POISONS

The toxins produced by the organisms in the marine environment and that are known to have cardiovascular toxicity are discussed in the following subsections.

7.2.1 Phycotoxins

Phycotoxins, produced by microalgae (e.g., dinoflagulates), are natural metabolites found commonly throughout the marine realm. Through bioaccumulation in the food chain, the toxins can concentrate in a variety of different marine organisms, from sponges through to carnivorous fish. Within the literature there are many examples of humans being poisoned by these compounds, usually by ingestion but also by dermal absorption. The various phycotoxins usually have very distinct molecular mechanisms of action; however, in many cases the exact mode of action is still unknown. These toxins are usually small to medium in size with molecular masses from 0.3 to more than 3 KDa; they include amino acids, alkaloids, and polyketides and are often actively bioaccummulated in marine food chains.

The typical pathway of bioaccumulation of phycotoxins proceeds from an algal bloom being ingested by filter-feeding shellfish that are then potentially ingested either by humans or by animals further up the trophic food chain; that is, ones that are not susceptible to the toxin. Distribution depends entirely on the presence of dinoflagellate algal blooms that produce the toxin; however, elimination time from hosts varies among species from days to years. Toxin profiles and uptake levels depend on a broad range of factors including geographic location, dinoflagellate species, prior exposure, and environmental conditions. Of these phycotoxins, saxitoxin, yessotoxin, palytoxin, azaspiracids, maitotoxin, and ciguatoxin are the main cardiotoxic groups that alter different ion channels and/or pumps at the cell membrane level.

Saxitoxin

Saxitoxin (STX) and its derivatives cause a poisoning commonly referred to as paralytic shellfish poisoning and paralytic pufferfish poisoning following bioaccumulation and ingestion. The toxins are produced in the marine environment by groups of dinoflagellates, namely *Alexandrium* spp., *Gymnodium* spp., and *Pyrodinium* spp.[1,2] A number of different bioaccumulation pathways have been shown, with typical vectors including bivalves and pufferfish and extending into atypical hosts (e.g., crustaceans, gastropods, a number of fish species, and a significant number of marine mammals and birds).

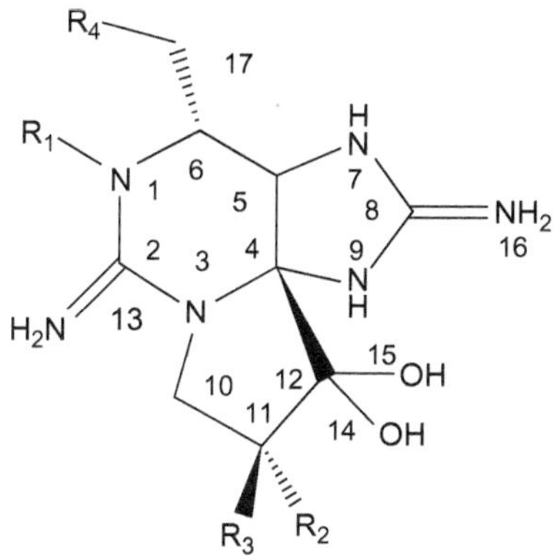

FIGURE 7.1 **The structure of saxitoxin.**[1]

The structure of saxitoxin, like most phycotoxins, is relatively complex (Figure 7.1) and has several different derivatives.[1] Saxitoxin toxicity is mediated through its effects on the Na^+, K^+, and Ca^{2+} channels. The primary target of saxitoxin has long been established as the Na^+ channel where it binds externally to receptor site 1 to halt inward Na^+ flow while also acting as a cationic substitute. In vitro saxitoxin has been shown to modify K^+ channel gating causing retardation of opening and activation while increasing the deactivation rate and subsequently resulting in decreased ion transmission through the channel.[3] Further, in vitro studies have demonstrated saxitoxins' effect on L-type Ca^{2+} channels where the toxin caused a channel blockade with a maximal effect of 49%, suggesting a mechanism other than a simple pore block as seen in Na^+ channels; as yet, this remains to be fully elucidated.[4]

Physiological effects of these channel blockades develop into a paralytic syndrome characterized by descending peripheral and central paralysis that begins with facial involvement, which results in paresthesia of the lips mouth, face, and neck, and then progresses to involve the diaphragmatic muscles wherein respiratory failure may ensue. Alterations in the central nervous system (CNS) are also apparent, with cardiac disturbances, ataxia, drowsiness, and lack of motor coordination also being reported.[1] Rapid cardiac failure in cats after administering high intravenous doses of saxitoxin with diverse arrhythmias, including ventricular and atrial fibrillation and hypotension that preceded rapid cardiac arrest within minutes, has also been shown.[5] In addition, it is known that excretion of the toxin is entirely renal and that no metabolic machinery exists to process saxitoxin in mammals.[5] It is also important to note that respiratory suppression has been shown to occur at much lower doses (3 μg/kg) than rapid cardiac failure (10 μg/kg); this reflects the emphasis on respiratory suppression described in paralytic shellfish poisoning syndrome. There is currently limited data available that details the presence or extent of long-term cardiac damage following intoxication.

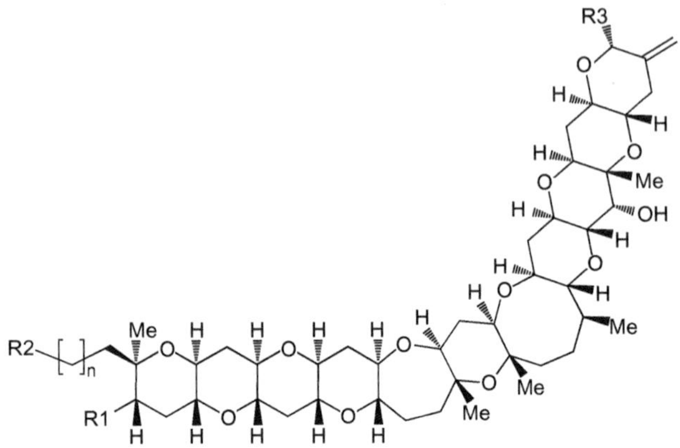

FIGURE 7.2 **The structure of yessotoxin.[6]**

Yessotoxin

Toxicological data detailing the effects of yessotoxin is largely incomplete and as yet there have been no reports of human toxicity. The toxin is produced by the dinoflagellate species *Gonyaulax grindley*, *G. polyedra*, and *G. spinifera* and typically bioaccumulates (Figure 7.2). A lack of reports about human toxicity have led to deregulation of the toxin and its analogs—of which more than 30 have been isolated and have had their structures described. However, there is opposition to this deregulation given the lack of toxicological knowledge and the toxin's propensity to accumulate at significant levels of up to several mg/kg.[6]

The toxicological properties of the yessotoxins have been described by several methods with varying ranges of toxicity. Intraperitoneal injection of the toxin into mice induced restlessness, dyspnea, shivering, jumping, ultrastructural cardiac changes, and death at doses between 0.1 and 0.5 mg/kg.[6,7] Oral administration showed significantly decreased toxicity and lethality has not been induced either by acute, repeated, or coexposure to okadaic acid—a toxin commonly found alongside yessotoxin.[8] Reports of ultrastructural changes following acute and repeated oral intoxication are conflicting but agree that the changes were characterized by mitochondrial swelling and myofibril disorganization in cardiomyocytes and resolve within three months postexposure.[9]

Direct in vitro studies of yessotoxin and its analogous effects on ion channels have excluded effects on Na^+ channels and have demonstrated a moderate effect on Ca^{2+} channels that is not linked to its cytotoxicity and that does not affect calcium homeostasis.[10,11] Inhibitory modification of the degradation pathway of E-cadherin, which is a protein important in cell–cell and cell–substrate adhesion, and a decrease in F-actin filaments has also been demonstrated; however, to date the mechanism and its physiological significance are not entirely understood.[6] Mitochondrial changes seen in cardiac tissue have been attributed to the stimulatory effects on the mitochondrial permeability transition pores that consequently lead to increased membrane depolarization and swelling.[9] Further studies have also revealed an apoptotic mechanism in a range of cell types but as yet no pathway has been described. At present the toxicological profile of the yessotoxins is poorly understood, and there is a distinct lack of chronic in vivo studies. Nevertheless, it appears likely that there is a potential risk to patients with preexisting cardiac conditions.

Palytoxins

Palytoxin and its analogs are considered to be among the most potent marine toxins, with the potential to cause rapid cardiac failure and death occurring within minutes of intoxication. The toxin was first isolated from the zoanthid *Palythoa toxica* and has since been found in various members of the *Palythoa* spp., two of the *Zoanthus* spp., a number of dinoflagellates originating from the *Ostreopsis* genus, the sea anemone *Radianthus macrodactylus*, the polychaete worm *Hermodice carunculata*, and several species of crab and fish wherein bioaccumulation presumably occurs.[12,13] The structure of palytoxin (Figure 7.3) and a number of its analogs have been determined and shown to exhibit heat stability and degradation in acidic and alkaline environments.[13]

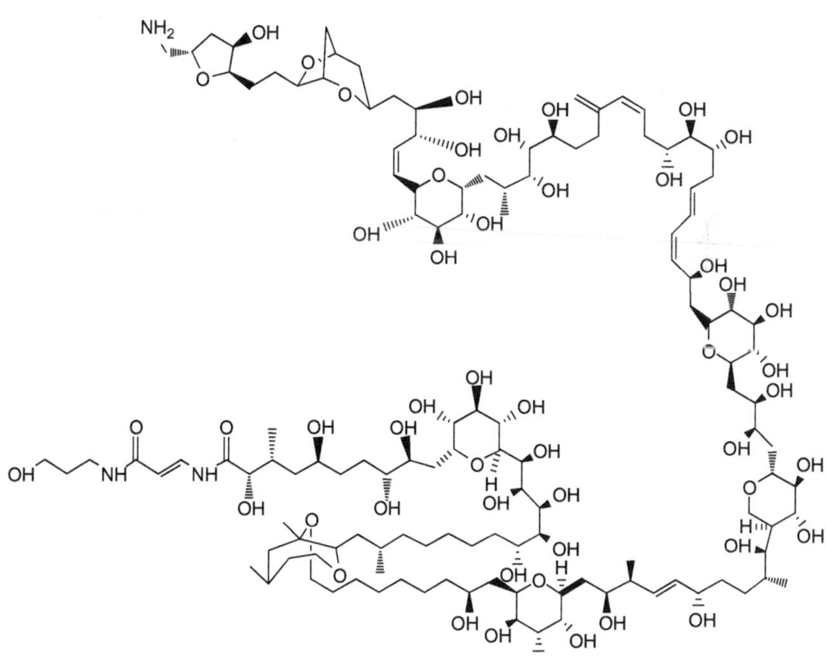

FIGURE 7.3 **The structure of palytoxin.**[14]

Toxicological and pharmacological analysis of the palytoxins has been extensive; in vitro the toxin has been shown to depolarize every cell type tested, an effect unequivocally attributed to the conversion of the Na/K-ATPase into an ion channel.[15] Binding occurs externally to the outer gate of the Na/K-ATPase, locking it in an open state and allowing only the inner gate to function; this transforms the pump into a channel and results in depolarization of the cell and influx of Ca^{2+}.[12,15] The depolarization induced by palytoxin subsequently causes contraction of cardiac smooth and skeletal muscles, stimulates excessive secretions, and causes cell death and subsequent secondary reactions depending on cell type.[12]

Further, in vitro analysis of palytoxins' effects on excitation–contraction coupling in ventricular and atrial cardiomyocytes revealed that the toxin inhibited sarcoplasmic reticulum Ca^{2+} uptake without exerting a direct action on the Ca^{2+} channels, depolarized the resting membrane potential, and caused delayed after-depolarization.[16] An inhibitory effect on the sarco/endoplasmic reticulum Ca^{2+}-ATPase (SERCA) has also been shown;[16] however, to date it has not been established whether the toxin can penetrate intact cells to reach the SERCA.[17]

Initial in vivo studies of palytoxin's toxicity largely focused on intraperitoneal and intravenous administration, which generally led to rapid cardiac arrest following extensive vascular and cardiac contraction.[18] More recent studies have focused on acute and repeated oral administration and have yielded interesting results. Del Favero et al.[19] found that the minimum lethal dose over 7 days was 30 μg/kg per day (previous oral LD50 studies found a median of 510–767 μg/kg), that death occurred on earlier days in animals subjected to higher dosages, and that death also occurred during the recovery period following palytoxin administration. This suggested that irreversible change resulted from repeated exposure and/or that the lipophilic portion of the toxin allows accumulation in tissues.[19]

Case studies of oral human intoxication are limited and few have been able to definitively identify palytoxin as the causative agent; this is generally due to the lack of available analysis equipment and trained staff, the physiological activity of the toxin at extremely low quantities, and the lack of a toxin source (e.g., leftover food). Intoxication has also been reported following inhalation of coral vapors after applying boiling water to *Palythoa* spp. and is characterized by mild respiratory symptoms, including wheezing and shortness of breath, low-grade fevers, some gastrointestinal involvement, and no cardiac changes.[20] Cutaneous intoxication also has been well documented, and case studies for both oral and dermal intoxications have described cardiac effects from both routes of contamination.[21]

Dermal and oral intoxication symptoms both vary significantly depending on the duration and extent of exposure but the principal concerns appear to be cardiac arrest, rhabdomyolysis, renal failure, and neurological alterations. A wide array of cardiac changes have been documented ranging from intermittent sinus bradycardia and sinus tachycardia, left and right bundle branch block, hypertension, T-wave inversions, and cardiac arrest.[21]

Azaspiracids

The azaspiracid toxins are poorly understood in terms of ecology and both acute and chronic toxicological effects. Production of azaspiracid-1 and azaspiracid-2—the two most common forms of the toxin—has only very recently been achieved in the marine algae *Azadinium spinosum*.[22] Isolation of the toxin and its analogs in shellfish and its presence in *Protoperidinium crassipes* blooms has been well documented in the coastal waters of northwestern Africa, western Europe, and eastern Canada; it is most prevelant from mid-summer to mid-winter.[23]

A vast number of azaspiracid analogs have been detected in shellfish tissue though azaspiracid-1 or azaspiracid-2 are typically predominant with azaspiracid-3 and other analogs forming a minority of total azaspiracids; a possible role of metabolic bioconversion in shellfish has been suggested given that only azaspiracid-1, azaspiracid-2, and azaspiracid-3 have been detected outside of these animals.[24] The current proposed structure for the azaspiracid toxin is shown in Figure 7.4.[25]

In vitro experiments have demonstrated that azaspiracid-1, azaspiracid-2, and azaspiracid-3 share a common molecular target in the hERG K^+ channel with the mechanism being proposed as a low-affinity occlusion of the cytoplasmic mouth of open pores.[26] Further studies have demonstrated cytotoxicity to almost

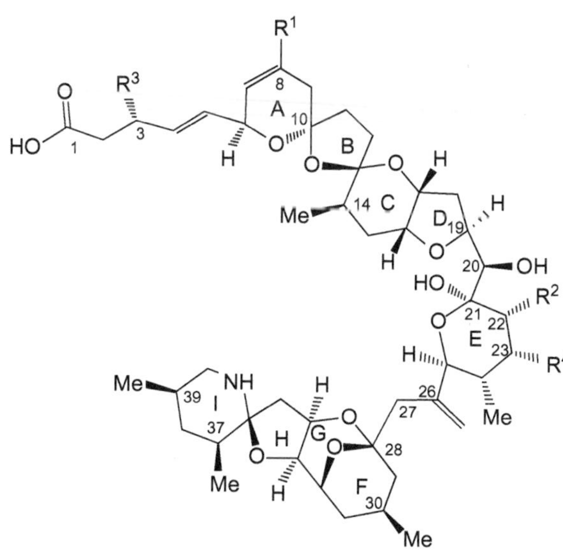

FIGURE 7.4 The structure of azaspiracid.[25]

all cell types tested in vitro, that azaspiracid toxins can enter cells directly in vitro, and that distribution is systemic following oral ingestion.[27–29] Oral intoxication results in changes to the gastrointestinal (GI) tract with no cardiac symptoms.[28–30]

However, a recent in vivo study into the cardiac effects of azaspiracid-2, which was conducted on the basis that other hERG K^+-channel modifiers (e.g., erythromycin) cause QT prolongation in ECGs and increase the risk of torsades de pointes (TdP), concluded that intravenous administration did not alter the QT complex. It, instead, caused premature ventricular contractions and prolongation of the PR interval that progressed to type 2 and 3 atrioventricular block and death with no changes in cardiac biomarkers.[31] At present, relatively little is known about the effects of azaspiracid intoxication on humans and the majority of the information leads toward the toxin primarily targeting the gastrointestinal system. There is, however, insufficient data to suggest that cardiac effects may not occur in high-risk patients with preexisting conditions.

Maitotoxin

Maitotoxin is widely considered the most potent of marine toxins discovered to date and is one of several implicated in ciguatera poisoning. Isolated from the dinoflagellate *Gambierdiscus toxicus*, maitotoxin is distributed in tropical and subtropical areas of the Pacific Ocean.[32] Although the structure of maitotoxin (Figure 7.5) has been known for some time, total synthesis has not yet been achieved and the majority of maitotoxin analogs have yet to be fully described.

Bioaccumulation in the food chain does not occur with maitotoxin due to its water solubility and thus isolation in fish has only been achieved from their digestive tracts.[33] The relative scarcity and the inability to synthesize maitotoxin has limited toxicological studies thus far. At present, in vitro studies on all cell types that have been examined have demonstrated drastic increases in cytosolic Ca^{2+} concentrations following conversion of the plasma membrane Ca^{2+}-ATPase pump into a nonselective cation channel.[34] The resultant influx of extracellular Ca^{2+} initiates a cell death cascade through membrane depolarization and activation of cytolytic pores.[35] In vivo oral studies and definitive human case studies are not available but, given the previously described mechanism of action, cardiac effects cannot be entirely excluded.

Ciguatera

Ciguatoxins are a heat-stabile toxin produced by the dinoflagellate species *Gambierdiscus* and induce a characteristic syndrome known as ciguatera fish poisoning. The distribution of *Gambierdiscus* is reported globally within tropical waters with sustainability of populations possible in waters between 21 and 31°C where turbulence is not excessive, light levels are $<10\%$ of

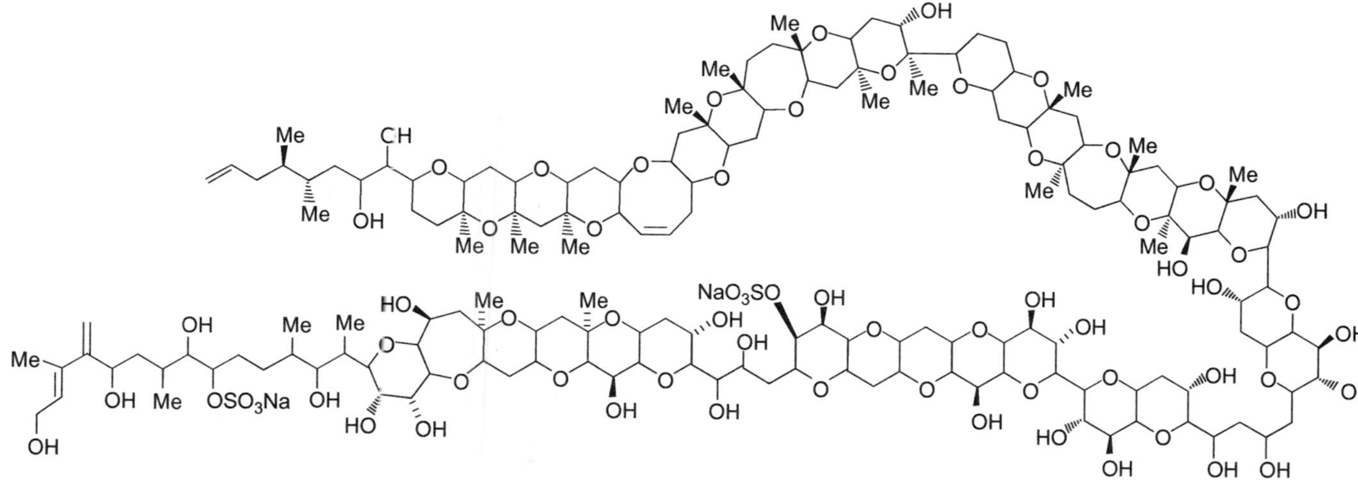

FIGURE 7.5 **The structure of maitotoxin.**[1]

incidence, and salinity levels are consistently high.[32] Ciguatoxin bioaccumulation is confined to herbivorous fish on coral reefs and their carnivorous predators and is typically reported in surgeonfish, triggerfish, red bass, snappers, coral trout, grouper, cods, emperors, moray eels, mackerel, tuna, and barracuda, though several other fish have been reported to contain the toxin.[32,36]

Ciguatoxins are distinct in that the toxin that is initially produced by *Gambierdiscus* spp. is not overly potent but ingestion by fish results in metabolic biotransformations that produce less polar and distinctly more potent analogs.[37] As a result of this biotransformation process, there is a veritable myriad analogs that are typically described broadly by region (e.g., Pacific, Caribbean, and Indian ciguatoxin) with toxicity profiles depending on the bioaccumulation and thus biotransformation route.[33]

The toxicological effects of ciguatoxins are attributed to the toxins' modulating effect on Na^+ channels, which result in increased membrane depolarization, which produces Na^+-dependent effects, while also inducing a Ca^{2+} influx and subsequent neurotransmitter release, both centrally and peripherally.[37] Further examination of this effect has demonstrated that ciguatoxins bind to the intramembrane segment at site 5 on the Na^+ channel to increase channel permeability and thus ion transmission.[38] Subsequent studies have revealed a potential role of a K^+-channel blockade, which may contribute to increased membrane excitability by limiting hyperpolarization and thus producing a lower action-potential threshold.[37]

Necrosis and marked swelling of myocytes has been demonstrated following acute and repeated oral administration of ciguatoxin to mice with a reversal of symptoms occurring within one month.[39,40] Indirect autonomic effects resulting from excessive neurotransmitter release are uncommon (1.2% of 3009 cases from the Pacific Islands between 1964 and 1967 and a reported 10–15% of U.S. cases) and typically manifest as bradycardia and hypotension.[41,42] The hemodynamic and cardiac effects are specifically attributed to ciguatoxins' inhibitory action on vasomotor centers innervated by the vagus nerve, thus preventing regulatory stimulation and removing sympathetic tone that establishes normal peripheral vascular resistance and cardiac rhythm.[42] Case studies have documented severe and prolonged cardiac symptoms lasting as long as four days despite corrective care.[43,44] The extended duration of symptoms—both the uncommon cardiac effects and more characteristic neurological symptoms—has been postulated to be due either to the lipophilic structure of the toxin allowing accumulation and preventing elimination or that ciguatoxin modulation of sodium channels is a permanent effect with relief relying on the development of new receptors.[45]

7.3 VENOMS

Although there are many venomous organisms in the marine environment, very few have cardiotoxic effects on envenomed humans and/or predators or prey. There are, however, several notable exceptions, mainly some species of box jellyfish and members of the *Scorpionid* fish family.

7.3.1 Jellyfish

The Cnidaria, or jellyfish, family contains the world's most venomous marine animal,[46] *Chironex fleckeri*, the multitentacled (or chirodropid) big box jellyfish, as well as a series of small, single tentacled box jellyfish (or carybdeids) that give rise to a debilitating syndrome known as Irukandji. Cnidarian venom is contained in millions of small stinging organelles, called nematocysts, that are present along the tentacles of jellyfish and are capable of injecting venom into prey items, predators, and/or human victims.

Chironex fleckeri is a large jellyfish that can grow to 30 cm with up to 15 tentacles on each corner of its body and a combined tentacle length potentially exceeding 60 m.[47] Due to its translucent body, it is seldom sighted before envenomation of a victim occurs.[48] Immediate onset of excruciating pain, however, and, in severe cases, rapid subsequent cardiovascular collapse[49] indubitably indicates an encounter with this highly venomous animal. *Chironex fleckeri* claims approximately one human life annually in Australia, with unsubstantiated reports of up to 100 deaths in the entire world from the chirodropid jellyfish.

The lethal agents that are in *C. fleckeri* and other related chirodropids are believed to exert immediate cardiotoxicity.[50–52] Further, cardiotoxic effects are present in the total absence of neural control;[52] for example, venom injected in isolated guinea pig hearts resulted in bradycardia similar to what was previously observed in live animals.[53]

Due to observed increases in calcium levels, initially, interference with transmembranic calcium channels of cardiac cells was suggested[54]; however, it is now believed the toxin has pore-building properties[55] and that massive influxes of Ca^{2+} are a secondary result of increased Na^{+} levels due to nonspecific cation leakage through the cell membrane.[56] In addition, it probably is the spontaneous excessive Na^{+}-mediated influx of Ca^{+} into the cytosol of cardiomyocytes that causes irregular contractions of the single cell, so that an overall asynchrony and flagging in communal contractions is observed, thus causing arrhythmias leading to eventual cardiac failure.[56]

Venom-induced contractile response on the aorta appear irreversible with prazosin treatment, an α1-adrenoceptor antagonist.[57] Similarly, ketanserin, a highly selective serotonin 5HT-receptor antagonist, appears ineffective in inhibiting hypertension and cardiovascular collapse following envenomation in anesthetized rats.[58] This suggests that the venom does not affect α1-helical adrenoreceptors of transmembrane proteins; this refutes a previous theory that envenomation symptoms originate from the provocation of a hyperadrenergic reaction[47] and supports an ionophoric, rather than a pore-forming mode of action.[57,58]

There is, however, an indication for an α-helical transmembrane-spanning region in *C. fleckeri* venom proteins.[59] This may potentially indicate that instead of binding to endogen receptors of transmembrane proteins of the victim, the venom uses its transmembrane-spanning region to anchor itself straight into the cell, creating pores in the process.

Research has determined and isolated the cardiotoxic fractions of the *C. fleckeri* venom that cause cardiovascular collapse.[60–63] It is now thought that two fractions, CTF-α and CTF-β, work in synergy, with CTF-α causing rapid short-term toxicity whereas CTF-β acts slower but instigates permanent damage to cardiac cells.[64] These results represent a valid rationalization for the observation that fish exposed to *C. fleckeri* venom did not result in significant heart rate changes until cardiac standstill,[65] while stroke volume decreased continuously (~50% of original volume at ~90% time from injection to death) prior to death.[65]

The mode of action for *Chiropsalmus quadrigatus*, another species of chirodropid, is thought to be slightly different from *C. fleckeri*. It is believed that cardiotoxicity manifests by involving L-type calcium channels,[66] a theory that has been refuted for *C. fleckeri* venom.[56,67] Increases in right-atrial contractile force, as well as increases in vasoconstriction in rats induced by *C. quadrigatus* venom, were significantly inhibited by diltiazem, which is an L-type calcium channel antagonist, supporting the calcium channel inference hypothesis.[66] Similarly, *C. quadrigatus* tentacle extract hemodynamic effects have been shown to be moderated by diltiazem.[68] Further, phentolamine, a nonselective α1-adrenoceptor blocking agent, did not prevent increases of venom-induced tension in rat aorta, contradicting the possibility that α1-adrenoceptors, which are a class of G-protein coupled receptors that bind catecholamines, are involved in the excessive Ca^{2+}-cytosolic influx.[66]

In addition, indomethacin, phentolamine, or atropine were unable to counter the hemodynamic effects of *Chiropsalmus quadrigatus*, thus suggesting it is unlikely that cyclooxygenase produces α-adrenergic receptors or that muscarinic receptors are involved in the mode of action.[68] Instead, the ability of diltiazem to attenuate the effects of the venom may suggest the involvement of voltage-dependent L-type calcium channels potentially causing an intracellular increase in calcium levels, eventually leading to a calcium overload that may lead to cardiac failure.[68]

7.3.2 Irukandji Syndrome

The expression *Irukandji syndrome*, a disease caused by envenomation from certain marine cnidarians, first appeared in the early 1950s to describe a set of debilitating symptoms that had been documented in Australian bathers since the 1920s. The definition of the syndrome has varied widely over time because of the diverse range of symptoms and severity of those experienced. Although this syndrome may be fairly innocuous to some individuals, others have experienced severe symptoms and complications as a result of envenoming toxins. In general, mild pain is felt at the sting site at the time of initial contact; the sensation fades with time. With some, especially individuals experiencing pain, other symptoms can be overshadowed by the delayed onset of severe systemic ones and hypertension; tachycardia; and, in extreme cases, pulmonary or cerebral edema. There have been two recorded deaths associated with Irukandji syndrome.[69–71]

Generally, Irukandji syndrome is thought to be a tropically based disease found around the coral reefs in northern Australia; however, no real research has been conducted into the true worldwide distribution of the syndrome. It has been recorded from many locations in Australia, the Torres Strait, Hawaii, Fiji, coastal Thailand, Puerto Rico, Manila Bay in the Philippines, the Gulf Sea, Key West in Florida, the French West Indies, Bonaire in the Caribbean, Timor-Leste and Papua New Guinea, Japan, North Wales in the United Kingdom, and throughout the Indonesian archipelago.[70]

Severe envenomings, which often result in various cardiac complications, do occur.[69,71–75] In two separate studies, one showed that 30% of patients experienced some degree of heart failure, while 22% experienced elevated troponin levels in the other.[69,72] In addition, severe cardiac dysfunction has been reported in several cases.[72–76]

Research using extracted venom has shown that it is quite complex. As many as 60 proteins were present in the venom extracted from nematocysts on mature bells of animals (SDS-PAGE gel analysis), and at least 45 different proteins were found in the venom extracted from tentacular nematocysts with proteins ranging in size from 25 to 250 kDa, the majority being less than 100 kDa.[77]

Laboratory studies into the components of Irukandji venom have predominantly revolved around the cardiac responses of extracted venom on both whole and isolated vertebrate models including pigs, guinea pigs, and rats.[78–82] Initial studies on the effects of *C. barnesi* venom have shown that serum levels of endogenous adrenaline increase, as well as pulse rate and blood pressure, with widening pulse pressure and a positive inotropic effect when injected into rats.[78] Similarly, studies utilizing crude blended whole specimens of *C. barnesi* injected into mechanically ventilated pigs saw a 200- and 100-fold increase in serum noradrenaline and adrenaline, respectively, with sustained tachycardia and systemic and pulmonary hypertension.[79,83] The venom's effects not only may cause the release of catecholamines but also direct vasoconstriction.[79]

Venom effects on isolated rat and guinea pig right atria have caused tachycardia in the presence of atropine.[79] However, this effect was almost abolished in an in vivo pig model by the prophylactic addition of tetrodotoxin restricted to peripheral postganglionic sympathetic sites and possibly the splanchnic nerve innervations and the adrenal medulla. This suggests that the venom extract functions as a neural sodium channel activator.[79] However, as crude venom extract and not pure extracted venom was used in these studies, and as subsequent research has found toxic components in tentacle extract devoid of nematocyst material, as well as cardiac response to this same material, these results must be viewed with some reservations.[79,80,82]

Subsequent laboratory studies have used the refined venom extraction technique.[77,84] These investigations showed some comparable cardiac effects, including severe pressor responses from the venom of *Carukia barnesi* and *Alatina nr mordens*, which is approximately 3 to 5 times less potent than that of *C. barnesi*, supporting the theory of a venom-induced catecholamine release after intravenous venom administration.[80,82] Interestingly, pressor responses do not appear to be dose-dependent, suggesting that the venom may be inducing a release of catecholamine into the circulation and does not actually contain a direct vasopressor itself.[80,82] In support of this, data exist that suggest that the venom from at least one species of Irukandji jellyfish (*Carukia barnesi*) is noncytotoxic.[85]

The administration of prazosin, which is an α-1 adrenoreceptor antagonist, in envenomed test animals both reduced the venom-induced pressor response and inhibited the tachycardia, supporting the hypothesis that this has an indirect rather than a direct effect on the peripheral vasculature and also not a direct β-adrenergic effect.[82] Conversely, cardiovascular collapse in envenomed animal models does appear to be dose-related, indicating the toxins may also be acting directly on the myocardium.[82] Salivation and urination with envenomed animals is also seen, which suggests parasympathetic stimulation resulting again from the venom-induced catecholaminemia.[80,82]

7.3.3 Stonefish

Stonefish have stout dorsal hollow spines with associated glands that are covered by thick overlapping skin supported by filamentous connective tissue;[86,87] a duct inside each spine transports venom to its tip and into the victim. Venom is produced in the glands between the fibers and spine.[86] When distressed, the first three dorsal spines stand vertically and appear more mobile while the other spines lie on a diagonal slope trailing down to the base of the tail, connected by a wide intermittent membrane between spines.[87,88]

Envenomation is triggered by pressure being applied to the integument sheath that encases the dorsal spine, which then compresses the dual venom sacs on either side of the spine.[86–91] Volume estimates from a single developed venom sac average 0.03 ml.[92,93] Severity of symptoms resulting from a sting is directly correlated with the amount of venom expressed.

Stonefish venom is a proteineous mixture,[94–99] while crude venom is opaque with a nitrogen content of 2% and a protein content of approximately 13%,[100] is stable at a pH range of 6 to 6.8, is antigenic,[92,95,98] and is said to be heat labile.[92,101–103] Multiple studies have detected a diversity of enzymes in stonefish venom; however, characterization of these enzymes to date is not significant. The prominent enzyme detected in the venom is hyaluronidase,[99,104–106] which contributes to the spreading of the venom and necrosis seen in envenomations.[107] Further, acetylcholinesterase, phosphodiesterase,[93] alkaline phosphomonoesterase, 5′-nucleotidase, proteinase, arginine, esterase and amidase,[93] esterase and aminopeptidase,[99] noradrenaline, dopamine, tryptophan, and acetylcholine may also be present.[108]

Stonefish envenomations show dose-dependent cardiovascular effects, which was the cause of much controversy in previous studies because results varied from positive to negative and biphasic cardiac activity.[109] This was specifically the case for one of the primary symptoms witnessed by clinicians—namely, marked hypotension. This complication has been well documented in envenomed mice and rabbits.[87,92,94,100–111] Peripheral vasodilatation also appears to contribute to the hypotensive effect because low dosages of venom injected into rabbits caused a decrease in blood pressure in the absence of, or prior to, a decrease in myocardial contractility.[100] This was later confirmed when decreases in heart rate were attributed to the endothelium-dependent relaxation in smooth muscles.[108] Comparatively, at high doses, a contraction rather than a relaxation occurred during cardiac examinations.[100]

This contractile response has also been observed in nonvascular smooth muscle, such as guinea pig ileum, believed to be through activation of acetylcholine and prostaglandin,[112] as well as in rat hindquarters.[110] Similarly, an increase in arterial blood pressure in anesthetized dogs,[92] rats,[112,113] guinea pigs, and rabbits has been observed.[113] Other studies have documented both an increase, followed by a marked decrease, of biphasic changes in blood pressure.[92,110,113] Research has yet to fully resolve whether cardiac activity is a consequence of specific cardiotoxic components within the venom or neuromuscular collapse, or whether the profound pain from the venom causes stress-induced complications within the cardiovascular system. Decreases in rate of cardiac contraction appear to be governed by muscarinic receptor activity, either through the presence of cholinomimetics or endogenous acetylcholine. Increases in contractibility have been described many times but are believed to involve the function of direct and indirect activity at both nonadrenergic and adrenergic receptors.[109]

Purified trachynilysin from these venoms appear to open Ca^{2+} channels in frog atrial heart muscle cells, reducing action potential duration and decreasing contracture times.[114] In contrast, verrucotoxin and p-verrucotoxin toxins that were isolated from stonefish, blocked Ca^{2+} channels and opened potassium channels.[115] The mechanism by which these venoms operate is yet to be elucidated, but it is known to decrease blood pressure through vasodilation via the release of endothelial nitric oxide (NO).[116]

7.4 CONCLUSION

Overall, within the marine environment there exist thousands of toxic compounds, many of which may cause cardiac complications in humans through either direct contact, ingestion, or delivery by venomous plants and/or animals. As a result, it is not surprising that our understanding of how these toxins operate and what their toxicity to humans may be is somewhat limited. What we do know, however, is that none of these toxins' origin are for use against humans. As a result, an effect seen in human intoxications is purely coincidental.

This then makes it difficult, if not impossible, to determine which compounds may be toxic without direct testing. What is required now is a systematic approach to studying these compounds, as well as extensive investigation into the mechanics of the toxins. Only then will we be in a position to determine the relative toxicity of marine cardiotoxins.

Key Learning Points

- There are many venomous and poisonous organisms in the marine environment; however, few of them produce cardiotoxic effects in humans. Notably, this is an area of ongoing research and much is still to be learned about these toxins' effects.
- The most venomous marine organisms include several species of box jellyfish and members of the scorpionid fish family.
- Irukandji syndrome is a disease caused by envenomation from certain marine cnidarians; its definition varies across the globe. It is known to cause fatal cardiac complications.
- Proper clothing and safety measures should be undertaken by all marine workers to prevent mortality and morbidity.

REFERENCES

1. Cusick KD, Sayler GS. An overview on the marine neurotoxin, saxitoxin: genetics, molecular targets, methods of detection and ecological functions. *Mar Drugs* 2013;**11**(4): 991–1018.
2. Araoz R, Molgo J, Tandeau de Marsac N. Neurotoxic cyanobacterial toxins. *Toxicon* 2010; **56**(5):813–28.
3. Wang J, Salata JJ, Bennett PB. Saxitoxin is a gating modifier of hERG K^+ channels. *J Gen Physiol* 2003;**121**(6):583–98.
4. Su Z, Sheets M, Ishida H, et al. Saxitoxin blocks L-type ICa. *J Pharmacol Exp Ther* 2004;**308**(1):324–9.
5. Andrinolo D, Michea LF, Lagos N. Toxic effects, pharmacokinetics and clearance of saxitoxin, a component of paralytic shellfish poison (PSP), in cats. *Toxicon* 1999;**37**(3): 447–64.
6. Tubaro A, Dell'ovo V, Sosa S, Florio C. Yessotoxins: a toxicological overview. *Toxicon* 2010;**56**(2):163–72.
7. Aune T, Sorby R, Yasumoto T, et al. Comparison of oral and intraperitoneal toxicity of yessotoxin towards mice. *Toxicon* 2002;**40**(1):77–82.
8. Sosa S, Ardizzone M, Beltramo D, et al. Repeated oral coexposure to yessotoxin and okadaic acid: a short term toxicity study in mice. *Toxicon* 2013;**76**:94–102.

9. Tubaro A, Giangaspero A, Ardizzone M, et al. Ultrastructural damage to heart tissue from repeated oral exposure to yessotoxin resolves in 3 months. *Toxicon* 2008;**51**(7): 1225–35.
10. Perez-Gomez A, Ferrero-Gutierrez A, Novelli A, et al. Potent neurotoxic action of the shellfish biotoxin yessotoxin on cultured cerebellar neurons. *Toxicol Sci* 2006;**90**(1): 168–77.
11. de la Rosa LA, Alfonso A, Vilarino N, et al. Modulation of cytosolic calcium levels of human lymphocytes by yessotoxin, a novel marine phycotoxin. *Biochem Pharmacol* 2001;**61**(7): 827–33.
12. Wu CH. Palytoxin: membrane mechanisms of action. *Toxicon* 2009;**54**(8):1183–9.
13. Ramos V, Vasconcelos V. Palytoxin and analogs: biological and ecological effects. *Mar Drugs* 2010;**8**(7):2021–37.
14. Usami M, Satake M, Ishida S, et al. Palytoxin analogs from the dinoflagellate *Ostreopsis siamensis*. *J Am Chem Soc* 1995;**117**(19):5389–90.
15. Artigas P, Gadsby DC. Large diameter of palytoxin-induced Na/K pump channels and modulation of palytoxin interaction by Na/K pump ligands. *J Gen Physiol* 2004;**123**(4):357–76.
16. Kockskamper J, Ahmmed GU, Zima AV, et al. Palytoxin disrupts cardiac excitation-contraction coupling through interactions with P-type ion pumps. *Am J Physiol Cell Physiol* 2004;**287**(2):C527–38.
17. Coca R, Soler F, Fernandez-Belda F. Characterization of the palytoxin effect on Ca^{2+}-ATPase from sarcoplasmic reticulum (SERCA). *Arch Biochem Biophys* 2008;**478** (1):36–42.
18. Munday R. Palytoxin toxicology: animal studies. *Toxicon* 2011;**57**(3):470–7.
19. Del Favero G, Beltramo D, Sciancalepore M, et al. Toxicity of palytoxin after repeated oral exposure in mice and in vitro effects on cardiomyocytes. *Toxicon* 2013;**75**:3–15.
20. Sud P, Su MK, Greller HA, et al. Case series: inhaled coral vapor—toxicity in a tank. *J Med Toxicol* 2013;**9**(3):282–6.
21. Tubaro A, Durando P, Del Favero G, et al. Case definitions for human poisonings postulated to palytoxins exposure. *Toxicon* 2011;**57**(3):478–95.
22. Jauffrais T, Kilcoyne J, Sechet V, et al. Production and isolation of azaspiracid-1 and -2 from *Azadinium spinosum* culture in pilot scale photobioreactors. *Mar Drugs* 2012;**10**(6): 1360–82.
23. Twiner MJ, Rehmann N, Hess P, Doucette GJ. Azaspiracid shellfish poisoning: a review on the chemistry, ecology, and toxicology with an emphasis on human health impacts. *Mar Drugs* 2008;**6**(2):39–72.
24. Rehmann N, Hess P, Quilliam MA. Discovery of new analogs of the marine biotoxin azaspiracid in blue mussels (*Mytilus edulis*) by ultra-performance liquid chromatography/ tandem mass spectrometry. *RCM* 2008;**22**(4):549–58.
25. Nicolaou KC, Pihko PM, Bernal F, et al. Total synthesis and structural elucidation of azaspiracid-1. Construction of key building blocks for originally proposed structure. *J Am Chem Soc* 2006;**128**(7):2244–57.
26. Twiner MJ, Doucette GJ, Rasky A, et al. Marine algal toxin azaspiracid is an open-state blocker of hERG potassium channels. *Chem Res Toxicol* 2012;**25**(9):1975–84.
27. Twiner MJ, Hanagriff JC, Butler S, et al. Induction of apoptosis pathways in several cell lines following exposure to the marine algal toxin azaspiracid. *Chem Res Toxicol* 2012;**25** (7):1493–501.
28. Aune T, Espenes A, Aasen JA, et al. Study of possible combined toxic effects of azaspiracid-1 and okadaic acid in mice via the oral route. *Toxicon* 2012;**60**(5):895–906.

29. Aasen JA, Espenes A, Hess P, Aune T. Sublethal dosing of azaspiracid-1 in female NMRI mice. *Toxicon* 2010;**56**(8):1419–25.
30. Furey A, O'Doherty S, O'Callaghan K, et al. Azaspiracid poisoning (AZP) toxins in shellfish: toxicological and health considerations. *Toxicon* 2010;**56**(2):173–90.
31. Ferreiro SF, Vilarino N, Carrera C, et al. In vivo arrhythmogenicity of the marine biotoxin azaspiracid-2 in rats. *Arch Toxicol* 2014;**88**(2):425–34.
32. Litaker RW, Vandersea MW, Faust MA, et al. Global distribution of ciguatera causing dinoflagellates in the genus *Gambierdiscus*. *Toxicon* 2010;**56**(5):711–30.
33. Lewis RJ. The changing face of ciguatera. *Toxicon* 2001;**39**(1):97–106.
34. Sinkins WG, Estacion M, Prasad V, et al. Maitotoxin converts the plasmalemmal Ca(2+) pump into a Ca(2+)-permeable nonselective cation channel. *Am J Physiol Cell Physiol* 2009; **297**(6):C1533–43.
35. Estacion M, Schilling WP. Maitotoxin-induced membrane blebbing and cell death in bovine aortic endothelial cells. *BMC Physiol* 2001;**1**:2.
36. Skinner MP, Brewer TD, Johnstone R, et al. Ciguatera fish poisoning in the Pacific Islands (1998 to 2008). *PLoS Negl Trop Dis* 2011;**5**(12):e1416.
37. Nicholson GM, Lewis RJ. Ciguatoxins: cyclic polyether modulators of voltage-gated ion channel function. *Mar Drugs* 2006;**4**(3):82–118.
38. Lombet A, Bidard J-N, Lazdunski M. Ciguatoxin and brevetoxins share a common receptor site on the neuronal voltage-dependent Na^+ channel. *FEBS Lett* 1987;**219**(2):355–9.
39. Terao K, Ito E, Oarada M, et al. Light and electron microscopic studies of pathologic changes induced in mice by ciguatoxin poisoning. *Toxicon* 1991;**29**(6):633–43.
40. Terao K, Ito E, Yasumoto T. Light and electron microscopic studies of the murine heart after repeated administrations of ciguatoxin or ciguatoxin-4c. *Nat Toxins* 1992;**1** (1):19–26.
41. Bagnis R, Kuberski T, Laugier S. Clinical observations on 3,009 cases of ciguatera (fish poisoning) in the South Pacific. *Am J Trop Med Hyg* 1979;**28**(6):1067–73.
42. Senthilkumaran S, Meenakshisundaram R, Michaels AD, et al. Cardiovascular complications in ciguatera fish poisoning: a wake-up call. *Heart* 2011;**12**(4):166–8.
43. Chan TY. Severe bradycardia and prolonged hypotension in ciguatera. *Singapore Med J* 2013;**54**(6):e120–2.
44. Pennotti R, Scallan E, Backer L, et al. Ciguatera and scombroid fish poisoning in the United States. *Foodborne Pathog Dis* 2013;**10**(12):1059–66.
45. Pearn J. Neurology of ciguatera. *J Neurol Neurosurg Psychiatry* 2001;**70**(1):4–8.
46. Endean R. Venom of Chironex, the world's most venomous animal. Venoms and Victims. *South Brisbane: Queensland Museum* 1988;15–24.
47. Bailey PM, Little M, Jelinek GA, Wilce JA. Jellyfish envenoming syndromes: unknown toxic mechanisms and unproven therapies. *Med J Aust* 2003;**178**(1):34–7.
48. Tibballs J. Australian venomous jellyfish, envenomation syndromes, toxins and therapy. *Toxicon* 2006;**48**(7):830–59.
49. Learmont SA. *Chironex fleckeri* (box jellyfish) envenomation: a case study. *Aus Emerg Nurs J* 2006;**9**(2):49–56.
50. Baxter EH. Marr AGM. Sea wasp (*Chironex fleckeri*) venom: Lethal, haemolytic and dermonecrotic properties. *Toxicon* 1969;**7**(3):195–210.
51. Endean R, Henderson L. Further studies of toxic material from nematocysts of the cubomedusan *Chironex fleckeri* Southcott. *Toxicon* 1969;**7**(4):303–14.
52. Turner R, Freeman SE. Effects of *Chironex fleckeri* toxin on the isolated perfused guinea pig heart. *Toxicon* 1969;**7**(4):277–86.

53. Freeman SE, Turner R. A pharmacological study of the toxin of a cnidarian, *Chironex fleckeri* Southcott. *Br J Pharmacol* 1969;**35**(3):510–20.
54. Burnett JW, Calton GJ. Response of the box-jellyfish (*Chironex fleckeri*) cardiotoxin to intravenous administration of verapamil. *Med J Aust* 1983;**2**(4):192.
55. Bailey PM, Bakker AJ, Seymour JE, et al. A functional comparison of the venom of three Australian jellyfish—*Chironex fleckeri*, *Chiropsalmus* spp., and *Carybdea xaymacana*—on cytosolic Ca^{2+}, haemolysis and *Artemia* spp. lethality. *Toxicon* 2005;**45**(2):233–42.
56. Mustafa MR, White E, Hongo K, et al. The mechanism underlying the cardiotoxic effect of the toxin from the jellyfish *Chironex fleckeri*. *Toxicol Appl Pharmacol* 1995;**133**(2):196–206.
57. Winter KL, Fernando R, Ramasamy S, et al. The in vitro vascular effects of two chirodropid (*Chironex fleckeri* and *Chiropsella bronzie*) venoms. *Toxicol Lett* 2007;**168**(1):13–20.
58. Ramasamy S, Isbister GK, Seymour JE, et al. Pharmacologically distinct cardiovascular effects of box jellyfish (*Chironex fleckeri*) venom and a tentacle-only extract in rats. *Toxicol Lett* 2005;**155**(2):219–26.
59. Brinkman D, Burnell J. Identification, cloning and sequencing of two major venom proteins from the box jellyfish, *Chironex fleckeri*. *Toxicon* 2007;**50**(6):850–60.
60. Winter KL, Isbister GK, McGowan S, et al. A pharmacological and biochemical examination of the geographical variation of *Chironex fleckeri* venom. *Toxicol Lett* 2010;**192**(3):419–24.
61. Brinkman DL, Aziz A, Loukas A, et al. Venom proteome of the box jellyfish *Chironex fleckeri*. *PloS one* 2012;**7**(12):e47866.
62. McClounan S, Seymour J. Venom and cnidome ontogeny of the cubomedusae *Chironex fleckeri*. *Toxicon* 2012;**60**(8):1335–41.
63. Saggiomo SL, Seymour JE. Cardiotoxic effects of venom fractions from the Australian box jellyfish *Chironex fleckeri* on human myocardiocytes. *Toxicon* 2012;**60**(3):391–5.
64. Chaouis S, Smout M, Wilson D, et al. Rapid short term and gradual permanent cardiotoxic effects of vertebrate toxins from *Chironex fleckeri* (Australian box jellyfish) venom. *Toxicon* 2014;**80**:17–26.
65. Kintner AH, Seymour JE, Edwards SL. Variation in lethality and effects of two Australian chirodropid jellyfish venoms in fish. *Toxicon* 2005;**46**(6):699–708.
66. Sakanashi M, Matsuzaki T, Nakasone J, et al. Effects of diltiazem on in vitro cardiovascular actions of crude venom obtained from Okinawan box-jellyfish (*Habu-kurage*), *Chiropsalmus quadrigatus*. *Anaesth Intensive Care* 2002;**30**(5):570–7.
67. Ramasamy S, Isbister GK, Seymour JE, et al. The in vivo cardiovascular effects of an Australasian box jellyfish (*Chiropsalmus* spp.) venom in rats. *Toxicon* 2005;**45**(3):321–7.
68. Koyama T, Noguchi K, Matsuzaki T, et al. Haemodynamic effects of the crude venom from nematocysts of the box-jellyfish *Chiropsalmus quadrigatus* (*Habu-kurage*) in anaesthetized rabbits. *Toxicon* 2003;**41**(5):621–31.
69. Fenner P, Carney I. The Irukandji syndrome. A devastating syndrome caused by a north Australian jellyfish. *Aust Fam Physician* 1999;**28**(11):1131.
70. Carrette TJ, Underwood AH, Seymour JE. Irukandji syndrome: a widely misunderstood and poorly researched tropical marine envenoming. Diving and hyperbaric medicine. *J South Pacific Underwater Med Soc* 2012;**42**(4):214–23.
71. Pereira P, Barry J, Corkeron M, et al. Intracerebral hemorrhage and death after envenoming by the jellyfish *Carukia barnesi*. *Clin Toxicol* 2010;**48**(4):390–2.
72. Huynh TT, Seymour J, Pereira P, et al. Severity of Irukandji syndrome and nematocyst identification from skin scrapings. *Med J Aust* 2003;**178**(1):38–42.

73. Fenner P, Williamson J, Burnett J, et al. The "Irukandji syndrome" and acute pulmonary oedema. *Med J Aust* 1988;**149**(3):150.
74. Little M, Pereira P, Mulcahy R, et al. Severe cardiac failure associated with presumed jellyfish sting. Irukandji syndrome? *Anaesth Intensive Care* 2003;**31**(6):642–7.
75. Little M, Mulcahy R, Wenck DJ. Life threatening cardiac failure in a healthy young female with Irukandji syndrome. *Anaesth Intensive Care* 2001;**29**:178–80.
76. Burnett JW, Calton GJ, Burnett HW, Mandojana RM. Local and systemic reactions from jellyfish stings. *Clin Dermatol* 1987;**5**(3):14–28.
77. Underwood AH, Seymour JE. Venom ontogeny, diet and morphology in *Carukia barnesi*, a species of Australian box jellyfish that causes Irukandji syndrome. *Toxicon* 2007;**49**(8):1073–82.
78. Burnett JW, Weinrich D, Williamson JA, et al. Autonomic neurotoxicity of jellyfish and marine animal venoms. *Clin Auton Res* 1998;**8**(2):125–30.
79. Winkel KD, Tibballs J, Molenaar P, et al. Cardiovascular actions of the venom from the Irukandji (*Carukia barnesi*) jellyfish: effects in human, rat and guinea-pig tissue in vitro and in pigs in vivo. *Clin Exp Pharmacol Physiol* 2005;**32**(9):777–88.
80. Winter KL, Isbister GK, Schneider JJ, et al. An examination of the cardiovascular effects of an 'Irukandji' jellyfish, *Alatina nr mordens*. *Toxicol Lett* 2008;**179**(3):118–23.
81. Wiltshire CJ, Sutherland SK, Fenner PJ, et al. Optimization and preliminary characterization of venom isolated from 3 medically important jellyfish: the box (*Chironex fleckeri*), Irukandji (*Carukia barnesi*), and blubber (*Catostylus mos*aicus) jellyfish. *Wilderness Environ Med* 2000;**11**(4):241–50.
82. Ramasamy S, Isbister GK, Seymour JE, et al. The in vivo cardiovascular effects of the Irukandji jellyfish (*Carukia barnesi*) nematocyst venom and a tentacle extract in rats. *Toxicol Lett* 2005;**155**(1):135–41.
83. Tibballs J, Hawdon G, Winkel K. Mechanism of cardiac failure in Irukandji syndrome and first aid treatment for stings. *Anaesth Intensive Care* 2001;**29**(5):552.
84. Carrette T, Seymour J. A rapid and repeatable method for venom extraction from Cubozoan nematocysts. *Toxicon* 2004;**44**(2):135–9.
85. Pereira P, Seymour JE. In vitro effects on human heart and skeletal cells of the venom from two cubozoans, *Chironex fleckeri* and *Carukia barnesi*. *Toxicon* 2013;**76**:310–5.
86. Endean R. A study of distribution, habitat, behaviour, venom apparatus, and venom of the stone-fish. *Mar Freshwater Res* 1961;**12**(2):177–90.
87. Gopalakrishnakone P, Gwee M. The structure of the venom gland of stonefish *Synanceja Horrida*. *Toxicon* 1993;**31**(8):979–88.
88. Halstead BW. *Poisonous and venomous marine animals of the world. Invertebrates.* Washington, DC: US Government Printing Office; 1965. 994p.
89. Sutherland SK, Tibballs J. *Australian animal toxins: the creatures, their toxins, and care of the poisoned patient.* Melbourne: Oxford University Press; 1983.
90. Halstead BW, Chitwood MJ, Modglin FR. Stonefish stings, and the venom apparatus of *Synanceja horrida (Linnaeus)*. *Trans Am Microsc Soc* 1956;**75**(4):381–97.
91. Maretic Z. Fish venoms. *Handbook of natural toxins: Marine toxins and venoms*, vol 3. New York: Marcel Dekker; 1983.
92. Wiener S. Observations on the venom of the stone fish (*Synanceja trachynis*). *Med J Aust* 1959;**46**(19):620.
93. Khoo H, Yuen R, Poh C, Tan C. Biological activities of *Synanceja horrida* (Stonefish) venom. *Nat Toxins* 1992;**1**(1):54–60.
94. Saunders PR. Venom of the stonefish *Synanceja horrida*. *Arch Int Pharmacodyn Ther* 1959;**123**:195–203.

95. Saunders PR, Tökés L. Purification and properties of the lethal fraction of the venom of the stonefish *Synanceja horrida* (Linnaeus). *Biochim Biophys Acta* 1961;**52**(3):527–32.
96. Deakins DE, Saunders PR. Purification of the lethal fraction of the venom of the stonefish *Synanceja horrida* (Linnaeus). *Toxicon* 1967;**4**(4):257–62.
97. Kreger AS. Detection of a cytolytic toxin in the venom of the stonefish (*Synanceia trachynis*). *Toxicon* 1991;**29**(6):733–43.
98. Poh CH, Yuen R, Khoo HE, et al. Purification and partial characterization of stonustoxin (lethal factor) from *Synanceja horrida* venom. *Comp Biochem Physiol, B: Comp Biochem* 1991;**99**(4):793–8.
99. Garnier P, Goudey-Perrière F, Breton P, et al. Enzymatic properties of the stonefish (*Synanceia verrucosa* Bloch and Schneider, 1801) venom and purification of a lethal, hypotensive and cytolytic factor. *Toxicon* 1995;**33**(2):143–55.
100. Saunders PR, Rothman S, Medrano VA, Chin H. Cardiovascular actions of venom of the stonefish. *Am J Physiol—Legacy Cont* 1962;**203**(3):429–32.
101. Phoon WO, Alfred ER. A study of stonefish (*Synanceja*) stings in Singapore with a review of the venomous fishes of Malaysia. *Singapore Med J* 1965;**6**(3):158–63.
102. Burnett JW. Aquatic adversaries: stonefish. *Cutis* 1998;**62**(6):269–70.
103. Isbister GK. Venomous fish stings in tropical northern Australia. *Am J Emerg Med* 2001;**19**(7):561–5.
104. Duhig J, Jones G. Haemotoxin of the venom of *Synanceja horrida*. *Aust J Exp Biol Med Sci* 1928;**5**:173–9.
105. Austin L, Gillis R, Youatt G. Stonefish venom: some biochemical and chemical observations. *Aust J Exp Biol Med Sci* 1965;**43**:79–90.
106. Poh C, Yuen R, Chung M, et al. Purification and partial characterization of hyaluronidase from stonefish (*Synanceja horrida*) venom. *Comp Biochem Physiol, B: Comp Biochem* 1992;**101**(1):159–63.
107. Khoo HE. Bioactive proteins from stonefish venom. *Clin Exp Pharmacol Physiol* 2002;**29**(9):802–6.
108. Church J, Hodgson W. Dose-dependent cardiovascular and neuromuscular effects of stonefish (*Synanceja trachynis*) venom. *Toxicon* 2000;**38**(3):391–407.
109. Church JE, Hodgson WC. The pharmacological activity of fish venoms. *Toxicon* 2002;**40**(8):1083–93.
110. Austin L, Cairncross K, Mccallum I. Some pharmacological actions of the venom of the stonefish (*Synanceja horrida*). *Archives Internationales de Pharmacodynamie et de Thérapie* 1961;**131**:339.
111. Low KS, Gwee MC, Yuen R, et al. Stonustoxin: effects on neuromuscular function in vitro and in vivo. *Toxicon* 1994;**32**(5):573–81.
112. Hopkins BJ, Hodgson WC, Sutherland SK. Pharmacological studies of stonefish (*Synanceja trachynis*) venom. *Toxicon* 1994;**32**(10):1197–210.
113. Hopkins BJ, Hodgson WC, Sutherland SK. Evidence for adrenergic and tachykinin activity in venom of the stonefish (*Synanceja trachynis*). *Toxicon* 1996;**34**(5):541–54.
114. Sauviat M-P, Meunier FA, Kreger A, et al. Effects of trachynilysin, a protein isolated from stonefish (*Synanceia trachynis*) venom, on frog atrial heart muscle. *Toxicon* 2000;**38**(7):945–59.
115. Garnier P, Grosclaude J-M, Goudey-Perrière F, et al. Presence of norepinephrine and other biogenic amines in stonefish venom. *J Chromatogr B Biomed Sci Appl* 1996;**685**(2):364–9.
116. Low KS, Gwee ME, Yuen R, et al. Stonustoxin: a highly potent endothelium-dependent vasorelaxant in the rat. *Toxicon* 1993;**31**(11):1471–8.

Chapter 8

Cardiovascular Toxicity of Cardiovascular Drugs

Ramachandran Meenakshisundaram,[1] Douglas Ewan Cannie,[1] P. Ravi Shankar,[2] Heidar Z. Zadeh,[3] Omi Bajracharya,[4] and Ponniah Thirumalaikolundusubramanian[5]

[1]*Department of Cardiology, Epsom and St. Helier University Hospitals NHS Trust, Epsom, United Kingdom*

[2]*Department of Pharmacology, Xavier University School of Medicine, Oranjestad, Aruba*

[3]*Department of Cardiology, Aintree University Hospital, Liverpool, United Kingdom*

[4]*Department of Pharmacology, KIST Medical College, Lalitpur, Nepal*

[5]*Department of Medicine, Chennai Medical College Hospital and Research Centre, Trichy, India*

8.1 INTRODUCTION

In 2008, the Global Health Observatory (GHO) of the World Health Organization (WHO) estimated that cardiovascular diseases were the leading cause of death among noncommunicable diseases (NCD), accounting for 17 million deaths during the year.[1] Cardiovascular disease (CVD) has assumed epidemic proportions in the developing world also. The Global Burden of Diseases (GBD) study reported the estimated mortality from coronary heart disease (CHD) in India was 1.6 million in the year 2010. A total of nearly 64 million cases of CVD are likely to occur in 2015, of which nearly 61 million will be CHD cases—the remaining will be stroke, rheumatic heart disease, and congenital heart diseases.

8.2 NORMAL CARDIAC CONTRACTION

Normal cardiac contraction is vital to maintain cardiovascular health and prevent CVDs. Cardiac contraction depends on the maintenance of a normal sinus rhythm and atrioventricular (AV) as well as inter- and intraventricular synchronization of activation, along with the integrity of the cardiac conduction pathway and well-organized excitation–contraction coupling. For adequate cardiac function, a rhythmic contraction and a forceful extrusion of blood is achieved through careful structural organization of sequential

The Heart and Toxins. DOI: http://dx.doi.org/10.1016/B978-0-12-416595-3.00008-6

contractions.[2] The contraction of cardiomyocytes is accomplished by a process that is called excitation–contraction coupling.

Action potential plays a key role in initiating, coordinating, and regulating this process, and therefore is at the core of cardiac function. This in turn is mediated by voltage-gated ion channels that selectively allow respective ion movements across the cellular membrane down their electrochemical gradient.[3] The cellular membrane is the key regulating element of the conductance of these voltage-gated ion channels and is thereby the driving force of the cardiac action potential. Subtle changes in transmembrane voltage can influence ion channel activation and inactivation ports and can rapidly change their confirmation, altering conductance of the ion channel and subsequently changing ionic currents and transmembrane voltage. This in turn influences the conductance of other ion channels and their respective ionic currents, giving rise to an electrical cascade known as an action potential.

In order to coordinate cardiomyocytes' contraction throughout the heart, action potentials need to be propagated between cells through gap junctions. Gap junctions are specialized structures consisting of collections of intercellular channels that connect adjacent cells in numerous tissues and organs and allow chemical and electrical communication. In the heart, they provide the path for intercellular current flow that is the result of rapid sodium (Na^+) influx during excitation and intracellular Na^+ rises. This passage of Na^+ ions depolarizes the cardiac myocytes beyond the excitation threshold and evokes excitation in the cell, thereby facilitating coordinated action potential propagation.

8.3 CARDIAC ARRHYTHMIAS AND CELLS

The word a *arrhythmia* refers to any change in the normal sequence and/or shape of electrical impulses during the cardiac cycle and is a significant health problem that places a great economic burden on health care systems all over the world. Arrhythmias can occur in up to 25% of patients treated with digitalis, 50% of anesthetized patients, and more than 80% of patients with acute myocardial infarction (AMI). These conditions of irregular heart rhythm may require treatment because the cardiac contractions may be too rapid, too slow, or asynchronous, resulting in altered cardiac output. The most common forms of arrhythmia leading to a high risk of cardiac morbidity and mortality are atrial fibrillation (AF) and ventricular tachycardia/fibrillation (VT/VF). Other rhythm disorders result mostly from unique mechanisms such as intranodal reentry, accessory pathways, or focal abnormal automaticity.

Arrhythmias are frequently asymptomatic but can be fatal. In the United States in 2008, 15,383 deaths were attributed to AF as the underlying cause and ventricular arrhythmias remain the most fatal arrhythmias worldwide.[4] Cardiac arrhythmias occur when there is a disturbance of initiation or conduction of cardiac impulse. To understand this we first have to understand the electrophysiology of the cardiac tissue.

The myocardial tissue is bound by a lipoprotein membrane that have criss-crossing receptor channels. These channels are ion-selective membrane proteins that permit passive movement of ions, such as sodium (Na^+), potassium (K^+), calcium (Ca^{2+}), and chloride (Cl^-), down their electrochemical gradients during the open but not during the closed state. In addition to these channels, there are two energy-requiring exchange pumps—that is, the Na^+/K^+ and the Na^+/Cl^-. During the resting phase, the concentration of Na^+ ions is more extracellular and there are more K^+ ions within the cell. Cardiac cells undergo depolarization and repolarization to create a cardiac action potential and these ion channels help in determining the amplitude and shape of the action potential in an individual cell.

8.3.1 Cardiac Cells at Rest

During the resting phase, the normal cardiac cell transmembrane potential is approximately 80 to 90 mV negative with respect to the exterior. In this stage, K^+ channels are in an open conformation. Thus, potassium can move through these channels across the cell membrane at negative potential in response to either electrical or concentration gradients. However, in this phase, due to the closure of Na^+ channels at negative transmembrane potentials, Na^+ does not enter normal resting cells.

8.3.2 Cardiac Cells When Depolarized

When an impulse is generated that is above the threshold potential, the atrial or ventricular cell at rest is depolarized and the Na^+ channels present in the cell membrane undergo conformational change from closed at resting stage to open at conducting stage, allowing loads of Na^+ ions to enter each cell. An increase in the concentration of Na^+ ions inside the cell allows the transmembrane potential to move toward the positive potential to a maximum of 65 mV. This indicates that phase 0 of the action potential with rapid depolarization of the cardiac cell membrane is associated with fast selective inflow of Na^+ ions. This opening of Na^+ channels lasts only for a millisecond, after which the channel again undergoes conformational change from the open to the inactive state. During the latter part of phase 0 of the cardiac action potential, Ca^{2+} also enters cells via Na^+/Ca^{2+} channels.

The entry of Ca^{2+} continues through phases 1 and 2 via the slow C_{av} channels. It stimulates further release of Ca^{2+} stores in the sarcoplasmic reticulum of the cardiac myocytes, raising the concentration of Ca^{2+} within these cells markedly. This intracellular free calcium interacts with the troponin-actin-myosin system and causes contraction of the heart. The P wave, the PR interval, and the QRS complex of the electrocardiogram (ECG) are inscribed during phases 0 and 1 of the action potential propagation in the atria, the atrioventricular nodal tissue and the ventricles, respectively.

After depolarization, repolarization takes place in several phases: phase 1 begins with the efflux of K^+ ions and Cl^- ions into the cells, resulting in a short and rapid repolarization. A prolonged plateau phase (phase 2) of the normal cardiac action potential occurs due to a balance between Ca^{2+} entering and K^+ leaving the cell. Rapid repolarization to the resting transmembrane potential is caused mainly by continuous extrusion of potassium via the K^+ channels (phase 3). In this phase the final ionic reconstitution of the cell is achieved by the Na^+/K^+ exchange, which actively pumps out sodium, exchanging it for K^+ (resting phase). This cycle is then repeated.

In the atria and ventricles, the membrane potential is steady throughout the diastole; in contrast, the SA node, AV node, and the His-Purkinje tissue depolarize spontaneously because the cell membrane is leaking Na^+. During the greater part of the action potential, the cardiac muscles are resistant to further stimulation; this condition is termed *refractoriness*. The earliest transient depolarization that can be produced (without propagation) marks the end of the absolute refractory period (ARP; phases 1 and 2), and the part of action potential during which a stimulus produces only a local, nonpropagated response is called the effective refractory period (ERP). The ERP is always longer and it includes the ARP. These concepts are important for understanding the mechanism of action of various antiarrhythmic drugs in terms of blockade of sodium, calcium, and potassium channels.

8.4 CLASSIFICATION OF ANTIARRHYTHMIC DRUGS

Most available antiarrhythmic drugs are potent compounds with a relatively narrow therapeutic index. If these agents are used cautiously, they can play a key role in enhancing or prolonging the lives of patients; however, if they are misused (e.g., selection of incorrect drug or inappropriate dosage regimen), they are capable of inducing arrhythmias. In addition, antiarrhythmic agents are, to a variable degree, negatively inotropic (except for digoxin). Vaughan Williams classified various drugs used in the treatment of arrhythmia,[5] which was subsequently modified by Harrison (Table 8.1).[6]

8.4.1 Mechanism of Action

This section examines the mechanism of action of the drug groups. Class I drugs primarily inhibit the rapid inward sodium current, thereby slowing the rate of rise of the cardiac action potential. Drugs of class I were later subclassified as IA, IB, and IC based on their electrophysiologic effects and on the duration of the cardiac action potential. Class IA drugs (e.g., quinidine, procainamide hydrochloride, and disopyramide) lengthen the repolarization and the refractoriness of isolated myocardial tissue in addition to blocking the rapid inward sodium current. As a result, they may increase the duration of the QRS and QT intervals. Drugs, such as lidocaine, tocainide, and mexiletine hydrochloride, of

TABLE 8.1 Classification of Antiarrhythmic Drugs with Examples

Class	Examples
IA	Quinidine, procainamide hydrochloride, disopyramide
IB	Lidocaine, tocainide, mexiletine hydrochloride
IC	Flecainide acetate, propafenone hydrochloride, moricizine
II	Propranolol hydrochloride and other β-adrenergic blocking agents
III	Sotalol hydrochloride, amiodarone, dronedarone, ibutilide fumarate, dofetilide
IV	Verapamil, diltiazem hydrochloride
Digitalis	Digoxin, digitoxin
Adenosine	Adenosine

class IB decrease the action potential duration (APD) but produce only modest inhibition of the rapid inward sodium current. Class IB differs from class IA drugs in that they have a minimal effect on the QRS complex and QT or PR intervals. Drugs of class IC (e.g., flecainide acetate, propafenone hydrochloride, and moricizine) slow the conduction velocity but have little effect on repolarization. These potent sodium channel inhibitors increase the QRS interval more than other class I drugs.

Propranolol hydrochloride and other β-adrenergic blocking agents belong to class II and are β-adrenergic antagonists that act by slowing the sinus rhythm and AV nodal conduction without substantially changing the QT or QRS interval. Drugs, such as N-acetylprocainamide, sotalol hydrochloride, amiodarone, and ibutilide fumarate, of class III block the outward potassium currents, thus prolonging the repolarization and refractory period. Class IV drugs (e.g., verapamil and diltiazem hydrochloride) are calcium channel blockers and act by blocking the slow, inward calcium current, thereby prolonging the PR interval without changing the QRS complex or QT interval. Details about the mechanism of action of digitalis and adenosine are discussed later in this chapter.

8.4.2 Caution While Using Antiarrhythmic Drugs

Some arrhythmias can precipitate more serious or even lethal rhythm disturbances; for example, early premature ventricular depolarization can precipitate VF. Treatment with antiarrhythmic drugs in such patients can be life-saving. That said, it should be kept in mind that such drugs used to treat patients can also precipitate lethal arrhythmias in some; this has led to a reevaluation of their relative risks and benefits. In general, it is very important to avoid the drugs for patients who have or have had asymptomatic nonfatal arrhythmias.

The pathophysiology of cardiac arrhythmias and treatment by drugs that are useful in preventing or controlling them is a complex process. Nevertheless, many patients with arrhythmias respond well to therapy with drugs and a working knowledge of their effects and indications pay dividends, as irregularity of the heartbeat is at the least inconvenient and at the worst fatal. Radiofrequency ablation and implantable defibrillators have often provided the best treatment options compared to antiarrhythmic drugs. However, there are numerous well-recognized limitations of implantable cardioverter defibrillator (ICD) therapy. These include the effects and the results of appropriate and inappropriate ICD shocks; cost of the devices; complications related both to the implantation procedure and to subsequent device function; device malfunction; and restricted efficacy despite normal device function in the presence of significant concomitant disease and, in particular, in the presence of severe left ventricular dysfunction. Due to these limitations, the development of novel approaches has been advantageous to minimize the impact of arrhythmias. Even though there are some drug options available for patients with cardiac arrhythmias, careful selection and use of them should be undertaken as antiarrhythmic drugs themselves are also capable of generating arrhythmias. Therefore, the drugs should only be used in the presence of clear indications of an arrhythmia.

Antiarrhythmic drugs are indicated to maintain the normal sinus rhythm and prevent rapid and irregular heartbeats, and they are used to relieve symptoms and complications associated with arrhythmias such as palpitation, syncope, difficulty breathing, heart failure, thromboembolic disease, and so on. In hypothetical terms, the goal of therapy is to reduce mortality directly related to arrhythmias. Currently available agents have limited efficacy and/or carry a risk of a variety of side effects such as drug toxicity or proarrhythmic potential.[7] Various cardiac toxicities that can be caused by antiarrhythmic drugs are listed in Table 8.2. Such antiarrhythmic agents, if not used properly, can lead to both drug-induced adverse effects outside the cardiovascular system and cardiovascular complications that can result in increased morbidity and mortality.

8.5 DRUG-INDUCED LONG QT SYNDROME

Drug-induced long QT syndrome is a common cardiovascular adverse effect described as a prolonged corrected QT interval (QTc) with an increased risk of torsades de pointes (TdP). TdP is a life-threatening polymorphic VT that occurs predominantly with the Vaughan Williams class I and class III antiarrhythmic agents. It was first described by Dessertenne as an ECG waveform in which a series of peaks are complicated by "torsade," meaning ornamental hair or threads on a column.[8,9] These are usually caused by prolongation of

TABLE 8.2 Cardiac Toxicity of Selected Antiarrhythmic Drugs

Class	Agents	Cardiac Side Effects
IA	Procainamide	Torsades de pointes (TdP)
	Disopyramide	TdP, CHF, negative inotropic effects
	Quinidine	TdP, increased AV node conduction, syncope
IC	Flecainide	VT, 2:1 AV conduction block, proischemic
	Propafenone	CHF, negative inotropic effects
III	Dofetilide	TdP
	Amiodarone	TdP (infrequent)
	Sotalol	TdP, CHF, bradycardia, hypotension, mild negative inotropic effects
	Dronedarone	Possible exacerbation of advanced heart failure
	Azimilide (phase 3 study)	TdP[10]
	Tedisamil (phase 3 study)	Proarrhythmic reactions[11]
Cardiac glycoside	Digoxin	Ventricular dysrhythmias, severe AV block, depression of myocardial contractility, proischemic effect
Adenosine		Premature ventricular complex, monomorphic sustained and nonsustained VT, TdP, premature atrial complex, AF, prolonged AV block

Source: Adapted from Camm AJ with permission.[11]

the QT interval, mostly by blockade of rapid outward movement of the potassium current during the repolarization phase of the action potential. This results in an inappropriate reentry current that may trigger and sustain the arrhythmia. If this is more rapid and prolonged, it can result in VF and sudden cardiac death.

8.5.1 Mechanism of QT Prolongation and Torsades de Pointes

The efflux of potassium ions is primarily mediated through myocardial repolarization. Delayed rectification is a time-dependent increase in cardiac K^+ conductance following depolarization. Noble and Tsien mentioned the underlying currents as I_{x1} and I_{x2}.[12,13] The rapid component I_{x1} showed delayed rectification, which has a greater ability to pass current in the inward or depolarizing direction compared to the outward or repolarizing direction, while the slower component I_{x2} showed a nearly linear conductance. Later, the rapid and slow delayed rectifier currents corresponding to I_{x1} and I_{x2} came to be known as I_{Kr} and I_{Ks} following the seminal work of the Sanguinetti laboratory.[14] These two currents are carried by distinct channels with differing ion-channel subunit

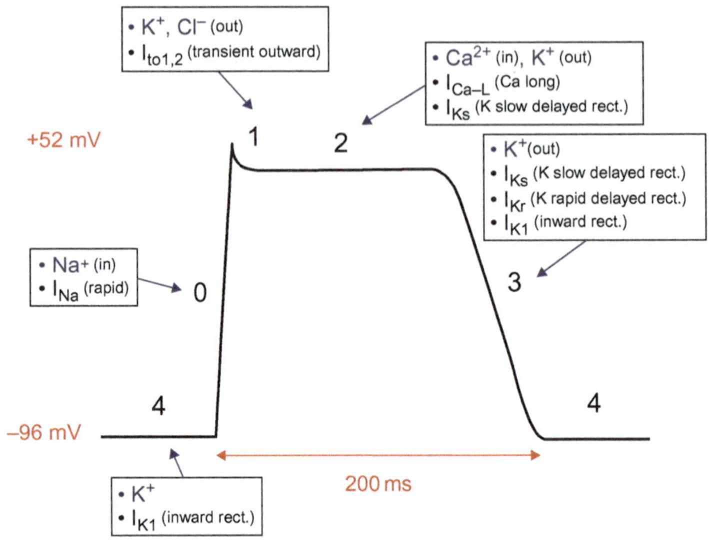

FIGURE 8.1 The myocardial action potential. Source: *From http://en.wikipedia.org/wiki/Cardiac_action_potential.*

compositions: the human Ether-à-go-go related (hERG) gene product for I_{Kr} versus the KCNQ1-encoded KvLQT1 α-subunit and KCNE1-encoded MinK β-subunit for I_{Ks}.[15,16]

Two important subtypes of the delayed rectifier K^+ current, I_{Kr} (rapid) and I_{Ks} (slow), are mainly responsible for repolarization. All drugs that prolong QTc practically block I_{Kr}.[17] This was demonstrated by De Bruin et al.[18] with the existence of a strong correlation between a drug's ability to block I_{Kr} and its potential to cause ventricular arrhythmias and sudden death. Blockade of I_{Kr} causes a delay in phase 3 rapid repolarization of the action potential (Figure 8.1). This increase in action potential duration (APD) is due to QT prolongation. Repetitive extra systoles (triggered activity), unidirectional block, and zones of slow conduction can lead to reentry and TdP.[19] Due to the activation of the inward depolarizing current, prolongation of repolarization may cause early after-depolarizations (EADs).[19] EADs appear as depolarizing fluctuations in the membrane voltage during phases 2 and 3 of the action potential (Figure 8.2). EADs that reach the threshold voltage cause ventricular extrasystoles.

Purkinje fibers and M cells (midmyocardial) are especially susceptible to drug-induced QT prolongation, early after-depolarizations, and ventricular extrasystoles.[20] Heterogeneity in ventricular repolarization (dispersion of refractoriness) can create zones of unidirectional block, which is required for the initiation of reentry.[21] Unidirectional block can be caused by preexisting heterogeneity in excitability, conduction, or refractoriness.[22] In the healthy heart, electrophysiological heterogeneities allow efficient contraction and relaxation.

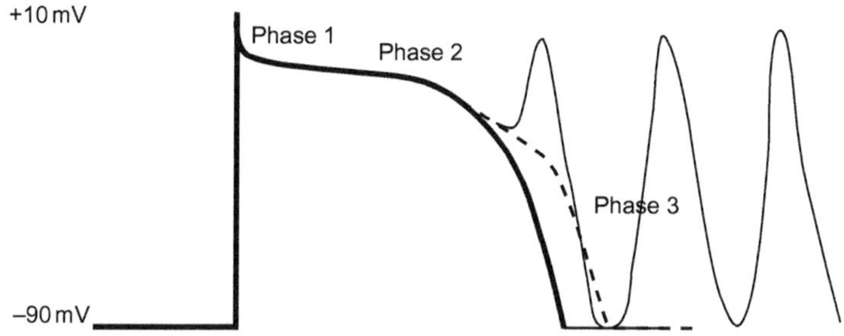

FIGURE 8.2 **Early after-depolarization fluctuations.**

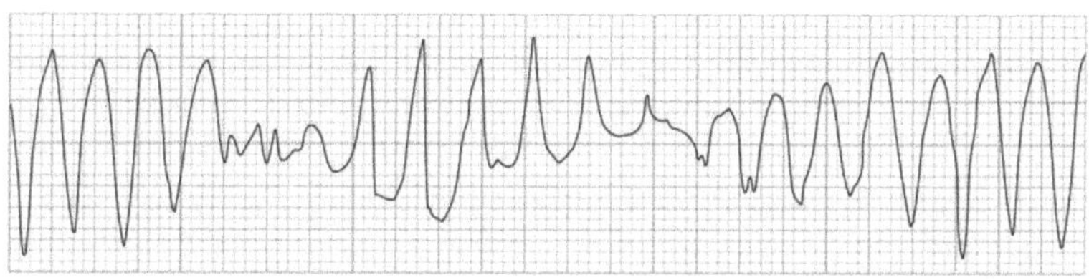

FIGURE 8.3 **Torsades de pointes showing short-long-short sequence.**

However, under pathological conditions, electrophysiological heterogeneities may underlie dysfunction and life-threatening cardiac arrhythmias.

Phase 0 rapid depolarization occurs and it is mediated by the influx of sodium ions into the cells. Phases 1 and 3 show the repolarization mediated by potassium efflux from cells. In phase 2, plateau occurs as a result of slow calcium influx and potassium efflux. Potassium reenters and sodium exits cells during phase 4 recovery.

Drug-induced TdP is usually preceded by a short-long-short ECG sequence (Figure 8.3).[23] This generally starts with one or more premature ventricular complexes followed by a compensatory pause. The subsequent sinus beat may have an especially long QT interval and T- or U-wave deformities.

The sinus beat is followed by another premature ventricular complex that precipitates TdP. Drugs that have a high risk of precipitating TdP include the class IA and class III antiarrhythmics, as shown earlier in Table 8.2. Potentially life-threatening polymorphic VT occurs predominantly with quinidine, disopyramide, and procainamide.[24–26] An increase in dose of the class IA drugs does not necessarily increase the incidence of TdP, which can occur even with a low or suboptimal dose.[23] This may be due to the QT prolongation caused by blockade of efflux potassium channels.[27,28]

Quinidine has a risk of TdP of approximately 1.5%.[29] Disopyramide has also been implicated in TdP.[30] Procainamide is less likely to cause TdP but its metabolite, N-acetylprocainamide, has effective I_{Kr}-blocking properties that may lead to QT prolongation and subsequent precipitation of TdP.[31] These conditions are usually seen in the patients with impaired renal function who may develop high N-acetylprocainamide levels.

Unlike class IA agents, class III antiarrhythmics (i.e., potent I_{Kr} blockers) prolong QTc in a dose-dependent manner. Dofetilide, ibutilide, and sotalol, which are class III drugs, pose the highest risk, progressively prolonging QTc as serum levels increase.[30,32] These drugs block I_{Kr} most effectively at low heart rates; this phenomenon is known as reverse-use dependence. Consequently, the risk of TdP increases with bradycardia. For sotalol, TdP risk ranges from 0.8 to 3.8%.[30] Similar rates (0.9–3.3%) have been reported with dofetilide.[33] Intravenous ibutilide has caused TdP in 3.6 to 8.3% of patients when it is administered for AF or atrial flutter.[34]

Amiodarone is unlikely to cause TdP despite significant QT prolongation.[35,36] In addition to blocking I_{Kr} without reverse-use dependence, amiodarone prolongs the action potential duration in a homogenous manner, reducing heterogeneity of refractoriness and making the myocardium less susceptible to reentry. Additional electrophysiologic effects that help explain its safety include a noncompetitive antagonism and inward L-calcium channel blockade, which may reduce EADs. Unlike other calcium channel blockers, bepridil prolongs QTc and has been associated with TdP.[37,38]

Antiarrhythmic drugs have implications for dental surgeons. Epinephrine and anticholinergic drugs, which have an excitatory effect on the heart, should be used with caution in patients on antiarrhythmic drugs.[39] Amiodarone inhibits lidocaine metabolism and increases serum lidocaine levels, which can predispose to lidocaine toxicity.

Procainamide, a class IA antiarrhythmic, can cause severe hypotension and dysrhthymias identical to those caused by quinidine. Hematologic abnormalities (e.g., agranulocytosis, thrombocytopenia, and hemolytic anemia) have been reported with long-term use of procainamide. Lidocaine is considered as the prototype of the class IB agents and can cause both neurologic and cardiovascular toxicity. The neurologic symptoms are usually the first to manifest and following this there is depression of the intrinsic cardiac pacemakers, depression of myocardial contractility, and delayed conduction. Large intravenous doses of lidocaine have resulted in asystole, complete heart block, and refractory hypotension. Many drugs, including cardiovascular, noncardiovascular, and antimicrobials, cause QT prolongation, as discussed in Chapter 10.

8.6 NEW ANTIARRHYTHMIC DRUGS

Because of problems that are associated with the currently available antiarrhthymic drugs, there has been an emphasis on the development of new drugs. Azimilide is a class III agent not related to the others in its class.

It has, however, been shown to cause concentration-dependent QTc prolongation.[40] Tedisamil is another newly developed class III agent that prolongs the duration of ventricular and atrial action potentials by acting on multiple potassium channels. The drug seems to be associated with the risk of QT prolongation and TdP.[41]

Dronedarone is an amiodarone analog without iodine in its structure; it is devoid of many of the life-threatening effects of amiodarone.[42] Most clinical studies have shown favorable effects of dronedarone in AF but a recent study, Antiarrhythmic Trial with Dronedarone in Moderate to Severe Congestive Heart Failure Evaluating Morbidity Decrease (ANDROMEDA), showed increased mortality in the dronedarone arm compared to the placebo arm.[43] The recent A Placebo-Controlled, Double-Blind Parallel Arm Trial to Assess the Efficacy of Dronedarone for the Prevention of Cardiovascular Hospitalization or Death from Any Cause in Patients with AF/Atrial Flutter (ATHENA) study, however, showed a significantly reduced risk of hospitalization and cardiovascular mortality in high-risk patients with AF treated with dronedarone compared to placebo.[44]

8.7 CARDIOVASCULAR TOXICITY OF BETA BLOCKERS

Beta-adrenoceptor antagonists, commonly known as beta blockers, are used to treat a wide variety of conditions including arrhythmias, hypertension, angina, heart failure, glaucoma, migraine, and anxiety. All of the beta blockers antagonize the effects of β-agonists at β-adrenergic receptors but vary according to their pharmacokinetics, receptor selectivity, and other properties.[45] This variation is useful in fulfilling the clinician's prescribing needs but leads to various toxicity profiles and clinical manifestations with overdose. The pharmacological and pharmacokinetic properties of beta blockers are listed in Table 8.3. The predominant cardiotoxic effects of the beta blockers are bradycardia and hypotension. These are best understood in the context of the pharmacological properties of beta blockers and their negative chronotropic, inotropic, and dromotropic characteristics.

There are three subtypes of β-adrenergic receptors, as follows. β1 receptors are seen in high concentration in cardiac tissue where they are acted on by catecholamines to increase heart rate and contractility. β2 receptors are prevalent in blood vessels, the lungs, and other sympathetic target organs where their stimulation leads to blood vessel dilatation, bronchodilatation, and relaxation of smooth muscle. β3 receptors are present in adipose tissue and are involved in the process of lipolysis. Beta blockers that exhibit β1 selectivity have therapeutic advantages over the nonselective beta blockers, particularly for those patients with reactive airway disease and diabetes. Selectivity does not persist in overdose situations.[46]

The influx of calcium across myocardial cell membranes triggers calcium release from intracellular storage organelles which in turn affects excitation–contraction coupling. Calcium influx is similarly required to maintain vascular

TABLE 8.3 Pharmacological and Pharmacokinetics Properties of Beta Blockers

β-Blocker	Receptor Selectivity	Partial Agonist Activity (ISA)	Protein Binding (%)	t½ (h)	Bioavailability (%)	Vd (L/kg)	Lipid Solubility	Metabolism/ Elimination	MSA
Acebutolol	β_1	Low	25	3–4	70	1.2	Low	Hepatic	Yes
Atenolol	β_1	No	5–10	6–9	50–60	1	Low	Renal	No
Betaxolol	β_1	No	50	14–22	90	5–13	Low	Hepatic	Low
Bisoprolol	β_1	No	30	9–12	80	3	Low	Renal	No
Cartelol	β_1, β_2	Yes	30	6	85		Low	Renal	Yes
Carvedilol	β_1, β_2, α_1	No	25–35	6–8	20	1–2	High	Hepatic	Yes
Esmolol	β_1	No	55	9 min	NA	2	Low	Erythrocyte esterases	No
Labetalol	β_1, β_2, α_1	No	50	6–8	90	9	Moderate	Renal	Yes
Metoprolol	β_1	No	10	3–4	90	4	Moderate	Hepatic	Low
Nadolol	β_1, β_2	No	25–30	20–24	30	2	Low	Renal	No
Oxprenolol	β_1, β_2	Yes	80	1–2	20–70	1.3	High	Hepatic	Yes
Penbutalol	β_1, β_2	Low	80–98	5	100	0.5	High	Hepatic	No
Pindolol	β_1, β_2	High	50	3–4	95	2	Moderate	Hepatic	Yes
Propranolol	β_1, β_2	No	90	3–5	90	4	High	Hepatic	High
Sotalol	β_1, β_2	No	0	7–15	80	2	Low	Renal	No
Timolol	β_1, β_2	No	10	4	90	2	Moderate	Hepatic	No

ISA, intrinsic sympathomimetic activity; MSA, membrane stabilising activity; NA, not applicable; t½, half-life; Vd, volume of distribution; h, hour

Source: Adapted from DeWitt and Waksman with permission.[45]

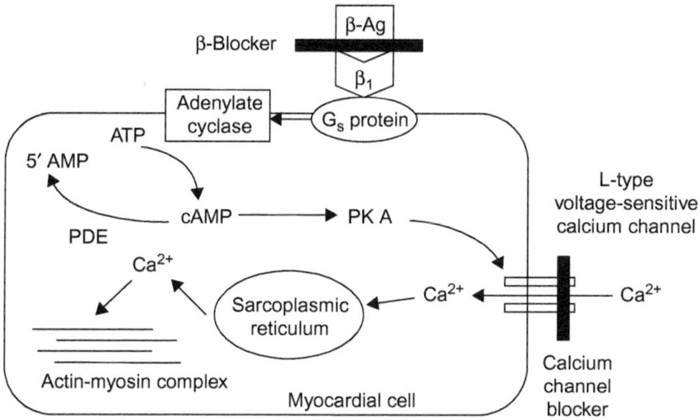

FIGURE 8.4 **The role of calcium in myocardial contraction and the effects of β-agonists, beta blockers, and calcium channel blockers on calcium influx.** 5′ AMP, 5′ adenosine monophosphate; ATP, adenosine triphosphate; β1, β1-receptor; β-Ag, β-agonist (catecholamine); cAMP, cyclic adenosine monophosphate; Gs, stimulatory G protein; PDE, phosphodiesterase; PK A, protein kinase A. **Source:** *Adapted from DeWitt and Waksman with permission.*[45]

tone.[47] Interference with this mechanism plays a key role in the cardiovascular dysfunction seen in beta-blocker toxicity. Beta blockers disrupt calcium flow into cells by competitive inhibition with catecholamines at the β1-adrenergic receptor. This prevents activation of the secondary messenger systems that normally lead to the opening of L-type calcium channels and increased calcium influx (Figure 8.4). Excessive inhibition of this pathway results in the hallmarks of β-blocker toxicity: bradycardia, conduction abnormalities, hypotension, and shock,[45] with 97% of symptomatic patients developing symptoms within 4 hours of ingestion.[48]

Some beta blockers exhibit structural similarity to isoproterenol, a β-receptor agonist. These beta blockers demonstrate varying degrees of agonist properties while also blocking the interaction of the catecholamines at the β-receptor.[49] This ability is termed *intrinsic sympathomimetic activity* (ISA), which influences the toxic nature of the drug. Theoretically, this should make the medication safer in overdose and this is corroborated by case studies and reviews of pindolol, the beta blocker with the greatest ISA. No deaths have been reported secondary to overdose with pindolol,[50–52] and there have been isolated reports of tachycardia and hypertension with pindolol overdose.[50,51] However, the possession of ISA has not prevented serious effects in overdose with other beta blockers, with acebutolol proving particularly dangerous.[50–54]

The most common ECG finding in beta-blocker toxicity is first degree heart block.[55] Bradycardia is not consistently seen in patients with symptomatic beta-blocker toxicity,[55] although the presence of bradycardia should not be underestimated.[50] This supports the supposition that profound hypotension sometimes seen in beta-blocker overdose is secondary to inotropic rather than chronotropic dysfunction and may explain why medication such as amrinone, which has limited effect on heart rate, is able to improve cardiac function with beta-blocker overdose.[56]

Propanolol in high doses has been reported to induce the typical coved ST segments and other ECG changes as seen in Brugada syndrome.[57,58] An explanation for this has been suggested by studies with canine models in which there was disparity in the APD with sodium channel blockage between the canine endocardium and epicardium.[59]

Beta blockers with membrane-stabilizing activity (MSA) have the ability to effect ventricular depolarization as a result of sodium channel blockade or altered calcium flux. It is of little significance at therapeutic doses; but in overdose, it can cause "quinidine-like" effects similar to those seen with tricyclic-antidepressant toxicity and in reference to activity demonstrated by quinidine and other class IA antidysrhythmic agents. Propranolol is widely acknowledged to possess the highest level of MSA with acebutolol, labetalol, carvedilol, and pindolol also possessing varying degrees of this characteristic. Widening of the QRS complex demonstrating intraventricular conduction delay is cited as a hallmark of beta-blocker overdose with MSA but is not invariably present. In a case series, only 8 out of 13 symptomatic patients of β-blocker overdose showed abnormal QRS morphology.[55] A prominent terminal R wave in lead aVR is sometimes seen on an ECG. Ventricular dysrhythmias are more common in overdose with beta blockers with MSA than those without.[60,61]

Beta blockers with MSA are responsible for the majority of morbidity and mortality in β-blocker toxicity. Love et al.[53] found that 94% of beta-blocker overdoses with cardiovascular compromise (i.e., hypotension, bradycardia, symptoms of hypoperfusion, or the need for therapeutic intervention) had taken beta blockers with MSA;[45] this has been consistently highlighted in a number of case studies where beta blockers with MSA were associated with higher mortality rates.[50,51,60] A review of fatal beta-blocker overdoses between 1975 and 1984 in England and Wales showed that 206 of 208 deaths involved beta blockers with MSA.[61]

Acebutolol is particularly cardiotoxic and potentially lethal in overdose. It is a partial agonist and possesses a high degree of MSA. Cases of overdose demonstrate prolongation of the QRS complex; this is consistent with that seen with other beta blockers possessing MSA. In addition, there are reports of significant QTc prolongation and VT including TdP.[54,62] This suggests an effect on ventricular repolarization that is not seen with propanolol intoxication,[63] which even more closely resembles the effects of the class IA antiarrhythmics. Acebutolol may, therefore, also act to block the outward-rectifying potassium channels in addition to sodium channels.

Sotalol is a nonselective beta blocker that does not possess ISA or MSA. As well as causing beta blockade, it possesses class III antiarrhythmic properties, blocking the delayed rectifier potassium current during repolarization.[45]

This leads to dose-dependent prolongation of the action potential and the QT interval, and this predisposes patients to ventricular dysrhythmias, including TdP.[64] Torsades has been described at both therapeutic dosing[64,65] and in overdose,[66,67] with polymorphic VT seen in up to 3.5% of patients.[64] Another case study described 6 patients presenting with bradycardia, hypotension, and considerable prolongation of the QTc interval after ingestion of 2.4 to 8 g of sotalol. Five of the 6 suffered severe ventricular tachyarrhythmias and the risk was greatest for up to 20 hours after ingestion. Peak QT prolongation was observed 4 to 15 hours after ingestion and it has been noted to take 3 to 4 days to normalize; this is consistent with the known long half-life of sotalol, which is about 10 to 20 hours[67]

The proarrhythmic effects secondary to sotalol's class III antiarrhythmic action have been well described. These effects are not usually seen in doses of less than 80 mg BD.[64] Less well documented are the proarrhythmic effects of the β-blocking properties of the drug. One case described polymorphic VT in a patient on a low dose of sotalol (40 mg TDS) and without prolongation of the QT interval.[68] When sotalol was stopped, symptoms attributed to the arrhythmia settled and no further ventricular tachycardia was seen. Coronary spasm was thought to be the underlying cause of the arrhythmia, supported by ergonovine-induced spasm at angiogram and complete resolution of symptoms with diltiazem.[68] A further case demonstrated sotalol-induced VT in a patient with dilated cardiomyopathy with coronary spasm again implicated as the underlying etiology.[69] In summary, sotalol has a narrow therapeutic index that needs close monitoring of the patient.

Beta blockers may exacerbate coronary vasospasm by leaving α-adrenergic vasoconstriction unopposed, although this has been disputed.[70] Beta blocker-induced coronary artery spasm has been implicated in patients with Prinzmetal's (variant) angina and in healthy individuals with cases demonstrating ST elevation on ECG with negative biochemical markers of myocardial cell necrosis.[71,72] The incidence of pulseless electrical activity (PEA) in cardiac arrest may be increasing with increased use of beta blockers. This corresponds with a decrease in VF as the presenting rhythm in out-of-hospital cardiac arrest. It is hypothesized that the antiarrhythmic properties of beta blockers may prevent earlier arrest with VF leading to later presentations with PEA. Additionally, beta blockade with a nonselective agent may provide physiological conditions more conducive to PEA with hypotension and low pulse pressure.[73]

8.7.1 Management of Beta-Blocker Overdose

The basic tenets of managing any critically ill patient apply to treating beta-blocker overdose, which should pay particular attention to airway, breathing, and circulation. Significant ingestions with signs of toxicity should prompt consideration of early intubation. A focused history should attempt to elicit the components of the overdose; it needs to include which beta blocker was taken and when, the amount, and the preparation that was ingested. Clinical effects, with nonsustained release formulations, manifest within 2 to 6 hours of ingestion.[53,54] Onset of symptoms beyond 6 hours has not been documented for beta blockers but is theoretically possible, with both metoprolol and propranolol having sustained release (SR) formulations.[74]

Investigations should include standard blood analysis and serial ECGs, particularly if sotalol or a beta blocker with MSA is involved. A chest x-ray may demonstrate pulmonary edema and arterial blood gases should be obtained to determine acid-base status, lactate, and glucose levels. The patient should be on a cardiac monitor. Bedside echocardiography or more invasive monitoring (e.g., pulmonary artery catheterization) is recommended to guide pharmacological therapy. IV crystalloids should be used for fluid resuscitation with ongoing monitoring and replacement of electrolytes. Fluid overload is a legitimate concern for patients with impaired inotropic function and a urinary catheter will aid in input and output monitoring.[45] As per standard guidelines, atropine should be given in cases of bradycardia with adverse features or in cases at risk of asystole. Atropine has not, however, been found to be effective in severe beta-blocker overdose.[45] Noncardiac complications of toxicity include bronchospasm secondary to $\beta 2$-antagonism and seizures, particularly related to propranolol overdose.[52] These should be managed with inhaled $\beta 2$-agonists and IV benzodiazepines, respectively.

Gastrointestinal decontamination has not been shown to improve overall outcomes with drug toxicity but can still play a role. Gastric lavage may be useful if a patient presents within 1 to 2 hours of a life-threatening ingestion as assessed on a case-by-case basis.[75] Consideration should be given to the risks of the lavage procedure itself.[74] Activated charcoal decreases systemic drug absorption and should be administered at a dose of 1 gm/kg within 1 to 2 hours of ingestion to maximize its effects.[76] Overdose with SR preparations extends the window in which gastrointestinal decontamination may be effective, best demonstrated by cases of SR calcium channel blocker (CCB) overdose.

The majority of beta blockers are highly lipid-soluble with large volumes of distribution making extracorporeal removal ineffective.[45] Atenolol, nadolol, and sotalol are more water-soluble, have smaller volumes of distribution, and are renally excreted. These properties make them more amenable to hemodialysis.[77] Acebutolol overdose has also been treated with hemodialysis, with extracorporeal membrane oxygenation initiated in the first instance to obtain hemodynamic stability.[78]

Beyond the basics, logic dictates two approaches to the management of beta-blocker toxicity: (1) to competitively overcome the agent at the blocked receptor or (2) to bypass the receptor in order to achieve a common physiological endpoint. While many case studies and reports exist, there are no clinical trials to guide therapeutic intervention due to β-blocker toxicity.[45] Additionally, patients are often managed with multiple agents, complicating accurate assessment of individual therapies.

Glucagon is a regulatory hormone that opposes the hypoglycemic action of insulin. It also has direct inotropic and chronotrophic effects on the myocardium; it achieves this via stimulation of adenylate cyclase through different receptors to the catecholamines. As a result, myocardial function can be maintained in the face of toxic beta blockade. This has been demonstrated in several animal models. Multiple case reports have shown a positive clinical response to glucagon therapy in beta-blocker toxicity in conjunction with other treatments, but fewer have demonstrated the efficacy of glucagon as the sole agent.[79]

Some other reports have described the failure of glucagon to adequately treat cardiac dysfunction in beta-blocker overdose.[80] In nonpoisoned individuals, onset of action of glucagon occurs at 1 to 3 min, peaks at 5 to 7 min, and effects persist for only 10 to 15 min. An initial dose of 50 to 150 μg/kg is recommended and this can be repeated within 3 to 5 min. A glucagon infusion can be initiated to counter its short duration of action; a reasonable approach is to establish the effective bolus dose and administer this every hour, titrating to cardiodynamic parameters.[80] Adverse effects include dose-related nausea, vomiting, and hyperglycemia. Despite lack of definitive evidence, glucagon therapy is recommended early in the course of beta-blocker toxicity and seems most effective for increasing heart rate.[80]

In light of the underlying pathophysiology of beta-blocker toxicity, calcium would seem an obvious treatment choice. There is, however, limited evidence to support this. Calcium was found to reverse negative inotropy in rodent and canine models but did not reverse bradycardia or conduction abnormalities.[81] Case reports have suggested some inotropic benefit but in no case was calcium therapy used alone. One isolated case of propranolol overdose demonstrated a dramatic improvement in hemodynamic stability and conduction in a temporal relationship to calcium boluses after other agents had failed.[82] Overall, clinical experience with calcium use in beta-blocker overdose is mixed. Calcium gluconate should be the preferred calcium compound,[80] giving 0.6 mL/kg bolus of 10% it followed by an infusion at 0.6 to 1.5 mL/kg per hour and titrating to improve blood pressure and contractility.

In theory, sodium bicarbonate may be of benefit for managing overdose with beta blockers that exhibit MSA; the pathophysiology of QRS widening and ventricular dysrhythmias involves the attenuation of sodium transport in myocardial cells. Animal models of propranolol toxicity provide mixed evidence for sodium bicarbonate therapy. A canine model showed no effect on QRS widening with sodium bicarbonate therapy[83] but raising the extracellular sodium concentration improved the ability to pace rat hearts poisoned with propranolol.[84] Sodium bicarbonate appeared to be of benefit in a case of acebutolol overdose with QRS widening and VT,[62] and it has been shown to improve hemodynamics in cases of drug toxicity-induced acidemia.[85] It should be given as an IV bolus at 1 to 2 mEq/kg in cases of QRS widening or ventricular dysrhythmia after overdose with a beta blocker with MSA. Repeat boluses may be given if they prove effective but fluid balance and respiratory function must be monitored because excessive sodium load can theoretically cause pulmonary edema.[45]

Sotalol overdose, in particular, predisposes patients to ventricular dysrhythmias; various methods have been used to manage these. Electrolyte disturbances should be replaced as a priority where the QT interval is prolonged. Magnesium infusions may be particularly beneficial for TdP.[65] Overdrive pacing isoprenaline and lidocaine may also be effective, although one case report indicated the increased frequency of torsades with lidocaine.[45,86]

Phosphodiesterase inhibitors (PDEI), such as amrinone and milrinone, slow the conversion of cyclic adenosine monophosphate (cAMP) to 5′ AMP, thus raising the intracellular levels of cAMP and, consequently, intracellular calcium. As previously described, this leads to a stronger inotropic response. This mechanism, much like that involving glucagon, bypasses the β-receptors.[45] Animal models with propranolol showed a good response with PDEI[87] but concerns have been raised about their effectiveness when used with glucagon. The combination of PDEI and glucagon caused significant tachycardia in one canine model of propranolol toxicity;[88] and, when used in animal models of CCB toxicity, showed either no improvement[89] or a worsening in mean arterial pressure.[90] Hypotension secondary to peripheral vasodilation can also occur with PDEI therapy. They are reserved as second-line agents.

Catecholamines are often used to support blood pressure in beta-blocker overdose. Their use is plagued by a lack of solid data because patients presenting with severe drug overdoses are often managed with multiple agents. Isoprenaline, a nonspecific β-agonist, has frequently been used in cases of beta-blocker overdose; its β-agonist characteristics are theoretically ideal to overcome beta blockade. It may, however, cause vasodilation and dysrhythmia, potentially exacerbating hypotension, and hard evidence supporting its use over other catecholamines is lacking. A case study of 39 beta-blocker overdoses showed improved heart rate with isoprenaline in only 11% of the cases and better blood pressure in only 22%. This contrasted with adrenaline in which heart rate improved in 67% of cases and blood pressure in 50% and dobutamine, which showed 25% improvement in each.

Selection of catecholamine should be considered on a case-by-case basis, depending on hemodynamic and cardiodynamic parameters as revealed by bedside echocardiography or more invasive monitoring. For example, adrenaline or noradrenaline should be preferred for the patient with depressed contractility and decreased peripheral resistance to benefit from both β- and α-agonist stimulation.[80] While the choice of agent remains unclear, it is apparent that whichever one is chosen, it will likely be needed in large quantities in order to obtain hemodynamic stability. Isoprenaline and dopamine were required at 15- and 5-fold dose increases to reverse propranolol-induced hemodynamic changes in canine models, while adrenaline became a lethal drug in large doses and did not increase cardiac output in standard doses.[91] Isoprenaline at 26 times the control dose was required to reverse labetalol-induced hypotension in human subjects,[92] and adrenaline was needed at high doses in a case of diltiazem and metoprolol overdose.[93]

Hyperinsulinemia-euglycemia (HIE) has been used to good effect in the treatment of beta-blocker overdose.[80] Insulin possesses inotropic properties, which may be beneficial for drug overdose; this is secondary to its role in carbohydrate metabolism. Under normal circumstances, free fatty acids are the heart's preferred energy substrate. This metabolic preference shifts to carbohydrates during drug-induced shock.[94] Insulin promotes carbohydrate metabolism and has been found to increase myocardial glucose uptake and cardiac function during beta blockade. Treatment with glucagon, adrenaline, and calcium promotes free fatty acid use with an increase in myocardial work, possibly explaining why these treatments can fail to work. In a canine-model, insulin was superior to both glucagon and adrenaline, significantly improving hemodynamic stability and reversing myocardial failure.[84,95] Clinical evidence for the use of HIE in beta-blocker overdose is restricted to case reports[96,97] and many of these involve overdose in conjunction with CCBs.

An insulin bolus of 1 U/kg is recommended, followed by a continuous infusion of 1 to 10 U/kg per hour initiated early in therapy.[97] Higher doses have been tolerated to good effect and may be necessary, where vasopressors have also been used to overcome the increased systemic vascular resistance and decreased cardiac output.[97] Studies with CCBs suggest no advantage or even a deleterious effect when HIE is used with vasopressors.[98] The adverse effects of HIE are predictable: hypokalemia and hypoglycemia. Potassium and glucose levels must be carefully monitored and a dextrose infusion should be run alongside the insulin therapy[97] with regular blood glucose (BM) checks.

Transvenous or transthoracic cardiac pacing can be used to maintain heart rate but often fails to achieve electrical capture. When capture is attained, improvement in blood pressure and contractility is not always observed[99] because of the calcium transport dysfunction into myocytes and the deficit in intracellular calcium. The recommended optimal pacing rate in both beta-blocker and CCB toxicity is therefore lower than that for other causes of hemodynamically significant bradycardia, at 50 to 60 bpm, to allow sufficient time for the myocardium to achieve forceful contraction.[80]

Where standard treatments have failed, extraordinary measures can be used in the intensive care setting. These include extracorporeal circulatory support,[78] aortic balloon pump,[100] and cardiopulmonary bypass.[101] Many of the cases suffered prolonged cardiac arrest (2.5–4 hours) but survived to obtain a good neurological outcome. This highlights the merits of persisting with cardiopulmonary resuscitation (CPR) in patients with cardiac drug toxicity.[80]

8.8 CALCIUM CHANNEL BLOCKERS

Calcium channel blockers (CCB) interfere with calcium flux across myocyte membranes by blocking flow through the L-type calcium channels. They are predominantly used in hypertension, arrhythmias, and angina. As with beta blockers, interruption of calcium flow results in decreased intracellular calcium, which can lead to cardiovascular dysfunction in toxicity. Beta-blocker and CCB toxicity present in similar ways, due to their related mechanisms of action. Therefore, the hallmarks of CCB overdose are also hypotension, bradycardia, and conduction abnormalities. As with beta blockers, the pharmacological profiles of the CCBs vary and a brief consideration of their individual properties will aid in understanding these drugs in overdose.

Calcium channel blockers can be divided into three main classes:[45]

- Phenylalkylamines, of which verapamil is a member
- Benzothiazepines, of which diltiazem is a member
- Dihydropyridines, which include the dipine groups (e.g., amlodipine and nifedipine)

The different classes vary according to their affinities for cardiac tissue and vascular smooth muscle. The dihydropyridines have little effect on contractility or on cardiac pacemaker cells, acting predominantly as vasodilators.[102] Verapamil and diltiazem affect both the heart and vascular smooth muscle. Of the two, verapamil has the more potent effects: greater suppression of contractility and AV node conduction and increased vasodilatation.[102] As with beta blockers, the selectivity of the CCB is lost in significant overdose.

With immediate release (IR) preparations, onset of symptoms is expected within 2 to 3 hours of ingestion. One year-long study showed all patients developed symptoms within 6 hours after taking IR preparations.[103] SR products present a different challenge, with symptoms delayed for 6 to 16 hours.[104] Recurrent toxicity after 24 hours has also been reported.[105] The risk of toxicity occurs at doses $<$10 mg/kg[106] and patients with a low body mass index (BMI) are at a greater risk.[107]

As well as the hallmark signs of hypotension, bradycardia, and conduction disturbances, CCB overdose can present with nausea, vomiting, diarrhea, abdominal pain, chest pain, shortness of breath, and/or lethargy.[108,109] Patients are also more likely to suffer acid-base disturbance than with β-blocker toxicity; a metabolic acidosis can occur, worsening myocardial depression. The mechanism underlying this is poorly understood but may be secondary to increased drug binding at the calcium channel. Metabolic acidosis may be related to the hyperglycemic effects of CCB; they impair pancreatic insulin secretion and induce systemic insulin resistance. As discussed in the management of beta-blocker overdose, HIE is therefore a particularly relevant therapy and blood glucose levels should be monitored.

In overdose, dihydropyridines can cause profound hypotension secondary to vasodilatation. There have been reports of bradycardia[110] but overdose is often associated with a reflex sinus tachycardia.[111] Conduction disturbances with dihydropyridine overdose is only rarely seen.[103] Overdoses of verapamil and diltiazem are associated with a greater incidence of bradycardia, AV nodal block, and escape rhythms. Verapamil is the most lethal CCB in overdose, as would be expected from its greater cardiosuppressive and vasodilatory effects. The incidence of hypotension is similar when comparing verapamil and nifedipine overdose.[103]

8.8.1 Management of CCB Toxicity

The management of acute CCB toxicity shares many of the principles of beta-blocker poisoning and readers should consult the section on the management of beta-blocker toxicity. Patients should receive the same basic assessment, investigations, and treatment to include airway and ventilator support, gut decontamination if appropriate, and fluid resuscitation. Metabolic acidosis, elevated lactate, hyperglycemia, and pulmonary edema are all more frequent in CCB overdose and investigations should be prioritized to reflect these.

As with beta-blocker overdose, calcium is a logical therapy for CCB toxicity. However, solid clinical evidence is sparse. Twenty-three patients were treated with calcium in a large case series of CCB overdose; 16 (70%) showed an increase in blood pressure in response.[103] Calcium treatment has failed to show an effect in other cases.[112,113] The beneficial effects of calcium are more likely to be seen in cases of impaired inotropy; therefore it is unlikely to be effective in patients with shock secondary to vasodilatation or in cases with severe bradycardia or AV block.

Glucagon therapy in CCB overdose is well supported by laboratory data. It has been shown to reverse bradycardia and hypotension induced by diltiazem, nifedipine, and verapamil in isolated heart preparations[114] and to increase the heart rate and contractility following verapamil infusion in intact animal studies.[115,116] Case reports are mixed; some support the use of glucagon[103,116] while others have demonstrated its failure to have an impact on blood pressure or heart rate with CCB overdose.[103]

Phosphodiesterase inhibitors have been discussed for beta-blocker overdose. Similar principles apply with regard to CCB overdose and the same concerns remain with regard to excessive tachycardia and hypotension. Animal models have suggested some success in verapamil overdose but PDEI remain a second-line treatment. Dopamine, dobutamine, noradrenaline, isoprenaline, and adrenaline have all been used to treat CCB poisoning[103,117–119] but none have demonstrated superiority. As advised in the case of beta-blocker overdose, many sources[80,119] support the use of cardiodynamic and hemodynamic monitoring to guide catecholamine use in CCB poisoning. Fundamentally, the agent of choice remains unclear.[45]

The mechanisms by which insulin therapy can be of benefit in drug-induced cardiotoxicity have been described for β-blocker toxicity. As with the beta blockers, there is good evidence from the laboratory of the efficacy of HIE for CCB overdose, but no trial data exist.[98] Insulin offers an additional advantage in the management of calcium channel blocker overdose. As previously mentioned, CCBs impair pancreatic insulin secretion in a way that beta blockers do not. Hypoinsulinemia, hyperglycemia, and poor tissue perfusion result and this can lead to a metabolic acidosis, exacerbating myocardial dysfunction.[102] Careful monitoring of acid-base status is therefore crucial in CCB overdose[80] and HIE is particularly relevant.

Current debate centers around whether the emphasis in management of CCB overdose should be on vasopressors or HIE. Levine et al. proposed an approach focusing on fluid resuscitation and vasopressors[117] while Holger et al. highlighted the inconsistent data supporting vasopressor use and promoted HIE as first-line treatment.[120] Catecholamine therapy can increase blood pressure and heart rate but may also increase systemic vascular resistance. This may in turn decrease cardiac output and perfusion of the vascular beds. Further studies have suggested that when HIE is used in conjunction with vasopressors, there may be a deleterious effect[97,98,120] with higher insulin doses required to overcome increased systemic vascular resistance and decreased cardiac output.

Debate over optimum management of CCB overdose will continue until the evidence base is fortified by well-designed prospective controlled trials. Until then, a sensible approach is outlined by Olson.[119] Where patients have shown evidence of shock secondary to vasodilation (i.e., tachycardia, low systemic vascular resistance on invasive monitoring, high output on ECG), a vasoconstrictor, such as noradrenaline, may be more effective. Patients with significant bradycardia or an AV block may require a pacemaker (noting that failure to capture can be problematic) or may benefit from isoprenaline, dopamine, or dobutamine. Atropine or glucagon may also be effective for the bradycardic patient. Patients with an inotropic deficit, as indicated by bedside ECG, may respond to dopamine or dobutamine as well as to calcium therapy. High-dose insulin therapy will likely produce the best results in those patients with depressed cardiac contractility.

Newer therapies also show some promise although, again, evidence is limited. Intravenous lipid emulsion has been used in cases of verapamil and diltiazem overdose, both of which have high lipid-solubility. IV lipid emulsion sequesters the toxic drug and removes it from the serum. It has been associated with favorable hemodynamic effects[121] but these are difficult to distinguish from other, simultaneously administered therapies. Levosimendan is a myocardial calcium sensitizer with positive inotropic effects. It binds to cardiac troponin C and improves calcium availability to actin and myosin fibers.[122] A rodent model of verapamil toxicity demonstrated increased cardiac output and survival but no improvement in mean arterial pressure. Cases have shown reversal of shock coinciding with levosimendan administration, but results are confounded by coadministration with other drugs.[122,123]

Methylene blue can theoretically counteract some of the vasodilatory effects of dihydropyridine CCB and beta blockers. One case reported reversal of hypotension in a case of atenolol and amlodipine overdose with methylene blue, after failure of standard therapy, including three vasopressors.[124] Again, evidence is confounded by concomitant use of multiple agents. Research also continues into the use of immunotherapy and vasopressin for CCB and beta-blocker overdose.[125] Extraordinary measures for use in critically ill patients have been described in the management of beta-blocker overdose and similar principles apply. Based on previously published literature and clinical evidence, DeWitt and Waksman formulated an algorithm for the management of beta-blocker and calcium channel blocker overdose (Figures 8.5 and 8.6).[45]

8.9 CARDIAC GLYCOSIDES

Cardiac glycosides act by inhibiting the sodium-potassium adenosine triphosphatase (Na^+/K^+-ATPase) pump. The subsections that follow describe a number of them.

8.9.1 Digoxin

Digoxin is the most widely used cardiac glycoside and has been used for centuries to aid ventricular rate control in AF and as a positive inotrope in heart failure; however, recently its use has been declining as newer, safer, and efficacious agents have been developed. Digoxin has high oral bioavailability and has a very narrow therapeutic index, which may be responsible for increased mortality and proarrhythmic action.

A study by Aronow has shown that 19% of nursing home residents were prescribed digoxin and 47% of those prescriptions were not necessary.[126] Digoxin controls the ventricular response in a patient with AF at rest but has reduced benefit in the presence of adrenergic stress.[127] In case of CHF, digoxin is considered as second-line therapy because it does not manage the excessive exercise-induced tachycardia that functionally limits patients.[128] Digoxin acts by inhibiting the active transport of Na^+ and K^+ across cell membranes by inhibiting the Na^+/K^+-ATPase, which results in an increase in extracellular potassium and intracellular Na^+. The increased intracellular concentration of Na^+ results in a reduction in the transmembrane Na^+ gradient and a subsequent increased activity of the Na^+/Ca^{2+} exchanger. This allows an increase in intracellular calcium ions[129] and creates a positive inotropic action by increased myofibril activity in the cardiac myocytes. Apart from this action, cardiac glycosides also increase vagal tone leading to a direct AV depression and conduction disturbances. The drug also produces a direct action on the refractory period in various regions of the heart.

The early signs and symptoms of digitalis toxicity appear to be premature ventricular contractions especially in a patient with AF whose ventricular response rate had been previously well controlled. Most patients with chronic, unintentional toxicity will complain first of anorexia and fatigue and will often have nausea and vomiting. Neurologic symptoms can begin as subtle visual changes, described as blurred vision, decreased visual acuity, or yellow halos, and progress to confusion, hallucinations, seizures, or coma.[130] In the case of acute poisoning with cardiac glycosides, there will be electrolyte disturbances

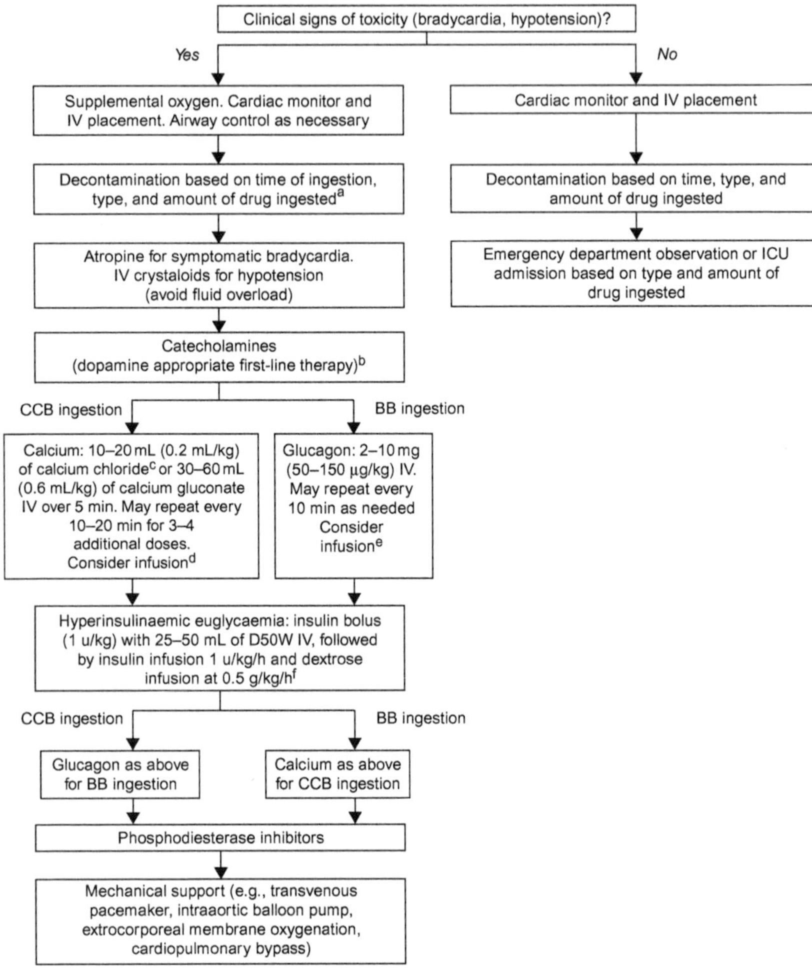

FIGURE 8.5 **Algorithm for treatment of beta-blocker and CCB overdose.** Definitions for superscript letters in figure follow: (a) maintenance of vital signs takes precedence over decontamination; (b) may need multiple pressors at very high doses; (c) administer calcium chloride via a central venous catheter; (d) calcium infusion: 0.4 mL/kg/hour of calcium chloride or 1.2 mL/kg/hour of calcium gluconate; (e) glucagon infusion: 2–10 mg/hour (50–100 μg/kg/h); and (f) dextrose infusion: 5 mL/kg/hour of D10W or 2 mL/kg/hour of D25W (solutions that are >10% dextrose should be administered via a central venous catheter). ICU, intensive care unit; IV, intravenous. **Source:** *Adapted from DeWitt and Waksman with permission from authors and publisher.*[45]

(e.g., hyperkalemia) that significantly contribute to dysrhythmias. Dysrhythmias can occur due to: (1) excitant activity (atrial, junctional, and ventricular premature beats and tachydysrhythmias); (2) suppressant activity (sinus bradycardia,

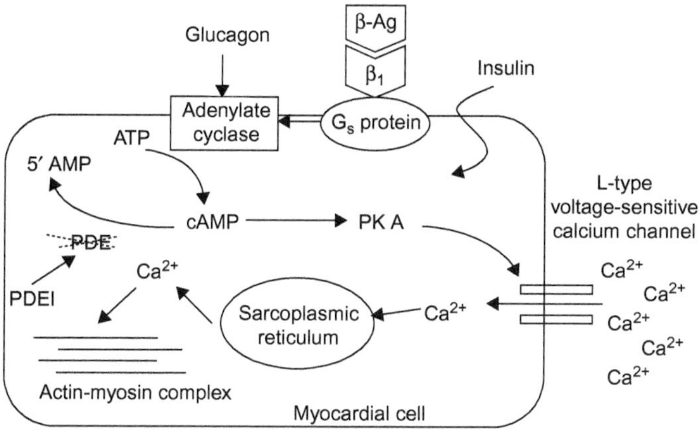

FIGURE 8.6 The mechanism of action of pharmacological therapies for calcium channel blockers and beta blockers and poisoning on cardiac myocytes.[45] 5′ AMP, 5′ adenosine monophosphate; ATP, adenosine triphosphate; β1, β1-receptor; β-Ag, β-agonist (catecholamine); cAMP, cyclic adenosine monophosphate; Gs, stimulatory G protein; PDE, phosphodiesterase; PDEI, phosphodiesterase inhibitors; PK A, protein kinase A. **Source:** *Adapted from DeWitt and Waksman with permission from authors and publisher.*[45]

bundle branch blocks, first-, second-, and third-degree blocks); and (3) a combination of excitant and suppressant activity (atrial tachycardia with AV block, second-degree block with junctional premature beats).

The most common dysrhythmia associated with toxicity induced by these agents is frequent premature ventricular beats. Bidirectional VT is specific for digitalis toxicity but is rarely seen. Bidirectional VT is a rare ventricular dysrhythmia characterized by a beat-to-beat alternation of the frontal QRS axis. Death due to digoxin poisoning usually results from cardiovascular collapse. The various conditions seen in toxicity are atrial tachycardia, ventricular dysrhythmias, severe AV block, and depression of myocardial contractility and may be refractory to most conventional therapies.

In dentistry, it is recommended that a baseline pulse rate be obtained and the cardiac rhythm recorded before commencing treatment. Macrolides and tetracyclines increase the risk of digoxin toxicity and should be used with caution. Also, digoxin toxicity is more common in renal failure because excretion is greatly affected. Digoxin-specific antibody fragments (Dig-Fab) are considered as first-line therapy for digoxin toxicity. In addition, gastric lavage, atropine, and cardiac pacing may be needed. Antiarrhythmics, such as lidocaine or phenytoin, may be needed if there is associated ventricular arrhythmias. If not contraindicated, oral administration of multidose activated charcoal prevents enterohepatic recycling. Hemodialysis is not effective in digoxin toxicity; however, hemoperfusion with charcoal or resin filters is efficient for such toxicity.[131,132] For more information, please refer to Chapter 5.

8.9.2 Diuretics

Diuretics are commonly used drugs for various conditions such as hypertension, cardiac failure, renal disease, and so on. These agents increase the excretion of sodium ions and water and are further classified according to their site of action and efficacy. The majority of drugs act by reducing sodium chloride reabsorption at different sites in the nephron, as a result of which there will be an increase in sodium and water loss in the urine (Figure 8.7).[133,134]

The drugs that cause cardiotoxicity are mostly loop and thiazide diuretics because use of these agents can cause serious complications such as hypokalemia. This is one of the important factors that influences the balance of intracellular and extracellular concentration of K^+ ions. Under normal circumstances, the concentration of K^+ ions is well maintained both intracellularly and extracellularly. But due to the use of drugs (e.g., diuretics) there will be disturbance in concentration due to the loss of K^+ ions in the urine. This is a very important factor that may precipitate dysrhythmias.[133]

The rate of fall in the extracellular K^+ ion concentration is of importance too, a more rapid fall being more liable to induce cardiac dysrhythmias. Also, the individual fall in extracellular K^+ concentration may influence the incidence of dysrhythmias even when the serum K^+ concentration is still within normal limits. A good correlation has been demonstrated between the individual change in serum K^+ level and the change in frequency of ventricular extrasystoles. Patients on prior diuretic therapy had a somewhat higher mean age than nondiuretic patients, they were more often treated with digitalis, and they developed larger infarctions and dyslipidemia.[135,136] These

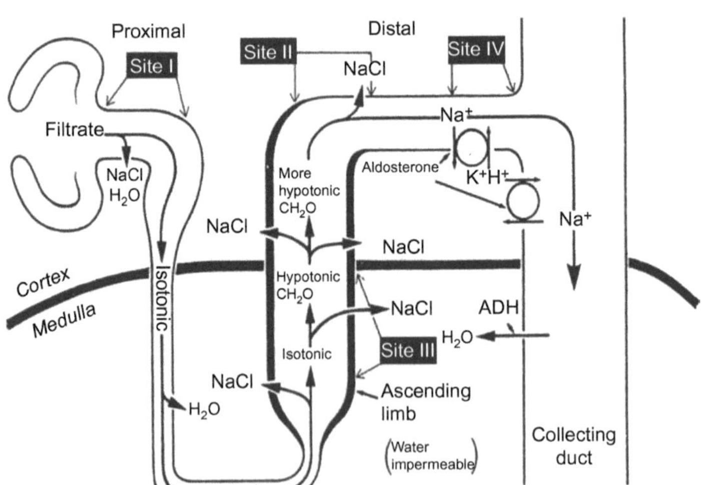

FIGURE 8.7 **Diagram of the renal tubules showing principal sites (I to IV) of diuretic action. Source:** *Adapted from Lant with permission.*[133]

facts would be expected to increase the incidence of cardiac dysrhythmias in the group on prior diuretic treatment.

8.9.3 Adenosine

Adenosine, the purine nucleoside, helps in regulating a variety of physiological functions in the heart. It is usually released in the body under conditions of stress such as hypoxia or ischemia, which increase the concentration of adenosine in the extracellular space dramatically.[137] The increased concentration of adenosine may protect the heart from these conditions. The effects are usually mediated by binding to receptors that regulate cellular function. These receptors have been subdivided into A1, A2A, A2B, and A3.[138]

Adenosine has become a very useful drug for the diagnosis and treatment of supraventricular arrhythmias because of its short half-life in humans—that is, <15 sec.[139] It causes transient AV nodal conduction block. Because of these properties, adenosine is currently used as a first-line agent for the differential diagnosis and therapy of narrow complex supraventricular tachycardias.[140] These drugs, when administered to some patients, may cause adverse effects such as arrhythmias including premature ventricular complex, monomorphic sustained and nonsustained VT, TdP, premature atrial complex, AF, and prolonged AV block.[141–146] These effects are due to the direct action on atrial, sinoatrial, and AV nodal cells.

In these cells, adenosine activates outward potassium current shortening the atrial action potentials and reducing the effective refractory periods, thus predisposing to AF. In the ventricular myocytes, adenosine blocks the effect of catecholamines that stimulate an inward calcium current through cAMP.[147] By doing so, adenosine suppresses most ventricular arrhythmias but adenosine administration is followed by an increase in sympathetic nerve traffic and increased catecholamine discharge that may be the cause of the proarrhythmic effect.[148] Another possible explanation for the ventricular proarrhythmic action of adenosine may be its effect on the purinergic-sensitive ventricular neurons.[149]

In 1929, an article was published by Drury and Szent Gyorgi; it described the potent action of purine nucleotides and nucleosides, ATP, and adenosine on the heart and blood vessels. Then, in 1970, evidence was presented for ATP as a neurotransmitter in nonadrenergic, noncholinergic (NANC) nerves supplying the gut, and in 1972 the word *purinergic* was coined and the purinergic neurotransmission hypothesis was proposed by Burnstock.[150] The action of purinergic substances occurs through the purinoceptors.[138] Burnstock proposed the nomenclature for these purinoceptor as P1 (adenosine) and P2 (ATP/ADP) receptors. This helped resolve some of the ambiguities in earlier reports, which were complicated by the breakdown of ATP to adenosine by enzymes so that some of the actions of ATP were directly on P2 receptors, whereas others were due to indirect action via P1 receptors.

Four subtypes of P1 receptors were cloned: namely, A1, A2A, A2B, and A3. In 1985, Burnstock and Kennedy[151] proposed a framework for distinguishing two types of P2 purinoceptors (i.e., P2X and P2Y) based largely on pharmacological criteria. In the early 1990s, studies of transduction mechanisms and cloning of both P2X and P2Y receptors were carried out, which led Abbracchio and Burnstock[152,153] to put forward, in 1994, a new nomenclature system, which is now widely accepted.[154] They proposed that there were two families of P2 receptors—namely, P2X ionotropic ligand-gated ion channel receptors and P2Y metabotropic G protein-coupled receptors.

When positive inotropic agents are used in many cardiovascular diseases, such as heart failure, arrhythmia, and so on, they cause upregulation of various types of receptors present in the heart and the blood vessels. This allows increased action of adenosine and ATP-induced increase in Ca^{2+} in cardiomyocytes, which may be impaired in heart failure due to myocardial infarction.[155] These drugs can cause cardiovascular toxicity by enhancing sympathetic nerve activity that can lead to cardiac dysfunction, arrhythmias, and sudden cardiac death due to myocardial ischemia. ATP is released along with noradrenaline (NA) and enhances NA release from sympathetic nerve terminals.[155]

8.10 PHOSPHODIESTERASE INHIBITORS

Phosphodiesterase is a ubiquitous enzyme that catalyzes the hydrolysis of phosphodiester bonds. Phosphodiesterase enzyme hydrolyzes cyclic adenosine monophosphate (cAMP) and cyclic guanosine monophosphate (cGMP), which play an important role in cardiovascular regulation by influencing function, gene expression, and morphology such as regulation of the inotropic mechanism, vasodilators effects, and so on. Thus far, 11 different isoforms of phosphodiesterase have been found in different human tissues. However, both cAMP and cGMP have numerous effects on the heart as well as other tissues and different phosphodiestrase isoenzymes are found in many tissues (Table 8.4).

In the heart, the inotropic mechanism increases either independently or depends on the activity of cAMP. Agents (e.g., PDEI) act via cAMP-dependent phosphorylation reaction-enhancing influx of Ca^{2+} through voltage-sensitive Ca^{2+} channels, helping in relaxation by enhancing uptake of Ca^{2+} by the sarcoplasmic reticulum (SR) and thus decreasing the sensitivity of contractile proteins.[156–159] cAMP-independent mechanisms are less clinically effective than those that are cAMP-dependent. Thus, an increase in cAMP in the cardiac tissue leads to a significant increase in inotropic effects (Table 8.5).

Apart from this mechanism, cAMP also activates cGMP, which is the mediator of nitric oxide (NO) and the atrial natriuretic peptide action that causes vasodilatation.[160–162] Among the phosphodiestrase (PDE) subtypes, PDE5 is responsible for the action in the heart and the vasculature. The primary effect of PDE5 activity is on cGMP levels, and this sequentially stimulates protein kinase G (PKG). Other subtypes of PDEs, such as PDE2 (shown

TABLE 8.4 Human PDEIs and the Cardiovascular System

PDE Isoenzyme	Tissue	Target	Inhibitors
1	Heart, brain, kidney, liver, skeletal muscle, smooth muscle	cGMP, cAMP	Vinpocetine, phenothiazines
2	Adrenal cortex, brain, corpus cavernosum, heart, liver, kidney, airway, smooth muscle, platelets	cGMP, cAMP	EHNA (erythro-9-2-hydroxy-3-nonyladenine)
3	Heart, corpus cavernosum, platelets, smooth muscle, liver, kidney, T lymphocytes, B lymphocytes, basophil, mast cell, monocyte, macrophage, endothelial cell	cAMP	Piroximone, olprinone, motapizone, cilostamide, cilostazol, milrinone, amrinone, pimobendan, imadazodan
4	Kidney, lung, heart, skeletal muscle, smooth muscle (i.e., vascular, visceral, airway), platelets, most inflammatory cells (i.e., T lymphocytes, B lymphocytes, basophil, mast cell, monocyte, macrophage, endothelial cell, eosinophil, neutrophil)	cAMP	Enoximone, rolipram (nanomolar), numerous others (piclamilast, tibenelast, benafentrine, zardaverine, and tolafentrine inhibit both PDE3 and PDE4)
5	Corpora cavernosum, platelets, skeletal muscle, smooth muscle kidney	cGMP	Sildenafil, zaprinast, dipyridamole
6	Retina	cGMP	PDE5 inhibitors
7	Skeletal muscle, heart, kidney, airways, T lymphocytes, B lymphocytes, monocytes, eosinophil	cAMP	Dipyridamole
8	8A: testis, ovary, ileum, colon; also heart, brain, kidney, pancreas, airways, monocyte 8B: thyroid	cAMP	Dipyridamole
9	Spleen, small intestine, brain	cGMP	Zaprinast (as for PDE5)
10	Brain (putamen, caudate nucleus)	cAMP, cGMP	Isobutylmethylxanthine
11	Skeletal muscle, prostate, kidney, liver, pituitary, salivary glands, testis	cAMP, cGMP	Zaprinast, dipyridamole

cAMP, cyclic adenosine monophosphate; cGMP, cyclic guanosine monophosphate.
Source: Adapted from Feneck with permission.[157]

TABLE 8.5 Mechanism of Inotropic Action

cAMP-Dependent Mechanisms	cAMP-Independent Mechanisms
Catecholamines	Direct activation of Ca channels
Phosphodiesterase inhibitors	α-1-adrenoceptor stimulation Direct inhibition of Na/Ca exchange α-1-adrenoceptor stimulation K-channel inhibition Ca salts Thyroid hormone Glucose/K/insulin Inhibition of ATPase-dependent Na/K pump Myofilament Ca sensitizer (e.g., Levosimendan)

cAMP, cyclic adenosine monophosphate.
Source: Adapted from Feneck with permission.[157]

in Table 8.4), also influence cGMP, which in turn regulates cAMP and thus cardiac and vascular function. Inhibition of PDE2 minimally altered basal or cAMP-stimulated L-type Ca^{2+} currents or sarcomere shortening, yet blocked suppression of cAMP-stimulated current and contraction by NO.[163] Smooth muscle relaxation occurs by several mechanisms including decreasing intracellular free calcium concentration, reduced calcium sensitization, and regulation of thin filament proteins. Phosphorylation occurs due to the increased concentration of calcium within the cell as a result of which interaction between actin and myosin occurs leading to smooth muscle contraction. Cyclic GMP can also stimulate PKG, and this is presently thought to be a major effector of PDE5 inhibition effects on blood vessels and the heart. Protein kinase G has an important role in regulating vascular tone and growth.[164]

Protein kinase G lowers calcium levels within the cell by activating calcium pumps, inhibiting voltage-gated Ca^{2+} channels and receptor/G protein-coupled Ca^{2+} activation; it is also associated with the inositol triphosphate (IP3) receptor to reduce Ca^{2+} release from the sarcoplasmic reticulum. It further reduces Ca^{2+} sensitivity by modulating myosin light-chain phosphorylation and through this pathway PDE5 inhibitors may inhibit RhoA−Rho-kinase (ROCK) signaling, thus leading to relaxation.[165−167] ROCKs are serine/threonine kinases. Their structure is comprised of a kinase domain located at the amino terminus of the protein, a coiled-coil region containing the Rho-binding domain and a pleckstrin homology domain with a cysteine-rich domain. They have various functions and one of the important ones is in vascular smooth muscle cells. The phosphorylation/dephosphorylation of the myosin light chain (MLC) is a major regulatory mechanism of smooth muscle contraction. A rise in $[Ca^{2+}]i$ causes activation of myosin light-chain kinase (MLCK) and consequent phosphorylation of MLC and smooth muscle contraction. However, MLC phosphorylation and vascular smooth muscle contraction can also be induced in the absence of significant increases in Ca^{2+} thereby causing Ca^{2+} sensitization of the contractile proteins, enhancing vascular smooth muscle contraction. Inhibition of stimulation of ROCK therefore leads to relaxation of vascular smooth muscles.

The drugs classified under PDEI are amrinone, enoximone, milrinone, olprinone, pimobendan, and vesnarinone. These are potent inotropes that also have a vasodilator affect and have been termed *inodilators*. These drugs are used for short-term treatment of severe heart failure. Long-term oral treatment is not recommended as PDEI drugs have the potential to cause hypotension and cardiac arrhythmias. Amrinone, when applied to the isolated canine Purkinje fibers[168] have no significant effect on the resting membrane potential, the rate of increase of phase 0 or phase 4 of the action potential, conduction velocity, effective refractory period, or automaticity. In slow fibers, amrinone may produce an increase in maximum velocity, overshoot, and duration of the action potential. In the intact canine heart, amrinone shortened the functional refractory period of the AV node and reduced conduction time through the AV node and the His-Purkinje system.[169]

The electrophysiological effects of milrinone on canine atrial cells have shown a significant increase in action potential amplitude and action potential duration but no significant effect on the resting membrane potential or the maximum velocity of upstroke of the action potential.[170] In addition to this, milrinone decreases the effective refractory period (ERP) and the functional refractory period (FRP) in atrial myocytes obtained from dogs or guinea pigs. In isolated ventricular myocytes, milrinone increases action potential amplitude and decreases APD and ERP. In isolated Purkinje fibers, milrinone has been shown to have no effect on action potential, conduction velocity, or ERP.[170,171] However, in ischemic preparations milrinone improves conduction and minimizes postrepolarization refractoriness. Milrinone also enhances automaticity in depressed Purkinje fibers.

Keister et al.[172] administered milrinone orally to 29 dogs with moderate to severe heart failure to assess the effects of the drug on left ventricular function. One of the dogs died because of refractory VT but the study could not distinguish whether the death was due to the drug or the disease because it was not placebo-controlled. This suggests that the PDEI milrinone, in the setting of myocardial ischemia, may increase ventricular extrasystoles (VEs) and increase the risk of development of VT.

A study of amrinone conducted among human patients with heart failure has characterized some with arrhythmogenic potential; however, it was difficult to determine whether the arrhythmias were drug-induced or caused by underlying ventricular dysfunction or other preexisting conditions.[173–177] In another study done by Leier et al., it was reported that there was an increase in VEs in 6 out of 24 patients, and new onset of supraventricular tachycardia in 2 out of 24 patients was observed during oral administration of amrinone (100 mg, 8 hours) therapy for the treatment of heart failure.[178] All these data suggest that the proarrhythmic potential of amrinone appears to be dose-related and primarily associated with IV administration, whereas milrinone, when used for long-term therapy, has the potential to cause arrhythmias.

In a randomized, double-blind, placebo-controlled study of 1088 patients with chronic cardiomyopathy, therapy with milrinone at 40 mg daily was associated with a 28% increase in all-cause mortality and a 34% increase in cardiovascular mortality.[179] Further, there was a significant increase in frequency of syncope in patients receiving milrinone than in the placebo group. Therefore, in summary, this study suggested that the use of PDEIs may produce either antiarrhythmic or proarrhythmic effects by improving impulse conduction in areas where it was depressed. These types of pharmacological effects primarily occur by elevating intracellular calcium concentration. An increased intracellular myocardial calcium concentration is known to promote the development of arrhythmias.

8.11 STATINS

Statins block the conversion of HMG-CoA to mevalonic acid and thus decrease the biosynthesis of cholesterol. They markedly decrease the level of low-density lipoprotein (LDL) cholesterol in the blood. The statins have recently been shown to exert direct cardiovascular effects independent of their cholesterol-lowering effects.[180] This group of drugs has been shown to have beneficial effects on cardiovascular mortality in many studies.[181] However, a recent systematic review has shown an increased risk of new-onset diabetes in patients treated with statins, and there is a link with the dose used; the more potent statins have an increased risk of causing diabetes.[182]

In an experimental study, atorvastatin has been shown to prevent development of diabetes mellitus in a rat model.[183] A recent article examined various studies on the diabetogenic potential of statins and recommends that the benefits of statin therapy appear to outweigh the risk of new-onset diabetes and that it would be inappropriate to discontinue statins for prevention of cardiovascular events because of the potential risk of incipient diabetes.[184]

Similarly, a recent meta-analysis of five clinical trials demonstrated that high-dose statin therapy was associated with an increased risk of new-onset diabetes compared with moderate-dose statin therapy.[185] The new onset of diabetes is not only associated with statins but also associated with other drugs such as beta blockers, thiazide diuretics, and glucocorticoids.[186,187] These drugs are often given along with statins, which further potentiate the risk of development of new cases of diabetes. The possible mechanism through which the statin induces diabetes may be via attenuation of the expression of glucose transporter (GLUT-4) in adipose tissue impairing glucose tolerance. In addition, dose-related suppression of glucose-induced elevation of intracellular Ca^{2+} levels may occur.[188]

8.12 ORAL ANTICOAGULANT THERAPY

Warfarin is the most common Coumadin derivative that produces an anticoagulant action by interfering with the interconversion of vitamin K and its epoxide. The coagulation factors II, VII, IX, and X require γ-carboxylation by vitamin K for their biologic activity. Warfarin causes the formation of partially carboxylated proteins with reduced coagulant activity. Both pharmacokinetic and pharmacodynamic factors influence the anticoagulant response to warfarin. Drug and food interactions affecting its absorption or clearance and genetic changes in the activity levels of the enzymes involved in metabolizing the drug can influence its level. The hemostatic response to a given drug concentration can also be altered by a variety of factors.

The main complication of oral anticoagulant therapy is bleeding, and the risk is related to the intensity of the anticoagulation. Age of more than 65 years, a history of stroke or gastrointestinal bleeding, comorbid conditions (e.g., anemia or renal insufficiency), and concomitant use of aspirin, other nonsteroidal antiinflammatory drugs (NSAIDs), and antiplatelet drugs increase the risk of major bleeding. The international normalized ratio (INR) serves as a measure of the anticoagulation achieved by warfarin. Bleeding that occurs when the INR is >3 is usually associated with trauma or an underlying gastrointestinal or urinary tract lesion.[189] In the elderly, a lower dose of warfarin may be advisable.

Insufficient prescription of warfarin may have profound clinical and economic consequences.[190] Overestimation of the risk of anticoagulation and the reduced likelihood of future use of warfarin by physicians, whose patients had developed adverse reactions to anticoagulation, have been reported.[191] Adverse events associated with underuse of warfarin—for example, a thromboembolic stroke in a patient of AF not on anticoagulation therapy—does not influence the odds that a physician will prescribe warfarin for subsequent patients.[192] Thus, adverse events due to warfarin use have had a significant impact on physicians' prescribing behavior.

It is very important to monitor the INR closely for the patients on warfarin. Numerous studies have demonstrated hemopericardium and hemorrhagic pericardial effusion associated with or without cardiac tamponade in patients on warfarin or other oral anticoagulant therapy when INR is above the normal limit.[193] Large pericardial effusion or cardiac tamponade is associated with ECG changes (electrical alternans).

8.13 ORGANIC NITRATES

The therapeutic use of nitroglycerine was discovered by William Murrell in 1879. These drugs are polyol esters of nitric acid having vasodilating effects and are used successfully in the treatment of angina pectoris. When used in the therapeutic dose, organic nitrates are reduced to organic nitrites, which are then converted into nitric oxide that directly acts as a potent vasodilator of smooth muscle and blood vessels, reducing the signs and symptoms of ischemic heart disease. Several adverse effects have been repeatedly described; among them headache, dizziness, flushing, and hypotension are well documented. Severe life-threatening reactions are rare, but among them are nitroglycerin-induced bradycardia, complete cardiac block, and asystole. The possible pathophysiologic mechanisms that are involved in producing these adverse effects may be neurocardiogenic syncope provoked by preload reduction. Apart from this, due to the administration of the nitrate, hypersensitivity of the AV node will occur, enhancing vagal tone. This enhanced vagal tone may contribute to a transient high-degree atrioventricular block.[194] The exaggerated reflex response in this phenomenon usually remains limited to the AV node.

In 1867 von Bezold (revived by Jarisch) described the Bezold–Jarisch reflex, which is a cardioinhibitory reflex characterized by bradycardia, vasodilatation, and hypotension in response to stimulation of cardiac C-fibers within the myocardium.[194] An excitation of these inhibitory fibers due to stretch, chemical substances, or drugs increases parasympathetic activity and inhibits sympathetic activity, with ensuing syncope. Similar effects are seen when glyceryl trinitrate (GTN) is used. The drug decreases myocardial filling due to a decrease in preload, which leads to deformation of afferent vagal fibers within the myocardium and enhances vagal tone.[195,196]

Another mechanism involved may be the result of selective baroreflex failure, which is characterized by phases of severe hypertension and tachycardia alternating with prolonged hypotension and bradycardia. This failure in reflex usually occurs after a complete loss of the afferent arc of the baroreflex and preservation of efferent nerves of sympathetic and parasympathetic nervous system, resulting in undamped sympathetic and parasympathetic activation and enhanced vagotonic response.[197] In similar studies, bradycardia and hypotension have also been documented in patients with acute myocardial infarction who received nitrate therapy.[198]

8.14 RANOLAZINE

Ranolazine is a new antianginal drug that has been approved for the treatment of chronic stable angina. The drug is preferred for patients who remain symptomatic while being on standard antianginal treatments. Ranolazine produces ion channel effects similar to those observed after chronic amiodarone treatment including reduced I_{Kr}, I_{Ks}, late I_{Na}, and I_{Ca}.[199] The drug suppresses EADs and reduces transmural dispersion of refractoriness, suggesting that it may possess antiarrhthymic activity in addition to its antianginal action. Ranolazine was tested for its ability to maintain sinus rhythm in 7 patients with resistant AF and was effective in maintaining sinus rhythm in 4.[200] The drug can increase the duration of the action potential and the QT interval.[201]

However, no increase in TdP and other significant arrhythmias has been reported with ranolazine.[202]

The drug also decreases the late sodium current, which is a depolarizing current that contributes to the plateau phase of the ventricular action potential.[203] Impaired inactivation of the sodium current causes increased intracellular concentration of Na leading to cellular calcium overload. This overload can precipitate arrhythmias and cause abnormal left ventricular relaxation and diastolic dysfunction. Ranolazine suppresses the late sodium current and reduces the abnormalities of ventricular repolarization and contractility associated with ischemia/reperfusion and heart failure.

8.15 IVABRADINE

Ivabradine is a drug used in the treatment of angina and heart failure when patients are in sinus rhythm and/or patients who are not able to tolerate beta blockers. Ivabradine causes direct inhibition of spontaneous depolarization of sinus node without interfering with the conduction of auricular and ventricular activities. Thus, the clinical effect is to cause a decrease in heart rate which is dose-dependent in therapeutic doses.[204] Nonetheless, owing to pharmacodynamic properties, some adverse events were noted observed both before- and after-marketing authorization such as AV block and syncope. Thus, ivabradine intoxication is thought to cause severe bradycardia and acute heart failure.[205,206] Ivabradine is metabolized by cytochrome P450 isoenzyme *CYP3A4*; there is thus a potentially high risk of pharmacokinetic interactions. Overall, no clinical studies or trials have been done with a large number of patients that have shown any serious cardiac adverse events from ivabradine.[206] However, ivabradine has been found to be safer and more effective for angina and chronic heart failure.

8.16 DOBUTAMINE

Dobutamine is an inodilator that stimulates β1- and β2 adrenergic receptors. The action on β1 receptors activates a guanine nucleotide regulatory cascade (via G proteins) that results in increased adenylate cyclase activity and increased conversion of ATP to the intracellular second messenger cAMP. Intracellular cAMP causes the release of calcium from the sarcoplasmic reticulum. The calcium is used by contractile proteins and results in increased stroke volume. In the vasculature, the α-adrenergic agonist effect appears to be counteracted by the partial agonism and the vasodilatory action caused by β2-receptor stimulation. This results in a modest decrease in systemic vascular resistances and venous filling pressures. This action on the heart leads to a decrease in heart rate due to reflex withdrawal of sympathetic tone in response to improved cardiovascular function. The major side effects experienced include tachycardia and ventricular arrhythmias. It has been hypothesized that the increased demands of the failing myocardium leads to a state of relative energy depletion through an initial compensatory phase of increased oxygen extraction. Further, drugs, such as dobutamine and dopamine, stimulate energy demand and ultimately accelerate myocardial cell death.[207,208]

8.17 DOPAMINE

Dopamine is an endogenous catecholamine having dose-dependent effects. At a dose of $\leq 2\ \mu g/kg/min$, it causes vasodilation by direct stimulation of the dopamine postsynaptic type 1 and the presynaptic type 2 receptors in the splanchnic and renal tubular epithelial beds. A major side effect of dopamine is tachycardia, which is more pronounced than with dobutamine. Apart from this, correct dosing of dopamine is also very important as its dose is based on lean body weight, and it can be difficult to estimate. A new or unexplained tachycardia or arrhythmia in a patient receiving low-dose dopamine should make a clinician suspect an inaccurate estimation of lean body weight resulting in an inappropriately high dopamine infusion rate.

8.18 ANGIOTENSIN-CONVERTING ENZYME INHIBITORS AND ANGIOTENSIN RECEPTOR BLOCKERS

Angiotensin-converting enzyme inhibitors (ACEIs) and angiotensin receptor blockers (ARBs) are widely used drugs that have been shown to be effective in reducing adverse cardiovascular outcomes in patients with cardiac failure or hypertension.[209,210] Angioedema is an uncommon but serious adverse event that can occur in a patient on ACEI therapy. The condition is associated with high levels of bradykinin in various organs and tissues and occurs more commonly among females and African-Americans. Management of this dangerous condition involves withdrawal of the ACEI and treatment of the reaction with various drugs depending on its severity. ARBs have been recommended as an alternative, though a small percentage of patients ($<10\%$) also develop episodes of angioedema after ARB intake. A recent study found decreased hemoglobin levels in patients with hypertension and diabetes mellitus treated with ACEIs. However, such an observation was not seen with ARBs.[211]

Orthostatic hypotension (OH) is defined as a reduction of systolic blood pressure of at least 20 mmHg, or diastolic blood pressure of at least 10 mmHg from a sitting to a standing position. ACEIs are one of the drugs prone to cause orthostatic hypotension and are one of the common causes of presyncope or syncope in hypertensive populations.[212,213] The mechanism of orthostatic hypotension is through inhibition of neurohumoral reflex, hormonal (i.e., inhibition of renin–angiotensin–aldosterone system), and vasodilatation.

Aliskiren is the first available direct inhibitor of renin, can be used orally, and is currently approved for the treatment of hypertension. Dual inhibition of the renin–angiotensin–aldosterone system with aliskiren and an ACEI or ARB may increase the risk of stroke in patients with diabetes and renal disease.[214] A recent review examined clinical trials with aliskiren and the newer ACEIs, telmisartan, olmesartan, and azilsartan.[215] The authors concluded that olmesartan is an effective and safe antihypertensive agent, but special attention should be paid to high-risk patients, such as those with coronary disease, because of the risk of an excessive reduction in blood pressure. Escobar and Barrios also noted that while azilsartan is probably the most potent ARB, there is still a lack of data regarding potential organ damage and the incidence of cardiovascular events.

8.19 ETHNIC DIFFERENCES IN ADVERSE EVENTS TO CARDIOVASCULAR DRUGS

A systematic review and meta-analysis published in 2006 examined differences in the risk of adverse events to cardiovascular drugs due to differences in ethnicity.[216] The authors concluded that their analysis confirmed that an adverse drug reaction (ADR) to ACEIs is consistent with those of studies in single ethnic populations that describe a higher incidence of angioedema among blacks.[217] Among patients on oral anticoagulants (e.g., warfarin), the risk of moderate-to-severe bleeding and intracranial hemorrhage was higher among black patients compared to those who were not black. A prospective study reported nonwhite race (i.e., black, Hispanic, or other) as a risk factor for admission to a hospital because of bleeding after oral anticoagulant treatment for deep vein thrombosis.[218] Black patients also had a higher risk of admission due to adverse events following digitalis treatment compared to white patients.[219] A small clinical study examined 58 patients who were treated with ibutilide and found 3 out of 20 black patients experienced TdP compared to 1 of 38 nonblack patients.

8.20 PREDICTING CARDIOTOXICITY DURING DRUG DEVELOPMENT

Drugs undergo both preclinical testing in animals and clinical trials among humans before they can be introduced into the market. There has been interest in developing preclinical models to predict cardiac toxicity. A recent review examines this important topic.[220] The authors concluded that embryonic stem cells can address the disconnect between the behavior of cultured cells and cells inside the human body. The differentiated cells can reflect the normal physiological image of their counterparts. The use of cardiac myocytes with a disease phenotype (e.g., hypertrophic cardiomyocytes) will provide a complete picture of the adverse effect profile of a compound on the heart.

Zebrafish assays have been recommended for detecting cardiotoxicity and other toxicities of drugs during development.[221] Rubinstein stated that the molecular mechanisms underlying cardiovascular development and function have been conserved during vertebrate evolution. For many years zebrafish heart rate has been used as an indicator of cardiac toxicity. Two studies have shown a correlation between drugs that cause prolongation of the QT interval in humans and those that cause changes in the heart rate of zebrafish.[222] Assay automation may be achieved through the use of a transgenic zebrafish with a fluorescent heart.

Regulators have recently emphasized cardiac safety during the clinical phase of drug development mainly because the effect of drugs on ventricular repolarization is now the commonest cause for drug withdrawals from the market and delay of marketing permission.[223] ECG indices of ventricular repolarization, which can be detected on a surface ECG, include the duration of the QT interval, displacement of the ST segment, morphological pattern of T and U waves, and their combination. The QTc interval is the only accepted surrogate marker of abnormal ventricular repolarization. It is mandatory that all candidates enrolled in clinical drug trials undergo a well-defined ECG during all phases and also during each clinic visit to ensure cardiac safety parameters.

8.21 INTERACTIONS BETWEEN HERBS, FRUIT JUICES, THE ENZYME *P450*, AND CARDIOVASCULAR DRUGS

There has been a significant increase in interest in alternative therapies and self-administration of herbal medicines.[224] Most of them are a mixture of more than one active ingredient and this increases the risk of herb–drug interactions. A review examined interactions between herbs and drugs used in cardiovascular pharmacotherapy.[225] The authors provided a comprehensive table that details interactions between herbs and various cardiovascular drugs. Digoxin, for example, interacts with guar gum, St. John's wort, Siberian ginseng, and wheat bran. Guar gum and wheat bran reduce the absorption of digoxin,[226] and Siberian ginseng reduces digoxin elimination.[227] A number of herbs and fruit juices interact with oral anticoagulant drugs and with antiplatelet drugs including aspirin.[228] There are numerous drug–drug interactions (DDIs) related to cardiovascular medications and many of these are mediated via the *cytochrome P450* (CYP) system. Thus, clinicians are encouraged to develop an awareness of CYP-mediated DDIs so that they can choose safe drug combination regimens, adjust drug dosages appropriately, and conduct therapeutic drug monitoring for drugs with narrow therapeutic indices.[229] In addition, please refer Chapter 6.

8.22 CONCLUSION

Cardiovascular drugs are widely used in therapeutics and can cause a number of cardiovascular abnormalities, which have been described in this chapter. Hypotension and arrhythmias are significant adverse effects. Cardiovascular adverse effects are a common cause for drug withdrawal from the market, and there needs to be an emphasis on predicting the cardiovascular toxicity of drugs during all phases of their development.

Key Learning Points

- Worldwide cardiovascular diseases are becoming a major cause of morbidity and mortality and drugs acting on this system are becoming important in therapeutics.
- Cardiovascular toxicity is, however, seen with many cardiovascular drugs, especially antiarrhythmics, digoxin, nitrates, diuretics, phosphodiesterase inhibitors, warfarin, ACE inhibitors, and dopamine and/or dobutamine.
- Arrhythmias, hypotension, myocardial ischemia, bleeding (i.e., pericardial effusion), and tachycardia are among the significant cardiovascular adverse effects.
- Cardiovascular adverse effects are a major cause of withdrawal of drugs from the market and newer approaches to predict cardiovascular toxicity during all stages of drug development are being investigated.

Acknowledgment We would like to thank all the authors and publishers who provided us with permission to include figures and tables from their articles.

REFERENCES

1. WHO. Deaths from CVD and diabetes. Geneva, Switzerland; 2008 [accessed 02-02-14]. Available from: www.who.int/gho/ncd/mortality_morbidity/cvd/en/.
2. Buckberg G, Hoffman JI, Mahajan A, et al. Cardiac mechanics revisited: the relationship of cardiac architecture to ventricular function. *Circulation* 2008;**118**(24):2571–87.
3. Roden DM, Balser JR, George Jr. AL, et al. Cardiac ion channels. *Annu Rev Physiol* 2002;**64**:431–75.
4. NIH. Morbidity and Mortality: 2012—Chart Book on Cardiovascular, Lung and Blood Diseases. National Institutes of Health, National Heart, Lung and Blood Institute; 2012 [accessed 02-01-14]. Available from: www.nhlbi.nih.gov/resources/docs/2012_ChartBook_508.pdf.
5. Vaughan Williams EM. Classifying antiarrhythmic actions: by facts or speculation. *J Clin Pharmacol* 1992;**32**(11):964–77.
6. Harrison DC. Is there a rational basis for the modified classification of antiarrhythmic drugs? In: Morganroth J, Moore EN, editors. *Cardiac arrhythmias: New therapeutic drugs and devices*. New York: Springer; 1985. p. 36–40.
7. Gramley F, Himmrich E, Mollnau H, et al. Recent advances in the pharmacological treatment of cardiac arrythmias. *Drug Today* 2009;**45**(11):807–24.
8. Dessertenne F. Ventricular tachycardia with two variable opposing foci. *Arch Mal Coeur Vaiss* 1966;**59**(2):263–72. [*La tachycardie ventriculaire a deux foyers opposes variables.*]
9. Dessertenne F, Fabiato A, Coumel P. A new chapter in electrocardiography: progressive variations in the amplitude of the electrocardiogram. *Actual Cardiol Angeiol Int (Paris)* 1966;**15**(4):241–58. [*Un chapitre nouveau d'electrocardiographie: les variations progressives de l'amplitude de l'electrocardiogramme.*]

10. Lombardi F, Borggrefe M, Ruzyllo W, et al. Azimilide vs. placebo and sotalol for persistent atrial fibrillation: the A-COMET-II (Azimilide-CardiOversion MaintEnance Trial-II) trial. *Eur Heart J* 2006;**27**(18):2224–31.
11. Camm AJ. Safety considerations in the pharmacological management of atrial fibrillation. *Int J Cardiol* 2008;**127**(3):299–306.
12. Noble D, Tsien RW. Outward membrane currents activated in the plateau range of potentials in cardiac Purkinje fibres. *J Physiol* 1969;**200**(1):205–31.
13. Noble D, Tsien RW. Reconstruction of the repolarization process in cardiac Purkinje fibres based on voltage clamp measurements of membrane current. *J Physiol* 1969;**200**(1): 233–54.
14. Sanguinetti MC, Jurkiewicz NK. Two components of cardiac delayed rectifier K + current. Differential sensitivity to block by class III antiarrhythmic agents. *J Gen Physiol* 1990;**96** (1):195–215.
15. Barhanin J, Lesage F, Guillemare E, et al. K(V)LQT1 and I_{Ks} (minK) proteins associate to form the IKs cardiac potassium current. *Nature* 1996;**384**(6604):78–80.
16. Sanguinetti MC, Curran ME, Zou A, et al. Coassembly of K(V)LQT1 and minK (I_{Ks}) proteins to form cardiac I_{Ks} potassium channel. *Nature* 1996;**384**(6604):80–3.
17. Mitcheson JS, Chen J, Lin M, et al. A structural basis for drug-induced long QT syndrome. *Proc Natl Acad Sci USA* 2000;**97**(22):12329–33.
18. De Bruin ML, Pettersson M, Meyboom RH, et al. Anti-HERG activity and the risk of drug-induced arrhythmias and sudden death. *Eur Heart J* 2005;**26**(6):590–7.
19. Roden DM, Lazzara R, Rosen M, et al. Multiple mechanisms in the long-QT syndrome. Current knowledge, gaps, and future directions. SADS Foundation Task Force on LQTS. *Circulation* 1996;**94**(8):1996–2012.
20. Antzelevitch C. Role of transmural dispersion of repolarization in the genesis of drug-induced torsades de pointes. *Heart Rhythm* 2005;**2**(Suppl. 2):S9–15 P.
21. Allessie MA, Bonke FI, Schopman FJ. Circus movement in rabbit atrial muscle as a mechanism of tachycardia. II. The role of nonuniform recovery of excitability in the occurrence of unidirectional block, as studied with multiple microelectrodes. *Circ Res* 1976;**39**(2): 168–77.
22. Vermeulen JT, Tan HL, Rademaker H, et al. Electrophysiologic and extracellular ionic changes during acute ischemia in failing and normal rabbit myocardium. *J Mol Cell Cardiol* 1996;**28**(1):123–31.
23. Yap YG, Camm AJ. Drug induced QT prolongation and torsades de pointes. *Heart* 2003;**89**(11):1363–72.
24. Chow MJ, Piergies AA, Bowsher DJ, et al. Torsade de pointes induced by N-acetylprocainamide. *J Am Coll Cardiol* 1984;**4**(3):621–4.
25. Roden DM, Thompson KA, Hoffman BF, Woosley RL. Clinical features and basic mechanisms of quinidine-induced arrhythmias. *J Am Coll Cardiol* 1986;**8**(1 Suppl A):73A–78AA.
26. Kimura Y, Takayanagi K, Sakai Y, et al. Torsades de pointes in paced patients with sick sinus syndrome after disopyramide administration. *Jpn Heart J* 1994;**35**(2):153–61.
27. Jackman WM, Friday KJ, Anderson JL, et al. The long QT syndromes: a critical review, new clinical observations and a unifying hypothesis. *Prog Cardiovasc Dis* 1988;**31**(2): 115–72.
28. Lazzara R. Antiarrhythmic drugs and torsade de pointes. *Eur Heart J* 1993;(14 Suppl H): 88–92.
29. Khan IA. Oral loading single dose flecainide for pharmacological cardioversion of recent-onset atrial fibrillation. *Int J Cardiol* 2003;**87**(2-3):121–8.

30. Bigger Jr JT, Sahar DI. Clinical types of proarrhythmic response to antiarrhythmic drugs. *Am J Cardiol* 1987;**59**(11):2E–9E.
31. Naccarelli GV, Wolbrette DL, Luck JC. Proarrhythmia. *Med Clin North Am* 2001;**85**(2):503–26 xii.
32. Soyka LF, Wirtz C, Spangenberg RB. Clinical safety profile of sotalol in patients with arrhythmias. *Am J Cardiol* 1990;**65**(2):74A–81A; discussion 82A-93A.
33. Kim SY, Benowitz NL. Poisoning due to class IA antiarrhythmic drugs. Quinidine, procainamide and disopyramide. *Drug Saf* 1990;**5**(6):393–420.
34. Woosley RL. Antiarrhythmic drugs. *Annu Rev Pharmacol Toxicol* 1991;**31**:427–55.
35. Cardiac Arrhythmia Suppression Trial (CAST) Investigators. Preliminary report: effect of encainide and flecainide on mortality in a randomized trial of arrhythmia suppression after myocardial infarction. *N Engl J Med* 1989;**321**(6):406–12.
36. Naccarelli GV, Wolbrette DL, Khan M, et al. Old and new antiarrhythmic drugs for converting and maintaining sinus rhythm in atrial fibrillation: comparative efficacy and results of trials. *Am J Cardiol* 2003;**91**(6A):15D–26D.
37. Torp-Pedersen C, Moller M, Bloch-Thomsen PE, et al. Dofetilide in patients with congestive heart failure and left ventricular dysfunction. Danish Investigations of Arrhythmia and Mortality on Dofetilide Study Group. *N Engl J Med* 1999;**341**(12):857–65.
38. Doval HC, Nul DR, Grancelli HO, et al. Randomised trial of low-dose amiodarone in severe congestive heart failure. Grupo de Estudio de la Sobrevida en la Insuficiencia Cardiaca en Argentina (GESICA). *Lancet* 1994;**344**(8921):493–8.
39. Becker DE. Cardiovascular drugs: implications for dental practice part 1—cardiotonics, diuretics, and vasodilators. *Anesth Prog* 2007;**54**(4):178–85; quiz 86-7.
40. Lombardi F, Terranova P. Pharmacological treatment of atrial fibrillation: mechanisms of action and efficacy of class III drugs. *Curr Med Chem* 2006;**13**(14):1635–53.
41. Levy S. Do we need pharmacological therapy for atrial fibrillation in the ablation era? *J Interv Card Electrophysiol* 2006;**17**(3):189–94.
42. Yalta K, Turgut OO, Yilmaz MB, et al. Dronedarone: a promising alternative for the management of atrial fibrillation. *Cardiovasc Drugs Ther* 2009;**23**(5):385–93.
43. Kober L, Torp-Pedersen C, McMurray JJ, et al. Increased mortality after dronedarone therapy for severe heart failure. *N Engl J Med* 2008;**358**(25):2678–87.
44. Hohnloser SH, Crijns HJ, van Eickels M, et al. Effect of dronedarone on cardiovascular events in atrial fibrillation. *N Engl J Med* 2009;**360**(7):668–78.
45. DeWitt CR, Waksman JC. Pharmacology, pathophysiology and management of calcium channel blocker and beta-blocker toxicity. *Toxicol Rev* 2004;**23**(4):223–38.
46. Wood AJ. Pharmacologic differences between beta blockers. *Am Heart J* 1984;**108**(4 Pt 2):1070–7.
47. Katz AM. Calcium channel diversity in the cardiovascular system. *J Am Coll Cardiol* 1996;**28**(2):522–9.
48. Love JN. Beta blocker toxicity after overdose: when do symptoms develop in adults? *J Emerg Med* 1994;**12**(6):799–802.
49. Jackson CD, Fishbein L. A toxicological review of beta-adrenergic blockers. *Fundam Appl Toxicol* 1986;**6**(3):395–422.
50. Frishman W, Jacob H, Eisenberg E, et al. Clinical pharmacology of the new beta-adrenergic blocking drugs. Part 8. Self-poisoning with beta-adrenoceptor blocking agents: recognition and management. *Am Heart J* 1979;**98**(6):798–811.
51. Prichard BN, Battersby LA, Cruickshank JM. Overdosage with beta-adrenergic blocking agents. *Adverse Drug React Acute Poisoning Rev* 1984;**3**(2):91–111.

52. Reith DM, Dawson AH, Epid D, et al. Relative toxicity of beta blockers in overdose. *J Toxicol Clin Toxicol* 1996;**34**(3):273–8.
53. Love JN, Howell JM, Litovitz TL, et al. Acute beta blocker overdose: factors associated with the development of cardiovascular morbidity. *J Toxicol Clin Toxicol* 2000;**38**(3): 275–81.
54. Love JN. Acebutolol overdose resulting in fatalities. *J Emerg Med* 2000;**18**(3):341–4.
55. Love JN, Enlow B, Howell JM, et al. Electrocardiographic changes associated with beta-blocker toxicity. *Ann Emerg Med* 2002;**40**(6):603–10.
56. Love JN, Leasure JA, Mundt DJ, et al. A comparison of amrinone and glucagon therapy for cardiovascular depression associated with propranolol toxicity in a canine model. *J Toxicol Clin Toxicol* 1992;**30**(3):399–412.
57. Rennyson SL, Littmann L. Brugada-pattern electrocardiogram in propranolol intoxication. *Am J Emerg Med* 2010;**28**(2):256.e7–8.
58. Aouate P, Clerc J, Viard P, et al. Propranolol intoxication revealing a Brugada syndrome. *J Cardiovasc Electrophysiol* 2005;**16**(3):348–51.
59. Krishnan SC, Antzelevitch C. Sodium channel block produces opposite electrophysiological effects in canine ventricular epicardium and endocardium. *Circ Res* 1991;**69**(2):277–91.
60. Love JN, Litovitz TL, Howell JM, et al. Characterization of fatal beta blocker ingestion: a review of the American Association of Poison Control Centers data from 1985 to 1995. *J Toxicol Clin Toxicol* 1997;**35**(4):353–9.
61. Henry JA, Cassidy SL. Membrane stabilising activity: a major cause of fatal poisoning. *Lancet* 1986;**1**(8495):1414–7.
62. Donovan KD, Gerace RV, Dreyer JF. Acebutolol-induced ventricular tachycardia reversed with sodium bicarbonate. *J Toxicol Clin Toxicol* 1999;**37**(4):481–4.
63. Love JN. The effect of propranolol intoxication on QTc interval in a canine model. *J Emerg Med* 1998;**16**(1):1–4 P.
64. Hohnloser SH, Woosley RL. Sotalol. *N Engl J Med* 1994;**331**(1):31–8.
65. Arstall MA, Hii JT, Lehman RG, et al. Sotalol-induced torsade de pointes: management with magnesium infusion. *Postgrad Med J* 1992;**68**(798):289–90.
66. Elonen E, Neuvonen PJ, Tarssanen L, et al. Sotalol intoxication with prolonged Q-T interval and severe tachyarrhythmias. *Br Med J* 1979;**1**(6172):1184.
67. Neuvonen PJ, Elonen E, Vuorenmaa T, et al. Prolonged Q-T interval and severe trachy-arrhythmias, common features of sotalol intoxication. *Eur J Clin Pharmacol* 1981;**20**(2): 85–9.
68. Lowe MD, Stone DL, Grace AA. Sotalol associated polymorphic ventricular tachycardia and coronary spasm. *Heart* 1998;**79**(5):518–20.
69. Muto S, Ashizawa N, Arakawa S, et al. Sotalol-induced coronary spasm in a patient with dilated cardiomyopathy associated with sustained ventricular tachycardia. *Intern Med* 2004;**43**(11):1051–5.
70. De Cesare N, Cozzi S, Apostolo A, et al. Facilitation of coronary spasm by propranolol in Prinzmetal's angina: fact or unproven extrapolation? *Coron Artery Dis* 1994;**5**(4):323–30.
71. Petrov D, Sardowski S, Gesheva M. "Silent" Prinzmetal's ST elevation related to atenolol overdose. *J Emerg Med* 2007;**33**(2):123–6.
72. Lin YJ, Liu YB, Wu CC, et al. Painless Prinzmetal's ST elevation related to propranolol: a case report. *Int J Cardiol* 1997;**60**(3):311–4.
73. Youngquist ST, Kaji AH, Niemann JT. Beta-blocker use and the changing epidemiology of out-of-hospital cardiac arrest rhythms. *Resuscitation* 2008;**76**(3):376–80.
74. Kerns 2nd W, Kline J, Ford MD. Beta-blocker and calcium channel blocker toxicity. *Emerg Med Clin North Am* 1994;**12**(2):365–90.

75. Vale JA, Kulig K. American Academy of Clinical Toxicology, and European Association of Poisons Centres and Clinical Toxicologists. Position paper: gastric lavage. *J Toxicol Clin Toxicol* 2004;**42**(7):933–43.
76. Chyka PA, Seger D. Position statement: single-dose activated charcoal. American Academy of Clinical Toxicology; European Association of Poisons Centres and Clinical Toxicologists. *J Toxicol Clin Toxicol* 1997;**35**(7):721–41.
77. Salhanick SD, Wax PM. Treatment of atenolol overdose in a patient with renal failure using serial hemodialysis and hemoperfusion and associated echocardiographic findings. *Vet Hum Toxicol* 2000;**42**(4):224–5.
78. Rooney M, Massey KL, Jamali F, et al. Acebutolol overdose treated with hemodialysis and extracorporeal membrane oxygenation. *J Clin Pharmacol* 1996;**36**(8):760–3.
79. Love JN, Sachdeva DK, Bessman ES, et al. A potential role for glucagon in the treatment of drug-induced symptomatic bradycardia. *Chest* 1998;**114**(1):323–6.
80. Kerns 2nd W. Management of beta-adrenergic blocker and calcium channel antagonist toxicity. *Emerg Med Clin North Am* 2007;**25**(2):309–31; viii.
81. Strubelt O. Evaluation of antidotes against the acute cardiovascular toxicity of propranolol. *Toxicol* 1984;**31**(3-4):261–70.
82. Brimacombe JR, Scully M, Swainston R. Propranolol overdose: a dramatic response to calcium chloride. *Med J Aust* 1991;**155**(4):267–8.
83. Love JN, Howell JM, Newsome JT, et al. The effect of sodium bicarbonate on propranolol-induced cardiovascular toxicity in a canine model. *J Toxicol Clin Toxicol* 2000;**38**(4):421–8.
84. Kerns 2nd W, Ransom M, Tomaszewski C, et al. The effects of extracellular ions on beta-blocker cardiotoxicity. *Toxicol Appl Pharmacol* 1996;**137**(1):1–7.
85. Tanen DA, Ruha AM, Curry SC, et al. Hypertonic sodium bicarbonate is effective in the acute management of verapamil toxicity in a swine model. *Ann Emerg Med* 2000;**36**(6):547–53.
86. Kontopoulos A, Filindris A, Manoudis F, et al. Sotalol-induced torsade de pointes. *Postgrad Med J* 1981;**57**(667):321–3.
87. Whitehurst VE, Vick JA, Alleva FR, et al. Reversal of propranolol blockade of adrenergic receptors and related toxicity with drugs that increase cyclic AMP. *Proceedings of the Society for Experimental Biology and Medicine Society for Experimental Biology and Medicine* 1999;**221**(4):382–5.
88. Sato S, Tsuji MH, Okubo N, et al. Combined use of glucagon and milrinone may not be preferable for severe propranolol poisoning in the canine model. *J Toxicol Clin Toxicol* 1995;**33**(4):337–42.
89. Tuncok Y, Apaydin S, Gidener S, et al. The effects of amrinone and glucagon on verapamil-induced myocardial depression in a rat isolated heart model. *Gen Pharmacol* 1997;**28**(5):773–6.
90. Tuncok Y, Apaydin S, Kalkan S, et al. The effects of amrinone and glucagon on verapamil-induced cardiovascular toxicity in anaesthetized rats. *Int J Exp Pathol* 1996;**77**(5):207–12.
91. Avery 2nd GJ, Spotnitz HM, Rose EA, et al. Pharmacologic antagonism of beta-adrenergic blockade in dogs. I. Hemodynamic effects of isoproterenol, dopamine, and epinephrine in acute propranolol administration. *J Thorac Cardiovasc Surg* 1979;**77**(2):267–76.
92. Richards DA, Prichard BN, Boakes AJ, et al. Pharmacological basis for antihypertensive effects of intravenous labetalol. *Br Heart J* 1977;**39**(1):99–106.
93. Anthony T, Jastremski M, Elliott W, et al. Charcoal hemoperfusion for the treatment of a combined diltiazem and metoprolol overdose. *Ann Emerg Med* 1986;**15**(11):1344–8.

94. Lheureux PE, Zahir S, Gris M, et al. Bench-to-bedside review: hyperinsulinaemia/euglycaemia therapy in the management of overdose of calcium-channel blockers. *Critical Care* 2006;**10**(3):212.
95. Kerns 2nd W, Schroeder D, Williams C, et al. Insulin improves survival in a canine model of acute beta-blocker toxicity. *Ann Emerg Med* 1997;**29**(6):748–57.
96. Hasin T, Leibowitz D, Antopolsky M, et al. The use of low-dose insulin in cardiogenic shock due to combined overdose of verapamil, enalapril and metoprolol. *Cardiol* 2006;**106**(4):233–6.
97. Engebretsen KM, Kaczmarek KM, Morgan J, et al. High-dose insulin therapy in beta-blocker and calcium channel-blocker poisoning. *Clin Toxicol* 2011;**49**(4):277–83.
98. Engebretsen KM, Morgan MW, Stellpflug SJ, et al. Addition of phenylephrine to high-dose insulin in dihydropyridine overdose does not improve outcome. *Clin Toxicol* 2010;**48**(8):806–12.
99. Weinstein RS. Recognition and management of poisoning with beta-adrenergic blocking agents. *Ann Emerg Med* 1984;**13**(12):1123–31.
100. Lane AS, Woodward AC, Goldman MR. Massive propranolol overdose poorly responsive to pharmacologic therapy: use of the intra-aortic balloon pump. *Ann Emerg Med* 1987;**16**(12):1381–3.
101. Holzer M, Sterz F, Schoerkhuber W, et al. Successful resuscitation of a verapamil-intoxicated patient with percutaneous cardiopulmonary bypass. *Crit Care Med* 1999;**27**(12):2818–23.
102. Salhanick SD, Shannon MW. Management of calcium channel antagonist overdose. *Drug Saf* 2003;**26**(2):65–79.
103. Ramoska EA, Spiller HA, Winter M, et al. A one-year evaluation of calcium channel blocker overdoses: toxicity and treatment. *Ann Emerg Med* 1993;**22**(2):196–200.
104. Spiller HA, Meyers A, Ziemba T, et al. Delayed onset of cardiac arrhythmias from sustained-release verapamil. *Ann Emerg Med* 1991;**20**(2):201–3.
105. Luscher TF, Noll G, Sturmer T, et al. Calcium gluconate in severe verapamil intoxication. *N Engl J Med* 1994;**330**(10):718–20.
106. Olson KR, Erdman AR, Woolf AD, et al. Calcium channel blocker ingestion: an evidence-based consensus guideline for out-of-hospital management. *Clin Toxicol* 2005;**43**(7):797–822.
107. Mihalcea M, Geiger A, Kopferschmitt J, et al. Calcium channel blocker sustained release: only three tablets can be life threatening. *J Emerg, Trauma, Shock* 2012;**5**(2):208–9.
108. Patel T, Tietze D, Mehta AN. Amlodipine overdose. *Proc (Bayl Univ Med Cent)* 2013;**26**(4):410–1.
109. Rizvi I, Ahmad A, Gupta A, et al. Life-threatening calcium channel blocker overdose and its management. *BMJ Case Rep.* 2012;**2012**. pii: bcr0120125643. Available from: http://dx.doi.org/10.1136/bcr.01.2012.5643.
110. Boyer EW, Shannon M. Treatment of calcium-channel-blocker intoxication with insulin infusion. *N Engl J Med* 2001;**344**(22):1721–2.
111. Howarth DM, Dawson AH, Smith AJ, et al. Calcium channel blocking drug overdose: an Australian series. *Hum Exp Toxicol* 1994;**13**(3):161–6.
112. Crump BJ, Holt DW, Vale JA. Lack of response to intravenous calcium in severe verapamil poisoning. *Lancet* 1982;**2**(8304):939–40.
113. Horowitz BZ, Rhee KJ. Massive verapamil ingestion: a report of two cases and a review of the literature. *Am J Emerg Med* 1989;**7**(6):624–31.

114. Zaritsky AL, Horowitz M, Chernow B. Glucagon antagonism of calcium channel blocker-induced myocardial dysfunction. *Crit Care Med* 1988;**16**(3):246–51.
115. Stone CK, May WA, Carroll R. Treatment of verapamil overdose with glucagon in dogs. *Ann Emerg Med* 1995;**25**(3):369–74.
116. Quezado Z, Lippmann M, Wertheimer J. Severe cardiac, respiratory, and metabolic complications of massive verapamil overdose. *Crit Care Med* 1991;**19**(3):436–8.
117. Levine M, Curry SC, Padilla-Jones A, et al. Critical care management of verapamil and diltiazem overdose with a focus on vasopressors: a 25-year experience at a single center. *Ann Emerg Med* 2013;**62**(3):252–8.
118. Megarbane B, Karyo S, Abidi K, et al. Predictors of mortality in verapamil overdose: usefulness of serum verapamil concentrations. *Basic Clin Pharmacol Toxicol* 2011;**108**(6):385–9.
119. Olson KR. What is the best treatment for acute calcium channel blocker overdose? *Ann Emerg Med* 2013;**62**(3):259–61.
120. Holger JS, Engebretsen KM, Stellpflug SJ, et al. Critical care management of verapamil and diltiazem overdose with a focus on vasopressors: a 25-year experience at a single center. *Ann Emerg Med* 2014;**63**(1):91–2.
121. French D, Armenian P, Ruan W, et al. Serum verapamil concentrations before and after Intralipid® therapy during treatment of an overdose. *Clin Toxicol* 2011;**49**(4):340–4.
122. Hasenfuss G, Pieske B, Castell M, et al. Influence of the novel inotropic agent levosimendan on isometric tension and calcium cycling in failing human myocardium. *Circulation* 1998;**98**(20):2141–7.
123. Varpula T, Rapola J, Sallisalmi M, et al. Treatment of serious calcium channel blocker overdose with levosimendan, a calcium sensitizer. *Anesth Analg* 2009;**108**(3):790–2.
124. Aggarwal N, Kupfer Y, Seneviratne C, et al. Methylene blue reverses recalcitrant shock in beta-blocker and calcium channel blocker overdose. *BMJ Case Rep* 2013;**2013** pii: bcr2012007402. Available from: http://dx.doi.org/10.1136/bcr-2012-007402.
125. Holger JS, Engebretsen KM, Fritzlar SJ, et al. Insulin versus vasopressin and epinephrine to treat beta-blocker toxicity. *Clin Toxicol* 2007;**45**(4):396–401.
126. Aronow WS. Prevalence of appropriate and inappropriate indications for use of digoxin in older patients at the time of admission to a nursing home. *J Am Geriatr Soc* 1996;**44**(5):588–90.
127. Arrigo M, Bettex D, Rudiger A. Management of atrial fibrillation in critically ill patients. *Crit Care Res Pract* 2014;**2014**:840615.
128. Jessup M, Abraham WT, Casey DE, et al. Focused update—Guidelines for the diagnosis and management of heart failure in adults: a report of the American College of Cardiology Foundation/AHA Task Force on Practice Guidelines: developed in collaboration with the International Society for Heart and Lung Transplantation. *Circulation* 2009;**119**(14): 1977–2016.
129. Wasserstrom JA, Aistrup GL. Digitalis: new actions for an old drug. *Am J Physiol Heart Circ Physiol* 2005;**289**(5):H1781–93.
130. Piltz JR, Wertenbaker C, Lance SE, et al. Digoxin toxicity. Recognizing the varied visual presentations. *J Clin Neuroophthalmol* 1993;**13**(4):275–80.
131. Warren SE, Fanestil DD. Digoxin overdose. Limitations of hemoperfusion-hemodialysis treatment. *JAMA* 1979;**242**(19):2100–1.
132. Smiley JW, March NM, Del Guercio ET. Hemoperfusion in the management of digoxin toxicity. *JAMA* 1978;**240**(25):2736–7.
133. Lant A. Diuretics. Clinical pharmacology and therapeutic use (Part I). *Drugs* 1985;**29**(1): 57–87.

134. Lant A. Diuretic drugs. Progress in clinical pharmacology. *Drugs* 1986;**31**(Suppl 4): 40–55.
135. Freis ED. The cardiotoxicity of thiazide diuretics: review of the evidence. *J Hypertens Suppl* 1990;**8**(2):S23–32.
136. Ames RP, Hill P. Raised serum lipid concentrations during diuretic treatment of hypertension: a study of predictive indexes. *Clin Sci Mol Med Suppl* 1978;**4**:311s–314ss.
137. Mubagwa K, Mullane K, Flameng W. Role of adenosine in the heart and circulation. *Cardiovasc Res* 1996;**32**(5):797–813.
138. Ralevic V, Burnstock G. Receptors for purines and pyrimidines. *Pharmacol Rev* 1998;**50**(3):413–92.
139. Till J, Shinebourne EA, Rigby ML, et al. Efficacy and safety of adenosine in the treatment of supraventricular tachycardia in infants and children. *Br Heart J* 1989;**62**(3): 204–11.
140. Blomstrom-Lundqvist C, Scheinman MM, Aliot EM, et al. Guidelines for the management of patients with supraventricular arrhythmias—executive summary: a report of the American College of Cardiology/AHA Task Force on Practice Guidelines and the European Society of Cardiology Committee for Practice Guidelines. *Circulation* 2003;**108**(15):1871–909.
141. Tan HL, Spekhorst HH, Peters RJ, et al. Adenosine-induced ventricular arrhythmias in the emergency room. *Pacing Clin Electrophysiol* 2001;**24**(4 Pt 1):450–5.
142. Gupta AK, Shah CP, Maheshwari A, et al. Adenosine-induced ventricular fibrillation in Wolff-Parkinson-White syndrome. *Pacing Clin Electrophysiol* 2002;**25**(4 Pt 1):477–80.
143. Misra D, Van Tosh A, Schweitzer P. Adenosine-induced monomorphic ventricular tachycardia. *Pacing Clin Electrophysiol* 2000;**23**(6):1044–6.
144. Romer M, Candinas R. Adenosine-induced non-sustained polymorphic ventricular tachycardia. *Eur Heart J* 1994;**15**(2):281–2.
145. Smith JR, Goldberger JJ, Kadish AH. Adenosine-induced polymorphic ventricular tachycardia in adults without structural heart disease. *Pacing Clin Electrophysiol* 1997;**20**(3 Pt 1): 743–5.
146. Kaplan IV, Kaplan AV, Fisher JD. Adenosine-induced atrial fibrillation precipitating polymorphic ventricular tachycardia. *Pacing Clin Electrophysiol* 2000;**23**(1):140–1.
147. Isenberg G, Belardinelli L. Ionic basis for the antagonism between adenosine and isoproterenol on isolated mammalian ventricular myocytes. *Circ Res* 1984;**55**(3):309–25.
148. Biaggioni I, Killian TJ, Mosqueda-Garcia R, et al. Adenosine increases sympathetic nerve traffic in humans. *Circulation* 1991;**83**(5):1668–75.
149. DiMarco JP, Miles W, Akhtar M, et al. Adenosine for paroxysmal supraventricular tachycardia: dose ranging and comparison with verapamil. Assessment in placebo-controlled, multicenter trials. Adenosine for PSVT Study Group. *Ann Intern Med* 1990;**113**(2):104–10.
150. Burnstock G. Purinergic nerves. *Pharmacol Rev* 1972;**24**(3):509–81.
151. Burnstock G, Kennedy C. Is there a basis for distinguishing two types of P2-purinoceptor? *Gen Pharmacol* 1985;**16**(5):433–40.
152. Abbracchio MP, Burnstock G. Purinoceptors: are there families of P2X and P2Y purinoceptors? *Pharmacol Ther* 1994;**64**(3):445–75.
153. Abbracchio MP, Burnstock G. Purinergic signalling: pathophysiological roles. *Jpn J Pharmacol* 1998;**78**(2):113–45.
154. Schwiebert EM, Zsembery A. Extracellular ATP as a signaling molecule for epithelial cells. *Biochim Biophys Acta* 2003;**1615**(1-2):7–32.

155. Burnstock G. Pathophysiology and therapeutic potential of purinergic signaling. *Pharmacol Rev* 2006;**58**(1):58–86.
156. Jones LR. Sarcolemmal enzymes mediating beta-adrenergic effects on the heart. In: Bronner FA, Shamoo AE, editors. *Current topics in membranes and transport: Regulaton of calcium transport across muscle membranes*. New York: Academic Press; 1985. p. 11–41.
157. Feneck R. Phosphodiesterase inhibitors and the cardiovascular system. *Contin Educ Anaesth Crit Care Pain* 2007;**7**(6):203–7.
158. Maurice DH. Cyclic nucleotide phosphodiesterase-mediated integration of cGMP and cAMP signaling in cells of the cardiovascular system. *Front Biosci* 2005;**10**:1221–8.
159. Eckly-Michel A, Martin V, Lugnier C. Involvement of cyclic nucleotide-dependent protein kinases in cyclic AMP-mediated vasorelaxation. *Br J Pharmacol* 1997;**122**(1): 158–64.
160. Hofmann F, Feil R, Kleppisch T, et al. Function of cGMP-dependent protein kinases as revealed by gene deletion. *Physiol Rev* 2006;**86**(1):1–23.
161. Potter LR, Abbey-Hosch S, Dickey DM. Natriuretic peptides, their receptors, and cyclic guanosine monophosphate-dependent signaling functions. *Endocr Rev* 2006;**27**(1):47–72.
162. Pilz RB, Casteel DE. Regulation of gene expression by cyclic GMP. *Circ Res* 2003;**93**(11): 1034–46.
163. Gustafsson AB, Brunton LL. Attenuation of cAMP accumulation in adult rat cardiac fibroblasts by IL-1beta and NO: role of cGMP-stimulated PDE2. *Am J Physiol Cell Physiol* 2002;**283**(2):C463–71.
164. Lincoln TM, Wu X, Sellak H, et al. Regulation of vascular smooth muscle cell phenotype by cyclic GMP and cyclic GMP-dependent protein kinase. *Front Biosci* 2006;**11**:356–67.
165. Murthy KS, Zhou H, Grider JR, et al. Inhibition of sustained smooth muscle contraction by PKA and PKG preferentially mediated by phosphorylation of RhoA. *Am J Physiol Gastrointest Liver Physiol* 2003;**284**(6):G1006–16.
166. Mack CP, Somlyo AV, Hautmann M, et al. Smooth muscle differentiation marker gene expression is regulated by RhoA-mediated actin polymerization. *J Biol Chem* 2001;**276** (1):341–7.
167. Liu HW, Halayko AJ, Fernandes DJ, et al. The RhoA/Rho kinase pathway regulates nuclear localization of serum response factor. *Am J Respir Cell Mol Biol* 2003;**29**(1):39–47.
168. Alousi AA, Farah AE, Lesher GY, et al. Cardiotonic activity of amrinone—Win 40680 [5-amino-3,4'-bipyridine-6(1H)-one]. *Circ Res* 1979;**45**(5):666–77.
169. Nusrat A, Tepper D, Hertzberg J, et al. Effects of amrinone on atrioventricular conduction in the intact canine heart. *J Clin Pharmacol* 1983;**23**(7):257–65.
170. Canniff PC, Farah AE, Sperelakis N, et al. The effect of milrinone (Win 47203) on the in vitro electropharmacological properties of mammalian cardiac tissue. *J Cardiovasc Pharmacol* 1985;**7**(5):813–21.
171. Iven H, Brasch H, Armah BI. Electrophysiologic effects of saterinone and milrinone in the isolated guinea pig myocardium. *Arzneimittelforschung* 1988;**38**(9):1298–302.
172. Keister DM, Kittleson MD, Bonagura JD, et al. Milrinone. A clinical trial in 29 dogs with moderate to severe congestive heart failure. *J Vet Intern Med* 1990;**4**(2):79–86.
173. Naccarelli GV, Goldstein RA. Electrophysiology of phosphodiesterase inhibitors. *Am J Cardiol* 1989;**63**(2):35A–40A.
174. Benotti JR, McCue JE, Alpert JS. Comparative vasoactive therapy for heart failure. *Am J Cardiol* 1985;**56**(3):19B–24B.

175. Guimond JG, Matuschak GM, Meyers F, et al. Augmentation of cardiac function in end-stage heart failure by combined use of dobutamine and amrinone. *Chest* 1986;**90**(2): 302–4.
176. Klein NA, Siskind SJ, Frishman WH, et al. Hemodynamic comparison of intravenous amrinone and dobutamine in patients with chronic congestive heart failure. *Am J Cardiol* 1981;**48**(1):170–5.
177. Benotti JR, Grossman W, Braunwald E, et al. Hemodynamic assessment of amrinone. A new inotropic agent. *N Engl J Med* 1978;**299**(25):1373–7.
178. Leier CV, Dalpiaz K, Huss P, et al. Amrinone therapy for congestive heart failure in outpatients with idiopathic dilated cardiomyopathy. *Am J Cardiol* 1983;**52**(3):304–8.
179. Packer M, Carver JR, Rodeheffer RJ, et al. Effect of oral milrinone on mortality in severe chronic heart failure. The PROMISE Study Research Group. *N Engl J Med* 1991;**325**(21):1468–75.
180. Lefer AM, Scalia R, Lefer DJ. Vascular effects of HMG CoA-reductase inhibitors (statins) unrelated to cholesterol lowering: new concepts for cardiovascular disease. *Cardiovasc Res* 2001;**49**(2):281–7.
181. Taylor FC, Huffman M, Ebrahim S. Statin therapy for primary prevention of cardiovascular disease. *JAMA* 2013;**310**(22):2451–2.
182. Bang CN, Okin PM. Statin treatment, new-onset diabetes, and other adverse effects: a systematic review. *Curr Cardiol Rep* 2014;**16**(3):461.
183. Madhu SV, Aslam M, Galav V, et al. Atorvastatin prevents type 2 diabetes mellitus-An experimental study. *Eur J Pharmacol* 2014;**728**:135–40.
184. Yoon JS, Lee HW. Diabetogenic effect of statins: a double-edged sword? *Diab Metab J* 2013;**37**(6):415–22.
185. Preiss D, Seshasai SR, Welsh P, et al. Risk of incident diabetes with intensive-dose compared with moderate-dose statin therapy: a meta-analysis. *JAMA* 2011;**305**(24):2556–64.
186. Gupta AK, Dahlof B, Dobson J, et al. Determinants of new-onset diabetes among 19,257 hypertensive patients randomized in the Anglo-Scandinavian Cardiac Outcomes Trial—Blood Pressure Lowering Arm and the relative influence of antihypertensive medication. *Diabetes Care* 2008;**31**(5):982–8.
187. Elliott WJ, Meyer PM. Incident diabetes in clinical trials of antihypertensive drugs: a network meta-analysis. *Lancet* 2007;**369**(9557):201–7.
188. Yada T, Nakata M, Shiraishi T, et al. Inhibition by simvastatin, but not pravastatin, of glucose-induced cytosolic Ca^{2+} signalling and insulin secretion due to blockade of L-type Ca^{2+} channels in rat islet beta-cells. *Br J Pharmacol* 1999;**126**(5):1205–13.
189. Landefeld CS, Goldman L. Major bleeding in outpatients treated with warfarin: incidence and prediction by factors known at the start of outpatient therapy. *Am J Med.* 1989;**87**(2):144–52.
190. Stewart S, Murphy NF, Walker A, et al. Cost of an emerging epidemic: an economic analysis of atrial fibrillation in the UK. *Heart* 2004;**90**(3):286–92.
191. Bungard TJ, Ghali WA, Teo KK, et al. Why do patients with atrial fibrillation not receive warfarin? *Arch Intern Med* 2000;**160**(1):41–6.
192. Choudhry NK, Anderson GM, Laupacis A, et al. Impact of adverse events on prescribing warfarin in patients with atrial fibrillation: matched pair analysis. *BMJ* 2006;**332**(7534): 141–5.
193. Levis JT, Delgado MC. Hemopericardium and cardiac tamponade in a patient with an elevated international normalized ratio. *West J Emerg Med* 2009;**10**(2):115–9.

194. Younas F, Janjua M, Badshah A, et al. Transient complete heart block and isolated ventricular asystole with nitroglycerin. *J Cardiovasc Med* 2012;**13**(8):533–5.
195. Brandes W, Santiago T, Limacher M. Nitroglycerin-induced hypotension, bradycardia, and asystole: report of a case and review of the literature. *Clin Cardiol* 1990;**13**(10): 741–4.
196. Mark AL. The Bezold–Jarisch reflex revisited: clinical implications of inhibitory reflexes originating in the heart. *J Am Coll Cardiol* 1983;**1**(1):90–102.
197. Jordan J, Shannon JR, Black BK, et al. Malignant vagotonia due to selective baroreflex failure. *Hypertension* 1997;**30**(5):1072–7.
198. Come PC, Pitt B. Nitroglycerin-induced severe hypotension and bradycardia in patients with acute myocardial infarction. *Circulation* 1976;**54**(4):624–8.
199. Antzelevitch C, Belardinelli L, Zygmunt AC, et al. Electrophysiological effects of ranolazine, a novel antianginal agent with antiarrhythmic properties. *Circulation* 2004;**110** (8):904–10.
200. Murdock DK, Overton N, Kersten M, et al. The effect of ranolazine on maintaining sinus rhythm in patients with resistant atrial fibrillation. *Indian Pacing Electrophysiol J* 2008;**8** (3):175–81.
201. Belardinelli L, Shryock JC, Wu L, et al. Use of preclinical assays to predict risk of drug-induced torsades de pointes. *Heart Rhythm* 2005;**2**(Suppl 2):S16–22.
202. Chaitman BR, Pepine CJ, Parker JO, et al. Effects of ranolazine with atenolol, amlodipine, or diltiazem on exercise tolerance and angina frequency in patients with severe chronic angina: a randomized controlled trial. *JAMA* 2004;**291**(3):309–16.
203. Belardinelli L, Shryock JC, Fraser H. Inhibition of the late sodium current as a potential cardioprotective principle: effects of the late sodium current inhibitor ranolazine. *Heart* 2006;**92**(Suppl 4):iv6–iv14.
204. Riccioni G. Ivabradine: from molecular basis to clinical effectiveness. *Adv Ther* 2010;**27** (3):160–7.
205. Deedwania P. Selective and specific inhibition of If with ivabradine for the treatment of coronary artery disease or heart failure. *Drugs* 2013;**73**(14):1569–86.
206. Mathiaux F, Dulaurent S, Julia F, et al. Case report of ivabradine intoxication. *J Anal Toxicol* 2014;**38**(4):231–2.
207. Katz AM. Cellular mechanisms in congestive heart failure. *Am J Cardiol* 1988;**62**(2): 3A–8A.
208. Katz AM. Is the failing heart energy depleted? *Cardiol Clin* 1998;**16**(4):633–44 viii.
209. Granger CB, McMurray JJ, Yusuf S, et al. Effects of candesartan in patients with chronic heart failure and reduced left-ventricular systolic function intolerant to angiotensin-converting-enzyme inhibitors: the CHARM-Alternative trial. *Lancet* 2003;**362**(9386): 772–6.
210. Turnbull F. Blood Pressure Lowering Treatment Trialists' Colloboration. Effects of different blood-pressure-lowering regimens on major cardiovascular events: results of prospectively-designed overviews of randomised trials. *Lancet* 2003;**362**(9395):1527–35.
211. Ajmal A, Gessert CE, Johnson BP, et al. Effect of angiotensin converting enzyme inhibitors and angiotensin receptor blockers on hemoglobin levels. *BMC Res Notes* 2013;**6**:443.
212. Poon IO, Braun U. High prevalence of orthostatic hypotension and its correlation with potentially causative medications among elderly veterans. *J Clin Pharm Ther* 2005;**30**(2): 173–8.
213. Hajjar I. Postural blood pressure changes and orthostatic hypotension in the elderly patient: impact of antihypertensive medications. *Drugs Aging* 2005;**22**(1):55–68.

214. Parving HH, Brenner BM, McMurray JJ, et al. Cardiorenal end points in a trial of aliskiren for type 2 diabetes. *N Engl J Med* 2012;**367**(23):2204–13.
215. Escobar C, Barrios V. An evaluation of the latest evidence relating to renin–angiotensin system inhibitors. *Expert Opin Drug Metab Toxicol* 2013;**9**(7):847–58.
216. McDowell SE, Coleman JJ, Ferner RE. Systematic review and meta-analysis of ethnic differences in risks of adverse reactions to drugs used in cardiovascular medicine. *BMJ* 2006; **332**(7551):1177–81.
217. Gibbs CR, Lip GY, Beevers DG. Angioedema due to ACE inhibitors: increased risk in patients of African origin. *Br J Clin Pharmacol* 1999;**48**(6):861–5.
218. Xie HG, Kim RB, Wood AJ, et al. Molecular basis of ethnic differences in drug disposition and response. *Annu Rev Pharmacol Toxicol* 2001;**41**:815–50.
219. Taylor AL, Ziesche S, Yancy C, et al. Combination of isosorbide dinitrate and hydralazine in blacks with heart failure. *N Engl J Med* 2004;**351**(20):2049–57.
220. Kettenhofen R, Bohlen H. Preclinical assessment of cardiac toxicity. *Drug Discov Today* 2008;**13**(15-16):702–7.
221. Rubinstein AL. Zebrafish assays for drug toxicity screening. *Expert Opin Drug Metab Toxicol* 2006;**2**(2):231–40.
222. Milan DJ, Peterson TA, Ruskin JN, et al. Drugs that induce repolarization abnormalities cause bradycardia in zebrafish. *Circulation* 2003;**107**(10):1355–8.
223. Gussak I, Litwin J, Kleiman R, et al. Drug-induced cardiac toxicity: emphasizing the role of electrocardiography in clinical research and drug development. *J Electrocardiol* 2004; **37**(1):19–24.
224. Eisenberg DM, Davis RB, Ettner SL, et al. Trends in alternative medicine use in the United States, 1990–1997: results of a follow-up national survey. *JAMA* 1998;**280**(18): 1569–75.
225. Izzo AA, Di Carlo G, Borrelli F, et al. Cardiovascular pharmacotherapy and herbal medicines: the risk of drug interaction. *Int J Cardiol* 2005;**98**(1):1–14.
226. Huupponen R, Seppala P, Iisalo E. Effect of guar gum, a fibre preparation, on digoxin and penicillin absorption in man. *Eur J Clin Pharmacol* 1984;**26**(2):279–81.
227. McRae S. Elevated serum digoxin levels in a patient taking digoxin and Siberian ginseng. *Can Med Assoc J* 1996;**155**(3):293–5.
228. Ge B, Zhang Z, Zuo Z. Updates on the Clinical Evidenced Herb-Warfarin Interactions. *Evid Based Complement Alternat Med* 2014;**2014**:957362. Available from: http://dx.doi.org/10.1155/2014/957362.
229. Scheen AJ. Cytochrome P450-mediated cardiovascular drug interactions. *Expert Opin Drug Metab Toxicol* 2011;**7**(9):1065–82.

Chapter 9

Cardiovascular Toxicity as a Result of Psychotropic Drugs

Mythily Subramaniam,[1] Amy See Yuning,[2] Janhavi Ajit Vaingankar,[2] and Siow Ann Chong[2]
[1]*Research Division, Institute of Mental Health and Saw Swee Hock School of Public Health, Singapore*
[2]*Research Division, Institute of Mental Health, Singapore*

9.1 INTRODUCTION

Cardiovascular disease (CVD) is one of the leading causes of death in the general population.[1,2] While heart disease, accounts for only 3% of all deaths for those aged 1 to 24, for persons 65 and over, it accounts for 27% of all deaths.[3] An overwhelming body of evidence suggests that patients with serious mental illness (SMI) are at significantly higher risk for cardiovascular morbidity and mortality than those in the general population[4,5]; and, in SMI patients, CVD is the most common cause of death.[6,7] The etiology of this excess CVD is multifactorial and includes the excessive accumulation of risk factors (e.g., smoking, obesity, physical inactivity, and dyslipidemia) in SMI patients; however, other mechanisms due to the mental illness itself, as well as the effects of psychotropic drugs, may also be influential in increasing risk in this group.

Sudden unexplained death of individuals with mental health problems was first described more than 40 years ago and a link with antipsychotic drugs was postulated.[8] Since the 1990s, regulatory authorities in several countries have expressed concerns about a relationship between antipsychotics and QTc interval prolongation, serious arrhythmias, and sudden death.[9,10] A recent meta-analysis reported that while there is no difference in cardiovascular risk assessed by weight or metabolic indices between individuals with an untreated first episode of psychosis and healthy controls, cardiovascular risk increases after a first exposure to any antipsychotic drug.[11] Similarly, the cardiovascular effects and cardiotoxicity of tricyclic antidepressants (TCAs) have been well documented in depressed patients without preexisting cardiac disease.[12,13]

The Heart and Toxins. DOI: http://dx.doi.org/10.1016/B978-0-12-416595-3.00009-8

Psychotropic drugs are prescription drugs that are used to treat or manage psychiatric symptoms or challenging behaviors; these include antipsychotic, antidepressant, and anxiolytic medications, as well as mood stabilizers. This research review for this chapter was performed using publications obtained through an electronic search of MEDLINE, PsycINFO, and ScienceDirect databases. There was no restriction with regard to the time period of the publications but the search was limited to those in the English language and included case reports and randomized controlled trials.

The following basic search terms were used, both alone and in combination: "cardiac," "cardiotoxic," "cardiovascular," "psychotropic," "antipsychotics," "neuroleptics," "antidepressant," "tricyclic," and "SSRI." In addition, the reference lists of retrieved articles to review were examined for cross-references. We focused predominantly on the effects of the available antipsychotic and antidepressant medications reported among clinical populations. However, a systematic evaluation was not possible given the breadth of the subject.

9.2 PSYCHOTROPIC DRUGS' MECHANISM OF ACTION

Psychotropic medications have various effects on the heart, ranging from hypotension and tachycardia, which may occur in many patients at the therapeutic range, to more severe effects such as bradycardia, arrhythmias, and sudden death. These effects on the cardiovascular system are either specific to the heart, by interfering with impulse conduction, cardiac rhythm, or myocardial contractility,[14] or via the vascular system by either blockade of adrenergic and cholinergic receptors or through central autonomic regulation and baroreceptor reflexes.[15]

9.3 PSYCHOTROPIC DRUG-INDUCED HYPOTENSION

Blood pressure changes leading to postural hypotension are a common vascular effect of certain antipsychotic drugs.[16] Orthostatic hypotension occurs when a blockade of α-1 adrenoceptors leads to vasodilation, resulting in postural hypotension, a $\geq$20 mmHg decrease in systolic blood pressure from the pooling of blood in the major veins after assuming an upright posture. Low-potency antipsychotics (e.g., chlorpromazine and thioridazine have a higher propensity to cause hypotension.[17] Silver et al.[18] reported the prevalence of postural hypotension in patients with schizophrenia ($n = 196$) on stable long-term antipsychotic treatment to be 77% at 1 min and 16.8% at 3 min after standing; this was compared with the absence of postural hypotension in unmedicated healthy controls at both time points.

A number of typical antipsychotics are associated with orthostatic hypotension: 29% of the schizophrenia patients ($n = 100$) on clozapine treatment developed hypotension, which usually resolved after the initial phase of treatment (i.e., 4–6 weeks) as tolerance developed with continued therapy.[19] Quetiapine can also cause some orthostasis, with one study reporting that 6 out of 12 elderly patients with psychosis had postural hypotension in an open-label, multiple-dose trial of quetiapine.[20] Furthermore, single case reports have documented postural hypotension among healthy volunteers prescribed a single low dose of olanzapine,[21] as well as a patient initiated on a low dose of risperidone.[22] Proportional reporting rates of orthostatic hypotension from the U.K.'s health care regulating agency, based on the Adverse Drug Reporting On Line Tracking (ADROIT) system ranged from 1 to 5% in patients given atypical antipsychotics; the highest rates were reported for sertindole and clozapine followed by quetiapine, risperidone, olanzapine, and lastly aripiprazole.[23]

TCAs are also antagonists of the α-1 adrenergic receptors that can lead to postural hypotension.[24] In a prospective study of 44 depressed patients given imipramine, an average fall in systolic pressure of 26 mmHg was observed when patients stood up; this was consistent over the four weeks of observation. The fall in blood pressure was independent of patients' age, pre-existing heart disease, and plasma level of the drug. A retrospective study was then conducted by the authors on 148 depressed patients treated with imipramine. They found that 20% of the patients had symptoms associated with orthostatic hypotension that were severe enough to interfere with their treatment and more than 4% sustained physical injuries.[25]

On the other hand, mirtazapine, a tetracyclic antidepressant classified as a *noradrenergic and specific serotonergic antidepressant* (NaSSA), has almost no adrenergic side effects. The clinical trial development analysis showed no changes in blood pressure, and the results were comparable to the placebo group ($n = 1378$) on mirtazapine versus the placebo group ($n = 348$).[26] In addition, a five-year review of 117 mirtazapine cases of overdose reported no significant toxicity, where toxicity was assessed on the basis of clinical data such as symptoms reported on arrival at the hospital, and on the Glasgow Coma Scale, heart rate, blood pressure, respiratory rate, temperature, and oxygen saturation. The laboratory data (e.g., serum urea, creatinine, electrolytes, creatinine kinase, and liver biochemistry), as well as the electrocardiographic intervals of PR, QRSD, QT, and QTc, were evaluated. Abnormal variables were identified as those outside the normal (95% of the population) reference values.[27]

Selective serotonin reuptake inhibitors (SSRIs) have little anticholinergic and noradrenergic activity and generally do not cause orthostatic hypotension.[28] Venlafaxine is not known to have α-1 adrenergic receptor activity, and therefore was not postulated to induce orthostatic hypotension. Johnson et al.,[29] however, found that of the 38 elderly patients who were not orthostatic at baseline, 50% developed orthostatic hypotension at least twice over the course of the study ($n = 62$). Of those who started treatment, 59 completed at least two weeks of the study, suggesting that older adults may have less tolerance for this drug. Conversely, 24% of initially normotensive participants and 54% of those with preexisting hypertension experienced an increase in blood pressure.[29]

Bupropion is an antidepressant of the aminoketone class and is chemically unrelated to tricyclic, tetracyclic, or the SSRI. It has not been reported to cause orthostatic hypotension even when taken by patients who had postural hypotension caused by tricyclic antidepressants.[30] However, a study evaluating the effects of medications used in the treatment of adults with attention deficit/hyperactivity disorder (ADHD) for blood pressure found that treatment with bupropion ($n = 16$) was associated with a statistically significant increase in mean systolic blood pressure in comparison to corresponding placebo values.[31]

Recent studies have focused on the newer psychotropic medications that have been approved for use among those with mental illnesses. Iloperidone, a recently approved atypical antipsychotic, has been significantly associated with orthostatic hypotension. A pooled analysis of 1908 patients, who entered three double-blind studies and received at least one dose of study medication, found that orthostatic hypotension was observed more frequently across all treatment groups than with the placebo—19.5% of iloperidone-treated patients versus 15.3% of haloperidol-treated patients, 12% of risperidone-treated patients, and 8.3% of placebo-treated patients. Sustained orthostatic hypotension was observed in 2.4% of patients treated with iloperidone and 0.8% of haloperidol-treated patients but in no patients treated with risperidone or placebo.[32]

Paliperidone extended release (ER), a 9-hydroxy metabolite of risperidone (9-hydroxyrisperidone), has a lower propensity to block alpha receptors than risperidone; it can, nevertheless, lead to orthostatic hypotension. In a pooled analysis of three pivotal studies, incidence of orthostatic hypotension for paliperidone-ER doses of 3 (2%), 6 (1%), and 9 mg per day (2%) was similar to that of the placebo (1%); however, the rates were higher among those who received 12 (4%) and 15 mg per day (3%).[33]

The selegiline transdermal system (STS) was developed and approved by the U.S. Food and Drug Administration (FDA) in 2006. The purpose was to overcome some of the limitations of oral monoamine oxidase inhibitors, specifically food–drug interactions. A postmarketing safety analysis conducted by Pae et al.[34] showed that cardiac and vascular adverse events (AEs) were reported by approximately 0.4% of the total exposed patients (number of AEs = 127/29,141), with hypotension (number of AEs = 25/127; 19.7%) being one of the most common among cardiovascular AEs. A total of 13 possible reports were considered to be hypertensive events. A review of the narrative summary for each case in the series led the investigators to categorize 5 of the 13 cases as possibly fitting the criteria for hypertensive crisis (0.017%). Thus, STS was associated with few cardiovascular AEs and shown to have a safety profile in clinical practice that was consistent with that observed in clinical trials.

9.4 CHRONOTROPIC EFFECTS

All antipsychotic drugs have antimuscarinic effects, and cardiovascular adverse effects are mainly the result of decreased vagal tone due to antagonism of type 2 muscarinic receptors. The type 2 receptors are responsible for vagal inhibition through increased potassium conductance and inhibition of calcium (Ca^{2+}) channels. Blockade of these receptors leads to tachycardia. Tachycardia as a result of the antimuscarinic effects is most common with low-potency typical antipsychotics (e.g., chlorpromazine and thioridazine). Warner et al.[35] reported that 111 patients on various typical antipsychotic drugs had a significantly higher heart rate (HR) compared to 42 unmedicated controls (83 ± 14 vs. 72 ± 14 beats/min), although largely within the normal range. Among the atypical antipsychotic medications, clozapine has potent antimuscarinic effects and tachycardia occurs in about 25% of patients receiving it.[36]

There are also individual case reports of subjects experiencing severe tachycardia in response to even low doses of clozapine.[37] A post hoc analysis of the risperidone (RIS)/paliperidone (PALI) clinical trials database of 64 studies of 11,096 RIS/PALI-treated patients revealed that risk for tachycardia was significantly increased (i.e., 2.4 times) in the RIS/PALI group versus the placebo one.[38]

A systematic literature review (comprising 102 reports and 185 patients with case-level data) described the cardiovascular effects consequent to overdose of five common atypical antipsychotic medications and found that among patients with single-substance ingestion, tachycardia occurred in the majority of quetiapine and olanzapine ingestions ($>50\%$), and was common following risperidone ingestion (37%); however, it was not reported following ziprasidone or aripiprazole ingestion. Tachycardia usually occurred within 8 hours of ingestion and rarely lasted more than 48 hours.[39] Typically, treatment with antipsychotics results in only mild tachycardia and tolerance develops over time, with clozapine being the exception wherein persistent tachycardia may require the use of beta adrenergic receptor antagonists in symptomatic patients.

Administration of TCAs leads to increased HR because of their anticholinergic effects.[40] Vohra et al.[41] reported a significant increase in the sinus rate in a study of 32 patients treated mostly with nortriptyline, while Bigger et al.[42] found a statistically significant increase in HR in seven patients treated with imipramine. Mirtazapine acts as an antagonist at central presynaptic α-2 adrenergic inhibitory autoreceptors and heteroreceptors, which results in an increase in central noradrenergic and serotonergic activity. Mirtazapine is a moderate antagonist at muscarinic receptors, resulting in a relatively low incidence of anticholinergic side effects associated with its use. Analysis of data from the clinical trial development program showed no changes in HR associated with the use of mirtazapine ($n = 1378$ on mirtazapine vs. $n = 348$ in placebo group).[26] However, Tulen et al.[43] reported that after four weeks of treatment with mirtazapine, 10 patients with unipolar affective disorder showed significant increases in HR. They postulated that the increase in HR may be attributed to the weak anticholinergic properties of mirtazapine, resulting in cardiac vagal inhibition.

Treatment with citalopram has been reported to cause mild bradycardia. The effect of citalopram on cardiac function was assessed by a retrospective analysis of 3746 electrocardiograms (ECGs) obtained in European clinical trials. Pretreatment ECGs were compared to those from treatment periods. Citalopram reduced the heart rate within the first weeks of treatment, without further reduction in the treatment period.[44] Some patients given venlafaxine reported an increased heart rate, while sinus tachycardia was observed in 2 of 14 reports of acute venlafaxine overdose.[45]

Newer atypical medications as well as rapid dosing have been associated with the development of chronotropic effects. The paliperidone ER tablet causes a similar profile of cardiac AEs as risperidone,[46,47] with tachycardia being the most commonly reported AE.[48] A meta-analysis from 15 randomized controlled trials (RCTs) with a total of 3779 patients found a greater incidence of tachycardia (9%) in the paliperidone group with an attributable risk of 4.[49] Iloperidone, in its therapeutic doses, is associated with tachycardia,[50,51] with the frequency of tachycardia being higher with higher doses.[52] It is contraindicated for treatment of psychotic symptoms in older adults with dementia due to its strong association with stroke during treatments. However, a recent RCT reported only mild tachycardia with iloperidone that did not lead to further sequelae.[53] An open-label randomized study on 169 in-patients with schizophrenia or schizoaffective disorder, given either rapid-dose quetiapine (800 mg/day by day 4) or a conventional dose, found higher number of AEs, including tachycardia, in the rapid-dose group within a week of the study.[54]

Literature on cardiotropic effects of antidepressants included a Cochrane review and meta-analysis of 39 RCTs of amitriptyline with a total of 3509 patients; it showed that, as compared to placebo or other treatments, amitriptyline was more likely to cause tachycardia.[55] Amitriptyline intoxication was also associated with tachycardia,[56,57] and more than 60% of these individuals experienced increased heart rate. Milnacipran, a selective serotonin and norepinephrine reuptake inhibitor (SNRIs), is associated with tachycardia.[58] Among 276 levomilnacipran (1S, 2R-milnacipran, an active enantiomer of milnacipran) sustained release (SR)-treated and 277 placebo-treated patients with major depressive disorder (MDD), levomilnacipran SR caused twice the number of AEs compared to a placebo, which included tachycardia.[59] Although more patients on levomilnacipran SR (9.4%) discontinued due to AEs, it was reportedly well tolerated throughout the 10-week trial.

9.5 ABNORMALITIES OF CARDIAC REPOLARIZATION AND ARRHYTHMIAS

The action potentials of the heart are caused by the interaction of the inward depolarizing and outward repolarizing ion currents. These are reflected as the QT interval in an electrocardiogram. The QT interval represents the time from the beginning of the Q wave to the end of the T wave, which manifests the ventricular depolarization and repolarization.[23] QTc is the QT interval corrected for heart rate (as a faster heart rate shortens the QT interval) and a rate-corrected QT interval is necessary to accurately assess the true length of the QT interval.

Ventricular depolarization depends on the fast sodium (Na^+) inward current. Drugs that inhibit the entry of Na^+ into cardiomyocytes result in disturbances of ventricular depolarization, the inhibition of which can result in reentry arrhythmias. The major ion channels involved in the repolarization are the potassium channels (K^+), which counterbalance the influx of Na^+ and Ca^{2+}. Drug-induced prolongation of the QT interval is mainly caused by the blockade of the delayed rectifier potassium channel Kv7.1. This channel, with its large central cavity containing many aromatic amino acids, allows binding of numerous drugs, which blocks the outflow of potassium ions resulting in delayed depolarization and prolongation of the QT interval. Studies have shown that patients with prolonged QT intervals are at greater risk of cardiac arrhythmias particularly torsades de pointes (TdP). The arrhythmia is usually self-limiting but may degenerate into ventricular fibrillation causing sudden cardiac death (SCD).[17] However, a direct cause−effect relationship between antipsychotic-induced QT prolongation and arrhythmogenic potential has not been established and a long QT interval does not always predict arrhythmias.

Both typical and atypical antipsychotics prolong QT interval. In a study of 495 patients,[60] there was a higher prevalence of QTc prolongation in patients receiving thioridazine (23%), droperidol (16%), and haloperidol (14%). Cross-sectional data and postmarketing analyses indicated that an intravenous form of haloperidol may carry a higher risk of QTc prolongation and TdP than the oral form.[61,62] Ozcki et al.[61] determined the potential for antipsychotic drugs to prolong the QTc interval by analyzing the electrocardiograms of 1017 patients suffering from schizophrenia. Using logistic regression analysis, the adjusted relative risk (RR) of intravenous haloperidol was 1.29 times (sample comprised of 47 patients who received IV haloperidol) for QTc prolongation. The FDA currently recommends cardiac monitoring for all patients receiving IV haloperidol. Pimozide has also been implicated in QT prolongation and TdP. Forty reports (16 deaths) of serious cardiac reactions were reported by the Committee on Safety of Medicines.[63] Prolongation of the QT interval in patients treated with atypical antipsychotics, such as olanzapine, risperidone, ziprasidone, and quetiapine, have been reported in various studies.[64,65]

A study evaluating the dose-dependent effect of olanzapine on QT interval among 26 patients with schizophrenia observed significant increases in QTc, interval after switching to "increased dose" olanzapine.[66] A recent study systematically analyzed all available case reports of risperidone, QTc interval prolongation, and/or TdP and identified 15 cases of QTc prolongation.[67] Data from 4306 adults who received ziprasidone in placebo- and active-comparator in Pfizer-sponsored phase II-IV RCTs was comprehensively analyzed to provide data on QTc changes associated with ziprasidone. Results revealed that while one subject had a QTc $\geq$480 milliseconds (ms), 33 (0.8%) had a QTc $\geq$450 ms. QTc prolongation of $\geq$30 ms was observed in 389 subjects (9.0%), $\geq$60 ms in 30 (0.7%), and $\geq$75 ms in 12 (0.3%). Mean change in QTc from baseline to end of study was 3.6 ms ($\pm$20.8) in the ziprasidone group; the corresponding QTc change in the pooled placebo group was -0.3 ms ($\pm$20.6). The final concentration-response analysis model, comprising 2966 data points from 1040 subjects, estimated an increase in QTc of 6 ms for each 100 ng/mL increase in ziprasidone concentration.[68]

Reilly et al.[60] suggested that even at therapeutic doses, TCAs prolong QTc interval independent of any cardiac problem. Thirteen out of 97 cases that prescribed TCA were classified to have a lengthened QTc interval. Overdoses of TCAs have also been associated with QT prolongation and ventricular arrhythmias.[69] Curtis et al.,[70] using the administrative claims database, conducted a retrospective cohort study with 4,825,345 subjects 18 years or older and reported that antidepressants were involved in nearly 50% of patients with prolonged QT interval. While SSRIs have a generally benign cardiovascular profile, case reports have linked most SSRIs to QT prolongation. Recently, the FDA has issued warnings implicating citalopram in drug-induced QTc interval prolongation and torsades de pointes when used in doses >40 mg per day.[71] A review of 235 cases of venlafaxine poisoning showed that QTc prolongation was highly prevalent in both men and women, irrespective of whether venlafaxine was coingested or ingested alone. Nonspecific electrocardiograph abnormalities including left and right bundle branch block patterns, T-wave inversion, and ST depression were also observed in nine cases.[72]

Lithium, which is used as a mood-stabilizer, is known to cause various cardiac disorders, including conduction abnormalities. The major cardiovascular side effects of lithium include unmasking of Brugada syndrome, sinus node dysfunction, atrioventricular (AV) conduction abnormalities, prolonged QT interval, and various arrhythmias.[73–77] The mechanisms underlying these conduction defects are not fully understood. Postulated mechanisms include evidence from animal studies that show that lithium decreases both the spontaneous rate of depolarization of the sinus node and the conduction velocity in the AV and intraventricular conduction system,[78] interference with Ca^{2+}-ion influx in pacing cells of the sinus node,[74] and the blocking of cardiac Na^{+} channels.[73] Although most side effects occur at the supratherapeutic levels, there are several case reports of conduction abnormalities induced by lithium in patients at levels within the therapeutic range.

Brugada syndrome is a familial condition that displays an autosomal dominant mode of transmission, characterized by coved ST elevation and J-point elevation of at least 2 mm in at least two of the right precordial ECG leads (V1–V3) that are unrelated to ischemia, electrolyte disturbances, or obvious structural heart disease.[79] It is characterized by ventricular arrhythmias, atrial arrhythmias, syncope, and sudden death. Brugada syndrome is usually caused by mutations in the cardiac Na^+ channel gene *SCN5A*, the gene encoding for the α-subunit of the Na^+ channel.[80] Sporadic cases of the Brugada ECG pattern without a family history of the condition have been increasingly reported following exposure to drugs with Na^+ channel blocking effects.[81] TCAs, fluoxetine, lithium, and trifluoperazine have all been documented in case reports/case series to induce Brugada ECG pattern.[73,82,83] Data shows that in asymptomatic patients without a family history of sudden death, the drug-induced Brugada ECG pattern is likely to be benign once the offending agent is discontinued.[84]

There are many risk factors that prolong QT interval and increase the risk of TdP. These include increasing age, female gender, preexisting cardiac disease, liver disease that impairs the metabolism for drugs, drug–drug interactions, electrolyte abnormalities (e.g., hypokalemia), and congenital long QT syndrome. In elderly patients, physiological decreases in hepatic and renal function, biotransformation, and/or elimination of drugs are affected and, therefore, adaptation of drug dosage may be needed. Estradiol can potentiate the effect of drugs that prolong QT interval and thus women have a higher risk of developing prolonged QT interval. Sex hormones have also been suggested to alter Ca^{2+} and/or K^+ currents, which may account for the gender differences in cardiac repolarization.[85] Preexisting bradycardia can prolong QT interval by delaying cellular repolarization in patients treated with antipsychotics.[86] Left ventricular hypertrophy and left ventricular dysfunction are also risk factors. Electrolyte imbalances, such as hypokalemia,[87] also increase the risk of prolonged QT intervals (Table 9.1).

The extent to which genetic factors contribute to psychotropic drug induced cardiotoxicity is not well defined. Existing data have reported that polymorphism in an ion channel gene increases the propensity of drug-induced arrhythmias. Abbott et al.[88] first identified a polymorphism of the *KCNE2* gene (subunit of the potassium channel, I_{Kr}), which was associated with an increased risk of quinidine-related TdP. The K897T polymorphism of the *KCNH2* gene is another common polymorphism that has been associated with the development of long QT syndrome and TdP.[89] Genetic polymorphisms encoding for enzymes involved in the metabolism of drugs can alter pharmacokinetics and cause fluctuations in plasma levels, leading to arrhythmia. One such enzyme is CYP2D6, which is involved in the metabolism of drugs that prolong QT interval; multiple polymorphisms of CYP2D6 reduce its function, resulting in increased vulnerability of individuals to psychotropic drug-induced cardiac toxicity.[90]

TABLE 9.1 Nonpsychotropic Drugs and CYP 450 Inhibitors that May Prolong the QT Interval*

Therapeutic Class	Drugs Involved
Class IA antiarrhythmics	Disopyramide Hydroquinidine Procainamide Quinidine
Class III antiarrhythmics	Amiodarone Ibutilide Sotalol
Class 1C antiarrhythmics	Flecainide
Class IV antiarrhythmics	Bepridil
Antibiotics (macrolis)	Erythromycin
Synthetic antimalarial agents	Halofantrine
Antileishmania agents and trypanocides	Diisethionate Pentamidine
H_1 antihistaminics	Loratadine Hydroxyzine
Antimicrobial and antimalarial drugs	Amantadine Clarithromycin Clindamycin Chloroquine Erythromycin Ketoconazole Pentamidine Quinine
Immunosuppressant	Tacrolimus
Antidiuretic hormone	Vasopressin
CYP 450 inhibitors	
CYP 450-1A2	Cimetidine Ciprofloxacin Fluvoxamine Norfloxacin Verapamil
CYP 450-3A4	Amiodarone Cimetidine Ciprofloxacin Clarithromycin Nefazodone Ritonavir Roxithromycin Erythromycin

	Fluvoxamine Fluoxetine Indinavir Isoniazid Itraconazole Ketoconazole Metronidazole
CYP 450-2D6	Amiodarone Cimetidine Chlorpheniramine Clomipramine Citalopram Fluoxetine Methadone Paroxetine Quinidine Ritonavir Sertraline

**This list is not exhaustive.*

Iloperidone and its metabolite P88 have a significant affinity for the human ether-a-go-go related gene (hERG) channel and thus the potential to prolong the QTc interval.[91] In short-term efficacy trials, iloperidone has demonstrated QTc interval duration changes from baseline ranging from 2.9 ms with iloperidone, 4 to 8 mg per day, to 11.4 ms with iloperidone of 20 to 24 mg per day.[51,92] A recent article reported on a "thorough QTc (TQT) study" with prolongation of the QTc serving as a surrogate end point for risk of TdP. Adults with schizophrenia or schizoaffective disorder and normal electrocardiograms at baseline ($n = 188$) were randomized to iloperidone doses of 8 mg twice daily (BID), 12 mg BID, 24 mg once daily (QD); quetiapine 375 mg BID; or ziprasidone 80 mg BID. Iloperidone BID produced mean changes in QTc Fridericia correction (QTcF) interval similar to those produced by ziprasidone (9.6 ms) and higher than those produced by quetiapine (1.3 ms). Iloperidone doses of 24 mg QD produced a mean QTcF change of 15.4 ms. However, the authors found that coadministration of metabolic inhibitors with iloperidone in subsequent periods of the treatment (paroxetine or paroxetine and ketoconazole) resulted in greater increases in the QTc interval. Increased QTc was observed in individuals with specific cytochrome P450 2D6 polymorphisms.[53]

A meta-analysis examining the efficacy and side effects of 15 antipsychotic drugs among schizophrenia patients reported that lurasidone, aripiprazole, paliperidone, and asenapine were not associated with significant QTc prolongation compared with placebo. The standardized mean differences of the other drugs compared with placebo ranged from 0·11 in haloperidol to 0·90 in sertindole. Clozapine and zotepine could not be included in the analysis.[93]

Studies have also examined the role of genetic variations in mediating antipsychotic-induced QT prolongation. Aberg et al.[94] examined data from 738 schizophrenia patients who participated in the Clinical Antipsychotic Trial of Intervention Effectiveness (CATIE) study. The candidate gene study suggested the involvement of the nitric oxide synthase 1 adaptor protein gene (NOS1AP) and the nucleotide binding protein-like (NUBPL). The top genome-wide association studies (GWAS) hit achieving genome-wide significance, located in SLC22A23, mediated the effects of quetiapine on prolongation. The authors concluded that genes mediating antipsychotic-induced QT prolongation partially overlap with the genes affecting normal QT-interval variation.

9.6 MYOCARDITIS AND CARDIOMYOPATHY

Antipsychotic medications have been associated with myocarditis. Haloperidol, quetiapine, and especially clozapine have been linked with myocarditis;[95–97] Reinders et al.[97] reported nine cases of myocarditis/cardiomyopathy among 94 patients initiated on clozapine. A study by Hill and Harrison-Woolrych,[98] who analyzed the Intensive Medicines Monitoring Programme's (IMMP) clozapine database, identified and reported 25 cases of myocarditis caused by clozapine and found that the majority of those developing myocarditis with clozapine did so within the first month of treatment and even very low doses of the drug were cardiotoxic. The cause for the myocarditis is unknown although it is hypothesized as being a result of type 1 IgE hypersensitivity.[17] Cardiomyopathy, which increases the risk of arrhythmias SCD, has also been reported with quetiapine; amisulpiride; and, most frequently, with clozapine.[97,99] The cases of cardiomyopathy were rare: Reinders et al.[97] reported one case out of 94 patients initiated on clozapine, Coffey and Williams[100] reported two cases with quetiapine, and Brakoulias et al.[101] reported a single case of cardiomyopathy with amisulpiride. The onset of cardiomyopathy has been found to be slower than that of myocarditis following commencement of clozapine therapy.

9.7 PSYCHOTROPIC DRUGS AND SUDDEN CARDIAC DEATH

Three main mechanisms have been suggested as causing sudden deaths of cardiac origin among patients on psychotropic drugs, and these include electrophysiological disturbances, myocardial ischemia, and myocarditis. Mehtonen et al.[102] reported 49 cases of sudden death; of these, 46 were treated with a therapeutic dose of a phenothiazine including thioridazine alone in 15 patients and combined with a tricyclic antidepressant in 5 patients. The authors concluded that the cause of death was TdP resulting from prolonged QT interval. Reilly et al.[103] conducted a case-control study of psychiatric in-patients dying suddenly in five hospitals in the northeast of England and surviving controls matched for age, gender, and mental disorder. Sixty-nine case-control clusters were identified. Probable sudden unexplained death was significantly associated with hypertension, ischemic heart disease, and current treatment with thioridazine. The authors concluded that the most likely explanation for the association was that thioridazine causes arrhythmias and SCD.

A more recent study by Ray et al.[104] analyzed sudden cardiac deaths among current users of antipsychotic drugs in a large retrospective cohort study of Medicaid enrollees in Tennessee. The primary analysis included 44,218 and 46,089 baseline users of single typical and atypical drugs, respectively, and 186,600 matched nonusers of antipsychotic drugs. During the 1,042,159 person-years of cohort follow-up, there were 1870 SCDs, or 17.9 per 10,000 person-years. Current users of typical antipsychotic drugs had an adjusted rate of SCD that was twice that for nonusers; a similar increased risk was also observed for current users of atypical antipsychotic drugs. Sertindole was shown to have a clear association with QTc prolongation in a prospective clinical study that investigated the effects of atypical neuroleptics on autonomic neurocardiac function—sertindole 12 mg per day ($n = 13$)[105] and the occurrence of arrhythmias and sudden death led to its suspension from the European markets in 1998.[106] The sertindole cohort study—that is, a multinational randomized, open-label, parallel-group, with a blind classification of outcomes, of 9858 patients with schizophrenia—showed that while sertindole did not increase all-cause mortality, it was associated with a greater than two-fold increased risk of SCD compared to risperidone.[107]

Another mechanism of sudden death associated with the use of neuroleptics is the development of myocarditis and cardiomyopathy. The Kilian et al.[108] study included 8000 patients from 1993 to 1999 and described structural lesions in 23 patients; of these 5 had cardiomyopathy of which 1 died after 36 months of treatment. The 18 remaining patients had myocarditis and 5 died (3 of sudden death and 2 of cardiac insufficiency) within the first 3 weeks of treatment.[108] La Grenade and colleagues identified 28 instances of myocarditis, including 18 deaths and 48 instances of cardiomyopathy, with 10 deaths reported to the U.S. FDA between September 1989 and December 1999.[109]

Thorogood et al.[110] observed an unexpected 17-fold increase in risk of total myocardial infarction associated with current use of psychotropic drugs while examining data on deaths among women aged 16 to 39 in England and Wales between January 1986 and December 1988, where the underlying cause of death was myocardial infarction. A prospective case-control study, the Finnish Genetic Study of Arrhythmic Events, compared the risk profiles of the victims of SCD with the survivors of an acute coronary event.[111] A total of 1814 SCD cases whose deaths were determined to be due to a coronary event, and where data on medications was available, were included in the study; the control group consisted of survivors of an acute MI ($n = 1256$). In the multivariate model, the use of antipsychotics emerged as a significant risk factor for SCD after adjustment for demographic parameters. The use of TCAs was also a significant predictor of SCD. The combined use of phenothiazines and antidepressants was significantly associated with an increased risk of SCD.

9.8 METABOLIC ABNORMALITIES

A major concern for patients on antipsychotics is the development of metabolic syndrome (MetS), encompassing abnormalities in weight, blood pressure, glucose metabolism, and lipids metabolism.[112] MetS is a constellation of different conditions, including abdominal obesity, insulin resistance, dyslipidemia, and elevated blood pressure, all of which have been recognized as independent risk factors for cardiovascular disease. Among typical antipsychotics, chlorpromazine and thioridazine are associated with greater risks of weight gain; most atypical antipsychotics are associated with weight gain.[113] A cross-sectional, multicenter study reported a weight gain >5 kg in 40.8% of patients after being treated with olanzapine ($n = 228$), 24.7% of patients on risperidone ($n = 232$), and 18.2% of patients on haloperidol ($n = 130$). However, none of the patients on quetiapine ($n = 43$) attained a weight gain of >5 kg (0.0%).[114] Because it has been hypothesized that the histamine H_1 receptors (H_1-Rs) regulate eating behavior (increased appetite), antipsychotics (e.g., aripiprazole, ziprasidone, and haloperidol), which have a low affinity for H_1-Rs, tend to lead to a minimal weight increment; while clozapine, olanzapine, thioridazine, and chlorpromazine, which have a high affinity for H_1-Rs, tend to cause clinically significant weight increments.[115]

Olanzapine and clozapine are also associated with higher lipid levels as compared to risperidone and quetiapine, while aripiprazole and ziprasidone produced minimal or no effects on lipid levels.[116] These changes are caused by both adiposity-dependent and possibly adiposity-independent changes in lipid metabolism.[117] Patients treated with atypical antipsychotics were found to be at a higher risk of developing type 2 diabetes than those on typical antipsychotics.[118] Among atypical antipsychotics, olanzapine and clozapine, followed by quetiapine and risperidone to a smaller degree, are clinically associated with increased risk of diabetes.[119] Current literature on antipsychotics and their effect on blood pressure are inconclusive. It has been reported that antipsychotics can either increase hypertension through weight gain[120] or have minimal to no influence on the development of hypertension.[121]

Use of antidepressants can also lead to weight gain. Amitriptyline has been reported to cause significantly higher weight gain as opposed to imipramine and desipramine.[122] The mechanism behind weight gain with TCAs is postulated to be a result of increased craving for carbohydrates.[123] Such a weight increase has been associated with risk of developing type 2 diabetes.[124] Less emphasis has been placed on the association between antidepressants and serum lipids, but TCAs have consistently been reported to significantly increase triglycerides and low-density lipoprotein cholesterol.[125] However, bupropion, venlafaxine, and duloxetine do not seem to consistently affect lipid levels. Long-term moderate to high doses (>2 years) of antidepressants can increase risk of development of diabetes, and this was found in both TCAs and SSRIs.[126]

Recent studies have focused on the differential risk of metabolic syndrome (and the consequent risk of cardiovascular morbidity and mortality) of antipsychotics and antidepressants.[127] Clozapine and olanzapine administration involves a high risk of weight gain, as well as lipid and glucose dysregulation. There is a medium risk of weight gain for risperidone, paliperidone, quetiapine, and iloperidone and aripiprazole, asenapine, ziprasidone have low risk.[128] The second-generation antidepressants have a markedly decreased risk for metabolic syndrome. Most are, in effect, weight-neutral with the exception of paroxetine and mirtazepine.[129] In addition, other studies have indicated that there is genetic vulnerability for antipsychotic-induced metabolic side effects. Genome-wide association studies have identified putative genomic regions that have certain susceptibility genes.[130] Polymorphisms associated with the 5-hydroxytryptamine (serotonin) receptor 1A (HTR2C) and the leptin gene have also been consistently associated with antipsychotic-induced weight gain.[131] These findings need to be replicated with larger samples and with different populations, however; such studies and other biomarker-identifying studies are important to pave the way for genetically informed, stratified medicine to manage the side effects of antipsychotics (Table 9.2).

9.9 STANDARD GUIDANCE

There are a number of reviews and meta-analyses on the cardiotropic effects of psychotropic medications, as highlighted in this chapter. However, few consensus statements or expert panel recommendations are available.

9.9.1 For the Use of Psychotropic Drugs and Chronic Cardiovascular Deterioration

A review of 47 adverse drug reaction reports found significant odds ratios for antipsychotics (i.e., clozapine, olanzapine), lithium, and antidepressants (i.e., clomipramine, amitriptyline, fluvoxamine)[132] associated with dilated cardiomyopathy. A review of 81 studies that reported clozapine-related adverse effects recommends discontinuation of clozapine without rechallenge for cardiomyopathy.[133] Incidence rates estimated from claims data of 2251 adult schizophrenic patients prescribed atypical and conventional antipsychotic medications found significantly lower odds of developing cardiomyopathy for aripiprazole (OR $= -3.45$; $P = 0.02$).[134] Among conventional antipsychotics, haloperidol, chlorpromazine, and fluphenazine were found to be associated with myocarditis and cardiomyopathy after review of more than two million adverse events reported in the WHO database.[135] The same review found fluphenazine and risperidone to be significantly associated with cardiomyopathy but not myocarditis.

Although myocardial infarction is linked to TCAs,[136,137] its association with the use of selective NRIs, SSRIs, or mixed serotonin/NRIs (SNRIs) is debatable.

TABLE 9.2 Cardiac Effect of Psychotropic Drugs

Cardiac Effects	Drug Class	Drug Name	Effect/Association
Hypotension	Antipsychotic drugs	Chlorpromazine Thioridazine Clozapine Quetiapine Olanzapine Risperidone Sertindole Aripiprazole	Hypotension Hypotension Orthostatic hypotension Orthostatic hypotension Postural hypotension Postural hypotension Orthostatic hypotension Orthostatic hypotension
	Tricyclic antidepressants	Imipramine Nortriptyline	Orthostatic hypotension Orthostatic hypotension
	SNRIs	Venlafaxine	Postural hypotension (among elderly patients)
Chronotropic effect	Antipsychotic drugs	Chlorpromazine Thioridazine Clozapine Quetiapine Olanzapine Risperidone Risperidone/ Paliperidone	Tachycardia Tachycardia Severe to persistent tachycardia, antimuscarinic Tachycardia Tachycardia Tachycardia Tachycardia
	Tricyclic antidepressants	Nortriptyline Imipramine	Increase in sinus rate Increase in heart rate
	SSRIs	Fluoxetine Fluvoxamine Paroxetine Citalopram	Mild bradycardia Mild bradycardia Mild bradycardia Mild bradycardia
	Tetracyclic antidepressants	Mirtazapine	Changes in heart rate (inconclusive)
	SNRIs	Venlafaxine	Sinus tachycardia, increase in heart rate
Abnormalities of cardiac repolarization and arrhythmias	Antipsychotic drugs	Droperidol Haloperidol Thioridazine Ziprasidone Quetiapine Risperidone Olanzapine Aripiprazole Pimozide Sertindole	QT prolongation QT prolongation QT prolongation QT prolongation QT prolongation QT prolongation QT prolongation QT prolongation QT prolongation, torsades de pointes QT prolongation, arrhythmias, sudden cardiac death

	Tricyclic antidepressants	–	QT prolongation
		–	QT prolongation, ventricular arrhythmias
	Antidepressants	–	QT prolongation
Cardiotoxicity	Antipsychotic drugs	–	Myocarditis
		Haloperidol	Myocarditis
		Quetiapine	Myocarditis
		Clozapine	Myocarditis
		Quetiapine	Cardiomyopathy
		Amisulpiride	Cardiomyopathy
Metabolic abnormalities	Antipsychotic drugs	–	Development of metabolic syndrome, encompassing abnormalities in weight, glucose metabolism, and lipid metabolism
			Increased hypertension through weight gain (inconclusive)
	Atypical antipsychotics	–	Type 2 diabetes compared to first-generation antipsychotics
		Chlorpromazine	Significant weight gain
		Thioridazine	Significant weight gain
		Olanzapine	Significant weight gain, hyperlipidemia, diabetes
		Clozapine	Significant weight gain, hyperlipidemia, diabetes
		Aripiprazole	Minimal weight gain
		Ziprasidone	Minimal weight gain
		Haloperidol	Minimal weight gain
		Risperidone	Diabetes
		Quetiapine	Diabetes
	Antidepressants	SSRIs and TCAs	Diabetes
		Amitriptyline	Weight gain
	Tricyclic antidepressants	–	Weight gain due to increased craving for carbohydrates
		–	Increase in triglycerides and low-density lipoprotein cholesterol

A study that reviewed the FDA's Adverse Event Reporting System and the WHO database[138] found an inverse relationship between cardiomyopathy and antidepressants that elevate norepinephrine levels and linked cardiomyopathy-related events with amitriptyline, imipramine, clomipramine, and venlafaxine use. Catecholamine-induced reversible cardiomyopathy was reported for milnacipran, which is a dual and equipotent inhibitor of norepinephrine and serotonin uptake.[139] Duloxetine[140] and venlafaxine[141] have been shown to be associated with takotsubo cardiomyopathy. However, to date no consensus statement has been issued on chronic cardiovascular deterioration.

9.9.2 For the Use of Psychotropic Drugs and Metabolic Their Side Effects

A number of consensus guidelines provide recommendations for psychiatrists when initiating and maintaining therapy with psychotropic medications.[142–144] Expert panels are in agreement that psychiatric disorders, such as bipolar disorder and schizophrenia, appear to be associated with a higher prevalence of type 2 diabetes. Experts generally agree that the relationship between schizophrenia and diabetes is not fully understood. An association between the two conditions has been recognized since the preantipsychotic era and studies have shown an association between the two disorders even for drug-naïve subjects.[145,146] The amount of literature on the association of second-generation antipsychotics (SGA) and metabolic disorders is much larger than for first-generation antipsychotics; and reports comprise abnormal glucose regulation, exacerbation of existing type 1 and type 2 diabetes, new-onset pseudo-type 1 or type 2 diabetes, diabetic ketoacidosis, coma, and death. Published evidence to date shows no convincing difference between agents, and impaired glucose metabolism is now considered to be a class effect.[143]

Expert consensus suggests a pragmatic approach wherein patients and family members are educated about the metabolic risk, and baseline screening and regular monitoring are carried out on patients on psychotropic medications. It is recommended that baseline screening measures be obtained before or as soon as clinically feasible after the initiation of any antipsychotic medication. These should include a personal and a family history of obesity, diabetes, dyslipidemia, hypertension, or cardiovascular disease; weight and height; waist circumference; fasting plasma glucose; and lipid profiles. Fasting plasma glucose, lipid levels, and blood pressure should also be assessed three months after initiation of antipsychotic medications. Thereafter, blood pressure and plasma glucose values should be obtained annually or more frequently for those who have a higher baseline risk for the development of diabetes or hypertension. It is also recommended that mental health teams take on some responsibility for managing general health issues in their patients (e.g., providing education about healthy living, ensuring that screening for diabetes is done, and that other services are involved when necessary).[143,147]

While choosing between second-generation antipsychotics that have equal efficacy, an SGA that has a low propensity for weight gain and diabetes, such as amisulpride, aripiprazole, or ziprasidone, should be chosen by psychiatrists for those patients who are at risk. If an SGA with a high risk for metabolic effects is indicated, a follow up should be done early on, and the treating general practitioner (GP)/internist should be informed.[144]

9.9.3 For the Use of Psychotropic Drugs and Cardiac Sudden Death

Following the study by Ray et al.,[104] who first suggested the mechanism of SCD among antipsychotic users, Schneeweiss and Avorn[148] proposed in their editorial that an ECG be obtained before and shortly after initiation of treatment with an antipsychotic drug. They suggested that, when QT interval prolongation is detected, the antipsychotic dose should be reduced or the drug discontinued, other risk factors for SCD should be addressed, and follow-up ECGs should be obtained.

However, the American Psychiatric Association[149] opined that instituting a policy of routine serial measurement of the QTc interval in all patients after initiating treatment with antipsychotic medications may be premature and, although prolongation of the QTc is the best available clinical surrogate for the development of TdP, it is an imperfect biomarker.[150,151] This is because it generally has low specificity for predicting arrhythmias, and for some drugs a dissociation exists between QTc prolongation and TdP risk.[151]

The APA guidance[149] states:

Given the methodological limitations of the study by Ray and colleagues, and the lack of data regarding the utility and cost-effectiveness of serial QTc measurement in antipsychotic-treated patients, clinicians should continue to observe extant practice guidelines for the work-up and management of psychotic patients. With regard to cardiac safety, these include obtaining a medical and medication history, a thorough physical exam, vital signs, and routine laboratory tests.[152] Thioridazine, mesoridazine, and pimozide should not be prescribed for patients with cardiac risk factors of known heart disease, a personal history of syncope, a family history of sudden death under age 40, or prolonged QTc syndrome.[153] If these agents are to be prescribed, serum potassium and an ECG should be checked before drug initiation; serum potassium and an ECG should also be checked in the presence of the ... cardiac risk factors prior to treatment with ziprasidone.[154] An ECG should be checked again following a significant change in dose of thioridazine, mesoridazine, pimozide, or, in the presence of cardiac risk factors, ziprasidone, or following the addition of another QTc-prolonging medication,[154] or if a patient presents with symptoms associated with a prolonged QTc interval (e.g., syncope).[153] An absolute QTc interval of >500 msec or an increase of 60 msec from baseline may be associated with an increased risk of TdP[155] and should prompt the reduction or discontinuation of the offending agent. When prescribing antipsychotic medication, clinicians are encouraged to use the lowest dose effective for any given patient in order to minimize dose-dependent adverse effect risks.

Similarly, the Royal College of Psychiatrists[156] in its consensus statement says this:

While a possible link has been postulated between antipsychotic drugs and ventricular tachycardia and sudden death, no consensus has been achieved on the frequency of these events, the contribution of high dosage, or even whether a true causal association exists. To reduce the risk of arrhythmia, all patients should be assessed (including electrocardiography) for cardiovascular disease prior to the institution of antipsychotic drug therapy. Periodic monitoring of the ECG and electrolytes during therapy is advocated when high-dose antipsychotic drug treatment is used.

9.10 CLINICAL IMPLICATIONS

Clinicians must obtain a detailed clinical history to establish the risk factors for cardiovascular disease as well as a family history of CVD or SCD in all patients with SMI. Although routine ECG monitoring has been recommended, it may not be cost effective to perform this for all patients. However, for patients with risk factors for QTc prolongation, QTc should be assessed at baseline and intermittently postinitiation when prescribing any antipsychotic drug. Consultation with a cardiologist must be considered for high-risk cases before and when needed during treatment. The Mount Sinai Conference[153] recommended that mental health care providers should *not* prescribe thioridazine, mesoridazine, or pimozide for patients with known heart disease, a personal history of syncope, a family history of sudden death at an early age (under 40, especially if both parents had sudden death), or congenital long QT syndrome. Antipsychotics should be used in monotherapy at recommended dosages and not in combination with other drugs that cause QT prolongation. Where clinically indicated, relevant investigations, especially ECGs and electrolytes, should be monitored.

One approach that has been suggested is to use psychotropic drugs that have a "relatively safe cardiac profile" for patients who are at high risk of QT prolongation (e.g., using sertraline for patients with cardiac disease).[157] Sertraline has few drug–drug interactions, has not been consistently linked to QTc prolongation, and has been the most studied drug for cardiac patients,[158] thus it has a "relatively safe cardiac profile" and should be considered for high-risk patients. However, metabolic side effects, such as increased body mass index (BMI), diabetes, and dyslipidemias, also need to be considered and monitored carefully as they all contribute to cardiovascular morbidity and mortality.

9.11 CONCLUSION

Serious mental illnesses are chronic conditions that impair the functioning and quality of life. Psychotropic drugs are effective in treatment of these conditions and are prescribed for considerable periods of time in the management of patients. An ideal psychotropic drug would be one that is devoid of side effects in the therapeutic dosage. However, there is no such existing drug and all drugs are associated with some adverse effects, including cardiovascular side effects such as hypotension, tachycardia, and arrhythmias.

Clinicians must keep the cardiac profile of psychotropic medications in mind and be especially mindful with vulnerable groups such as the elderly and those with preexisting cardiac disorders. However, it is also important to exercise caution in seemingly healthy patients receiving psychotropic medications because SCD can occur in this patient population as well.

All clinicians—not just psychiatrists—should be aware of the increased risk of cardiac events in patients with SMI as they also have an increased risk of comorbid physical illness and are likely to present in a primary or medical specialist health-care setting and where care must be exercised while prescribing nonpsychotropic drugs.

Clinicians must carefully consider the risk–benefit ratio of prescribing various psychotropic drugs and monitoring of cardiovascular health markers, such as blood pressure, heart rate, ECG, and metabolic parameters (e.g., weight, BMI, and blood lipids), regularly, which is vital for these patients. However, physicians should continue to be able to prescribe antipsychotic drugs when there is clear evidence of a benefit for conditions such as schizophrenia and bipolar disorders. In patients for whom the drug is truly indicated, a small risk of rare but fatal side effects may be acceptable until new medications with a safer cardiac risk profile are developed.

Key Learning Points

- There is an excess of mortality among people with serious mental illnesses and the main contributor is cardiovascular disease.
- The use of psychotropic medications, in particular the antipsychotics (both first- and second-generation) and tricyclic antidepressants, has the concomitant risk of cardiovascular events.
- Clinicians must be cognizant of the risk, which is not only related to the type and dose of the psychotropic medication but also determined by patient characteristics such as age, coexisting medical conditions, and the use of other medications that may lead to a drug–drug interaction. This calls for the judicious and individualized use of medications as well as close clinical monitoring including regular ECGs.

REFERENCES

1. Fuster V, Voute J. MDGs: chronic diseases are not on the agenda. *Lancet* 2005;**366**: 1512–4.
2. Lozano R, Naghavi M, Foreman K, et al. Global and regional mortality from 235 causes of death for 20 age groups in 1990 and 2010: a systematic analysis for the Global Burden of Disease Study 2010. *Lancet* 2012;**380**(9859):2095–128.
3. Miniño AM. *Death in the United States, 2009*. NCHS data brief, no 64. Hyattsville, MD: National Center for Health Statistics; 2011.
4. Colton CW, Manderscheid RW. Congruencies in increased mortality rates, years of potential life lost, and causes of death among public mental health clients in eight states. *Prev Chronic Dis* 2006;**3**:A42.

5. Saha S, Chant D, McGrath J. A systematic review of mortality in schizophrenia. *Arch Gen Psychiatry* 2007;**64**:1123–31.
6. Laursen TM, Munk-Olsen T, Agerbo E. Somatic hospital contacts, invasive cardiac procedures, and mortality from heart disease in patients with severe mental disorder. *Arch Gen Psychiatry* 2009;**66**:713–20.
7. Bouza C, López-Cuadrado T, Amate JM. Hospital admissions due to physical disease in people with schizophrenia: a national population-based study. *Gen Hosp Psychiatry* 2010;**32**: 156–63.
8. Reinert RE, Hermann CG. Unexplained deaths during chlorpromazine therapy. *J Nerv Ment Dis* 1960;**131**:435–42.
9. Committee on Safety of Medicines. Drug-induced prolongation of the QT interval. *Curr Probl Pharmacovig* 1996;**22**:2.
10. Royal College of Psychiatrists. *The association between anti-psychotic drugs and sudden death.* Report of the Working Group of the Royal College of Psychiatrists' Psychopharmacology Subgroup (Council Report CR57). London: Royal College of Psychiatrists; 1997.
11. Foley DL, Morley KI. Systematic review of early cardiometabolic outcomes of the first treated episode of psychosis. *Arch Gen Psychiatry* 2011;**68**(6):609–16.
12. Pacher P, Ungvari Z, Nanasi PP, et al. Speculations on difference between tricyclic and selective serotonin reuptake inhibitor antidepressants on their cardiac effects. Is there any? *Curr Med Chem* 1999;**6**:469–80.
13. Glassman AH, Bigger Jr JT. Cardiovascular effects of therapeutic doses of tricyclic antidepressants. A review. *Arch Gen Psychiatry* 1981;**38**:815–20.
14. Sala M, Coppa F, Cappucciati C, et al. Antidepressants: their effects on cardiac channels, QT prolongation and torsade de pointes. *Curr Opin Investig Drugs* 2006;**7**(3):256–63.
15. Sugiyama A, Satoh Y, Hashimoto K. In vivo canine model comparison of cardiohemodynamic and electrophysiological effects of a new antipsychotic drug aripiprazole (OPC-14597) to haloperidol. *Toxicol Appl Pharmacol* 2001;**173**(2):120–8.
16. Collaborative Working Group on Clinical Trials Evaluations. Adverse effects of the atypical antipsychotics. *J Clin Psychiatry* 1998;**59**(Suppl. 12):17–22.
17. Mackin P. Cardiac side effects of psychiatric drugs. *Hum Psychopharmacol* 2008;**23** (Suppl. 1):3–14.
18. Silver H, Kogan H, Zlotogorski D. Postural hypotension in chronically medicated schizophrenics. *J Clin Psychiatry* 1990;**51**(11):459–62.
19. Marinkovic D, Timotijevic I, Babinski T, et al. The side-effects of clozapine: a four year follow-up study. *Prog Neuropsychopharmacol Biol Psychiatry* 1994;**18**(3):537–44.
20. Jaskiw GE, Thyrum PT, Fuller MA, et al. Pharmacokinetics of quetiapine in elderly patients with selected psychotic disorders. *Clin Pharmacokinet* 2004;**43**(14):1025–35.
21. Markowitz JS, DeVane CL, Boulton DW, et al. Hypotension and bradycardia in a healthy volunteer following a single 5-mg dose of olanzapine. *J Clin Pharmacol* 2002;**42**(1): 104–6.
22. Ravin DS, Levenson JW. Fatal cardiac event following initiation of risperidone therapy. *Ann Pharmacother* 1997;**31**(7/8):867–70.
23. Drici MD, Priori S. Cardiovascular risks of atypical antipsychotic drug treatment. *Pharmacoepidemiol Drug Saf* 2007;**16**(8):882–90.
24. Warrington SJ, Padgham C, Lader M. The cardiovascular effects of antidepressants. *Psychol Med Monogr Suppl* 1989;**16**(i-iii):1–40.
25. Glassman AH, Bigger Jr JT, Giardina EV, et al. Clinical characteristics of imipramine-induced orthostatic hypotension. *Lancet* 1979;**1**(8114):468–72.

26. Montgomery SA. Safety of mirtazapine: a review. *Int Clin Psychopharmacol* 1995;**10**(Suppl 4):37–45.
27. Waring WS, Good AM, Bateman DN. Lack of significant toxicity after mirtazapine overdose: a five-year review of cases admitted to a regional toxicology unit. *Clin Toxicol (Phila)* 2007;**45**(1):45–50.
28. Roose SP. Considerations for the use of antidepressants in patients with cardiovascular disease. *Am Heart J* 2000;**140**(Suppl 4):S84–8.
29. Johnson EM, Whyte E, Mulsant BH, et al. Cardiovascular changes associated with venlafaxine in the treatment of late-life depression. *Am J Geriatr Psychiatry* 2006;**14**(9):796–802.
30. Farid FF, Wenger TL, Tsai SY, et al. Use of bupropion in patients who exhibit orthostatic hypotension on tricyclic antidepressants. *J Clin Psychiatry* 1983;**44**(5 Pt 2):170–3.
31. Wilens TE, Hammerness PG, Biederman J, et al. Blood pressure changes associated with medication treatment of adults with attention-deficit/hyperactivity disorder. *J Clin Psychiatry* 2005;**66**(2):253–9.
32. Crabtree BL, Montgomery J. Iloperidone for the management of adults with schizophrenia. *Clin Ther* 2011;**33**(3):330–45.
33. Meltzer HY, Bobo WV, Nuamah IF, et al. Efficacy and tolerability of oral paliperidone extended-release tablets in the treatment of acute schizophrenia: pooled data from three 6-week, placebo-controlled studies. *J Clin Psychiatry* 2008;**69**:817–29.
34. Pae CU, Bodkin JA. Portland KB, et al. Safety of selegiline transdermal system in clinical practice: analysis of adverse events from postmarketing exposures. *J Clin Psychiatry* 2012;**73**(5):661–8.
35. Warner JP, Barnes TR, Henry JA. Electrocardiographic changes in patients receiving neuroleptic medication. *Acta Psychiatr Scand* 1996;**93**:311–3.
36. Lieberman JA, Safferman AZ. Clinical profile of clozapine: adverse reactions and agranulocytosis. *Psychiatr Q* 1992;**63**(1):51–70.
37. Stampfer H, Swanepoel P. Severe tachycardia following low-dose clozapine treatment. *Aust Psychiatry* 2005;**13**:80–2.
38. Gopal S, Hough D, Karcher K, et al. Risk of cardiovascular morbidity with risperidone or paliperidone treatment: analysis of 64 randomized, double-blind trials. *J Clin Psychopharmacol* 2013;**33**(2):157–61.
39. Tan HH, Hoppe J, Heard K. A systematic review of cardiovascular effects after atypical antipsychotic medication overdose. *Am J Emerg Med* 2009;**27**(5):607–16.
40. Roos JC. Cardiac effects of antidepressant drugs. A comparison of the tricyclic antidepressants and fluvoxamine. *B J Clin Pharmacol* 1983;**15**(Suppl 3):439S–45S.
41. Vohra J, Burrows GD, Sloman G. Assessment of cardiovascular side effects of therapeutic doses of tricyclic anti-depressant drugs. *Aust NZ J Med* 1975;**5**:7–11.
42. Bigger Jr JTH, Kantor SJ, Glassman AH, Perel JM. Cardiovascular effects of tricyclic antidepressant drugs. In: Lipton MA, DiMascio A, Killam KF, editors. *Psychopharmacology: a generation of progress*. New York: Raven Press; 1978. p. 1033–46.
43. Tulen JH, Bruijn JA, de Man KJ, et al. Cardiovascular variability in major depressive disorder and effects of imipramine or mirtazapine (Org 3770). *J Clin Psychopharmacol* 1996;**16**(2):135–45.
44. Enemark B. The importance of ECG monitoring in antidepressant treatment. *Nord J Psychiatry* 1993;**47**:57–65.
45. Rudolph RL, Derivan AT. The safety and tolerability of venlafaxine hydrochloride: analysis of the clinical trials database. *J Clin Psychopharmacol* 1996;**16**(3 Suppl 2):54S–9S.

46. Nussbaum AM, Stroup TS. Paliperidone for treatment of schizophrenia. *Schizophr Bull* 2008;**34**(3):419–22.
47. Nussbaum A, Stroup TS. Paliperidone for schizophrenia. *Cochrane Database Syst Rev* 2008;(2):CD006369.
48. Turkoz I, Bossie CA, Lindenmayer JP, et al. Paliperidone ER and oral risperidone in patients with schizophrenia: a comparative database analysis. *BMC Psychiatry* 2011;**11**:21.
49. Harrington CA, English C. Tolerability of paliperidone: a meta-analysis of randomized, controlled trials. *Int Clin Psychopharmacol* 2010;**25**(6):334–41.
50. Arif SA, Mitchell MM. Iloperidone: a new drug for the treatment of schizophrenia. *Am J Health Syst Pharm* 2011;**68**(4):301–8.
51. Cutler AJ, Kalali AH, Weiden PJ, et al. Four-week, double-blind, placebo- and ziprasidone-controlled trial of iloperidone in patients with acute exacerbations of schizophrenia. *J Clin Psychopharmacol* 2008;**28**(2 Suppl. 1):S20–8.
52. Citrome L, Meng X, Hochfeld M, Stahl SM. Efficacy of iloperidone in the short-term treatment of schizophrenia: a post hoc analysis of pooled patient data from four phase III, placebo- and active-controlled trials. *Hum Psychopharmacol* 2012;**27**(1): 24–32.
53. Potkin SG, Preskorn S, Hochfeld M, Meng X. A thorough QTc study of 3 doses of iloperidone including metabolic inhibition via CYP2D6 and/or CYP3A4 and a comparison to quetiapine and ziprasidone. *J Clin Psychopharmacol* 2013;**33**(1):3–10.
54. Boidi G, Ferro M. Rapid dose initiation of quetiapine for the treatment of acute schizophrenia and schizoaffective disorder: a randomised, multicentre, parallel-group, open study. *Hum Psychopharmacol* 2007;**22**(5):299–306.
55. Leucht C, Huhn M, Leucht S. Amitriptyline versus placebo for major depressive disorder. *Cochrane Database Syst Rev* 2012;**12**:CD009138.
56. Bebarta VS, Maddry J, Borys DJ, Morgan DL. Incidence of tricyclic antidepressant-like complications after cyclobenzaprine overdose. *Am J Emerg Med* 2011;**29**(6):645–9.
57. Güloglu C, Orak M, Ustündag M, Altunci YA. Analysis of amitriptyline overdose in emergency medicine. *Emerg Med J* 2011;**28**(4):296–9.
58. Forman MB, Sutej PG, Jackson EK. Hypertension, tachycardia, and reversible cardiomyopathy temporally associated with milnacipran use. *Tex Heart Inst J* 2011;**38**(6): 714–8.
59. Sambunaris A, Bose A, Gommoll CP, et al. A phase III, double-blind, placebo-controlled, flexible-dose study of levomilnacipran extended-release in patients with major depressive disorder. *J Clin Psychopharmacol* 2014;**34**(1):47–56.
60. Reilly JG, Ayis SA, Ferrier IN, et al. QTc-interval abnormalities and psychotropic drug therapy in psychiatric patients. *Lancet* 2000;**355**(9209):1048–52.
61. Ozeki Y, Fujii K, Kurimoto N, et al. QTc prolongation and antipsychotic medications in a sample of 1017 patients with schizophrenia. *Prog Neuropsychopharmacol Biol Psychiatry* 2010;**34**:401–5.
62. U.S. Food and Drug Administration. Information for healthcare professionals: haloperidol. FDA Alert 2007:9. Available at: <*www.fda.gov/Drugs/DrugSafety/ucm085203.htm*> [accessed 08.25.13].
63. Committee on Safety of Medicines. Cardiotoxic effects of pimozide. *Curr Probl Pharmacovigilance* 1990;**29**:1.
64. Pacher P, Kecskemeti V. Cardiovascular side effects of new antidepressants and antipsychotics: new drugs, old concerns? *Curr Pharm Des* 2004;**10**(20):2463–75.

65. Raviña T, Gutierrez J, Raviña P. Acquired long QT syndrome: long-term electrocardiographic (Holter) recording of torsades de pointes ending in asystole: II. *Int J Cardiol* 2007; **116**(2):272–5.
66. Suzuki Y, Ono S, Sugai T, et al. Dose-dependent effects of olanzapine on QT intervals and plasma prolactin levels in Japanese patients with stable schizophrenia. *Hum Psychopharmacol* 2011;**26**(6):440–3.
67. Vieweg WV, Hasnain M, Hancox JC, et al. Risperidone, QTc interval prolongation, and torsade de pointes: a systematic review of case reports. *Psychopharmacol (Berl)* 2013;**228**(4):515–24.
68. Camm AJ, Karayal ON, Meltzer H, et al. Ziprasidone and the corrected QT interval: a comprehensive summary of clinical data. *CNS Drugs* 2012;**26**(4):351–65.
69. Yap YG, Camm J. Risk of torsades de pointes with non-cardiac drugs. Doctors need to be aware that many drugs can cause QT prolongation. *BMJ* 2000;**320**(7243):1158–9.
70. Curtis LH, Østbye T, Sendersky V, et al. Prescription of QT-prolonging drugs in a cohort of about 5 million outpatients. *Am J Med* 2003;**114**(2):135–41.
71. Vieweg WV, Hasnain M, Howland RH, et al. Citalopram, QTc interval prolongation, and torsade de pointes. How should we apply the recent FDA ruling? *Am J Med* 2012;**125**(9): 859–68.
72. Howell C, Wilson AD, Waring WS. Cardiovascular toxicity due to venlafaxine poisoning in adults: a review of 235 consecutive cases. *Br J Clin Pharmacol* 2007;**64**(2):192–7.
73. Darbar D, Yang T, Churchwell K, et al. Unmasking of brugada syndrome by lithium. *Circulation* 2005;**112**:1527–31.
74. Rosenqvist M, Bergfeldt L, Aili H, Mathé AA. Sinus node dysfunction during long-term lithium treatment. *Br Heart J* 1993;**70**(4):371–5.
75. Serinken M, Karcioglu O, Korkmaz A. Rarely seen cardiotoxicity of lithium overdose: complete heart block. *Int J Cardiol* 2009;**132**:276–8.
76. Kayrak M, Duman C, Gul EE, et al. A bizarre electrocardiographic pattern due to chronic lithium therapy. *Ann Noninvasive Electrocardiol* 2010;**15**:289–92.
77. Mamiya K, Sadanaga T, Sekita A, et al. Lithium concentration correlates with QTc in patients with psychosis. *J Electrocardiol* 2005;**38**:148–51.
78. Ricciutti MA, Lisi KR, Damato AN. A metabolic basis of the electrophysiologic effects of lithium. *Circulation* 1971;**43**(Suppl. 2):217.
79. Antzelevitch C, Brugada P, Borggrefe M, et al. Brugada syndrome: report of the Second Consensus Conference. Endorsed by the Heart Rhythm Society and the European Heart Rhythm Association. *Circulation* 2005;**111**(5):659–70.
80. Chen Q, Kirsch GE, Zhang D, et al. Genetic basis and molecular mechanisms for idiopathic ventricular fibrillation. *Nature* 1998;**392**:293–6.
81. Hermida JS, Jandaud S, Lemoine JL, et al. Prevalence of drug-induced electrocardiographic pattern of the Brugada syndrome in a healthy population. *Am J Cardiol* 2004;**94**:230–3.
82. Bebarta VS, Phillips S, Eberhardt A, et al. Incidence of Brugada electrocardiographic pattern and outcomes of these patients after intentional tricyclic antidepressant ingestion. *Am J Cardiol* 2007;**100**:656–60.
83. Rouleau F, Asfar P, Boulet S, et al. Transient ST segment elevation in right precordial leads induced by psychotropic drugs: relationship to the Brugada syndrome. *J Cardiovasc Electrophysiol* 2001;**12**:61–5.
84. Akhtar M, Goldschlager NF. Brugada electrocardiographic pattern due to tricyclic antidepressant overdose. *J Electrocardiol* 2006;**39**:336–9.

85. Coker SJ. Drugs for men and women: how important is gender as a risk factor for TdP? *Pharmacol Ther* 2008;**119**(2):186–94.
86. Sadanaga T, Sadanaga F, Yao H, Fujishima M. Abnormal QT prolongation and psychotropic drug therapy in psychiatric patients: significance of bradycardia-dependent QT prolongation. *J Electrocardiol* 2004;**37**(4):267–73.
87. Trojak B, Astruc K, Pinoit JM, et al. Hypokalemia is associated with lengthening of QT interval in psychiatric patients on admission. *Psychiatry Res* 2009;**169**(3):257–60.
88. Abbott GW, Sesti F, Splawski I, et al. MiRP1 forms I_{Kr} potassium channels with HERG and is associated with cardiac arrhythmia. *Cell* 1999;**97**(2):175–87.
89. Hu D, Viskin S, Oliva A, et al. Genetic predisposition and cellular basis for ischemia-induced ST-segment changes and arrhythmias. *J Electrocardiol* 2007;**40**(Suppl. 6):S26–9.
90. Glue P, Banfield C. Psychiatry, psychoparmacoloy and P-450s. *Hum Psychopharmacol* 1996; **11**:97–114.
91. FDA Center for Drug Evaluation and Research. Fanapt (iloperidone) approval package: Pharmacology review. Part 1. May 2009. Available from: <*www.accessdata.fda.gov/drugsatfda_docs/nda/2009/022192s000_PharmR_P1.pdf*> [accessed 01.12.13].
92. Weiden PJ, Cutler AJ, Polymeropoulos MH, Wolfgang CD. Safety profile of iloperidone. A pooled analysis of 6-week acute-phase pivotal trials. *J Clin Psychopharmacol* 2008;**28** (2 Suppl 1):S12–9.
93. Leucht S, Cipriani A, Spineli L, et al. Comparative efficacy and tolerability of 15 antipsychotic drugs in schizophrenia: a multiple-treatments meta-analysis. *Lancet* 2013;**382**(9896): 951–62.
94. Aberg K, Adkins DE, Liu Y, et al. Genome-wide association study of antipsychotic-induced QTc interval prolongation. *Pharmacogen J* 2012;**12**(2):165–72.
95. Bhatia MS, Gupta R, Dhawan J. Myocarditis after overdose of conventional antipsychotics. *World J Biol Psychiatry* 2009;**10**:606–8.
96. Roesch-Ely D, Van Einsiedel R, Kathöfer S, et al. Myocarditis with quetiapine. *Am J Psychiatry* 2002;**159**(9):1607–8.
97. Reinders J, Parsonage W, Lange D, et al. Clozapine-related myocarditis and cardiomyopathy in an Australian metropolitan psychiatric service. *Aust NZ J Psychiatry* 2004;**38**(11-12):915–22.
98. Hill GR, Harrison-Woolrych M. Clozapine and myocarditis: a case series from the New Zealand intensive medicines monitoring programme. *NZ Med J* 2008;**121**(1283):68–75.
99. Bhushan M, Asirvatham SJ. The conundrum of ventricular arrhythmia and cardiomyopathy: which abnormality came first? *Curr Heart Fail Rep* 2009;**6**(1):7–13.
100. Coffey S, Williams M. Quetiapine-associated cardiomyopathy. *NZ Med J* 2011;**124** (1337):105–7.
101. Brakoulias V, Bannan E, Cohen P, Geary G. Amisulpride and cardiomyopathy. *Aust NZ J Psychiatry* 2005;**39**(8):738.
102. Mehtonen OP, Aranko K, Malkanen L, Vapaatalo H. A survey of sudden death associated with the use of antipsychotic or antidepressant drugs: 49 cases in Finland. *Acta Psychiatr Scand* 1991;**84**:58–64.
103. Reilly JG, Ayis SA, Ferrier IN, et al. Thioridazine and sudden unexplained death in psychiatric in-patients. *Br J Psychiatry* 2002;**180**:515–22.
104. Ray WA, Chung CP, Murray KT, et al. Atypical antipsychotic drugs and the risk of sudden cardiac death. *N Engl J Med* 2009;**360**(3):225–35.

105. Agelink MW, Majewski T, Wurthmann C, et al. Effects of newer atypical antipsychotics on autonomic neurocardiac function: a comparison between amisulpride, olanzapine, sertindole, and clozapine. *J Clin Psychopharmacol* 2001;**21**(1):8–13.
106. Cincotta SL, Rodefer JS. Emerging role of sertindole in the management of schizophrenia. *Neuropsychiatr Dis Treat* 2010;**6**:429–41.
107. Thomas SH, Drici MD, Hall GC, et al. Safety of sertindole versus risperidone in schizophrenia: principal results of the sertindole cohort prospective study (SCoP). *Acta Psychiatr Scand* 2010;**122**(5):345–55.
108. Kilian JG, Kerr K, Lawrence C, Celermajer DS. Myocarditis and cardiomyopathy associated with clozapine. *Lancet* 1999;**354**:1841–5.
109. La Grenade L, Graham D, Trontell A. Myocarditis and cardiomyopathy associated with clozapine use in the United States. *N Engl J Med* 2001;**345**:224.
110. Thorogood M, Cowen P, Mann J, et al. Fatal myocardial infarction and use of psychotropic drugs in young women. *Lancet* 1992;**340**(8827):1067–8.
111. Honkola J, Hookana E, Malinen S, et al. Psychotropic medications and the risk of sudden cardiac death during an acute coronary event. *Eur Heart J* 2012;**33**(6):745–51.
112. Pramyothin P, Khaodhiar L. Metabolic syndrome with the atypical antipsychotics. *Curr Opin Endocrinol Diabetes Obes* 2010;**17**(5):460–6.
113. Allison DB, Mentore JL, Heo M, et al. Antipsychotic-induced weight gain: a comprehensive research synthesis. *Am J Psychiatry* 1999;**156**:1686–96.
114. Bobes J, Rejas J, Garcia-Garcia M, et al. Weight gain in patients with schizophrenia treated with risperidone, olanzapine, quetiapine or haloperidol: results of the EIRE study. *Schizophr Res* 2003;**62**(1-2):77–88.
115. Newcomer JW. Second-generation (atypical) antipsychotics and metabolic effects: a comprehensive literature review. *CNS Drugs* 2005;**19**(Suppl. 1):1–93.
116. Casey DE. Dyslipidemia and atypical antipsychotic drugs. *J Clin Psychiatry* 2004;**65** (Suppl 18):27–35.
117. Correll CU, Manu P, Olshanskiy V, et al. Cardiometabolic risk of second-generation antipsychotic medications during first-time use in children and adolescents. *JAMA* 2009;**302** (16):1765–73.
118. Smith M, Hopkins D, Peveler RC, et al. First- versus second-generation antipsychotics and risk for diabetes in schizophrenia: systematic review and meta-analysis. *Br J Psychiatry* 2008;**192**:406–11.
119. Ramaswamy K, Masand PS, Nasrallah HA. Do certain atypical antipsychotics increase the risk of diabetes? A critical review of 17 pharmacoepidemiologic studies. *Ann Clin Psychiatry* 2006;**18**:183–94.
120. Henderson DC, Daley TB, Kunkel L, et al. Clozapine and hypertension: a chart review of 82 patients. *J Clin Psychiatry* 2004;**65**(5):686–9.
121. Saddichha S, Vishnuvardhan G, Akhtar S. Obesity, diabetes and hypertension associated with antipsychotic use in remitted schizophrenia. *Int J Risk Saf Med* 2011;**23**(3):181–5.
122. Garland EJ, Remick RA, Zis AP. Weight gain with antidepressants and lithium. *J Clin Psychopharmacol* 1988;**8**(5):323–30.
123. Laimer M, Kramer-Reinstadler K, Rauchenzauner M, et al. Effect of mirtazapine treatment on body composition and metabolism. *J Clin Psychiatry* 2006;**67**:421–4.
124. Chokka P, Tancer M, Yeragani VK. Metabolic syndrome: relevance to antidepressant treatment. *J Psychiatry Neurosci* 2006;**31**(6):414.

125. McIntyre RS, Soczynska JK, Konarski JZ, Kennedy SH. The effect of antidepressants on lipid homeostasis: a cardiac safety concern? *Expert Opin Drug Saf* 2006;**5**(4):523–37.
126. Andersohn F, Schade R, Suissa S, Garbe E. Long-term use of antidepressants for depressive disorders and the risk of diabetes mellitus. *Am J Psychiatry* 2009;**166**(5):591–8.
127. Darbà J, Kaskens L, Aranda P, et al. A simulation model to estimate 10-year risk of coronary heart disease events in patients with schizophrenia spectrum disorders treated with second-generation antipsychotic drugs. *Ann Clin Psychiatry* 2013;**25**(1):17–26.
128. Pappadopulos E, Newcomer JW, Kolluri S. Changes in weight, plasma lipids, and glucose in adults treated with ziprasidone: a comprehensive analysis of Pfizer-initiated clinical trials. *J Clin Psychiatry* 2012;**73**(6):e742–8.
129. Hasnain M, Vieweg WV, Hollett B. Weight gain and glucose dysregulation with second-generation antipsychotics and antidepressants: a review for primary care physicians. *Postgrad Med* 2012;**124**(4):154–67.
130. Athanasiu L, Brown AA, Birkenaes AB, et al. Genome-wide association study identifies genetic loci associated with body mass index and high density lipoprotein-chloestrol levels during pspychopharmacological treatment—a cross-sectional naturalistic study. *Psychiatry Res* 2012;**197**(3):327–36.
131. Lett TA, Wallaace TJ, Chowdhury NI, et al. Pharmacogenetics of antipsychotic-induced weight gain: review and clinical implications. *Mol Psychiatry* 2012;**17**(3):242–66.
132. Montastruc G, Favreliere S, Sommet A, et al. French Association of Regional Pharmacovigilance Centres. Drugs and dilated cardiomyopathies: a case/noncase study in the French pharmacovigilance database. *Br J Clin Pharmacol* 2010;**69**(3):287–94.
133. Nielsen J, Correll CU, Manu P, Kane JM. Termination of clozapine treatment due to medical reasons: when is it warranted and how can it be avoided? *J Clin Psychiatry* 2013;**74**(6):603–13.
134. Jerrell JM, McIntyre RS. Cerebro- and cardiovascular conditions in adults with schizophrenia treated with antipsychotic medications. *Hum Psychopharmacol* 2007;**22**(6):361–4.
135. Coulter DM, Bate A, Meyboom RH, et al. Antipsychotic drugs and heart muscle disorder in international pharmacovigilance: data mining study. *BMJ* 2001;**322**(7296):1207–9.
136. Cohen HW, Gibson G, Alderman MH. Excess risk of myocardial infarction in patients treated with antidepressant medications: association with use of tricyclic agents. *Am J Med* 2000;**108**:2–8.
137. Hippisley-Cox J, Fielding K, Pringle M. Depression as a risk factor for ischaemic heart disease in men: population based case control study. *BMJ* 1998;**316**:1714–9.
138. Ratcliffe S, Younus M, Hauben M, Reich L. Antidepressants that inhibit neuronal norepinephrine reuptake are not associated with increased spontaneous reporting of cardiomyopathy. *J Psychopharmacol* 2010;**24**(4):503–11.
139. Forman MB, Sutej PG, Jackson EK. Hypertension, tachycardia, and reversible cardiomyopathy temporally associated with milnacipran use. *Tex Heart Inst J* 2011;**38**(6):714–8.
140. Selke KJ, Dhar G, Cohn JM. Takotsubo cardiomyopathy associated with titration of duloxetine. *Tex Heart Inst J* 2011;**38**(5):573–6.
141. Christoph M, Ebner B, Stolte D, et al. Broken heart syndrome: Tako Tsubo cardiomyopathy associated with an overdose of the serotonin-norepinephrine reuptake inhibitor Venlafaxine. *Eur Neuropsychopharmacol* 2010;**20**(8):594–7.
142. American Diabetes Association, American Psychiatric Association, American Association of Clinical Endocrinologists, and North American Association for the Study of Obesity. Consensus Development Conference on Antipsychotic Drugs and Obesity and Diabetes. *Diabetes Care* 2004;**27**(2):596–601.

143. Expert Consensus Meeting. Schizophrenia and diabetes, Dublin 2003: consensus summary. *Br J Psychiatry* 2004;**47**(Suppl):S112.
144. De Nayer A, De Hert M, Scheen A, et al. Conference report: Belgian consensus on metabolic problems associated with atypical antipsychotics. *Int J Psychiatry Clin Pract* 2005;**9**(2):130–7.
145. Kirkpatrick B, Miller BJ, Garcia-Rizo C, et al. Is abnormal glucose tolerance in antipsychotic-naive patients with nonaffective psychosis confounded by poor health habits? *Schizophr Bull* 2012;**38**(2):280–4.
146. Verma SK, Subramaniam M, Liew A, Poon LY. Metabolic risk factors in drug-naive patients with first-episode psychosis. *J Clin Psychiatry* 2009;**70**(7):997–1000.
147. Lester H, Shiers DE, Rafi I, et al. Positive Cardiometabolic Health Resource: an intervention framework for patients with psychosis on antipsychotic medication. *London: Royal College of Psychiatrists,* 2012.
148. Schneeweiss S, Avorn J. Antipsychotic agents and sudden cardiac death: how should we manage the risk? *N Engl J Med* 2009;**360**:294–6.
149. Lieberman JA, Merrill D, Parameswaran S. *APA guidance on the use of antipsychotic drugs and cardiac sudden death.* Prepared for the New York State Office of Mental Health. Available from: <www.omh.ny.gov/omhweb/advisories/adult_antipsychotic_use_attachement.html> [accessed 05.25.14].
150. Shah RR. Drug-induced QT dispersion: does it predict the risk of torsades de pointes? *J Electrocardiol* 2005;**38**:10–8.
151. Sager PT. Key clinical considerations for demonstrating the utility of preclinical models to predict clinical drug-induced torsades de pointe. *Br J Pharmacol* 2008;**154**:1544–9.
152. American Psychiatric Association. Practice guidelines. *Psychiatric evaluation of adults.* 2nd Ed. Arlington, VA: APA, 2006.
153. Marder SR, Essock SM, Miller AL, et al. Physical health monitoring of patients with schizophrenia. *Am J Psychiatry* 2004;**161**:1334–49.
154. American Psychiatric Association. Practice guidelines. *Treatment of patients with schizophrenia.* 2nd Ed. Arlington, VA: APA, 2004.
155. Haddad PM, Anderson IM. Antipsychotic-related QTc prolongation, torsades de pointes and sudden death. *Drugs* 2002;**62**:1649–71.
156. Royal College of Psychiatrists. Consensus statement on high dose anti-psychotic medication, Council Report CR138, May 2006. Available from: *<www.rcpsych.ac.uk/files/pdfversion/cr138.pdf>* [accessed 12.25.13].
157. Beach SR, Celano CM, Noseworthy PA, et al. QTc prolongation, torsades de pointes, and psychotropic medications. *Psychosomatics* 2013;**54**(1):1–13.
158. Glassman AH, O'Connor CM, Califf RM, et al. Sertraline treatment of major depression in patients with acute MI or unstable angina. *JAMA* 2002;**288**:701–9.

Chapter 10

Cardiovascular Toxicity of Noncardiovascular Drugs

Omi Bajracharya,[1] P. Ravi Shankar,[2] and Nisha Jha[3]
[1,3]Department of Pharmacology, KIST Medical College, Lalitpur, Nepal
[2]Department of Pharmacology, Xavier University School of Medicine, Oranjestad, Aruba

10.1 INTRODUCTION

This chapter discusses the cardiovascular toxicity of noncardiovascular drugs. Cardiovascular toxicity from cardiovascular, psychotropic, anticancer, and recreational drugs are discussed in other chapters. Cardiotoxicity is a significant adverse reaction to a drug, and it can even lead to life-threatening effects. In various studies, it has been documented that, due to the abundance of mitochondria in cardiac muscle cells and the close relationship between oxidative metabolism and myocardial function and viability, mitochondrial dysfunction should always be suspected in cardiotoxicity.

Drug-induced cardiac toxicity has been broadly classified into the following three categories:

1. Structural damage
2. Functional deficits that may or may not be associated with histopathologic changes
3. Altered cell or tissue homeostasis in the absence of obvious structural or functional deficits[1]

The following is a partial list of drug-induced cardiovascular diseases; descriptions of each can be found in the sections that follow.

- Myocardial ischemia and acute coronary syndrome
- Heart failure
- Arrhythmias
- Valvular heart disease
- Hypertension
- Hypotension

The Heart and Toxins. DOI: http://dx.doi.org/10.1016/B978-0-12-416595-3.00010-4

10.2 MYOCARDIAL ISCHEMIA AND ACUTE CORONARY SYNDROME

Myocardial ischemia generally occurs as a result of an increase in myocardial oxygen demand, decreased myocardial oxygen supply, or both. It is very common in patients with atherosclerotic coronary artery disease. Acute coronary syndrome (ACS) is defined as one of three syndromes:

1. ST-segment elevation myocardial infarction (MI)
2. Non-ST-segment elevation myocardial infarction
3. Unstable angina

A working group set up by the British Cardiology Society proposed ACS to be divided into three categories:

1. ACS with unstable angina
2. ACS with myocyte necrosis
3. ACS with clinical MI[2]

Drugs or medicines used in the treatment of various disease conditions have been reported to cause ACS, angina pectoris, and myocardial ischemia. In a report obtained from The Netherlands Centre for Monitoring of Adverse Reaction to Drugs, data collected over a period of 20 years found several classes of drugs that can produce chest pain and myocardial ischemia. Therefore it is very important to consider drugs as a possible cause for these conditions, particularly in patients with normal coronary arteries.[3]

Myocardial ischemia or acute coronary syndrome can occur in association with short-term medicine use or long-term therapy. Determination of the prevalence of drug-induced ischemia and acute coronary syndromes are difficult in the general population as, with the exception of a small number of drugs, data on drug-induced ischemia may not be available. A recent article examined both cardiac and vascular postapproval adverse drug events reported to the U.S. Food and Drug Administration (FDA).[4] Cardiac arrhythmias, coronary artery disorders, cardiac disorder signs and symptoms, and heart failure were found to be among the most common cardiac adverse effects, while changes in blood pressure, embolism, thrombosis, and hemorrhagic disorders were the most common vascular ones.

10.2.1 Mechanism of Drug-Induced Myocardial Ischemia

An imbalance between coronary blood flow, myocardial oxygen supply, and oxygen demand results in myocardial ischemia, which occurs when myocardial oxygen demand exceeds supply. There are three factors that increase myocardial oxygen demand, namely increased heart rate, contractility, and left ventricular systolic wall tension.[5] In the healthy human heart, there is a linear relationship between myocardial oxygen demand, oxygen consumption, and coronary blood flow, and adjustments are made on a beat-by-beat basis.

Among these factors, the easiest to determine clinically is heart rate; an increase in heart rate leads to an increase in oxygen demand. Myocardial contractility increases when there is a rise in intraventricular pressure during isovolumetric contraction, and this is influenced by various other factors such as functioning of the autonomic nervous system, heart rate, blood calcium level, and body temperature. The third determining factor that increases myocardial oxygen demand is systolic wall tension, which is directly related to systolic pressure and the wall radius of ventricles and inversely related to wall thickness.

Preload and afterload are important determinants of these components. A reduction in systolic blood pressure is required to decrease the myocardial oxygen demand, and this will ultimately reduce the afterload. Even at rest, the myocardium's oxygen use is high relative to the blood flow. About 70 to 80% of the oxygen is extracted from each unit of blood crossing the myocardial capillaries.[6] This leaves little margin for increasing oxygen extraction in conditions of increased oxygen demand.

A drug that modifies any of these factors may cause myocardial ischemia. For example, abrupt discontinuation of beta blockers may directly increase the heart rate; and drugs (e.g., nifedipine, minoxidil, and so on), which reduce the blood pressure, indirectly increase heart rate through reflex baroreceptor stimulation; tachycardia can precipitate myocardial ischemia.

Factors that decrease myocardial oxygen supply can also aggravate myocardial ischemia. Among the factors that determine myocardial oxygen supply are coronary blood flow, enhanced platelet activation, coronary thrombosis, and the oxygen-carrying capacity of the blood. Clinically, the most important is coronary blood flow.[5] The coronary system contains arteries, arterioles, capillaries, venules, and veins. The right and left main coronary arteries exit the ascending aorta just above the aortic valve (coronary ostia); these two branches again divide and course over the surface of the heart (epicardium). They gradually divide into smaller branches and progress inward to penetrate the epicardium and supply blood to the transmural myocardium. Eventually coronary arteries branch into arterioles, which then branch into innumerable capillaries through which the blood flows back to the cardiac chambers.

The coronary arteries supply blood to the myocardium and reduced blood flow results in coronary ischemia, angina, reduced performance, and/or infarction. Formation of atherosclerotic plaques in branches of the coronary arteries can result in a regional reduction in blood flow. Normally, the arteriolar resistance vessels are the most important regulators of coronary blood flow, whereas large epicardial vessels are low-resistance. Enhanced platelet activation leads to formation of platelet plaques, which can further increase the blockage caused by atherosclerotic deposits in the coronary arteries.

Changes in the electrocardiogram can be seen when there is insufficient blood flow to a region of the heart. The isoelectric period (ST segment) elevation can suggest myocardial injury. Acute transmural ischemia typically causes ST segment elevation, which is often associated with high-amplitude T waves. These changes are most often seen in the precordial leads, but may also be seen in leads I, II, III, and aVF if the inferior wall is ischemic. ST segment depression is often seen with myocardial ischemia, but there can be many false positives. The condition could happen because myocardial ischemia is delaying the process of ventricular repolarization, especially if the depression is horizontal or downsloping.[7]

Whenever coronary blood flow decreases below that necessary to meet metabolic requirements, the myocardium is considered ischemic; the pumping capability of the heart is impaired and there are allied changes in electrical activity. Complex factors that determine coronary blood flow include duration of diastole and coronary vascular resistance. Coronary vascular resistance is modified by metabolic control, autoregulation, extravascular compressive forces, and humoral and neural factors. Modification of any of these factors by a drug can influence the myocardial oxygen supply. Conditions (e.g., anemia) can affect oxygen-carrying capacity[8] and has been recognized as a risk factor for ischemic heart disease.[9]

The role of reduced oxygenation as a result of reduced hemoglobin and hematocrit on myocardial performance has been studied.[10,11] Inadequate oxygen-carrying capacity due to diminished oxygenation can have serious clinical implications, chiefly because of ischemic effects on the myocardium and brain. Increased stroke volume[12] and decreased peripheral vascular resistance[13] to maintain adequate cardiac output and oxygen supply to the tissues in anemia can precipitate myocardial ischemia. When hemoglobin concentration is reduced to less than half the normal level, ventricular function is impaired, presumably because the coronary flow has approached its maximum and oxygen supply cannot be increased further.[14] Angina pectoris usually occurs in patients with underlying coronary disease[15]; however, in anemic patients, it may occur even without coronary disease because of the low hemoglobin level.[16,17]

10.2.2 Cardiovascular Adverse Effects Due to Drugs

A study published in 2002 concluded that of the 548 new chemical entities approved during the period from 1975 to 1999 by the FDA, 45 drugs (8.2%) now have one or more black-box warnings that were not present when the drugs were approved, while 16 approved drugs (2.9%) have been withdrawn from the market and, of these, 5 had acquired a black-box warning prior to withdrawal.[18] Among the drugs that acquired a new warning for cardiovascular adverse effects during the period 1975 to 2000 were the noncardiovascular drugs danazole, ketoconazole, cyclosporine, terfenadine, ribavirin, cisapride, and astemizole. Among the noncardiovascular drugs withdrawn from the market due to cardiovascular toxicity were azaribine due to thromboembolism; terfenadine, astemizole, and cisapride due to drug interactions causing cardiotoxicity; and grapafloxacin due to cardiovascular events.[18]

More recently the cyclo-oxygenase-2 (COX-2) inhibitors have attracted attention as a result of the risk of cardiovascular events. Rofecoxib was precipitously withdrawn from the market and a black-box warning was added to valdecoxib for life-threatening skin reactions and cardiovascular risk.[19] The atypical antipsychotics sertindole and mesoridazine received black-box warnings in the early 2000s.[20] In February 2006, the FDA's Drug Safety and Risk Management Advisory Committee voted to include boxed warnings on methylphenidate formulations used to treat attention deficit hyperactivity disorder due to their possible cardiovascular side effects.[21] However, this was not implemented and research is ongoing in this important area. On November 14, 2007, the FDA added a boxed warning to the diabetes medication rosiglitazone, citing the risk of heart failure or heart attack in patients with underlying heart disease, or at a high heart attack risk,[22] while pioglitazone acquired a black-box warning of risk of congestive heart failure. Table 10.1 shows various mechanisms involved in drug-induced myocardial ischemia and acute coronary syndrome.

10.3 DRUG-INDUCED HEART FAILURE

Heart failure occurs when the heart is not able to deliver enough oxygen and nutrients to meet the needs of the tissues.[24] It is a complex condition arising from defects in left ventricular filling and/or emptying and symptoms include shortness of breath and fatigue. The symptoms of heart failure are a result of inadequate tissue perfusion, venous congestion, and disturbed water and electrolyte balance.

Conditions, such as valvular heart disease, hypertension, and coronary heart disease, can result in heart failure, which is a major cause of morbidity and mortality in developed countries. In the United States alone, around 5.7 million people experience heart failure and each year it is estimated that 55,000 deaths occur due to heart failure. In 2008, heart failure was mentioned as a contributing cause in more than 280,000 patients and half the people who were diagnosed with it died within 5 years.[25,26]

Drugs can induce heart failure by a variety of mechanisms. They can reduce myocardial contractility (e.g., anthracyclines), increase preload (e.g., nonsteroidal antiinflammatory drugs, NSAIDs), glitazones, COX-2 inhibitors, corticosteroids), and increase afterload (e.g., sympathomimetic agents).[27] Table 10.2 shows various types of drugs that can induce or worsen heart failure.

10.3.1 Cardiac Toxicity Due to Biologics, Anorexients, and Anthracyclines

Biologics, such as trastuzumab, induce cardiac toxicity by binding to specific receptors. The toxicity may be related to its mechanism of therapeutic action. Trastuzumab binds to the HER2 receptors on cardiac myocytes and this

TABLE 10.1 Various Mechanisms of Drug-Induced Myocardial Ischemia and Acute Coronary Syndrome

Mechanism	Examples
Drug-induced myocardial ischemia	This can occur because of increased oxygen demand or decreased oxygen supply, or both.
Increased oxygen demand	
Increasing myocardial oxygen demand and consumption and increased myocardial contractility	*Directly by stimulating the myocardium:* adrenergic β-agonists, sympathomimetics, withdrawal of β-antagonists (due to receptor upregulation and increased sensitivity) *Indirectly through changes in blood pressure:* potent vasodilators (e.g., nifedipine, minoxidil, hydralazine)—that is, through baroreceptor stimulation and reflex tachycardia
Increased left ventricular systolic wall tension	
Vasoconstrictors (increasing venous return and/or peripheral resistance)	Amphetamines, antihistamines, caffeine, oxymetazoline, phenylephrine, pseudoephedrine, tetrahydrozoline
Decreased myocardial oxygen supply (temporary, nontotal occlusion)	
Increased coronary vascular resistance (vasospasm, thrombosis)	Cocaine, antimigraine agents (e.g., ergot alkaloid, triptans)
Decreased coronary diastolic perfusion pressure	Enalapril, nifedipine, minoxidil, hydralazine, nitroprusside, adenosine, dipyridamole
Drug-induced acute coronary syndromes	
Associated with acute drug therapy	
Coronary artery thrombosis superimposed on a damaged atherosclerotic plaque	Sympathomimetics (e.g., cocaine, ephedrine, amphetamines, phenylpropranolaime) oral contraceptives, estrogens, COX-2 inhibitors, NSAIDs (n.b., may increase risk of thrombosis), bromocriptine
Coronary artery vasospasm (with or without a superimposed coronary artery thrombosis)	Cocaine, antimigraine agents (e.g., ergot alkaloids, triptans), marijuana, amphetamines, alcohol and ephedrine-based weight-loss products
Associated with chronic drug therapy, increased cardiovascular risk	Cocaine, estrogens, COX-2 inhibitors, NSAIDs, HIV agents and/or protease inhibitors,* oral contraceptives, rosiglitazone

**Protease inhibitors can lead to effects on the coronary circulation through dyslipidemia caused by inhibition of cellular enzymes involved in lipid metabolism.*[23]

Source: Adapted and reprinted with permission from Tisdale and Miller, p. 406.[17]

TABLE 10.2 Noncardiac Drugs Known to Induce or Make Heart Failure Worse According to the Mechanism(s) Implicated

Cardiac Problem and Class	Drug
Cardiomyopathy	Trastuzumab, interferon-α-2, interleukin-2, infliximab, etanercept
	Itraconazole, amphotericin B
	Anthracyclines (doxorubicin and others)
Pulmonary hypertension	
Antimigraine drugs	Methysergide, ergotamine
Appetite suppressants	Fenfluramine, dexfenfluramine, phentermine
Heart-valve abnormalities	
Antimigraine drugs	Methysergide, ergotamine
Appetite suppressants	Fenfluramine, dexfenfluramine, phentermine
Fluid overload	
NSAIDs, including cyclo-oxygenase-2-inhibitors	All
Antidiabetic drugs	Rosiglitazone, pioglitazone*
Glucocorticoids	All
Herbal drugs	Those containing liquorice or adulterated with NSAIDs

**Available evidence suggests that edema is a class effect of the thiazolidinediones and is multifactorial in origin.[28] Thiazolidinedione-associated edema seems to be dose related and occurs most frequently when these drugs are used in combination with insulin.*

Source: Adapted from Slordal and Spigset, and reprinted with permission from Springer Publishers, 2006.[27]

receptor is likely to be involved in the survival of normal healthy myocytes under conditions of stress.[29] Based on studies carried out with conditionally mutated mice with a deficiency of ERbB2 (HER2) only in the ventricle,[30] it was postulated that the ERbB2/ERbB4 complex provides a survival mechanism for cardiac myocytes in the presence of stresses that would otherwise cause cellular death. Cardiomyocytes isolated from the mutated mice were also more susceptible to anthracycline toxicity. Bevacizumab is an antibody against the human vascular endothelial growth factor (VEGF). Cardiovascular adverse events associated with this medicine include arterial thromboembolic events, hemorrhage, and hypertension.[29] The potential mechanisms underlying these events may be either the endothelial cells lining the blood vessels becoming prothrombotic or the platelets becoming directly activated.

In mammals serotonergic drugs, many of which are sympathomimetic anorexients, can cause pulmonary hypertension.[31] Possible mechanisms may involve the release of serotonin from platelets and prolonging its vasoconstrictor influence. Certain anorexients may cause pulmonary hypertension by inhibiting potassium current in pulmonary vascular smooth muscle, an effect similar to that induced by hypoxia.[32] The mechanism of heart valve abnormalities caused by antimigraine drugs and appetite suppressants are discussed in detail later in this chapter.

The anthracycline anticancer drugs (e.g., doxorubicin) are well known to cause cardiotoxicity. A variety of mechanisms including increased oxidative stress,[33] decreased levels of antioxidants,[34] inhibition of nucleic acid and protein synthesis, release of vasoactive amines, altered adrenergic function, and decreased expression of cardiac-specific genes[35,36] have been postulated to be responsible. Cardiomyocyte death through apoptosis, necrosis, and other mechanisms is a primary mechanism for the progression of doxorubicin-induced cardiac toxicity.[37]

10.4 DRUG-INDUCED ARRHYTHMIAS

In the last few years, a wide variety of drugs belonging to various therapeutic classes have shown rare but serious cardiotoxic events. These have been a major cause of concern among patients and clinicians, as well as drug developers and regulators. To increase the safety profile of a pharmaceutical, a group of drug regulatory agencies from Europe, the United States, and Japan developed the International Conference on Harmonization of Technical Requirements for Registration of Pharmaceuticals for Human Use (ICH). The ICH has developed guidelines and recommendations that drugs having arrhythmogenic potential should be recognized at the stages of both preclinical and clinical drug development to improve the safety of newly introduced drugs. Drugs can induce various types of arrhythmias ranging from sinus bradycardia, atrioventricular (AV) block, atrial fibrillation, monomorphic ventricular tachycardia, and paroxysmal atrial tachycardia, among others. This chapter focuses mainly on QT interval prolongation and atrial fibrillation.

Electrical depolarization in myocardial tissue results in the mechanical process of atrial and ventricular contraction. The sinoatrial (SA) node, which is considered as the heart's intrinsic pacemaker, is normally the source of electrical impulses and determines the frequency of depolarization. In resting adults, the SA node's normal intrinsic depolarization rate varies between 60 and 100/min, whereas both the AV node and the ventricular tissue possess intrinsic depolarization rates ranging between 40 and 60/min and 30 and 40/min, respectively. When the SA node fails to generate depolarization, often the AV node will take over the role of pacemaker.

In the cardiac myocytes, movement of the ions results in depolarization. During the resting phase, myocytes have a resting potential of -70 to -90 mV as a result of the sodium–potassium adenosine triphosphatase (ATPase) pump, which maintains high extracellular concentrations of sodium and low extracellular concentrations of potassium.[38] In these cells the action potential is shaped by the sequential opening of distinct classes of ion channels. The electrical activity of the normal Purkinje cell is subdivided into five phases from 0 to 4; where phases 0 to 3 correspond to systole and phase 4 corresponds to diastole (Figure 10.1).

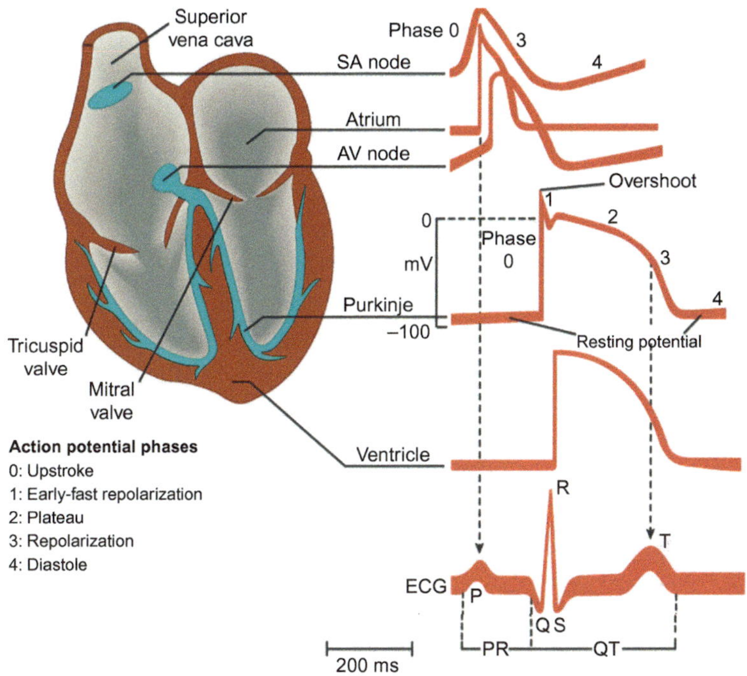

FIGURE 10.1 **Schematic representation of the heart and normal cardiac electrical activity.** The ECG is the body surface manifestation of the depolarization and repolarization waves of the heart. The P wave is generated by atrial depolarization, the QRS by ventricular muscle depolarization, and the T wave by ventricular repolarization. Thus, the PR interval is a measure of conduction time from atrium to ventricle, and the QRS duration indicates the time required for all of the ventricular cells to be activated (i.e., the intraventricular conduction time). The QT interval reflects the duration of the ventricular action potential. **Source**: *Reproduced with permission from Katzung BG, Master SB, Trevor AJ,* Basic and Clinical Pharmacology, *11th edition. (For further information, see reference 140.)*

When the membrane potential reaches a threshold, which is usually −60 to −80 mV, there will be a rapid influx of sodium ions through voltage-gated sodium channels (phase 0 of the action potential), thereby depolarizing the cell to a point at which it overshoots the membrane potential to between +20 and +30 mV.[38] This depolarization is subsequently followed by the activation of rapidly activating and inactivating potassium channels, which generates the initial repolarization of phase 1. In phase 2, consisting of repolarization there will be continued movement of potassium ions from the cell, but the membrane potential remains at a plateau, near 0 mV, due to intracellular influx of calcium and sodium via slow channels. During phase 3 of repolarization and T wave in the electrocardiogram (ECG), potassium efflux occurs at a higher rate than calcium and sodium influx, further lowering the membrane potential, which returns toward negativity.

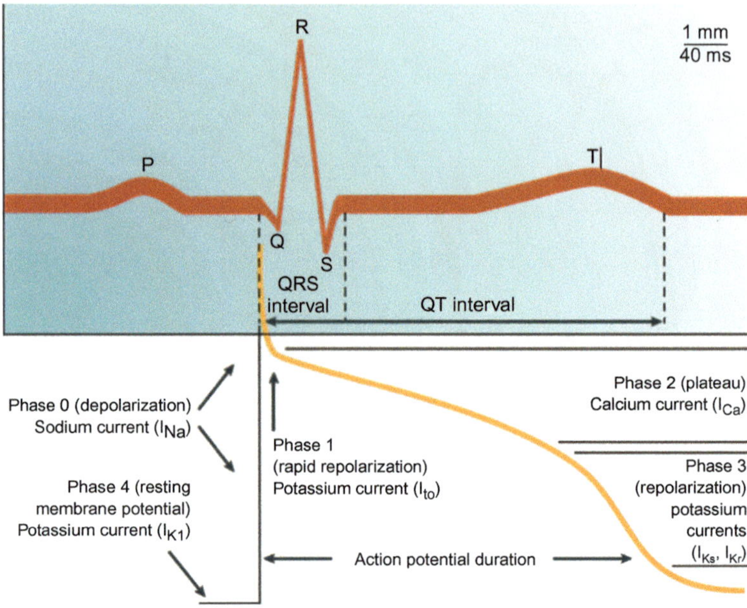

FIGURE 10.2 **Correlation between electrocardiogram and QT.** **Source**: *Reprinted with permission of Macmillan Publishers Ltd., copyright 2003:* Nature Reviews Drug Discovery *(Fermini B, Fossa AA. The impact of drug-induced QT interval prolongation on drug discovery and development.)*

Phase 4 completes the action potential during which the ATPase pump promotes the efflux of sodium and returns the cell to a resting membrane potential that is between −70 and −90 mV. Abnormal impulse generation or conduction, or a combination of both, can precipitate arrhythmias. Usually arrhythmias develop when there is generation of an abnormal impulse immediately following the absolute refractory period of cardiac tissues. During this phase, depolarization of myocytes does not occur because the impulse generated does not trigger it as the cells are in the refractory phase thereby resulting in abnormal conduction of the heart and thus precipitating arrhythmias.[39]

Many factors increase automaticity and among them drugs are an important one. Drug-induced arrhythmias are usually associated with QT interval prolongation and development of polymorphic ventricular tachycardias. The main mechanism of drug-induced long QT syndrome involves blockade of specific potassium channels, which ultimately prolongs repolarization. Other mechanisms involved in prolonging the QT interval may be enhancing inward sodium current or reducing cell surface expression of functional channels.

The morphology and duration of ventricular action potential is determined by numerous overlapping ionic currents (Figure 10.2). Rapid entry of sodium ions through selective sodium channels initiates the depolarization of the ventricles (phase 0), followed by a rapid repolarization through transiently activating and inactivating outward potassium channels (I_{to}, phase 1). This is followed by a plateau phase (phase 2), which is mainly determined by the entry of calcium ions through L-type calcium channels. Repolarization, or phase 3, results from the inactivation of calcium channels and increase in net outward potassium currents carried mainly by the slow (I_{Ks}) and rapid (I_{Kr}) components of the delayed rectifier potassium channels. Inwardly rectifying potassium channels (I_{K1}) contributes to phase 3 repolarization and to the maintenance of the resting membrane potential (phase 4). In humans, I_{Kr} seems to play a significant role in determining ventricular action potential duration (APD) and repolarization, as congenital mutations of this channel are associated with a decrease in current amplitude and prolongation of the QT interval.

The balance between inward and outward currents determines the morphology and duration of the action potential and the QT interval. A major problem in measuring the QT interval is determining the end of the T wave.[40] A recent article mentions that to get an accurate measure of the QT interval, several beats in each lead of the ECG should be considered and then a median obtained of the values from six or more leads.[41] A manual approach may be more accurate but may be time consuming. The authors suggest using a manual measurement where more than one lead of the 12-lead ECG is measured and then the median QT interval is calculated. The end of the T wave is determined visually as the point where it returns to the baseline.

The authors of this paper[41] recommend that only the absolute QT interval be used and not the corrected one. The QT interval is plotted against the heart rate on the QT nomogram and if the plot is above the specified line on the nomogram then there is increased risk of torsades de pointes (TdP).[41] This was exemplified by the drug amisulpiride where the greater the orthogonal distance of the pair of the HR and QT interval above the QT nomogram line, the greater was the risk of TdP.[42] The QT interval is usually corrected for heart rate (QTc). QTc intervals are usually around 400 msec in duration, and values lower than 440 are considered normal. The greater the duration, the more likely TdP becomes, but 500 msec has frequently been used as a cutoff because longer QTc intervals are associated with substantially higher risk. However, other factors can influence this risk.[20]

Drug-induced inhibition of the I_{Kr} current can delay repolarization and lengthen the action potential duration and the QT interval. Lengthening repolarization further delays the inactivation of calcium channels as shown in Figure 10.3. The resulting late inflow of calcium contributes to the propensity of developing early afterdepolarizations (EADs). Prolongation of the action potential also increases the amplitude of the intracellular calcium concentration transient, thereby promoting EADs. When generated in the presence of transmural heterogeneity in ventricular repolarization, EADs are thought to contribute to the generation of extrasystoles that can cause TdP.

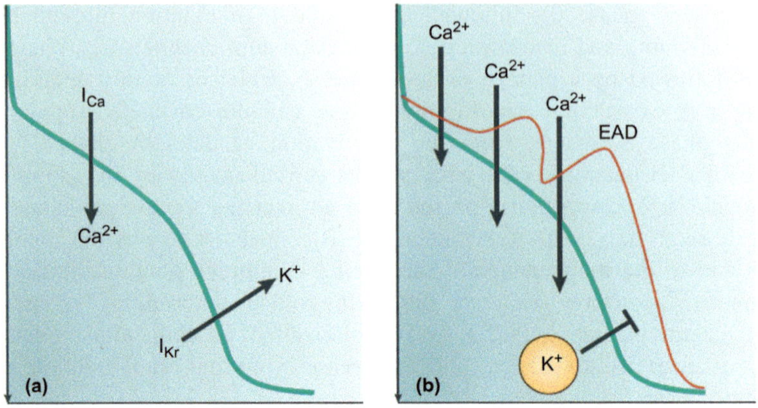

FIGURE 10.3 **Graphic representation of inward and outward movement of calcium (a) and potassium (b) ions and how drugs delay repolarization precipitating arrhythmia.** **Source**: *Reprinted with permission of Macmillan Publishers Ltd., copyright 2003:* Nature Reviews Drug Discovery *(Fermini B, Fossa AA. The impact of drug-induced QT interval prolongation on drug discovery and development.)*

The blockade of the ion channel coded by the human Ether-á-go-go-related gene (hERG) has been correlated to a prolongation of the QT interval in the ECG, which increases the potential risk of TdP.[43] The hERG potassium channels conduct the rapid component of the delayed rectifier potassium current, I_{Kr}, which is crucial for repolarization of cardiac action potentials. A reduction in hERG currents either due to genetic defects or adverse drug effects can lead to hereditary or acquired long QT syndromes characterized by action potential prolongation, lengthening of the QT interval on the surface ECG, and an increased risk for TdP arrhythmias and sudden death.[43,44]

Between June 1990 and March 2001, 8 noncardiovascular pharmaceuticals were withdrawn from the market in the United States and elsewhere because of their propensity to delay cardiac repolarization, prolong the QT interval on the ECG, and cause TdP. At present, more than 50 noncardiac drugs have been reported to significantly prolong the corrected QT interval and/or induce TdP. Table 10.3 shows a list of drugs that can cause QT prolongation. The potential for noncardiac drugs to alter cardiac repolarization by prolonging the QT interval and to induce potentially fatal arrhythmias such as TdP is now the second leading cause for withdrawing approved drugs from the market.[45]

Possible risk factors that increase the likelihood that a person receiving a particular medicine will develop TdP have been studied. A recent case series analysis has shown that increasing age, female gender, concomitant drug use, and other comorbid conditions were common risk factors in TdP induced by macrolides.[46] At least one cardiac abnormality was noted in 42% of all the reports and cardiomyopathy and congestive heart failure were most

TABLE 10.3 List of Drugs that Can Produce QT Prolongation

Category	Individual Drugs
Anticancer drugs	Arsenic trioxide*, eribulin△, lapatinib△, nilotinib△, vorinostat△
Antiemetic drugs	Cisapride*, dolasetron△, domperidone*, granisetron△, ondansetron*
Anesthetic agents	Cocaine*, droperidol*, sevoflurane*
Antiinfectives	Amantadine▽, amphotericin, atazanavir△, azithromycin*, bedaquiline△, ciprofloxacin▽, chloroquine*, clarithromycin*, clindamycin, erythromycin*, fluconazole△, foscarnet△, gatifloxacin△, gemifloxacin△, grepafloxacin*, halofantrine*, itraconazole▽, ketoconazole▽, levofloxacin△, mefloquine, miconazole, moxifloxacin*, norfloxacin△, ofloxacin△, pentamidine*, quinine sulfate▽, rilpivirine△, ritonavir▽, roxithromycin△, saquinavir△, sparfloxacin*, spiramycin, telavancin△, telithromycin△, trimethoprim sulfamethoxazole▽ troleandomycin, voriconazole△
Antihistamines	Astemizole*, azelastine, clemastine, diphenhydramine▽, ebastine, hydroxyzine, promethazine△ and terfenadine*
Antiseizure drugs	Felbamate△, fosphenytoin△
Drug for peptic ulcer	Famotidine△
Drugs used for psychiatric disorders	Amisulpiride▽, amitriptyline▽, amoxapine▽, chlorpromazine*, citalopram*, clomipramine▽, clozapine△, desipramine▽, doxepin▽, escitalopram*, fluoxetine▽, haloperidol*, iloperidone△, imipramine▽, lithium▽, mesridazine*, mirtazapine△, nortriptyline▽, olanzapine△, paroxetine▽, pimozide*, protriptyline▽, quetiapine△, risperidone△, sertindole△, sertraline▽, thioridazine*, trazodone▽, trimipramine▽, venlafaxine△, ziprasidone△
Diuretics	Indapamide△, triamterene, moexipril/HCTZ△
Hormones	Octreotide, vasopressin
Immunosuppressive	Tacrolimus△
Migraine–serotonin	Naratriptan, sumatriptan, zolmitriptan
Muscle relaxant	Tizanidine△
Narcotic detoxification, narcotics	Levomethadyl*, methadone*
Respiratory	Salmeterol, sympathomimetics
Miscellaneous agents	Cisapride*, budipine, furosemide, probucol*, terodiline

**Drug with known risk of TdP.*
△Drug with possible risk of TdP.
▽Drug with conditional risk of TdP.

Source: Compiled from Arizona Center for Education and Research on Therapeutics' CredibleMeds. Available from http://crediblemeds.org/.

commonly noted with cisapride, accounting for the majority of the drug interactions. Another report of four cases found risk factors for QT interval prolongation induced by drugs were female gender, increasing age, presence of underlying heart disease (including congestive heart failure), hypokalemia, hypomagnesemia, and use of other drugs known to prolong the QT interval.[47]

The CredibleMeds website provides a list of different medicines that prolong the QT interval and/or cause TdP.[48] A recent article proposes a scheme for risk stratification for cardiac events among patients with long-QT syndrome according to genotype and gender.[49] Mutations in potassium-channel genes KCNQ1 (LQT1 locus) and KCNH2 (LQT2 locus) and the sodium-channel gene SCN5A (LQT3 locus) are the most common causes of the inherited long-QT syndrome. Low-risk patients have a QTc of <500 msec, male gender, and mutations at the LQT2 and LQT1 locuses. Patients at moderate risk have a QTc of <500 msec and could be female with an LQT2 or LQT3 mutation or males with a mutation at LQT3.

The authors concluded that prophylactic treatment is warranted in male and female patients with a mutation at the LQT1 locus who have a QTc of 500 msec or more, male patients with a mutation at the LQT2 locus who have a QTc of 500 msec or more, all female patients with a mutation at the LQT2 locus irrespective of the QTc, and all patients with a mutation at the LQT3 locus.[49] Thirteen different variants of the long QT syndrome and the genetic mutations associated with the same have been elucidated recently.[50]

Electrolyte disorders can alter cardiac ion currents and can promote or reduce arrhythmogenesis depending on the changes.[51] Hypokalemia can result in prolonged action potential duration and QT interval, early afterdepolarizations, and TdP. Hypocalcemia can result in prolonged ST segment and QT interval. Magnesium disorders are frequently associated with other electrolyte abnormalities rendering its clinical significance difficult to determine. Intravenous magnesium by blocking the L-type Ca^{2+} current can successfully terminate TdP without affecting the prolonged QT interval. Also, despite the frequency of sodium abnormalities, particularly hyponatremia, its effects on cardiac electrophysiology are rarely clinically significant.[51] The potassium currents are modulated by the potassium gradient itself and the concentration of other electrolytes (e.g., sodium, magnesium, and calcium).

Fever and hyperthermia have been reported to cause QT interval prolongation and increase the risk of TdP in patients with type 2 congenital long QT syndrome.[52,53] This has been postulated to be due to a weak increase in the hERG current density.[53] Hypothermia also prolongs the QTc interval and increases the potential for developing TdP.[54] However, further investigations are required.

Predicting compounds that are likely to prolong the QT interval and have a risk of inducing TdP is of great importance to the pharmaceutical industry and new drug development. Measuring the activity of the compound at the hERG channel is an important first step as, in the majority of cases, a drug inactive at this channel may not prolong the QT interval.[55] Plotting the QT interval against the heart rate in a QT nomogram as previously described[41] is a good risk assessment tool for QT prolongation and the occurrence of TdP.

A study published in 2009 examined the magnitude of QT interval prolongation of 30 nonantiarrhythmic drugs.[56] The authors concluded that prolongation of the QTc interval greater than 12 msec when patients were on monotherapy, or 25 msec in the presence of metabolic inhibition, favored a stronger association with TdP. A recent article states drug effects on QTc interval do not always explain the observed QTc values in the population.[57] The prevalence of high QTc values in the real-life population can be influenced by comorbidities and concomitant medications. These factors should be accounted for when evaluating cardiovascular risk of medicinal products. Drugs that show blockage of the hERG potassium channel at concentrations close to those achieved during clinical use are at risk of causing TdP.[58]

The authors suggested that a 30-fold margin between IC_{50} and C_{max} at the hERG channel may be enough to ensure safety from arrhythmogenesis. The margin can be reduced for drugs used in diseases that are lethal if not treated and for which there is a paucity of effective drugs. The hERG channel shows promiscuous binding across a wide range of chemical structures and class characteristics may not be predictive of arrhythmogenic potential.

10.5 DRUG-INDUCED ATRIAL FIBRILLATION

Several classes of drugs can induce atrial fibrillation (AF) through very different mechanisms.[59] Sympathomimetics, parasympathomimetics, adenosine, and theophylline exert direct electrophysiological effects such as increasing focal activity, shortening action potential duration, and/or slowing conduction. Alcohol, acetylcholine, adenosine, and sympathomimetics change the cardiac autonomic tone (increase in sympathetic and/or vagal tone) while amifostine, apomorphine, calcium channel blockers, docetaxel, interleukin-2, paclitaxel, and phosphodiesterase inhibitors cause drug-induced hypotension leading to changes in cardiac autonomic tone.

Certain cancer chemotherapeutic drugs can cause direct myocardial damage while some drugs (e.g., acetylcholine, chemotherapy agents, ondansetron, and sumatriptan) can cause coronary vasoconstriction/thrombosis leading to myocardial ischemia. Other mechanisms, including electrolyte disturbances (e.g., diuretics, glucocorticoids), abnormalities in Ca^{2+} handling (alemtuzumab, inotropic agents), release of cytokines (interleukin-2), and increased oxidative stress (chemotherapy agents) have also been implicated in drug-induced atrial fibrillation. Bisphosphonate-induced AF is described in Section 10.7.5.[60]

10.6 VALVULAR HEART DISEASE

Stedman's electronic medical dictionary defines valvular disease as any disease process involving one or more of the valves of the heart (the aortic and mitral valves on the left and the pulmonary and tricuspid valves on the right). This condition can also be precipitated by the use of drugs such as the combination of fenfluramine and phenteramine (appetite suppressants), ergot alkaloids used for migraines, and ergot-derived dopamine agonists (e.g., cabergoline and pergolide) used for Parkinson's disease. These have been associated with fibrotic reaction in the heart with consequent valvular regurgitation.

Serotonin may exert proliferative effects on fibroblasts and smooth muscle cells[61–63] through regulation of transforming growth factor β, which in turn stimulates the production of glycosaminoglycan.[64] The serotonin receptor 5HT2B found on both aortic and mitral valves is known to have proliferative action, thus contributing to the development of valvulopathy. In the patient with drug-induced valvular heart disease, it has been found that when drugs such as appetite suppressants are used, high concentrations of glycosaminoglycan are detected. Studies done in animal models also suggest that long-term administration of serotonin induces valvular lesions similar to those seen in drug-induced and carcinoid heart disease.[65,66]

Drugs, such as norfenfluramine (a metabolite of fenfluramine), methylergonovine (an active metabolite of ergotamine and methysergide), and pergolide, have a high affinity for 5HT2B receptors. By contrast, fluoxetine and trazodone (drugs not known to cause valvular heart disease) have a low affinity for this receptor. Activation of 5HT2B receptor causes dissociation of the G-protein, which leads to activation of protein kinases and potentiation of the effects of transforming growth factor β, thought to be involved in cell proliferation.[67] The development of drug-induced valvular heart disease is probably a manifestation of the complex interaction between various factors including serotonin, 5HT2B receptors, and the serotonin transporter. Individual susceptibility to drug-induced valvular disease may be attributed partially to the interplay of these factors or to polymorphisms in gene expression.[68]

10.6.1 Ergot Alkaloids

The ergot alkaloid derivatives methysergide and ergotamine are two classical drugs used in the prophylaxis and treatment of migraine headaches. Serotonin is the agent responsible for valve disease in the carcinoid syndrome, involving endocardial fibrosis. In the mid-1960s, Graham reported the development of valvular insufficiency in patients on these drugs; 36 patients taking methysergide developed cardiac murmurs of mitral or aortic insufficiency, or both, during treatment.[69] On discontinuation of therapy, these murmurs regressed either completely or partially in more than a third of the patients.[69]

During those days it was thought that only left-sided valves were affected; however, in the 1970s a group of physicians during an autopsy examination found that the tricuspid valves were also affected.[70,71] Ergotamine-induced valvular disease was first reported in 1974[72] and patients that have ergotamine-related and methysergide-related disease are still being reported.[73–75] Although rarely used now because of their adverse effects, methysergide and ergotamine remain licensed for migraine prophylaxis.[76–78]

It has been reported that chronic ingestion of methysergide or ergotamine can induce endocardial thickening that results in valve dysfunction.[76] The endocardial involvement comprises a fibrotic reaction that coats valves, chordae, papillary muscles, and the endomyocardial surface. Fibrosis causes valve and chordae retraction that results in either stenosis or regurgitation. The process is similar to that described in the carcinoid syndrome, but while carcinoid-associated valve disease is restricted to the right-sided valves (except in the case of bronchial carcinoid), in ergot alkaloid-associated valve disease; although all four valves can be involved, the aortic and mitral valves are most often damaged.

Having examined common heart diseases, let us now look at a selection of individual drugs that can cause cardiovascular toxicity.

10.7 INDIVIDUAL NONCARDIOVASCULAR DRUGS CAUSING CARDIOVASCULAR TOXICITY

This section examines cardiovascular toxicity caused by various classes of drugs, including antimicrobial agents; antihistaminics; centrally acting appetite suppressants; gastrointestinal tract drugs; drugs acting on the endocrine system; drugs that act on the musculoskeletal system, nervous system, and respiratory system; and miscellaneous drugs.

10.7.1 Antimicrobial Chemotherapeutic Agents: Erythromycin and Others

Large doses of erythromycin are known to increase the risk of the heart disturbance called QT interval prolongation. However, there are limited studies regarding the effect of commonly used oral doses of the antibiotic on the QT interval. QT interval prolongation can result in chaotic heartbeat, which can cause fainting, seizures, and even sudden cardiac death. Researchers found the rate of sudden cardiac death was twice as high in people who were currently taking erythromycin and five times as high among those taking erythromycin and a strong CYP3A inhibitor. Other examples of CYP3A inhibitors are ketoconazole, itraconazole, fluconazole, cimetidine, clarithromycin, troleandomycin, and grapefruit juice.[79]

Some antibiotics, such as macrolides and fluoroquinolones, antimalarials, and imidazole antifungal agents, can also cause QT prolongation and TdP. Torsades de pointes is rare, and it does not occur with the entire range of antimicrobials that prolong the QT interval. Intravenous erythromycin lengthens the QT interval, causes dispersion of recovery across the ventricular wall, and can induce torsades de pointes. The fluoroquinolones sparfloxacin, levofloxacin, and grepafloxacin (now withdrawn in most countries) cause prolongation of the QT interval; however, it does not occur with ofloxacin. Quinine prolongs the QT interval at standard doses, as does halofantrine, particularly when combined with mefloquine. Ketoconazole prolongs the QT interval by directly blocking I_{Kr} and by delaying the cytochrome P450-dependent metabolism of other drugs that prolong the QT interval.

10.7.2 Antihistaminics

In the early 1990s two nonsedating antihistamines, terfenadine and astemizole, captured the attention of regulatory authorities because of their unexpected association with QT prolongation, TdP, and sudden death. The adverse effects of terfenadine appeared to be concentration-dependent, occurring at supraclinical doses or at normal doses in patients also taking other drugs that inhibit cytochrome P450 drug metabolizing enzymes (e.g., imidazole antifungals and some macrolide antibiotics) and in patients with congenital long QT syndrome. Adverse effects of most QT interval prolonging drugs may be concentration-dependent.

The cardiac safety of newer nonsedating antihistamines (e.g., ebastine, loratadine, cetirizine, acrivastine, fexofenadine, and mizolastine) required confirmation, but only some and not others block I_{Kr} and may well provoke clinical arrhythmia. The likelihood of causing cardiac arrhythmia must be assessed carefully because these drugs are liberally prescribed for self-limiting, nonfatal diseases.[80] Terfenadine and astemizole prolonged the QT interval in laboratory animals but this was not seen with loratidine or cetrizine.[81] In vitro studies showed that terfenadine and astemizole block the cardiac K^+ channels, leading to delayed ventricular repolarization and QT-interval prolongation; in contrast, neither loratadine nor its metabolite desloratadine significantly inhibited cardiac K^+ channels at clinically achievable blood levels. Studies confirm the absence of electrocardiographic effects of azelastine, cetirizine, fexofenadine, and loratadine at several times the recommended dose or when administered concomitantly with agents that inhibit their metabolism and elimination.[81] The authors concluded that loratadine, cetirizine, azelastine, and fexofenadine are not associated with torsades de pointes or other ventricular arrhythmias.

10.7.3 Centrally Acting Appetite Suppressants

Centrally acting appetite suppressants can cause pulmonary hypertension by inducing vasoconstriction or may cause abnormalities in cardiac valvular morphology.

Dexfenfluramine-Induced Pulmonary Hypertension and Valvular Heart Disease

Dexfenfluramine and fenfluramine cause pulmonary vasoconstriction by inhibiting voltage-gated potassium channels in the smooth muscle cells of resistance-level pulmonary arteries. Although pulmonary vasoconstriction due to these anorexigens was observed in all normal rats, the degree of the response was small and tended to occur when the dose of the drug was increased. Dexfenfluramine causes marked pulmonary vasoconstriction at concentrations similar to those used in humans for weight reduction.[82–84] The anorexic agent fenfluramine considerably increases the risk of primary pulmonary hypertension. The mechanism of pulmonary hypertension's effect is unknown. The action of fenfluramine that reduces the appetite is mediated by its interaction with the metabolism of serotonin—5-hydroxytryptamine (5-HT)—in the brain. Preexisting K_V-channel insufficiency may predispose some patients to the development of pulmonary hypertension during fenfluramine treatment.[85]

Exposure to fenfluramine, alone or in combination with phentermine, a noradrenergic central nervous system stimulant, has been related to abnormal cardiac valvular morphology and resulting regurgitation of the left and right heart valves. The prevalence of significant valvular disease associated with the use of these anorectic drugs is reported to be as high as 23%.[84]

10.7.4 Gastrointestinal Tract Drugs

Cisapride has attracted much recent attention because of reports of QT prolongation and torsades de pointes. Thirty-four cases of TdP and 23 cases of QT prolongation associated with cisapride have been reported to the Food and Drug Administration. Cisapride has been withdrawn from the market in many countries including the United States.[86] Proton pump inhibitors (PPIs) have been correlated with lower magnesium levels and the risk of arrhythmias.[60] Patients receiving these drugs should be followed closely for magnesium deficiency, especially if they experience acute cardiovascular events.

A review examined the increased risk of myocardial infarction, stroke, and cardiovascular events associated with coprescription of PPI and clopidogrel.[87] The coprescription of these two drugs was recommended to be avoided. However, no increase in cardiovascular events was noted in major trials, and there have been observational studies on the use of these two groups of drugs together, thus undermining the clinical relevance of avoiding their coprescription.[88,89]

10.7.5 Drugs that Act on the Endocrine System

A variety of drugs that act on the endocrine system (e.g., bisphosphonates, corticosteroids, thyroid hormone, growth hormone, and oral contraceptives) may exert cardiovascular adverse effects.

Bisphosphonate-Induced Atrial Fibrillation

Altered intracellular electrolyte homeostasis and proinflammatory, profibrotic, and antiangiogenic effects provide potential mechanisms by which atrial conduction could be affected in patients treated with bisphosphonates. A recent thorough review on this topic examined the potential relationship between atrial arrhythmogenesis and the pharmacological action of bisphosphonates. The authors concluded that evidence that bisphosphonates affect atrial conduction is currently lacking but might be an important area for further investigation.[90]

Corticosteroids

High doses of corticosteroids are standard treatment for a wide array of medical disorders. There are several case reports about atrial fibrillation after pulse methylprednisolone therapy of two patients with multiple sclerosis, of two children with nephrotic syndrome (one with systemic lupus erythematosus [SLE]), of a man with SLE, and of a woman with rheumatoid arthritis. It has been postulated that methylprednisolone mediates potassium efflux via a direct effect on the cell membrane. The local potassium efflux in turn may influence arrhythmogenesis.[91]

Thyroid Hormone

Thyroid hormone is an important regulator of cardiac gene expression and may be responsible for many of the cardiac manifestations of thyroid dysfunction.[92,93] The thyroid hormone affects the repolarization currents and duration of the action potential in cardiac myocytes through both genomic and nongenomic mechanisms.[94] Sinus tachycardia is the rhythm disturbance most commonly associated with the heart in thyrotoxicosis. The prevalence of atrial fibrillation in this disease varies between 2% and 20%.

Growth Hormone

Excess of growth hormone (GH) exerts adverse effects on the cardiovascular system.[95] Patients with a GH excess develop acromegalic cardiomyopathy, biventricular hypertrophy, diastolic dysfunction, hypertension, endothelial dysfunction, mitral and aortic valve abnormalities, and dysrhythmias; while growth hormone deficiency may predispose one to increased risk of atherosclerosis, decreased left ventricular mass, and dyslipidemia.

Oral Contraceptives

In women taking high doses of oral contraceptive pills (OCPs), the incidence of hypertension is 5%. Family history of hypertension, preexisting gestational hypertension, cigarette smoking, diabetes, obesity, and age of more than 35 years increase the risk. The mechanism thought to be involved in OCP-induced hypertension is increased synthesis of angiotensinogen in the liver, which causes an increased level of aldosterone secretion, plasma volume, and exchangeable sodium.[96]

10.7.6 Drugs that Act on the Musculoskeletal System

NSAIDs are an important group of drugs acting on the musculoskeletal system that may exert cardiovascular adverse effects.

Nonsteroidal Antiinflammatory Drug-Induced Hypertension

The nonsteroidal antiinflammatory drugs (NSAIDs) have potentially adverse effects on blood pressure. NSAIDs inhibit both cyclooxygenase-1 (COX-1) and cyclooxygenase-2 (COX-2) enzymes, thus resulting in a reduction in the formation of prostaglandin (PG). These drugs can have a wide range of beneficial and harmful effects, depending on the situation of the patient. Drug-induced hypertension related to NSAIDs is due to the effect of the drug on the renal system. For example, dose-related sodium and water retention usually occur with the use of NSAIDs. This action is also seen with COX-2 selective agents such as celecoxib.

The COX-1 and COX-2 isoforms are both expressed within the normal adult kidney, with COX-1 in the glomerulus and afferent arteriole and COX-2 in the afferent arteriole, the podocytes, and macula densa. The different locations of each of these isoenzymes in the kidney translate into notably different effects on renal function. The prostaglandins produced by COX-1 affect renal homeostasis mainly by promoting vasodilation in the renal vascular bed, reducing renal vascular resistance, consequently increasing renal perfusion. Prostaglandins produced by the COX-2 isoenzyme have diuretic and natriuretic effects. In patients who are hemodynamically compromised, the two isoenzymes are essential for the maintenance of renal perfusion because of their vasodilatory effects. The production of COX-1 and COX-2 prostaglandins is inhibited by NSAIDs, and renal side effects are not uncommon, occurring in approximately 1 to 5% of NSAID users.

Arachidonic acid released from phospholipids is converted by cycleoxygenase in the kidney to PGE_2, $PGF_{2\alpha}$, PGD_2, and, probably, to PGI_2 and thromboxane A_2. PGE_2 and $PGF_{2\alpha}$ production is predominantly, but not exclusively, in the medulla, whereas degradative enzymes are present in both the cortex and the medulla. Entry of prostaglandin into the lumen takes place by facilitated transport and they are partially reabsorbed from the urine in the distal nephron. Urine prostaglandins probably reflect renal synthesis. In response to ischemia, vasoconstriction, or angiotensin II, the synthesis of prostaglandin in the kidney will increase to modulate renal vascular resistance. In conscious animals or humans, no role has been recognized for prostaglandins in the maintenance of basal renal blood flow or renal sodium excretion. By inhibiting the action of vasopressin, prostaglandin E

(PGE) influences renal water excretion. There is evidence that renal prostaglandins are involved either primarily or secondarily in many types of hypertension.[97]

NSAID-induced hypertension occurs due to the inhibition of prostaglandin synthesis. Prostaglandins play an important role in regulating blood pressure by exerting vasodilatory and natriuretic effects. Inhibition of the prostaglandins by NSAIDs causes renal vasoconstriction with sodium and water retention, weight gain, and blood volume expansion, increasing blood pressure (BP).

10.7.7 Drugs that Act on the Nervous System

With regard to drugs acting on the nervous system, anesthetics, antimigraine drugs, dopamine agonists, anticholinergics, antiepileptics, and hypnotics exert cardiovascular effects.

Anesthetics

Atrial fibrillation (AF) has been described during epidural anesthesia using bupivacaine in a person with a history of stable angina pectoris but no arrhythmias. The atrial fibrillation persisted despite treatment during the period of epidural blockade. Reports suggest that bupivacaine inhibits the Na^+/K^+ pump, thus reducing the potential of the resting cell membrane. Resultant changes may increase cardiac automaticity, particularly in cardiac fibers previously partially depolarized because of ischemic heart disease. It has been postulated that bupivacaine acted as a trigger for the onset of atrial fibrillation.[98]

Antimigraine Drugs

The antimigraine drug sumatriptan, which is a serotonin-1 agonist, has commonly been associated with chest pain and myocardial infarction presumably due to vasoconstriction of the coronary arteries. Sumatriptan-induced atrial fibrillation is uncommon, but several cases have been reported with positive rechallenge, which means occurrence of an anticipated adverse drug reaction after administering sumatriptan. In vivo sumatriptan causes rapid constriction of dural and meningeal vessels. During migraine attack, it does not modify cerebral blood flow but constricts arteriovenous anastamoses that may be dilated.

This evidence recommends that, with its antimigraine activity, sumatriptan has a direct, dose-related, vasoconstrictor action on certain intracranial blood vessels. However, sumatriptan may act directly on the trigeminal sensory nerve terminals within the cranial blood vessel, thus inhibiting the release of sensory neuropeptides. Myocardial ischemia secondary to coronary vasospasm could be a trigger for atrial fibrillation.[99]

Dopamine Agonists

Use of apomorphine, a dopamine receptor agonist in a patient devoid of cardiovascular disease, has been shown to have an association with AF. The man developed AF within 5 min of apomorphine administration and this was converted to sinus rhythm (SR) after medical treatment. In this person, other possible causes for atrial fibrillation were excluded. In 10% of patients with Parkinsonism, subcutaneous use of apomorphine injection has been reported to cause postural hypotension and vasovagal response. It has been shown that AF in this patient may have been induced by an imbalance of autonomic tone associated with increased vagal activity. The risk of cardiac valve regurgitation may increase with the use of dopamine agonists (e.g., pergolide and cabergoline) for the treatment of Parkinson's disease and restless leg syndrome.[100]

Anticholinergics

Cardiac dysrhythmias are among the major adverse reactions of the anticholinergic and vagal inhibitory agent atropine. It has shown that there is an association of AF with use of ophthalmic atropine eye drops after glaucoma surgery. Two cases have been described, and AF resolved after antiarrhythmic treatment.[101] A recent review mentioned that atropine may induce AF through the cholinergic muscarinic receptors, leaving the nicotinic receptors in the autonomic ganglia free; however, details were not provided.[102] A possible mechanism may be blockade of muscarinic receptors on the heart, leaving the nicotinic receptors in the ganglia to be stimulated, resulting in ganglionic activation.

A large number of drugs cause central anticholinergic syndrome and post-anesthetic depression, and the drug physostigmine, a cholinesterase inhibitor and vagal stimulator, is commonly used to treat such conditions. One case report describes the occurrence of AF directly after the administration of physostigmine. The mechanism is thought to be due to severe depression of both the sinoatrial and atrioventricular nodes as a result of increased vagal tone. Also, donepezil, a cholinesterase inhibitor used in Alzheimer's disease, has been associated with atrial fibrillation in clinical trials.[91]

Antiepileptic Drugs

Antiepipleptic drugs (e.g., phenytoin, fosphenytoin, carbamazepine, oxcarbazepine, and lamotrigine) should not be prescribed for elderly patients with abnormalities of cardiac conduction or a history of ventricular arrhythmia.[103] The occurrence of bradycardia, hypotension, respiratory depression, and cardiopulmonary arrest has been reported in persons receiving intravenous injections of phenytoin and fosphenytoin.[104,105] This reaction has been postulated to be due to the solvent, propylene glycol in which the drugs are dissolved. Many serologic markers that are relevant to cardiovascular risk have been shown to be elevated in patients on enzyme-inducing antiepileptics.[106]

Phenytoin and carbamazepine especially have been implicated and the risk may be less with newer antiepileptics, which do not induce microsomal enzymes. Children with epilepsy may be at an increased risk of developing atherosclerosis.[107] The newer antiepileptic drugs may be a safer option for children with a higher risk of developing atherosclerotic disease but further, long-term studies are needed.

Hypnotics

Various medications used to induce sedation, including benzodiazepines and propofol, can cause hypotension. The causes could be decreased environmental stimulation and sympathetic tone and vasodilation and hypotension; it more commonly occurs in persons who are hemodynamically unstable and hypovolemic.[108] Bradycardia and QT interval prolongation and arrhythmias are also noted and may be increased by electrolyte abnormalities, structural abnormalities of the heart, genetic predisposition, and drug–drug interactions.

A recent review examined the cardiovascular manifestations of sedatives and hypnotics in a critical care unit.[109] Benzodiazepines provide the greatest cardiovascular safety but could cause significant hypotension in hemodynamically unstable patients. Propofol may cause bradycardia, lactic acidosis, and increased risk of cardiac arrest, especially in patients with head injury and those suffering from septic shock, but is otherwise a drug with cardioprotective features. Etomidate shows good hemodynamic stability even in patients suffering from shock and hypovolemia.

10.7.8 Drugs that Act on the Respiratory System

Sympathomimetics and xanthines are among the group of drugs acting on the respiratory system that can have cardiovascular adverse effects.

Sympathomimetic Inhalants

Alpha and beta sympathomimetic inhalants are prescribed to induce bronchodilation in lung disease patients. They are known for their potential to cause cardiovascular adverse effects such as sinus tachycardia and exacerbation of existing arrhythmias. Drug-induced AF has been reported in infants <1 year of age after excessive therapeutic doses of pseudoephedrine (>4 mg/kg/day). Albuterol treatment using a spacer device was reported to induce atrial fibrillation with a positive rechallenge in a healthy young man.[110] The arrhythmia was not seen when the spacer was discontinued and was postulated to be due to increased deposition of the drug when used with a spacer device.

Xanthines

The arrhythmogenic, positive inotropic and chronotropic effects of xanthines are well known. Atrial fibrillation associated with intravenous aminophylline has been reported in three patients without any underlying cardiac disease. Rechallenge was positive in one patient. Conversion to SR occurred 9 to 14 h after cessation of the drug in all three patients.[111]

10.7.9 Miscellaneous Drugs

It is suggested that sildenafil caused profound hypotension leading to syncope and reflex tachycardia via catecholamine excess. Hypotension occurs due to vasodilation causing reflex tachycardia. It is speculated that atrial fibrillation caused by increased sympathetic activity due to hypotension may be provoked by sildenafil.[112]

There are several noncardiovascular drugs that can have an effect on blood vessels leading either to hypotension or hypertension, which sometimes can blunt the effect of antihypertensive treatment. This leads to exacerbation of previously well-controlled hypertension and/or turns it into resistant to therapy. However, drugs that are used for the treatment of erectile dysfunction, such as phosphodiesterase type-5 inhibitors (e.g., sildenafil, tadalafil, and vardenafil), have notable systemic hemodynamic effects that can cause modest reduction in systolic and diastolic blood pressure without significant effect on heart rate.

A study done by Arruda-Olson et al. found that when sildenafil is used in combination with nitrates, there is a dramatic decrease in blood pressure.[113] Therefore a combination of these drugs is contraindicated.[112,113] In the patient with benign prostatic hypertrophy, alpha-1 adrenergic receptor antagonists are commonly used and when used together with phosphodiesterase type-5 inhibitors, it causes or worsens orthostatic hypotension and syncope especially in elderly patients.[114–116]

Resistant hypertension is a significant condition defined as the failure to reach the targeted blood pressure in patients who are following lifestyle modifications and taking full doses of an appropriate 3-drug regimen that includes a diuretic.[117,118] Many conditions can cause resistant hypertension, and drugs may be responsible in nearly 58% of the cases.[119] Examples of drugs that may cause resistant hypertension are NSAIDs, cocaine, amphetamines, sympathomimetics, oral contraceptives, adrenal steroids, and so on.[120]

Corticosteroids are an important class of drugs having antiinflammatory action and are indicated in a wide variety of disease conditions. The prevalence rate of hypertension is 20% of patients on synthetic corticosteroids in a dose-dependent manner. The mechanism involved in the drug-induced hypertension is due to increased synthesis of angiotensinogen, sympathetic nervous system activation, and the mineralocorticoid effect causing salt and water retention.[121] Two-thirds of patients were reported to have developed drug-induced hypertension due to the chronic use of erythropoietin.[122] The mechanism involved is complex and includes an increase in cytosolic Ca^{2+} content in vascular smooth muscles,[123] activation of the renin–angiotensin system (RAS),[124] and an increased level of endothelin-1 production,[125] among others.

Cyclosporine has been shown to induce hypertension in human subjects.[126] Several mechanisms including activation of the sympathetic nervous system, endothelin-mediated systemic vasoconstriction, impaired vasodilatation secondary to reduction in prostaglandin and nitric oxide, altered cytosolic calcium translocation, and activation of the renin–angiotensin system have been suggested.[127,128] A recent study proposes that the drug represses HNF4-alpha activity through calcineurin inhibitor-mediated inhibition of the nuclear factor of activation of T cells (NFAT) which in turn represses HNF4-alpha, leading to a disturbed balance of the RAS.[129]

10.8 RECENT ADVANCES

Drugs may have beneficial as well as hazardous effects. The study done by Keller et al. showed that cardiovascular effects were common adverse drug reactions often caused by noncardiovascular drugs and highlighted the importance of systematic evaluation of adverse drug reactions.[130] A number of noncardiovascular drugs have been withdrawn from the market due to unacceptable adverse cardiac side effects such as sudden cardiac death, arrhythmias, and QT-interval prolongation. Terfenadine is a nonsedating antihistaminic agent that was first marketed during the 1990s for the treatment of various allergic conditions. Later it was shown to be associated with fatal cardiac arrhythmias when administered along with cytochrome P450 enzyme inhibitors such as erythromycin. Therefore, it was withdrawn in 1997 after the approval of fexofenadine.[131,132]

In the same year, fenfluramine and dexfenfluramine were also withdrawn as the data suggested that these drugs can cause cardiac valvular disease.[133] Cisapride, a drug that increases gastrointestinal motility, was withdrawn from the market in 2000 following reports of cardiac arrhythmias and deaths.[134] However, it has been reported that the use of drugs, such as older tricyclic antidepressants (e.g., amitriptyline and imipramine), fluoroquinolone antibiotics (e.g., moxifloxacin and grepafloxacin), and some opioids (e.g., methadone), can cause QT prolongation.

Clinical trials test the safety and efficacy of a drug in a limited number of individuals and under controlled conditions of use. Thus, postmarketing surveillance of drugs is important as information is obtained about adverse effects under conditions of routine use in a diverse patient population. Cardiovascular events can be determined by direct clinical observation; however, potentially important pathophysiologic changes can be measured objectively to predict events as well. At present, the FDA and the European Medicines Agency (EMA) have mandated measurement of the electrocardiographic QT interval during the development of drugs for those that are capable of causing potentially lethal arrhythmias.[135,136] Similarly, for drugs having an effect on cardiac repolarization, ICH documents E14 and S7B emphasize the need for a thorough evaluation of the QT interval during preclinical and clinical trials,[137,138] supported by in vivo cardiovascular safety studies and in vitro electrophysiological studies.[139] The ECG-bedside and quick test, which is important to obtain information about the possible cardiotoxicity of drugs including fatal arrhythmias, was shown earlier in Figure 10.1. This is a schematical presentation of the action potential and ECG in the heart. (The figure is reproduced with permission from Katzung et al.[140])

10.9 CONCLUSION

There are many noncardiovascular drugs that can have off-target effects on the heart. Some effects are common and their mechanisms are understood, others are rare, uncommon, unproven, and/or not well understood. Various cases of withdrawal of a drug due to the harmful cardiovascular effects suggest that there is a need for new approaches to evaluate potential cardiovascular risk in the context of noncardiovascular therapy.

Key Learning Points

- Due to extensive use of drugs during the course of treatment, patients may suffer from minor to severe cardiovascular ADRs, which can have unexpected and serious consequences.
- Cardiotoxicity is a well-known side effect of several drugs, especially noncardiovascular ones such as antimicrobial agents, antihistamines, antimigraine drugs, NSAIDs, and so on.
- Several types of cardiac complications, such as myocardial infarction, acute coronary syndrome, heart failure, cardiac arrhythmias, and valvular heart disease, have been identified that may be caused by the use of noncardiovascular drugs.
- There is a need for new approaches to evaluate potential cardiovascular risk in the context of noncardiovascular therapy.

GLOSSARY

Arrhythmias Also called cardiac dysrhythmia. A condition where there is abnormal electrical activity in the heart.

Cardiotoxicity Occurrence of heart electrophysiology dysfunction and/or muscle damage.

Drug Also referred to as medicine or medication. Can be loosely defined as any chemical substance (or product comprising a chemical substance) intended for use in medical diagnosis, cure, treatment, or prevention of disease.

Drug-induced disease A disease that is due to the use of a drug or drugs.

Heart failure Often called congestive heart failure (CHF) or congestive cardiac failure (CCF). Occurs when the heart is unable to provide sufficient pump action to maintain blood flow to meet the needs of the body.

Myocardial ischemia An imbalance between myocardial oxygen supply and demand. Left untreated, it results in angina pectoris; myocardial stunning; myocardial hibernation; ischemic preconditioning or postconditioning; or, under the most severe instances, acute coronary syndrome and myocardial infarction.

Noncardiovascular drugs Medicines used to treat a disease other than a cardiovascular one.

Valvular heart disease Any disease process involving one or more of the heart valves (the aortic and mitral valves on the left and the pulmonary and tricuspid valves on the right).

REFERENCES

1. Wallace KB, Hausner E, Herman E, et al. Serum troponins as biomarkers of drug-induced cardiac toxicity. *Toxicol Pathol* 2004;**32**:106–21.
2. Fox KA, Birkhead J, Wilcox R, et al. British Cardiac Society Working Group on the definition of myocardial infarction. *Heart* 2004;**90**:603–9.
3. Ottervanger JP, Wilson JH, Stricker BH. Drug-induced chest pain and myocardial infarction. Reports to a National Centre and Review of the Literature. *Eur J Clin Pharmacol* 1997; **53**:105–10.
4. Laverty H, Benson C, Cartwright E, et al. How can we improve our understanding of cardiovascular safety liabilities to develop safer medicines? *Br J Pharmacol* 2011;**163**:675–93.
5. Ardehall A, Port TA. Myocardial oxygen supply and demands. *Chest* 1990;**98**:699–705.
6. CDC. Chapter 3: Physiological responses and long-term adaptations to exercise In: Blair SN, Franks AL, Shelton DM, Livengood JR, Hull FL, Breedlove B, editors. Physical activity and health. A report of the Surgeon General. U.S. Department of Health and Human Services, Centers for Disease Control and Prevention, National Center for Chronic Disease Prevention and Health Promotion, 1996. p. 61–80.
7. Best B. Electrocardiogram (ECG, EKG) interpretation emphasizing ST-segment and T wave deviation (ischemic changes). Available from: <www.benbest.com/health/ECG.html#ST> [accessed 08.21.13].
8. Zeidman A, Fradin Z, Blecher A, et al. Anemia as a risk factor for ischemic heart disease. *Isr Med Assoc J* 2004;**6**:16–8.
9. Shulman LN, Braunwald E, Rosenthal DS. Hematological-oncological disorders and heart disease. In: Braunwald E, editor. *Heart disease: A textbook of cardiovascular medicine*. 5th ed. Phildadelphia: WB Saunders; 1997. p. 1786–808.
10. Brannon ES, Merrill AJ, Warren JV, et al. The cardiac output in patients with chronic anemia as measured by the technique of right atrial catheterization. *J Clin Invest* 1945;**24**:332–6.
11. Florezano F, Diaz G, Regonesi C, et al. Left ventricular function in chronic anemia: evidence of noncathecholamine positive inotropic factor in the serum. *Am J Cardiol* 1984;**54**:638–45.
12. Duke M, Abelmann WH. The hemodynamic response to chronic anemia. *Circulation* 1969; **39**:503–13.

13. Roy SB, Bhatia ML, Mathur VS, et al. Hemodynamic effects of chronic severe anemia. *Circulation* 1963;**28**:346–56.
14. Case RB, Berglund E, Sarnoff SJ. Ventricular function. VII. Changes in coronary resistance and ventricular function resulting from acutely induced anemia and the effect thereon of coronary stenosis. *Am J Med* 1955;**18**:397–405.
15. Varat MA, Adolph RJ, Fowler NO. Cardiovascular effect of anemia. *Am Heart J* 1972; **83**:415–26.
16. Zeidman A, Fradin Z, Blecher A, et al. Anemia as a risk factor for ischemic heart disease. *Isr Med Assoc J* 2004;**6**:16–8.
17. Tisdale JE, Miller DA, editors. *Drug-induced diseases: Prevention, detection and management.* 2nd ed. American Society of Health System Pharmacist; 2010.
18. Lasser KE, Allen PD, Woolhandler SJ, et al. Timing of new black box warnings and withdrawals for prescription medications. *JAMA* 2002;**287**:2215–20.
19. Topol EJ. Arthritis medicines and cardiovascular events—"house of coxibs." *JAMA* 2005; **293**:366–8.
20. Glassman AH, Bigger Jr. JT. Antipsychotic drugs: prolonged QTc interval, torsade de pointes, and sudden death. *Am J Psychiatry* 2001;**158**:1774–82.
21. Nissen SE. ADHD drugs and cardiovascular risk. *N Engl J Med* 2006;**354**:1445–8.
22. Starner CI, Schafer JA, Heaton AH, et al. Rosiglitazone and pioglitazone utilization from January 2007 through May 2008 associated with five risk-warning events. *J Manag Care Pharm* 2008;**14**:523–31.
23. Garg H, Joshi A, Mukherjee D. Cardiovascular complications of HIV infection and treatment. *Cardiovasc Hematol Agents Med Chem* 2013;**11**:58–66.
24. Ardehall A, Ports AT. Myocardial oxygen supply and demand. *Chest* 1990;**98**: 699–705.
25. Roger VL, Go AS, Lloyd-Jones DM, et al. Heart disease and stroke statistics—2012 update: a report from the American Heart Association. *Circulation* 2012;**125**:e2–220.
26. Kochanek KD, Xu JQ, Murphy SL, et al. Deaths: final data for 2009. *Natl Vital Stat Rep* 2011;**59**(4):117. Available from: <http://198.246.112.54/pub/Health_Statistics/NCHS/Publications/DVD/DVD_3/National_Vital_Statistics_Reports/nvsr60/nvsr60_03.pdf> [accessed 06.27.13].
27. Slordal L, Spigset O. Heart failure induced by non-cardiovascular drugs. *Drug Saf* 2006; **29**:567–89.
28. Mudaliar S, Chang AR, Henry RR. Thiazolidinediones, peripheral edema, and type 2 diabetes: incidence, pathophysiology, and clinical implications. *Endocr Pract* 2003;**9**:406–16.
29. Clarke JB. Mechanism of adverse drug reactions to biologics. *Handb Exp Pharmacol* 2010; **196**:453–74.
30. Crone SA, Zhao YY, Fan L, et al. ErbB2 is essential in the prevention of dilated cardiomyopathy. *Nat Med* 2002;**8**:459–65.
31. Egermayer P, Town GI, Peacock AJ. Role of serotonin in the pathogenesis of acute and chronic pulmonary hypertension. *Thorax* 1999;**54**:161–8.
32. Weir EK, Reeve HL, Huang JM, et al. Anorexic agents aminorex, fenfluramine, and dexfenfluramine inhibit potassium current in rat pulmonary smooth muscle and cause pulmonary vasoconstriction. *Circulation* 1996;**94**:2216–20.
33. Doroshow JH. Effect of anthracycline antibiotics on oxygen radical formation in rat heart. *Cancer Res* 1983;**43**:460–72.
34. Odom AL, Hatwig CA, Stanley JS, et al. Biochemical determinants of adriamycin toxicity in mouse liver, heart and intestine. *Biochem Pharmacol* 1992;**43**:831–6.

35. Takemura G, Fujiwara H. Doxorubicin-induced cardiomyopathy from the cardiotoxic mechanisms to management. *Prog Cardiovasc Dis* 2007;**49**:330–52.
36. Chatterjee K, Zhang J, Honbo N, et al. Doxorubicin cardiomyopathy. *Cardiol* 2010;**115**: 155–62.
37. Zhang YW, Shi J, Li YJ, et al. Cardiomyocyte death in doxorubicin-induced cardiotoxicity. *Arch Immunol Ther Exp (Warsz)* 2009;**57**:435–45.
38. Sanoski CA, Schoen MD, Bauman JL. The arrhythmias. In: Dipiro JT, Talbert RL, Yee GC, Matzke GR, Wells BG, Posey LM, editors. *Pharmacotherapy: A pathophysiologic approach.* 7th Ed. New York: McGraw-Hill; 2008. p. 279–314.
39. Mines GR. On circulating excitations in heart muscle and their possible relation to tachycardia and fibrillation. *Trans R Soc Canada* 1914;**8**:43–52.
40. Malik M, Camm AJ. Evaluation of drug-induced QT interval prolongation: implications for drug approval and labelling. *Drug Saf* 2001;**24**:323–51.
41. Isbister GK, Page CB, Drug-induced QT. prolongation: the measurement and assessment of the QT interval in clinical practice. *Br J Clin Pharmacol* 2013;**76**:48–57.
42. Joy JP, Coulter CV, Duffull SB, et al. Prediction of torsade de pointes from the QT interval: analysis of a case series of amisulpride overdoses. *Clin Pharmacol Ther* 2011;**90**:243–5.
43. Möller C. Keeping the rhythm: hERG and beyond in cardiovascular safety pharmacology. *Expert Rev Clin Pharmacol* 2010;**3**:321–9.
44. Thomas D, Karle CA, Kiehn J. The cardiac hERG/IKr potassium channel as pharmacological target: structure, function, regulation, and clinical applications. *Curr Pharm Des* 2006; **12**:2271–83.
45. Fermini B, Fossa AA. The impact of drug-induced QT interval prolongation on drug discovery and development. *Nat Rev Drug Discov* 2003;**2**:439–47.
46. Shaffer D, Singer S, Korvick J, et al. Concomitant risk factors in reports of torsades de pointes associated with macrolide use: review of the United States Food and Drug Administration Adverse Event Reporting System. *Clin Infect Dis* 2002;**35**:197–200.
47. Bertino Jr JS, Owens Jr RC, Carnes TD, et al. Gatifloxacin-associated corrected QT interval prolongation, torsades de pointes, and ventricular fibrillation in patients with known risk factors. *Clin Infect Dis* 2002;**34**:861–3.
48. CredibleMeds. Combined List of All QT Drugs and the List of Drugs to Avoid for Patients with Congenital Long QT Syndrome; May 2014. Available from: <www.crediblemeds.org/everyone/composite-list-all-qtdrugs/> [accessed 08.22.13].
49. Priori SG, Schwartz PJ, Napolitano C, et al. Risk stratification in the long-QT syndrome. *N Engl J Med* 2003;**348**:1866–74.
50. Monteforte N, Napolitano C, Priori SG. Genetics and arrhythmias: diagnostic and prognostic applications. *Rev Esp Cardiol (Engl Ed)* 2012;**65**:278–86.
51. El-Sherif N, Turitto G. Electrolyte disorders and arrhythmogenesis. *Cardiol J* 2011; **18**:233–45.
52. Lim SM, Pak HN, Lee MH, et al. Fever-induced QTc prolongation and ventricular fibrillation in a healthy young man. *Yonsei Med J* 2011;**52**:1025–7.
53. Amin AS, Herfst LJ, Delisle BP, et al. Fever-induced QTc prolongation and ventricular arrhythmias in individuals with type 2 congenital long QT syndrome. *J Clin Invest* 2008; **118**:2552–61.
54. Khan JN, Prasad N, Glancy JM. QTc prolongation during therapeutic hypothermia: are we giving-it the attention it deserves? *Europace* 2010;**12**:266–70.

55. Pollard CE, Valentin JP, Hammond TG. Strategies to reduce the risk of drug-induced QT interval prolongation: a pharmaceutical company perspective. *Br J Pharmacol* 2008; **154**:1538–43.
56. Lin YL, Kung MF. Magnitude of QT prolongation associated with a higher risk of torsades de pointes. *Pharmacoepidemiol Drug Saf* 2009;**18**:235–9.
57. Chain A, Dieleman J, van Noord C, et al. Not-in-trial simulation I: bridging cardiovascular risk from clinical trials to real life conditions. *Br J Clin Pharmacol* 2013;**76**(6): 964–72.
58. Redfern WS, Carlsson L, Davis AS, et al. Relationships between preclinical cardiac electrophysiology, clinical QT interval prolongation and torsade de pointes for a broad range of drugs: evidence for a provisional safety margin in drug development. *Cardiovasc Res* 2003;**58**:32–45.
59. Tamargo J, Caballero R, Delpón E. Drug-induced atrial fibrillation: does it matter? *Discov Med* 2012;**14**:295–9.
60. Erichsen R, Christiansen CF, Frøslev T, et al. Intravenous bisphosphonate therapy and atrial fibrillation/flutter risk in cancer patients: a nationwide cohort study. *Br J Cancer* 2011; **105**:881–3.
61. Dmitrienko A, Smith BP, Electrocardiogram reference ranges for Lilly clinical trials. Eli Lilly and Company, March 2002. Available from: <www.fda.gov/ohrms/dockets/ac/03/briefing/pubs/dmitrien.pdf> [accessed 06.25.13].
62. Pakala R, Willerson JT, Benedict CR. Mitogenic effect of serotonin on vascular endothelial cells. *Circulation* 1994;**90**:1919–26.
63. Seuwen K, Magnaldo I, Pouyssegur J. Serotonin stimulates DNA synthesis in fibroblasts acting through 5-HT1B receptors coupled to a Gi-protein. *Nature* 1988;**335**:254–6.
64. Nemecek GM, Coughlin SR, Handley DA, et al. Stimulation of aortic smooth muscles cell mitogenesis by serotonin. *Proc Natl Acad Sci USA* 1986;**83**:674–8.
65. Jian B, Xu J, Connolly J, et al. Serotonin mechanisms in heart valve disease I: serotonin-induced up-regulation of transferring growth factor-β1 via G-protein signal transduction in aortic valve interstitial calls. *Am J Pathol* 2002;**161**:2111–21.
66. Gustafsson BI, Tommeras K, Nordrum I. Long-term serotonin administration induces heart valve disease in rats. *Circulation* 2005;**111**:1517–22.
67. Droogmans S, Franken PR, Garbar C, et al. In vivo model of drug-induced valvular heart disease in rats: pergolide-induced valvular heart disease demonstrated with echocardiography and correlation with pathology. *Eur Heart J* 2007;**28**:2156–62.
68. Roth BL. Drugs and valvular heart disease. *N Engl J Med* 2007;**356**:6–9.
69. Ferrari MD. The economic burden of migraine to society. *Pharmacoeconomics* 1998;**13**; 667–76.
70. Michel P, Dartigues JF, Lindousli A, et al. Loss of productivity and quality of life in migraineurs among French workers: results from the GAZEL cohort. *Headache* 1997;**37**:71–8.
71. Graham JR. Cardiac and pulmonary fibrosis during methysergide therapy for headache. *Am J Med Sci* 1967;**254**:1–12.
72. Misch KA. Development of heart valve lesions during methysergide therapy. *BMJ* 1974; **2**:365–6.
73. Bana DS, MacNeal PS, LeCompte PM, et al. Cardiac murmurs and endocardial fibrosis associated with methysergide therapy. *Am Heart J* 1974;**88**:640–55.
74. Seiler K, Jenzer HR, Kacl J. Disease course of ergotism with femoral artery occlusion and mitral insufficiency. *Vasa* 1973;**2**:366–71.

75. Hauck AJ, Edwards WD, Danielson GK, et al. Mitral and aortic valve disease associated with ergotamine therapy for migraine. Report of two cases and review of literature. *Arch Pathol Lab Med* 1990;**114**:62–4.
76. Martínez Quintana E, Llorens León R, Redondo Martínez E, et al. Valvular heart disease associated with ergotamine. *Rev Esp Cardiol* 2005;**58**:97–9.
77. Goujon-Dubois J, El Allaf D. Clinical case of the month. Triple valvular dysfunction associated with long-term ergotamine therapy. *Rev Med Liege* 2003;**58**:70–2.
78. Redfield MM, Nicholson WJ, Edwards WD, et al. Valve disease associated with ergot alkaloid use: echocardiographic and pathologic correlations. *Ann Intern Med* 1992;**117**: 50–2.
79. Ray WA, Murray KT, Meredith S, et al. Oral erythromycin and the risk of sudden death from cardiac causes. *N Engl J Med* 2004;**351**:1089–96.
80. Yap YG, Camm J. Risk of torsades de pointes with non-cardiac drugs. *BMJ* 2000;**320**: 1158–9.
81. DuBuske LM. Second-generation antihistamines: the risk of ventricular arrhythmias. *Clin Ther* 1999;**21**:281–95.
82. Belohlávková S, Simák J, Kokesová A, et al. Fenfluramine-induced pulmonary vasoconstriction: role of serotonin receptors and potassium channels. *J Appl Physiol* 2001;**91**:755–61.
83. Weir EK, Reeve HL, Huang J, et al. Anorexic agents aminorex, fenfluramine and dexfenfluramine inhibit potassium current in rat pulmonary vascular smooth muscle and cause pulmonary vasoconstriction. *Circulation* 1996;**94**:2216–20.
84. Mlczoch J, Weir E, Reeves J, et al. Long-term effects of the anorectic agent fenfluramine alone and in combination with aminorex on pulmonary and systemic circulation of the pig. *Basic Res Cardiol* 1979;**74**:313–20.
85. Naeije R, Wauthy P, Maggiorini M, et al. Effects of dexfenfluramine on hypoxic pulmonary vasoconstriction and embolic pulmonary hypertension in dogs. *Am J Respir Crit Care Med* 1995;**151**:692–7.
86. Quigley EM. Cisapride: what can we learn from the rise and fall of a prokinetic? *J Dig Dis* 2011;**12**:147–56.
87. Abraham NS. Prescribing proton pump inhibitor and clopidogrel together: current state of recommendations. *Curr Opin Gastroenterol* 2011;**27**:558–64.
88. Bhatt DL, Cryer BL, Contant CF, et al. Clopidogrel with or without omeprazole in coronary artery disease. *N Engl J Med* 2010;**363**:1909–17.
89. Depta JP, Bhatt DL. Antiplatelet therapy and proton pump inhibition: cause for concern? *Curr Opin Cardiol* 2012;**27**:642–50.
90. Pazianas M, Compston J, Huang CL. Atrial fibrillation and bisphosphonate therapy. *J Bone Miner Res* 2010;**25**:2–10.
91. van der Hooft CS, Heeringa J, van Herpen G, et al. Drug-induced atrial fibrillation. *J Am Coll Cardiol* 2004;**44**:2117–24.
92. Lowes BD, Minobe W, Abraham WT, et al. Changes in gene expression in the intact human heart. Downregulation of alpha-myosin heavy chain in hypertrophied, failing ventricular myocardium. *J Clin Invest* 1997;**100**:2315–24.
93. Danzi S, Klein I, Portman M. Effect of triiodothyronine on gene transcription during cardiopulmonary bypass in infants with ventricular septal defect. *Am J Cardiol* 2005;**95**: 787–9.
94. Klein I, Danzi S. Thyroid disease and the heart. *Circulation* 2007;**116**:1725–35.
95. Palmeiro CR, Anand R, Dardi IK, et al. Growth hormone and the cardiovascular system. *Cardiol Rev* 2012;**20**:197–207.

96. August P, Oparil S. Hypertension in women. *J Clin Endocrinol Metab* 1999;**84**:1862–6.
97. Ito S. Kidney and hypertension: role of the juxtaglomerular apparatus. *Tohoku J Exp Med* 1997;**181**:411–29.
98. Pratila MG, Pratilas V. Dysrhythmia occurring during epidural anesthesia with bupivacaine. *Mt Sinai J Med* 1982;**49**:130–2.
99. Morgan DR, Trimble M, McVeigh GE. Atrial fibrillation associated with sumatriptan. *BMJ* 2000;**321**:275.
100. Steiger M, Jost W, Grandas F, et al. Risk of valvular heart disease associated with the use of dopamine agonists in Parkinson's disease: a systematic review. *J Neural Transm* 2009;**116**:179–91.
101. Merli GJ, Weitz H, Martin JH, et al. Cardiac dysrhythmias associated with ophthalmic atropine. *Arch Intern Med* 1986;**146**:45–7.
102. van der Hooft CS, Heeringa J, van Herpen G, et al. Drug-induced atrial fibrillation. *J Am Coll Cardiol* 2004;**44**:2117–24.
103. Jankovic SM, Dostic M. Choice of antiepileptic drugs for the elderly: possible drug interactions and adverse effects. *Expert Opin Drug Metab Toxicol* 2012;**8**:81–91.
104. York RC, Coleridge ST. Cardiopulmonary arrest following intravenous phenytoin loading. *Am J Emerg Med* 1988;**6**:255–9.
105. Adams BD, Buckley NH, Kim JY, et al. Fosphenytoin may cause hemodynamically unstable bradydysrhythmias. *J Emerg Med* 2006;**30**:75–9.
106. Mintzer S. Metabolic consequences of antiepileptic drugs. *Curr Opin Neurol* 2010;**23**:164–9.
107. Jakubus T, Michalska-Jakubus M, Lukawski K, et al. Atherosclerotic risk among children taking antiepileptic drugs. *Pharmacol Rep* 2009;**61**:411–23.
108. Devlin JW, Mallow-Corbett S, Riker RR. Adverse drug events associated with the use of analgesics, sedatives, and antipsychotics in the intensive care unit. *Crit Care Med* 2010;**38**(Suppl 6):S231–43.
109. Darrouj J, Karma L, Arora R. Cardiovascular manifestations of sedatives and analgesics in the critical care unit. *Am J Ther* 2009;**16**:339–53.
110. Breeden CC, Safirstein BH. Albuterol and spacer-induced atrial fibrillation. *Chest* 1990;**98**:762–3.
111. Varriale P, Ramaprasad S. Aminophylline induced atrial fibrillation. *Pacing Clin Electrophysiol* 1993;**16**:1953–5.
112. Cheitlin MD, Hutter Jr AM, Brindis RG, et al. ACC/AHA expert consensus document. Use of sildenafil (Viagra) in patients with cardiovascular disease. *J Am Coll Cardiol* 1999;**33**:273–82.
113. Arruda-Olson AM, Mahoney DW, Nehra A, et al. Cardiovascular effects of sildenafil during exercise in men with known or probable coronary artery disease: a randomized crossover trial. *JAMA* 2002;**287**:719–25.
114. Santillo VM, Lowe FC. Treatment of benign prostatic hyperplasia in patients with cardiovascular disease. *Drugs Aging* 2006;**23**:795–805.
115. Sica DA. Alpha1-adrenergic blockers: current usage considerations. *J Clin Hypertens (Greenwich)* 2005;**7**:757–62.
116. Nieminen T, Koobi T, Tammela TL, et al. Hypotensive potential of sildenafil and tamsulosin during orthostasis. *Clin Drug-investig* 2006;**26**:667–71.
117. Chobanian AV, Bakris GL, Black HR, et al. The seventh report of the Joint National Committee on Prevention, Detection, Evaluation, and Treatment of High Blood Pressure. *JAMA* 2003;**289**:2560–72.

118. Calhoun DA, Jones D, Textor S, et al. Resistant hypertension: diagnosis, evaluation, and treatment: a scientific statement from the American Heart Association Professional Education Committee of the Council for High Blood Pressure Research. *Circulation* 2008;**117**: e510–26.
119. Kaplan NM. Resistant hypertension. *J Hypertens* 2005;**23**:1441–4.
120. Rossi GP, Seccia TM, Maniero C, et al. Drug-related hypertension and resistance to antihypertensive treatment: a call for action. *J Hypertens* 2011;**29**:2296–309.
121. Hammer F, Stewart PM. Cortisol metabolism in hypertension. *Best Pract Res Clin Endocrinol Metab* 2006;**20**:337–53.
122. Schaefer RM, Leschke M, Strauer BE, et al. Blood rheology and hypertension in hemodialysis patients treated with erythropoietin. *Am J Nephrol* 1988;**8**:449–53.
123. Morakkabati N, Gollnick F, Meyer R, et al. Erythropoietin induces Ca2þ mobilization and contraction in rat mesangial and aortic smooth muscle cultures. *Exp Hematol* 1996; **24**:392–7.
124. Eggena P, Willsey P, Jamgotchian N, et al. Influence of recombinant human erythropoietin on blood pressure and tissue renin-angiotensin systems. *Am J Physiol* 1991;**261**: E642–6.
125. Carlini RG, Dusso AS, Obialo CI, et al. Recombinant human erythropoietin (rHuEPO) increases endothelin-1 release by endothelial cells. *Kidney Int* 1993;**43**:1010–4.
126. Singer DR, Jenkins GH. Hypertension in transplant recipients. *J Hum Hypertens* 1996; **10**:395–402.
127. Cauduro RL, Costa C, Lhulier F, et al. Endothelin-1 plasma levels and hypertension in cyclosporine-treated renal transplant patients. *Clin Transplant* 2005;**19**:470–4.
128. Lassila M. Interaction of cyclosporine A and the renin-angiotensin system; new perspectives. *Curr Drug Metab* 2002;**3**:61–71.
129. Niehof M, Borlak J. HNF4alpha dysfunction as a molecular rational for cyclosporine induced hypertension. *PLoS One* 2011;**6**:e16319.
130. Keller AG, Girolamo GD, Alvarez PA. Pharmacovigilance and the cardiovascular system: two sides to every story. *Curr Drug Saf* 2011;**6**:224–9.
131. Honig PK, Wortham DC, Zamani K, et al. Terfenadine-ketoconazole interaction. Pharmacokinetic and electrocardiographic consequences. *JAMA* 1993;**269**:1513–8.
132. Honig PK, Woosley RL, Zamani K, et al. Changes in the pharmacokinetics and electrocardiographic pharmacodynamics of terfenadine with concomitant administration of erythromycin. *Clin Pharmacol Ther* 1992;**52**:231–8.
133. Devereux RB. Appetite suppressants and valvular heart disease. *N Engl J Med* 1998; **339**:765–6.
134. Wysowski DK, Corken A, Gallo-Torres H, et al. Postmarketing reports of QT prolongation and ventricular arrhythmia in association with cisapride and Food and Drug Administration regulatory actions. *Am J Gastroenterol* 2001;**96**:1698–703.
135. European Medicines Agency. Clinical evaluation of QT/QTc interval prolongation and proarrhythmic potential for non-antiarrhythmic drugs, ICH E14; 2012. Available from: <www.ema.europa.eu/docs/en_GB/document_library/Scientific_guideline/2009/09/WC500002879.pdf> [accessed 06.25.13].
136. The European Agency for the Evaluation of Medicinal Products. Committee for Proprietary Medicinal Products. Points to consider: the assessment of the potential for QT interval prolongation by noncardiovascular medicinal products; 1997. Available from: <www.emea.europa.eu/docs/en_GB/document_library/Annual_report/2009/12/WC500016809.pdf> [accessed 06.25.13].

137. ICH E14, Step 2: The clinical evaluation of QT/QTc interval prolongation and proarrhythmic potential for non-antiarrhythmic drugs; 2012. Available from: <www.fda.gov/ohrms/dockets/dockets/04d0377/04D-0377-EC7-Attach-1.pdf> [accessed 06.29.13].
138. Cavero I, Crumb W. ICH S7B draft guideline on the non-clinical strategy for testing delayed cardiac repolarisation risk of drugs: a critical analysis. *Expert Opin Drug Saf* 2005; **4**:509–30.
139. Finlayson K, Witchel HJ, McCulloch J, et al. Acquired QT interval prolongation and hERG: implications for drug discovery and development. *Eur J Pharmacol* 2004;**500**: 129–42.
140. Hune JR, Grant AO. Agents used in cardiac arrhythmias. In: Katzung BG, Masters SB, Trevor AJ, editors. *Basic and clinical pharmacology*. 11th ed. New York: McGraw-Hill/Lange;2009. Chapter 14.

Chapter 11

Cardiovascular Toxicity from Chemotherapy and Anticancer Treatment

Angela Esposito, Lucia Gelao, Carmen Criscitiello, and Giuseppe Curigliano
Division of Early Drug Development for Innovative Therapies, Istituto Europeo di Oncologia (European Institute of Oncology), Milan, Italy

11.1 INTRODUCTION

Cardiovascular toxicity is a potential short- or long-term complication of anticancer therapy involving the heart and circulation, as well as exacerbating and/or unmasking existing heart disease. Moreover, many cancer patients have multiple risk factors for both cardiac and coronary disease such as cigarette smoking, diabetes, alcohol consumption, obesity, and advanced age. In patients with either cancer or cardiovascular disease, the ultimate goal of therapy is to maximize survival and quality of life. Preexisting cardiovascular disease may significantly limit the diagnosis, staging, and therapy of the cancer patient. This is a particularly common problem in the older patient because of the higher prevalence of cardiovascular and other comorbidities. The management of cardiac disease in the cancer patient is unique and in many ways unlike managing cardiac disease in patients without cancer. An understanding of the potential cardiac toxicities of the chemotherapeutic regimen used is essential.

11.2 CARDIOVASCULAR TOXICITY INDUCED BY ANTICANCER AGENTS

The use of specific chemotherapeutic agents and molecular targeted therapies can affect the cardiovascular system, either through a direct effect on heart and endothelial function, or peripherally through hemodynamic flow alteration (hypertension and/or thrombotic events). The use of many targeted agents (e.g., signaling inhibitors) induces a cardiac dysfunction that resolves for most patients over time. Ewer et al.[1] proposed a system to identify drugs

The Heart and Toxins. DOI: http://dx.doi.org/10.1016/B978-0-12-416595-3.00011-6

TABLE 11.1 Potential Cardiac Toxicity Induced by Anticancer Chemotherapeutic Agents

Drug	Toxic Dose Range	Cardiac Toxicity	Percent
Doxorubicin Epirubicin Idarubicin	>450 mg/m^2 >900 mg/m^2	Left ventricular dysfunction	3–12% 0.9–3.3%
Paclitaxel Docetaxel	Conventional dose	Left ventricular dysfunction	5–15% 2.3–8%
Cyclophosphamide Ifosfamide	>100–120 mg/kg	Left ventricular dysfunction	3–5%
Capecitabine Fluorouracil	Conventional dose	Cardiac ischemia	3–9% 1–68%
Paclitaxel Docetaxel	Conventional dose	Cardiac ischemia	<1–5% 1.7%
Trabectedin	Conventional dose	Cardiac ischemia	1%
Arsenic trioxide	Conventional dose	QTc prolongation	26–93%
Paclitaxel	Conventional dose	QTc prolongation	0.1–31%

that have the potential to cause irreversible damage (type 1) versus drugs that predominantly induce reversible dysfunction (type 2). This classification system does have limitations; for example, trastuzumab, a type 2 drug, can trigger irreversible cardiac damage in patients with severe preexisting cardiac disease, or potentiate anthracycline type 1 cardiotoxicity. In type 1 cardiotoxicity, usually pathophysiology is related to cell loss; in type 2, cellular dysfunction (mitochondrial and protein alterations) underlies the reversible damage.

Although nonreversible damage can induce a progressive cardiovascular disease, a reversible dysfunction is usually temporary, with no injury marker release, and will be recovered from with normalization of cardiovascular function. This is a seminal concept that can impact the "go-no-go" decision in case of left ventricular dysfunction (LVD) induced by targeted agents and that identifies type 1 (e.g., anthracyclines) and type 2 (e.g., trastuzumab) agents.[2] The literature describes several kinds of cardiotoxicities: left ventricular dysfunction, cardiac ischemia, hypertension, QT prolongation, and venous thromboembolism (Table 11.1).

11.2.1 Left Ventricular Dysfunction

One of the most common manifestations of cardiotoxicity associated with exposure to anticancer therapies is the development of LVD and overt heart failure.

One definition of left ventricular dysfunction has been provided by the Cardiac Review and Evaluation Committee.[3] According to the committee, LVD is characterized by

(1) decrease in cardiac left ventricular ejection fraction (LVEF) that was either global or more severe in the septum; (2) symptoms of congestive heart failure (CHF); (3) associated signs of CHF, including but not limited to S3 gallop, tachycardia, or both; and (4) decline in LVEF of at least 5% to less than 55% with accompanying signs or symptoms of CHF, or a decline in LVEF of at least 10% to below 55% without accompanying signs or symptoms.

Left ventricular dysfunction and heart failure (HF) have been defined by the Common Terminology Criteria for Adverse Events (CTCAE) for the purposes of uniform reporting.[3] The CTCAE criteria, however, have changed over the years; and in the recently updated version 4, HF and more specialized testing were introduced (e.g., echocardiography and biomarker testing) to provide a framework for a more sophisticated detection of toxicity with newer chemotherapeutic and targeted agents. Recent definitions have varied and include a larger change in LVEF to below the lower limit of normal (LLN) or LVEF <50%. As a consequence, at present, a consensus definition for cardiotoxicity is still lacking.

Chemotherapeutic Drugs

Several chemotherapeutic agents may cause LVD, among them antimetabolites, alkylating agents, antitumor antibiotics, and anthracyclines.

Anthracyclines, including doxorubicin, epirubicin, and daunorubicin, are a class of chemotherapeutics widely used in the management of multiple malignancies, and the cardiotoxicity induced by them can be categorized into acute, early-onset chronic progressive, and late-onset chronic progressive. Acute cardiotoxicity occurs in <1% of patients and it is usually observed within 14 days from the beginning of treatment. It manifests as an acute, transient decline in myocardial contractility, which is usually reversible. The early-onset chronic progressive form occurs in 1.6 to 2.1% of patients during therapy or within the first year after treatment. Late-onset chronic progressive cardiotoxicity occurs at least one year after completion of therapy in 1.6 to 5% of patients.

Early- and late-onset chronic progressive cardiotoxicity typically present as dilated cardiomyopathy in adults and can be progressive. Risk factors for anthracycline toxicity include cumulative dose; intravenous bolus administration; higher single doses; history of prior mediastinal irradiation; the use of other concomitant agents known to have cardiotoxic effects including cyclophosphamide, trastuzumab, and paclitaxel; female gender; underlying cardiovascular disease; age (i.e., young and elderly); increased length of time since completion of chemotherapy; and an increase in cardiac biomarkers, as troponins and natriuretic peptides, during and after administration.[4–7]

The main hypothesis to explain the anthracyclines-related cardiotoxicities is based on the generation of reactive oxygen species (ROS), which causes damage to DNA, proteins, and lipids and can result in myocyte death.[8] A study by Zhang et al. suggested that doxorubicin-induced cardiotoxicity is mediated by topoisomerase IIβ in cardiomyocytes and that cardiomyocyte-specific deletion of *Top2β* (encoding topoisomerase-IIβ) protects cardiomyocytes from doxorubicin-induced DNA double-strand breaks and transcriptome changes that are responsible for defective mitochondrial biogenesis and ROS formation.[9]

Left ventricular dysfunction has also been described for two other classes of cytotoxic agents: alkylating agents and inhibitors of microtubule polymerization. In the first case, LVD has been reported in 7 to 28% of patients receiving a cyclophosphamide-based therapy.[10,11] The risk of cardiotoxicity appears to be dose-related (≥150 mg/kg and 1.5 g/m^2/day). Another alkylating agent, ifosfamide, can induce HF with a dose−response trend (doses ≥12.5 g/m^2).[12] The incidence of heart failure associated with the inhibitors of microtubule polymerization is relatively low. In the Breast Cancer International Research Group Trial 001, the overall incidence of congestive heart failure (including those during follow-up) was 1.6% among patients treated with a TAC regimen (e.g., docetaxel, doxorubicin, cyclophosphamide) and 0.7% for those treated with an FAC regimen (e.g., 5-fluorouracil, doxorubicin, cyclophosphamide) ($P = 0.09$).[13]

Targeted Agents

Within the past decade, the advent of biologic targeted agents has brought into the clinic additional therapies with cardiotoxic concerns. Whereas the cardiac toxicity of anthracyclines is fairly well described, cardiac dysfunction related to targeted agents is just beginning to be elucidated.

Use of the humanized monoclonal antibody trastuzumab, directed against the HER2 receptor, has revolutionized the treatment for HER2-positive breast cancer, with landmark adjuvant phase 3 trials demonstrating a 50% reduction in recurrence of disease and a 33% improvement in survival.[14−17] Rates of cardiac toxicity reported in the adjuvant trials of trastuzumab are variable and reflect differences in trial design, chemotherapy administration, and definitions of cardiac events.

In the trastuzumab adjuvant trials[18−22] (Table 11.2), the highest reported incidence of symptomatic or severe congestive heart failure with trastuzumab was 4%, which occurred when trastuzumab was administered with paclitaxel after anthracycline exposure. A low rate of 0.4% CHF was reported in the BCIRG 006 Adjuvant Trial with the trastuzumab/docetaxel/carboplatin combination regimen in the absence of previous anthracycline therapy.[20]

The exact pathogenesis of trastuzumab-induced cardiac damage has still not been determined. Trastuzumab is thought to cause cardiac dysfunction through

TABLE 11.2 Cardiac Toxicity Induced by Trastuzumab

Trial	Design	Asymptomatic Drop in LVEF (≥10 Percentage Points to <55%)	Severe CHF/ Cardiac Events (NYHA Class III/IV CHF or Death)	Discontinued for Cardiac Reasons
NSABP B31[18] *n* = 2043	AC + TH + H vs. AC + T	34% vs. 17%	4.1% vs. 0.8%	19%
NCCTG N9831[19] *n* = 2766	AC + TH + H vs. AC + T + H vs. AC + T	5.8–10.4% vs. 4.0–7.8% vs. 4.0–5.1%	3.3% vs. 2.8% vs. 0.3%	n/a
BCIRG 006[20] *n* = 3222	AC + T vs. AC + TH + H vs. TCaH[2]	11% vs. 19% vs. 9%	0.7% vs. 2.0% vs. 0.4%	n/a
HERA[21] *n* = 5102	Adj chemo[3] → H vs. adj chemo alone	7.1% vs. 2.2%	0.6% vs. 0.06%	4.3%
FinHER[22] *n* = 232	V or T + H vs. V or T[4] → FEC × 3	3.5% vs. 8.6%	0% vs. 3.4%	n/a

Note that 6.7% did not receive H after A due to unacceptable drops in LVEF; included a nonanthracycline arm. In addition, 96% of the chemotherapy was A-containing. No prior anthracycline before H exposure; H exposure limited to 9 weeks.
Abbreviations: A, anthracycline; C, cyclophosphamide; T, taxane; H, trastuzumab; Ca, carboplatin; V, vinorelbine; F, 5-flourouracil; E, epirubicin; n/a, information not available

the interruption of the HER2/ErbB2 signaling pathway in myocardium, thus interfering with normal growth, repair, and survival of cardiomyocytes. Preclinical studies have shown that normal HER2 signaling plays a critical role in myocardiogenesis and cellular response to stress.[23] Loss of the HER2/ErbB2-mediated pathways, which normally mitigate the effects of acute stressors after trastuzumab, might lead to cardiac dysfunction associated with exposure to chemotherapeutic agents.[24] Recent preclinical data demonstrated that trastuzumab cardiotoxicity is more complex and does not involve only HER2 inhibition.[25]

Some studies have revealed low toxicity after the use of lapatinib, a tyrosine kinase inhibitor with dual action against HER2 and epidermal growth factor (EGF).[26] Therefore another suggested mechanism of trastuzumab cardiotoxicity is related to the effect on cardiomyocytes of cytotoxic immune reactions triggered by the IgG1 domain of trastuzumab[27] and the modulation of the mitochondrial integrity via the Bcl-X family proteins that lead to adenosine triphosphate (ATP) depletion and to contractile dysfunction.[28]

Lapatinib, an oral receptor tyrosine kinase inhibitor of HER2 and EGFR, has an approved role in combination with capecitabine chemotherapy in the treatment of patients with trastuzumab-resistant breast cancer. Risk of cardiac toxicity is estimated to be very low (rates of symptomatic CF = 1.4%), specifically in a population with prior exposure to anthracycline and trastuzumab.[29] Some studies indicate that lapatinib may reduce trastuzumab's cytotoxicity when the two drugs are administered together. This is due to their different affect on adenosine monophosphate-activated protein kinase (AMPK); while AMPK is reduced by trastuzumab, reducing intracellular ATP enhances lapatinib, leading to increased adenosine triphosphate production by oxidative pathways.[30]

Bevacizumab, a humanized monoclonal antibody directed against vascular endothelial growth factor (VEGF), has demonstrated significant antitumor activity when combined with chemotherapy, leading to regulatory approval for several advanced solid tumors including breast, lung, colorectal, and renal carcinomas. With greater use of bevacizumab, data are emerging regarding potential cardiac toxicity. To date, rates of cardiac toxicity associated with bevacizumab therapy appear to be relatively low. In the major phase 3 trials in metastatic breast cancer, reported rates of CTCAE grade 3/4 CHF were 0.8 to 2.2% in a mostly anthracycline pretreated population.[31] To date, clinical trial data do not suggest significant increases in cardiac toxicity during treatment with bevacizumab, even in the setting of concurrent treatment with other cardiotoxic agents. However, long-term monitoring of patients who have completed bevacizumab therapy has not been performed; therefore, the safety of adjuvant bevacizumab in the setting of cancer survivorship is unknown.

Multiple small molecule tyrosine kinase inhibitors of the VEGF receptor (VEGFR) have been developed, including sunitinib and sorafenib, with cross activity against other growth factor receptors including platelet-derived growth factor (PDGF), c-Kit, and BRAF. In contrast to bevacizumab, VEGFR antagonists appear to have a more profound effect on cardiac function. Initial reports of sunitinib in renal cell carcinoma suggested a 10% incidence of asymptomatic drop in LVEF, with full recovery when treatment was completed.[32]

Screening and Diagnosis

At present, the most frequently used modality for detecting left ventricular dysfunction is the periodic measurement of LVEF by using either echocardiography or multigated acquisition scanning. To date, however, there are no evidence-based guidelines for cardiotoxicity monitoring. Although several guidelines are available, none specify how often, by what means, nor how long cardiac function should be monitored during and after cancer treatment.[33,34] Serial evaluation of left ventricular ejection fraction is recommended for patients treated with trastuzumab (Figure 11.1).[35,36] However,

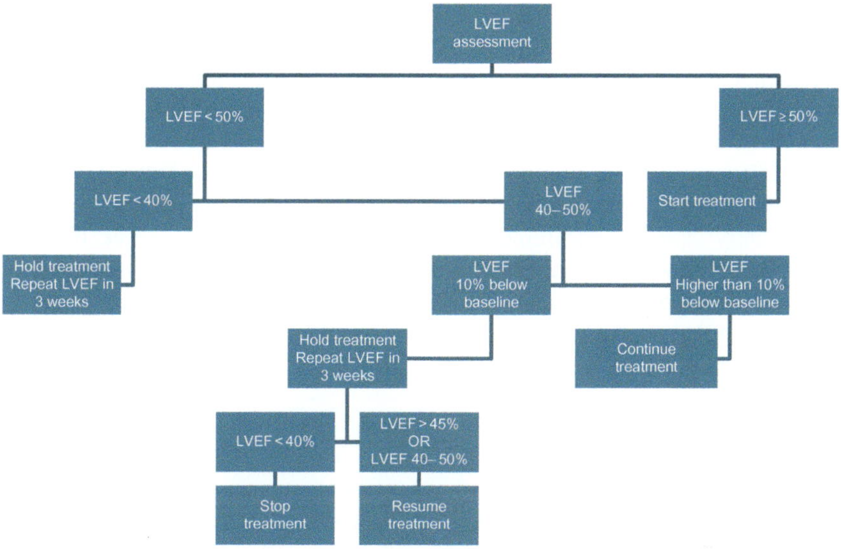

FIGURE 11.1 Algorithm for continuation and discontinuation of trastuzumab based on left ventricular ejection fraction assessments.

LVEF measurement is a relatively insensitive tool for detecting cardiotoxicity at an early stage because no considerable change in LVEF occurs until a critical amount of myocardial damage has taken place, and it only comes to the forefront after compensatory mechanisms are exhausted. In addition, the measurement of LVEF presents a number of challenges related to image quality, assumption of left ventricular geometry, load dependency, and expertise.

Multiple gated acquisition (MUGA) scan can reduce interobserver variability with the disadvantages of including the exposure to radioactivity and the limited information that can be obtained about cardiac structure and diastolic function. Magnetic resonance imaging (MRI) is considered the gold standard for the evaluation of LV volumes, mass, and function. However, its lack of availability and high cost limit its routine use. Novel ultrasound imaging techniques, such as contrast echocardiography and real-time 3D echocardiography, are under investigation. Small studies examining tissue Doppler and strain rate imaging appear promising in detecting early subclinical changes in cardiac performance, even if long-term data on large populations that confirm the clinical relevance of such changes are not available yet.[37]

During the last decade, a new approach, based on the use of cardiac biomarkers, in particular troponins (TNI), has emerged. It has proved to be a more sensitive and more specific tool for early, real-time identification, assessment, and monitoring of anticancer drug-induced cardiac injury.[6,7] Strong data indicate that TNI detects anticancer drug-induced cardiotoxicity in its earliest phase, long before any reduction in LVEF has occurred. Its evaluation during high-dose chemotherapy allows for the early identification of patients at risk of developing cardiac dysfunction, the stratification of risk of cardiac events after chemotherapy, and the opportunity for a preventive therapy in selected high-risk patients.[6,7] For patients treated with trastuzumab, TNI may help distinguish between reversible and irreversible cardiac injury by identifying myocardial cell necrosis.[38] Measurement of troponin immediately before and immediately after each cycle of cancer therapy seems to be effective enough and is also transferable from clinical research to real-world routine assessment.[39,40] Therefore more information from ongoing clinical trials is required before we can establish the practical benefits of TNI in routine clinical activity.

Treatment

Regarding the treatment of LVD induced by anticancer drugs, the use of angiotensin-converting-enzyme (ACE) inhibitors (ACE-I) and beta blockers (BB) may be highly effective. Recent findings reported, in patients with anthracycline-induced cardiomyopathy, that the time elapsed from the end of chemotherapy to the start of therapy, with ACE-I and with BB, is a crucial variable for recovery of cardiac dysfunction.[41,42]

No evidence-based recommendations for the treatment of patients developing cardiac dysfunction after trastuzumab therapy have been proposed. The evidence that support the use of ACE-Is and BB in this setting is limited to case series. Management of trastuzumab-related cardiotoxicity has two distinct aspects: withdrawal of trastuzumab therapy and treatment of cardiac dysfunction. The "stopping/restarting" rules used in the adjuvant trials were effective and are recommended, with some modifications regarding recommendations for a cardiology consult or treatment of cardiac dysfunction (or both) when appropriate. Many medications have been investigated for a preventive activity in the group of patients at increased risk of LVD. Carvedilol may prevent cardiac damage induced by doxorubicin due to its antioxidant activity. The effect of carvedilol was confirmed in a randomized study in which prophylactic use of carvedilol in a small population of patients treated with anthracycline prevented LVD and reduced mortality.[43]

Nakamae et al. demonstrated that valsartan, an angiotensin receptor blocker given concurrently with anthracycline-containing regimens, prevents cardiac damage.[44] Dexrazoxane, an iron-chelating agent, significantly reduces anthracycline-related cardiotoxicity in adults with various solid tumors and in children with acute lymphoblastic leukemia and Ewing's sarcoma.[45] Dexrazoxane is not routinely used in clinical practice; it is recommended as a cardioprotectant by the American Society of Clinical Oncology only for patients with metastatic breast cancer who have already received more than 300 mg/m^2 of doxorubicin.

A randomized clinical trial evaluated the prophylactic role of ACE inhibitors and beta blockers in the prevention of chemotherapy-induced LVD.[46] The OVERCOME trial showed that the combined treatment with enalapril and carvedilol may prevent heart failure in patients treated for hematologic malignancies.[46] The MANTICORE study is still ongoing and it is evaluating the efficacy of perindopril and bisoprolol in the prevention of trastuzumab-mediated left ventricular remodeling in HER2-positive breast cancer.[47]

11.2.2 Cardiac Ischemia

Cardiac ischemia that is related to chemotherapy administration is an unusual occurrence, but an increased risk of acute coronary syndrome has been associated with administration of cytotoxic, hormonal, and targeted agents for cancer treatment. Antimetabolites and inhibitors of microtubule polymerization agents are most frequently responsible for ischemic heart disease.

The most commonly cardiotoxic effect attributed to 5-fluorouracil (FU) is the ischemic syndrome, which varies clinically from angina pectoris to acute myocardial infarction.[48] Ischemia can take place in patients without underlying coronary artery disease (CAD) (1.1%), but the incidence is higher in patients with CAD (4.5%).[49] Cardiac toxicity typically occurs with early onset (within 2–5 days of starting therapy) and is usually reversible on cessation of 5-FU and implementation of antiischemic medical therapy. Coronary artery thrombosis, arteritis, or vasospasm, secondary to drug exposure, have been proposed as the most likely underlying mechanisms. Capecitabine, an oral 5-FU analog, is currently used for treatment of breast and gastrointestinal cancers and is believed to be less toxic than 5-FU.[50] Paclitaxel administration has been associated with cases of myocardial ischemia and infarction.

In a large study of approximately 1000 patients, the incidence of cardiac toxicity was 14%.[51] Vinca alkaloids-related cardiac events are more likely to occur in women than in men.[52] These agents may cause angina with ECG changes,[53] myocardial ischemia, and myocardial infarction.[54] Moreover, endocrine agents, such as tamoxifen,[55] and aromatase inhibitors,[56] which are widely used in the treatment of hormone receptor-positive breast cancer, are associated with cardiac ischemia risk. Cardiac events, including myocardial infarction and cardiac failure, have been reported at low frequency in major adjuvant trials comparing use of AIs to a control arm of five years of tamoxifen.[57]

Differential changes in lipid profile has been proposed as an etiology for these observations, however a strong signal linking AIs and relevant changes in lipid levels is lacking. Finally, whether novel molecular therapies also increase risk of cardiac ischemia is not clear. Some studies described cardiac complications and arterial thromboembolism risk with the use of tyrosine-kinase inhibitor (TKI). Schmidinger et al. reported symptomatic cardiotoxicity in 18% of 74 renal cancer patients treated with sunitinib or sorafenib. All patients recovered and anticancer treatment was continued.[58]

A meta-analysis by Choueiri et al., including 10 trials and 9387 patients taking sunitinib or sorafenib, showed an increase in risk for arterial thromboembolism with an incidence of 4% and a relative risk (RR) of 3.3%.[59] Another meta-analysis by Schutz et al.,[60] which included 4679 patients using pazopanib, sorafenib, or sunitinib, reported that the RR of fatal adverse events was 2.23 compared to controls, with bleeding being the most common etiology (45% of fatal adverse events), followed by myocardial infarction (15% of fatal adverse events). The increased incidence of thromboembolic events with antiangiogenic target therapies can be explained with various mechanisms. VEGF inhibition blocks the capability of endothelial cells to regenerate, causing endothelial dysfunction.[61,62]

Recently ponatinib, a new oral TKI was approved for therapy of chronic myeloid leukemia (CML). It was proposed as first-line therapy in patients with CML (EPIC clinical trial [ClinicalTrials.gov number NCT01650805]) or in pretreated patients to overcome imatinib resistance, which occurs in about 20% of cases.[63] Despite that in a phase 2 ponatinib Ph-positive acute lymphoblastic leukemia and CML evaluation (PACE) clinical trial ponatinib showed clinically significant antileukemic activity, serious-grade arterial thrombotic events (including cardiovascular, cerebrovascular, and peripheral vascular events) were seen in 11.8% of the patients, whereas the incidence of all arterial thrombotic events, serious or not, was 17.1%.[64] The exact mechanism of TKI-induced arterial thrombotic events is yet to be elucidated. It was hypothesized that vascular toxicities associated with ponatinib are mediated by direct on-target ABL kinase inhibition or by the targeting of non-ABL tyrosine kinases such as vascular endothelial growth factor receptor 2.[63]

In light of these findings, the EPIC trial was recently stopped because of high rates of thrombotic events in ponatinib-exposed patients and the U.S. Food and Drug Administration (FDA) asked the manufacturer to suspend marketing and sales of the drug. More data are needed to determine the cardiovascular risk that can be attributed to ponatinib and the mechanism of action underlying these events. European Society for Medical Oncology (ESMO) guidelines recommend baseline ECG evaluation and frequent vital signs monitoring during chemotherapeutic agent infusion, particularly with 5-FU or paclitaxel.[65]

The management of cardiac ischemia and coronary heart disease is similar to management of patients with coronary artery disease without cancer, with an emphasis after intervention on platelet inhibition.[66] Placing coronary stents requires antiplatelet therapy for prolonged periods of time, and chemotherapy may lead to thrombocytopenia, which makes it challenging to manage antiplatelet therapy.[67] Moreover, antineoplastic and antiinflammatory therapies may have a direct effect on stent reendothelialization. Dual antiaggregation (antiplatelet) therapy is mandatory in patients stented and usually consists of acetylsalicylic acid (100 mg/day) and clopidogrel (75 mg/day) for at least 6 to 12 months depending on the type of stent. Such therapy has been shown to reduce cardiac clinical events significantly, although it increases slightly the risk of bleeding, especially in patients with thrombocytopenia. In this case, it is possible to adapt and personalize therapy with antiplatelet agents to reduce the risk of bleeding. Thus, the management of coronary disease in patients with active cancer or undergoing chemotherapy has never been evaluated, and this must be considered by the cardio-oncologist in planning therapy.

11.2.3 Venous Thromboembolism

Venous thromboembolism (VTE) represents one of the most important causes of morbidity and mortality in cancer patients. According to population-based case control studies, the two-year cumulative incidence of VTE is between 0.8% and 8%. The increased risk of recurrent VTE in cancer patients is greatest in the first few months after malignancy is diagnosed and can persist for many years after an initial episode of symptomatic VTE.[68–72] The risk for VTE varies widely between various subgroups of cancer patients and even in the same cancer patient over time. Because the natural history of cancer is dynamic, the risk for VTE may increase and subside over time as a result of hospitalization, chemotherapy, metastasis, remission, and many other factors. Therefore, risk factor assessment is an ongoing process throughout the course of care for the cancer patient.

It is essential to identify risk factors predictive of VTE to better assess the potential for thromboprophylaxis therapy. Anticancer agents correlated with an increased risk of VTE include antiangiogenic agents, thalidomide, lenalidomide, cisplatin, and tamoxifen. It was seen that the incidence of VTE with thalidomide in monotherapy was 5%.[73] The risk increases when thalidomide is administered in combination with dexamethasone or chemotherapy, and particularly doxorubicin, in the absence of thromboprophylaxis. Lenalidomide, a thalidomide analog, is generally less toxic, but the risk of VTE is high and varies from 3 to 75%.[74]

The thrombogenic mechanism of these drugs involves direct action on endothelial cells and increased platelet aggregation.[75] The antivascular endothelial growth factor agent bevacizumab has been associated with an increased risk of venous thromboembolism in one meta-analysis,[76] in contrast with three other analyses that showed no increased risk of venous thromboembolism in patients treated with bevacizumab.[77–79] Hurwitz showed that the risk of grade 3 to 5 bleeding in patients treated with anticoagulant after a VTE was low and was not increased by bevacizumab treatment.[79] These results are concordant with other reports indicating that patients with cancer can be safely treated with anticoagulation therapy while on chemotherapy[80] and that the risk of serious bleeding is not increased by bevacizumab.[77,81]

Many studies have shown evidence of a mild increase in the incidence of thromboembolic events with VEGF signaling pathways (VSP) inhibitor TKIs compared to placebo. A meta-analysis, that included 7441 patients enrolled in 17 trials, did not demonstrated a significant overall increase in the risk of all-grade or grade ≥ 3 VTE with or without one of the VEGFR tyrosine-kinase inhibitors (sunitinib, sorafenib, pazopanib, vandetanib, and axitinib). The RR of all-grade and high-grade VTEs for the TKI versus no TKI arms was 1.10 (95%, CI 0.73–1.66, $P = 0.64$) and 0.85 (95%, CI 0.58–1.25, $P = 0.41$), respectively.[82] Based on the results of several studies, extensive, routine prophylaxis for advanced cancer patients receiving chemotherapy is not recommended but may be considered and discussed with high-risk cancer patients,[83] while prophylaxis with low-molecular-weight heparin (LMWH) or fondaparinux in hospitalized cancer patients confined to bed with an acute medical complication is recommended.

The American Society of Clinical Oncology guidelines suggest that, with the exception of patients with active bleeding or with contraindications to anticoagulation, antithrombotic prophylaxis should be considered in hospitalized cancer patients when platelet counts are >50,000/μl. In patients with intermediate to high risk of cardiovascular events (previous cardiovascular event or recent stent implantation) hospitalized for a medical condition (e.g., ischemic stroke, acute medical illness, congestive heart failure, respiratory acute disease, or sepsis), maintaining antiplatelet therapy and adding LMWH or fondaparinux for VTE prophylaxis is recommended as long as the additional risk factor for thrombosis is present.

In patients at low cardiovascular risk or with a high tendency to bleed, the risk–benefit ratio between cardiovascular recurrence and VTE prevention should be carefully evaluated. In these patients, withdrawing aspirin during VTE prophylaxis with LMWH or fondaparinux (i.e., during the time of hospitalization or at the time of exposure to newer VTE risk conditions) may be considered. The standard initial treatment of an acute episode of VTE in cancer and noncancer patients consists of the administration of subcutaneous LMWH at a dose adjusted to body weight: 200 U/kg once daily (200 units of antiXa activity per kg of body weight) administered once daily (e.g., dalteparin), 100 U/kg (100 units of antiXa activity per kg of body weight) administered twice daily (e.g., enoxaparin) or unfractioned heparin (UFH) intravenously (IV) in continuous infusion.

11.2.4 Hypertension

Hypertension is another possible cardiovascular complication during chemotherapy. Aggressive management of hypertension beginning from the initiation of therapy is important to avoid stress of the myocardium. An understanding of the potential cardiac toxicities of the chemotherapeutic regimen used is essential, giving further support to the concept of a multidisciplinary strategy for management. Rates of significant hypertension appear to depend on the agent used, the tumor type, and patient-related factors including age and comorbidity. It is an adverse effect of many VEGFR inhibitors.

The overall incidence of hypertension reported following bevacizumab therapy is 4 to 35%.[31,84–87] About one quarter of patients (19–24%) treated with a TKI will develop secondary hypertension.[88,89] The incidence of sunitinib-induced hypertension varies from 5 to 24%.[90–94] The incidence of sorafenib-induced hypertension has been reported to be between 17 and 43%.[88] The exact mechanism underlying the hypertension associated with angiogenesis inhibitors has not been well defined. However, several theories have been proposed, including an imbalance in neurohumoral factors, the development of vascular rarefaction, or an alteration in vascular nitric oxide balance.[95,96]

The alternative hypothesis of vascular rarefaction, the process of reduced microvascular density, has been proposed to result in hypertension through increases in systemic vascular resistance.[97,98] Patients with greater risk are individuals who have a systolic blood pressure (BP) ≥160 mmHg or a diastolic BP ≥100 mmHg; diabetes mellitus; established cardiovascular disease including any history of ischemic stroke, cerebral hemorrhage, or transient ischemic attack; myocardial infarction; angina; coronary revascularization; peripheral artery disease; subclinical organ damage previously documented by ECG or echocardiogram revealing left ventricular hypertrophy; cigarette smoking; and dyslipidemia. High blood pressure has been proposed as a surrogate biomarker of antitumoral activity. A huge variety of studies tested the relationship between hypertension and outcome.[31,88,93,99–101] No definitive data are available on this topic.

According to the recommendations put forth in Common Terminology Criteria for Adverse Events (CTCAE), version 4.0,[102] of the National Cancer Institute (NCI) and the Joint National Committee on Prevention, Detection, Evaluation, and Treatment of High Blood Pressure (JNC7) guidelines,[103] the treatment of hypertension should be started as soon as prehypertension is recorded, if cardiovascular risk factors such as diabetes mellitus and obesity are present, or if there is evidence of organ damage, such as left ventricular hypertrophy, chronic kidney disease, and/or peripheral arterial disease.

In 2010, the NCI published the recommendations for the management of VSP inhibitor induced hypertension.[104] In individuals considered at risk, ESMO guidelines recommend repeated BP measurements,[59] while there is no evidence-based guideline for follow-up echocardiograms in asymptomatic patients. Preferred antihypertensive agents for angiogenesis inhibitor-associated hypertension include angiotensin-converting enzyme inhibitors and dihydropyridine calcium channel blockers, although there are minimal data to suggest superiority of a single class of agents. Early and aggressive initiation of antihypertensive therapy appears to help maintain treatment schedule and reduce the risk of significant complications, including malignant hypertension and reversible posterior leukoencephalopathy.

11.2.5 QT Prolongation

Prolongation of the QT interval can lead to life-threatening cardiac arrhythmias, including torsades de pointes (TdP).[105] Although prolongation of the QT interval is not the best predictor of proarrhythmic risk, it represents the principal clinical surrogate marker by which to evaluate the arrhythmic risk of a drug, and it has led to withdrawal of several anticancer drugs from the market. Patients with a history of QT interval prolongation, patients who are taking antiarrhythmics, or patients with relevant cardiovascular disease, bradycardia, thyroid dysfunction, or electrolyte disturbances should be screened and monitored. Periodic monitoring with on-treatment ECGs and electrolytes should be considered.

11.3 CARDIAC TOXICITY INDUCED BY RADIOTHERAPY

Radiotherapy (RT) also plays an important role in cardiotoxicity. The RT of the chest and mediastinum exposes the heart irradiation and this is one of the most critical dose-limiting organs in RT. Several clinical studies have identified adverse clinical consequences of radiation-induced heart disease on the outcome of long-term cancer survivors.[106] In fact, advances in the treatment of cancer have reduced the mortality rates from this disorder and have led to an ever-increasing population of long-term survivors, some with premature cardiac disease related to chemotherapy or RT. The main heart complications include pericardial disease, valvular disorders, conduction abnormalities, ventricular dysfunction, and coronary disease.[107,108]

The physiopathological and clinical consequences of heart irradiation have been mostly studied in patients with Hodgkin lymphoma and breast cancer. Cardiac complications after RT are thought to be due to radiation-induced inflammation and fibrosis of individual cardiac structures, with signs and symptoms typically becoming manifest 5 to 10 years after completion of treatment. Preexisting factors of cardiovascular risk, including chemotherapy, can increase risk of cardiotoxicity. Introduction of conformational radiotherapy and new technologies allowed a drastic reduction in cardiac mortality.[109]

For patients with breast cancer, it is recommended that the irradiated heart volume be minimized to the greatest possible degree without compromising the target coverage. Cardiotoxicity related to RT is somewhat dose dependent and the relative risk is increased even at doses less than 30 Gy.[110] In many cases, conformal blocking and breath-hold techniques can essentially eliminate the heart from the primary beams. For the vast majority of lymphoma patients who are receiving chemotherapy (particularly doxorubicin) and RT, it seems prudent to limit whole heart doses to $\sim$15 Gy, with field reductions as appropriate in the given clinical situation, to areas of persistent (postchemotherapy) residual tumor or to areas of previous bulky involvement.

Although currently there is no direct evidence that successful treatment of traditional cardiac risk factors will alter the natural history of radiation-associated cardiac disease, it is prudent to optimize patient cardiovascular risk profiles. For survivors of Hodgkin lymphoma, ESMO guidelines recommend annual blood pressure, serum glucose, lipid screening, aggressive management of cardiovascular risk factors, and baseline stress test/echocardiogram at 10 years after treatment.

11.4 CONCLUSION

Therapeutic options for patients with cancer now include increasingly complex combinations of medications, radiation therapy, and surgical intervention. Many highly effective agents in contemporary oncology, including anthracyclines, trastuzumab, and anti-VEGF, are associated with potential adverse cardiac effects and are likely to have significant effects on patient outcomes. The development of cardiovascular disease during the course of cancer treatment can adversely impact management of the underlying malignancy. Therefore, understanding cardiac toxicity is crucial to their effective management.

Given the growth of novel biologic therapies, efforts are needed to promote strategies for risk detection and management and to avoid dangerous toxicities that may impede development of patient access to new agents. Progress in understanding of treatment-related cardiac toxicity requires the development of standardized measurement approaches for defining cardiac toxicity that must present high predictive validity for clinical outcomes. Studies are also needed to more accurately predict which patients are at highest risk of developing treatment-related cardiotoxicity; these studies should include genomic testing to explain the variability of incidence and onset. Thus, we can conclude that more research is needed to assess and manage patients with heart disease and cancer, and it is necessary to forge a dynamic partnership between oncologists and cardiologists with the development of a new generation of "cardio-oncology" investigators with the aim of obtaining the best results for patients' wellness.

REFERENCES

1. Ewer MS, Vooletich MT, Durand JB, et al. Reversibility of trastuzumab-related cardiotoxicity: new insights based on clinical course and response to medical treatment. *J Clin Oncol* 2005;**23**:7820–6.
2. Valero V, Gill E, Paton VE, et al. Normal cardiac biopsy results following coadministration of doxorubicin(A), cyclophosphamide (C) and trastuzumab (H) to women with HER2-positive metastatic breast cancer. *J Clin Oncol* 2004;**23**:20(Suppl; abstr 572).
3. Seidman A, Hudis C, Pierri MK, et al. Cardiac dysfunction in the trastuzumab clinical trials experience. *J Clin Oncol* 2002;**20**:1215–21.

4. Jones RL, Swanton C, Ewer MS. Anthracycline cardiotoxicity. *Expert Opin Drug Saf* 2006; **5**:791–809.
5. Raschi E, Vasina V, Ursino MG, et al. Anticancer drugs and cardiotoxicity:insights and perspectives in the era of targeted therapy. *Pharmacol Ther* 2010;**125**:196–218.
6. Cardinale D, Sandri MT, Martinoni A, et al. Left ventricular dysfunction predicted by early troponin I release after high-dose chemotherapy. *J Am Coll Cardiol* 2000;**36**:517–22.
7. Cardinale D, Sandri MT, Martinoni A, et al. Myocardial injury revealed by plasma troponin I in breast cancer treatment with high dose chemotherapy. *Ann Oncol* 2002;**13**:710–5.
8. Ewer MS, Vooletich MT, Benjamin RS. A mathematical model for doxorubicin cardiotoxicity: added evidence for the concept of sequential stress. *Proc Am Soc Clin Oncol* 2004; **23**:148(Abstract 2086).
9. Zhang S, Liu X, Bawa-Khalfe T, et al. Identification of the molecular basis of doxorubicin-induced cardiotoxicity. *Nat Med* 2012;**18**:1639–42.
10. Braverman AC, Antin JH, Plappert MT, et al. Cyclophosphamide cardiotoxicity in bone marrow transplantation: a prospective evaluation of new dosing regimens. *J Clin Oncol* 1991;**9**:1215–23.
11. Goldberg MA, Antin JH, Guinan EC, et al. Cyclophosphamide cardiotoxicity: an analysis of dosing as a risk factor. *Blood* 1986;**68**:1114–8.
12. Quezado ZM, Wilson WH, Cunnion RE, et al. High-dose ifosfamide is associated with severe, reversible cardiac dysfunction. *Ann Intern Med* 1993;**118**:31–6.
13. Martin M, Pienkowski T, Mackey J, et al. Adjuvant docetaxel for node-positive breast cancer. *N Engl J Med* 2005;**352**:2302–13.
14. Slamon D, Eiermann W, Robert N, et al. Adjuvant trastuzumab in HER2-positive breast cancer. *N Engl J Med* 2011;**365**(14):1273–83.
15. Romond EH, Perez EA, Bryant J, et al. Trastuzumab plus adjuvant chemotherapy for operable HER2-positive breast cancer. *N Engl J Med* 2005;**353**:1673–84.
16. Piccart-Gebhart MJ, Procter M, Leyland-Jones B, et al. Trastuzumab after adjuvant chemotherapy in HER2-positive breast cancer. *N Engl J Med* 2005;**353**:1659–72.
17. Slamon DJ, Leyland-Jones B, Shak S, et al. Use of chemotherapy plus a monoclonal antibody against HER2 for metastatic breast cancer that overexpresses HER2. *N Engl J Med* 2001; **344**:783–92.
18. Romond EH, Perez EA, Bryant J, et al. Trastuzumab plus adjuvant chemotherapy for operable HER2-positive breast cancer. *N Engl J Med* 2005;**353**:1673–84.
19. Tan-Chiu E, Yothers G, Romond E, et al. Assessment of cardiac dysfunction in a randomized trial comparing doxorubicin and cyclophosphamide followed by paclitaxel, with or without trastuzumab as adjuvant therapy in node-positive, human epidermal growth factor receptor 2-overexpressing breast cancer: NSABP B-31. *J Clin Oncol* 2005;**23**:7811–9.
20. Slamon D, Eiermann W, Robert N. Adjuvant trastuzumab in HER2-positive breast cancer. *N Engl J Med* 2011;**365**(14):1273–83.
21. Piccart-Gebhart MJ, Procter M, Leyland-Jones B, et al. Herceptin Adjuvant (HERA) Trial Study Team. Trastuzumab after adjuvant chemotherapy in HER2-positive breast cancer. *N Engl J Med* 2005;**353**(16):1659–72.
22. Joensuu H, Kellokumpu-Lehtinen PL, Bono P, et al. FinHer Study Investigators Adjuvant docetaxel or vinorelbine with or without trastuzumab for breast cancer. *N Engl J Med* 2006; **354**(8):809–20.
23. Chien KR. Stress pathways and heart failure. *Cell* 1999;**98**:555–8.
24. Chien KR. Herceptin and the heart: a molecular modifier of cardiac failure. *N Engl J Med* 2006;**354**:789–90.

25. De Keulenaer GW, Doggen K, Lemmens K. The vulnerability of the heart as a pluricellular paracrine organ: lessons from unexpected triggers of heart failure in targeted ErbB2 anticancer therapy. *Circ Res* 2010;**106**:35–46.
26. Geyer CE, Forster J, Lindquist D, et al. Lapatinib plus capecitabine for HER2-positive advanced breast cancer. *N Engl J Med* 2006;**355**:2733–43.
27. Sliwkowski MX, Lofgren JA, Lewis GD, et al. Nonclinical studies addressing the mechanism of action of trastuzumab (herceptin). *Semin Oncol* 1999;**26**:60–70.
28. Shell SA, Lyass L, Trusk PB, et al. Activation of AMPK is necessary for killing cancer cells and sparing cardiac cells. *Cell Cycle* 2008;**7**:1769–75.
29. Perez EA, Koehler M, Byrne J, et al. Cardiac safety of lapatinib: pooled analysis of 3689 patients enrolled in clinical trials. *Mayo Clin Proc* 2008;**83**:679–86.
30. Spector NL, Yarden Y, Smith B, et al. Activation of AMP-activated protein kinase by human EGF receptor 2/EGF receptor tyrosine kinase inhibitor protects cardiac cells. *Proc Natl Acad Sci USA* 2007;**104**:10607–12.
31. Miller K, Wang M, Gralow J, et al. Paclitaxel plus bevacizumab versus paclitaxel alone for metastatic breast cancer. *N Engl J Med* 2007;**357**:2666–76.
32. Motzer RJ, Hutson TE, Tomczak P, et al. Sunitinib versus interferon alfa in metastatic renal-cell carcinoma. *N Engl J Med* 2007;**356**:115–24.
33. Hunt SA, Abraham WT, Chin MH, et al. Focused update incorporated into the ACC/AHA 2005 guidelines for the diagnosis and management of heart failure in adults. A report of the American College of Cardiology Foundation/American Heart Association Task Force on Practice Guidelines developed in collaboration with the International Society for Heart and Lung Transplantation. *J Am Coll Cardiol* 2009;**53**:e1–90.
34. Eschenhagen T, Force T, Ewer M, et al. Cardiovascular side effects of cancer therapies: a position statement from the Heart Failure Association of the European Society of Cardiology. *Eur J Heart Fail* 2011;**13**:1–10.
35. Jones AL, Barlow M, Barrett-Lee PJ, et al. Management of cardiac health in trastuzumab-treated patients with breast cancer: updated United Kingdom National Cancer Research Institute recommendations for monitoring. *Br J Cancer* 2009;**100**:684–92.
36. Martin M, Esteva FJ, Alba E, et al. Minimizing cardiotoxicity while optimizing treatment efficacy with trastuzumab: review and expert recommendations. *Oncologist* 2009;**14**:1–11.
37. Geisberg CA, Sawyer DB. Mechanisms of anthracycline cardiotoxicity and strategies to decrease cardiac damage. *Curr Hypertens Rep* 2010;**12**:404–10.
38. Cardinale D, Colombo A, Torrisi R, et al. Trastuzumab-induced cardiotoxicity: clinical and prognostic implications of troponin I evaluation. *J Clin Oncol* 2010;**28**:3910–6.
39. Kelley WE, Januzzi JL, Christenson RH. Increases of cardiac troponin in conditions other than acute coronary syndrome and heart failure. *Clin Chem* 2009;**55**:2098–112.
40. Horwich TB, Patel J, MacLellan WR, et al. Cardiac troponin I is associated with impaired hemodynamics, progressive left ventricular dysfunction, and increased mortality rates in advanced heart failure. *Circulation* 2003;**108**:833–8.
41. Cardinale D, Colombo A, Sandri MT, et al. Prevention of high-dose chemotherapy-induced cardiotoxicity in high-risk patients by angiotensin-converting enzyme inhibition. *Circulation* 2006;**114**:2474–81.
42. Cardinale D, Colombo A, Lamantia G, et al. Anthracycline-induced cardiomyopathy. Clinical relevance and response to pharmacologic therapy. *J Am Coll Cardiol* 2010;**55**:213–20.
43. Kalay N, Basar E, Ozdogru I, et al. Protective effects of carvedilol against anthracyclines-induced cardiomyopathy. *J Am Coll Cardiol* 2006;**48**:2258–62.

44. Nakamae H, Tsumura K, Terada Y, et al. Notable effects of angiotensin II receptor blocker, valsartan, on acute cardiotoxic changes after standard chemotherapy with cyclophosphamide, doxorubicin, vincristine, and prednisolone. *Cancer* 2005;**104**:2492–8.
45. Huh WW, Jaffe N, Durand JB, et al. Comparison of doxorubicin cardiotoxicity in pediatric sarcoma patients when given with dexrazoxane versus continuous infusion. *Pediatr Hematol Oncol* 2010;**27**:546–57.
46. Bosch X, Esteve J, Sitges M, et al. Prevention of chemotherapy-induced left ventricular dysfunction with enalapril and carvedilol: rationale and design of the OVERCOME trial. *J Card Fail* 2011;**17**:643–8.
47. Pituskin E, Haykowsky M, Mackey JR, et al. Rationale and design of the Multidisciplinary Approach to Novel Therapies in Cardiology Oncology Research Trial (MANTICORE 101-Breast): a randomized, placebo-controlled trial to determine if conventional heart failure pharmacotherapy can prevent trastuzumab-mediated left ventricular remodelling among patients with HER^{2+} early breast cancer using cardiac MRI. *BMC Cancer* 2011;**11**:318.
48. Gradishar WJ, Vokes EE. 5-Fluorouracil cardiotoxicity: a critical review. *Ann Oncol* 1990;**1**:409–14.
49. Labianca R, Beretta G, Clerici M, et al. Cardiac toxicity of 5-fluorouracil: a study on 1083 patients. *Tumori* 1982;**68**:505–10.
50. Frickhofen N, Beck FJ, Jung B, et al. Capecitabine can induce acute coronary syndrome similar to 5-fluorouracil. *Ann Oncol* 2002;**13**:797–801.
51. Trimble EL, Adams JD, Vena D, et al. Paclitaxel for platinum-refractory ovarian cancer: results from the first 1000 patients registered to National Cancer Institute Treatment Referral Center 9103. *J Clin Oncol* 1993;**11**:2405–10.
52. Lapeyre-Mestre M, Gregoire N, Bugat R, et al. Vinorelbine-related cardiac events: a meta-analysis of randomized clinical trials. *Fundam Clin Pharmacol* 2004;**18**:97–105.
53. Yancey RS, Talpaz M. Vindesine-associated angina and ECG changes. *Cancer Treat Rep* 1982;**66**:587–9.
54. Lejonc JL, Vernant JP, Macquin J, et al. Myocardial infarction following vinblastine treatment. *Lancet* 1980;**2**:692.
55. Braithwaite RS, Chlebowski RT, Lau J, et al. Meta-analysis of vascular and neoplastic events associated with tamoxifen. *J Gen Intern Med* 2003;**18**:937–47.
56. Nabholtz JM, Gligorov J. Cardiovascular safety profiles of aromatase inhibitors: a comparative review. *Drug Saf* 2006;**29**:785–801.
57. Cuppone F, Bria E, Verma S, et al. Do adjuvant aromatase inhibitors increase the cardiovascular risk in postmenopausal women with early breast cancer? Meta-analysis of randomized trials. *Cancer* 2008;**112**:260–7.
58. Schmidinger M, Zielinski CC, Vogl UM, et al. Cardiac toxicity of sunitinib and sorafenib in patients with metastatic renal cell carcinoma. *J Clin Oncol* 2008;**26**:5204–12.
59. Choueiri TK, Schutz FA, Je Y, et al. Risk of arterial thromboembolic events with sunitinib and sorafenib: a systematic review and meta-analysis of clinical trials. *J Clinl Oncol* 2010;**28**:2280–5.
60. Schutz FA, Je Y, Richards CJ, et al. Meta-analysis of randomized controlled trials for the incidence and risk of treatment-related mortality in patients with cancer treated with vascular endothelial growth factor tyrosine kinase inhibitors. *J Clinl Oncol* 2012;**30**:871–7.
61. Kilickap S, Abali H, Celik I. Bevacizumab, bleeding, thrombosis, and warfarin. *J Clin Oncol* 2003;**21**(18):3542–3; author reply 3543.
62. Tam BY, Wei K, Rudge JS, et al. VEGF modulates erythropoiesis through regulation of adult hepatic erythropoietin synthesis. *Nat Med* 2006;**12**:793–800.

63. Groarke JD, Cheng S, Moslehi J. Cancer-drug discovery and cardiovascular surveillance. *N Engl J Med* 2013;**369**(19):1779–81.
64. Cortes JE, Kim D-W, Pinilla-Ibarz J, et al. A phase 2 trial of ponatinib in Philadelphia chromosome-positive leukemias. *N Engl J Med* 2013;**369**(19):1783–96.
65. Curigliano G, Cardinale D, Suter T, et al. Cardiovascular toxicity induced by chemotherapy, targeted agents and radiotherapy: ESMO Clinical Practice Guidelines. *Ann Oncol* 2012;**23** (Suppl. 7):155–66.
66. Chen CL, Parameswaran R. Managing the risks of cardiac therapy in cancer patients. *Semin Oncol* 2013;**40**(2):210–7.
67. Yusuf SW, Iliescu C, Bathina JD, et al. Antiplatelet therapy and percutaneous coronary intervention in patients with acute coronary syndrome and thrombocytopenia. *Tex Heart Inst J* 2010;**37**:336–40.
68. Winter PC. The pathogenesis of venous thromboembolism in cancer: emerging links with tumour biology. *Hematol Oncol* 2006;**24**:126–33.
69. Petralia GA, Lemoine NR, Kakkar AK. Mechanisms of disease: the impact of antithrombotic therapy in cancer patients. *Nat Clin Pract Oncol* 2005;**2**:356–63.
70. Lip GY, Chin BS, Blann AD. Cancer and the prothrombotic state. *Lancet Oncol* 2002;**3**:27–34.
71. Yusuf SW, Razeghi P, Yeh ET. The diagnosis and management of cardiovascular disease in cancer patients. *Curr Probl Cardiol* 2008;**33**:163–96.
72. Khorana AA. Risk assessment for cancer-associated thrombosis: What is the best approach? *Thromb Res* 2012;**129**:S10–5.
73. Rodeghiero F, Elice F. Thalidomide and thrombosis. *Pathophysiol Haemost Thromb* 2003; **33**:15–8.
74. Hirsh J. Risk of thrombosis with lenalidomide and its prevention with aspirin. *Chest* 2007; **131**:275–7.
75. Yeh ET, Bickford CL. Cardiovascular complications of cancer therapy: incidence, pathogenesis, diagnosis, and management. *J Am Coll Cardiol* 2009;**53**:2231–47.
76. Nalluri SR, Chu D, Keresztes R, et al. Risk of venous thromboembolism with the angiogenesis inhibitor bevacizumab in cancer patients: A metaanalysis. *JAMA* 2008;**300**: 2277–85.
77. Scappaticci FA, Skillings JR, Holden SN, et al. Arterial thromboembolic events in patients with metastatic carcinoma treated with chemotherapy and bevacizumab. *J Natl Cancer Inst* 2007;**99**:1232–9.
78. Calvo V, Ramirez N, Saura C, et al. Risk of venous and arterial thromboembolic events in patients with metastatic breast cancer treated with bevacizumab: a meta-analysis. *J Clin Oncol* 2010;**28**:15s(Suppl, abstr 1043).
79. Hurwitz HI, Saltz LB, Van Cutsem E, et al. Venous thromboembolic events with chemotherapy plus bevacizumab: a pooled analysis of patients in randomized phase II and III studies. *J Clin Oncol* 2011;**29**(13):1757–64.
80. Khorana AA, Connolly GC. Assessing risk of venous thromboembolism in patients with cancer. *J Clin Oncol* 2009;**27**:4839–47.
81. Leighl N, Bennouna J, Kuo H. Safety of bevacizumab treatment in non-small cell lung cancer (NSCLC) subjects receiving full-dose anticoagulation (FDAC) treated on protocol BO17704. *Eur J Cancer* 2007;**391**(Suppl. 5, abstr 6610).
82. Sonpavde G, Je Y, Schutz F, et al. Venous thromboembolic events with vascular endothelial growth factor receptor tyrosine kinase inhibitors: a systematic review and meta-analysis of randomized clinical trials. *Crit Rev Oncol Hematol* 2013;**87** (1):80–9.

83. Agnelli G, Gussoni G, Bianchini C, et al. Nadroparin for the prevention of thromboembolic events in ambulatory patients with metastatic or locally advanced solid cancer receiving chemotherapy: a randomised, placebo-controlled, double-blind study. *Lancet Oncol* 2009;**10**:943–9.
84. Miller KD, Chap LI, Holmes FA, et al. Randomized phase III trial of capecitabine compared with bevacizumab plus capecitabine in patients with previously treated metastatic breast cancer. *J Clin Oncol* 2005;**23**:792–9.
85. Hurwitz H, Fehrenbacher L, Novotny W, et al. Bevacizumab plus irinotecan, fluorouracil, and leucovorin for metastatic colorectal cancer. *N Engl J Med* 2004;**350**:2335–42.
86. Kabbinavar FF, Schulz J, McCleod M, et al. Addition of bevacizumab to bolus fluorouracil and leucovorin in first-line metastatic colorectal cancer: results of a randomized phase II trial. *J Clin Oncol* 2005;**23**:3697–705.
87. Pande A, Lombardo J, Spangenthal E, et al. Hypertension secondary to antiangiogenic therapy: experience with bevacizumab. *Anticancer Res* 2007;**27**:3465–70.
88. Escudier B, Eisen T, Stadler WM, et al. Sorafenib in advanced clear-cell renal-cell carcinoma. *N Engl J Med* 2007;**356**:125–34.
89. Rini BI, Cohen DP, Lu DR, et al. Hypertension as a biomarker of efficacy in patients with metastatic renal cell carcinoma treated with sunitinib. *J Natl Cancer Inst* 2011;**103**:763–73.
90. Burstein HJ, Elias AD, Rugo HS, et al. Phase II study of sunitinib malate, an oral multitargeted tyrosine kinase inhibitor, in patients with metastatic breast cancer previously treated with an anthracycline and a taxane. *J Clin Oncol* 2008;**26**:1810–6.
91. Motzer RJ, Michaelson MD, Redman BG, et al. Activity of SU11248, a multitargeted inhibitor of vascular endothelial growth factor receptor and platelet-derived growth factor receptor, in patients with metastatic renal cell carcinoma. *J Clin Oncol* 2006;**24**:16–24.
92. Demetri GD, van Oosterom AT, Garrett CR, et al. Efficacy and safety of sunitinib in patients with advanced gastrointestinal stromal tumour after failure of imatinib: a randomised controlled trial. *Lancet* 2006;**368**:1329–38.
93. Motzer RJ, Hutson TE, Tomczak P, et al. Sunitinib versus interferon alfa in metastatic renal-cell carcinoma. *N Engl J Med* 2007;**356**:115–24.
94. Motzer RJ, Rini BI, Bukowski RM, et al. Sunitinib in patients with metastatic renal cell carcinoma. *JAMA* 2006;**295**:2516–24.
95. Kamba T, McDonald DM. Mechanisms of adverse effects of anti-VEGF therapy for cancer. *Br J Cancer* 2007;**96**:1788–95.
96. Verheul HM, Pinedo HM. Possible molecular mechanisms involved in the toxicity of angiogenesis inhibition. *Nat Rev Cancer* 2007;**7**:475–85.
97. Yang JC. A randomized trial of bevacizumab, an antivascular endothelial growth factor antibody, for metastatic renal cancer. *N Engl J Med* 2003;**349**:427–34.
98. Maitland ML. Blood pressure (BP) as a biomarker for sorafenib (s.), an inhibitor of the vascular endothelial growth factor (veGF) signalling pathway. *ASCO Meeting Abstracts* 2006;**24**:2035.
99. Escudier B, Pluzanska A, Koralewski P, et al. Bevacizumab plus interferon alfa-2a for treatment of metastatic renal cell carcinoma: a randomised, double-blind phase III trial. *Lancet* 2007;**370**(9605):2103–11.
100. Motzer RJ, Hutson TE, Tomczak P, et al. Overall survival and updated results for sunitinib compared with interferon alfa in patients with metastatic renal cell carcinoma. *J Clin Oncol* 2009;**27**(22):3584–90.

101. Floyd JD, Nguyen DT, Lobins RL, et al. Cardiotoxicity of cancer therapy. *J Clin Oncol* 2005;**23**:7685–96.
102. National Cancer Institute. Common Terminology Criteria for Adverse Events (CTCAE), version 4.0. Bethesda, MD: NCI, 2009. Available at <http://evs.nci.nih.gov/ftp1/CTCAE/About.html> [accessed 08.27.13].
103. Chobanian AV, Bakris GL, Black HR, et al. Seventh report of the Joint National Committee on Prevention, Detection, Evaluation. *Treat High Blood Press Hypertens* 2003;**42**:1206–52.
104. Maitland ML, Bakris GL, Black HR, et al. Initial assessment, surveillance, and management of blood pressure in patients receiving vascular endothelial growth factor signaling pathway inhibitors. *J Natl Cancer Inst* 2010;**102**:596–604.
105. Suter TM, Ewer MS. Cancer drugs and the heart: importance and management. *Eur Heart J* 2013;**34**(15):1102–11.
106. Lancellotti P, Nkomo VT, Badano LP, et al. Expert consensus for multi-modality imaging evaluation of cardiovascular complications of radiotherapy in adults: a report from the European Association of Cardiovascular Imaging and the American Society of Echocardiography. *Eur Heart J Cardiovasc Imaging* 2013;**14**(8):721–40.
107. Heidenreich PA, Hancock SL, Lee BK, et al. Asymptomatic cardiac disease following mediastinal irradiation. *J Am Coll Cardiol* 2003;**42**:743–9.
108. Heidenreich PA, Hancock SL, Vagelos RH, et al. Diastolic dysfunction after mediastinal irradiation. *Am Heart J* 2005;**150**:977–82.
109. Petersen C, Würschmidt F. Late toxicity of radiotherapy: a problem or a challenge for the radiation oncologist? *Breast Care (Basel)* 2011;**6**(5):369–74.
110. Darby SC, Ewertz M, McGale P, et al. Risk of ischemic heart disease in women after radiotherapy for breast cancer. *N Engl J Med* 2013;**368**(11):987–98.

Chapter 12

Association of Human Immunodeficiency Virus Infection with Exposure to Highly Active Antiretroviral Therapy and Its Adverse Cardiovascular Effects

Steven E. Lipshultz,[1] Tracie L. Miller,[2] James D. Wilkinson,[2] Gwendolyn B. Scott,[2] Gabriel Somarriba,[2] Thomas R. Cochran,[3] Catherine M. Mas,[1] and Stacy D. Fisher[4]

[1]*Carmen and Ann Adams Department of Pediatrics, Wayne State University School of Medicine, University Pediatricians, Children's Hospital of Michigan, Detroit*

[2]*Department of Pediatrics, University of Miami Miller School of Medicine, Holtz Children's Hospital of the University of Miami, Jackson Memorial Medical Center, and the Sylvester Comprehensive Cancer Center, Miami*

[3]*University of Kansas School of Medicine, University of Kansas Medical Center, Kansas City*

[4]*Departments of Medicine and Pediatrics, University of Maryland School of Medicine, University of Maryland Comprehensive Heart Center, Baltimore*

12.1 INTRODUCTION

Human immunodeficiency virus (HIV) infection is a primary cause of acquired heart disease, particularly of accelerated atherosclerosis, symptomatic heart failure, and pulmonary arterial hypertension.[1–17] Cardiac complications often occur in the later stages of prolonged HIV infection and are therefore more relevant as longevity improves.[1–17] Multiagent HIV therapies that help sustain life may also directly increase the risk of cardiovascular events and accelerated atherosclerosis.[1,18–23]

By 2011, between 31 and 36 million people worldwide were living with HIV,[24] an estimated 0.8% of all people aged 15 to 49. Globally, the treatment and burden of this epidemic varies greatly from one region to the next. One of

The Heart and Toxins. DOI: http://dx.doi.org/10.1016/B978-0-12-416595-3.00012-8

the most severely affected regions is Sub-Saharan Africa, where 69% of all people living with HIV worldwide reside and nearly 1 in every 20 adults is infected.[24] In 2011, 330,000 children acquired an HIV infection (90% of whom are in Sub-Saharan Africa). Compared to the rate of new infections in children in 2003, this is a 43% reduction.[24]

In the early 1990s, most HIV-infected children did not receive antiretroviral therapy (ART) or only received monotherapy with zidovudine. These children often had abnormal echocardiographic measures of the left ventricular (LV) structure and function, predictors of early mortality.[25] Although the cardiovascular effects of HIV and antiretroviral therapy exposures are not fully understood, HIV-infected children are routinely exposed to ART or highly active antiretroviral therapy (HAART) while the cardiovascular system is still developing. Subclinical cardiac abnormalities may develop into symptomatic cardiomyopathy in adulthood.

Cardiac abnormalities associated with HIV infection (Table 12.1) include premature myocardial infarction (MI) or stroke, pericardial effusion, lymphocytic interstitial myocarditis, LV diastolic dysfunction, dilated cardiomyopathy (frequently with myocarditis), infective endocarditis, and malignancy (myocardial Kaposi's sarcoma and B-cell immunoblastic lymphoma).[3,4,26,27] Multidrug class ART regimens, including nonnucleoside reverse transcriptase inhibitors (NNRTI), protease inhibitors (PIs), and integrase inhibitors, are associated with clinically significant drug–drug interactions with cardiovascular agents (e.g., statins, calcium channel blockers, and milrinone, among others).[18] Antiretroviral therapies may also change repolarization or prolong the QT interval, increasing the risk of serious arrhythmias and sudden cardiac death.[19]

12.1.1 Historical and Ethical Considerations

In 1987, the first ART agent, zidovudine, became available. By the late 1990s, clinicians began treating HIV-infected patients with a comprehensive drug regimen combining three or more drugs from the nucleoside reverse transcriptase inhibitor (NRTI), NNRTI, and PI drug classes, known as highly active antiretroviral therapy. As of 2011, more than 8,000,000 people worldwide were receiving ART.[24] People living with HIV who receive ART now have a near-normal life expectancy.[28] Increased use of HAART has transformed HIV/AIDS from a rapidly fatal disease into a chronic and manageable illness. However, in the HAART era, concerns have been raised about treatment-related cardiovascular and other morbidities associated with long-term exposure to potent antiretroviral regimens[29] and HIV infection.

Zidovudine treatment, introduced as prenatal prophylaxis in the early 1990s, reduced the incidence of maternal–infant transmission from about 25% to 8%, giving the world hope for eradication of vertical transmission of

TABLE 12.1 Summary of HIV-Associated Cardiovascular Diseases

Disease	Possible Causes	Incidence/Prevalence	Diagnosis	Treatment
Accelerated atherosclerosis	Protease inhibitors, atherogenesis with virus-infected macrophages, chronic inflammation, glucose intolerance, dyslipidemia, endothelial dysfunction	Up to 8% prevalence	ECG, stress testing, echocardiography, lipid profile, CT angiography, and calcium scoring	Smoking cessation, low-fat diet, aerobic exercise, blood pressure control, guideline-based statin use, percutaneous coronary intervention, coronary artery bypass surgery
Dilated cardiomyopathy Systolic dysfunction	Coronary artery disease *Drug related*: cocaine, AZT, IL-2, doxorubicin, interferon *Infectious*: HIV, toxoplasma, coxsackievirus group B, EBV, CMV, adenovirus *Metabolic or endocrine*: selenium or carnitine deficiency, anemia, hypocalcemia, hypophosphatemia, hyponatremia, hypokalemia, hypoalbuminemia, hypothyroidism, growth hormone deficiency, adrenal insufficiency, hyperinsulinemia *Cytokines*: TNF-α, nitric oxide, TGF-β, endothelin-I, interleukins *Immunodeficiency*: CD4 < 100 Autoimmune	Up to 8% of asymptomatic patients Up to 25% of autopsy cases	*Chest radiograph findings ECG*: Nonspecific conduction abnormalities, PVCs, PACs *Echocardiographic findings*: low–normal LV wall thickness, increased LV mass, dilated LV, systolic LV dysfunction *Possible laboratory studies*: Troponin T, brain natriuretic peptide level, CD4 count, viral load, viral PCR, toxoplasma serology, thyroid-stimulating hormone, cortisol, carnitine, selenium, serum ACE, stress testing, myocardial biopsy, cardiac catheterization	Diuretics, digoxin, ACE inhibitors, beta blockers Adjunctive treatment in HIV patients Treatment of infection with nutritional replacement IVIg Intensify antiretroviral therapy Follow-up serial echocardiograms

(*Continued*)

TABLE 12.1 Summary of HIV-Associated Cardiovascular Diseases (Continued)

Disease	Possible Causes	Incidence/Prevalence	Diagnosis	Treatment
LV diastolic dysfunction	TNF, interleukin-6 Hypertension Chronic viral infection	Up to 37% prevalence asymptomatic	Echocardiography Tissue Doppler imaging	Treat hypertension Intensify antiretroviral therapy
Pulmonary hypertension	Plexogenic pulmonary arteriopathy	0.5% prevalence	ECG, echocardiography, right heart catheterization	Anticoagulation, vasodilators, prostacyclin analogs, endothelin antagonists, PDE-5 inhibitors
Pericardial disease	*Bacteria*: Staphylococcus, streptococcus, proteus, klebsiella, enterococcus, listeria, nocardia, mycobacterium *Viral pathogens*: HIV, HSV, CMV, adenovirus, echovirus *Other pathogens*: Cryptococcus, toxoplasma, histoplasma *Malignancy*: Kaposi's sarcoma, lymphoma, capillary leak/ wasting/malnutrition Hypothyroidism Immunodeficiency Uremia	11%/year and markedly reduced in post-HAART era studies Spontaneous resolution in 42% of affected patients Approximately 30% increase in 6-month mortality	Pericardial rub on examination Echocardiography Fluid analysis for gram stain and culture, cytology; ECG-low voltage/PR depression Associated pleural and peritoneal fluid analysis Pericardial biopsy	Treat the cause *Follow up*: Serial echocardiography Intensify antiretroviral therapy Pericardiocentesis or window histoplasma

Infective endocarditis	*Autoimmune bacteria*: *Staphylococcus aureus* or *Staphylococcus epidermidis*, salmonella, streptococcus, *Hemophilus parainfluenzae*, *Pseudallescheria boydii*, HACEK organisms *Fungal*: *Aspergillus fumigatus*, candida, *Cryptococcus neoformans*	Increased incidence in IVDA, regardless of HIV status	Blood cultures; echo-cardiography	IV antibiotics, valve replacements
Nonbacterial thrombotic endocarditis	Valvular damage, vitamin C deficiency, malnutrition, wasting, DIC, hypercoagulable state, prolonged acquired immunodeficiency	Rare, but clinically relevant emboli in 42% of cases when it occurs	Echocardiography	Anticoagulation, treat vasculitis or underlying illness
Malignancy	Kaposi's sarcoma, non-Hodgkin lymphoma, leiomyosarcoma, low CD4 count, prolonged immunodeficiency HHV-8, EBV	Approximately 1% incidence Usually metastatic in HIV + patients	Echocardiography, biopsy	Chemotherapy possible
Right ventricle Disease	Recurrent pulmonary infections, pulmonary arteritis, microvascular pulmonary emboli, COPD		ECG, echocardiography, right heart catheterization	Diuretics, treat underlying lung infection or disease, anticoagulation as clinically indicated
Vasculitis	Drug therapy with antibiotics and antivirals	Increasing incidence	Clinical diagnosis	Systemic corticosteroids, withdrawal of drug

(*Continued*)

TABLE 12.1 Summary of HIV-Associated Cardiovascular Diseases (Continued)

Disease	Possible Causes	Incidence/Prevalence	Diagnosis	Treatment
Autonomic dysfunction	CNS disease, drug therapy, prolonged immunodeficiency, malnutrition, sedentary lifestyle	Increased in patients, with CNS disease	Tilt-table test, Holter or event monitoring	Procedural precautions
Arrhythmias	Drug therapy, pentamidine, autonomic dysfunction, acidosis electrolyte abnormalities		*ECG*: Long QT, Holter monitoring, exercise stress testing	Discontinue drug, procedural precautions, electrolyte replacement
Lipodystrophy	Drug therapy: NRTI, protease inhibitors		Echocardiography, lipid profile, cardiac catheterization, coronary calcium score	Hyperlipidemia therapy (beware of drug interactions), aerobic exercise, altered antiretroviral, therapy, cosmetic surgery/fat implantation

ACE, angiotensin-converting enzyme; AZT, azidothymidine; CMV, cytomegalovirus; CNS, central nervous system; DIC, disseminated intravascular coagulation; EBV, Epstein–Barr virus; ECG, electrocardiogram; HACEK, *Cardiobacterium* sp., *Actinobacillus* sp., *Cardiobacterium hominis, Eikenella corrodens, Kingella kingae*; HHV, human herpesvirus; HIV, human immunodeficiency virus; HSV, herpes simplex virus; HTN, hypertension; IL-2, interleukin-2; IVDA, intravenous drug abuse; IVIg, intravenous immunoglobulin; LV, left ventricular; NRTI, nonnucleoside reverse transcriptase inhibitor; PAC, premature atrial complex; PCR, polymerase chain reaction; PVC, premature ventricular complex; TGF, transforming growth factor; TNF, tumor necrosis factor.

Source: *Modified with permission from Fisher and Lipshultz, 2011.*[141]

HIV in the foreseeable future.[30] Moreover, the use of prenatal multidrug ART prophylaxis during pregnancy has reduced the rate of maternal–infant transmission to below 2%.[31] Successful reduction in vertical transmission has been greatest in developed countries where there is widespread maternal access to multidrug ART prophylaxis. These successes have stimulated efforts to provide wider access to effective vertical-transmission preventive and therapeutic regimens for HIV-infected persons in countries with the highest prevalence of the virus and that also have the most limited health resources. In 2012, the Joint United Nations Program on HIV/AIDS adopted the goal of eliminating maternal–infant transmission by 2015.[24] While international organizations and governmental programs seek to provide wider access to antiretroviral treatment in low- and middle-income countries, scientists and clinicians continue to encourage identification and management of treatment-related toxicities.

With the advent of wider access to prenatal prophylactic therapies for pregnant HIV-infected women, important ethical questions have been raised. In 1994, Ronald Bayer presented a commentary, published in the same issue as an article presenting the results of the new prophylactic zidovudine regimen, urging the medical and public health community to consider the social impact of these findings.[31,32] He raised concerns about the physical, social, and economic consequences of mandatory testing and imposed treatment on mothers.[32] Is it ethical to compel a mother to undergo testing for HIV when she does not have access to prophylactic therapy if she tests positive? What might be the effects of being stigmatized by HIV infection, and what might be the unknown long-term effects on her unborn child of prenatal zidovudine exposure? With current guidelines recommending a more effective multidrug ART or HAART prenatal prophylaxis regimen, what additional ethical issues are raised regarding the counterpoise between the clear benefits of prenatal HAART prophylaxis to prevent vertical transmission of HIV versus the potential long-term cardiovascular and other risks to the HIV-exposed but uninfected infants?

Longitudinal prospective studies of ART's cumulative effects have found increased risk for cardiovascular dysfunction.[21] Studies of in utero exposure to ART on uninfected infants born to HIV-infected mothers have also found that such exposure is associated with abnormal cardiac structure and function in the first years of life.[33] Specifically, these children had smaller hearts but increased LV function. These effects were more pronounced in girls than in boys. Clinicians need to understand that HIV-infected patients experience higher cardiac risks because of their treatment, and that the associated cumulative and delayed effects from exposure to antiretroviral drugs must be managed carefully. Longitudinal follow up of HIV-uninfected infants exposed to prophylactic multidrug ART in utero is needed to determine whether there are any long-term cardiac effects and risks as they grow into adulthood.

12.2 CARDIAC CONSIDERATIONS OF HIV EXPOSURE AND ANTIRETROVIRAL THERAPY

12.2.1 Accelerated Atherosclerosis

Since the advent of ART, patients with HIV infection live longer, but chronic conditions, including atherosclerotic and metabolic disease, are becoming more prevalent in this population.[34] HAART causes metabolic syndrome, which is well characterized in adults by unfavorable body composition (reduced subcutaneous and increased visceral fat), insulin resistance and abnormal glucose metabolism, and dyslipidemia.[35,36] Metabolic syndrome places patients at risk for atherosclerotic cardiovascular disorders. In fact, several studies have suggested that HAART-related fat redistribution in adults increases their risk of MI secondary to the level of viral control (increased inflammation) or to ART exposure (including PIs and certain NRTIs).[37–39] Acute MI can be the primary presentation of atherosclerotic disease.[40] However, whether metabolic syndrome in HIV-infected patients is exclusively related to ART exposure or to the HIV infection itself is debatable. Synergistic causes may include traditional risk factors, such as family history, high low-density lipoprotein (LDL) cholesterol, low high-density lipoprotein (HDL) cholesterol, diabetes, hypertension, age greater than 55 years, HIV viral load, and medication-specific ART exposure. Studies with children show similar, although not identical, findings with the use of ART, including abnormal body composition, insulin resistance, and dyslipidemia. Further, the risk appears to increase at older ages and with longer duration of HAART treatment.[41–47] The onset of puberty has been proposed as another factor associated with accelerating these changes.[48]

Early studies in HIV-infected children showed that PI therapy improved weight, weight-to-height ratio, and mid-arm muscle circumference independent of concurrent decreases in the HIV viral load and improved CD4 T-lymphocyte counts.[49] The immediate treatment effects were most apparent as improved weight and mid-arm muscle circumference and, to some extent, increased height and lean body mass. However, in addition to positive improvements in growth and body mass, HAART is also associated with abnormalities in fat distribution in children, although some studies reported similar lean mass in HIV-infected and uninfected children.[50] Jacobson et al. found that limb-to-trunk fat ratios were lower in HIV-infected children than in HIV-exposed but uninfected (HEU) children.[42] These findings suggest that both peripheral lipoatrophy and central obesity occur in such children. Other studies have shown that most children experience fat redistribution within three years of initiating a PI-containing regimen and that this redistribution progresses over time. Still other studies have identified metabolic abnormalities induced by other specific classes of drugs; for example, stavudine has been associated with lipoatrophy, possibly by altering mitochondrial number and function.[51–54]

After exposure to ART, LDL cholesterol, HDL cholesterol, and total cholesterol, concentrations increase in both adults and children. Children newly exposed to ART experienced a rapid rise in LDL cholesterol concentration during the first six months of treatment and that continued through 12 months.[55] In a cohort of 449 children in the United Kingdom, 10% had LDL cholesterol concentrations greater than the 95th percentile for age, and PIs increased total cholesterol concentrations more than NNRTIs. The authors concluded that dietary and exercise interventions and a change in ART might help address these metabolic abnormalities.[56]

For children with incident hypercholesterolemia, Jacobson et al. found that changing the ART regimen was associated with cholesterol concentrations returning to normal.[57] Statistical power to detect effects of switching to specific ARTs was low; however, a higher viral load at baseline was associated with the return to normal concentrations. According to the Department of Health and Human Services Panel on Antiretroviral Guidelines for Adults and Adolescents, switching from one PI to another or to the same PI at a lower dosing frequency may reduce dyslipidemia.[58] Evaluating metabolic changes in children as they start or change ART can help to determine specific effects of ART because children have fewer confounding lifestyle factors (e.g., smoking, alcohol use, and obesity) that can independently affect metabolism.

Atherosclerotic cardiovascular disease (CVD) often results from an environment hostile to the endothelium, which may occur from a complex interaction of HIV, the adverse effects of ART, traditional risk factors for CVD, inflammation, and coinfections.[59] Autopsies of HIV-infected patients 23 to 32 years of age who died unexpectedly revealed atherosclerotic plaque with features common to both coronary atherosclerosis and transplant vasculopathy, histologic characteristics more often seen with single-vessel disease in which plaque rupture caused the MI.[60,61] Imaging data suggest that inflammation also causes such premature cardiovascular events (Table 12.2).

Endothelial dysfunction is one possible causal link between HIV infection and atherosclerosis. HIV-infected patients have increased expression of vascular adhesion molecules (e.g., E-selectin, intercellular adhesion molecule 1, and vascular cell adhesion molecule 1) and inflammatory cytokines such as interleukin-6 (IL-6) and tumor necrosis factor (TNF)-α.[62,63] That there is an endothelial response to injury is supported by the correlation between viral load and higher plasma TNF-α, IL-6, and von Willebrand factor concentrations.[44,62,64] The risk of MI increases with exposure to combination ART (Figure 12.1).[64] Inflammation and immune activation associated with untreated HIV infection may increase the risk for cardiovascular disease as imaging, such as carotid intimal medial thickness and brachial flow reactivity, suggest subclinical vascular disease.

A recent review of numerous adult studies of HIV-infected individuals have associated elevated biomarkers, including C-reactive protein, fibrinogen,

TABLE 12.2 Evidence of Atherosclerosis in HIV-Infected Individuals

Imaging Modality	HIV vs. Matched Controls	Associations
Carotid ultrasound Carotid intimal-medial thickness	First to show higher rates of atherosclerosis; 0.04 mm thicker in HIV (from meta-analysis)	Smoking, dyslipidemia, low nadir CD4 T-cell count and increased lymphocyte activation that is correlated with higher IMT and progression
Computed tomography Calcium scores CT angiography	HIV-infected patients have higher mean Agatston scores and proportion of scores >0 Higher prevalence of noncalcified plaque	Framingham risk, metabolic syndrome, higher levels of asymmetric dimethylarginine, and fatty liver CD4/CD8 ratio and HIV duration independently predict plaque burden
Magnetic resonance angiography		Association of HIV viremia and atherosclerotic plaque burden in the aorta; extensively used in cerebral and peripheral vascular beds
Flow-mediated brachial artery dilation	Impaired in HIV infected	Degree of HIV viremia, injection drug use, periodontal disease, and Vitamin D deficiency Statins, niacin, and pentoxifylline have improved flow-mediated dilation
Future potential imaging: Intravascular ultrasound Intracoronary optical coherence tomography Future PET imaging of 18 FDG uptake Molecular targeted MRI		

Source: Modified with permission from Fisher and Lipshultz, 2011.[141]

D-dimer, and IL-6, with progression of HIV disease and attendant morbidity and mortality.[65] However, it is not clear that these biomarkers are specific for heart disease in this population. Indeed, a randomized study comparing abacavir/lamivudine to tenofovir/emtricitabine found no significant differences in a large panel of inflammatory biomarkers.[66] Another randomized switch study comparing abacavir regimens to tenofovir regimens found no differences in

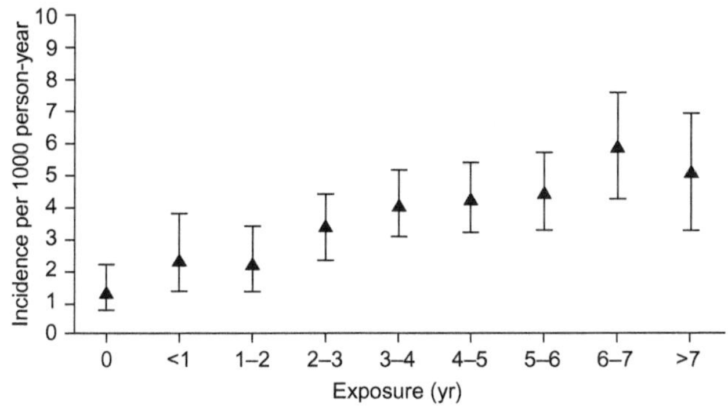

	0	<1	1–2	2–3	3–4	4–5	5–6	6–7	>7	Total
No. of events	16	17	20	41	61	62	51	47	30	345
No. of person-years	11,815	7105	9027	12,098	14,892	14,394	11,351	7935	5853	94,469

FIGURE 12.1 **Risk of myocardial infarction according to exposure to combination ART.** The adjusted relative rate of MI, according to cumulative exposure to combination antiretroviral therapy, was 1.16 per year of exposure (95% CI 1.09–1.23). Arrows, mean values; bars, 95% CIs. **Source:** *Reprinted with permission from DAD Study Group, 2007.*[64]

biomarkers of inflammation, coagulation, platelet function, or endothelial function.[67] In the setting of newer antiretroviral agents, consideration is being given to treating at higher CD4 counts in large part hoping to limit inflammation and subsequent clinical CAD in the host environment.[68]

Premature cerebrovascular disease is also prevalent in HIV-infected adults, and providers should be aware of its risks for young adults. A review of autopsies from AIDS patients between 1983 and 1987 estimated the prevalence of stroke to be 13 (8%). Evidence of cerebral emboli was found in 4 of these 13 patients, and the embolus had a clear cardiac source in 3 of these 4 patients. These patients were treated in the pre-HAART era either without directed therapy or with azidothymidine (AZT), so the cause of cerebral emboli may be different.

As HIV-infected children age, the most common causes of stroke should be considered and atherosclerotic cardiovascular disease should be suspected. Premature atherosclerosis is generally found in children treated with antiretroviral therapy, although it is not clear whether it is ART-related or related to long-standing immune suppression, concomitant infection, inflammation, or the HIV infection itself. Acute stroke investigation in HIV-infected individuals should be somewhat different from the general population as a result of infectious and immune-mediated vasculopathy, tumors, opportunistic infections, and cardioembolism.[69]

Prevention of premature atherosclerotic disease should be directed at identifying and decreasing known risk factors. Low-cholesterol diets can reduce the incidence of dyslipidemia.[70] In addition, patients should be encouraged to engage in regular aerobic exercise, follow heart-healthy diets, avoid secondhand smoke, and/or stop smoking, all of which also lower lipid levels and help decrease atherosclerotic risk.[71,72] Glucose and lipid concentrations should be monitored regularly for those with HIV infection.[70,71] Current treatment guidelines should be followed for patients with dyslipidemia for primary and secondary risk prevention. Known drug interactions should be avoided, such as those between certain statins and protease inhibitors, which can increase protease inhibitor concentrations 400-fold.[73,74]

12.2.2 Left Ventricular Systolic Dysfunction

Progressive abnormalities in left ventricular size and function have been found in children infected with HIV before the use of HAART. These children generally presented with dilated cardiomyopathy resulting in depressed LV contractility and dilation.

Incidence

Before HAART therapy, the 2- to 5-year incidence of symptomatic heart failure ranged from 4 to 28% in HIV patients, suggesting a prevalence of symptomatic HIV-related heart failure of between 4 million and 5 million cases worldwide.[5,6] The incidence of clinically important cardiac disease in HIV-infected patients has been markedly reduced by HAART. However, HAART is only available to a few of those in need.[6,75–77]

Patients receiving HAART often have asymptomatic abnormalities in cardiac structure.[78] Echocardiographic measurements indicate that 18% have LV systolic dysfunction, 6.5% have LV hypertrophy, and 40% have left atrial dilation.[76] A history of MI, current tobacco smoking, and elevated highly sensitive C-reactive protein concentrations were associated with LV systolic dysfunction.[76]

Among pre-HAART-era HIV-infected children 10 years of age or younger, 25% died from chronic cardiac disease[5,6] and 28% experienced serious cardiac events after an AIDS-defining illness.[1,5,6,77] Only in children has a mild decrease in LV systolic function or an increase in LV mass been predictive of mortality.[25] The National Heart, Lung, and Blood Institute (NHLBI) funded the Highly Active Antiretroviral Therapy-Associated Cardiotoxicity (CHAART-II) study, which collected longitudinal echocardiographic measurements of HIV-infected children and adolescents exposed to HAART or multidrug ART. Compared to HIV-infected but relatively less ART-exposed children in the HIV-infected cohort from the Pediatric Pulmonary and Cardiovascular Complications of Vertically Transmitted HIV Infection (P^2C^2 HIV) study, the CHAART-II patients had persistently lower LV mass. Although in infancy, LV contractility was substantially better in the CHAART-II patients than in the P^2C^2 HIV patients, at 10 years of follow up, LV contractility had declined significantly in the CHAART-II group compared to those in the P^2C^2 HIV group during infancy.[79] These findings suggest that long-term HAART exposure may be cardioprotective for a finite period early in life but this protection decreases as this HIV-infected population ages into adolescence and early adulthood. Further longitudinal follow-up studies are needed with adolescents and young adults who were perinatally infected with HIV to better characterize their future cardiac risk.

After 11 years of HAART exposure, LV function in CHAART-II patients was the same as that of the HAART-unexposed P^2C^2 HIV-infected cohort. The conclusion was that the protective effects of HAART on cardiac function diminish 11 years after exposure. A larger, but otherwise similar, HIV-infected cohort from the National Institutes of Health (NIH)-funded Pediatric HIV/AIDS Cohort Study's Adolescent Master Protocol collected only a single echocardiogram from each patient. Generally, measures of LV structure and function were better in this long-term HAART-exposed cohort than in the relatively HAART-unexposed P^2C^2 HIV cohort, but these measures were not as normal as those in an HEU control group.[80] Although a conclusion was that HAART exposure in HIV-infected children appeared to be cardioprotective, the cross-sectional study did not show the same long-term trajectories of cardiac health or dysfunction. Serial echocardiographic and other cardiovascular risk screenings in this cohort could better define the long-term cardiovascular risk in perinatally HIV-infected children treated during the HAART era.

The CHAART-I study collected serial echocardiograms in a cohort of HEU children exposed perinatally to either multidrug ART or HAART.[33] At age 2, these children had below-normal LV mass, LV dimension, and septal wall thickness, indicating smaller hearts. In contrast, LV function was increased. These differences were more pronounced in girls.[33] In a larger cohort of HEU, perinatally HAART-exposed and slightly older (ages 3–5) children from the Pediatric HIV/AIDS Cohort Study (PHACS) Surveillance Monitoring for ART Toxicities Study in the HIV-Uninfected Children Born to HIV-Infected Women (SMARTT) protocol, preliminary results from a single echocardiogram showed that 16% of children had at least one abnormal echocardiographic measure. First-trimester exposure to various ART agents was associated with specific echocardiographic abnormalities. For instance, first-trimester exposure to abacavir was associated with decreased LV wall thickness. In a separate study of the PHACS SMARTT HEU cohort, serum cardiac biomarker measurements suggested that HEU children perinatally exposed to multiple ART agents may have subclinical myocardial inflammation. Specifically, abacavir exposure was potentially associated with deleterious cardiac effects.[80]

The results of cardiac biomarkers in the PHACS AMP HIV-infected cohort are still being analyzed and could provide further insights into both the long-term pathophysiologic effects of HAART exposure and how to best evaluate long-term cardiovascular risk. Currently, additional analyses are comparing the cross-sectional echocardiographic measures in this PHACS SMARTT cohort to the relatively ART-unexposed P^2C^2 HIV cohort and the smaller but longitudinally followed CHAART-I perinatally HAART-exposed cohort. The results of these analyses may clarify the effects of prenatal HIV and ART exposures on cardiac measures of structure and function in HEU children.

Recent data show a marked decline in the incidence of both clinical cardiomyopathy and structural abnormalities and an apparent cardioprotective effect of HAART for children and adolescents.[33,69,75,77] HIV-infected patients post-MI have similar short-term outcomes but increased readmission for congestive heart failure at a one-year follow up.[81]

Clinical Presentation

Concurrent pulmonary infections, anemia, pulmonary hypertension, malnutrition, portal hypertension, and malignancy can change or confuse the distinctive signs in HIV-infected patients that otherwise define heart failure in other populations. Patients can present with LV systolic dysfunction ranging anywhere from asymptomatic to New York Heart Association Class III (marked functional limitations) or IV heart failure (severe functional limitations).[74]

Echocardiography, including strain measurements, and cardiac magnetic resonance imaging are useful for assessing LV function, in addition to diagnosing LV dysfunction. Images often reveal LV hypertrophy, dilation, or low-to-normal wall thickness, as well as left atrial dilation.[5,25,40,74,76] Echocardiographic assessment is recommended at baseline and every 1 to 2 years thereafter, or as indicated, in any patient at elevated cardiovascular risk who has unexplained or persistent pulmonary symptoms or viral coinfections or with any clinical manifestations of CVD.[5,61,74,76]

Electrocardiography often reveals nonspecific conduction defects or repolarization changes in ART-naïve patients. Chest radiography has low sensitivity and specificity for diagnosing heart failure in HIV-infected patients treated or untreated with ART due to concomitant pulmonary disease. Several small studies of HIV-infected individuals revealed that serum concentrations of brain natriuretic peptide were inversely correlated with LV ejection fraction. This inverse correlation can be useful in the differential diagnosis of congestive cardiomyopathy in HIV-infected patients and may help to identify those in need of further diagnostic testing including echocardiography.[40,80,82]

Progressive LV dilation is common in children infected with HIV. Left ventricular dilation may precede heart failure (5-year cumulative incidence, 12.3%) and is associated with elevated LV afterload, LV hypertrophy, and reduced LV function.[77] Early and continuous treatment with HAART for at least five years prevented clinically important heart failure in HIV-infected children better than in earlier groups and preserved cardiac structure and function, indicating that HAART may be cardioprotective.[77]

Both the PHACS and CHAART studies suggest that any cardiac changes in the HAART era are generally subclinical in children. Further, in addition to characterizing lifetime ART exposure, traditional non-HIV cardiovascular risk factors will be needed to best determine differences in global cardiovascular risk between perinatally HIV-infected and HEU children and that in the general population.

Pathogenesis in Children

Two mechanisms of pathogenesis have been described for children with perinatally transmitted HIV infection treated with ART in the pre-HAART era: dilation of the LV with a reduced ratio of the LV wall thickness to end-systolic dimension and concentric hypertrophy of the muscle and dilation, in which the ratio of LV thickness to end-systolic dimension remains normal or is increased.[5]

Pathogenesis in Young Adults

Traditional atherosclerosis risk factors are often less favorable and have more of an immediate clinical risk as HIV-infected children enter adolescence and adulthood. Several causes have been postulated for HIV-related cardiomyopathy in children and adults treated in the pre-HAART era (see Table 12.1).[33,74,77]

Myocarditis

Dilated cardiomyopathy can be related to the direct action of HIV on myocardial tissue or to proteolytic enzymes or cytokine mediators induced by HIV alone or with coinfecting viruses.[83] Endomyocardial biopsy specimens have revealed *Toxoplasma gondii*, coxsackievirus group B, Epstein–Barr virus, cytomegalovirus, adenovirus, and HIV in cardiomyocytes. Whether these coinfecting agents are also problematic in the post-HAART era needs to be determined.

Only scant and patchy inflammatory cell infiltrates in the myocardium have been identified in autopsy and biopsy findings,[5,74,83,84] indicating that HIV can infect myocardial interstitial cells and is rarely found in cardiomyocytes. Patients with confirmed myocarditis have an increased number of infected interstitial cells in which proteolytic enzymes or increased concentrations of TNF-α or interleukin may injure the cardiomyocytes. These patients have increased concentrations of TNF-α, inducible nitric oxide synthase, and IL-6.[5,74,83,85] About 40% of patients with HIV-related cardiomyopathy have had no opportunistic infection prior to the onset of cardiac symptoms.[5,6]

Cytokine Alterations

Increased TNF-α production induced by HIV infection can increase nitric oxide production and alter intracellular calcium homeostasis, transforming growth factor-β and endothelin-1 activity.[86] When nitric oxide concentrations were increased experimentally, cardiomyocytes were killed or injured, causing negative inotropic effects.[86] Clinical trials are needed to determine the effect of cytokine alterations in the current post-HAART era. Two recent studies, one with HIV-infected adults and the other with HIV-infected children, found the highest levels of the cardiac isoform of α2-macroglobulin (CA2M) in those with cardiac disease compared to those without, as well as compared to non-HIV infected groups with or without heart disease.[87,88] Although CA2M is presented as a reliable early indicator of heart disease in both HIV-infected and uninfected individuals, large prospective studies are needed to determine its clinical utility as a biomarker in terms of diagnosis and prognosis.

Nutritional Deficiencies

Nutritional deficiencies are common in HIV-infected individuals, particularly in late-stage HIV infection and in infants. Electrolyte imbalances and deficiencies in elemental nutrients are often a result of diarrhea and poor absorption of nutrition. Deficiencies of trace elements have been associated with cardiomyopathy. For example, coxsackievirus is more virulent in selenium-deficient cardiac tissue.[61] Selenium replacement restores left ventricular function and reverses cardiomyopathy. Concentrations of vitamin B12, carnitine, growth hormone, and thyroid hormone can be altered in HIV disease; all have been associated with LV dysfunction.[80,89]

Course of Disease

Patients with asymptomatic LV dysfunction, which is defined as a LV fractional shortening less than 28% with global LV hypokinesis, may have echocardiographically defined transient disease. One serial echocardiographic study reported that 3 of 6 patients with abnormal LV fractional shortening had normal readings after a mean of 9 months. The three patients with persistently depressed LV function all died within a year after LV systolic dysfunction was diagnosed.[5]

Prognosis

Mortality is increased in HIV-infected patients with cardiomyopathy, independent of CD4 count, sex, age, or HIV risk group. In the pre-HAART era, median survival from diagnosis to AIDS-related death was 101 days in patients with LV dysfunction. Patients with normal hearts had a median survival of 472 days at a similar stage of infection.[1,5] Neither isolated right ventricular dysfunction nor borderline LV dysfunction increased the risk of AIDS-related death.

In the P^2C^2 HIV study, the median survival was 2.1 years, and the five-year cumulative survival was 64%.[5] Children with depressed LV fractional shortening or increased LV dimension, mass, thickness, heart rate, blood pressure, or wall stress at baseline had higher mortality. Increased LV wall thickness and decreased LV fractional shortening also predicted adjusted survival (Figure 12.2).[5] Although increased LV wall thickness identified a

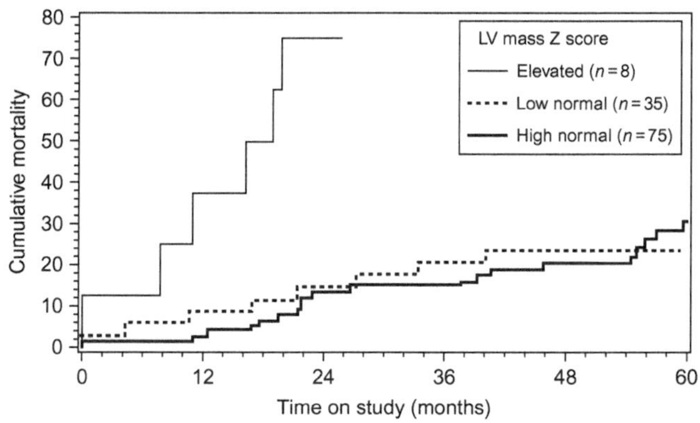

FIGURE 12.2 **Mildly increased left ventricular mass is a risk marker for early HIV mortality. Source:** *Reprinted with permission from Fisher et al., 2005.*[25]

population at risk only 18 to 24 months before death, LV fractional shortening was abnormal for three years before death. Despite the fact that most patients received zidovudine at some time during the P^2C^2 HIV study, a separate report found that zidovudine was not associated with cardiac complications.[90] Thus, LV fractional shortening may be a useful long-term predictor of mortality, and LV wall thickness a useful short-term predictor for children receiving ART during the pre-HAART era.[5,25,77,91]

In the P^2C^2 HIV-infected cohort, echocardiographic evidence of increased left ventricular mass was associated with postmortem cardiomegaly and chronically increased heart rate before death but not with anemia, HIV viral load, or encephalopathy.[5] Mild persistent depressed LV function and mildly elevated LV mass were associated with higher all-cause mortality in children infected with HIV.[25,33,77] A reduction in LV fractional shortening from 34 to 30% in a 10-year-old patient (a reduction equivalent to 2 z-scores) was associated with an increase in five-year mortality, from 15 to 55%.[33,77] Left ventricular fractional shortening was higher in HIV-uninfected children of HIV-infected mothers with in utero exposure to ART than in HIV-uninfected children of HIV-infected mothers unexposed to ART. However, exposure to ART was associated with decreased LV mass, LV dimension, and septal wall thickness.[33] Any exposure to HAART in perinatally infected children with HIV markedly affects LV mass, LV contractility, and LV afterload.[86] Rapid-onset heart failure has a grim prognosis in both HIV-infected children and adults. More than half of patients die from primary cardiac failure within a year of presentation.[1,5,74]

Therapy

As with nonischemic cardiomyopathy, therapy for dilated cardiomyopathy that is associated with HIV infection includes diuretics, digoxin, aldosterone antagonists, beta blockers, and angiotensin-converting enzyme inhibitors. The efficacy of specific cardiac therapeutic regimens other than intravenous immunoglobulin is unknown.[2] Patients with low systemic vascular resistance may be very sensitive to angiotensin-converting enzyme inhibitors. Preventing most cardiomyopathy in the first place using HAART remains the best treatment.[70,71,74] Infections should be treated to improve or resolve related cardiomyopathy. Right ventricular biopsy may help target therapy, in addition to identifying infectious causes of failure.[72] Right ventricular biopsy may be underused.[6,74,76,83]

Serial echocardiography should be performed at clinically relevant intervals (e.g., four months after medical therapy is begun). Monitoring recommendations for testing and timing of follow up are based on studies relating impaired LV fractional shortening to a worse prognosis. A biopsy should be considered if cardiac function continues to deteriorate or if the clinical course worsens. Patients with heart failure who have not responded to two weeks of medical therapy may benefit from cardiac catheterization and endomyocardial biopsy, which may reveal lymphocytic infiltrates, suggesting myocarditis or treatable opportunistic infections (by use of special stains) and permitting aggressive therapy of an underlying pathogen.[5,33,59,74,83,84] Angiography should be performed selectively if there are risk factors for atherosclerotic disease or suggestive clinical symptoms (Figure 12.3).[40,51]

In HIV-uninfected children, intravenous immunoglobulin helps treat acute congestive cardiomyopathy and nonspecific myocarditis. Monthly immunoglobulin infusions have minimized LV dysfunction, increased LV wall thickness, and reduced peak LV wall stress in HIV-infected children, suggesting that both impaired myocardial growth and LV dysfunction can be immunologically mediated.[2]

Although cardiac transplantation is not widely available, it remains an area of active research and has been successful in patients infected with HIV.[82]

Animal Models

Exposure to heat-killed *Mycobacterium avium* complex, a ubiquitous environmental agent, results in exaggerated myocardial pathology in rhesus macaques infected with simian immunodeficiency virus. In this model, enternacept (a TNF antagonist) prevented left ventricular dysfunction, suggesting a TNF-α-dependent pathway in the development of cardiomyopathy in HIV infection.[86]

12.2.3 Left Ventricular Diastolic Dysfunction

Diastolic dysfunction is relatively common in long-term survivors of HIV infection, as suggested by clinical and echocardiographic data. Such left ventricular dysfunction may precede LV systolic dysfunction and mark an early manifestation of HIV-associated cardiac disease.[18,89,92–95] However, LV

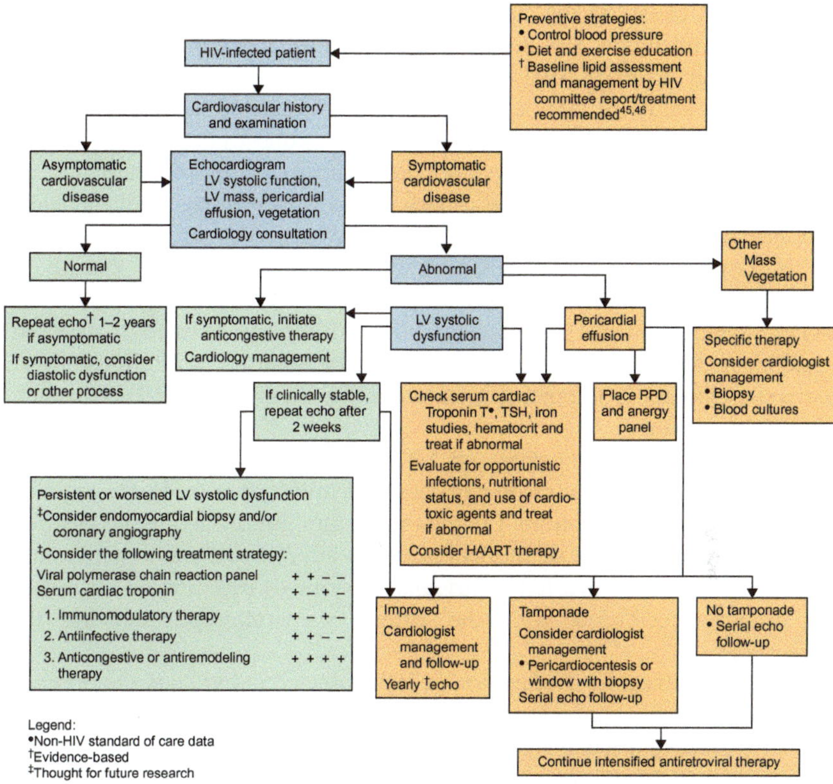

FIGURE 12.3 **Algorithm for treating cardiac dysfunction for patients infected with HIV.** HAART, highly active antiretroviral therapy; LV, left ventricular; PPD, purified protein derivative; TSH, thyroid-stimulating hormone. **Source:** *Reprinted with permission from Fisher and Lipshultz, 2011.*[141]

diastolic function is not well characterized in HIV-uninfected children exposed in utero to ART. Slower LV relaxation during diastole decreases early diastolic filling. Left ventricular compliance decreases as LV diastolic dysfunction worsens and left atrial pressure increases. Moderate-to-severe LV diastolic dysfunction independently predicts mortality, regardless of normal LV systolic function.[96] The clinical impact of LV diastolic function has been studied in children with cardiomyopathy and with other comorbidities such as obesity, generalized autoimmune disease, and diabetes.[97–102]

In one cross-sectional study, early diastolic mitral valve annular velocity was lower in HEU children born to HIV-infected mothers who were exposed in utero to ART than in a group of HIV-uninfected children born to HIV-uninfected mothers with no perinatal ART exposure. In addition, lower early diastolic mitral valve annular velocity was associated with lower maternal CD4 counts in the final trimester.[103] The longitudinal CHAART-I study found subclinical LV diastolic abnormalities in both LV compliance and relaxation among HEU children exposed perinatally to multidrug ART.[97] In a study of 656 asymptomatic HIV-infected adults, 26% had screening echocardiographic evidence of LV diastolic dysfunction.[76] Adults with left ventricular diastolic dysfunction, compared to those without, were older, tended to have higher body mass indexes, were more likely to have hypertension, and had been infected longer.[95] Whether LV diastolic dysfunction is associated with an increased risk of early coronary disease is unknown.[89,95] Also unknown is the clinical importance of LV systolic versus diastolic dysfunction in HIV-infected and HEU children perinatally exposed to multidrug ART. The time between LV systolic and LV diastolic echocardiographic changes is important in determining the effects of HIV exposure and ART exposure.

Uncontrolled HIV replication and ART increase IL-6 concentrations.[104] In animal models, viral proteins or replication of myocardial macrophages may cause LV diastolic dysfunction. Longitudinal mitral inflow and tissue Doppler echocardiographic studies of rhesus macaques infected with simian immunodeficiency virus found that left ventricular diastolic dysfunction was common and strongly correlated with the extent of viral replication in the myocardium.[104]

Therapy for diastolic dysfunction in HIV-infected individuals should optimize blood pressure, intravascular volume status, and include negative inotropy when left ventricular hypertrophy (LVH) is significant. Treatments are not specific to HIV and may include beta blockers or calcium channel blockers, diuretics, and ACE inhibitors for hypertension. HIV-infected patients may have a lower systemic vascular resistance than anticipated and agents should be initiated at low doses.

12.2.4 Pulmonary Hypertension

Pulmonary arterial hypertension (PAH) occurs in about 0.5% of HIV-infected patients. This percentage does not include cases of elevated pulmonary pressure secondary to interstitial lung disease or chronic obstructive pulmonary disease, in which the pathophysiology and response to therapies differ. The introduction of HAART has not changed the prevalence of pulmonary arterial hypertension.[3,105–108] In HIV-infected patients, normal endothelial structure is replaced by plexogenic pulmonary arteriopathy, which is characterized by remodeling of the pulmonary vasculature with intimal fibrosis.[106,107] Perfusion scans are normal and lung fields may be clear on examination and on chest radiographs.[106] Pulmonary arterial hypertension has been reported in HIV-infected patients without a history of thromboembolic disease, intravenous drug use, or pulmonary infections associated with HIV.[3,107,108]

Primary pulmonary hypertension has been found in patients with hemophilia receiving lyophilized factor VIII, intravenous drug users, and patients with LV dysfunction, all of which obscure any relationship with HIV.[3,107] Whether PAH is associated with human herpesvirus 8 is unclear. HIV or a coinfection might cause endothelial damage and mediator-related vasoconstriction of the pulmonary arteries.

Two recent studies found that CD4 count was independently associated with survival in 154 patients with HIV and PAH; pulmonary hypertension was the direct cause of death in 72% of those affected. Survival rates at 1, 2, and 3 years were 73%, 60%, and 47%, respectively. Survival rates in New York Heart Association functional Classes III and IV patients at the time of diagnosis were 60%, 45%, and 28% at 1, 2, and 3 years.[107,108] In one year, 52% of 549 patients with HIV and PAH died, 51% from right heart failure.[108]

In the setting of PAH, morbidity and mortality seem to be related to the PAH more than to the HIV infection. Patients do respond to PAH-directed therapies summarized in the American College of Cardiology expert consensus document on pulmonary hypertension.[106] Standard treatments for PAH, such as phosphodiesterase type 5 (PDE5) inhibitors, endothelin antagonists, and prostacyclin analogs, have been effective in HIV-infected patients. Therapy may also include anticoagulation. Affected patients should continue or start HAART. In patients with HIV, PAH needs to be treated aggressively because it is life-threatening.[106]

12.2.5 Pericardial Effusion

Children infected with HIV who have not been treated with HAART may also present with pericardial effusions. Pericardial effusions have been found to be one of the more common cardiac abnormalities present in patients infected with HIV.

Incidence

Pericardial effusions were found in up to 11% of patients with AIDS before the HAART era. The prevalence of effusion in asymptomatic AIDS patients reaches a mean of about 22% after 25 months, rising over time.[109] In a recent study, only 2 of 802 HAART-treated patients had clinically important effusions, indicating the greatly reduced incidence with treatment of HIV.[109]

Clinical Presentation

HIV-infected patients with pericardial effusions generally have lower CD4 counts than those of patients without effusions.[106] Effusions are generally small and asymptomatic. An HIV infection should be suspected whenever a patient presents with unexplained pericardial effusion or tamponade. In a retrospective series from a city hospital, 13 of 37 (35%) patients with cardiac tamponade had HIV infection.[109] Although rare, tuberculosis has been a presenting infection for pericardial effusions in underdeveloped geographic areas where tuberculosis is prevalent.[109,110] These cases have therapeutic implications and deserve special attention.[111]

Pathogenesis

Pericardial effusion is often part of a generalized serous effusive process also involving pleural and peritoneal surfaces. Enhanced cytokine production in AIDS may be associated with this "capillary leak." Other well-described associations (see Table 12.1) include uremia from HIV-associated nephropathy or drug nephrotoxicity. Effusion nearly triples the risk of death among AIDS patients.[109] Immune reconstitution inflammatory syndrome can cause pericardial effusions and pericarditis in patients coinfected with HIV and tuberculosis.[112]

Monitoring and Therapy

Baseline echocardiographic and electrocardiographic measurements should be obtained from all HIV-infected patients with evidence of heart failure, Kaposi's sarcoma, tuberculosis, and/or unexplained shortness of breath. Pericardiocentesis is indicated for pericardial effusion when there are clinical signs of tamponade (e.g., elevated jugular venous pressure, dyspnea, hypotension, persistent tachycardia, or pulsus paradoxus), or echocardiographic signs of tamponade (e.g., continuous-wave Doppler echocardiographic evidence of respiratory variation in valvular inflow, septal bounce, right ventricular diastolic collapse, and a large effusion) or as needed for diagnostic purposes. Pericardiocentesis is a safe and effective treatment of tuberculosis-related pericardial effusions in HIV-infected patients combined with directed antimicrobial therapy.[113]

Patients with pericardial effusion without tamponade should be evaluated for malignancy and opportunistic infections such as tuberculosis. HAART should be considered if it has not already been started. Repeat echocardiography is recommended after a month or sooner if clinically indicated (see Figure 12.3).[21,74]

12.2.6 Acquired Valvular Disease: Infective Endocarditis

Infective endocarditis has been reported in adults with HIV infection, often right-sided, and most commonly in intravenous drug users. Endocarditis has the same incidence in IV drug users with or without HIV. For HIV-infected adults, the most common organism that is associated with endocarditis is *Staphylococcus aureus*. Endocarditis caused by *Aspergillus fumigatus*,

Candida species, and *Cryptococcus neoformans* are more common in intravenous drug users with HIV than in those without it. HIV influences the response to the infection with certain organisms where T-cell mediated injury to the valve may be less in those infected with it. This is the case in *salmonella endocarditis*, which is rare and not generally responsive to medical therapy in the general population, but successfully medically treated in several HIV-infected individuals, most likely due to a lower background immune response (see Table 12.1).[21]

Late-stage AIDS patients with poor nutritional status and severely compromised immune systems may experience a more fulminant course and a higher mortality; however, several patients have been successfully treated with antibiotics. Surgical indications in HIV-infected patients with endocarditis include bacteremia that persists, despite intravenous antibiotics to which the organism is sensitive; hemodynamic instability; persistent embolization; and severe valvular destruction in patients with a reasonable life expectancy after surgery.

Endocarditis in HIV-infected children rarely has been reported. In one report, a two-month-old HIV-infected Ugandan boy presented with disseminated *Staphylococcus aureus* infection and a large obstructing vegetation on the free wall of the right ventricle in association with a purulent pericardial effusion and an empyema. Echocardiograms showed no structural abnormalities other than a patent foramen ovale.[114]

12.2.7 Nonbacterial Thrombotic Endocarditis

Marantic or nonbacterial thrombotic endocarditis involves the deposition of large, friable, sterile vegetations predominantly on the cardiac valves. These vegetations have been associated with disseminated intravascular coagulation and systemic embolization. Vegetations are rarely diagnosed before death, but when they are, clinically important emboli are likely.[21] Marantic endocarditis is not common in children, but it has been described in a child with HIV diagnosed at age 14 months. The child had pneumonia and staphylococcus sepsis and later developed acute cardiac failure with valvular dysfunction, hepatosplenomegaly, ascites, and failure to thrive, and eventual related death. An echocardiogram showed bright echoes within the chordea of the tricuspid valve and the leaflet tips.[115] Marantic endocarditis is more likely to be identified in patients with delayed HIV diagnosis, limited or no access to antiretroviral therapy, and in those with progressive disease. In the early HIV epidemic, several case series in adults suggested a high incidence of this uncommon disorder; however, few cases have since been reported.

12.2.8 Cardiovascular Malignancy

Malignancy affects many adult AIDS patients, generally in the later stages of the disease. Cardiac malignancy may be a primary tumor or a metastatic secondary site. Although lymphomas have been associated with malignancy in HIV-infected children, the incidence is low, and cardiac malignancy is rare. The Children's Cancer Group and the Pediatric HIV Clinic at the National Cancer Institute diagnosed 65 tumors between 1982 and 1997 in 64 HIV-infected children,[116] although these patients were not on treatment. Non-Hodgkin lymphoma accounted for 42 of these tumors. In this study, almost one-third of the children with this disease had normal or moderate immune suppression. Leiomyosarcoma occurred in 11 and Kaposi's sarcoma in 3.

Kaposi's sarcoma (angiosarcoma) affected up to 35% of AIDS patients early in the HIV epidemic and is associated with human herpesvirus 8. Its incidence is inversely related to CD4 count. Although sarcoma is not often described as a primary cardiac tumor, autopsy studies have found that 28% of HIV-infected patients with widespread Kaposi's sarcoma had cardiac involvement.[4] Kaposi's sarcoma is often an endothelial cell neoplasm with a predilection in the heart for subpericardial fat around the coronary arteries.[4,21] Combination antiretroviral therapy has markedly decreased the incidence of Kaposi's sarcoma from that in the pre-HAART era.[21]

Children with HIV infection may harbor human herpesvirus 8, the virus associated with Kaposi's sarcoma, which is endemic in eastern equatorial Africa. It can cause a lymphadenopathic type of Kaposi's sarcoma that is found mainly in children, which may have a fulminant course and ultimately also invade organ systems. Two children were reported with it in the United States early in the epidemic; both had progressive HIV infection with severe immune deficiency and died before one. Kaposi's sarcoma lesions were in the lymph glands and spleen and in the thymus in one child.[117]

Primary cardiac malignancy associated with HIV infection is generally caused by cardiac lymphoma. Lymphoma, an AIDS-defining illness, has a higher incidence in HIV-infected populations. Non-Hodgkin lymphomas are 25 to 60 times more common in HIV-infected individuals. They are the first manifestation of AIDS in up to 4% of new cases.[4] This disease is not specifically associated with severe immune suppression. Patients with primary cardiac lymphoma can present with signs of heart failure, chest pain, or arrhythmias. Cardiac lymphoma can cause rapid progression to cardiac tamponade, heart failure, myocardial infarction, tachyarrhythmias, conduction abnormalities, or superior vena cava syndrome. Malignant cells may be found in the pericardial fluid. Systemic multiagent chemotherapy, with and without concomitant radiation or surgery, has benefitted some patients, but overall, the prognosis is poor.[4] Treatment with HAART has not substantially affected the incidence of HIV-related non-Hodgkin lymphomas.[21,118]

12.2.9 Isolated Right Ventricular Disease

Isolated right ventricular hypertrophy is rare in HIV-infected individuals, with or without right ventricular dilation. It is generally related to pulmonary disease that increases pulmonary vascular resistance. Possible causes include pulmonary arteritis from the immunological effects of HIV disease, multiple bronchopulmonary infections, or microvascular pulmonary emboli caused by thrombus or contaminants in illicit injected drugs such as talc.[3] Right ventricular diastolic dysfunction has been reported in asymptomatic patients studied with tissue Doppler imaging.[119]

12.2.10 Vasculitis

Vasculitis may occur in patients with fever of unknown origin, unexplained arthritis or myositis, unexplained multisystem disease, glomerulonephritis, or peripheral neuropathy (especially mononeuritis multiplex) and in unexplained gastrointestinal, cardiac, or central nervous system ischemia. Several types have been described in HIV-infected patients, but all show diffuse inflammation of the vessel walls.[120] Successful immunomodulatory therapy has been reported, chiefly with systemic corticosteroid.[120] The HIV protein, transactivator of transcription (Tat), has been implicated in the pathogenesis of vasculitis.[120]

12.2.11 Sudden Cardiac Death

Sudden cardiac death is becoming increasingly common as the HIV-infected population ages. In one study, sudden cardiac death accounted for 86% of all cardiac-related deaths (30 of 35). The mean rate of sudden cardiac death was 2.6 per 1000 person-years (95% confidence interval: 1.8–3.8), which was 4.5-fold as high as that expected in an age-matched uninfected population.[121] One report found that patients dying from sudden cardiac death were older than those dying from AIDS (mean age at death, 49 vs. 45 years, $P = 0.02$) and had a higher prevalence of positive histories for MI (17% vs. 1%, $P < 0.001$), cardiomyopathy (23% vs. 3%, $P < 0.001$), heart failure (30% vs. 9%, $P = 0.004$), and arrhythmias (20% vs. 3%, $P = 0.003$).[121]

12.2.12 Cardiac Arrhythmias: QT Interval, PR Prolongation, and Atrial Fibrillation

HIV infection is associated with QT prolongation and torsades de pointes ventricular tachycardia. The risk of sudden death from these conditions increases late in HIV infection and specifically in AIDS. The incidence of QT prolongation also increases as the disease progresses to AIDS.[122] Hepatitis C is independently associated with increased QT duration. One study found that the risk of QT prolongation (i.e., QTc values of 470 msec or higher) was 16% with HIV alone and 30% with both HIV and hepatitis C infections.[118] The risk of increased QT duration is also higher in patients treated with antituberculosis medications such as levofloxacin, moxifloxacin, and bedaquiline.[58] The mechanism of prolonged QT interval and PR interval may be related to influences on vulnerable cell membrane structure exaggerated in the setting of electrolyte imbalance, mild underlying systolic or diastolic cardiomyopathy, and multiple drug interactions.

Different protease inhibitor-based regimens have similar, minimal effects on the QT interval, but they do markedly prolong the PR interval, from 3 msec in nonboosted proteaseinhibitor regimens to 5.11 msec in boosted proteaseinhibitor regimens. The interval returns to normal when protease inhibitor therapy is withdrawn, and prolongation is not associated with NNRTIs. The clinical importance of a prolonged PR interval has not been well established.[123] Some investigators believe that PR prolongation may increase the likelihood of complete heart block in the presence of immune reconstitution inflammatory syndrome during initiation of ART.

A recent report from the U.S. Veterans Administration of HIV-infected individuals from 1996–2011 found a cumulative incidence of atrial fibrillation of 2.6% (median follow up of 6.8 years).[124] Patients with CD4 counts <100 cell/mm^3 (compared to those with CD4 count >350 cells/mm^3) or a viral load >100,000 copies (compared to those with VL <500 copies) had a 1.4 to 1.7 times increased risk of atrial fibrillation, respectively.

12.2.13 Autonomic Dysfunction

Preliminary clinical signs of autonomic dysfunction in HIV-infected patients include syncope and presyncope, diarrhea, diminished sweating, bladder dysfunction, and impotence. In one study, heart rate variability, Valsalva ratio, cold pressor testing, hemodynamic responses to isometric exercise, tilt-table testing, and standing showed that autonomic dysfunction occurred in HIV-infected individuals and was pronounced in AIDS patients, although AIDS patients receiving HAART were relatively protected. Patients with HIV-associated nervous system disease had the greatest abnormalities in autonomic function, and procedural precautions, such as advanced airway management and expectant use of atropine, are advised (Figure 12.4).[125]

12.2.14 Complications of Therapy

Antiretroviral medications have greatly reduced mortality by delaying the progression to AIDS and have increased the quality of life of those who are HIV infected.[59] However, these same therapies have been shown to have several complications.[21,64,94,126,127]

As previously detailed, altered body composition and hyperlipidemia are associated with PIs and NRTIs and may increase the risk of cardiometabolic dysfunction in children.[126,127] Lipid abnormalities vary between protease inhibitors. Ritonavir has the most adverse effects on lipids, with a mean increase in total cholesterol concentration of 2.0 mmol/L and a mean increase

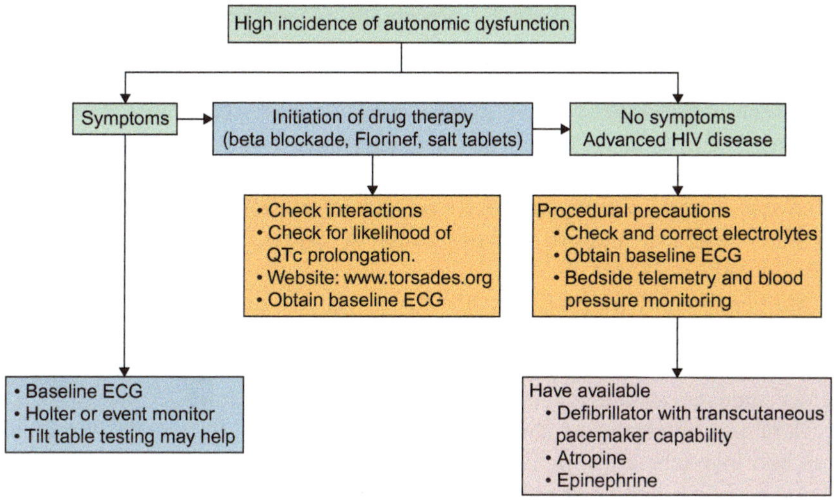

FIGURE 12.4 Algorithm for evaluating and treating dysautonomia. ECG, electrocardiography. **Source:** *Reprinted with permission from Fisher and Lipshultz, 2011.*[141]

in triglyceride concentration of 1.83 mmol/L.[64,94,128] More modest increases of total cholesterol concentration without marked triglyceride increases were found in patients taking indinavir and nelfinavir. In combination with saquinavir (including atanazavir and saquinavir in salvage therapy), ritonavir did not further elevate total cholesterol concentrations. Protease inhibitors significantly increase lipoprotein(a) concentrations in patients who had elevated pretreatment values (>20 mg/dL), which is another risk marker for atherosclerotic cardiovascular disease.[64,94,128] In some cases, switching protease inhibitors may reverse both elevations in triglyceride concentrations and abnormal fat deposition.

Low-level aerobic exercise may also help reverse lipid abnormalities.[21,60] Zidovudine or AZT has been implicated in skeletal muscle myopathies. In culture, AZT causes a dose-dependent destruction of human myotubes. Human cultured cardiac muscle cells treated with AZT developed mitochondrial abnormalities, and NRTIs in general have been associated with altered mitochondrial DNA replication and cardiac structure.[21,129] Whether altered mitochondrial DNA replication is the cause of cardiomyopathy is unknown; however, cardiac myopathies have not been evident in clinical data.

Some patients with LV dysfunction may improve when AZT therapy is stopped.[21] Some evidence associates ARTs with mitochondrial toxicity.[130] This toxicity and additional factors may reduce aerobic capacity in children infected with HIV. HIV-infected children and adolescents have lower cardiorespiratory fitness, lower extremity strength, and less flexibility than do their uninfected counterparts. Additionally, HAART exposure for more than five years and a higher total body fat percentage independently and negatively affected aerobic capacity.[71]

Intravenous pentamidine, used to treat *Pneumocystis jirovecii* pneumonia in patients who do not tolerate trimethoprim-sulfamethoxazole, has been associated with torsades de pointes and refractory ventricular tachycardia.[21] Pentamidine should be reserved for patients with a QTc interval below 480 msec. Several medication reactions and interactions have occurred during HIV treatment and are a major cause of cardiac emergencies in HIV-infected patients (Table 12.3).[74,77,131]

Mother-to-child transmission has been reduced in the United States to approximately 2%.[30] Intrauterine exposures to these potent ARTs have some effects on the child.[132] At birth, children exposed to HIV and ARTs were lighter than a comparison group with no exposure to ARTs and showed accelerated growth during the first two years of life. Additionally, the children had less subcutaneous fat and decreasing mid-upper arm circumference over time when compared to national standards.[133]

12.2.15 Perinatal Transmission of HIV Infection

Most children with HIV are infected perinatally.[134–135] Current therapies and guidelines for prophylaxis of infants born to women with HIV infection recommend six weeks of neonatal AZT, which can decrease perinatal transmission of HIV to less than 2%. A worldwide UNAIDS goal is to eliminate perinatal transmission by the end of 2014.

Rates of congenital cardiovascular malformations ranged from 5.6 to 8.9% in cohorts of HIV-uninfected and HIV-infected children born to HIV-infected mothers. Although these rates were not higher than those in similarly screened normal populations, they were 5 to 10 times as high as those reported in population-based epidemiological studies.[134]

In the same cohorts, serial echocardiography that was performed every 4 to 6 months showed subclinical cardiac abnormalities to be common, persistent, and often progressive.[5,33,74] Some patients had dilated cardiomyopathy (LV contractility of 2 standard deviation points or more below the mean of a normative population and a LV end-diastolic dimension of 2 standard deviation points or more above the mean), whereas others had mildly increased cardiac mass for height and weight. Depressed LV function was correlated with immune dysfunction at baseline but not over time. This correlation suggests that the CD4 cell count may not be a useful surrogate marker of HIV-associated LV dysfunction.

Disease can progress rapidly or slowly in children with perinatally transmitted HIV-1 infection.[74] Rapid progressors have higher heart rates, higher respiratory rates, and lower fractional shortening on serial examinations than do those who are nonrapid progressors and HIV-uninfected children who are similarly screened. Rapid progressors also have higher HIV-1 viral loads,

TABLE 12.3 Cardiac Interactions and Side Effects of Drugs Commonly Used in HIV Therapy

Class	Cardiac Drug Interactions	Cardiac Side Effects
Antiretroviral: Nucleoside (and nucleotide) reverse transcriptase inhibitors Abacavir (ABC), didanosine (ddI), emtricitabine (FTC) lamivudine (3TC), stavudine (d4T), tenofovir (TDF), zalcitabine (ddC), zidovudine (ZDV, AZT)	Zidovudine and dipyridamole Stavudine and DDI	*Rare*: Lactic acidosis, hypotension; accelerated risk with cardiopulmonary bypass *Zidovudine*: skeletal muscle myopathy, myocarditis Mitochondrial toxicity with lipodystrophy
Nonnucleoside reverse transcriptase inhibitors: Delavirdine (DLV), efavirenz (EFV), nevirapine (NVP), rilpivirine (RPV), etravirine (ETR)	Calcium-channel blockers, warfarin, beta blockers, nifedipine, quinidine, steroids, theophylline Delavirdine can cause serious toxic effects if given with antiarrhythmic drugs and calcium channel blockers	Arrhythmia
Protease inhibitors: Amprenavir (APV), atazanavir (ATV), darunavir (DRV), fosamprenavir (FPV) indinavir (IDV), lopinavir/ritonavir (LPV/r), nelfinavir (NFV), ritonavir (RTV), saquinavir (SQV), tipranavir (TPV)	Metabolized by cytochrome P450 and interact with other drugs metabolized through this pathway such as selected antimicrobials, antidepressant and antihistamine agents, cisapride, HMG CoA reductase inhibitors (lovastatin, simvastatin), and sildenafil	Implicated in premature atherosclerosis, dyslipidemia, insulin resistance, diabetes mellitus, fat wasting, and redistribution Abacavir may be associated with increased risk of MI13

(*Continued*)

TABLE 12.3 Cardiac Interactions and Side Effects of Drugs Commonly Used in HIV Therapy (Continued)

Class	Cardiac Drug Interactions	Cardiac Side Effects
	Potentially dangerous interactions that require close monitoring or dose adjustment can occur with amiodarone, disopyramide, flecainide, lidocaine, mexiletine, propafenone, quinidine, and ranolazine (1.8–2.3 × increase in ranolazine level) Ritonavir is the most potent cytochrome activator (CYP3A) and P-glycoprotein inhibitor is most likely to interact; indinavir, amprenavir, and nelfinavir are moderate activators Saquinavir has the lowest probability of interaction Calcium-channel blockers, prednisone, quinine, beta blockers (1.5- to 3-fold increase) Decreases theophylline concentrations	
Integrase strand transfer inhibitors (INSTIs): Elvitegravir (EVG), raltegravir (RAL)		
CCR5 antagonists: Maraviroc		
Fusion inhibitor: Enfuvirtide		

Antiinfective antibiotics	*Rifampin*: Reduces therapeutic effect of digoxin by inducing intestinal P-glycoprotein; reduces protease inhibitor concentration and effect *Erythromycin*: Cytochrome P450 metabolism and drug interactions *Trimethoprim-sulfamethoxazole*: (Bactrim) increases warfarin effects	*Erythromycin*: Orthostatic hypotension, ventricular tachycardia, bradycardia, torsades (with drug interactions) *Clarithromycin*: QT prolongation and torsades de pointes *Trimethoprim-sulfamethoxazole*: Orthostatic hypotension, anaphylaxis, QT prolongation, torsades de pointes, hypokalemia *Sparfloxacin (fluoroquinolones)*: QT prolongation
Antifungal agents	*Amphotericin B*: Digoxin toxicity *Ketoconazole or itraconazole*: Cytochrome P450 metabolism and drug interactions—increases levels of sildenafil, warfarin, HMG CoA reductase inhibitors, nifedipine, digoxin	*Amphotericin B*: Hypertension, arrhythmia, renal failure, hypokalemia, thrombophlebitis, bradycardia, angioedema, dilated cardiomyopathy; liposomal formulations still have the potential for electrolyte imbalance and QT prolongation *Ketoconazole, fluconazole, itraconazole*: QT prolongation and torsades de pointes
Antiviral agents	*Ganciclovir*: Zidovudine	*Foscarnet*: Reversible cardiac failure, electrolyte abnormalities *Ganciclovir*: Ventricular tachycardia, hypotension
Antiparasitic		*Pentamidine*: Hypotension, QT prolongation, arrhythmias (torsades de pointes), ventricular tachycardia, hyperglycemia, hypoglycemia, sudden death; these effects are enhanced by hypomagnesemia and hypokalemia
Chemotherapy agents	*Vincristine, doxorubicin*: Decrease digoxin level	*Recombinant human interferon-α*: Hypertension, hypotension, tachycardia, acute coronary events, dilated cardiomyopathy, arrhythmias, sudden death, atrioventricular block, and peripheral vasodilation; contraindicated in patients with unstable angina or recent myocardial infarction

(*Continued*)

TABLE 12.3 Cardiac Interactions and Side Effects of Drugs Commonly Used in HIV Therapy (Continued)

Class	Cardiac Drug Interactions	Cardiac Side Effects
		Interleukin-2: Hypotension, arrhythmia, sudden death, myocardial infarction, dilated cardiomyopathy, capillary leak, thyroid alterations *Anthracyclines (doxorubicin, daunorubicin, mitoxantrone)*: Myocarditis, cardiomyopathy *Liposomal anthracyclines*: As above for doxorubicin and also vasculitis
Pentoxifylline		*Pentoxifylline*: Decreased triglyceride levels, arrhythmias, chest pain *Megace*: Edema, thrombophlebitis, hyperglycemia
Megestrol acetate (Megace)		*Epoetin α (erythropoietin)*: Hypertension, ventricular dysfunction
Methadone		Prolonged QT interval
Amphetamines		Increased heart rate and blood pressure

Source: Modified with permission from Fisher and Lipshultz, 2011.[141]

higher five-year cumulative mortality, and lower $CD8^+$ (cytotoxic) T-cell counts. With the capability of early diagnosis and treatment of HIV-infected infants, rapid progression of the disease and early mortality can be prevented.

Studies of HIV-uninfected infants born to HIV-infected mothers have reported that fetal exposure to ART is associated with reduced left ventricular dimension, LV mass, and septal wall thickness along with higher LV fractional shortening and contractility during the first two years of life.[129] In utero exposure to ART may initially improve LV function while impairing myocardial growth. Although LV function is improved, it is still below normal.[33] These effects are more pronounced in girls.[33]

12.2.16 Cardiac Monitoring Recommendations

Routine, systematic cardiac evaluation, including a comprehensive history and a thorough cardiac examination, is essential care for HIV-infected children and adults. The history should include traditional risk factors, environmental exposures, prior opportunistic infections, and therapeutic and illicit drug use. Laboratory tests should include a lipid profile, fasting glucose concentrations, and HIV viral load (Figure 12.5). Routine blood pressure monitoring is important because HIV-infected individuals can experience hypertension at a younger age and this occurs more often than in the general population.[21,40,59]

Unless patients have symptoms, such as palpitations, syncope, stroke, or dysautonomia, routine electrocardiographic and Holter monitoring are not warranted. However, these tests can be useful for baseline and monitoring before, during, and after therapies with drugs, such as pentamidine, methadone, or antibiotics, that may prolong the QT interval.[122]

Asymptomatic cardiac disease related to HIV can be fatal. When present, cardiac symptoms are often disguised by the secondary effects of HIV infection. Thus, systematic echocardiographic monitoring is warranted.[76,77,136,137] An international consensus panel recommended echocardiographic monitoring, with a baseline, for any patient at high risk or with any clinical manifestation of CVD, in addition to studies every 1 to 2 years or as clinically indicated. Patients with cardiac symptoms should begin directed therapy and receive a formal cardiac assessment, including baseline electrocardiogram, echocardiogram, and Holter monitoring.[138] Brain natriuretic peptide concentrations may help distinguish a cardiac etiology from the other etiologies of dyspnea.[139,140]

Serum troponin assays are indicated in patients who have LV dysfunction. Persistent elevated concentrations of serum troponin warrant consideration of endomyocardial biopsy and cardiac catheterization. Therapy with intravenous immunoglobulin should be considered for biopsy-proven myocarditis.[2] Echo-cardiography should be repeated after two weeks of therapy.

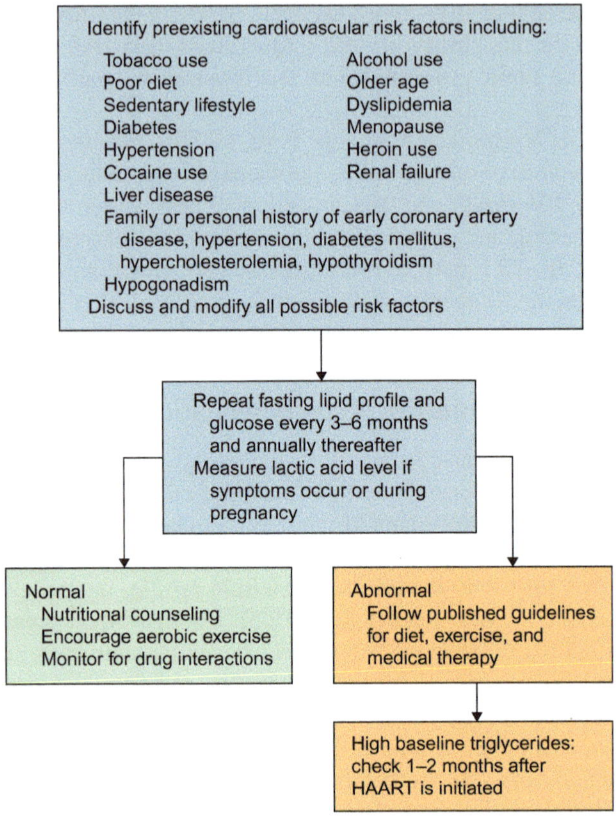

FIGURE 12.5 **Cardiovascular considerations when initiating highly active antiretroviral therapy. Source:** *Reprinted with permission from Fisher and Lipshultz, 2011.*[141]

Therapy should be continued if improvement has occurred and should become more aggressive if LV dysfunction persists or becomes worse.

Stress testing and coronary assessment (e.g., CT angiography or cardiac catheterization) should be considered when appropriate.[40,51,61,74] Guidelines for using implantable cardioverter-defibrillators should be followed in this population, especially for patients after an MI who are being treated for HIV infection and have a reasonable functional status.[40]

12.3 CONCLUSION

As a chronic disease, HIV-related CVD is a vital area of research. If HIV can be used as a model of chronic immunosuppression in a large population, findings may translate to other populations. Already cardiomyopathy is less frequent due to advances in therapy in this group. Understanding genetic predisposition to QT prolongation may guide therapy regimes. Understanding the causes of cardiomyopathy may benefit diverse research studies, such as the effects of cytokines, mitochondria, and neurohormonal pathways. Observations, such as increased mortality related to LV mass and very mild LV dysfunction, may enhance diagnostic testing and identify at-risk populations affected by other poorly understood cardiomyopathies.

REFERENCES

1. Currie PF, Jacob AJ, Foreman AR, et al. Heart muscle disease related to HIV infection: prognostic implications. *BMJ* 1994;**309**:1605.
2. Lipshultz SE, Orav EJ, Sanders SP, et al. Immunoglobulins and left ventricular structure and function in pediatric HIV infection. *Circulation* 1995;**92**:2220–5.
3. Saidi A, Bricker JT. Pulmonary hypertension in patients infected with HIV. In: Lipshultz SE, editor. *Cardiology in AIDS*. New York: Chapman & Hall; 1998. p. 187–94.
4. Jenson HB, Pollock BH. Cardiac cancers in HIV-infected patients. In: Lipshultz SE, editor. *Cardiology in AIDS*. New York: Chapman & Hall; 1998. p. 255–63.
5. Lipshultz SE, Easley KA, Orav EJ, et al. Cardiac dysfunction and mortality in HIV-infected children: the Prospective P^2C^2 HIV Multicenter Study. Pediatric Pulmonary and Cardiac Complications of Vertically Transmitted HIV Infection (P^2C^2 HIV) Study Group. *Circulation* 2000;**102**:1542–8.
6. Morse CG, Kovacs JA. Metabolic and skeletal complications of HIV infection: the price of success. *JAMA* 2006;**296**:844–54.
7. Lipshultz SE, Fisher SD, Miller TL, et al. The cardiovascular manifestations of HIV infection. *Dialog Cardiovasc Med* 2007;**12**(1):5–23.
8. Zareba KM, Miller TL, Lipshultz SE. Cardiovascular disease and toxicities related to HIV infection and its therapies. *Expert Opin Drug Saf* 2005;**4**(6):1017–25.
9. Zareba KM, Lipshultz SE. Cardiovascular complications in patients with HIV infection. *Curr Infect Dis Rep* 2003;**5**(6):513–20.
10. Barbaro G, Fisher SD, Lipshultz SE. Pathogenesis of HIV-associated cardiovascular complications. *Lancet Infect Dis* 2001;**1**(2):115–24.
11. Keesler MJ, Fisher SD, Lipshultz SE. Cardiac manifestations of HIV infection in infants and children. *Ann N Y Acad Sci* 2001;**946**:169–78.
12. Langston C, Cooper ER, Goldfarb J, et al. Human immunodeficiency virus-related mortality in infants and children: data from the Pediatric Pulmonary and Cardiovascular Complications of Vertically Transmitted HIV P^2C^2 Study. *Pediatrics* 2001;**107**(2):328–38.
13. Starc TJ, Lipshultz SE, Kaplan S, et al. Cardiac complications in children with human immunodeficiency virus infection. Pediatric Pulmonary and Cardiac Complications of Vertically Transmitted HIV Infection (P^2C^2 HIV) Study Group, National Heart, Lung, and Blood Institute. *Pediatrics* 1999;**104**(2):e14.
14. Lipshultz SE, Easley KA, Orav EJ, et al. Left ventricular structure and function in children infected with human immunodeficiency virus: the Prospective P^2C^2 HIV Multicenter Study. Pediatric Pulmonary and Cardiac Complications of Vertically Transmitted HIV Infection (P^2C^2 HIV) Study Group. *Circulation* 1998;**97**(13):1246–56.
15. Epstein JE, Eichbaum QG, Lipshultz SE. Cardiovascular manifestations of HIV infection. *Compr Ther* 1996;**22**(8):485–91.

16. Lane-McAuliffe EM, Lipshultz SE. Cardiovascular manifestations of pediatric HIV infection. *Nurs Clin North Am* 1995;**30**(2):291–316.
17. Luginbuhl LM, Orav EJ, McIntosh K, et al. Cardiac morbidity and related mortality in children with HIV infection. *JAMA* 1993;**269**(22):2869–75.
18. Patel N, Abdelsayed S, Veve M, et al. Predictors of clinically significant drug–drug interactions among patients treated with nonnucleoside reverse transcriptase inhibitor-, protease inhibitor-, and raltegravir-based antiretroviral regimens. *Ann Pharmacother* 2011;**45**(3): 317–24.
19. Chinello P, Lisena FP, Angeletti C, et al. Role of antiretroviral treatment in prolonging QTc interval in HIV-positive patients. *J Infect* 2007;**54**(6):597–602.
20. Mas CM, Miller TL, Cordero C, et al. The effects of fetal and childhood exposure to antiretroviral agents. *J AIDS Clin Res* 2011;**S2**:001.
21. Fisher SD, Kanda BS, Miller TL, et al. Cardiovascular disease and therapeutic drug-related cardiovascular consequences in HIV-infected patients. *Am J Cardiovasc Drugs* 2011;**11** (6):383–94.
22. Dubé MP, Lipshultz SE, Fichtenbaum CJ, et al. Effects of HIV infection and antiretroviral therapy on the heart and vasculature. *Circulation* 2008;**118**(2):e36–40.
23. Fisher SD, Miller TL, Lipshultz SE. Impact of HIV and highly active antiretroviral therapy on leukocyte adhesion molecules, arterial inflammation, dyslipidemia, and atherosclerosis. *Atherosclerosis* 2006;**185**(1):1–11.
24. UNAIDS: 2012 Report on the Global AIDS epidemic: <www.unaids.org>.
25. Fisher SD, Easley KA, Orav EJ, et al. Mild dilated cardiomyopathy and increased left ventricular mass predict mortality: the Prospective P^2C^2 HIV Multicenter Study. *Am Heart J* 2005;**150**(3):439–47.
26. Lipshultz SE. Dilated cardiomyopathy in HIV-infected patients. *N Engl J Med* 1998;**339** (16):1153–5.
27. Harmon WG, Dadlani GH, Fisher SD, et al. Myocardial and pericardial disease in HIV. *Curr Treat Options Cardiovasc Med* 2002;**4**(6):497–509.
28. WHO, UNICEF, UNAIDS. *Global update on HIV treatment 2013: results, impact, and opportunities*. Geneva, Switzerland: World Health Organization; 2013. Available from: <www.who.int/hiv/pub/progressreports/update2013/en/>.
29. Brady MT, Oleske JM, Williams PL, et al. Declines in mortality rates and changes in causes of death in HIV-1-infected children during the HAART era. *J Acquir Immune Defic Syndr* 2010;**53**(1):86–94.
30. Connor EM, Sperling RS, Gelber R, et al. Reduction of maternal-infant transmission of human immunodeficiency virus type 1. *N Engl J Med* 1994;**331**(18):1173–80.
31. Cooper ER, Charurat M, Mofenson L, et al. Combination antiretroviral strategies for the treatment of pregnant HIV-1-infected women and prevention of perinatal HIV-1 transmission. *J Acquir Immune Defic Syndr* 2002;**29**(5):484–94.
32. Bayer R. Ethical Challenges posed by zidovudine treatment to reduce vertical transmission of HIV. *N Engl J Med* 1994;**331**:1223–15.
33. Lipshultz SE, Shearer WT, Thompson B, et al. Cardiac effects of antiretroviral therapy in HIV-negative infants born to HIV-positive mothers: NHLBI CHAART-1 (National Heart, Lung, and Blood Institute Cardiovascular Status of HAART Therapy in HIV-Exposed Infants and Children cohort study). *J Am Coll Cardiol* 2011;**57**(1):76–85.
34. Fitch K, Grinspoon S. Nutritional and metabolic correlates of cardiovascular and bone disease in HIV-infected patients. *Am J Clin Nutr* 2011;**94**(6):1721S–8S.

35. Carr A, Samaras K, Chisholm DJ, et al. Abnormal fat distribution and use of protease inhibitors. *Lancet* 1998;**351**(9117):1736.
36. Grinspoon S, Carr A. Cardiovascular risk and body-fat abnormalities in HIV-infected adults. *N Engl J Med* 2005;**352**(1):48–62.
37. Palella Jr FJ, Phair JP. Cardiovascular disease in HIV infection. *Curr Opin HIV AIDS* 2011;**6**(4):266–71.
38. Friis-Møller N, Weber R, Reiss P, et al. Cardiovascular disease risk factors in HIV patients: association with antiretroviral therapy. Results from the DAD study. *AIDS* 2003; **17**(8):1179–93.
39. Mary-Krause M, Cotte L, Simon A, et al. Increased risk of myocardial infarction with duration of protease inhibitor therapy in HIV-infected men. *AIDS* 2003;**17**(17):2479–86.
40. Fihn SD, Gardin JM, Abrams J, et al. Practice Guideline: 2012. Guideline for the diagnosis and management of patients with stable ischemic heart disease. A report of the American College of Cardiology Foundation/American Heart Association Task Force on Practice Guidelines, and the American College of Physicians, American Association for Thoracic Surgery, Preventive Cardiovascular Nurses Association, Society for Cardiovascular Angiography and Interventions, and Society of Thoracic Surgeons. *J Am Coll Cardiol* 2012;**60**:e44–164.
41. Arpadi SM, Bethel J, Horlick M, et al. Longitudinal changes in regional fat content in HIV-infected children and adolescents. *AIDS* 2009;**23**(12):1501–9.
42. Jacobson DL, Patel K, Siberry GK, et al. Body fat distribution in perinatally HIV-infected and HIV-exposed but uninfected children in the era of highly active antiretroviral therapy: outcomes from the Pediatric HIV/AIDS Cohort Study. *Am J Clin Nutr* 2011; **94**(6):1485–95.
43. Geffner ME, Patel K, Miller TL, et al. Factors associated with insulin resistance among children and adolescents perinatally infected with HIV-1 in the Pediatric HIV/AIDS Cohort Study. *Horm Res Paediatr* 2011;**76**(6):386–91.
44. Miller TL, Borkowsky W, DiMeglio LA, et al. Metabolic abnormalities and viral replication are associated with biomarkers of vascular dysfunction in HIV-infected children. *HIV Med* 2012;**13**(5):264–75.
45. Miller TL, Orav EJ, Lipshultz SE, et al. Risk factors for cardiovascular disease in children infected with human immunodeficiency virus-1. *J Pediatr* 2008;**153**(4):491–7.
46. Sanchez-Torres AM, Munoz-Muniz R, Madero R, et al. Prevalence of fat redistribution and metabolic disorders in human immunodeficiency virus-infected children. *Eur J Pediatr* 2005;**164**(5):271–6.
47. Brambilla P, Bricalli D, Sala N, et al. Highly active antiretroviral-treated HIV-infected children show fat distribution changes even in absence of lipodystrophy. *AIDS* 2001; **15**(18):2415–22.
48. Moscicki AB, Ellenberg JH, Murphy DA, et al. Associations among body composition, androgen levels, and human immunodeficiency virus status in adolescents. *J Adolesc Health* 2006;**39**(2):164–73.
49. Miller TL, Mawn BE, Orav EJ, et al. The effect of protease inhibitor therapy on growth and body composition in human immunodeficiency virus type 1-infected children. *Pediatrics* 2001;**107**(5):E77.
50. Aldrovandi GM, Lindsey JC, Jacobson DL, et al. Morphologic and metabolic abnormalities in vertically HIV-infected children and youth. *AIDS* 2009;**23**(6):661–72.
51. Boccara F, Teiger E, Cohen A, et al. Percutaneous coronary intervention in HIV infected patients: immediate results and long-term prognosis. *Heart* 2006;**92**:543–4.

52. Vigano A, Mora S, Testolin C, et al. Increased lipodystrophy is associated with increased exposure to highly active antiretroviral therapy in HIV-infected children. *J Acquir Immune Defic Syndr* 2003;**32**(5):482–9.
53. Arpadi SM, Cuff PA, Horlick M, et al. Lipodystrophy in HIV-infected children is associated with high viral load and low $CD4^+$-lymphocyte count and $CD4^+$-lymphocyte percentage at baseline and use of protease inhibitors and stavudine. *J Acquir Immune Defic Syndr* 2001;**27**(1):30–4.
54. Gerschenson M, Shiramizu B, LiButti DE, et al. Mitochondrial DNA levels of peripheral blood mononuclear cells and subcutaneous adipose tissue from thigh, fat and abdomen of HIV-1 seropositive and negative individuals. *Antivir Ther* 2005;**10**(Suppl. 2):M83–9.
55. Sztam KA, Jiang H, Jurgrau A, et al. Early increases in concentrations of total, LDL, and HDL cholesterol in HIV-infected children following new exposure to antiretroviral therapy. *J Pediatr Gastroenterol Nutr* 2011;**52**(4):495–8.
56. Rhoads MP, Lanigan J, Smith CJ, et al. Effect of specific ART drugs on lipid changes and the need for lipid management in children with HIV. *J Acquir Immune Defic Syndr* 2011;**57**(5):404–12.
57. Jacobson DL, Williams P, Tassiopoulos K, et al. Clinical management and follow-up of hypercholesterolemia among perinatally HIV-infected children enrolled in the PACTG 219C study. *J Acquir Immune Defic Syndr* 2011;**57**(5):413–20.
58. Panel on Antiretroviral Guidelines for Adults and Adolescents. Guidelines for the use of antiretroviral agents in HIV-1-infected adults and adolescents; 2012. Available from: <http://aidsinfo.nih.gov/contentfiles/lvguidelines/adultandadolescentgl.pdf>.
59. NHLBI, AIDS working group. <www.nhlbi.nih.gov/meetings/workshops/AIDSworking.htm>. September 6–7, 2012, Bethesda, MD.
60. Libby P, Ridker PM, Hansson GK. Inflammation in atherosclerosis: from pathophysiology to practice. *J Am Coll Cardiol* 2009;**54**:2129–38.
61. Longenecker C, Hoit B. Imaging atherosclerosis in HIV: carotid intima-media thickness and beyond. *Transl Res* 2012;**159**(3):127–39.
62. Miller TL, Somarriba G, Orav EJ, et al. Biomarkers of vascular dysfunction in children infected with human immunodeficiency virus-1. *J Acquir Immune Defic Syndr* 2010;**55**(2):182–8.
63. Foster SB, Lu M, Glaze DG, et al. Associations of cytokines, sleep patterns, and neurocognitive function in youth with HIV infection. *Clin Immunol* 2012;**144**(1):13–23.
64. DAD Study Group and Friis-Møller N, Reiss P, Sabin CA, et al. Class of antiretroviral drugs and the risk of myocardial infarction. *N Engl J Med* 2007;**356**(17):1723–35.
65. Fichtenbaum CJ. Inflammatory markers associated with coronary heart disease in persons with HIV infection. *Curr Infect Dis Rep* 2011;**13**(1):94–101.
66. Martinez E, Larroussea M, Podzamczer D, et al. Abacavir-based therapy does not affect biological mechanisms associated with cardiovascular dysfunction. *AIDS* 2010;**24**(3):F1–9.
67. Martin A, Amin J, Cooper DA, et al. Abacavir does not affect circulating levels of inflammatory or coagulopathic biomarkers in suppressed HIV: a randomized clinical trial. *AIDS* 2010;**24**(17):2657–63.
68. Longenecker CT, Triant VA. Initiation of antiretroviral therapy at high CD4 cell counts: does it reduce the risk of cardiovascular disease? *Curr Opin HIV AIDS* 2014;**9**(1):54–62.
69. Cruse B, Cysique LA, Markus R, et al. Cerebrovascular disease in HIV-infected individuals in the era of highly active antiretroviral therapy. *J Neurovirol* 2012;**18**(4):264–76.

70. Lazzaretti RK, Kuhmmer R, Sprinz E, et al. Clinical research: dietary intervention prevents dyslipidemia associated with highly active antiretroviral therapy in human immunodeficiency virus type 1-infected individuals. A randomized trial. *J Am Coll Cardiol* 2012; **59**:979–88.
71. Somarriba G, Lopez-Mitnik G, Ludwig DA, et al. Physical fitness in children infected with the human immunodeficiency virus: associations with highly active antiretroviral therapy. *AIDS Res Hum Retroviruses* 2013;**29**(1):112–20.
72. Miller TL, Somarriba G, Kinnamon DD, et al. The effect of a structured exercise program on nutrition and fitness outcomes in human immunodeficiency virus-infected children. *AIDS Res Hum Retroviruses* 2010;**26**(3):313–9.
73. Dubé MP, Cadden JJ. Lipid metabolism in treated HIV infection. *Best Practice Res Clin Endocrinol Metab* 2011;**25**(3):429–42.
74. Lipshultz SE, Mas CM, Henkel JM, et al. HAART to heart: highly active antiretroviral therapy and the risk of cardiovascular disease in HIV-infected or exposed children and adults. *Expert Rev Anti Infect Ther* 2012;**10**(6):661–74.
75. Butt AA, Chang CC, Kuller L, et al. Risk of heart failure with human immunodeficiency virus in the absence of prior diagnosis of coronary heart disease. *Arch Intern Med* 2011; **171**(8):737–43.
76. Mondy KE, Gottdiener J, Overton ET, et al. High prevalence of echocardiographic abnormalities among HIV-infected persons in the era of highly active antiretroviral therapy. *Clin Infect Dis* 2011;**52**(3):378–86.
77. Lipshultz SE, Williams PL, Wilkinson JD, et al. Cardiac status of children infected with human immunodeficiency virus who are receiving long-term combination antiretroviral therapy: results from the Adolescent Master Protocol of the Multicenter Pediatric HIV/AIDS Cohort Study. *JAMA Pediatr* 2013;**167**(6):520–7.
78. Zareba KM, Lavigne JE, Lipshultz SE. Cardiovascular effects of HAART in infants and children of HIV-infected mothers. *Cardiovasc Toxicol* 2004;**4**(3):271–9.
79. Lipshultz SE, Easley KA, Orav EJ, et al. Cardiovascular status of infants and children of women infected with HIV-1 (P^2C^2 HIV): a cohort study. *Lancet* 2002;**360**(9330):368–73.
80. Wilkinson JD, Williams PL, Leister E, et al. Cardiac biomarkers in HIV-exposed uninfected children. *AIDS* 2013;**27**(7):1099–108.
81. Lorgis L, Cottenet J, Molins G, et al. Outcomes after acute myocardial infarction in HIV-infected patients: analysis of data from a French nationwide hospital medical information database. *Circulation* 2013;**127**(17):1767–74.
82. Grossi PA. Update in HIV infection in organ transplantation. *Curr Opin Organ Transplant* 2012;**17**(6):586–93.
83. Pozzan G, Pagliari C, Tuon FF, et al. Diffuse-regressive alterations and apoptosis of myocytes: possible causes of myocardial dysfunction in HIV-related cardiomyopathy. *Int J Cardiol* 2009;**132**(1):90–5.
84. Bowles NE, Kearney DL, Ni J, et al. The detection of viral genomes by polymerase chain reaction in the myocardium of pediatric patients with advanced HIV disease. *J Am Coll Cardiol* 1999;**34**(3):857–65.
85. Rogers JS, Zakaria S, Thom KA, et al. Immune reconstitution inflammatory syndrome and human immunodeficiency virus-associated myocarditis. *Mayo Clin Proc* 2008;**83**(11):1275–9.
86. Yearley JH, Mansfield KG, Carville AAL, et al. Antigenic stimulation in the simian model of HIV infection yields dilated cardiomyopathy through effects of TNF alpha. *AIDS* 2008; **22**(5):585–94.

87. Subbiah R, Changat V, Clifton JD, et al. Cardiac isoform of alpha 2 macroglobulin and its reliability as a cardiac marker in HIV patients. *Heart Lung Circ* 2010;**19**(2):93–5.
88. Meenakshisundaram R, Sweni S, Thirumalaikolundusubramanian P. Cardiac isoform of alpha 2 macroglobulin: a marker of cardiac involvement in pediatric HIV and AIDS. *Pediatr Cardiol* 2010;**31**(2):203–7.
89. Reinsch N, Neuhaus K, Esser S, et al. Prevalence of cardiac diastolic dysfunction in HIV-infected patients: results of the HIV-HEART study. HIV. *Clin Trials* 2010;**11**(3):156–62.
90. Lipshultz SE, Orav EJ, Sanders SP, et al. Cardiac structure and function in children with human immunodeficiency virus infection treated with zidovudine. *N Engl J Med* 1992;
327(18):1260–5.
91. Lipshultz SE, Orav EJ, Sanders SP, et al. Limitations of fractional shortening as an index of contractility in pediatric patients infected with human immunodeficiency virus. *J Pediatr* 1994;**125**(4):563–70.
92. Lipshultz SE, Shearer WT, Thompson B, et al. Antiretroviral therapy (ART) cardiac effects in HIV-infected children: the multicenter NHLBI Cardiac Highly Active Antiretroviral Therapy (CHAART-II) study. *Circulation* 2009;**120**:S909–10.
93. Hulten E, Mitchell J, Scally J, et al. HIV positivity, protease inhibitor exposure and subclinical atherosclerosis: a systematic review and meta-analysis of observational studies. *Heart* 2009;**95**:1826–35.
94. Worm SW, Sabin C, Weber R, et al. Risk of myocardial infarction in patients with HIV infection exposed to specific individual antiretroviral drugs from the 3 major drug classes: the data collection on adverse events of anti-HIV drugs (D:A:D) study. *J Infect Dis* 2010; **201**:318–30.
95. Nayak G, Ferguson M, Tribble DR, et al. Cardiac diastolic dysfunction is prevalent in HIV-infected patients. *AIDS Patient Care & STDs* 2009;**23**(4):231–8.
96. O'Brien S, Sasaki N, Eidem BW, et al. Left ventricular diastolic dysfunction in HIV-negative infants exposed in utero to antiretroviral therapy from HIV-positive mothers: The Prospective NHLBI CHAART-I Study. *Circulation* 2011;**124**:A10808.
97. Halley CM, Houghtaling PL, Khalil MK, et al. Mortality rate in patients with diastolic dysfunction and normal systolic function. *Arch Intern Med* 2011;**171**(12):1082–7.
98. McMahon CJ, Nagueh SF, Pignatelli RH, et al. Characterization of left ventricular diastolic function by tissue Doppler imaging and clinical status in children with hypertrophic cardiomyopathy. *Circulation* 2004;**109**(14):1756–62.
99. McMahon CJ, Nagueh SF, Eapen RS, et al. Echocardiographic predictors of adverse clinical events in children with dilated cardiomyopathy: a prospective clinical study. *Heart* 2004;**90**(8):908–15.
100. Ingul CB, Tjonna AE, Stolen TO, et al. Impaired cardiac function among obese adolescents: effect of aerobic interval training. *Arch Pediatr Adolesc Med* 2010;**164**(9):852–9.
101. Wojcik M, Rudzinski A, Starzyk J. Left ventricular diastolic dysfunction in adolescents with type 1 diabetes reflects the long- but not short-term metabolic control. *J Pediatr Endocrinol Metab* 2010;**23**(10):1055–64.
102. Plazak W, Kopec G, Tomkiewicz-Pajak L, et al. Heart structure and function in patients with generalized autoimmune diseases: echocardiography with tissue Doppler study. *Acta Cardiol* 2011;**66**(2):159–65.
103. Cade WT, Waggoner AD, Hubert S, et al. Reduced diastolic function and left ventricular mass in HIV-negative preadolescent children exposed to antiretroviral therapy in utero. *AIDS* 2012;**26**(16):2053–8.

104. Kelly KM, Tarwater PM, Karper JM, et al. Diastolic dysfunction is associated with myocardial viral load in simian immunodeficiency virus-infected macaques. *AIDS* 2012; **26**(7):815–23.
105. Opravil M, Sereni D. Natural history of HIV-associated pulmonary arterial hypertension: trends in the HAART era. *AIDS* 2008;**22**(Suppl 3):S35.
106. McLaughlin VV, Archer SL, Badesch DB, et al. ACCF/AHA 2009 expert consensus document on pulmonary hypertension: a report of the American College of Cardiology Foundation Task Force on Expert Consensus Documents and the American Heart Association, developed in collaboration with the American College of Chest Physicians, American Thoracic Society, Inc., and the Pulmonary Hypertension Association. *Circulation* 2009;**119**(16):2250–94.
107. Cicalini S, Almodovar S, Grilli E, et al. Pulmonary hypertension and human immunodeficiency virus infection: epidemiology, pathogenesis, and clinical approach. *Clin Microbiol Infect* 2011;**17**(1):25–33.
108. Janda S, Quon B, Swiston J. HIV and pulmonary arterial hypertension: a systematic review. HIV and pulmonary arterial hypertension. *HIV Med* 2011;**11**(10):620–34.
109. Lindl A, Reinschl N, Nehausl K, et al. Pericardial effusion of HIV-infected patients: results of a prospective multicenter cohort study in the era of antiretroviral therapy. *Eur J Med Res* 2011;**16**:480–3.
110. Bolt RJ, Rammeloo LA, van Furth AM, van Well GT. A 15-year-old girl with a large pericardial effusion. *Eur J Pediatr* 2008;**167**(7):811–2.
111. Yoon SA, Hahn YS, Hong JM, et al. Tuberculous pericarditis presenting as multiple free floating masses in pericardial effusion. *J Korean Med Sci* 2012;**27**(3):325–8.
112. Rapose A, Sarvat B, Sarria JC. Immune reconstitution inflammatory syndrome presenting as pericarditis and pericardial effusion. *Cardiology* 2008;**110**(2):142–4.
113. Reuter H, Burgess LJ, Louw VJ, et al. The management of tuberculous pericardial effusion: experience in 233 consecutive patients. *Cardiovasc J S Afr* 2007;**18**(1):20–5.
114. Van Doorn CA, Yates R, Tsang VT. Endocarditis as the first presentation of AIDS in infancy. *Arch Dis Child* 1998;**79**(2):179–80.
115. Steinherz LJ, Brochstein JA, Robins J. Cardiac involvement in congenital acquired immunodeficiency syndrome. *Am J Dis Child* 1986;**140**(12):1241–4.
116. Granovsky MO, Mueller BU, Nicholson HS, et al. Cancer in human immunodeficiency virus-infected children: a case series from the Children's Cancer Group and the National Cancer Institute. *J Clin Oncol* 1998;**16**(5):1729–35.
117. Buck BE, Scott GB, Valdes-Dapena M, et al. Kaposi sarcoma in two infants with acquired immune deficiency syndrome. *J Pediatr* 1983;**103**(6):911–3.
118. Zoufaly A, Stellbrink HJ, Heiden MA, et al. Cumulative HIV viremia during highly active antiretroviral therapy is a strong predictor of AIDS-related lymphoma. *J Infect Dis* 2009;**200**(1):79–87.
119. Karavidas A, Tsiachris D, Lazaros G, et al. Doppler tissue imaging unmasks right ventricular function abnormalities in HIV-infected patients. *Cardiol J* 2010;**17**(6):587–93.
120. Guillevin L. Vasculitides in the context of HIV infection. *AIDS* 2008;**22**(3):S27–33.
121. Tseng ZH, Secemsky EA, Dowdy D, et al. Sudden cardiac death in patients with human immunodeficiency virus infection. *J Am Coll Cardiol* 2012;**59**:1891–6.
122. Sani MU, Okeahialam BN. QTc interval prolongation in patients with HIV and AIDS. *J Natl Med Assoc* 2005;**97**:1657–61.
123. Soliman EZ, Lundgren JD, Roediger MP, et al. Boosted protease inhibitors and the electrocardiographic measures of QT and PR durations. *AIDS* 2011;**25**(3):367–77.

124. Hsu JC, Li Y, Marcus GM, Hsue PY, et al. Atrial fibrillation and atrial flutter in human immunodeficiency virus-infected persons: incidence, risk factors, and association with markers of HIV disease severity. *J Am Coll Cardiol* 2013;**61**(22):2288–95.
125. Correia D, Rodrigues de Resende LA, Molina RJ, et al. Power spectral analysis of heart rate variability in HIV-infected and AIDS patients. *Pacing Clin Electrophysiol* 2006; **29**:53–8.
126. Dapena M, Jiménez B, Noguera-Julian A, et al. Metabolic disorders in vertically HIV-infected children: future adults at risk for cardiovascular disease. *J Pediatr Endocrinol Metab* 2012;**25**(5–6):529–35.
127. Piloya T, Bakeera-Kitaka S, Kekitiinwa A, et al. Lipodystrophy among HIV-infected children and adolescents on highly active antiretroviral therapy in Uganda: a cross sectional study. *J Int AIDS Soc* 2012;**15**(2):17427.
128. Lang S, Mary-Krause M, Cotte L, et al. Impact of individual antiretroviral drugs on the risk of myocardial infarction in human immunodeficiency virus-infected patients: a case-control study nested within the French Hospital Database on HIV ANRS cohort CO4. *Arch Intern Med* 2010;**170**:1228–38.
129. Lipshultz SE, Orav EJ, Sanders SP, et al. Cardiac structure and function in children with human immunodeficiency virus infection treated with zidovudine. *N Engl J Med* 1992; **327**(18):1260–5.
130. Crain MJ, Chernoff MC, Oleske JM, et al. Possible mitochondrial dysfunction and its association with antiretroviral therapy use in children perinatally infected with HIV. *J Infect Dis* 2010;**202**(2):291–301.
131. Chanock SJ, Luginbuhl LM, McIntosh K, et al. Life-threatening reaction to trimethoprim/sulfamethoxazole in pediatric human immunodeficiency virus infection. *Pediatrics* 1994; **93**(3):519–21.
132. Centers of Disease Control and Prevention (CDC). Achievements in public health. Reduction in perinatal transmission of HIV infection—United States, 1985–2005. *MMWR Morb Mortal Wkly Rep* 2006;**55**(21):592–7.
133. Neri D, Somarriba GA, Schaefer NN, et al. Growth and body composition of uninfected children exposed to human immunodeficiency virus: comparison with a contemporary cohort and United States national standards. *J Pediatr* 2013;**163**(10):249–54.
134. Mofenson LM, Brady MT, Danner SP, et al. Guidelines for the prevention and treatment of opportunistic infections among HIV-exposed and HIV-infected children: recommendations from CDC, the National Institutes of Health, the HIV Medicine Association of the Infectious Diseases Society of America, the Pediatric Infectious Diseases Society, and the American Academy of Pediatrics. *MMWR Recomm Rep* 2009;**58**(RR-11):1–166.
135. Hornberger LK, Lipshultz SE, Easley KA, et al. Cardiac structure and function in fetuses of mothers infected with HIV: the Prospective P^2C^2 HIV Multicenter Study. *Am Heart J* 2000;**140**(4):575–84.
136. Lipshultz SE, Fisher SD, Lai WW, et al. Cardiovascular risk factors, monitoring, and therapy for HIV-infected patients. *AIDS* 2003;**17**(Suppl 1):S96–122.
137. Lipshultz SE, Fisher SD, Lai WW, et al. Cardiovascular monitoring and therapy for HIV-infected patients. *Ann NY Acad Sci* 2001;**946**:236–73.
138. Saidi AS, Moodie DS, Garson Jr A, et al. Electrocardiography and 24-hour electrocardiographic ambulatory recording (Holter monitor) studies in children infected with human immunodeficiency virus type 1. The Pediatric Pulmonary and Cardiac Complications of Vertically Transmitted HIV-1 Infection Study Group. *Pediatr Cardiol* 2000;**21**(3): 189–96.

139. Ratnasamy C, Kinnamon DD, Lipshultz SE, et al. Associations between neurohormonal and inflammatory activation and heart failure in children. *Am Heart J* 2008;**155**:527–33.
140. Rusconi PG, Ludwig DA, Ratnasamy C, et al. Serial measurements of serum NT-proBNP as markers of left ventricular systolic function and remodeling in children with heart failure. *Am Heart J* 2010;**160**:776–83.
141. Fisher SD, Lipshultz SE. Cardiovascular abnormalities in HIV-infected individuals. In: Bonow RO, Mann DL, Zipes DP, Libby P, editors. *Braunwald's heart disease: a textbook of cardiovascular medicine*. 9th ed. Philadelphia: Elsevier Saunders; 2011. p. 1618–27.

Chapter 13

Toxic Effects of Alcohol on the Heart

Subroto Acharjee,[1] Bhaskar Purushottam,[1] and Vincent M. Figueredo[2]
[1]*Einstein Institute for Heart and Vascular Health, Einstein Medical Center, Philadelphia*
[2]*Einstein Institute for Heart and Vascular Health, Einstein Medical Center, and Jefferson Medical College, Philadelphia*

13.1 INTRODUCTION

Dietary alcohol has been a part of human culture dating back almost to the beginning of recorded history. It has served many roles over time—for example, in religion for rituals of worship, in medicine for antiseptic and analgesic properties, as a social lubricant enhancing relaxation and merry-making, and nutritionally as a thirst-quencher and source of calories.

Although discovery and consumption of the earliest alcoholic beverage may have been a "fortuitous accident," purposeful fermentation took place as early as the Neolithic period (10,000 B.C.), as evidenced by the discovery of beer jugs from the late Stone Age.[1] The earliest evidence of the existence of a fermented beverage itself comes from biochemical analysis of residue from Chinese artifacts dating to around 7000 B.C.[2] The art of using fermented grapes for winemaking, however, may have actually originated in the Middle East, with the earliest chemical evidence of a tartaric acid-rich sediment found in storage jars from Northern Iran dating from 5400 to 5000 B.C.[3]

Wine appeared as a finished product in Egyptian pictographs as early as 4000 B.C.[4] In Sumer (modern day Iraq), it may have been used for medicinal purposes as early as 2000 B.C.[5] An alcoholic beverage was being consumed between 3000 B.C. and 2000 B.C. in India.[6] In Babylon, wine deities were worshiped as early as 2700 B.C.,[7] and the art of winemaking had reached the Greek civilization by 2000 B.C.[8] Alcohol consumption has been practiced in Rome since its founding in 753 B.C., and the Romans, subsequently, facilitated its spread throughout Europe.[9] Although beer and wine arrived in North America with the first colonial settlers in the early seventeenth century, several Native American fermented beverages were already in existence during the pre-Columbian era.[10]

The Heart and Toxins. DOI: http://dx.doi.org/10.1016/B978-0-12-416595-3.00013-X

With the widespread popularity of alcohol, knowledge of its toxic effects and efforts at regulation also increased. Indeed, in many of these time periods, there is evidence of specific guidance advocating moderation and warning against the dangers of intoxication.[11,12] Abraham Lincoln, sixteenth president of the United States, famously stated of alcohol in his address to the Springfield Washington Temperance Society in 1842:[13]

It is true, that even then, it was known and acknowledged, that many were greatly injured by it; but none seemed to think the injury arose from the use of a bad thing, but from the abuse of a very good thing.

Although the specific detrimental effects of alcohol on the heart were reported as early as the nineteenth century, understanding the relationship between alcohol consumption and cardiovascular diseases remains complex to date. While light-to-moderate drinking may have a favorable effect on morbidity and mortality from coronary heart disease (CHD) and ischemic stroke, this cardiovascular protective effect is neutralized by toxic effects in the setting of excessive overall volume of intake or by a consumption pattern consisting of episodic heavy drinking (binge drinking). This chapter discusses patterns of alcohol consumption around the world and notes its epidemiological impact on morbidity, mortality, and health economics. We outline the pharmacology and metabolism of alcohol, and focus on its toxic effects on various components of the cardiovascular system, with a special emphasis on evidence-based observations and conclusions.

13.2 GLOBAL HEALTH BURDEN, PATTERNS OF USE, AND ECONOMIC CONSEQUENCES

Excessive consumption of alcohol is one of the world's leading health problems, contributing to more than 60 major types of disorders and accounting for 3.8% of all deaths worldwide: 6.2% for men and 1.1% for women.[14] Approximately 2.5 million deaths each year are attributable to alcohol, with a net loss of life of 2.25 million, even after accounting for the estimated favorable impact of moderate drinking in selected circumstances. Alcohol consumption is a necessary cause for more than 30 International Classification of Diseases (ICD)-10 codes, and may be a component of more than 200 other ICD-10 disease codes. Alcohol ranks eighth among global risk factors for death.

Cardiovascular disease (CVD) and diabetes mellitus account for 14% of all alcohol-attributable deaths, the third-largest disease category after cancers and liver cirrhosis. The overall burden of alcohol-related morbidity can be measured in disability-adjusted life years. In 2004, 4.5% of the global burden of disease and injury was attributable to alcohol: 7.4% for men and 1.4% for women. This makes alcohol the world's third largest risk factor for disease and disability.[14] Interestingly, while chronic diseases account for the majority of alcohol-attributable deaths, neuropsychiatric disorders related to alcohol consumption account for the major portion of alcohol-related disability.

Globally, alcohol use is prevalent in cultures throughout the world. In 2005, worldwide per-capita consumption was 6.13 liters of pure alcohol consumed per person aged 15 or older, although almost half of all men and two thirds of women classify themselves as nondrinkers (i.e., they had not consumed alcohol in the past year).[14] Heavy episodic drinking, a harmful pattern particularly associated with intoxication and injury, was noted in 11.5% of drinkers. The highest consumption levels can be found in the developed world, mostly the Northern Hemisphere (excluding North America), but also in Argentina, Australia, and New Zealand. Countries of North Africa and Sub-Saharan Africa, the eastern Mediterranean region, and Southern Asia and the Indian Ocean report low consumption. Southern Africa and North and South America are classified as medium-consumption regions.

Interestingly, although income levels are directly related to consumption patterns, the same does not hold true for alcohol-attributed mortality. The lower the economic development of a country, the higher the alcohol-attributable mortality and burden of disease per liter of pure alcohol consumed. For example, while Western and Eastern Europe are both high-consumption regions, deaths due to alcohol consumption are not at the same level. In fact, the Russian Federation and neighboring Eastern European countries have a "perfect storm" combination of high consumption levels and risky drinking patterns, leading to the world's highest incidence of alcohol-related death and disability. Within each country, consequences of alcohol use are also influenced by income, with those that are impoverished and marginalized from society being disproportionately affected. Over time, these patterns of global alcohol consumption have remained relatively stable although consumption is rapidly increasing in the most populous countries of China and India.

Besides well-known chronic and acute health effects, alcohol consumption is also associated with far-reaching psychosocial consequences, including absenteeism and loss of productivity at the workplace and domestic violence and child neglect at home.[14] Thus, use of alcohol is associated with an overall cost to society, which includes health care costs, law enforcement costs, other direct social costs, direct administrative or social work services costs, and other indirect costs. In one meta-analysis, compiling cost studies from four high-income countries and two middle-income countries, the total costs attributable to alcohol ranged from 1.3 to 3.3% of gross domestic product,[15] with the greatest contributor being the cost of productivity loss.

In another study from Australia,[16] three-quarters of the adult population surveyed reported having been affected by someone else's drinking within the past year. More than $13 billion were lost directly by these members of society due to the heavy drinking of people around them, including costs such as out-of-pocket expenses and time away from work. Hospital and child protection costs added a further $765 million, leading to a total tangible cost to others of \$14.2 billion. In addition, there were large intangible costs (due to loss of quality of life of others), estimated at a minimum of \$6.4 billion dollars. Thus, the real cost of excessive alcohol consumption far exceeds (and is roughly double) the sum total of measured costs to the drinker and to society, largely due to the magnitude of its effect on others within the same social grouping as the consumer.

13.3 CHEMISTRY AND PHARMACOLOGY

Following oral ingestion, alcohol is readily absorbed by passive diffusion across the stomach wall (approximately 20%) and the small bowel (80%) into the bloodstream. Peak plasma levels occur half an hour after consumption. Only 10% of alcohol consumed is eliminated in the urine, breath, or perspiration, with the majority being metabolized.[17,18] Ethanol metabolism occurs primarily through the oxidative pathway using three different enzyme systems: alcohol dehydrogenase (ADH), cytochrome P450 2E1 (CYP2E1), and catalase. The major oxidative pathway involves hepatic oxidation to acetaldehyde by ADH. Toxic acetaldehyde is then eliminated by aldehyde dehydrogenase (ALDH) via nonreversible oxidation to acetate. Acetyl CoA, the activated form of acetate, is then further metabolized into ketone bodies, amino acids, fatty acids, and steroids, which enter the Krebs cycle, leading to end products of water and CO_2.

Both ADH and ALDH enzymes use the cofactor nicotinamide adenine dinucleotide (NAD^+) which is reduced to NADH. As a result, ethanol oxidation by this enzyme system can raise the $NADH/NAD^+$ ratio, altering the cellular redox state and leaving hepatocytes prone to damage from acetaldehyde and free radicals. At higher ethanol concentrations and with chronic use, the hepatic cytochrome P450 isozyme CYP2E1 is induced, which metabolizes ethanol leading to production of several reactive oxygen species, including hydroxyethyl, superoxide anion, and hydroxyl radicals, capable of causing oxidative tissue damage. A minor pathway of ethanol oxidation involves the peroxisomal enzyme catalase.[19,20] The reversible enzyme-based oxidation of ethanol to acetaldehyde by ADH is the rate-limiting step in ethanol metabolism. Due to high enzyme affinity and low capacity, ADH can quickly become saturated. This may in turn lead to a rapid accumulation of ethanol in the body after only a few drinks.[21]

The various metabolites of ethanol have several toxic effects on biological systems. Acetaldehyde produced by hepatic oxidation of alcohol can bind proteins, such as enzymes, microsomal proteins, and microtubules, causing impaired hepatic protein secretion and contributing to hepatomegaly. Acetaldehyde adducts in the plasma membrane of hepatocytes may contribute to immune-mediated liver damage by generating antibodies that eventually destroy hepatocytes—a process called antibody-dependent cell-mediated cytotoxicity. Adducts formed by the interaction of acetaldehyde with erythrocyte membranes may be associated with ethanol-induced macrocytosis. Finally, acetaldehyde can also form adducts with the neurotransmitter dopamine to form salsolinol, which may contribute to alcohol dependence, and with DNA to form adducts, which may predispose to carcinogenesis.[19]

Chronic alcohol consumption leads to greater exposure to acetaldehyde, which can cause mitochondrial injury and cell apoptosis. Acetate, the other important ethanol metabolite, increases hepatic blood flow while causing depressed central nervous system activity along with affecting several metabolic processes.[20] Interestingly, because of metabolism, in part by the cytochrome P450 enzyme system, alcohol consumption may cause decreased clearance of other CYP2E1 substrates such as phenytoin and warfarin.[20,21]

Significant gender- and population-based differences exist in alcohol metabolism. Experimental evidence shows that when men and women consume the same amount of alcohol, women experience higher blood alcohol concentrations and are more prone to its toxicity.[22] Plausible explanations include decreased "first-pass" metabolism in women because of lower levels of gastric ADH, interaction of ADH with female sex hormones, lower lean mass, and lower body water leading to a smaller volume of distribution and higher blood levels, and more rapid metabolism of alcohol in the liver by women causing ADH to become saturated.[22]

Several polymorphisms occur at the ADH1B and the ADH1C gene loci that are associated with varying levels of enzymatic activity and affinity to ethanol. In particular, allelic variations in ADH1B occur at various frequencies in different populations. For example, some blacks and Native Americans with the ADH1B*3 allele can metabolize alcohol at a faster rate than those with ADH1B*1. The ADH1B*1 form is predominant in Caucasian and Black populations, whereas ADH1B*2 is more frequent among Chinese and Japanese populations.[19]

13.4 EVIDENCE OF CARDIOVASCULAR BENEFIT

Interestingly, the first description of the beneficial effects of alcohol on longevity dates back to the Prohibition era in the United States when Raymond Pearl, one of the premier epidemiologists of the twentieth century, published a collection of case histories in 1926 demonstrating that moderate drinkers had the greatest survival rate, followed by abstainers, and then heavy drinkers.[23] Since then, a variety of studies done with a number of population groups have reported a J-shaped or U-shaped relationship between alcohol consumption and all-cause mortality (Figure 13.1),[24] as well as cardiovascular morbidity and mortality (Figure 13.2).[25] The pleiotropic mechanisms underlying this apparent cardioprotection by alcohol are summarized in Figure 13.3.[26]

Ethanol consumption demonstrated a strong U-shaped relationship with coronary heart disease (CHD) mortality for male nonsmokers and heavy smokers at

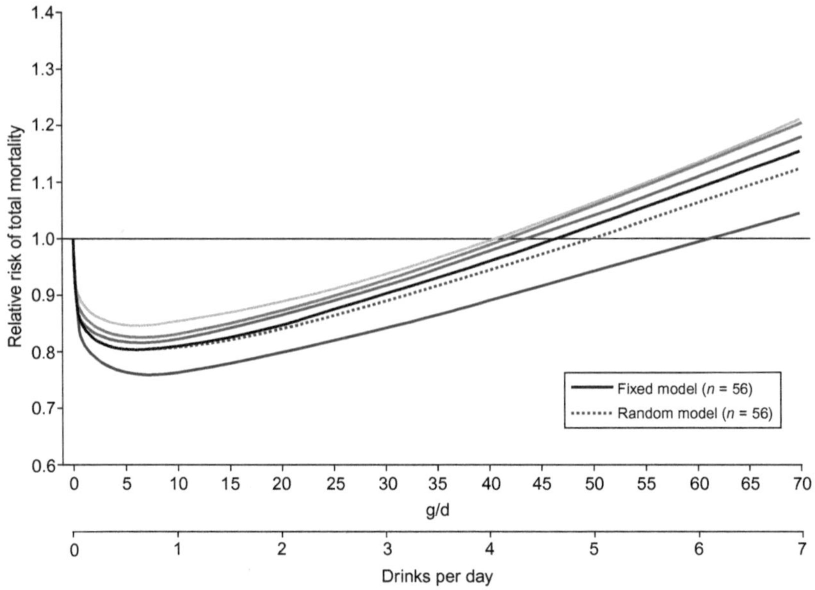

FIGURE 13.1 Relationship between total mortality and alcohol intake extracted from 56 curves using fixed- and random-effects models. Source: *Used with permission from Di Castelnuovo et al., 2006.*[24]

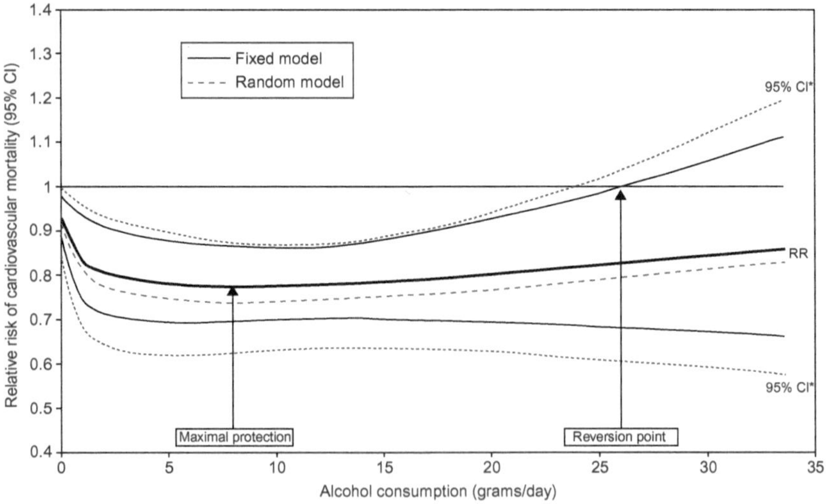

FIGURE 13.2 Alcohol consumption in relation to cardiovascular mortality in CVD patients. Source: *Used with permission from Costanzo et al., 2010.*[25]

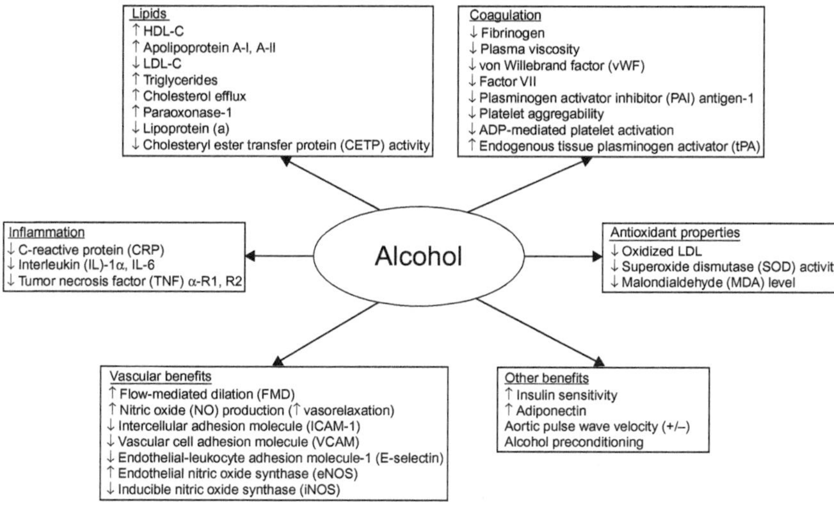

FIGURE 13.3 **Potential beneficial mechanisms of light-to-moderate alcohol consumption. Source:** *Used with permission from Movva and Figueredo, 2013.*[26]

24 years of follow up in the Framingham Heart Study.[27] In the Physicians' Health Study, 89,299 U.S. men, 40 to 84 years old and free of CVD at enrollment in 1982, were followed for 5.5 years. A U-shaped relationship was observed between alcohol intake and total mortality, with no overall benefit or harm at the ≥2 drinks per day level. The curve for CVD mortality was L-shaped, with risk reductions noted even with ≥2 drinks per day.[28] Similar U-shaped dose–response curves for alcohol and all-cause mortality have been reported among British male physicians, French wine drinkers, and German beer drinkers.[29–31]

In an analysis done using nine samples of the National Health Interview Survey, light and moderate alcohol consumption was associated with reduced cardiovascular mortality over a 25-year period, even when compared with lifetime abstainers.[32] However, those consuming ≥3 drinks per day had consistently higher mortality risk compared with those consuming 2 drinks per day. The Cancer Prevention Study II, an American prospective mortality study involving 490,000 men and women followed for nine years, demonstrated that rates of CVD death were 30 to 40% lower among those reporting at least one drink daily than among nondrinkers. Overall risk of death in middle age was slightly reduced with alcohol intake, although all-cause mortality increased with heavier drinking, particularly among adults under age 60 with a lower risk of cardiovascular disease.[33]

Alcohol intake has also been reported to affect the risk of developing CHD. An inverse association was demonstrated between alcohol and CHD risk in an analysis of pooled data from eight prospective North American and European studies involving 192,067 women and 74,919 men.[34] Among 8867 physicians from the Health Professionals Follow-up Study who were already low risk based on body mass index, physical activity, smoking, and diet, moderate alcohol intake was associated with lower risk for myocardial infarction (MI) during 16 years of follow up.[35]

Another analysis among 11,711 hypertensive men from the same study demonstrated an almost linear reduction in risk of MI with increasing alcohol intake, but no reductions in total or cardiovascular mortality.[36] Among the hypertensive subgroup in the Physicians' Health Study, an inverse relationship was noted between moderate alcohol consumption and incident CHD.[37]

Specific epidemiologic data focusing on the effects of alcohol on women have been published. In the Swedish Women's Lifestyle and Health Study of 47,921 women aged 30 to 49, during a 713,295 person-year follow up, light-to-moderate drinking yielded a significant inverse association with all-cause mortality, predominantly related to reduced CVD mortality.[38] Among 87,526 middle-aged female nurses in the Nurses' Health Study, moderate alcohol consumption reduced CHD risk during 334,382 person-years of follow up.[39] Among diabetic women in the same study, moderate alcohol consumption was also associated with reduced CHD risk during 39,092 person-years of follow up.[40]

Among 10,576 black and 105,610 white postmenopausal women free of CHD in the Women's Health Initiative (WHI), alcohol consumption in moderate amounts was associated with reduced risk of total mortality among Caucasian women during a mean follow up of eight years, regardless of hypertensive status, but only in hypertensive black women.[41] The effect of moderate alcohol intake among postmenopausal women with diabetes was assessed in the Women's Health Initiative observational study, where 3198 women without prior CHD were followed for 22,546 person-years. Both frequency and quantity of alcohol intake were inversely associated with the risk of developing CHD.[42]

The effects of alcohol on non-Western populations have also been previously described. In one study, 64,597 Chinese men aged ≥40 years and free of clinical CHD were followed for 494,084 person-years. A linear inverse relationship was noted between alcohol consumption and CHD events, which remained consistent even among heavy drinkers.[43] In another study from China that enrolled 18,244 men aged 45 to 64, intake of 1 to 14 drinks a week was associated with a 19% reduction in overall mortality, whereas drinking 28 or fewer drinks per week was associated with a 36% reduction in CHD mortality.[44] From Japan, a pooled analysis of six ongoing large-scale cohort studies involving 309,082 subjects revealed a J- or U-shaped association for the risk of total and major causes of mortality in men, and the risk of total and heart disease mortality in women.[45]

A second Japanese study included 34,776 men and 48,906 women aged 40 to 79 who were followed for a median of 14.2 years. While light-to-moderate drinking was associated with reduced mortality from CVD for both sexes, heavy alcohol consumption (≥46.0 g ethanol/day) was linked to greater mortality from stroke and CVD in men and from CHD in women.[46]

Similar associations have been reported even among patients with known CVD. In the Second Manifestations of Arterial Disease study, 5447 patients with known vascular disease or diabetes were followed for 4.7 years. Moderate alcohol consumption (1–2 drinks/day) was associated with a reduced risk of vascular and all-cause death and nonfatal events from CHD, stroke, and amputations. A U-shaped relationship was noted between alcohol intake and total mortality, vascular mortality, and amputations.[47] In one meta-analysis of 16,351 patients with CVD from eight prospective studies, light-to-moderate alcohol consumption (5–25 g/day) was associated with a lower incidence of CVD events, with a J-shaped association of alcohol intake with all-cause and CVD mortality.[25]

In the Optimal Trial in Myocardial Infarction with Angiotensin II Antagonist Losartan study, there was a strong positive association between moderate alcohol use and survival in a cohort of 5477 patients for more than 2.7 years following heart failure and/or evidence of left ventricular dysfunction after a complicated MI.[48] During a three-year follow up, no adverse effects were noted among 325 moderate drinkers who continued consuming alcohol after an MI. In the Stockholm Heart Epidemiology Program study, among 1346 consecutive patients between 45 years and 70 years with a first nonfatal MI followed for more than eight years, moderate alcohol consumption in the year prior to acute MI was associated with reduced total and cardiac mortality, along with reductions in hospitalization for MI, stroke, or heart failure.[49]

Another study with 437 survivors of a recent MI revealed a significant reduction in the risk of subsequent complications over four years with moderate wine drinking.[50] Finally, moderate alcohol consumption was protective against coronary atherosclerosis progression as demonstrated by quantitative coronary angiography in the Stockholm Female Coronary Risk Angiographic study.[51]

13.5 EVIDENCE OF LACK OF BENEFIT OR HARM

The effects of different alcohol intake patterns on cardiovascular health have been well described in several population-based studies. In one case-control study, 427 Caucasian males aged 35 to 69 with incident MI were compared to 905 healthy controls. Men who drank mainly without food in the prior 12 to 24 months had approximately 50% higher odds of MI compared to lifetime abstainers. In addition, participants who drank only on weekends had a 90% greater risk of MI compared to those who drank less than once a week.[52] Heavy episodic drinking was associated with a higher risk of MI or coronary death in a case-control study involving 11,511 cases and 6077 randomly selected controls from Australia.[53]

In a Swedish case-control study, women who reported drinking enough to be intoxicated or drinking at least a half bottle of spirits on occasion had an 80% incidence of MI compared to those without this drinking pattern.[54] In a Canadian longitudinal study among 1154 participants aged 18 to 64, consumption of eight or more drinks at one sitting increased the risk of having a CHD event in both men (hazard ratio (HR) = 2.26) and women (HR = 1.10).[55] The Prospective Epidemiological Study of Myocardial Infarction evaluated the effect of alcohol intake patterns on incident CHD risk in Northern Ireland and France. Although regular and moderate alcohol intake (prevalent in France) was associated with a low risk of CHD, the binge-drinking pattern (prevalent in Northern Ireland) conferred a higher risk.[56]

The pattern of drinking appears to have more impact on cardiovascular mortality than the volume of alcohol consumed among survivors of acute MI. In one prospective study of men aged 45 to 64, heavy drinking (≥3 drinks/day) was associated with a 50% higher risk of all-cause death among 455 survivors of MI over a mean follow up of 12.8 years.[57] Another study among 1935 survivors of MI demonstrated that binge drinking (≥3 drinks within 1–2 hours), even among light drinkers, was found to be associated with a two-fold higher all-cause and cardiovascular mortality.[58] Binge drinking completely attenuated the decrease in cardiovascular mortality associated with moderate drinking.

Several meta-analyses have examined the relationship between alcohol consumption patterns and CHD risk. In one paper combining data from 28 cohort studies, consumption of ≥89 g per day of alcohol was associated with a greater relative risk of CHD.[59] In another meta-analysis of six studies, regular heavy drinkers experienced a protective effect, while heavy irregular or binge drinkers showed a statistically significant increase in risk of incident CHD. Further, while the dose–response curve for regular drinkers was L-shaped (suggesting a continued protective effect even at higher doses), for binge drinkers, the curve was J-shaped (with higher doses having a deleterious effect on CHD risk).[60]

A pooled analysis of 34 studies (1,015,835 subjects and 94,533 deaths) found that although total mortality was reduced in women who consumed one drink daily and men who had one to two drinks daily compared to nondrinkers, intake of more than two drinks a day for women and three drinks a day for men was associated with increased mortality.[24] Finally, a meta-analysis of 14 studies including 4718 CHD events showed that the beneficial effect of moderate alcohol consumption was neutralized when an overall light-to-moderate drinking pattern was mixed with episodes of binge drinking (>60 g of pure alcohol or ≥5 drinks per occasion at least monthly). In fact, such a pattern was associated with a 45% higher CHD risk.[61]

Similar data are available from population groups known to have a higher frequency of heavy episodic drinking. Data from the National Health and Nutrition Examination Survey Epidemiologic Follow-up Study examined the relationship between average volume of alcohol intake and all-cause mortality among 2054 blacks aged 25 to 75, who were followed over 19 years. No J-shaped curve was found, and there was no evidence of any beneficial effect. Instead, mortality appeared to increase with increasing average consumption of more than one drink a day, particularly at levels of >28 drinks per week.[62]

In a prospective cohort study including 6502 Russian men aged 25 to 64, frequent heavy drinkers had a >60% greater risk for total mortality, while deaths from CVD increased more than two-fold.[63] In the Lipid Research Clinics Prevalence and Follow-up Studies, among middle-aged men and women followed for mortality over 13 years, alcohol consumption in moderation had beneficial effects on all-cause and CVD mortality among U.S. adults. However, for Russian men and women, this protective influence could not be demonstrated.[64]

In a prospective cohort study among 5766 Scottish men followed over 21 years, there was no beneficial effect of alcohol consumption on mortality from CHD. However, the association between alcohol consumption and mortality was unfavorable for men drinking more than 22 units a week, with a 22 to 49% higher relative risk of all-cause mortality.[65]

Among populations representing more than a third of the human race, epidemiologic studies contradict the purported cardiovascular benefits of alcohol intake. In one study involving 1732 cases with MI and 2204 matched controls among South Asians from the INTERHEART study, alcohol consumption did not provide protection from CHD.[66] Among Indian participants, there was a 64% higher rate of MI for alcohol users compared with nonusers.

Similar findings were noted in a survey among 11,898 Asian-Indian men, including 4465 self-reported alcohol users. No inverse association was found between alcohol intake and the prevalence of CHD. In fact, alcohol intake was associated with 40% higher odds for having CHD.[67] More than half of those who reported drinking behavior were binge drinkers. Comparable data also exist among a Chinese population. In a prospective, population-based cohort study of 56,167 Chinese men and women aged 65 or over, followed for 4.2 years, no overall association was noted between moderate alcohol use and lower CHD mortality in either men or women.[68]

In another case-control study among 1476 consecutive Chinese men who underwent coronary angiography, both moderate and heavy intake of alcohol was associated with an 80 to 100% higher risk of having coronary artery disease (CAD). Drinking alcohol 4 to 5 days per week, or more frequently, independently predicted greater risk of CAD, as did a history of drinking for >15 years. Further, greater frequency, duration, or volume of alcohol intake was directly related to a higher incidence of CAD.[69]

Even when not considering specific consumption patterns, several studies have found a lack of benefit and evidence of significant cardiovascular harm associated with alcohol intake. The Multi-Ethnic Study of Atherosclerosis examined the relationship between alcohol intake and subclinical CHD. Over a two- to four-year period, no protective association was found between alcohol intake and the primary outcome measure, the presence, amount, or progression of coronary calcium as studied by computed tomography. Light-to-moderate alcohol consumption did show some beneficial effect on clinical CHD risk. Heavy alcohol consumption (>2 drinks/day) was, however, associated with greater coronary calcium at baseline and faster progression throughout the study period.[70]

Even in the Coronary Artery Risk Development in Young Adults study, alcohol consumption was directly associated with coronary calcification in 3037 subjects aged 33 to 45, suggesting a potential proatherogenic effect of alcohol.[71] Data on alcohol and subclinical atherosclerosis are also available from the Cardiovascular Risk in Young Finns study, where alcohol consumption, carotid intima–media thickness (IMT), and conventional cardiovascular risk factors were investigated with 2074 young subjects without CHD. A linear and independent relationship was noted between increasing frequency of alcohol consumption and greater carotid IMT, suggesting that alcohol consumption may have proatherogenic effects in a young healthy population.[72]

In another prospective population-based study among 7169 British men age 45 to 64 with CHD, regular light alcohol consumption (1–14 units/week) was not associated with any significant effect on CHD, CVD mortality, or all-cause mortality compared with occasional drinkers. Former drinkers had the highest risk of these endpoints even after adjustment for lifestyle characteristics and preexisting disease. Among 455 men with prior MI, ex-drinkers showed a significant 75% increase in risk of all-cause mortality, and moderate-to-heavy drinking (≥3 drinks/day) was associated with a 50% increase in CVD mortality and all-cause mortality.[73]

The study by Shaper and Wannamethee highlighted a critical limitation of several prior studies reporting alcohol-related benefits—that of using "current nondrinkers" as a baseline group for comparison.[73] The authors hypothesized that as drinkers age, they may become frail or ill with comorbid conditions, leading to abstention from alcohol and greatly enhancing their mortality risk (the "sick quitter" hypothesis). Combining such people with lifelong teetotalers into a "current nondrinkers" group raises the risk profile for the entire group. This in turn exaggerates the benefits of alcohol intake when comparisons are made using this group as referent. Interestingly, one meta-analysis attempted to test this hypothesis by excluding studies where the "nondrinkers" group had been contaminated by former drinkers.

Out of 54 all-cause mortality studies and 35 CHD mortality studies, only 7 studies for all-cause and 2 studies for CHD mortality were judged error free. No evidence of alcohol-related all-cause or cardiac protection was found, confirming that prior studies with the "abstainer error" and meta-analyses including those studies may have overestimated the cardioprotective effects of alcohol. Exclusion of such a large number of studies, however, reduced the overall statistical power of this particular meta-analysis.[74]

As is evident here, data regarding the cardiovascular effects of alcohol in humans are derived exclusively from observational studies, which typically rely on self-reported alcohol intake. Potential limitations include uncertain verification of alcohol use, varying definitions of a standard drink, evolving drinking behavior over the study period, and association of confounding lifestyle factors (e.g., educational status, socioeconomic status, smoking, exercise, diet) with patterns of alcohol use, which cannot be completely accounted for by statistical methods. Although a randomized controlled trial that assigns participants to specific doses of alcohol could provide definitive answers, such a study is unlikely to ever take place, given the risk of addiction to alcohol and consequent ethical considerations.

13.6 SPECIFIC TOXIC EFFECTS OF ALCOHOL ON THE CARDIOVASCULAR SYSTEM

Although the epidemiological studies just reviewed largely provide evidence of the toxic effects of alcohol on CHD events, other specific harmful sequelae of alcohol intake have also been described. Further, the mechanisms of the adverse impact of alcohol on myocardial function, electrical conduction, blood pressure regulation, and cerebrovascular health have also been elucidated.

13.6.1 Alcoholic Cardiomyopathy

Descriptions abound of a heart failure syndrome related to excessive alcohol intake, going as far back as the nineteenth century.[75] German neurologist and pathologist Nikolaus Friedreich described cardiac hypertrophy related to alcohol in 1861.[76] His compatriot, pathologist Otto Bollinger, described a similar condition he named "Munich beer heart," characterized by cardiac dilatation and hypertrophy in many citizens of Munich, where the average yearly consumption of beer was a reported 432 liters.[77]

Other examples include "patchy cirrhosis of the heart," described by W.H. Walsche in 1873, and the "arsenic-beer drinkers' disease" reported in England in 1900.[78,79] In 1902, in his textbook *The Study of the Pulse*, William Mackenzie used the term "alcoholic heart disease" to describe heart failure from alcohol.[80] After the description of classic high-output cardiovascular beriberi by Aalsmeer and Wenckebach in 1929,[81] nutritional deficiency remained the dominant presumed etiology of alcohol-related heart failure syndromes. However, in 1957, Brigden described "alcoholic cardiomyopathy" as a separate entity that does not respond to thiamine.[82] In contrast, beriberi heart disease is rapidly (within 1–2 weeks) reversible after thiamine therapy, manifests predominantly as right heart failure, is associated with increased cardiac output (high output failure), and exhibits normal myocardial histology.[83]

Alcoholic cardiomyopathy is the most common cause of nonischemic cardiomyopathy in Western societies; it is responsible for up to 45% of cases.[84] Some reports peg the prevalence of alcoholic cardiomyopathy at 21 to 32% of all dilated cardiomyopathies at referral centers, although others attribute at least half of dilated cardiomyopathy cases to alcohol.[85] Among alcohol users, the exact prevalence of alcoholic cardiomyopathy is unclear. In one cross-sectional study among asymptomatic alcoholics (50 women and 100 men), approximately one-third had evidence of cardiomyopathy.[86]

Many alcoholics, however, exhibit subclinical depression of cardiac function, with autopsy studies revealing enlarged hearts and other signs of cardiomyopathy in alcoholics who did not show overt symptoms.[87] There also appears to be a correlation between cirrhosis and left ventricular dysfunction, with one study revealing that 43% of 30 alcoholics with cardiomyopathy had cirrhosis compared to only 6% of 30 alcoholics without cardiomyopathy.[88]

Conversely, of the 20 alcoholics with cirrhosis, half had evidence of dilated cardiomyopathy. Among men and women alcoholics, the prevalence of dilated cardiomyopathy is similar, although women require a lower total lifetime dose of ethanol (~60% of that in men) to develop the condition. Women also experience a steeper decline in myocardial contractile function with increasing doses of alcohol.[86,89] Thus, women appear to be more sensitive to alcohol-related cardiac toxicity, largely due to the metabolic differences mentioned previously.

Among healthy young volunteers, even light intoxication (serum ethanol 23 mmol/liter) lowered median left ventricular ejection fraction at rest by 5%, while the decline with heavy intoxication (serum ethanol 45 mmol/liter) was 11%. This was likely due to a dose-dependent impairment of cardiac contractility.[90] Interestingly, despite concerns about alcoholic-induced myocardial depression, light-to-moderate drinking appears to be associated with a 10 to 20% lower risk for heart failure.[91]

Heavy drinking (≥3 drinks/day), however, is associated with a significantly increased risk of nonischemic heart failure.[92] Over the long term, risk of alcoholic cardiomyopathy is related to both the mean daily alcohol dose and the duration of usage.[93–95] Thus, the aggregate of clinical studies suggests that most patients who develop alcoholic cardiomyopathy report an intake level of more than 80 to 90 g of ethanol per day for ≥5 years.

Multiple mechanisms have been proposed to explain the toxic effects of alcohol on myocardial function (Figure 13.4).[96] Both animal and human studies have reported cardiomyocyte apoptosis induced by ethanol.[97] Ethanol and its metabolites (chiefly acetaldehyde) cause direct myocardial depression through a variety of noxious effects on the myocyte sarcoplasm and the mitochondria, including oxidative damage, deposition of triglycerides, altered fatty acid extraction, decreased myofilament calcium sensitivity,

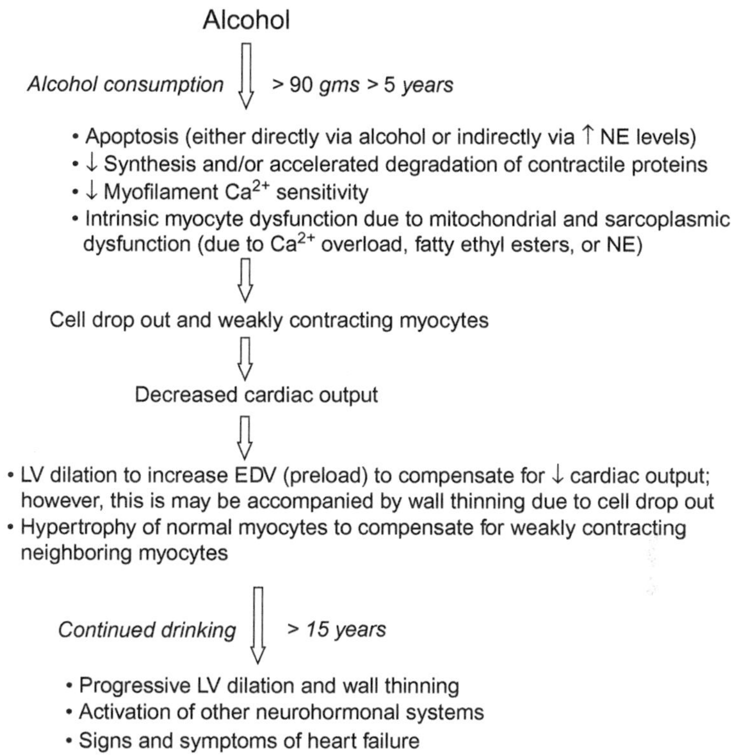

FIGURE 13.4 **Hypothetical schema for the pathogenesis of alcoholic cardiomyopathy.** NE, norepinephrine; LV, left ventricular; EDV, end-diastolic volume. **Source:** *Reprinted with permission from Piano, 2002.*[96]

defective myofibrillar protein turnover, and impaired protein synthesis due to reduced amounts of heat shock proteins.[98,99]

Alcohol-induced hypertension in humans is associated with the activation of the renin–angiotensin system, a pathway that can contribute to a decline in myocardial function.[100] The nutritional deficiencies and toxic effects of additives to alcoholic beverages may act synergistically with the mechanisms just mentioned to exacerbate myocardial damage in selected situations.[101] Finally, genetic factors may also play a role. Plasma and cardiac levels of the angiotensin converting enzyme (ACE) are 1.5- to 3-fold higher in patients with the DD polymorphism, which has been noted to predispose alcoholics to the development of left ventricular dysfunction.[102]

In chronic asymptomatic alcoholics, left ventricular dilation with preserved ejection fraction and impaired relaxation are often noted.[93] Thus, diastolic damage may precede systolic dysfunction. With established alcoholic cardiomyopathy, progressive dilatation, increased left ventricular mass, thin (or normal thickness) walls, and systolic impairment are common.[103] Echocardiographic abnormalities are usually noted prior to onset of clinical signs and symptoms.

13.6.2 Alcohol and Arrhythmias

The association between alcohol and electrical abnormalities in the heart was first recognized in the mid-twentieth century. In 1959, Evans described the electrocardiogram of patients with alcoholic cardiomyopathy, reporting characteristic T-wave changes (i.e., dimple, cloven, spinous, and blunt) along with the presence of atrial fibrillation, paroxysmal atrial tachycardia (PAT), and bundle branch blocks.[104] Others confirmed higher frequencies of atrial fibrillation, nodal rhythms, ventricular extrasystoles, complete heart block, and left and right bundle branch blocks, as well as abnormal P and T waves.[105]

The term "holiday heart" was coined in 1978 by Ettinger et al. to describe mostly supraventricular rhythms "associated with heavy ethanol consumption in a person without other clinical evidence of heart disease and disappearing without evident residual, with abstinence." A temporal pattern was noted, with episodes usually following heavy weekend or holiday drinking sprees, resulting in hospitalization on Monday or close to the year-end holidays. Atrial fibrillation was the most common arrhythmia, followed by atrial flutter.[106]

The role of alcohol as a precipitant in atrial fibrillation was examined in one case-control study, where 15 to 30% of idiopathic atrial fibrillation was judged to be alcohol-related and 5 to 10% of new episodes of atrial fibrillation were triggered by alcohol.[107] Among 10,333 participants from the Framingham Heart study, intake of ≥36 g per day of alcohol increased the risk of atrial fibrillation by 34%, after adjusting for potential confounders.[108] Among 47,949 individuals in the Danish Diet, Cancer and Health study, an average intake of alcohol of ≥20 g per day was associated with an increased risk of atrial fibrillation and flutter in men but not in women.[109]

In another cohort study among 16,415 Danes from the Copenhagen City Heart study, consuming ≥35 drinks per week was associated with greater risk of atrial fibrillation in men.[110] Among middle-aged women in the Women's Health Study, subjects consuming ≥2 drinks per day had a 60% higher risk of atrial fibrillation compared with nondrinkers.[111] Lastly, in a recent meta-analysis of 14 studies involving 130,820 participants and 7558 cases, a direct linear relationship was found between atrial fibrillation risk and alcohol consumption, suggesting that abstinence from alcohol was most favorable in terms of atrial fibrillation risk reduction.[112]

The association between alcohol and atrial flutter was confirmed in a case-control study involving 195 consecutive cases of patients with atrial fibrillation and/or flutter. Among those ≤60 years of age, subjects with atrial flutter were significantly more likely to be daily drinkers, and increasing frequency of alcohol intake was associated with greater odds of atrial flutter. Right atrial effective refractory periods were shortened significantly

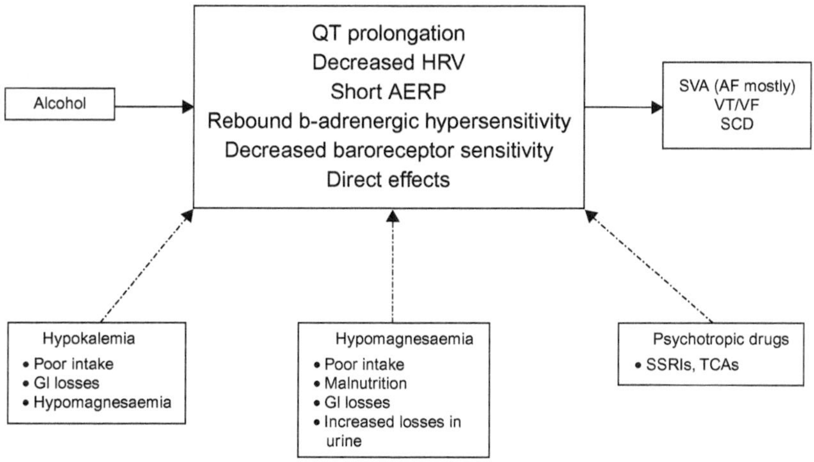

FIGURE 13.5 **Potential mediators of alcohol-induced cardiac arrhythmias.** TCA, tricyclic antidepressants; SSRI, selective serotonin reuptake inhibitors. **Source:** *Reprinted with permission from George and Figueredo, 2010.*[116]

with increasing amounts of alcohol intake, which could hypothetically allow the right atrium to sustain a rapid rate or allow a critically timed premature atrial beat to initiate the arrhythmia.[113]

Although some studies support an association between heavier drinking ($\geq$6 drinks/day) and all supraventricular arrhythmias, others indicate that the association with alcohol may not be as clear for nonfibrillation atrial arrhythmias.[114,115] Potential mediators of alcohol-induced cardiac arrhythmias are summarized in Figure 13.5.[116]

13.6.3 Alcohol and Sudden Cardiac Death

That heavy drinking can induce cardiac arrhythmias and ventricular tachycardia (VT) was reported as early as 1972.[117] Among 14 patients with prior rhythm disturbances and a known history of alcohol consumption, who were studied electrophysiologically, 10 of them developed sustained or nonsustained atrial or ventricular tachyarrhythmias after ingestion of 90 mL of 80-proof whiskey.[118]

The earliest epidemiological demonstration of an association between heavy alcohol use and sudden cardiac death (SCD) was published in 1977, when "problem drinkers" from the Chicago Peoples Gas Company were shown to have significantly higher 15-year rates of sudden death.[119] In the prospective British Regional Heart Study among 7735 men between 40 years and 59 years, heavy drinkers had a nearly two-fold increase in SCD risk when compared to nondrinkers.[120] Alcohol intemperance was strongly associated with an increased risk of SCD after myocardial infarction in a prospective 10-year study done in Sweden.[121] In that study, subjects registered with the Swedish Temperance Board (for heavy alcohol abuse, drinking offenses, or driving under the influence) were almost 2.5 times more likely to experience SCD.

In one study from New Zealand, a higher proportion of heavy drinkers had acute coronary events manifest as SCD,[122] whereas in the Yugoslavia Cardiovascular Disease Study, recent drunkenness was positively correlated with SCD.[123] The Framingham study found that sudden death, in the absence of CHD, was more common among men drinking more than 90 ounces of ethanol per month.[124]

However, as with CAD and heart failure, the relationship between alcohol intake and VT/SCD is more complex. Data from a 12-year follow up of 22,071 apparently healthy male physicians participating in the Physicians' Health Study demonstrated that the risk of VT/SCD is lower for individuals with low alcohol intake (2–6 drinks/week) compared with those who rarely or never consumed alcohol. The relationship between alcohol consumption and SCD was U-shaped, with the risk approaching unity at approximately two drinks per day.[125]

Similarly, among 85,067 healthy women from the Nurses' Health Study, a U-shaped association between alcohol consumption and risk of SCD was reported after 26 years of follow up. While the lowest risk was at the 5 to 14.9 g per day alcohol (0.5–1 drink/day) consumption level, intake of >30 g per day of alcohol was associated with a higher risk of SCD.[126]

Acute alcohol intoxication, acute withdrawal, and chronic alcohol abuse can all lead to SCD as a result of several overlapping mechanisms. In acute intoxication, there is evidence of increasing catecholamine levels, decreased heart rate variability due to lower vagal modulation and reduced baroreceptor sensitivity, and QT interval prolongation, all of which can portend development of fatal arrhythmias (see Figure 13.5).[127–129] Severe coronary vasospasm may be triggered by heavier doses of alcohol.[130]

During the acute withdrawal period, there is a state of autonomic hyperactivity with elevated catecholamine levels and increased β-adrenoceptor density and activity.[131] Mechanisms similar to acute intoxication are also at play here: decreased heart rate variability and baroreceptor sensitivity, along with QT interval prolongation.[132] Additionally, electrolyte abnormalities (e.g., hypokalemia, hypomagnesemia, and hyponatremia) can be present during withdrawal from chronic abuse, which can contribute to a proarrhythmogenic milieu.[133]

In chronic heavy drinkers, besides autonomic dysfunction due to vagal neuropathy, prolonged QTc, and dyselectrolytemia, increased QTc dispersion, steeper QT/RR slope, and the development of cardiomyopathy may all contribute to SCD.[134–137]

13.6.4 Alcohol and Hypertension

It is known that a well-established relationship exists between alcohol consumption and hypertension. A plethora of cross-sectional studies among the full spectrum of races and ethnicities consistently demonstrate higher mean blood pressure and/or greater prevalence of hypertension with increasing alcohol consumption, independent of salt intake, smoking, adiposity, education, and other potential confounders.

Among 83,947 men and women of three races participating in the Kaiser-Permanente Multiphasic Health Examination, consumption of ≥3 drinks per day was associated with higher systolic and diastolic pressures, and substantially higher prevalence of pressures ≥160/95 mmHg.[138] Among 1310 normotensive middle-aged Japanese men, the risk for incident hypertension increased in a dose-dependent manner as alcohol intake increased.[139] Among 70,891 healthy young women, a J-shaped association of alcohol intake and incident hypertension was demonstrated over eight years of follow up. While lower levels of intake appeared protective, women consuming >2 drinks per day had a 31% higher risk of developing chronic hypertension.[140]

A meta-analysis of 12 cohort studies from the United States, Japan, and Korea revealed that the risk for hypertension increased linearly with alcohol consumption for both men and women. The review revealed a linear dose–response relationship with a relative risk of 1.57 at 50 g pure alcohol per day and 2.47 at 100 g per day for men and 1.81 at 50 g per day and 2.81 at 100 g pure alcohol per day for women.[141]

A cohort analysis involving 8334 participants from the Atherosclerosis Risk in Communities Study, who were free of hypertension and coronary heart disease at baseline, showed that higher levels of consumption of all types of alcoholic beverages were associated with higher risk of hypertension for all race-gender strata at six years of follow up.[142] The consumption of alcohol in amounts ≥210 g per week was an independent risk factor for hypertension in North American populations. In the black population, consumption of even light-to-moderate amounts of alcohol was also associated with a higher risk of hypertension.

Another study followed 28,848 women from the Women's Health Study and 13,455 men from the Physicians' Health Study for 10.9 and 21.8 years, respectively, for incident hypertension. In women, a J-shaped association between alcohol intake and hypertension was noted, whereas in men, even light-to-moderate alcohol consumption increased hypertension risk. The threshold above which alcohol became deleterious for hypertension risk was ≥4 drinks per day for women versus a moderate level of ≥1 drink per day for men.[143] A Mediterranean population study with a median follow up of 4.2 years revealed that the consumption of beer or spirits, but not wine, was associated with a higher risk of progression to hypertension, irrespective of the pattern of consumption.[144]

Results from some interesting intervention trials support a causal relationship of alcohol to hypertension. Using a crossover design among 16 men with hypertension and a daily alcohol intake of 80 g, one study successfully demonstrated a clear pressor effect of alcohol.[145] Blood pressure remained high in eight patients who continued their regular alcohol consumption up to the fourth day after study initiation, and then fell significantly after alcohol intake was halted. The other eight patients had no alcohol for the first three days after admission, but resumed alcohol consumption from day 4 to 7. In these patients, blood pressure fell slightly after admission and significantly increased with reintroduction of alcohol.

Similar crossover trials from Australia demonstrated declining blood pressures accompanying protocol-mandated reductions in alcohol consumption among normotensive and ambulatory men, with blood pressure rising when usual drinking habits were resumed.[146,147] In one meta-analysis of 15 randomized controlled trials involving 2234 participants, alcohol reduction as the only intervention for patients with hypertension was associated with a significant reduction in mean systolic (−3.32 mmHg) and diastolic blood pressure (−2.04 mmHg), with a clear dose−response relationship between mean percentage of alcohol reduction and mean blood pressure reduction.[148] As the effects of intervention were enhanced for those with higher baseline blood pressure, this study highlighted the importance of alcohol intake reduction among those with diagnosed hypertension.

The exact mechanisms of alcohol-induced hypertension remain unclear. Although prior studies suggested an overactive and hyperresponsive sympathetic nervous system, one study assessing high blood pressures in drinkers found no differences in the sympathoadrenal or renal pressor mechanisms after stresses as compared to nondrinkers.[149] A Japanese study showed that while low-to-moderate alcohol intake appeared to have a possible vasculoprotective effect (measured by pulse wave velocity), heavy drinking caused a direct detrimental effect on the vasculature.[150]

Polymorphisms of alcohol-metabolizing enzymes may influence the effects of alcohol on blood pressure as well. For example, genotypic variations in catechol-O-methyltransferase (*COMT*) mRNA, and the consequent reduced inactivation of catecholamines, may exacerbate alcohol-induced blood pressure elevation.[151] Conversely, certain polymorphisms of alcohol dehydrogenase—that is, the ADH2(1)/2(1) genotype—are associated with lower levels of systolic blood pressure.[152]

13.6.5 Alcohol and Stroke

Although heavy alcohol consumption appears to increase the risk of all strokes, the effect of moderate alcohol intake on stroke risk appears discordant for two major stroke types. Most studies of moderate drinking support a decreased risk of ischemic stroke, which is similar to the observed decreases in ischemic heart disease events, but a neutral effect or increase in the risk of hemorrhagic stroke.[153] In a prospective cohort study of 38,156 male health professionals followed for 14 years, light and moderate average alcohol use was generally not associated with an increased risk for ischemic stroke, although a statistically significant trend for increased risk was reported for heavier drinkers (≥30 g or ≥3 drinks daily).[154]

In a case-control study involving 677 subjects with first ischemic stroke from an urban multiethnic population in northern Manhattan, a three-fold increase in risk of ischemic stroke was noted among those consuming ≥7 drinks per day.[155] A meta-analysis of 19 cohort and 16 case-control studies found that heavy alcohol use (>60 g or four drinks/day) increased the risk for all strokes by 64%, including a 69% increase in risk of ischemic stroke and a more than doubling in risk of hemorrhagic strokes.[153]

Four-year follow-up data from 87,526 female nurses aged 34 to 59 in the Nurses' Health Study revealed that even light drinkers had twice the risk of subarachnoid hemorrhage compared to abstainers. Although only 28 cases of subarachnoid hemorrhage were noted, the relative risk for 5 to 14 g per day of alcohol intake was 3.7.[156]

In a study from Finland involving 212 consecutive patients with acute ischemic stroke, recent heavy drinking was an independent risk factor for stroke, increasing the risk for cardioembolic and cryptogenic stroke several fold, particularly among those with a known high-risk source or prominent large-artery atherosclerosis.[157] In another Finnish study with a 10-year follow up, among 15,965 men and women aged 25 to 64, binge drinking (≥6 drinks in one session for men, or ≥4 drinks for women) was associated with an 85% higher risk for all strokes, and an almost two-fold increase in ischemic strokes.[158] Thus, drinking patterns also seem to affect stroke risk.

Mechanisms for this apparent dichotomous relationship are speculative at this time. While alcohol-induced hypertension and antithrombotic effects could account for the increased hemorrhagic strokes, cardiomyopathy and atrial fibrillation associated with alcohol use may explain some of the increased embolic events. Additionally, the same antiatherosclerotic effects that protect moderate drinkers from CAD could be responsible for the decrease in ischemic stroke noted among some populations with controlled alcohol consumption (refer to Figure 13.3).

13.7 CONCLUSION

For people who choose to drink, current U.S. dietary guidelines recommend restricting alcohol consumption to one drink per day for women and two drinks per day for men.[159] Similar recommendations advocating moderation exist in several other countries.[160] One standard drink in the United States has 14 g or 0.6 ounces of alcohol, equivalent to 12 ounces of regular beer (5% alcohol),

5 ounces of wine (12% alcohol), or 1.5 ounces of 80 proof (40% alcohol) distilled spirits.[159]

The available evidence from epidemiological studies supports the association between light-to-moderate regular alcohol consumption and a reduction in cardiovascular and all-cause mortality, although several variables such as age, sex, genetic and ethnic profiles, and comorbid medical conditions modify this relationship. Furthermore, randomized controlled trials, the "gold-standard" applied to all other population-based health initiatives, do not exist with regard to these purported benefits of drinking in moderation.

Finally, it appears that no definite "safe limit" exists for alcohol intake, with the threshold of benefit and harm varying from one study to the other. However, heavy or binge drinking has well-substantiated harmful cardiovascular effects, including cardiomyopathy, hypertension, arrhythmias, sudden death, and stroke. Cirrhosis of the liver, type 2 diabetes, cancer, cognitive decline, traumatic injury, and traffic accidents are all known adverse consequences of excessive drinking, in addition to other clinical and social problems such as sexually transmitted disease, unintended pregnancy, violent crime, and the risk of addiction. In view of the preceding, nondrinkers should not be routinely advised to initiate alcohol consumption for potential cardiovascular protection, although selected individuals who drink responsibly and in moderation may be reassured.

Key Learning Points

- At low-to-moderate levels of consumption, alcohol may have an overall beneficial effect on cardiovascular health for many, but not all, individuals. Heavy regular drinking or heavy episodic drinking (binge drinking) has definite harmful effects, many manifested as cardiovascular events.
- Cardiomyopathy, arrhythmias, sudden cardiac death, hypertension, and stroke (particularly hemorrhagic stroke) are the major known adverse cardiovascular effects of alcohol.
- In view of the impact of alcohol on other aspects of human health (e.g., cirrhosis, cancer, neuropathy), as well as the economic and psychosocial sequelae of alcohol dependency and abuse, nondrinkers should not be routinely advised to initiate alcohol consumption in anticipation of potential cardiovascular health benefits.

REFERENCES

1. Patrick CH. *Alcohol, culture, and society*. Durham: Duke University Press; 1952.
2. McGovern PE, Zhang J, Tang J, et al. Fermented beverages of pre- and proto-historic China. *Proc Natl Acad Sci USA* 2004;**101**(51):17593–8.
3. McGovern PE, Glusker DL, Exner LJ, et al. Neolithic resinated wine. *Nature* 1996;**381**(6582):480–1.
4. Lucia SP. *A history of wine as therapy*. Philadelphia: J.B. Lippincott; 1963.
5. Babor T. *Alcohol: customs and rituals*. New York: Chelsea House; 1986.

6. Dasgupta A. *The science of drinking: how alcohol affects your body and mind.* Lanham: Rowman & Littlefield; 2011.
7. Hyams E. *Dionysus: a social history of the wine vine.* New York: Macmillan; 1965.
8. Younger WA. *Gods, men, and wine. Wine and food society.* London: Michael Joseph/ Penguin; 1966.
9. Austin GA. *Alcohol in Western society from antiquity to 1800: a chronological history.* Santa Barbara: ABC-Clio Information Services; 1985.
10. MacAndrew C, Edgerton RB. *Drunken comportment: a social explanation.* Chicago: Aldine; 1969.
11. Darby WJ, Ghalioungui P, Grivetti L. *Food: the gift of osiris*, vols. 1 and 2. London: Academic Press; 1977.
12. Lutz HF. *Viticulture and brewing in the ancient orient.* New York: J.C. Heinrichs; 1922.
13. Lincoln A, Basler RP. *The collected works of Abraham Lincoln.* New Brunswick, NJ: Rutgers University Press; 1953.
14. World Health Organization. Global status report on alcohol and health. Geneva; 2011.
15. Room R, Jernigan D, Carlini-Cotrim B, et al. *Alcohol in developing societies: a public health approach.* Finnish Foundation for Alcohol Studies; 2002.
16. Laslett AM, Catalano P, Chikritzhs T, et al. The range and magnitude of alcohol's harm to others. AER Centre for Alcohol Policy Research, Turning Point Alcohol and Drug Centre, Victoria: Eastern Health; 2010.
17. Norberg A, Jones AW, Hahn RG, et al. Role of variability in explaining ethanol pharmacokinetics: research and forensic applications. *Clin Pharmacokine* 2003;**42**(1):1–31.
18. Holford NHG. Clinical pharmacokinetics of ethanol. *Clin Pharmacokinet* 1987;**13**(5): 273–92.
19. Zakhari S. Overview: how is alcohol metabolized by the body? *Alcohol Res Health* 2006;**29**(4):245–54.
20. Agarwal DP. Genetic polymorphisms of alcohol metabolizing enzymes. *Pathol Biol (Paris)* 2001;**49**(9):703–9.
21. Gemma S, Vichi S, Testai E. Individual susceptibility and alcohol effects:biochemical and genetic aspects. *Ann Ist Super Sanita* 2006;**42**(1):8–16.
22. Thomasson HR. Gender differences in alcohol metabolism. Physiological responses to ethanol. *Recent Dev Alcohol* 1995;**12**:163–79.
23. Pearl R. *Alcohol and longevity.* New York: Alfred A. Knopf; 1926.
24. Di Castelnuovo A, Costanzo S, Bagnardi V, et al. Alcohol dosing and total mortality in men and women: an updated meta-analysis of 34 prospective studies. *Arch Intern Med* 2006;**166**(22):2437–45.
25. Costanzo S, Di Castelnuovo A, Donati MB, et al. Alcohol consumption and mortality in patients with cardiovascular disease: a meta-analysis. *J Am Coll Cardiol* 2010;**55**(13): 1339–47.
26. Movva R, Figueredo VM. Alcohol and the heart: to abstain or not to abstain? *Int J Cardiol* 2013;**164**(3):267–76.
27. Friedman LA, Kimball AW. Coronary heart disease mortality and alcohol consumption in Framingham. *Am J Epidemiol* 1986;**124**(3):481–9.
28. Gaziano JM, Gaziano TA, Glynn RJ, et al. Light-to-moderate alcohol consumption and mortality in the Physicians' Health Study enrollment cohort. *J Am Coll Cardiol* 2000;**35**(1):96–105.
29. Doll R, Peto R, Hall E, et al. Mortality in relation to consumption of alcohol: 13 years' observations on male British doctors. *BMJ* 1994;**309**(6959):911–8.

30. Renaud SC, Guéguen R, Schenker J, et al. Alcohol and mortality in middle-aged men from eastern France. *Epidemiol* 1998;**9**(2):184–8.
31. Keil U, Chambless LE, Döring A, et al. The relation of alcohol intake to coronary heart disease and all-cause mortality in a beer-drinking population. *Epidemiol* 1997;**8**(2):150–6.
32. Mukamal KJ, Chen CM, Rao SR, et al. Alcohol consumption and cardiovascular mortality among U.S. adults, 1987 to 2002. *J Am Coll Cardiol* 2010;**55**(13):1328–35.
33. Thun MJ, Peto R, Lopez AD, et al. Alcohol consumption and mortality among middle-aged and elderly U.S. adults. *N Engl J Med* 1997;**337**(24):1705–14.
34. Hvidtfeldt UA, Tolstrup JS, Jakobsen MU, et al. Alcohol intake and risk of coronary heart disease in younger, middle-aged, and older adults. *Circulation* 2010;**121**(14):1589–97.
35. Mukamal KJ, Chiuve SE, Rimm EB. Alcohol consumption and risk for coronary heart disease in men with healthy lifestyles. *Arch Intern Med* 2006;**166**(19):2145–50.
36. Beulens JW, Rimm EB, Ascherio A, et al. Alcohol consumption and risk for coronary heart disease among men with hypertension. *Ann Intern Med* 2007;**146**(1):10–9.
37. Britton KA, Gaziano JM, Sesso HD, et al. Relation of alcohol consumption and coronary heart disease in hypertensive male physicians (from the Physicians' Health Study). *Am J Cardiol* 2009;**104**(7):932–5.
38. Behrens G, Leitzmann MF, Sandin S, et al. The association between alcohol consumption and mortality: the Swedish women's lifestyle and health study. *Eur J Epidemiol* 2011;**26**(2): 81–90.
39. Stampfer MJ, Colditz GA, Willett WC, et al. A prospective study of moderate alcohol consumption and the risk of coronary disease and stroke in women. *N Engl J Med* 1988; **319**(5):267–73.
40. Solomon CG, Hu FB, Stampfer MJ, et al. Moderate alcohol consumption and risk of coronary heart disease among women with type 2 diabetes mellitus. *Circulation* 2000; **102**(5):494–9.
41. Freiberg MS, Chang YF, Kraemer KL, et al. Alcohol consumption, hypertension, and total mortality among women. *Am J Hypertens* 2009;**22**(11):1212–8.
42. Rajpathak SN, Freiberg MS, Wang C, et al. Alcohol consumption and the risk of coronary heart disease in postmenopausal women with diabetes: Women's Health Initiative Observational Study. *Eur J Nutr* 2010;**49**(4):211–8.
43. Bazzano LA, Gu D, Reynolds K, et al. Alcohol consumption and risk of coronary heart disease among Chinese men. *Int J Cardiol* 2009;**135**(1):78–85.
44. Yuan JM, Ross RK, Gao YT, et al. Follow up study of moderate alcohol intake and mortality among middle aged men in Shanghai, China. *BMJ* 1997;**314**(7073):18–23.
45. Inoue M, Nagata C, Tsuji I, et al. Research Group for the Development and Evaluation of Cancer Prevention Strategies in Japan. Impact of alcohol intake on total mortality and mortality from major causes in Japan: a pooled analysis of six large-scale cohort studies. *J Epidemiol Comm Health* 2012;**66**(5):448–56.
46. Ikehara S, Iso H, Toyoshima H, et al. Japan Collaborative Cohort Study Group Alcohol consumption and mortality from stroke and coronary heart disease among Japanese men and women: the Japan Collaborative Cohort Study. *Stroke* 2008;**39**(11):2936–42.
47. Beulens JW, Algra A, Soedamah-Muthu SS, et al. Alcohol consumption and risk of recurrent cardiovascular events and mortality in patients with clinically manifest vascular disease and diabetes mellitus: the Second Manifestations of ARTerial (SMART) Disease study. *Atherosclerosis* 2010;**212**(1):281–6.
48. Brügger-Andersen T, Pönitz V, Snapinn S, et al. OPTIMAAL study group. Moderate alcohol consumption is associated with reduced long-term cardiovascular risk in patients

following a complicated acute myocardial infarction. *Int J Cardiol* 2009;**133**(2):229–32.
49. Janszky I, Ljung R, Ahnve S, et al. Alcohol and long-term prognosis after a first acute myocardial infarction: the SHEEP study. *Eur Heart J* 2008;**29**(1):45–53.
50. de Lorgeril M, Salen P, Martin JL, et al. Wine drinking and risks of cardiovascular complications after recent acute myocardial infarction. *Circulation* 2002;**106**(12):1465–9.
51. Janszky I, Mukamal KJ, Orth-Gomér K, et al. Alcohol consumption and coronary atherosclerosis progression: the Stockholm Female Coronary Risk Angiographic Study. *Atherosclerosis* 2004;**176**(2):311–9.
52. Trevisan M, Dorn J, Falkner K, et al. Drinking pattern and risk of non-fatal myocardial infarction: a population-based case-control study. *Addiction* 2004;**99**(3):313–22.
53. McElduff P, Dobson AJ. How much alcohol and how often? Population based case-control study of alcohol consumption and risk of a major coronary event. *BMJ* 1997;**314**(7088):1159–64.
54. Hammar N, Romelsjö A, Alfredsson L. Alcohol consumption, drinking pattern and acute myocardial infarction. A case referent study based on the Swedish Twin Register. *J Intern Med* 1997;**241**(2):125–31.
55. Snow WM, Murray R, Ekuma O, et al. Alcohol use and cardiovascular health outcomes: a comparison across age and gender in the Winnipeg Health and Drinking Survey Cohort. *Age Ageing* 2009;**38**(2):206–12.
56. Ruidavets JB, Ducimetière P, Evans A, et al. Patterns of alcohol consumption and ischaemic heart disease in culturally divergent countries: the Prospective Epidemiological Study of Myocardial Infarction (PRIME). *BMJ* 2010;**341**:c6077.
57. Shaper AG, Wannamethee SG. Alcohol intake and mortality in middle aged men with diagnosed coronary heart disease. *Heart* 2000;**83**(4):394–9.
58. Mukamal KJ, Maclure M, Muller JE, et al. Binge drinking and mortality after acute myocardial infarction. *Circulation* 2005;**112**(25):3839–45.
59. Corrao G, Rubbiati L, Bagnardi V, et al. Alcohol and coronary heart disease: a meta-analysis. *Addiction* 2000;**95**(10):1505–23.
60. Bagnardi V, Zatonski W, Scotti L, et al. Does drinking pattern modify the effect of alcohol on the risk of coronary heart disease? Evidence from a meta-analysis. *J Epidemiol Comm Health* 2008;**62**(7):615–9.
61. Roerecke M, Rehm J. Irregular heavy drinking occasions and risk of ischemic heart disease: a systematic review and meta-analysis. *Am J Epidemiol* 2010;**171**(6):633–44.
62. Sempos CT, Rehm J, Wu T, et al. Average volume of alcohol consumption and all-cause mortality in African Americans: the NHEFS cohort. *Alcohol Clin Exp Res* 2003;**27**(1):88–92.
63. Malyutina S, Bobak M, Kurilovitch S, et al. Relation between heavy and binge drinking and all-cause and cardiovascular mortality in Novosibirsk, Russia: a prospective cohort study. *Lancet* 2002;**360**(9344):1448–54.
64. Deev A, Shestov D, Abernathy J, et al. Association of alcohol consumption to morality in middle-aged U.S. and Russian men and women. *Ann Epidemiol* 1998;**8**(3):147–53.
65. Hart CL, Smith GD, Hole DJ, et al. Alcohol consumption and mortality from all causes, coronary heart disease, and stroke: results from a prospective cohort study of Scottish men with 21 years of follow up. *BMJ* 1999;**318**(7200):1725–9.
66. Joshi P, Islam S, Pais P, et al. Risk factors for early myocardial infarction in South Asians compared with individuals in other countries. *JAMA* 2007;**297**(3):286–94.

67. Roy A, Prabhakaran D, Jeemon P, et al. Sentinel Surveillance in Industrial Populations Study Group. Impact of alcohol on coronary heart disease in Indian men. *Atherosclerosis* 2010;**210**(2):531–5.
68. Sun W, Schooling CM, Chan WM, et al. Moderate alcohol use, health status, and mortality in a prospective Chinese elderly cohort. *Ann Epidemiol* 2009;**19**(6):396–403.
69. Zhou X, Li C, Xu W, et al. Relation of alcohol consumption to angiographically proved coronary artery disease in chinese men. *Am J Cardiol* 2010;**106**(8):1101–3.
70. McClelland RL, Bild DE, Burke GL, et al. Alcohol and coronary artery calcium prevalence, incidence, and progression: results from the Multi-Ethnic Study of Atherosclerosis (MESA). *Am J Clin Nutr* 2008;**88**(6):1593–601.
71. Pletcher MJ, Varosy P, Kiefe CI, et al. Alcohol consumption, binge drinking, and early coronary calcification: findings from the Coronary Artery Risk Development in Young Adults (CARDIA) Study. *Am J Epidemiol* 2005;**161**(5):423–33.
72. Juonala M, Viikari JS, Kähönen M, et al. Alcohol consumption is directly associated with carotid intima-media thickness in Finnish young adults: the Cardiovascular Risk in Young Finns Study. *Atherosclerosis* 2009;**204**(2):e93–8.
73. Shaper AG, Wannamethee SG. Alcohol intake and mortality in middle aged men with diagnosed coronary heart disease. *Heart* 2000;**83**(4):394–9.
74. Fillmore KM, Stockwell T, Chikritzhs T, et al. Moderate alcohol use and reduced mortality risk: systematic error in prospective studies and new hypotheses. *Ann Epidemiol* 2007; **17**(Suppl 5):S16–23.
75. Wood GB. *A treatise on the practice of medicine*. 4th ed. Philadelphia: Lippincott, Grambo and Co; 1885.
76. Friedreich N. *Handbuch der speziellen pathologie und therapie, fifth section: die krankheiten des herzens*. Erlangen: Verlag von Ferdinand Enke; 1861.
77. Bollinger O. *Ueber die haufigkeit und ursachen der idiopathischen herzhypertrophie in munchen. Deutsch Med Wchnschr* 1884;**10**:180–1.
78. Walshe WH. *Diseases of the heart and great vessels*. 4th ed. London: Smith, Elder and Co; 1873.
79. Royal Commission Appointed to Inquire into Arsenical Poisoning from the consumption of beer and other articles of food or drink. Final report. Part I. Wyman and Sons; 1903.
80. MacKenzie J. *The study of the pulse*. Edinburgh and London: Y. J. Pentland; 1902.
81. Aalsmeer WC, Wenckebach KF. *Herz und Kreislauf bei der Beri-Beri Krankheit. Wien Arch Inn Med* 1929;**16**:193–272.
82. Brigden W. Uncommon myocardial diseases; the non-coronary cardiomyopathies. *Lancet* 1957;**273**(7007):1179–84.
83. Gubbay ER. Beri-Beri heart disease. *Can Med Assoc J* 1966;**95**(1):21–7.
84. Kasper EK, Agema WR, Hutchins GM, et al. The causes of dilated cardiomyopathy: a clinicopathologic review of 673 consecutive patients. *J Am Coll Cardiol* 1994;**23**(3): 586–90.
85. Segel LD, Klausner SC, Gnadt JT, et al. Alcohol and the heart. *Med Clin North Am* 1984;**68**(1):147–61.
86. Urbano-Márquez A, Estruch R, Fernández-Solà J, et al. The greater risk of alcoholic cardiomyopathy and myopathy in women compared with men. *JAMA* 1995;**274** (2):149–54.
87. Davidson DM. Cardiovascular effects of alcohol. *West J Med* 1989;**151**(4):430–9.
88. Estruch R, Fernández-Solà J, Sacanella E, et al. Relationship between cardiomyopathy and liver disease in chronic alcoholism. *Hepatol* 1995;**22**(2):532–8.

89. Fernández-Solà J, Estruch R, Nicolás JM, et al. Comparison of alcoholic cardiomyopathy in women versus men. *Am J Cardiol* 1997;**80**(4):481–5.
90. Kelbaek H, Gjørup T, Brynjolf I, et al. Acute effects of alcohol on left ventricular function in healthy subjects at rest and during upright exercise. *Am J Cardiol* 1985;**55**(1):164–7.
91. Padilla H, Michael Gaziano J, Djoussé L. Alcohol consumption and risk of heart failure: a meta-analysis. *Phys Sportsmed* 2010;**38**(3):84–9.
92. Klatsky AL, Chartier D, Udaltsova N, et al. Alcohol drinking and risk of hospitalization for heart failure with and without associated coronary artery disease. *Am J Cardiol* 2005;**96**(3):346–51.
93. Lazarević AM, Nakatani S, Nesković AN, et al. Early changes in left ventricular function in chronic asymptomatic alcoholics: relation to the duration of heavy drinking. *J Am Coll Cardiol* 2000;**35**(6):1599–606.
94. Wilke A, Kaiser A, Ferency I, et al. Alcohol and myocarditis. *Herz* 1996;**21**(4):248–57.
95. Lee WK, Regan TJ. Alcoholic cardiomyopathy: is it dose-dependent? *Congest Heart Fail* 2002;**8**(6):303–6.
96. Piano MR. Alcoholic cardiomyopathy: incidence, clinical characteristics, and pathophysiology. *Chest* 2002;**121**(5):1638–50.
97. Fernández-Solà J, Fatjó F, Sacanella E, et al. Evidence of apoptosis in alcoholic cardiomyopathy. *Hum Pathol* 2006;**37**(8):1100–10.
98. Ren J, Wold LE. Mechanisms of alcoholic heart disease. *Ther Adv Cardiovasc Dis* 2008;**2**(6):497–506.
99. Zhang X, Li SY, Brown RA, Ren J. Ethanol and acetaldehyde in alcoholic cardiomyopathy: from bad to ugly en route to oxidative stress. *Alcohol* 2004;**32**(3):175–86.
100. Cheng CP, Cheng HJ, Cunningham C, et al. Angiotensin II type 1 receptor blockade prevents alcoholic cardiomyopathy. *Circulation* 2006;**114**(3):226–36.
101. Kesteloot H, Roelandt J, Willems J, et al. An enquiry into the role of cobalt in the heart disease of chronic beer drinkers. *Circulation* 1968;**37**(5):854–64.
102. Fernández-Solà J, Nicolás JM, Oriola J, et al. Angiotensin-converting enzyme gene polymorphism is associated with vulnerability to alcoholic cardiomyopathy. *Ann Intern Med* 2002;**137**(5 Part 1):321–6.
103. Mathews Jr EC, Gardin JM, Henry WL, et al. Echocardiographic abnormalities in chronic alcoholics with and without overt congestive heart failure. *Am J Cardiol* 1981;**47**(3):570–8.
104. Evans W. The electrocardiogram of alcoholic cardiomyopathy. *Br Heart J* 1959;**21**(4):445–56.
105. Brigden W, Robinson J. Alcoholic heart disease. *Br Med J* 1964;**2**(5420):1283–9.
106. Ettinger PO, Wu CF, De La Cruz Jr C, et al. Arrhythmias and the "holiday heart": alcohol-associated cardiac rhythm disorders. *Am Heart J* 1978;**95**(5):555–62.
107. Koskinen P, Kupari M, Leinonen H, et al. Alcohol and new onset atrial fibrillation: a case-control study of a current series. *Br Heart J* 1987;**57**(5):468–73.
108. Djoussé L, Levy D, Benjamin EJ, et al. Long-term alcohol consumption and the risk of atrial fibrillation in the Framingham Study. *Am J Cardiol* 2004;**93**(6):710–3.
109. Frost L, Vestergaard P. Alcohol consumption and the risk of atrial fibrillation or flutter: secondary publication. a cohort study. *Ugeskr Laeger* 2005;**167**(35):3308–10.
110. Mukamal KJ, Tolstrup JS, Friberg J, et al. Alcohol consumption and risk of atrial fibrillation in men and women: the Copenhagen City Heart Study. *Circulation* 2005;**112**(12):1736–42.

111. Conen D, Tedrow UB, Cook NR, et al. Alcohol consumption and risk of incident atrial fibrillation in women. *JAMA* 2008;**300**(21):2489–96.
112. Kodama S, Saito K, Tanaka S, et al. Alcohol consumption and risk of atrial fibrillation: a meta-analysis. *J Am Coll Cardiol* 2011;**57**(4):427–36.
113. Marcus GM, Smith LM, Whiteman D, et al. Alcohol intake is significantly associated with atrial flutter in patients under 60 years of age and a shorter right atrial effective refractory period. *Pacing Clin Electrophysiol* 2008;**31**(3):266–72.
114. Cohen EJ, Klatsky AL, Armstrong MA. Alcohol use and supraventricular arrhythmia. *Am J Cardiol* 1988;**62**(13):971–3.
115. Koskinen P, Kupari M. Alcohol consumption of patients with supraventricular tachyarrhythmias other than atrial fibrillation. *Alcohol* 1991;**26**(2):199–206.
116. George A, Figueredo VM. Alcohol and arrhythmias: a comprehensive review. *J Cardiovasc Med (Hagerstown)* 2010;**11**(4):221–8.
117. Singer K, Lundberg WB. Ventricular arrhythmias associated with the ingestion of alcohol. *Ann Intern Med* 1972;**77**(2):247–8.
118. Greenspon AJ, Schaal SF. The "holiday heart": electrophysiologic studies of alcohol effects in alcoholics. *Ann Intern Med* 1983;**98**(2):135–9.
119. Dyer AR, Stamler J, Paul O, et al. Alcohol consumption, cardiovascular risk factors, and mortality in two Chicago epidemiologic studies. *Circulation* 1977;**56**(6):1067–74.
120. Wannamethee G, Shaper AG. Alcohol and sudden cardiac death. *Br Heart J* 1992; **68**(5):443–8.
121. Lithell H, Aberg H, Selinus I, Hedstrand H. Alcohol intemperance and sudden death. *Br Med J (Clin Res Ed)* 1987;**294**(6585):1456–8.
122. Fraser GE, Upsdell M. Alcohol and other discriminants between cases of sudden death and myocardial infarction. *Am J Epidemiol* 1981;**114**(4):462–76.
123. Kozarevic D, Vojvodic N, Gordon T, et al. Drinking habits and death. The Yugoslavia cardiovascular disease study. *Int J Epidemiol* 1983;**12**(2):145–50.
124. Gordon T, Kannel WB. Drinking habits and cardiovascular disease: the Framingham study. *Am Heart J* 1983;**105**(4):667–73.
125. Albert CM, Manson JE, Cook NR, et al. Moderate alcohol consumption and the risk of sudden cardiac death among U.S. male physicians. *Circulation* 1999;**100**(9):944–50.
126. Chiuve SE, Rimm EB, Mukamal KJ, et al. Light-to-moderate alcohol consumption and risk of sudden cardiac death in women. *Heart Rhythm* 2010;**7**(10):1374–80.
127. Perman ES. The effect of ethyl alcohol on the secretion from the adrenal medulla in man. *Acta Physiol Scand* 1958;**44**(3-4):241–7.
128. Koskinen P, Virolainen J, Kupari M. Acute alcohol intake decreases short-term heart rate variability in healthy subjects. *Clin Sci (Lond)* 1994;**87**(2):225–30.
129. Aasebø W, Erikssen J, Jonsbu J, et al. ECG changes in patients with acute ethanol intoxication. *Scand Cardiovasc J* 2007;**41**(2):79–84.
130. Altura BM, Altura BT, Carella A. Ethanol produces coronary vasospasm: evidence for a direct action of ethanol on vascular muscle. *Br J Pharmacol* 1983;**78**(2):260–2.
131. Mäki T, Heikkonen E, Härkönen T, et al. Reduction of lymphocytic beta-adrenoceptor level in chronic alcoholics and rapid reversal after ethanol withdrawal. *Eur J Clin Invest* 1990;**20**(3):313–6.
132. Bär KJ, Boettger MK, Boettger S, et al. Reduced baroreflex sensitivity in acute alcohol withdrawal syndrome and in abstained alcoholics. *Drug Alcohol Depend* 2006;**85**(1): 66–74.

133. Cunha DF, Monteiro JP, Ortega LS, et al. Serum electrolytes in hospitalized pellagra alcoholics. *Eur J Clin Nutr* 2000;**54**(5):440–2.
134. Yokoyama A, Ishii H, Takagi T, et al. Prolonged QT interval in alcoholic autonomic nervous dysfunction. *Alcohol Clin Exp Res* 1992;**16**(6):1090–2.
135. Genovesi S, Prata Pizzala DM, Pozzi M, et al. QT interval prolongation and decreased heart rate variability in cirrhotic patients: relevance of hepatic venous pressure gradient and serum calcium. *Clin Sci (Lond)* 2009;**116**(12):851–9.
136. Malpas SC, Whiteside EA, Maling TJ. Heart rate variability and cardiac autonomic function in men with chronic alcohol dependence. *Br Heart J* 1991;**65**(2):84–8.
137. Fauchier L. Alcoholic cardiomyopathy and ventricular arrhythmias. *Chest* 2003;**123**(4): 1320.
138. Klatsky AL, Friedman GD, Siegelaub AB, et al. Alcohol consumption and blood pressure Kaiser-Permanente Multiphasic Health Examination data. *N Engl J Med* 1977;**296**(21): 1194–200.
139. Nakanishi N, Yoshida H, Nakamura K, et al. Alcohol consumption and risk for hypertension in middle-aged Japanese men. *J Hypertens* 2001;**19**(5):851–5.
140. Thadhani R, Camargo Jr CA, Stampfer MJ, et al. Prospective study of moderate alcohol consumption and risk of hypertension in young women. *Arch Intern Med* 2002;**162**(5): 569–74.
141. Taylor B, Irving HM, Baliunas D, et al. Alcohol and hypertension: gender differences in dose-response relationships determined through systematic review and meta-analysis. *Addiction* 2009;**104**(12):1981–90.
142. Fuchs FD, Chambless LE, Whelton PK, et al. Alcohol consumption and the incidence of hypertension: the Atherosclerosis Risk in Communities Study. *Hypertension* 2001;**37**(5): 1242–50.
143. Sesso HD, Cook NR, Buring JE, et al. Alcohol consumption and the risk of hypertension in women and men. *Hypertension* 2008;**51**(4):1080–7.
144. Núñez-Córdoba JM, Martínez-González MA, Bes-Rastrollo M, et al. Alcohol consumption and the incidence of hypertension in a Mediterranean cohort: the SUN study. *Rev Esp Cardiol* 2009;**62**(6):633–41.
145. Potter JF, Beevers DG. Pressor effect of alcohol in hypertension. *Lancet* 1984; **1**(8369):119–22.
146. Puddey IB, Beilin LJ, Vandongen R, et al. Evidence for a direct effect of alcohol consumption on blood pressure in normotensive men. A randomized controlled trial. *Hypertension* 1985;**7**(5):707–13.
147. Rakic V, Puddey IB, Burke V, et al. Influence of pattern of alcohol intake on blood pressure in regular drinkers: a controlled trial. *J Hypertens* 1998;**16**(2):165–74.
148. Xin X, He J, Frontini MG, et al. Effects of alcohol reduction on blood pressure: a meta-analysis of randomized controlled trials. *Hypertension* 2001;**38**(5):1112–7.
149. Arkwright PD, Beilin LJ, Vandongen R, et al. The pressor effect of moderate alcohol consumption in man: a search for mechanisms. *Circulation* 1982;**66**(3):515–9.
150. Matsumoto C, Tomiyama H, Yamada J, et al. Association of blood pressure levels with the effects of alcohol intake on the vasculature in Japanese men. *Hypertens Res* 2009; **32**(2):127–32.
151. Stewart SH, Oroszi G, Randall PK, et al. COMT genotype influences the effect of alcohol on blood pressure: results from the COMBINE study. *Am J Hypertens* 2009; **22**(1):87–91.

152. Hashimoto Y, Nakayama T, Futamura A, et al. Relationship between genetic polymorphisms of alcohol-metabolizing enzymes and changes in risk factors for coronary heart disease associated with alcohol consumption. *Clin Chem* 2002;**48**(7):1043–8.
153. Reynolds K, Lewis B, Nolen JD, et al. Alcohol consumption and risk of stroke: a meta-analysis. *JAMA* 2003;**289**(5):579–88.
154. Mukamal KJ, Ascherio A, Mittleman MA, et al. Alcohol and risk for ischemic stroke in men: the role of drinking patterns and usual beverage. *Ann Intern Med* 2005;**142**(1):11–9.
155. Sacco RL, Elkind M, Boden-Albala B, et al. The protective effect of moderate alcohol consumption on ischemic stroke. *JAMA* 1999;**281**(1):53–60.
156. Stampfer MJ, Colditz GA, Willett WC, et al. A prospective study of moderate alcohol consumption and the risk of coronary disease and stroke in women. *N Engl J Med* 1988; **319**(5):267–73.
157. Hillbom M, Numminen H, Juvela S. Recent heavy drinking of alcohol and embolic stroke. *Stroke* 1999;**30**(11):2307–12.
158. Sundell L, Salomaa V, Vartiainen E, et al. Increased stroke risk is related to a binge-drinking habit. *Stroke* 2008;**39**(12):3179–84.
159. U.S. Department of Agriculture and U.S. Department of Health and Human Services. *Dietary Guidelines for Americans*. 7th Ed. U.S. Government Printing Office; 2010.
160. National Health and Medical Research Council (NHMRC). *Australian Guidelines to Reduce Health Risks from Drinking Alcohol*. Commonwealth of Australia; 2009.

Chapter 14

The Effects of Active and Passive Smoking and Cardiovascular Disease

Mahmood Ahmad,[1,2] Emmanuel Selvaraj,[3] and Ramachandran Meenakshisundaram[4]
[1]*Tahir Heart Institute, Rabwah, Pakistan,*
[2]*Darent Valley Hospital, Dartford and Gravesham NHS Trust, Dartford, United Kingdom*
[3]*London Chest Hospital, Barts Health NHS Trust, London, United Kingdom*
[4]*Epsom and St. Helier University Hospitals NHS Trust, Epsom, United Kingdom*

14.1 INTRODUCTION

Tobacco smoking is prevalent throughout the world; however, there has been a rapid rise in the prevalence of tobacco smoking because of industrialization and modernization. Among the variety of ways to use tobacco, cigarette smoking is the most common preferred method. Smoking is one of the major modifiable risk factors for coronary artery disease and a significant percentage of it worldwide is due to smoking. One in 10 deaths from cardiovascular disease can be attributed to smoking.[1] This effect persists even if the patient's cholesterol is well under control.[2]

Smoking is the largest risk factor for coronary artery disease for those below the age of 45 and in women more than men.[3] It is not clear why it becomes less of a risk factor as patient's age, but this may be due to the fact that smokers die much younger than nonsmokers. The number of cigarettes smoked per day is significantly correlated to the risk of cardiovascular events.[4] Tobacco smoke contains many chemicals and carcinogens, thus it affects almost all the organs. Also, cigarette smoking imposes a heavy economic burden. However, considering the scope of this chapter and book, we have concentrated only on tobacco smoke and cardiovascular health. Smoking cessation significantly reduces the number of cardiovascular events that occur over time[5] and the benefit is greater for women than it is for men.[3]

The Heart and Toxins. DOI: http://dx.doi.org/10.1016/B978-0-12-416595-3.00014-1

14.2 SMOKING AND ENDOTHELIAL DYSFUNCTION

Endothelium lines all the blood vessels and lymphatics with the capillaries being made up entirely from endothelial cells. Smokers have an impaired ability to regenerate endothelium, as they have a reduced number of circulating endothelial progenitor cells.[6] Areas of disturbed flow, such as bifurcations, tend to have higher rates of endothelial turnover and the damage caused by smoking may affect these areas more significantly.[7] Endothelial dysfunction can also be caused by dyslipidemia, hypertension, and diabetes through similar mechanisms. The endothelium produces a number of factors that cause vascular hemostasis, including small molecules, proteins, and lipids that cause activation or relaxation of the underlying endothelium.[7]

Tobacco smoking affects the vasodilator substances produced by endothelial cells. These include nitric oxide (NO), prostacyclin, atrial natriuretic peptide (ANP),[8] adrenomedullin,[9] and the endothelium-deprived hyperpolarizing factor.[10] These vasodilator substances also oppose processes such as inflammation, cell proliferation and atherogenesis. Instead, there is increased production of endothelin and thromboxane.[11] The endothelium is significantly affected by decreased NO levels, increased level of the inhibitors of NO synthase (NOS), decreased NOS expression, increased production of oxygen radicals, and inflammatory reactions caused by cigarette smoking.[12,13]

Nitric oxide itself is partially reduced by free oxygen radicals, large numbers of which are found in both the tar and gas phase of smoking. Nitric oxide causes vasodilation and the inhibition of the adhesion of platelets and leucocytes to vessel walls.[14] This in turn reduces the risk of thrombosis and atherosclerosis. The endothelium-deprived relaxation factor (EDRF), which was previously thought to be an unknown factor causing vascular relaxation, is now known to be NO. This is reduced by free radicals at a very rapid rate. NO is stabilized by superoxide dismutase. The effect of NO on vasodilation is through the stimulation of guanylate cyclase to cyclic guanosine monophosphate.[15] Smoking causes impairment of endothelial NOS (eNOS), which results in atherosclerosis, hypertension, and impairment of blood flow.[16] There was a significant increase in the vascular resistance with animal models if they were exposed to antagonists of eNOS. There are increased adherence and infiltration of mononuclear cells into the intima, which leads to rapid formation of atherosclerotic lesions and aneurysmal changes in animal models with low NOS levels.[17]

There may be a lack of substrate or cofactors for eNOS; thus, it causes decreased expression of eNOS. Glu298Asp is one of the eNOS polymorphisms that makes patients even more sensitive to cigarette smoking.[18] Also, it has been linked with the reduced ability to form collaterals in patients with chronic coronary artery occlusions.[19] Patients with homozygous eNOS alleles have a greater risk of congestive heart disease if they smoke.[20] Smokers have high levels of thiocyanate (SCN^-), which is a myeloperoxidase (MPO) substrate that causes hypothiocyanous acid (HOSCN) formation. This is turn targets eNOS and thus causes lower levels of NO. Rat aortic rings exposed to HOSCN had both lower levels of NO and cyclic GMP guanosine monophosphate.[21] It is notable that eNOS$^{-/-}$ mice who have been exposed to cigarette smoking for six months also develop pulmonary hypertension secondary to the radical nitrogen species from cigarette smoking.[22]

Smoking enhances the increased blunting of the flow-mediated vasodilation found in a patient's brachial arteries as measured by vasodilation of them. This phenomenon has been found in both smokers and passive smokers.[23] The effect on endothelial dysfunction caused by a reduction in NO levels is also found in patients with heart failure.[24] Low levels of nitric oxide appear to decrease the blood flow in coronary arteries.[25] It has also been found that mice that lack all isoforms of NO synthase have a significant risk of having a spontaneous myocardial infarction (MI),[26] and in animal models increased NO in vessel walls prevents further atherosclerosis.[27]

14.3 SMOKING AND OXIDATIVE STRESS

Increased creation of the reactive oxygen species disturbs the equilibrium within the body and causes increased oxidation through a phenomenon called oxidative stress. Smoking increases the stress by following pathways. The uncoupling of eNOS, nicotinamide adenine dinucleotide (NAD) plus hydrogen (NADH/NADPH) oxidase, and xanthine oxidase are all causes of this increased reactive oxygen species production. These reactive oxygen species are free radicals; they include hydrogen peroxide, lipid radicals, hydroxyl radicals, superoxide anion, hypochlorous acid, and peroxinitrate (formed by the superoxide anion combining with NO).[28] The oxygen radicals can attack polyunsaturated fatty acids causing a fatty acid peroxyl radical; this molecule further creates unsaturated aldehydes with cytotoxic effects. Also, lipid radicals can react with NO to form less reactive secondary nitrogen-containing products (LONO and LOONO).

Oxidized low-density lipoprotein (LDL) itself causes impaired endothelial relaxation. Asymmetric dimethylarginine (ADMA), which is an endogenous competitive inhibitor of eNOS, is produced during degradation of methylated nuclear proteins and is also a strong inhibitor of NO.[29] ADMA is degraded by the enzyme dimethylarginine dimethylaminohydrolase, which is rendered less effective by oxidative stress. Antioxidant vitamins are also reduced by smoking because they are consumed in response to excess oxidative stress induced by smoking. Moreover, nicotine causes injury to endothelial cells and the number of endothelial cells in the blood, which is a strong indicator of endothelial damage and is increased in smokers.[30] Oral nicotine alone has been found to cause increased excessive thickening of the aortic myointima after injury in rats, which also indicates damage to the endothelium,[31] and

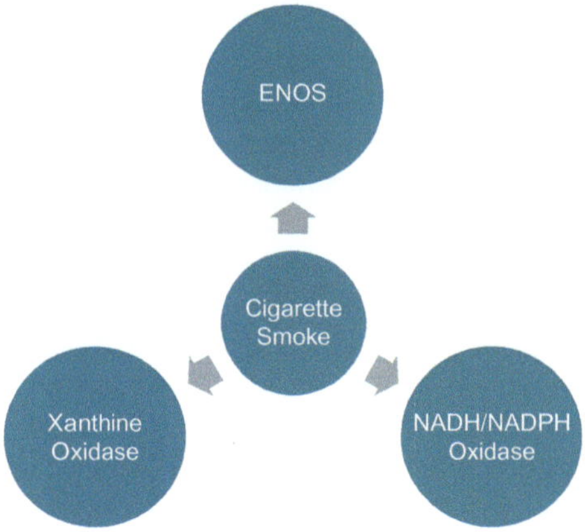

FIGURE 14.1 **Graphic of smoking and oxidative stress.** ENOS, endothelial nitric oxide synthase.

nicotine also increases the levels of transforming growth factor (TGF) β1 and the fibroblast growth factor in in vitro studies.[32]

Xanthine oxidoreductase exists in two interconvertible forms—xanthine oxidase, which increases production of O_2^- and H_2O_2 from molecular oxygen, and santhine dehydrogenase, which reduces NAD^+. Xanthine oxidase has been found to be upregulated in smokers. Xanthine oxidoreductase metabolizes the oxidation of hypoxanthine and xanthine. O_2^- deprived from xanthine oxidase reduces NO bioavailability. Its inhibitor decreases the level of free radicals in rats. Xanthine oxidase can also exist in a molybdenum-deficient state. It uses NADH rather than xanthine as a substrate, which causes the creation of O_2^-. NADH and NADPH oxidase are the main producers of superoxides. Cytokines, hormones, and mechanical forces all have an effect on their expression.[33] Figure 14.1 illustrates the effect on smoking.

14.4 ATHEROSCLEROSIS AND SMOKING

Smoking significantly accelerates the onset of coronary artery disease presentation in smokers, with the average person developing coronary artery calcification, a sensitive marker for the early development of coronary artery disease, 10 years earlier than a nonsmoker.[34] Smoking mainly causes endothelial damage through a number of mechanisms that predispose to increased atherosclerosis. Smoking causes damage to the endothelial cells caused by low levels of NO. This damage is further increased by more expression of macrophages and platelets, which adhere to the damaged endothelium. This in turn provokes inflammation and thrombosis in the area of the damaged endothelium. Low-density lipoprotein cholesterol (LDLc) is oxidized by the radicals from smoking. It is deposited in plaque and causes increased migration of neutrophils. The macrophages that migrate then absorb the oxidized

LDLc turning it into foam cells. Finally, smoking also causes remodeling of the tissues and has extensive prothrombotic effects in itself.[35]

Smoking causes increased inflammation as evidenced by higher levels of leucocytes, C-reactive protein (CRP), fibrinogen, and monocyte activation. CRP is an important marker for accelerated atherosclerosis and indicates increased oxidative stress, damage to DNA, and fibrogenic changes in the atherosclerotic plaques.[36,37] CRP levels slowly decrease over time after the cessation of smoking.[37]

Smoking causes increased levels of oxidized LDL and lipid peroxidation. This oxidized LDL leads to significantly increased atherogenecity. Peroxidized lipids increase the platelet-activating factor and also increases the migration and degranulation of neutrophils. In addition, the reduction of NO level increases the binding of monocytes to the vessel wall, which in turn activates the dendritic cells. Butadiene, which is another vapor-phase constituent of cigarettes, and polycyclic aromatic hydrocarbons (PAH) during smoking accelerate the atherosclerotic lesions.[38,39] Nicotine also causes increased angiogenesis through expression of cytokines and endothelial growth factors. Nicotine has effects on the endothelium mediated through the nicotinic acetylcholine receptor (nACh), which allows all these angiogenic factors to express their effects.[40] Thus, the atherosclerotic process is accelerated through multiple pathways.[41]

14.5 SMOKING AND THROMBOSIS

Smoking causes increased vascular adhesion molecules such as the intercellular adhesion molecule-1 (ICAM-1), the vascular cell adhesion molecule (VCAM), and E-selectin in circulation.[42] These molecules increase the adhesion of leucocytes and their infiltration into the vessel wall. Increased levels of ICAM and VCAM are associated with a higher cardiovascular risk.[43] Smoking increases the secretion of the von Willebrand factor (VWF) and tissue factor (TF), which in turn cause platelet adhesion.[44] Increased levels of VWF and TF have been observed in atherosclerotic plaques in smokers.[45,46] In smokers, the tissue plasminogen activator (tPA) level is decreased and the plasminogen activator inhibitor-1 (PAI-1) level is increased, which impairs the ability to lyse thrombus and raises the risk of a cardiovascular event.[47]

Smoking causes an increased risk of sudden death and myocardial infarction (MI) rather than angina. The risk of sudden cardiac death is increased by 2.5 times in smokers, and smokers present with a coronary event, such as an MI, in younger age groups than nonsmokers.[48] Thrombolysis also works better in these patients, which indicates that the cause of myocardial infarction in this group is more likely to be a thrombus that has expanded quickly.[49] In a primary angioplasty in myocardial infarction (PAMI) study, it has been demonstrated that mortality and recurrent acute MI are higher in smokers than nonsmokers who were treated with primary angioplasty.[49] Thrombosis is caused by the conversion of fibrinogen to fibrin. This process is dependent on endothelium, platelets, proteins in the plasma, and vascular tissue. Smoking increases the replacement of platelets and increases the aggregation and activation of platelets. Increased levels of thromboxane and its byproducts are noted in smokers, which indicates the platelet activation; however, thromboxane related to the cyclooxygenase-1 (COX-1) pathway is suppressed by aspirin.[50]

Nitric oxide decreases platelet recruitment to thrombus, and thus its suppression enhances thrombus formation in smokers. The levels of nitric oxide are reduced in smokers who have myocardial infarction compared to patients who have had a myocardial infarction and do not smoke.[51] Increased tissue factor, which is significantly thrombogenic, is secreted from the vascular smooth muscle in smokers. Platelets are oxidized themselves and show increased aggregation under shear stress,[52] and oxidized platelets decrease the expression of glycoprotein 1b, a receptor for VWF.[53]

Fibrinogen levels are significantly increased in smokers and higher fibrinogen levels are associated with a rise in cardiovascular risk.[54] Tyrosine residues are nitrated in smokers, which indicates NO-dependent damage. Fibrinogen itself is nitrated by tyrosine residues, which causes increased thrombogenicity.[55]

14.6 DYSLIPIDEMIA AND SMOKING

Smoking has a vast array of effects on lipid metabolism, including increasing adipose tissue lipid breakdown. Cigarette smoking can affect the postprandial lipid metabolism and increase triglyceride levels after meals in comparison to nonsmokers.[56] It decreases plasma lipoprotein lipase (LPL) levels. LPL is involved in the clearance of total triglyceride from triglyceride rich lipoprotein (TGRL). The cholesteryl ester transfer protein (CETP) is another enzyme that may be affected in smokers and causes the transfer of cholesterol esters from high-density lipoprotein (HDL) to other lipoproteins.[57] Smoking also decreases apolipoprotein A-1 (APO A-1) (a reduction of 5–6%), HDL (lowered by 6–9%), and LDL APO-B levels. Smoking causes an increase in triglyceride levels, very low-density lipoprotein (VLDL), LDL, and APO-B levels.

In a previous study among smokers, there was a rise in total cholesterol (TC), triglycerides (TGL), LDL, and APO-B, and a fall in HDL and APO-A; these changes were significant ($P < 0.05$).[58] The mechanisms for an altered lipid profile among smokers are[58]:

- Nicotine stimulates the release of adrenaline from the adrenal cortex leading to an increased serum concentration of free fatty acids (FFA), which further stimulates hepatic synthesis and secretion of cholesterol as well as hepatic secretion of VLDL, thus increased TGL.

- Smoking decreases estrogen levels and further leads to a decreased HDL cholesterol concentration. Also, HDL concentration has been shown to be inversely related to VLDL concentration in serum.
- Smoking increases insulin resistance and, thus, causes hyperinsulinemia. LDL, VLDL, and TGL are elevated during hyperinsulinemic conditions due to the decreased activity of lipoprotein lipase.

14.7 SMOKING AND THE BLOOD VESSELS

Cigarette smoking delivers a decreased level of antioxidants (e.g., vitamin C) and causes increased oxidation. The increased oxidation is due to nitrogen oxides and free radicals in cigarette smoke. In turn, these increase lipid peroxidation products, which are found both in urine and blood.[59] Acrolein is a reactive aldehyde produced through endogenous lipid peroxidation that damages APO A-1, which is the major protein in HDL. Acrolein causes thioredoxin 1 and 2 oxidation, and the byproducts formed have a significant antioxidant effect, which causes endothelial cell dysfunction.[60] Also, acrolein increases the production of COX-2, which is highly expressed during the atherosclerotic process.[61]

Most cigarettes smoked contain multiple metals, such as aluminum, copper, lead, mercury, nickel, zinc, and cadmium; this, especially in combination with other metals, increases the oxidation of cell proteins.[62] Cadmium deposits in the aortic wall and predisposes a person to MI and hypertension.[63] Smoking also causes inhibition of bradykinin-mediated endothelial venodilation.[64] Smoking causes increased red cell mass and hematocrit due to increased levels of carbon monoxide in the blood. It also causes increased viscosity and deformability of the blood. These hematological changes have been observed in babies born to women who were smokers,[65] and they persist for a long period even after cessation of smoking.[66]

Smoking promotes inflammatory and thromobogenic processes in the blood vessels through various proinflammatory, inflammatory, and hematological factors, as discussed previously. In addition, smoking decreases nitric oxide, which is a potent dilator and relaxer in blood vessels. Overall, smoking increases the stiffness of the blood vessels, and this stiffness decreases with smoking cessation.[67] It is a well-known fact that smoking is an independent significant risk factor for aneurysmal changes in the blood vessels. Multiple studies have confirmed abdominal aortic aneurysm (AAA) in smokers, which is directly correlated with number of packs smoked over years, and smoking increases the mortality in such patients.[68]

14.8 THE INFLUENCE OF SMOKING ON HYPERTENSION

Smoking is in itself a strong risk factor for hypertension, an effect likely mediated by its multiple effects on endothelial function, its prothrombotic effects, its inflammatory effects, and its effects on arterial stiffness (as previously discussed).[69] Nicotine stimulates the sympathetic nervous system, which in turn causes increased epinephrine and norepinephrine release from the adrenal medulla and from nerve endings. These increase heart rate, blood pressure, and myocardial contractibility (through myocardial β2 receptors). The β2 and α2 coronary receptors are affected. The α receptors are effected peripherally causing peripheral vasoconstriction. Nicotine causes constriction in the skin arterioles and dilatation in the skeletal artery beds. Also, smoking activates the renin aldosterone angiotensin (RAAS) pathway and nACh receptors, which all cause blood pressure to increase.[70]

14.9 SMOKING AND MYOCARDIAL WORKLOAD

Each cigarette contains around 1 to 2 mg of nicotine, and the half-life of the nicotine is around 2 hours. Nicotine levels go down to a trough level in about 6 to 8 hours. There is, however, a basal level of nicotine in smokers.[71] Nicotine itself causes increased catecholamine release, especially from the adrenal gland. The levels of epinephrine, norepinephrine, and enhanced 24-hour urinary excretion of catecholamines are all higher in smokers.[70] Thus, tachycardia ensues in both short-term and chronic smokers. However, the response of blood vessels to a requirement for more blood flow is impaired in smokers. As a result, the myocardial workload is increased as a compensatory process.[72]

14.10 SMOKING AS IT RELATES TO OXYGEN-CARRYING CAPACITY

In addition to the effect of smoking on blood vessels, the levels of carboxyhemoglobin are increased in smokers. Heavy smokers have 10% carboxyhemoglobin, average smokers have 5% carboxyhemoglobin, and nonsmokers have less than 2% carboxyhemoglobin in their blood.[73] Carboxyhemoglobin in the blood impairs oxygen binding to the red cells, and to overcome the effect, red cell mass and hematocrit increases. In turn, viscosity and coagulopathy are increased as a result of reduced oxygen-carrying capacity.[74]

14.11 ARRHYTHMIAS CAUSED BY SMOKING

Through multiple mechanisms described previously (i.e., atherogenesis, hypoxia, increased catecholamine release, thrombotic and inflammatory effects), smoking causes arrhythmias. Patients with chronic obstructive pulmonary disease (COPD) have high levels of carbon monoxide, which causes hypoxia, and these patients have an increased propensity to ventricular arrhythmia and left ventricular (LV) dysfunction during exercise.[75] Most COPD patients are smokers or these patients have similar carbon monoxide levels to smokers.

Carbon monoxide inhalation increases ventricular fibrillation (VF) in animal models.[76] Patients with LV dysfunction and implantable cardiac defibrillators (ICD) have been found to have much higher rates of ventricular tachycardia if they smoked rather than if they did not; this is despite the fact that patients who smoked were much younger.[77] Nicotine causes increased catecholamine release, which in turn leads to higher rates of arrhythmias. Carbon monoxide and oxygen radicals released from cigarette smoking cause hypoxia-induced arrhythmias.

The Cardiac Arrhythmia Suppression Trial (CAST), which involved 2752 patients, looked at antiarrhythmic drugs and their efficacy in the reduction of ectopic ventricular activity. A subgroup analysis of patients who continued smoking as compared to those who stopped showed that those who no longer smoked had reduced arrhythmia-induced mortality at one year; however, this finding was of marginal statistical significance. In the subgroup of patients who had untreated coronary artery disease, those who had LV dysfunction had the greatest statistically significant reduction in arrhythmias on the cessation of smoking.[78] Smoking is a risk factor for nocturnal ventricular tachycardia in patients who have poor LV function.[79]

Tsai et al. found in a large study group of 375,000 patients that areca nut, which is chewed with smokeless tobacco, was independently associated with higher levels of atrial fibrillation (AF), with an adjusted odds ratio for 1.02 (95% CI 1.00–1.04) in risk of AF prevalence.[80] The Atherosclerosis Risk in Communities (ARIC) study of 15,000 patients showed that the rate of smoking was a significant risk factor for AF, with hazard ratios (HR) of 1.32 (95% CI 1.10–1.57) in former smokers, 2.05 (95% CI 1.71–2.47) in current smokers, and 1.58 (95% CI 1.35–1.85) in people who had never smoked.[81] The consortium Cohorts for Heart and Aging Research in Genomic Epidemiology–Atrial Fibrillation (CHARGE-AF) looked at individual data for three large studies, which had 26,000 participants, and found an increased risk of AF in smokers; this was adjusted into their model to predict AF over five years.[82] Patients with AF who smoked were also found to have higher mortality rates over time in the Atrial Fibrillation Follow-Up Investigation of Rhythm Management (AFFIRM) dataset of 4060 patients, with 21% of smokers as compared to nonsmokers dying over five years.[83] Among 7115 patients undergoing coronary bypass grafting (CABG), 2270 patients were found to have a much higher risk for postoperative atrial fibrillation if they were smokers prior to surgery.[84]

14.12 CORONARY ARTERY DISEASE AND SPASM AS A RESULT OF SMOKING

As already discussed, multiple mechanisms are considered to be coronary artery disease pathogenetic factors (e.g., atherogenesis, thrombogenesis, hypertension, dyslipidemia, insulin resistance, the inflammatory process, free radical damage, effects on blood vessels, and oxidative stress). Smoking associated with a low inflammatory effect is more prone to cause angina than a MI.[85] The glutathione S-transferase M1 (GSTM1) polymorphism null genotype and certain 9p21 polymorphisms, when found in smokers, cause a significantly increased risk of coronary artery disease.[86,87] The Greek Atorvastatin and Coronary-Heart-Disease Evaluation (GREACE) study showed that the risk of atherosclerosis can be reduced in smokers by statins, but not fully counteracted.[88] It is well known that atherosclerosis is a systemic process and, thus, patients with triple vessel disease often have AAA.[89]

Many previous studies have confirmed that acute coronary syndrome occurs earlier (in an average of ~10 years) in tobacco users than in nonusers.[90–92] This is despite previous studies showing a "smoker's paradox" (smokers were found to have an improved survival rate after a myocardial infarction and a better response to clot-busting medications), a possible explanation of which is that coronary events occur in the younger age group of smokers and, thus, the likelihood of concurrent comorbidities are lower.[93]

De Luca et al. found that the infarct size of smokers was similar to that of nonsmokers in patients undergoing primary percutaneous coronary intervention (PCI); however, the ages of the patients who smoked were significantly lower than the ages of nonsmokers.[92] Also, there were increased rates of reintervention needed in the active smokers who already had revascularization therapy for coronary artery disease treated with CABG.[94] Multiple studies have confirmed that during PCI, smokers carry a high risk of in-stent thrombosis and increased mortality. These observations have been noted in the Platelet Inhibition and Patient Outcomes (PLATO)[95] and the Prevention of Restenosis with Tranilast and Its Outcomes (PRESTO) trials.[96] Notably, tobacco decreases the efficacy of antiplatelet drugs through platelet inhibition.[95,96]

Tobacco smoking is one of the most significant risk factors for coronary artery spasm in patients having normal coronary arteries confirmed by angiograms.[97] Smoking decreases the blood flow in the epicardial arteries and increases total coronary artery vascular resistance (this is mediated through α adrenergic receptors). Smoking increases the cardiac output by up to 40% acutely but reduces the blood flow response to exercise because of endothelial dysfunction. In fact, smoking causes endothelial dysfunction through multiple pathways, as previously described. In addition, all other factors related to smoking (e.g., proinflammatory, thromobogenic, free radical damage, oxidative stress) involved in the atherosclerotic process are likely to cause coronary artery spasm.[98] Smokers found significant improvement in vasospastic angina after cessation of smoking.[98]

14.13 SMOKING AND HEART FAILURE/CARDIOMYOPATHY

Hartz et al. found that smoking is a significant risk factor for developing ventricular wall motion abnormalities independent of coronary artery disease, which suggests that it can lead to left ventricular dysfunction.[99] In rats, smoking leads to ventricular dilatation in both systole and diastole, which results in LV dysfunction. Cigarette smoking causes left ventricular remodeling and is often multifactorial (e.g., inflammatory, oxidative stress, and nutritional deficiency, especially vitamin- and mineral-deficient states).[100] There was a significantly increased risk of ventricular tachycardia in patients with LV dysfunction as observed in the Multicenter Automatic Defibrillator Implantation Trial—Cardiac Resynchronization Therapy (MADIT-CRT) study, which was carried out across the patients with both ischemic and nonischemic cardiomyopathy.[77]

14.14 PERIPHERAL VASCULAR DISEASE, SUDDEN CARDIAC DEATH, AND SMOKING

Cigarette (tobacco) smoking has a significant association with peripheral vascular disease (PVD), as consistently proved in many large studies.[101,102] The risk of PVD directly correlates with smoking pack years. Arterial reocclusion rates drop significantly after PVD surgery for patients who stopped smoking. Also, such patients did not develop critical limb ischemia.[103] Smoking is also implicated in nonatherosclerotic processes such as thromboangiitis obliterans and Raynaud's phenomenon.[104,105]

As previously discussed, smoking accelerates many of the cardiovascular insults, including fatal arrhythmias. Sudden cardiac death is a major cause of cardiovascular mortality, and it has been found to be associated with cigarette smoking.[106] Smoking was also found to be a risk factor ($P = 0.06$) for sudden cardiac death (albeit not as significant as high cholesterol or preexisting heart disease) in a British study of 7735 patients.[107] A similar prospective study in Paris of 7746 men also found that smoking was an independent risk factor for sudden cardiac death.[108]

14.15 SMOKING CESSATION AND CARDIOVASCULAR RISK

Smoking cessation remains one of the most cost-effective and best methods of reducing cardiovascular risk. A number of medications such as nicotine subsitutes (e.g., nicotine gum, transdermal patches, inhalers, or sprays) or bupropion have been approved. Vareniciline, which also stimulates nicotine receptors, has also been approved but data for cardiovascular patients are not yet available.[109] Other medications, such as nortriptyline and clonidine, have also been used but are not approved by the U.S. Food and Drug Administration (FDA). However, the success rates for smoking cessation remain only about one-third. The majority of patients relapse within the first week due to strong flavor or withdrawal symptoms such as anxiety, unrelenting craving, lack of sleep, decreased concentration, and weight gain. In the Lung Health Study, nicotine replacement therapy (NRT) was found to be safe and to have no significant negative cardiovascular events compared to smokers over five years,[110] and similar observations have been noted in different trials.[111] In addition, NRT is safe for post-MI patients.[112] Nicotine replacement is widely available and all of the substitutes have equal efficacy.

Bupropion is thought to block nACh receptors and stop norepinephrine and dopamine uptake. It does, however, increase the seizure threshold in epileptics. Bupropion does not increase cardiovascular risk in smoking cessation patients.[113,114] The effectiveness of bupropion has been found in one trial to be even higher than NRT for smoking cessation, with rates of 30.3% as compared to 16.4%.[114] A second trial has shown a benefit in short-term effectiveness in improving smoking cessation (37–27%) but not long term.[113] Tonstad et al. showed no increase in cardiovascular risk but showed significant smoking cessation rates (27% with bupropion to 12% for the control group).[115]

Overall, nicotine replacement therapy has been found to double the rates of smoking cessation in a number of systematic reviews.[116] There is a growing body of evidence for the use of electronic cigarettes for smoking cessation, but their long-term effects still need to be studied.[116,117] Patients who were on NRT but stopped smoking, experience the following effects:[116]

- Significant improvement in HDL cholesterol
- Rise in the HDL to LDL ratio
- Improvement in plasma fibrinogen and hemostatic factors
- Reduction in inflammatory markers

14.16 PASSIVE SMOKING AND SMOKELESS TOBACCO

Passive smoking causes increased fibrinogen, factor VII, plasminogen activator inhibitor-1 (PAI-1), thromboxane, and thrombomodulin levels, and it has been noted in the children of smokers.[116] Passive smoking has been found to be an important risk factor for coronary artery disease.[118] The mechanism involved in passive smoking is similar to the neurohormonal, hemotological, metabolic, genetic, and biochemical processes and damages caused by active smoking. The World Health Organization (WHO) has estimated that passive smoking causes the death of 600,000 people each year, including 165,000 children,[119] including an increase in the risk of sudden infant death syndrome (SIDS).[120]

The EUROASPIRE III study of 8729 coronary artery disease patients showed that 6060 (69.4%) were nonsmokers and that 24.2% of these were passive smokers.[121] Also, it implied that exposure to passive smoking significantly reduced the rates of smoking cessation in active smokers, probably because of social and emotional support.[121] Le-Ha et al., in a study of 1754 young adults, found a significant lowering of HDL cholesterol in passive adolescent female smokers than male smokers.[122]

Public health campaigns to reduce the exposure of people to passive smoking have had some effect, with Spain's Registre Gironí del Cor (REGICOR) study showing that even partial smoke-free legislation introduced in 2006 significantly decreased acute MI incidence and mortality rates (RR = 0.89, 95% CI 0.81–0.97, and RR = 0.82, 95% CI 0.71–0.94, respectively). This effect was especially significant for women and for those ages 65 to 74.[123] Also, Germany passed laws against passive smoking in 2007 and 2008 that caused a significant reduction in rates of acute MI and angina in an insurance cohort of 3.7 million patients.[124] Smoke-free legislation in the United Kingdom in 2007 has been shown to have significantly reduced exposure to second-hand smoking by the reduction in cotinine (a constituent of cigarette smoke) levels found in saliva in National Health Survey data.[125] A meta-analysis that included 45 published studies showed that smoke-free legislation led to significantly reduced rates of admissions for cardiac events in the 33 different regions or countries in which they were enacted.[126]

Smokeless tobacco (called *snus*) is commonly used in countries such as Pakistan and Sweden. It contains many of the same chemicals as are in cigarettes including nicotine, nitrosamines, aldehydes, and PAHs; however, it can also contain betel juice, lime, areca nut, and tobacco. The evidence for smokeless tobacco causing MI is conflicting. A meta-analysis of eight Swedish studies showed no increase in MI after snus use, but did show a short-term increase in mortality if the patient had an acute myocardial infarction.[127] The amount of nicotine in cigarettes and smokeless tobacco is similar. In the INTERHEART study, the risk of an acute MI was significantly increased in patients who smoked both cigarettes and smokeless tobacco.[128] The risk of cardiovascular disease increased with smokeless tobacco and the risk correlates directly with amount used and years of smoking. The INTERHEART study also revealed that the risk of an acute MI with smokeless tobacco is significantly increased depending on the varieties of content used in it.[128]

14.17 BIOMARKERS IN SMOKING AND CARDIOVASCULAR DISEASE

As a result of the multiple pathological processes involved in smoking (as discussed previously), certain biomarkers have been released, as follows:

- Due to inflammatory processes, markers, such as interleukin-6, fibrinogen, CRP, total white cell count, and neutrophils, are elevated.
- Smoking affects coagulation factors and thus increases PAI-1, homocystiene, tPA, red cell mass, blood viscosity, P-selectin, and thromboxane A2; some of them can be measured in urine as well.

- Markers of oxidative stress include oxidized LDL, fibrinogen and F2 isoprostanes (found in blood and urine), and urine thromboxane A2 metabolites.
- Vitamin B6, vitamin C, vitamin B12, folic acid, and β-carotenes are decreased as a result of excessive consumption in response to oxidative stress.[129]
- PAH, a compound in cigarettes, is thought to contribute to atherosclerosis, and the metabolites of PAH (1-hydroxypyrene) are excreted in the urine.
- Smoking also causes significantly reduced blood levels of uric acid, which is likely secondary to oxidative stress.[130]

14.18 CONCLUSION

Smoking has significant effects through a vast array of factors released from the tobacco, and these harmful materials have many toxic effects in human beings. It needs to be kept in mind that the mechanisms behind atherosclerotic damage caused by smoking are similar to those of other cardiovascular risk factors such as hypertension, high cholesterol, and diabetes. This may be due to common pathways through which they act, such as endothelial damage (diabetes) or dysfunction, or through increased stiffening of the arteries. Smoking is an independent cause of hypertension, dyslipidemia, insulin resistance, and atherosclerosis.

There is a significant improvement in mortality and morbidity especially with regard to cardiovascular events for people who stop smoking. The safer and effective techniques of smoking cessation include NRT and use of bupropion. Like active smoking, passive smoking and smokeless tobacco also carry similar risks of cardiovascular diseases. Thus, there is a need for health campaigns to create an awareness about active and passive smoking among the public.

Key Learning Points

- Smoking causes endothelial dysfunction through multiple mechanisms such as reduction of nitric oxide level and endothelial cell damage.
- Smoking has significant prothrombotic effects through increased platelet adhesion and its effect on prothrombotic factors.
- Smoking contributes to increased atherosclerosis, and biomarkers of atherosclerosis are raised because of smoking.
- Smoking causes dyslipidemia, hypertension, and insulin resistance.
- Smoking cessation is a powerful tool to reduce the cardiovascular risk caused by it over time. Different methods, such as nicotine replacement therapy and bupropion, are effective.
- Passive smoking and smokeless tobacco carry similar risks for cardiovascular events to active smoking.

REFERENCES

1. Ezzati M, Henley SJ, Thun MJ, et al. Role of smoking in global and regional cardiovascular mortality. *Circulation* 2005;**112**(4):489–97.
2. Blanco-Cedres L, Daviglus ML, Garside DB, et al. Relation of cigarette smoking to 25-year mortality in middle-aged men with low baseline serum cholesterol: the Chicago Heart Association Detection Project in Industry. *Am J Epidemiol* 2002;**155**(4): 354–60.
3. Kawachi I, Colditz GA, Stampfer MJ, et al. Smoking cessation and time course of decreased risks of coronary heart disease in middle-aged women. *Arch Intern Med* 1994; **154**(2):169–75.
4. Whincup PH, Gilg JA, Emberson JR, et al. Passive smoking and risk of coronary heart disease and stroke: prospective study with cotinine measurement. *BMJ* 2004;**329**(7459): 200–5.
5. Suskin N, Sheth T, Negassa A, et al. Relationship of current and past smoking to mortality and morbidity in patients with left ventricular dysfunction. *J Am Coll Cardiol* 2001;**37** (6):1677–82.
6. Vasa M, Fichtlscherer S, Aicher A, et al. Number and migratory activity of circulating endothelial progenitor cells inversely correlate with risk factors for coronary artery disease. *Circ Res* 2001;**89**(1):E1–7.
7. Chang E, Harley CB. Telomere length and replicative aging in human vascular tissues. *Proc Natl Acad Sci USA* 1995;**92**(24):11190–4.
8. Chen HH, Burnett Jr. JC. C-type natriuretic peptide: the endothelial component of the natriuretic peptide system. *J Cardiovasc Pharmacol* 1998;**32**(Suppl 3):S22–8.
9. Brain SD, Grant AD. Vascular actions of calcitonin gene-related peptide and adrenomedullin. *Physiol Rev* 2004;**84**(3):903–34.
10. Busse R, Fleming I. Regulation of endothelium-derived vasoactive autacoid production by hemodynamic forces. *Trends Pharmacol Sci* 2003;**24**(1):24–9.
11. Vanhoutte PM, Feletou M, Taddei S. Endothelium-dependent contractions in hypertension. *Br J Pharmacol* 2005;**144**(4):449–58.
12. Toda N, Tanabe S, Nakanishi S. Nitric oxide-mediated coronary flow regulation in patients with coronary artery disease: recent advances. *Int J Angiol* 2011;**20**(3):121–34.
13. Sitia S, Tomasoni L, Atzeni F, et al. From endothelial dysfunction to atherosclerosis. *Autoimmun Rev* 2010;**9**(12):830–4.
14. Kubes P, Suzuki M, Granger DN. Nitric oxide: an endogenous modulator of leukocyte adhesion. *Proc Natl Acad Sci USA* 1991;**88**(11):4651–5.
15. Smolenski A, Burkhardt AM, Eigenthaler M, et al. Functional analysis of cGMP-dependent protein kinases I and II as mediators of NO/cGMP effects. *Naunyn Schmiedebergs Arch Pharmacol* 1998;**358**(1):134–9.
16. Toda N, Ayajiki K, Okamura T. Control of systemic and pulmonary blood pressure by nitric oxide formed through neuronal nitric oxide synthase. *J Hypertens* 2009;**27**(10): 1929–40.
17. Kuhlencordt PJ, Gyurko R, Han F, et al. Accelerated atherosclerosis, aortic aneurysm formation, and ischemic heart disease in apolipoprotein E/endothelial nitric oxide synthase double-knockout mice. *Circulation* 2001;**104**(4):448–54.
18. Leeson CP, Hingorani AD, Mullen MJ, et al. Glu298Asp endothelial nitric oxide synthase gene polymorphism interacts with environmental and dietary factors to influence endothelial function. *Circ Res* 2002;**90**(11):1153–8.

19. Lamblin N, Cuilleret FJ, Helbecque N, et al. A common variant of endothelial nitric oxide synthase (Glu298Asp) is associated with collateral development in patients with chronic coronary occlusions. *BMC Cardiovasc Disord* 2005;**5**:27.
20. Ichihara S, Yamada Y, Fujimura T, et al. Association of a polymorphism of the endothelial constitutive nitric oxide synthase gene with myocardial infarction in the Japanese population. *Am J Cardiol* 1998;**81**(1):83–6.
21. Talib J, Kwan J, Suryo Rahmanto A, et al. The smoking-associated oxidant hypothiocyanous acid induces endothelial nitric oxide synthase dysfunction. *Biochem J* 2014;**457**(1): 89–97.
22. Wright JL, Zhou S, Churg A. Pulmonary hypertension and vascular oxidative damage in cigarette smoke exposed eNOS$^{-/-}$ mice and human smokers. *Inhal Toxicol* 2012;**24**(11): 732–40.
23. Celermajer DS, Sorensen KE, Georgakopoulos D, et al. Cigarette smoking is associated with dose-related and potentially reversible impairment of endothelium-dependent dilation in healthy young adults. *Circulation* 1993;**88**(5 Pt 1):2149–55.
24. Ishibashi Y, Takahashi N, Tokumaru A, et al. Activation of inducible NOS in peripheral vessels and outcomes in heart failure patients. *J Card Fail* 2008;**14**(9):724–31.
25. Sezgin N, Barutcu I, Sezgin AT, et al. Plasma nitric oxide level and its role in slow coronary flow phenomenon. *Int Heart J* 2005;**46**(3):373–82.
26. Nakata S, Tsutsui M, Shimokawa H, et al. Spontaneous myocardial infarction in mice lacking all nitric oxide synthase isoforms. *Circulation* 2008;**117**(17):2211–23.
27. Cooke JP, Singer AH, Tsao P, et al. Antiatherogenic effects of L-arginine in the hypercholesterolemic rabbit. *J Clin Invest* 1992;**90**(3):1168–72.
28. Beckman JS, Koppenol WH. Nitric oxide, superoxide, and peroxynitrite: the good, the bad, and ugly. *Am J Physiol* 1996;**271**(5 Pt 1):C1424–37.
29. Vallance P, Leiper J. Cardiovascular biology of the asymmetric dimethylarginine:dimethylarginine dimethylaminohydrolase pathway. *Arterioscler Thromb Vasc Biol* 2004;**24**(6): 1023–30.
30. Davis JW, Shelton L, Eigenberg DA, et al. Effects of tobacco and non-tobacco cigarette smoking on endothelium and platelets. *Clin Pharmacol Ther* 1985;**37**(5):529–33.
31. Krupski WC, Olive GC, Weber CA, et al. Comparative effects of hypertension and nicotine on injury-induced myointimal thickening. *Surgery* 1987;**102**(2):409–15.
32. Villablanca AC. Nicotine stimulates DNA synthesis and proliferation in vascular endothelial cells in vitro. *J Appl Physiol* 1998;**84**(6):2089–98.
33. Cai H, Harrison DG. Endothelial dysfunction in cardiovascular diseases: the role of oxidant stress. *Circ Res* 2000;**87**(10):840–4.
34. Lehmann N, Mohlenkamp S, Mahabadi AA, et al. Effect of smoking and other traditional risk factors on the onset of coronary artery calcification: results of the Heinz Nixdorf Recall Study. *Atherosclerosis* 2014;**232**(2):339–45.
35. Messner B, Bernhard D. Smoking and cardiovascular disease: mechanisms of endothelial dysfunction and early atherogenesis. *Arterioscler Thromb Vasc Biol* 2014;**34**(3):509–15.
36. Koenig W, Sund M, Frohlich M, et al. C-Reactive protein, a sensitive marker of inflammation, predicts future risk of coronary heart disease in initially healthy middle-aged men: results from the MONICA (Monitoring Trends and Determinants in Cardiovascular Disease) Augsburg Cohort Study, 1984 to 1992. *Circulation* 1999;**99**(2):237–42.
37. Bakhru A, Erlinger TP. Smoking cessation and cardiovascular disease risk factors: results from the Third National Health and Nutrition Examination Survey. *PLoS Med* 2005;**2**(6):e160.
38. Penn A, Snyder CA. 1,3 Butadiene, a vapor phase component of environmental tobacco smoke, accelerates arteriosclerotic plaque development. *Circulation* 1996;**93**(3):552–7.

39. Penn A, Snyder C. Arteriosclerotic plaque development is "promoted" by polynuclear aromatic hydrocarbons. *Carcinogenesis* 1988;**9**(12):2185–9.
40. Heeschen C, Weis M, Aicher A, et al. A novel angiogenic pathway mediated by non-neuronal nicotinic acetylcholine receptors. *J Clin Invest* 2002;**110**(4):527–36.
41. Hioki H, Aoki N, Kawano K, et al. Acute effects of cigarette smoking on platelet-dependent thrombin generation. *Eur Heart J* 2001;**22**(1):56–61.
42. Mazzone A, Cusa C, Mazzucchelli I, et al. Cigarette smoking and hypertension influence nitric oxide release and plasma levels of adhesion molecules. *Clin Chem Lab Med* 2001;**39**(9):822–6.
43. Blankenberg S, Rupprecht HJ, Bickel C, et al. Circulating cell adhesion molecules and death in patients with coronary artery disease. *Circulation* 2001;**104**(12):1336–42.
44. Blann AD, McCollum CN. Adverse influence of cigarette smoking on the endothelium. *Thromb Haemost* 1993;**70**(4):707–11.
45. Prisco D, Fedi S, Brunelli T, et al. The influence of smoking on von Willebrand factor is already manifest in healthy adolescent females: the Floren-teen (Florence Teenager) Study. *Int J Clin Lab Res* 1999;**29**(4):150–4.
46. Matetzky S, Tani S, Kangavari S, et al. Smoking increases tissue factor expression in atherosclerotic plaques: implications for plaque thrombogenicity. *Circulation* 2000;**102**(6): 602–4.
47. Barua RS, Ambrose JA, Saha DC, et al. Smoking is associated with altered endothelial-derived fibrinolytic and antithrombotic factors: an in vitro demonstration. *Circulation* 2002;**106**(8):905–8.
48. Metz L, Waters DD. Implications of cigarette smoking for the management of patients with acute coronary syndromes. *Prog Cardiovasc Dis* 2003;**46**(1):1–9.
49. Bowers TR, Terrien EF, O'Neill WW, et al. Effect of reperfusion modality on outcome in nonsmokers and smokers with acute myocardial infarction (a Primary Angioplasty in Myocardial Infarction [PAMI] Substudy). *Am J Cardiol* 1996;**78**(5):511–5.
50. Nowak J, Murray JJ, Oates JA, et al. Biochemical evidence of a chronic abnormality in platelet and vascular function in healthy individuals who smoke cigarettes. *Circulation* 1987;**76**(1):6–14.
51. Freedman JE, Ting B, Hankin B, et al. Impaired platelet production of nitric oxide predicts presence of acute coronary syndromes. *Circulation* 1998;**98**(15):1481–6.
52. Chung TW, Tyan YC, Hsieh JH, et al. Shear stress-induced aggregation of oxidized platelets. *Thromb Res* 2002;**105**(4):325–9.
53. Escolar G, White JG. Changes in glycoprotein expression after platelet activation: differences between in vitro and in vivo studies. *Thromb Haemost* 2000;**83**(3):371–86.
54. Thomas AE, Green FR, Lamlum H, et al. The association of combined alpha and beta fibrinogen genotype on plasma fibrinogen levels in smokers and non-smokers. *J Med Genet* 1995;**32**(8):585–9.
55. Petruzzelli S, Puntoni R, Mimotti P, et al. Plasma 3-nitrotyrosine in cigarette smokers. *Am J Respir Crit Care Med* 1997;**156**(6):1902–7.
56. Mero N, Van Tol A, Scheek LM, et al. Decreased postprandial high density lipoprotein cholesterol and apolipoproteins A-I and E in normolipidemic smoking men: relations with lipid transfer proteins and LCAT activities. *J Lipid Res* 1998;**39**(7):1493–502.
57. Dullaart RP, Hoogenberg K, Dikkeschei BD, et al. Higher plasma lipid transfer protein activities and unfavorable lipoprotein changes in cigarette-smoking men. *Arterioscler Thromb* 1994;**14**(10):1581–5.
58. Meenakshisundaram R, Rajendiran C, Thirumalaikolundusubramanian P. Lipid and lipoprotein profiles among middle aged male smokers: a study from southern India. *Tob Induc Dis* 2010;**8**:11.

59. Morrow JD, Frei B, Longmire AW, et al. Increase in circulating products of lipid peroxidation (F2-isoprostanes) in smokers. Smoking as a cause of oxidative damage. *N Engl J Med* 1995;**332**(18):1198–203.
60. Conklin DJ, Bhatnagar A, Cowley HR, et al. Acrolein generation stimulates hypercontraction in isolated human blood vessels. *Toxicol Appl Pharmacol* 2006;**217**(3):277–88.
61. Park YS, Kim J, Misonou Y, et al. Acrolein induces cyclooxygenase-2 and prostaglandin production in human umbilical vein endothelial cells: roles of p38 MAP kinase. *Arterioscler Thromb Vasc Biol* 2007;**27**(6):1319–25.
62. Bernhard D, Csordas A, Henderson B, et al. Cigarette smoke metal-catalyzed protein oxidation leads to vascular endothelial cell contraction by depolymerization of microtubules. *FASEB J* 2005;**19**(9):1096–107.
63. Abu-Hayyeh S, Sian M, Jones KG, et al. Cadmium accumulation in aortas of smokers. *Arterioscler Thromb Vasc Biol* 2001;**21**(5):863–7.
64. Chalon S, Moreno Jr. H, Hoffman BB, et al. Angiotensin-converting enzyme inhibition improves venous endothelial dysfunction in chronic smokers. *Clin Pharmacol Ther* 1999;**65**(3):295–303.
65. al-Alawi E, Jenkins D. Does maternal smoking increase the risk of neonatal polycythaemia? *Ir Med J* 2000;**93**(6):175–6.
66. Haustein KO, Krause J, Haustein H, et al. Changes in hemorheological and biochemical parameters following short-term and long-term smoking cessation induced by nicotine replacement therapy (NRT). *Int J Clin Pharmacol Ther* 2004;**42**(2):83–92.
67. Yujie W, Huiliang L, Zhigeng J. Smoking cessation can improve arterial stiffness. *Angiol* 2014;**65**(1):88.
68. Vardulaki KA, Walker NM, Day NE, et al. Quantifying the risks of hypertension, age, sex and smoking in patients with abdominal aortic aneurysm. *Br J Surg* 2000;**87**(2):195–200.
69. Virdis A, Giannarelli C, Neves MF, et al. Cigarette smoking and hypertension. *Curr Pharm Des* 2010;**16**(23):2518–25.
70. Benowitz NL, Gourlay SG. Cardiovascular toxicity of nicotine: implications for nicotine replacement therapy. *J Am Coll Cardiol* 1997;**29**(7):1422–31.
71. Benowitz NL, Jacob 3rd P, Jones RT, et al. Interindividual variability in the metabolism and cardiovascular effects of nicotine in man. *J Pharmacol Exp Ther* 1982;**221**(2): 368–72.
72. Czernin J, Waldherr C. Cigarette smoking and coronary blood flow. *Prog Cardiovasc Dis* 2003;**45**(5):395–404.
73. Benowitz NL, Kuyt F, Jacob 3rd P. Circadian blood nicotine concentrations during cigarette smoking. *Clin Pharmacol Ther* 1982;**32**(6):758–64.
74. Lowe GD, Lee AJ, Rumley A, et al. Blood viscosity and risk of cardiovascular events: the Edinburgh Artery Study. *Br J Haematol* 1997;**96**(1):168–73.
75. Sheps DS, Herbst MC, Hinderliter AL, et al. Production of arrhythmias by elevated carboxyhemoglobin in patients with coronary artery disease. *Ann Intern Med* 1990;**113**(5): 343–51.
76. DeBias DA, Banerjee CM, Birkhead NC, et al. Effects of carbon monoxide inhalation on ventricular fibrillation. *Arch Environ Health* 1976;**31**(1):42–6.
77. Plank B, Kutyifa V, Moss AJ, et al. Smoking is associated with an increased risk of first and recurrent ventricular tachyarrhythmias in patients with mild heart failure with ischemic and nonischemic cardiomyopathy: A MADIT-CRT Substudy. *Heart Rhythm* 2014;**11**(5): 822–7.
78. Peters RW, Brooks MM, Todd L, et al. Smoking cessation and arrhythmic death: the CAST experience. The Cardiac Arrhythmia Suppression Trial (CAST) Investigators. *J Am Coll Cardiol* 1995;**26**(5):1287–92.

79. Javaheri S, Shukla R, Wexler L. Association of smoking, sleep apnea, and plasma alkalosis with nocturnal ventricular arrhythmias in men with systolic heart failure. *Chest* 2012;**141**(6):1449–56.
80. Tsai WC, Chen CY, Kuo HF, et al. Areca nut chewing and risk of atrial fibrillation in Taiwanese men: a nationwide ecological study. *Int J Med Sci* 2013;**10**(7):804–11.
81. Chamberlain AM, Agarwal SK, Folsom AR, et al. Smoking and incidence of atrial fibrillation: results from the Atherosclerosis Risk in Communities (ARIC) Study. *Heart Rhythm* 2011;**8**(8):1160–6.
82. Alonso A, Krijthe BP, Aspelund T, et al. Simple risk model predicts incidence of atrial fibrillation in a racially and geographically diverse population: the CHARGE-AF Consortium. *J Am Heart Assoc* 2013;**2**(2):e000102.
83. Pawar PP, Jones LG, Feller M, et al. Association between smoking and outcomes in older adults with atrial fibrillation. *Arch Gerontol Geriatr* 2012;**55**(1):85–90.
84. Thoren E, Hellgren L, Jideus L, et al. Prediction of postoperative atrial fibrillation in a large coronary artery bypass grafting cohort. *Interact Cardiovasc Thorac Surg* 2012;**14**(5):588–93.
85. Larsen SB, Grove EL, Hvas AM, et al. Platelet turnover in stable coronary artery disease: influence of thrombopoietin and low-grade inflammation. *PloS One* 2014;**9**(1):e85566.
86. Zhou D, Hu W, Wang Q, et al. Glutathione S-transferase M1 polymorphism and coronary heart disease susceptibility: a meta-analysis involving 47,596 subjects*Heart Lung Circ* 2014. S1443-9506(14)00032-8. Available from: http://dx.doi.org/10.1016/j.hlc.2014.01.003.
87. Hamrefors V, Hedblad B, Hindy G, et al. Smoking modifies the associated increased risk of future cardiovascular disease by genetic variation on chromosome 9p21. *PloS One* 2014;**9**(1):e85893.
88. Athyros VG, Tziomalos K, Katsiki N, et al. The impact of smoking on cardiovascular outcomes and comorbidities in statin-treated patients with coronary artery disease: a post hoc analysis of the GREACE study. *Curr Vasc Pharmacol* 2013;**11**(5):779–84.
89. Durieux R, Van Damme H, Labropoulos N, et al. High prevalence of abdominal aortic aneurysm in patients with three-vessel coronary artery disease. *Eur J Vasc Endovasc Surg* 2014;**47**(3):273–8.
90. Perk J, De Backer G, Gohlke H, et al. European guidelines on cardiovascular disease prevention in clinical practice (version 2012). Fifth Joint Task Force of the European Society of Cardiology and other societies on cardiovascular disease prevention in clinical practice (constituted of representatives of nine societies and invited experts). *Eur Heart J* 2012;**33**(13):1635–701.
91. Piano MR, Benowitz NL, Fitzgerald GA, et al. Impact of smokeless tobacco products on cardiovascular disease: implications for policy, prevention, and treatment: a policy statement from the American Heart Association. *Circulation* 2010;**122**(15):1520–44.
92. De Luca G, Parodi G, Sciagra R, et al. Smoking and infarct size among STEMI patients undergoing primary angioplasty. *Atherosclerosis* 2014;**233**(1):145–8.
93. Aune E, Roislien J, Mathisen M, et al. The "smoker's paradox" in patients with acute coronary syndrome: a systematic review. *BMC Med* 2011;**9**:97.
94. Inci S, Arslan S, Bakirci EM, et al. Predictors of reintervention after coronary artery bypass grafting. *Eur Rev Med Pharmacol Sci* 2014;**18**(1):66–70.
95. Cornel JH, Becker RC, Goodman SG, et al. Prior smoking status, clinical outcomes, and the comparison of ticagrelor with clopidogrel in acute coronary syndromes-insights from the PLATelet inhibition and patient Outcomes (PLATO) trial. *Am Heart J* 2012;**164**(3):334–42.

96. Assali AR, Moustapha A, Sdringola S, et al. Acute coronary syndrome may occur with in-stent restenosis and is associated with adverse outcomes (the PRESTO trial). *Am J Cardiol* 2006;**98**(6):729–33.
97. Mahemuti A, Abudureheman K, Schiele F, et al. Association between inflammatory markers, hemostatic markers, and traditional risk factors on coronary artery spasm in patients with normal coronary angiography. *J Interv Cardiol* 2014;**27**(1):29–35.
98. Yasue H, Nakagawa H, Itoh T, et al. Coronary artery spasm: clinical features, diagnosis, pathogenesis, and treatment. *J Cardiol* 2008;**51**(1):2–17.
99. Hartz AJ, Anderson AJ, Brooks HL, et al. The association of smoking with cardiomyopathy. *N Engl J Med* 1984;**311**(19):1201–6.
100. Gu L, Pandey V, Geenen DL, et al. Cigarette smoke-induced left ventricular remodelling is associated with activation of mitogen-activated protein kinases. *Eur J Heart Fail* 2008;**10**(11):1057–64.
101. Freund KM, Belanger AJ, D'Agostino RB, et al. The health risks of smoking. The Framingham Study: 34 years of follow-up. *Ann Epidemiol* 1993;**3**(4):417–24.
102. Murabito JM, Evans JC, Nieto K, et al. Prevalence and clinical correlates of peripheral arterial disease in the Framingham Offspring Study. *Am Heart J* 2002;**143**(6):961–5.
103. Jonason T, Bergstrom R. Cessation of smoking in patients with intermittent claudication. Effects on the risk of peripheral vascular complications, myocardial infarction and mortality. *Acta Med Scand* 1987;**221**(3):253–60.
104. Malecki R, Zdrojowy K, Adamiec R. Thromboangiitis obliterans in the 21st century: a new face of disease. *Atherosclerosis* 2009;**206**(2):328–34.
105. Pope JE. The diagnosis and treatment of Raynaud's phenomenon: a practical approach. *Drugs* 2007;**67**(4):517–25.
106. Kannel WB, Schatzkin A. Sudden death: lessons from subsets in population studies. *J Am Coll Cardiol* 1985;**5**(Suppl 6):141B–149BB.
107. Wannamethee G, Shaper AG, Macfarlane PW, et al. Risk factors for sudden cardiac death in middle-aged British men. *Circulation* 1995;**91**(6):1749–56.
108. Jouven X, Desnos M, Guerot C, et al. Predicting sudden death in the population: the Paris Prospective Study I. *Circulation* 1999;**99**(15):1978–83.
109. Le Houezec J, Aubin HJ. Pharmacotherapies and harm-reduction options for the treatment of tobacco dependence. *Expert Opin Pharmacother* 2013;**14**(14):1959–67.
110. Murray RP, Bailey WC, Daniels K, et al. Safety of nicotine polacrilex gum used by 3094 participants in the Lung Health Study. *Chest* 1996;**109**(2):438–45.
111. Working Group for the Study of Transdermal Nicotine in Patients with Coronary Artery Disease. Nicotine replacement therapy for patients with coronary artery disease. *Arch Intern Med* 1994;**154**(9):989–95.
112. Meine TJ, Patel MR, Washam JB, et al. Safety and effectiveness of transdermal nicotine patch in smokers admitted with acute coronary syndromes. *Am J Cardiol* 2005;**95**(8):976–8.
113. Rigotti NA, Thorndike AN, Regan S, et al. Bupropion for smokers hospitalized with acute cardiovascular disease. *Am J Med* 2006;**119**(12):1080–7.
114. Cahill K, Stevens S, Perera R, et al. Pharmacological interventions for smoking cessation: an overview and network meta-analysis. *Cochrane Database Syst Rev* 2013;**5**:CD009329.
115. Tonstad S, Johnston JA. Does bupropion have advantages over other medical therapies in the cessation of smoking? *Expert Opin Pharmacother* 2004;**5**(4):727–34.
116. Bonetti PO, Lardi E, Geissmann C, et al. Effect of brief secondhand smoke exposure on endothelial function and circulating markers of inflammation. *Atherosclerosis* 2011;**215**(1):218–22.

117. Saitta D, Ferro GA, Polosa R. Achieving appropriate regulations for electronic cigarettes. *Ther Adv Chronic Dis* 2014;**5**(2):50–61.
118. Taylor BV, Oudit GY, Kalman PG, et al. Clinical and pathophysiological effects of active and passive smoking on the cardiovascular system. *Can J Cardiol* 1998;**14**(9): 1129–39.
119. Palavra IR, Franelic IP, Milanovic SM, et al. *Pasivno pusenje—aktivni ubojica.* [Passive smoking–active killer]. *Lijec Vjesn* 2013;**135**(11-12):326–9.
120. Mitchell EA, Milerad J. Smoking and the sudden infant death syndrome. *Rev Environ Health* 2006;**21**(2):81–103.
121. Prugger C, Wellmann J, Heidrich J, et al. Passive smoking and smoking cessation among patients with coronary heart disease across Europe: results from the EUROASPIRE III survey. *Eur Heart J* 2014;**35**(9):590–8.
122. Le-Ha C, Beilin LJ, Burrows S, et al. Gender difference in the relationship between passive smoking exposure and HDL-cholesterol levels in late adolescence. *J Clin Endocrinol Metab* 2013;**98**(5):2126–35.
123. Aguero F, Degano IR, Subirana I, et al. Impact of a partial smoke-free legislation on myocardial infarction incidence, mortality and case-fatality in a population-based registry: the REGICOR Study. *PloS One* 2013;**8**(1):e53722.
124. Sargent JD, Demidenko E, Malenka DJ, et al. Smoking restrictions and hospitalization for acute coronary events in Germany. *Clin Res Cardiol* 2012;**101**(3):227–35.
125. Sims M, Mindell JS, Jarvis MJ, et al. Did smokefree legislation in England reduce exposure to secondhand smoke among nonsmoking adults? Cotinine analysis from the Health Survey for England. *Environ Health Perspect* 2012;**120**(3):425–30.
126. Tan CE, Glantz SA. Association between smoke-free legislation and hospitalizations for cardiac, cerebrovascular, and respiratory diseases: a meta-analysis. *Circulation* 2012;**126** (18):2177–83.
127. Hansson J, Galanti MR, Hergens MP, et al. Use of snus and acute myocardial infarction: pooled analysis of eight prospective observational studies. *Eur J Epidemiol* 2012;**27**(10): 771–9.
128. Teo KK, Ounpuu S, Hawken S, et al. Tobacco use and risk of myocardial infarction in 52 countries in the INTERHEART Study: a case-control study. *Lancet* 2006;**368**(9536): 647–58.
129. Cogswell ME, Weisberg P, Spong C. Cigarette smoking, alcohol use and adverse pregnancy outcomes: implications for micronutrient supplementation. *J Nutr* 2003;**133**(5 Suppl 2): 1722S–31S.
130. Haj Mouhamed D, Ezzaher A, Neffati F, et al. Effect of cigarette smoking on plasma uric acid concentrations. *Environ Health Prev Med* 2011;**16**(5):307–12.

Chapter 15

Cardiovascular Toxicity as a Result of Recreational Drugs

Rohin Francis[1] **and Azad Ghuran**[2]
[1]*University of Cambridge, Cambridge, United Kingdom*
[2]*East and North Hertfordshire NHS Trust, Lister Hospital, Stevenage, United Kingdom*

15.1 INTRODUCTION

Recreational drug use remains common worldwide. In the United Kingdom, between 2011 and 2012, 8.9% of people reported using illegal drugs. Among those aged 16 to 24, this figure was 19.3%.[1] An estimated 1 in 3 people in the United Kingdom have taken illegal drugs at some point.[2] The majority of recreational drugs currently available have effects on the cardiovascular system, although for many of these agents, data concerning the cardiovascular repercussions are scarce. In adults under 45 years of age, a quarter of myocardial infarctions have been attributed to cocaine use alone.[3] However, the wide range of illicit drugs presents a challenge to the physician and this chapter attempts to summarize the cardiovascular complications associated with the major recreational drugs in use today and up-to-date guidance on management.

One danger inherent in the use of illicit drugs is a lack of quality control, such that users risk consuming a wide range of inert or potentially toxic substances along with the desired narcotic. As a result, the clinical picture can vary between patients even if they report using the same drug. The varying levels of active drug in a "dose" also results in the risk of accidental overdose being greatly increased in comparison to the abuse of prescription medication. It should also be noted that polydrug use is extremely common, with alcohol and tobacco being most frequently taken along with one or more illegal drugs. Although the interactions can be complex and unpredictable, the addition of tobacco or alcohol to use of an illicit drug tends to increase cardiotoxicity.[4] Many of the recreational drugs can be administered intravenously and the risk of infective endocarditis is considerable.[4]

The Heart and Toxins. DOI: http://dx.doi.org/10.1016/B978-0-12-416595-3.00015-3

Specific treatments for each narcotic agent are considered in the relevant subsections that follow, but many supportive measures are common to all recreational drugs. Neurological management is not covered, but the potential exists for conscious levels to be reduced and a patent airway should be maintained. Patients should have heart rate, blood pressure, oxygen saturation, temperature, and Glasgow Coma Scale monitored as well as an electrocardiogram (ECG) performed.[5] In many cases blood tests including arterial blood sampling may be appropriate. Short-term sedation and mechanical ventilation may be required with specialist anesthetic input. In all cases, a specialist poisons service should be consulted for agent-specific guidance.

15.2 COCAINE

Derived from the *Erythroxylum coca* plant, cocaine has been ingested for thousands of years, but its purification in the late nineteenth century ushered in the era of modern cocaine abuse. Between 2011 and 2012, 4.2% of 16 to 24-year-olds used powder cocaine in the United Kingdom. For several decades it has remained the second most commonly used illegal drug in the United Kingdom, behind cannabis, and its use increased between 1996 and 2011.[2] However, it is overrepresented when examining the causes of death in young people and accounts for more emergency department presentations than any other illegal drug.[6] In the United States, recent data have shown that 56% of cocaine-using patients attending the emergency department presented with cardiovascular complications and 1 in 20 of those presenting with chest pain suffered myocardial infarction.[7]

Cocaine is a crystalline tropane alkaloid that is used in two forms. Cocaine hydrochloride ("coke") is a water-soluble powder form of the drug, normally administered intranasally, intravenously, or via the buccal membranes. "Crack" freebase cocaine comprises crystals of cocaine without the hydrochloride moiety and can be ingested orally, smoked, inhaled, or injected.[5]

15.2.1 Pharmacology

Cocaine exerts its cardiovascular effects via two opposing mechanisms: sympathomimetic overdrive and direct action on sodium cardiac ion channels.[8] At low doses the sympathomimetic actions predominate but at high doses the class I effects are more pronounced.[9] In fact, the mechanism by which cocaine causes fatal arrhythmias remains somewhat open to debate in that it is not clear whether one of the pathophysiological processes is more important than the others. In histological studies, cocaine has been shown to cause myocardial interstitial fibrosis; perivascular fibrosis; small vessel disease; and, in cases of sudden death, contraction band necrosis.[10] These histological changes are consistent with excess catecholamine exposure.

15.2.2 Adverse Effects

Sympathomimetic Activation and Vasospasm: Promotion of Ischemia

The sympathetic nervous system's effects are driven by cocaine causing inhibition of noradrenaline and dopamine reuptake at preganglionic sympathetic nerve terminals and by increasing the sensitivity of adrenergic neuroreceptors to noradrenaline.[11] Heart rate and blood pressure rise in a dose-dependent manner,[12] increasing myocardial oxygen demand. Intranasal use of cocaine results in an increase in heart rate ($18 \pm 17\%$ beats/min), an increase in mean systemic arterial pressure ($8 \pm 7\%$ mmHg), an increase in cardiac index ($18 \pm 18\%$ L/min/m^2), and an increase in dP/dt (18 ± 20 mmHg/s) (Figure 15.1).[13] At low-to-moderate doses, cocaine can produce an increase in contractility via a sympathetically mediated positive inotropic effect, whereas at higher doses it can reduce left ventricular (LV) function and increase end-systolic wall stress, secondary to sodium channel blockade.[14]

Cocaine is a potent vasoconstrictor, acting by stimulating the release of endothelin-1 and inhibiting nitric oxide production.[15,16] The increased levels

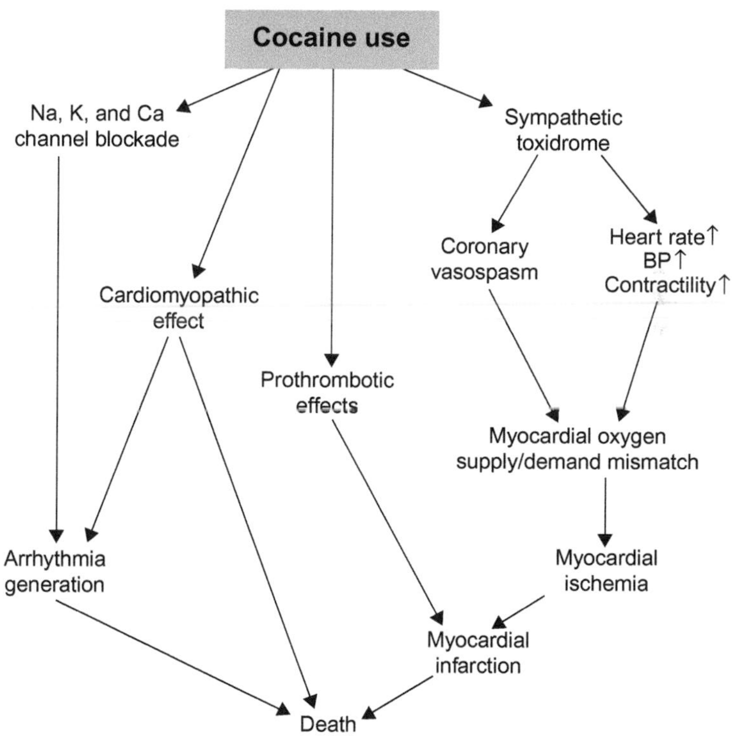

FIGURE 15.1 **The multifaceted way in which cocaine promotes morbidity and mortality.**

of catecholamines also act as coronary vasoconstrictors themselves. Cocaine can rapidly reduce arterial pH.[17] Acidosis exacerbates ischemia, slows cardiac electrical conduction, and can stabilize drug binding, leading to slower recovery from intoxication.[18]

Generation of Arrhythmias

Cocaine is capable of producing any possible arrhythmia in both canine and human hearts.[19] It interferes with normal cardiac depolarization resulting in increases in PR, QRS, and QTc intervals on the surface ECG,[20] which is thought to result from cocaine's inhibition of L-type Ca currents, delayed rectifier K currents, and cardiomyocyte Na currents. In particular, voltage-gated Na channels are thought to be of key importance. These are responsible for the rapid upstroke of cardiac action potential, and cocaine induces sodium channel blockade, which can lead to impaired cardiac conduction. Inhibition of L-type Ca currents and K currents can lead to elongated refractory periods.[18]

Cocaine can exacerbate arrhythmias generated by a preexisting cardiac disease;[21] however, postmortem examinations of patients who died from cocaine abuse revealed that most cases did not show any preexisting myocardial abnormality.[22]

The coconsumption of alcohol causes an increase in heart rate and blood pressure above what is observed for either drug in isolation[23] and the risk of adverse cardiac events also rises.[24] This is, at least in part, contributed to by cocaethylene, a metabolite of cocaine and alcohol that additionally slows cardiac conduction and delays repolarization. Cocaine causes a localized increase in the concentration of extracellular K, which indirectly reduces the availability of Na channels, further slowing cardiac conduction and promoting re-entrant arrhythmias.[18]

The most common arrhythmia associated with cocaine toxicity is monomorphic ventricular tachycardia,[8] but it can also cause torsades de pointes (TdP), probably in a similar manner to other QT-prolonging medications.[18] Indeed TdP has been observed in cocaine users with a preexisting prolonged QTc interval.[25] Several studies have shown QTc prolongation in association with both acute and chronic cocaine use, with the prevalence markedly elevated for patients who died from its use.[25] Due to the Na channel-blocking action, many cases have been reported of cocaine producing a type I Brugada syndrome ECG appearance, and subsequent pharmacological challenge tests have been both positive and negative.[26]

Prothrombotic Effects

Cocaine increases the risk of acute coronary thrombosis.[27] Platelet activation occurs, encouraging platelet aggregation and additional release of platelet α-granule,[28] plasminogen inhibitor activity,[29] fibrinogen, and von

Willebrand's factor.[30] Aneurysm and calcifications occur in the coronary arteries of cocaine users.[31,32]

"Conventional" myocardial infarction, occurring in patients who have not used cocaine, is typically produced by rupture of a cholesterol-rich plaque that promotes thrombosis. However, in acute cocaine-induced myocardial infarction, an intracoronary thrombosis occurs in association with fibrous plaques, without plaque rupture.[33] The fibrous plaques are rich in smooth muscle cells and show medial or intimal inflammation, normally plasma cells or lymphocytes.[33]

Chronic cocaine use can accelerate coronary atherosclerosis.[33] Focal coronary vasospasm and occlusion has been proposed as one mechanism for cocaine-induced myocardial infarction.[34,35] For patients who have already undergone percutaneous angioplasty with stent insertion, cocaine can increase the risk of stent thrombosis.[36] Cocaine users also had higher circulating levels of IL-1β, TNF-α, and IL-10, which may suggest cocaine-induced inflammation contributes to plaque rupture.[36]

Cardiomyopathic Effects

Cocaine can depress myocardial function in both the acute and chronic setting, in part related to its sodium channel blockade effect. It has been shown to transiently reduce left ventricular function, increase LV dimensions, and create a toxic cardiomyopathy analogous to that seen in takotsubo or pheochromocytoma-associated cardiomyopathies.[37,38]

In patients who use cocaine regularly, left ventricular ejection fraction can be reduced even if asymptomatic. In a small study of cocaine users undergoing coronary angiography, ejection fraction was abnormal in 55% of patients and less than or equal to 30% in 18% of patients.[39] Cessation of its use produces a very favorable outcome in patients with cocaine-induced cardiomyopathy.[40]

15.2.3 Clinical Presentation and Diagnostic Testing

Cocaine has a fairly short serum half-life of 30 to 90 min.[4] The U.K. National Poisons Information Service lists the numerous cardiac (and extracardiac) clinical features of cocaine use,[5] which include tachycardia, hypertension, chest pain, pyrexia, arrhythmias, syncope, coronary artery dissection,[41] aortic dissection, metabolic acidosis, and cardiorespiratory arrest.

Two thirds of patients present within 3 hours, but the median time to presentation is 18 hours.[42] The risk of myocardial infarction (MI) in the first hour after use of cocaine is said to be increased 24-fold,[43] but the temporal relationship between cocaine use and chest pain is variable. The phenomenon of cocaine-associated chest pain is poorly understood. The presentation is often atypical and risk stratification can be difficult, although the rate of MI is low, at between 0.7% and 6%.[7] A small proportion of patients' chest pain can be attributed to myocardial ischemia, myocardial infarction, and cocaine-related barotrauma causing pneumothorax or pneumomediastinum, but the majority of them have an unclear etiology.[4]

As with all illegal drugs, patients may not report use and if cocaine is suspected, they should be questioned directly. Several methods of testing for its presence exist, with the most commonly used being an assay for urinary benzoylecgonine, a metabolite of cocaine.[12] This is detectable in urine for 24 to 48 hours. A cocaine user presenting with chest pain should be worked up in a similar manner to a patient presenting with an acute coronary syndrome, with serial ECGs and cardiac enzymes; the American Heart Association also recommends some form of stress testing.[12]

Electrocardiogram

The ECG has been reported as abnormal in 56 to 84% of patients presenting with cocaine-associated chest pain; however, many of them are young males and frequently have the normal variant of early repolarization present on their ECG, which is thought to account for a significant proportion of abnormal findings.[12] A single study showed 42% of patients presenting with cocaine-associated chest pain had ST elevation, although none of them were found to have positive cardiac enzymes.[44] QTc prolongation and dispersion have also been reported more frequently in patients with cocaine-associated chest pain.[45] Other abnormalities seen in various studies, to varying degrees, include LV hypertrophy by voltage criteria and nonspecific changes including T-wave abnormalities.[12]

Other Investigations

Left ventricular hypertrophy has been reported more commonly in chronic cocaine abusers and, indeed, this has been postulated as yet another mechanism of promoting myocardial ischemia or arrhythmias.[46] As a result of this increased left ventricular mass, echocardiography is thought to offer limited information in the acute setting.[12] Limited guidance exists on which modality of stress testing is most useful, but because baseline ECGs may be abnormal, imaging-based stress tests are likely to be preferable. It should be noted that all noninvasive cardiac investigations, including computed tomography, exercise-testing, and myocardial perfusion scans, have a very low yield among patients with cocaine-associated chest pain.[8]

Several studies have examined coronary angiographic findings in patients presenting with cocaine-associated chest pain. A single study demonstrated that patients presenting with cocaine-associated myocardial infarction, as proven by positive cardiac enzymes, are likely to have obstructive coronary artery disease, whereas those with negative cardiac enzymes are not.[45] Of the patients deemed high enough risk to undergo coronary angiography,

77% had significant coronary artery disease; however, this study examined patients with a mean age of 43 ± 7 years. In smaller studies of younger patients, rates of significant coronary disease have been reported in 35 to 55% of patients undergoing coronary angiography after presenting with cocaine-associated chest pain.[47]

15.2.4 Treatment

Very few complications from cocaine occur more than 12 hours after use.[7] Therefore, patients who do not present with high-risk features (ST segment elevation or depression, elevated cardiac enzymes, ongoing chest pain, or hemodynamic instability) will normally not need admission but can be observed for up to 12 hours.[48] However, in asymptomatic cases, patients need only be observed for a maximum of four hours after cocaine use.[5]

Observation

In general, patients with cocaine-associated chest pain are managed in broadly the same way as any other patient presenting with chest pain and a suspected acute coronary syndrome, although there are a few differences such as the roles of benzodiazepines and beta blockers (Figure 15.2). No randomized, placebo-controlled trials have shown survival benefit for any specific treatment[12] and most management guidelines in existence are based on observational or animal studies. No guidance exists on the newer high-sensitivity troponin I assays in the current patient population.

Chest Pain and Hypertension

Many anecdotal reports exist for antiplatelet and antithrombin agents but their role in cocaine-mediated acute coronary syndromes is unclear. Aspirin is highly likely to be safe[7] and standard loading and maintenance doses of it and other antiplatelet agents (e.g., clopidogrel, ticagrelor, prasugrel), low-molecular-weight heparin, or fondaparinux should be administered according to the clinical presentation and in line with the current acute coronary syndrome guidelines.

Benzodiazepines and nitrates form the mainstay of treatment for chest pain and hypertension, along with second-line agents including calcium channel blockers, phentolamine, and sodium nitroprusside.[5]

ST-Segment Elevation Myocardial Infarction

If a patient presents with an ST-segment elevation myocardial infarction (STEMI), the treatment of choice is percutaneous coronary intervention (PCI) over thrombolytic therapy, which should be avoided unless PCI is unavailable.[4,12,49,50]

Again, no data exist to guide the choice of bare-metal stent versus drug-eluting stent, if one is required. However, cocaine abusers have extremely

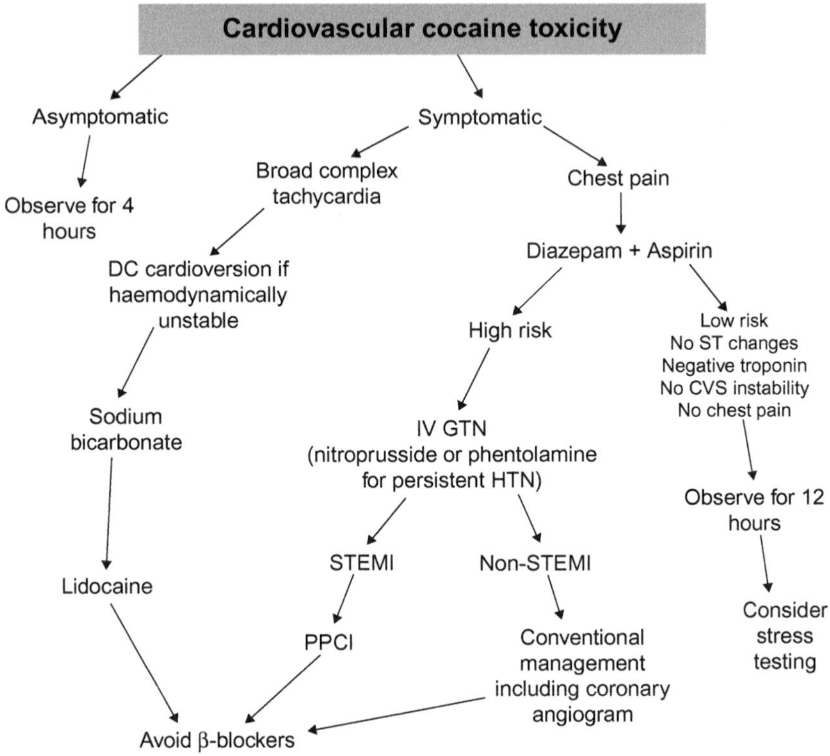

FIGURE 15.2 Simplified management of cocaine cardiovascular toxicity.

high rates of further cocaine use and compliance with antiplatelet regimens may be poor; as such, the shorter course of dual antiplatelet therapy required after implantation of a bare metal stent may be preferable.

Benzodiazepines

A major difference between the management of conventional acute coronary syndromes and cocaine-associated chest pain is the early use of benzodiazepines, normally intravenous diazepam. Large doses may be required, with the U.K. National Poisons Information Service advocating up to 100 mg in incremental doses via a slow intravenous infusion.[5] Diazepam acts in several ways in that it directly reduces the neuropsychiatric complications, which can drive the cardiovascular effects. It reduces agitation and can centrally reduce the stimulatory potency of cocaine, along with acting favorably on elevated blood pressure.[37] Of note, convulsions are not uncommon with cocaine toxicity and diazepam is also the first-line choice for management of seizures.

Nitroglycerine

Nitrates can be used for the management of chest pain or hypertension that does not settle with diazepam.[51,52] Nitroglycerine can be administered rapidly via sublingual or buccal routes, followed by an intravenous infusion.[5] It is thought to alleviate cocaine-associated chest pain by opposing coronary vasospasm and causing a decrease in blood pressure by reducing preload, afterload, and left ventricular end diastolic pressure.[52]

Calcium Channel Antagonists

Calcium channel blockers are described as a second-line therapy for persistent chest pain and hypertension, as their role remains unclear. Although there is some angiographic evidence that intravenous verapamil alleviates cocaine-induced coronary vasospasm,[4] cocaine can have a complex and highly variable effect on myocyte calcium metabolism, causing an unpredictable response to calcium channel antagonists.[53] The American Heart Association suggests calcium channel antagonists be considered only for patients who have ongoing chest pain despite nitroglycerin and diazepam, and it advises against the use of short-acting nifedipine.[12]

Phentolamine

Phentolamine is an α-antagonist with supportive anecdotal evidence of utility in cocaine-induced acute coronary syndrome.[50] Experimental evidence has shown it can reduce the hemodynamic effects of cocaine,[54] and it may be of use in treating ischemia driven by cocaine. Both phentolamine (5–10 mg IV) and sodium nitroprusside (0.5–1.5 μg/kg/min to a maximum of 8 μg/kg/min) are listed by the U.K. National Poisons Information Service as possible choices for resistant hypertension.

Beta Blockers

The role of β-blockade in managing cocaine-mediated cardiac complications remains complex. Cocaine-mediated smooth muscle vasospasm is driven by α-adrenergic stimulation and use of a selective β-adrenergic blocking drug may lead to unopposed α-adrenergic vasospasm and make both systemic hypertension and coronary artery spasm worse.[53] In animal models, propranolol worsened the seizure threshold and expedited death, whereas in humans, coronary artery vasospasm got worse.[55,56]

Labetalol has been proposed as preferable to other beta blockers due to its combined α- and β-blocking properties, but it is predominantly β-blocking and does not appear to be significantly different from other beta blockers.[57] Some retrospective data exist to suggest that beta blocker use may reduce complications when used for cocaine-associated myocardial infarction,[58,59] but no randomized studies exist. Beta blockers currently remain contraindicated in the acute phase of a cocaine-mediated acute coronary syndrome.[12,19]

Arrhythmias

Narrow complex tachycardias rarely require treatment in themselves and respond to supportive measures and benzodiazepines, as has been previously described. Electrolyte abnormalities should be sought and corrected. For sustained supraventricular tachycardias, adenosine can be used safely. It should be noted that the hyperadrenergic state caused by stimulant drugs will reduce the efficacy of adenosine, meaning reentrant tachycardias can persist.[53] Intravenous verapamil can also be considered if a patient is very symptomatic.[5]

Ventricular arrhythmias caused by cocaine can be treated differently based on their time of onset in relation to the cocaine use. In all cases, the use of oxygenation, cardioversion, and defibrillation should follow Advanced Life Support guidelines as is clinically indicated. Ventricular arrhythmias that develop immediately after use are caused by sodium channel inhibitory effects, and animal models have shown a response to hypertonic sodium bicarbonate solution and lidocaine, a class IB antiarrhythmic.[19]

For a broad complex tachycardia that develops immediately after cocaine use, sodium bicarbonate (50 mmol IV) should be administered first, via a central line. If the tachycardia does not respond to a second dose of up to 50 mmol, lidocaine should be given as a bolus, followed by an infusion of 4 mg/min for 30 min, 2 mg/min for 2 hours, and then 1 mg/min for up to 24 hours.[5] No data exist for the efficacy or safety of amiodarone in this population.

In cases of TdP caused by cocaine use, overdrive pacing, magnesium, and lidocaine have been advocated but no data exist on their relative efficacies. Magnesium is advised for patients with a QTc in excess of 500 ms.[19]

Ventricular arrhythmias that occur several hours after cocaine use may be induced by ischemia and alleviating this should be a priority. In terms of therapy choices, it is advisable to use lidocaine for persistent arrhythmias.[60] Magnesium and sodium bicarbonate can also be considered.[19] Again, no data exist regarding the beneficial nor the deleterious effects of amiodarone in this setting. In all cases, beta blockers and class IA and IC antiarrhythmics are absolutely contraindicated.[19]

15.2.5 Other Complications

Endocarditis is associated with intravenous cocaine abuse, with an apparent preponderance to left-sided valves, although this has not been found in all studies.[4,61] Aortic dissection is a rare complication of cocaine use.[62] Both endocarditis and aortic dissection are thought to be caused by the direct effect cocaine has on the vascular endothelium, that is mediated by increased endothelin-1 and decreased NO.[8] The sympathetic stimulation caused by cocaine is also thought to be relevant in aortic dissection as a rapid spike in blood pressure and shear-stress forces acting on the arterial wall.[62,63]

15.2.6 Discharge Planning

After discharge, the most important measure patients should accomplish is cessation of cocaine use. Among those who do abstain, the rates of further chest pain are low, but unfortunately 60% of patients admit to further cocaine use over the 12 months following discharge.[64] Conventional cardiac risk factors should be addressed. In terms of drug regimens, no specific guidance exists and patients who have suffered a cocaine-induced myocardial infarction are generally treated in line with acute coronary syndrome secondary prevention; however, beta blockers are also contraindicated in this follow-up phase due to the risk of further cocaine use.[8,12,53] Angiotensin-converting enzyme inhibitors, angiotensin receptor blockers, statins, and diuretics have not been studied in this population but are unlikely to adversely interact with cocaine.[8]

15.3 AMPHETAMINE-RELATED DRUGS

Amphetamine refers to a group of related compounds including amphetamine itself ("speed"); methamphetamine ("crystal meth"); and, the most commonly used one in the United Kingdom, 3,4-methylenedioxy-N-methylamphetamine, which is also known as MDMA or ecstasy. Amphetamine-related compounds occur naturally, in both *Ephedra* and khat plants. Amphetamines can be ingested orally; absorbed intranasally; or, less often, injected intravenously. Amphetamine and its substituted derivatives are classified as stimulants, hallucinogens, and enactogens.

In the United Kingdom, MDMA and amphetamine represent the third and fourth most popular illegal drugs among young people and overall an estimated 800,000 people used an amphetamine-like drug between 2011 and 2012.[2] Ecstasy was directly implicated in an estimated 200 deaths in the United Kingdom between the years 2003 and 2007, although often in combination with another drug.[65]

15.3.1 Pharmacology

The metabolism of amphetamines is variable and up to 30% of the drug can be excreted unchanged.[53] The plasma half-life is also subject to variation, ranging from 7 to 34 hours depending on urinary pH (a higher pH gives a longer half-life).[5] The clinical effects typically commence an hour after oral ingestion and last 4 to 6 hours.

Amphetamines cause an acute increase in serotonin, noradrenaline, and dopamine at synaptic junctions, via both the inhibition of reuptake as well as increased release.[66] This elevated catecholamine phase is then followed by a period of depletion. Predominantly enactogenic drugs appear to preferentially affect serotonin (i.e., 5-hydroxytryptamine or 5-HT) reuptake, whereas stimulant drugs exert a greater effect on dopamine reuptake. However, amphetamines invariably produce their effects via a combination of both pathways.[67]

Serum concentration does not appear to correlate with severity of systemic effects. These neurotransmitters are integral to control of homeostatic mechanisms such as thermoregulation, sleep, reward, appetite, and the autonomic nervous system. Toxicity appears to be generated by oxidative stress and mitochondrial dysfunction.[68]

15.3.2 Adverse Effects

Amphetamines have a host of acute and chronic neuropsychiatric effects including euphoria, tremor, agitation, and hallucinations. The adverse cardiovascular effects of amphetamines are predominantly mediated via its sympathomimetic properties, in a similar manner to cocaine.[53] The common cardiovascular effects include tachyarrhythmias and hypo- or hypertension.

Rare complications are diverse, with hyperpyrexia, rhabdomyolysis, heart failure, myocardial infarction, takotsubo cardiomyopathy, pneumopericardium, multiorgan failure, and sudden cardiac death having been reported, secondary to MDMA use.[66–69] There is a paucity of information as to why amphetamines cause sudden death, but it is believed that a sympathetic toxidrome precipitates a malignant ventricular arrhythmia. Many fatal cases of MDMA use have been found to have pulmonary edema. MDMA has also been associated with a prolonged QTc, which can predispose toward TdP ventricular tachycardia and death.[69]

MDMA-induced myocardial infarction has been reported on several occasions, presumably secondary to an interplay between coronary artery vasospasm, increased myocardial oxygen demand, and thrombus formation.[53] Myocardial ischemia can act as a substrate for arrhythmias even in the absence of coronary artery occlusion. Cases of myocardial infarction and ventricular tachycardia have been reported in MDMA addicts, even with normal coronary angiography.[70]

Prolonged exposure to serotonin can lead to pulmonary hypertension, through activation of constrictor 5-HT1B receptors and proliferative 5-HT2B receptors and possibly through direct intracellular effects. MDMA and metamphetamine have been shown to produce proliferative thickening of cardiac valve tissue in vitro and may cause pulmonary hypertension with chronic use.[71] A possible link between MDMA and drug-induced valvular heart disease, with increased valvular strands on echocardiography, has been described in vivo. This is an abnormal thickening of valve leaflets resulting in restricted valve movement in the absence of any other causes.[72]

Hyperpyrexia appears to drive cases of rhabdomyolysis and subsequent multiorgan failure. The genesis of increased body temperature is multifactorial, caused not only by the role of serotonin and dopamine in the thermoregulatory process, but also by the environment in which amphetamine-like drugs are normally used. MDMA, in particular, has a significant association with nightclubs and electronic music, and users can spend hours dancing in crowded, hot environments. The euphoria caused by the drug can lead to inadequate fluid replacement, exacerbating the problem.[66] This is of relevance to the cardiovascular system because sympathetic stresses can be potentiated by hypovolemia and hyperpyrexia. Complex direct effects of MDMA on skeletal muscle have been reported, although no information exists about the effects on the cardiac muscle. Cases of acute cardiomyopathy, some of which were not recovered from, can occur from MDMA use.[73]

Users of amphetamines often try to counteract the hyperpyrexia by drinking large amounts of water, which can create a hyponatremic state and an elevated level of antidiuretic hormone (ADH). This has resulted in death secondary to cerebral edema.[69] Although the vast majority of amphetamine and MDMA users are young, the theoretical risk exists of pulmonary edema and other complications of fluid overload. Methamphetamine has been associated with life-threatening pulmonary edema,[5] and MDMA has been implicated in deaths of patients with preexisting cardiomyopathies.[74]

Postmortem examination of amphetamine users revealed similar myocardial histological changes caused by excess catecholaminergic activity, also seen with cocaine use. Left ventricular hypertrophy, interstitial fibrosis, and microvascular disease have been described in MDMA users.[10] These changes can act as foci for arrhythmias.

15.3.3 Clinical Presentation

Amphetamine-like drugs can cause a wide range of neuropsychiatric symptoms, which typically dominate the clinical picture. Cardiovascular signs and symptoms include chest pain, which is of unclear origin, palpitations, and either extremes of blood pressure.[5] In a similar manner to cocaine, the catecholamine excess can create any arrhythmia, with narrow complex and ventricular tachycardias having been reported. It should also be noted that heart rate and blood pressure can continue to be more labile than normal for several days after drug use.

As with many illegal drugs, use of amphetamines with alcohol increases its toxicity.[75] The serotonin syndrome is more likely to occur if a patient also has exposure to another drug that potentiates the effect of serotonin, including selective serotonin reuptake inhibitors, cocaine, tramadol, and tricyclic antidepressants, among others.[5]

15.3.4 Treatment

Observations and supportive management are in line with that for cocaine toxicity, as previously described, including the therapies for myocardial ischemia and hypertension and the prominent role for diazepam. Activated charcoal administered within one hour can again be of use if the drug has been taken orally, but owing to the nature of amphetamines' duration of onset, patients rarely present in the first hour after ingestion. Along with observations and investigations suggested in the management of cocaine toxicity, serum creatine kinase should be included.

The National Poisons Information Service suggests resuscitation should continue for a minimum of one hour after cardiac arrest in the context of amphetamines use. Owing to the lower incidence of cardiovascular complications in comparison to cocaine, the management of cardiovascular effects is more permissive. A narrow complex tachycardia with adequate cardiac output should be treated with intravenous fluids and observed, as most arrhythmias will spontaneously terminate as the drug is metabolized. As with cocaine, adenosine can be safely administered for sustained supraventricular tachycardia, but the shorter dromotropic effect caused by the sympathetic toxidrome will reduce its efficacy.

A further corollary to cocaine management is the risk of administering beta blockers because this can produce unopposed α-adrenergic activity exacerbating hypertension and vasospasm. A small double-blind, placebo-controlled trial showed that the nonselective beta blocker pindolol reduced the tachycardia caused by MDMA but not blood pressure or any noncardiac adverse effects.[75]

Fluid losses can be considerable, but fluid replacement should not be overly aggressive as fluid status can fluctuate rapidly. Hyponatremia may indicate fluid overload and should be managed initially with fluid restriction until paired urine and plasma osmolality suggest otherwise.[69] Hypotension and tachycardia are treated with adequate intravenous crystalloid replacement. Fluid-resistant hypotension should be treated aggressively in a critical care environment with inotropes and vasopressors, if indicated.[5]

15.4 HEROIN

Diacetylmorphine (heroin) is a semisynthetic opioid analgesic that is normally injected intravenously but can be smoked in its freebase form. Heroin is metabolized to morphine and its cardiovascular pharmacology and management are in line with other opioid drugs discussed elsewhere in this book. However, it is somewhat surprising how little evidence exists concerning the optimum management of a patient compromised by opioid overdose.

Heroin use in the United Kingdom has remained stable over the last decade and although less prevalent than cannabis, amphetamines, MDMA, and cocaine, heroin abusers present health care providers with significant pathology. Opioid toxicity can produce profound hypotension, bradycardia, and noncardiogenic pulmonary edema (NCPE); and intravenous injection significantly increases the risk of infective endocarditis.

15.4.1 Adverse Effects

All opioid analgesics can precipitate catastrophic circulatory collapse with hypotension and profound bradycardia, additionally complicated by respiratory depression.

Heroin users may be taking methadone simultaneously, in spite of the fact methadone is intended to help in withdrawing from heroin. Methadone appears to cause ECG abnormalities more frequently than heroin, but more than half of patients using either agent will have an abnormal ECG. Both heroin and methadone can prolong the QTc, cause bradycardia, and produce U waves. Bradycardia and enhanced automaticity can precipitate an increase in ectopic activity, atrial fibrillation, idioventricular rhythm, or potentially lethal ventricular tachyarrhythmias.[76]

Both left- and right-sided heart valves are affected by endocarditis in heroin use, although many classical descriptions of right-sided valve endocarditis exist. Postmortem examinations of heroin users have revealed increased inflammatory cells in the myocardium and it is thought that frequently noted hypertrophic findings are likely to be due to additional use of stimulant drugs, such as cocaine or amphetamines, as opposed to heroin itself.[10] Like all intravenous drug use, heroin is associated with a higher rate of HIV infection, which can cause myocarditis and cardiomyopathy.

An esoteric but well-described effect of heroin is acute cardiomyopathy and noncardiogenic pulmonary edema, which may present up to 24 hours after drug use.[53] The pathophysiology is unclear and focal myocardial myolysis has been observed in postmortem specimens from heroin-induced acute cardiomyopathic deaths. Other mechanisms that may be responsible for pulmonary edema include an anaphylactic reaction to the drug, an increase in pulmonary capillary hydrostatic pressure due to hypoxia-induced pulmonary smooth muscle vasoconstriction, or a disruption in alveolar capillary membrane integrity.[53] The administration of narcotic analgesic receptor blockers may theoretically precipitate the development of NCPE by reversing opioid-induced vasodilatation and venous pooling, thereby increasing venous return to the lungs. Heroin cardiomyopathy is thought to be irreversible.[77]

15.4.2 Management

Heroin toxicity is managed in the same way as morphine toxicity, as discussed previously.

15.5 CANNABIS

Cannabis is by far the most popular recreational drug in the United Kingdom and probably the world. Between 2011 and 2012, 2.3 million 16- to 59-year-olds in the United Kingdom used cannabis. It is normally smoked mixed with tobacco, but can also be ingested orally. It is a flowering plant that actually exists in three normally used varieties, all of which produce their pleasurable effects via tetrahydrocannabinol (THC), a psychoactive aromatic terpenoid. Modern strains of cannabis plants have been genetically modified such that THC concentrations can be extremely high and toxic effects are not well correlated with amount smoked or consumed. Toxic symptoms are also not linearly related to serum levels.[66] As cannabis is normally smoked with tobacco, many of the adverse effects may be related to tobacco, which is not discussed in this chapter.

15.5.1 Pharmacology

Cannabis, or specifically the cannabinoids contained therein, act via the endocannabinoid system (ECS). It is comprised of two G-protein-coupled cannabinoid receptors, CB1 and CB2; endogenous ligands for these receptors; and proteins that regulate endocannabinoid levels and distribution. These proteins also control the activity of cannabinoid receptors.[78] Endogenous and exogenous cannabinoids for CB1 and CB2 demonstrate differing selectivities, which underlie their varied effects. CB1 receptors are predominantly expressed in the brain, heart, vasculature, liver, gastrointestinal system, and kidneys. CB2 receptors are predominantly expressed in immune cells.[79]

Cannabis has a biphasic effect on the cardiovascular system, depending on dose. Cannabis joints vary widely in strength and size, but a low dose could be considered to be less than 0.5 g of cannabis.[78] At low doses it causes an increase in heart rate and blood pressure by stimulating increased sympathetic nervous system activity and reducing parasympathetic activity. Heart rate can rapidly increase by between 20% and 100%, supine blood pressure can rise, and cardiac output can also be increased by approximately 30%. At higher doses, the reverse is true and bradycardia and hypotension result.[78] Cannabis has a plasma half-life of 20 to 30 hours and is normally detected with urinary testing, which is positive for several days.

15.5.2 Adverse Effects

Although the ECS has been implicated in several clinical conditions of relevance to the cardiovascular system, many of the potential effects of cannabis are inferred; there are few published studies demonstrating a causative link from cannabis itself. Stimulation of CB1 receptors located in coronary artery endothelial cells causes increased production of reactive oxygen species and mitogen-activated protein kinases that cause endothelial cell damage. However, CB2 receptor stimulation appears to have a protective effect on endothelium. This antagonistic relationship is mirrored in the effect of cannabinoids on macrophage activity, where CB1 agonism is proatherogenic and CB2 agonism prevents atherogenesis. CB1 receptors appear to have a beneficial effect on the lipid profile, decreasing triglycerides and increasing high-density lipoproteins. The opposing actions of CB1 and CB2 receptors is not duplicated in ischemia-reperfusion injury, where both appear to have a cardioprotective effect.[79]

As with many of the other drugs discussed in this chapter, an increase in heart rate and sympathetic activity can lead to increased myocardial oxygen demand. Cannabis exerts hemodynamic effects that could cause plaque rupture and promote thrombosis. Marijuana smoking may be associated with acute coronary syndromes, with a 4.8-fold increase in myocardial infarction risk in the first hour after smoking a "joint."[80] The duration of exercise required to bring on angina is reduced in comparison to tobacco.[81] Rare cases of acute coronary vasospasm after cannabis use have been reported, which is likely to be mediated via sympathomimetic activity.[82]

The mechanism by which cannabis precipitates an acute coronary syndrome is not clear. Cannabis use may destabilize preexisting plaques simply through its hemodynamic effects. THC activates platelets via CB1 and CB2 receptors, causing increased GPIIb/IIIa and P-selectin expression and activation of factor VII.[83] However, a separate mechanism of platelet activation has been proposed, which makes use of a cyclooxygenase pathway that produces thromboxane A2.[79]

Life-threatening arrhythmias are not thought to be caused by cannabis, but the potential exists for the development of a multitude of arrhythmias because sympathomimetic activity may be significant. Cannabis may mediate its arrhythmic properties by vagal stimulation; action potential duration is reduced and atrial cells have a shorter refractory period and are hyperpolarized. The wavelength of atrial activation is reduced and this can predispose a person to the re-entrant mechanism of atrial fibrillation.[84,85] A cannabis-mediated arteritis affecting peripheral circulation has been well described, although this has not been observed in coronary circulation.

Long-term cardiovascular effects of cannabis are difficult to delineate as there are invariably a significant number of confounding factors present among cannabis users, chiefly tobacco smoking. However, the increase in blood carboxyhemoglobin is greater than from tobacco alone, which may have a significant impact on long-term cardiorespiratory health.

15.5.3 Clinical Presentation

Neuropsychiatric symptoms predominate the clinical picture, in addition to sympathetic or parasympathetic activity, depending on the dose. ECG abnormalities reported have included increased P-wave width and amplitude and ST-segment and T-wave changes, although it is not clear whether these changes are related to drug ingestion independent of the effects on heart rate.[84] Many explanations have been proposed for the ECG changes, including myocardial ischemia, catecholaminergic effects, reduced oxygen delivery secondary to increased blood carboxyhemoglobin, and a cellular response to damage. Scant reports have associated paroxysmal atrial fibrillation, paroxysmal atrial flutter, and AV block to cannabis use.[81,84,85] Atrial and ventricular ectopics may also increase in frequency[53] and a pseudo-Brugada appearance may be produced.[81]

15.5.4 Treatment

Cardiovascular compromise caused by cannabis is exceedingly rare; however, cases of acute coronary syndrome or arrhythmias, potentially provoked by cannabis in isolation, have been reported. However, most acute complications occur when cannabis is taken in conjunction with cocaine or amphetamines as it can potentiate their effects. In the absence of preexisting heart disease, the cardiovascular effects of cannabis on its own are generally well tolerated.

All patients require a 12-lead ECG recording and if any abnormalities are present, serial ECG monitoring. Profound bradycardia or hypotension may require treatment with atropine or fluids and vasopressors, respectively. Acute coronary syndromes and arrhythmias are managed conventionally, in line with current guidelines.

15.6 LYSERGIC ACID DIETHYLAMIDE AND PSILOCYBIN MUSHROOMS

Normally referred to as "acid," lysergic acid diethylamide (LSD) is a semisynthetic ergoline drug with potent psychedelic effects that began its life as a failed psychiatric medication during the 1940s. Originally isolated from the parasitic rye fungus *Claviceps purpurea*, it enjoyed peak popularity in the 1960s and 1970s; its use has waned in the United Kingdom over the last few decades, but it remains in widespread use around the world. Psilocybin-containing or "magic" mushrooms are also potent hallucinogens, but markedly less so than LSD. Psilocybin is a substituted indolealkylamine that is converted to psilocin in the body, which produces a reaction similar to LSD. Both of these drugs are normally ingested orally and typically do not cause significant cardiovascular toxicity but can result in diverse psychiatric complications.[86]

15.6.1 Pharmacology

Both drugs exert their effects via serotonergic, dopaminergic, and adrenergic pathways, to varying degrees. LSD's psychological effects present between

15 to 45 min after oral ingestion and last 4 to 6 hours. It is metabolized by the liver and has a plasma half-life of 100 min.[53] LSD interacts with serotonin receptors in a complex way. It appears to preferentially inhibit serotenergic cell firing and prevent postsynaptic serotonin receptors from being upregulated or downregulated.[86] A label given to describe LSD's actions is a mixed 5-HT2/5-HT1 receptor partial agonist.

LSD also affects the dopaminergic system, acting both as an agonist and antagonist at central dopamine D1 and D2 receptors.[86] In terms of the cardiovascular properties, a sympathomimetic response is produced but to a lesser extent than amphetamines or cocaine. Animal subjects receiving very high doses of LSD have developed tachycardia but no significant cardiovascular complications. In humans, moderate LSD doses have produced modest heart rate and blood pressure increases, in the region of 15 to 19 bpm and 5 to 10 mmHg, respectively.[86] It should be noted that some users exhibit a parasympathetic-dominant picture, with mild bradycardia and hypotension.

Psilocybin also acts predominantly via serotonergic pathways. Its metabolite, psilocin, appears in the plasma 30 min after oral ingestion and psychological effects peak at 70 to 90 min. Similar to LSD, it produces a mild sympathetic stimulus, but in contrast it does not have any affinity for dopamine D2 receptors. Human subjects have shown moderate increases in heart rate, but blood pressure is often unaffected.[87] A small placebo-controlled, double-blind trial showed no effect on ambulatory holter ECG recordings from any dose of psilocybin.[88]

15.6.2 Adverse Effects

Case reports available have associated LSD with arrhythmia and subsequent myocardial infarction, circulatory complications including lower limb arterial vasospasm requiring percutaneous angioplasty, and carotid artery occlusion.[89–91] Anecdotal reports also exist of psilocybin causing rhabdomyolysis, severe renal failure, and encephalopathy.[92]

The mechanism by which psilocybin may promote abnormal clotting is not known but may relate to the ability of LSD to inhibit serotonin-mediated platelet aggregation. Disordered coagulation has been observed in vitro and in vivo.[93] Sympathetic vasospasm may also be involved.[53]

15.6.3 Treatment

Treatment required for LSD or psilocybin toxicity generally deals with the neuropsychiatric complications, which can include psychosis and convulsions. Agitation can be treated with benzodiazepines and overall treatment is supportive because the majority of symptoms resolve within 12 hours.[5] Activated charcoal is unlikely to be of benefit as users normally present greater than 1 hour after ingestion. Supraventricular tachyarrhythmias can be treated with adenosine and verapamil. Severe hypertension and myocardial ischemia should follow conventional recommended treatment.

15.7 INHALED VOLATILE SUBSTANCE ABUSE

Recreational abuse of volatile substances is widespread and appears to be increasing, especially in the developing world and among children.[94] Many of the substances are legal and easy to obtain such as nail varnish remover, glue, paint, shoe polish, fuels, nitrous oxide, and assorted cleaning products. Medicinal inhalants are also utilized such as volatile nitrites normally used for the treatment of angina, commonly known as "poppers." The morbidity and mortality associated with this diverse group of substances is alarming, and many cases of sudden death of healthy young people have been reported.

Users may employ techniques such as spraying the substance into a plastic bag and hyperventilating to maximize the quantity inhaled. Onset of action is rapid, ranging from seconds to minutes and although dependent on the agent being abused, effects typically wear off quickly, causing the user to take another dose to maintain a "high." Agents that fall into the category of volatile substance or solvent abuse include toluene, propane, butane, acetone, haloalkanes, and alkyl nitrites.

15.7.1 Pharmacology

Volatile agents are diverse and interact with the body in different ways. They are rapidly absorbed through the lungs and exert their neuropharmacological effects via dopaminergic, glutaminergic, GABAergic, and serotonergic pathways and have an effect on cell membranes and ion channels.[94] Some molecules are structurally similar to volatile anesthetic agents and produce similar degrees of cardiac and respiratory depression.[95] Metabolism is also rapid and predominantly through the cytochrome P450 system of the liver.

15.7.2 Adverse Effects

Few data exist concerning the effects of recreationally abused volatile agents. The main cause of mortality is "sudden sniffing death syndrome." This is poorly understood, but it is possible that the mechanism of death involves inhalants sensitizing the myocardium to adrenaline.[96] Even normal amounts of adrenaline produced as a physiological response to stress can then induce a malignant arrhythmia.

Volatile substances actually stabilize the membranes of myocardial cells to depolarization, inhibiting electrical impulse conduction. Death often results rapidly after inhalation but can be delayed by hours as volatile substances tend to dissipate slowly from lipid-rich membranes.[96] The theory is supported by documented ventricular fibrillation after solvent inhalation, reports of sudden death during sympathetic stimulation (e.g., running or sexual intercourse), and ventricular arrhythmias observed in animal studies following adrenaline administration after inhalation of volatile hydrocarbons.[97] Toluene can inhibit myocyte sodium channels.[98] Hexane, a hydrocarbon, demonstrated cardiotoxicity in rats, lowering the threshold for ventricular fibrillation and reducing myocardial magnesium and potassium levels.[99]

Some ECG changes have been noted, with toluene abuse in children increasing P wave, QRS, and QT durations along with QT dispersion, over controls.[100] Echocardiography also revealed increased diameters of the left ventricle, left atrium, ascending aorta, and coronary arteries in children abusing toluene in comparison to normal subjects.[101] Although rare, complications (e.g., dilated cardiomyopathy, myocarditis, and myocardial infarctions) have been reported with toluene.[102,103]

Fatal bradycardia can also result, normally from a direct effect on the cardiac conduction system. Sinoatrial node automaticity is reduced and an atrioventricular block can be induced.[104] Less commonly, bradycardia can result from profound vagal stimulation, particularly when agents are sprayed directly into the mouth.[53]

More than half of all deaths from volatile substance abuse are due to sudden sniffing death syndrome.[94] From among the numerous noncardiovascular complications, anoxia is another key cause of death. This can be caused by aspiration of vomitus, respiratory depression, laryngeal edema, or asphyxiation. Some users wrap a plastic bag around their head to ensure minimal atmospheric escape of the substance. When the neuropsychiatric effect of the drug causes impaired consciousness, this can result in a fatal outcome.

Nitrites are somewhat different from hydrocarbons (e.g., toluene, propane, butane, xylene) and ketones (e.g., acetone) in that they have marked direct effects on the vasculature, causing smooth muscle relaxation and vasodilation.[94] Adverse effects can include those normally described for medical nitrate use, such as flushing, tachycardia, hypotension, penile engorgement, and syncope. Nitrites can also cause methemoglobinemia.[94]

15.7.3 Treatment

Airway damage may compromise breathing and anesthetic support should be sought as appropriate. As with all other drugs described, neuropsychiatric complications may be the chief concern and should be promptly addressed. Much of the time this simply means supportive measures in a calm, quiet environment. If systemic features develop, patients should be observed for 12 hours after inhalation. If they are asymptomatic after this period, they may be discharged safely.[5]

ECG monitoring should be continuous and serial 12-lead ECGs should be recorded. Metabolic status, including the presence of rhabdomyolysis, should be thoroughly assessed, as a metabolic acidosis can occur, further promoting cardiac arrhythmias. QRS prolongation may respond to sodium bicarbonate and QTc prolongation should be treated with magnesium sulfate.[5]

Hypotension should be treated with intravenous fluids. Patients exhibiting fluid-refractory hypotension can deteriorate rapidly and should be managed in a critical care environment. Inotropic medications pose a theoretical risk in the context of a sensitized myocardium and should be avoided. Cardiac arrhythmias and myocardial ischemia should be treated conventionally, with hemodynamically significant tachyarrhythmias being cardioverted promptly. Sustained tachyarrhythmias may respond to beta blockers or amiodarone. Significant bradyarrhythmias, conversely, may require atropine and temporary pacing.[53]

If pulmonary edema occurs, it is unlikely to be of cardiac origin and should be treated with respiratory support, such as continuous positive airway pressure or mechanical ventilation, because this is similar to acute respiratory distress syndrome.

15.8 KETAMINE AND PHENCYCLIDINE

Ketamine is well known to medical professionals because it has been used in anesthesia for several decades. Over recent years is has also grown in popularity as an illegal drug of abuse, taken intravenously, smoked, or via nasal inhalation. 1-(1-phenylcyclohexyl) piperidine, or phencyclidine or simply PCP, belongs to the same family as ketamine and can be administered via the same routes; however, its use in the United Kingdom is extremely rare and limited data are available.

15.8.1 Pharmacology

Ketamine and PCP are antagonists of the N-methyl-D-aspartate (NDMA) receptor, part of the glutaminergic pathway, which accounts for its analgesic properties. However, they also exhibit activity affecting opioid receptors, cholinergic neurotransmission, act as serotonin and noradrenaline reuptake inhibitors, and inhibit some sodium and voltage-gated calcium channels.[105] Ketamine has a half-life of 2.5 to 3 hours and effects typically develop minutes after drug use and last under 2 hours.

Ketamine produces an increase in blood pressure, heart rate, and stroke volume, but systemic vascular resistance is unchanged. Myocardial ischemia may result, especially in patients with preexisting heart disease for whom ketamine has been shown to reduce systolic and diastolic function. These effects are, in keeping with many recreational drugs, thought to be mediated by the adrenergic properties of ketamine.[105]

15.8.2 Adverse Effects

The vast majority of published data pertains to medicinal ketamine, where doses are closely monitored. Cardiovascular complications are exceedingly rare, despite the theoretical risk of a sympathetic toxidrome that could result in similar complications to amphetamine or cocaine toxicity. Even when ketamine is used illegally and taken in a far less regulated manner, scant information exists to suggest it poses a significant cardiovascular risk.[106]

Chronic ketamine use has been examined in mice, where after six weeks of ketamine administration, ST elevation and elevated troponin levels were observed in one-third of the subjects, suggesting myocardial damage. When combined with alcohol, the damage was even greater.[107]

15.8.3 Treatment

Management of cardiovascular complications would be in line with cocaine and amphetamines, but it is unlikely that ketamine alone would result in significant cardiovascular embarrassment. Standard measures are required, namely blood pressure and ECG monitoring, metabolic assessment, and supportive management for neurological complications. Benzodiazepines can prove useful not only for agitation but for reducing any excess sympathomimetic drive. Beta blockers are safe in ketamine toxicity.

15.9 NOVEL AND UNCOMMON DRUGS

The "designer drug" market evolves rapidly to meet demand and to attempt to evade legislation. As a result, very few data exist on the effects and risks of newer agents and what are sometimes referred to as "legal highs." For the clinician; fortunately, many are chemically similar to existing compounds and in the absence of evidence to suggest otherwise, management of potential cardiovascular toxicity would be in line with preexisting guidance for the more familiar drugs. (For a summary of them and treatment recommendations, see Table 15.1.)

Khat has been used at least since the seventh century, whereas mephedrone is a relatively new drug. However, many more enactogenic and hallucinogenic agents exist that exploit similar pharmacological pathways, and management of toxicity broadly follows the same route. These two drugs in particular produce their effects via the monoamine alkaloid cathinone, which is chemically similar to amphetamine. Khat is a plant that is chewed, whereas mephedrone is a synthetic compound that can be inhaled, absorbed transnasally, ingested orally, or injected intravenously. Their effects are similar to the amphetamine-like compounds previously described.

Khat has been associated with incidences of elevated blood pressure and heart rate, coronary vasospasm, myocardial infarction, arrhythmias, and

TABLE 15.1 Summary of Recreational Drugs, Interactions with the Cardiovascular System, Complications, and Treatment

Drug	Mechanism	Complication	Management
Cocaine	Na, K, and Ca channel blockade	Bradyarrhythmias	Atropine, pacing
		Supraventricular tachycardias	Observe, adenosine, DCCV, avoid beta blockers
		Ventricular arrhythmias	DCCV, sodium bicarbonate, lidocaine, magnesium sulfate
		Acute heart failure/cardiomyopathy	Diuresis, inotropes, ventilation, IABP, conventional
	Prothrombotic effect via platelet activation, plaque disruption, and increased clotting factors	Myocardial ischemia	Diazepam, nitrates, primary PCI for STEMI, conventional ACS management (DAPT, fondaparinux)
	Premature atherosclerosis		
	Catecholamine reuptake inhibition: sympathetic toxidrome causing coronary vasospasm and increased myocardial O_2 demand		*Consider*: Calcium channel antagonists, phentolamine
		Hypertensive emergencies	Diazepam, nitrates, calcium channel antagonists, phentolamine
MDMA and amphetamines	Catecholamine reuptake inhibition: sympathetic toxidrome	Hypertension, arrhythmias	As for cocaine
Heroin	Sympathetic inhibition, mast cell degranulation	Bradycardia	Atropine, pacing
		Tachyarrhythmias	As for cocaine
		Hypotension	IV fluids, vasopressors, inotropes
		Endocarditis (*as with any injected drug*)	Conventional

Cannabis	Via endocannabinoid system *Low dose*: sympathetic effects *High dose*: parasympathetic effects	Tachy- and bradyarrhythmias ACS	Observation As for cocaine if treatment required (*rare*) Conventional
LSD, magic mushrooms	Serotonergic, dopaminergic, and adrenergic pathways	Arrhythmias Vasospasm ACS	Observe As for cocaine if treatment required Conventional
Inhaled volatile substances	*Diverse mechanisms*: e.g., affecting ion channels, cell membranes, or glutaminergic, dopaminergic, GABAergic, and serotonergic pathways	Sudden death Bradyarrhythmias Supraventricular arrhythmias Ventricular arrhythmias Hypotension ACS, cardiomyopathy	ALS guidelines Atropine, pacing Adenosine, DCCV, beta blockers DCCV, lidocaine, amiodarone IV fluids, vasopressors, *avoid inotropes* As for cocaine
Ketamine	Catecholamine reuptake inhibition, also affect Na and Ca channels	Hypo- or hypertension, arrhythmias Myocardial ischemia	Observe As for cocaine if treatment required Conventional
Mephedrone and khat	Via cathinone; similar to amphetamine	Arrhythmias, hypertension, cardiomyopathy, and ACS	As for cocaine

Na, sodium; K, potassium; Ca, calcium; DCCV, direct current cardioversion; IABP, intra-aortic balloon pump; PCI, percutaneous intervention; STEMI, ST-segment elevation infarct; ACS, acute coronary syndrome; DAPT, dual antiplatelet therapy; MDMA, 3,4-methylenedioxy-N-methylamphetamine; LSD, lysergic acid diethylamide; GABA, γ-aminobutyric acid dilated cardiomyopathy.[108] It enjoys popularity in Africa and the Middle East and data from Yemen have shown a peak daily incidence of myocardial infarction at "chewing time" and a premature onset of cardiovascular disease among khat chewers. However, tobacco smoking is a major confounder as this is more common among those who chew khat.[109]

Mephedrone has been directly implicated in more than 100 deaths in the United Kingdom, although little information is available regarding its effects.[110] Worrisome is that between 2011 and 2012 the number of deaths in the United Kingdom tripled from 6 to 18.[2] Again, it produces a similar clinical profile to khat or amphetamines and has specifically been associated with acute myocarditis.[111] 2,5-dimethoxy-4-iodoamphetamine (DOI) falls into the substituted amphetamine class and, again, very sparse data exist.

Methoxetamine is a ketamine derivative that also inhibits NDMA receptors and acts as a selective serotonin inhibitor. It has caused hospitalizations with complaints similar to ketamine. During a four-month period in 2012, 47 telephone inquiries were made to the U.K. National Poisons Information Service, with the most common presenting complaint being that of sympathetic nervous system symptoms.[112] The death of a young man has been reported; he was found to have high serum levels of methoxetamine and synthetic cannabinoids postmortem.[113] Another NDMA receptor antagonist is 3-methoxyphencyclidine (3-MeO-PCP), which also shares similarities with PCP.

Synthetic cannabinoids are normally collectively referred to as "spice." These have gained in popularity in recent years and are easily purchased, often legally, over the Internet. Preliminary information suggests they produce their effects via the endocannabinoid system in a similar manner to THC. Reports of fatal intoxication from spice use are increasing; calls for advice regarding it to the American Association of Poison Control Centers rose from 53 calls in 2009 to more than 6000 in 2011. Synthetic cannabinoids may have a stronger affinity to and greater potency at cannabinoid receptors, in addition to having a longer half-life.[114]

15.10 BODY PACKERS

A scenario commonly encountered by health care professionals is that of "body packing," which was first reported in 1973 and now forms a major method of transporting illegal drugs across national and international borders.[115] Body packers are individuals who swallow sealed packets of drugs, often a condom coated with wax, to transport them between countries. Body packers usually carry approximately 1 kg of a drug, divided into packets of around 10 g each, but case reports of patients having ingested 200 packets have been described.[116] The most common narcotics transported in this way are cocaine, heroin, and amphetamines. Each packet of these agents contains a potentially lethal dose.[115] Body packers can suffer a range of medical complications, however, in terms of cardiovascular effects, the practice simply represents a method of rapidly introducing a large dose into the body.

Treatment is normally conservative, allowing the packets to pass. Those who have consumed packets of cocaine are at the highest risk of any group because no known drug effectively antagonizes the toxic effects.[115] Urgent surgical removal is indicated if the packets are suspected to have leaked. Supportive measures can be used to counteract neurological and cardiovascular complications. Benzodiazepines are administered for seizures and agitation, which can also reduce heart rate as a secondary effect. Ventricular arrhythmias and hypertension are treated as previously detailed.

15.11 CONCLUSION

A vast and ever-developing range of recreational drugs exists and the majority have cardiovascular effects. While management is often supportive, the signs and symptoms of cardiovascular toxicity from recreational drugs must be identified to guide treatment. The burden on health services is created not only by acute toxicity, but by longer-term effects. While cardiorespiratory effects are the most common cause of death from recreational drug use, their adverse effects are wide-ranging and require a multifaceted approach with involvement from several specialties. In particular, acute toxicity is frequently typified by neuropsychiatric symptoms, which have not been discussed in this chapter but clearly have the potential to complicate management of a hemodynamically unstable patient.[117]

Users of recreational substances invariably present having used more than one drug, with alcohol and tobacco frequently taken along with something illegal. Even considering the most studied recreational drug, cocaine, overall data surrounding cardiovascular effects are scarce.

Key Learning Points

- Cocaine produces cardiovascular toxicity via multiple routes:
  - Sympathetic toxidrome promoting myocardial ischemia
  - Coronary vasospasm
  - Promotion of arrhythmias by acting directly on Na, K, and Ca channels
  - Prothrombotic effects
  - Cardiomyopathic effects
  - Direct action on the endothelium
- The mainstay of treatment for chest pain associated with stimulant drugs, such as cocaine or amphetamine-related drugs, centers around benzodiazepine use and nitrates if hypertensive, along with conventional antiplatelets. However, the rate of myocardial infarction in patients presenting with cocaine-associated chest pain is low.
- Beta blockers are contraindicated in cocaine and amphetamine toxicity as they can result in unopposed α-adrenergic activity, which can exacerbate hypertension and vasospasm.

- Recreational drug-induced ventricular arrhythmias should be treated with sodium bicarbonate or lidocaine; no data exist for amiodarone.
- Volatile substance inhalation can result in sudden sniffing death syndrome, thought to be caused by a myocardium that has been sensitized to adrenaline by the volatile substance, resulting in a ventricular arrhythmia.
- Management of recreational drug toxicity is predominantly supportive and neuropsychiatric symptoms often dominate the clinical picture so a multidisciplinary approach is essential.

REFERENCES

1. Department of Health [Internet]. Reducing Drugs Misuse and Dependence. [accessed 4.20.13]. Available from: <www.gov.uk/government/policies/reducing-drugs-misuse-and-dependence>.
2. Home Office [Internet]. Drug misuse declared: findings from the 2011 to 2012 Crime Survey for England and Wales (CSEW). 2nd ed. [accessed 5.29.13]. Available from: <www.gov.uk/government/publications/drug-misuse-declared-findings-from-the-2011-to-2012-crime-survey-for-england-and-wales-csew-second-edition/drug-misuse-declared-findings-from-the-2011-to-2012-crime-survey-for-england-and-wales-csew-second-edition>.
3. Qureshi AI, Suri MF, Guterman LR, et al. Clinical investigation and reports cocaine use and the likelihood of nonfatal myocardial infarction and stroke data from the Third National Health and Nutrition Examination survey. *Circulation* 2001;**103**(4):502–6.
4. Afonso L, Mohammad T, Thatai D. Crack whips the heart: a review of the cardiovascular toxicity of cocaine. *Am J Cardio* 2007;**100**(6):1040–3.
5. National Poisons Information Service [Internet]. TOXBASE [accessed April–June 2013]. Available from: <www.toxbase.org>.
6. Substance Abuse and Mental Health Services Administration, Office of Applied Studies. Emergency Department Trends From the Drug Abuse Warning Network, Final Estimates 1995–2002. DAWN Series: D-24, DHHS Publication No. SMA-03-3780. U.S. Department of Health and Human Services; 2003.
7. Weber JE, Chudnofsky CR, Boczar M, et al. Cocaine-associated chest pain: how common is myocardial infarction? *Acad Emerg Med* 2000;**7**:873–7.
8. Turillazzi E, Bello S, Neri M, et al. Cardiovascular effects of cocaine: cellular, ionic and molecular mechanisms. *Curr Med Chem* 2012;**19**(33):5664–76.
9. Schwartz BG, Rezkella S, Kloner RA. Cardiovascular effects of cocaine. *Circulation* 2010; **122**:2558–69.
10. Milroy CM, Parai JL. The histopathology of drugs of abuse. *Histopathol* 2011;**59**(4):579–93.
11. Vongpatanasin W, Mansour Y, Chavoshan B, et al. Cocaine stimulates the human cardiovascular system via a central mechanism of action. *Circulation* 1999;**100**:497–502.
12. McCord J, Jneid H, Hollander J, et al. Management of cocaine-associated chest pain and myocardial infarction: a scientific statement from the American Heart Association. *Circulation* 2008;**117**(14):1897–907.
13. Boehrer JD, Moliterno DJ, Willard JE, et al. Hemodynamic effects of intranasal cocaine in humans. *J Am Coll Cardiol* 1992;**20**:90–3.
14. Mehta PM, Grainger TA, Lust RM, et al. Effect of cocaine on left ventricular function: relation to increased wall stress and persistence after treatment. *Circulation* 1995;**91**:3002–9.

15. Wilbert-Lampen U, Seliger C, Zilker T, et al. Cocaine increases the endothelial release of immunoreactive endothelin and its concentrations in human plasma and urine: reversal by coincubation with sigma-receptor antagonists. *Circulation* 1998;**98**:385–90.
16. Mo W, Singh AK, Arruda JA, et al. Role of nitric oxide in cocaine-induced acute hypertension. *Am J Hypertens* 1998;**11**:708–14.
17. Knopf H, Theising R, Moon CH, et al. Continuous determination of extracellular space and changes of K^+, Na^+, Ca^{2+}, and H^+ during global ischaemia in isolated rat hearts. *J Mol Cell Cardiol* 1990;**22**:1259–72.
18. O'Leary ME, Hancox JC. Role of voltage-gated sodium, potassium and calcium channels in the development of cocaine-associated cardiac arrhythmias. *Br J Clin Pharm* 2010; **69**(5):427–42.
19. Hoffman R. Treatment of patients with cocaine-induced arrhythmias: bringing the bench to the bedside. *Br J Clin Pharmacol* 2010;**69**(5):448–57.
20. Schwartz AB, Janzen D, Jones RT, et al. Electrocardiographic and hemodynamic effects of intravenous cocaine in awake and anesthetized dogs. *J Electrocardiol* 1989;**22**:159–66.
21. Gussak I, Antzelevitch C, Bjerregaard P, et al. The Brugada syndrome: clinical, electrophysiologic and genetic aspects. *J Am Coll Cardiol* 1999;**33**:5–15.
22. Bauman JL, DiDomenico RJ. Cocaine-induced channelopathies: emerging evidence on the multiple mechanisms of sudden death. *J Cardiovasc Pharmacol Ther* 2002;**7**:195–202.
23. Foltin RW, Fischman MW. Ethanol and cocaine interactions in humans: cardiovascular consequences. *Pharmacol Biochem Behav* 1988;**31**:877–83.
24. McCance-Katz EF, Price LH, McDougle CJ, et al. Concurrent cocaine-ethanol ingestion in humans: pharmacology, physiology, behavior, and the role of cocaethylene. *Psychopharmacol* 1993;**111**:39–46.
25. Schrem SS, Belsky P, Schwartzman D, et al. Cocaine-induced torsades de pointes in a patient with the idiopathic long QT syndrome. *Am Heart J* 1990;**120**:980–4.
26. Yap TG, Behr ER, Camm AJ. Brugada. *Europace* 2009;**11**(8):989–94.
27. Stenberg RG, Winniford MD, Hillis LD, et al. Simultaneous acute thrombosis of two major coronary arteries following intravenous cocaine use. *Arch Pathol Lab Med* 1989;**113**: 521–4.
28. Heesch CM, Wilhelm CR, Ristich J, et al. Cocaine activates platelets and increases the formation of circulating platelet containing microaggregates in humans. *Heart* 2000;**83**: 688–95.
29. Moliterno DJ, Lange RA, Gerard RD, et al. Influence of intranasal cocaine on plasma constituents associated with endogenous thrombosis and thrombolysis. *Am J Med* 1994; **96**:492–6.
30. Siegel AJ, Mendelson JH, Sholar MB, et al. Effect of cocaine usage on C-reactive protein, von Willebrand factor, and fibrinogen. *Am J Cardiol* 2002;**89**:1133–5.
31. Satran A, Bart BA, Henry CR, et al. Increased prevalence of coronary artery aneurysms among cocaine users. *Circulation* 2005;**111**:2424–9.
32. Lai S, Lima JA, Lai H, et al. Human immunodeficiency virus 1 infection, cocaine, and coronary calcification. *Arch Intern Med* 2005;**165**:690–5.
33. Kolodgie FD, Virmani R, Cornhill JF, et al. Increase in atherosclerosis and adventitial mast cells in cocaine abusers: an alternative mechanism of cocaine-associated coronary vasospasm and thrombosis. *J Am Coll Cardiol* 1991;**17**:1553–60.
34. Zimmerman FH, Gustafson GM, Kemp Jr HG. Recurrent myocardial infarction associated with cocaine abuse in a young man with normal coronary arteries: evidence for coronary artery spasm culminating in thrombosis. *J Am Coll Cardiol* 1987;**9**(4):964–8.

35. Ascher EK, Stauffer JC, Gaasch WH. Coronary artery spasm, cardiac arrest, transient electrocardiographic Q waves and stunned myocardium in cocaine-associated acute myocardial infarction. *Am J Cardiol* 1988;**61**(11):939–41.
36. Ibrahim M, Hasan R, Awan M. Cocaine-induced coronary stent thrombosis. *Exp Clin Cardiol* 2013;**18**(1):e57–9.
37. Maraj S, Figueredo VM, Morris DL. Cocaine and the heart. *Clin Cardiol* 2010;**33**(5): 264–9.
38. Chokshi SK, Moore R, Pandian NG, et al. Reversible cardiomyopathy associated with cocaine intoxication. *Ann Intern Med* 1989;**111**(12):1039–40.
39. Om A, Warner M, Sabri N, et al. Frequency of coronary artery disease and left ventricle dysfunction in cocaine users. *Am J Cardiol* 1992;**69**:1549–52.
40. Willens HJ, Chakko SC, Kessler KM. Cardiovascular manifestations of cocaine abuse: a case of recurrent dilated cardiomyopathy. *Chest* 1994;**106**:594–600.
41. Eskander KE, Brass NS, Gelfand ET. Cocaine abuse and coronary artery dissection. *Ann Thorac Surg* 2001;**71**:340–1.
42. Amin M, Gabelman G, Karpel J, et al. Acute myocardial infarction and chest pain syndromes after cocaine use. *Am J Cardiol* 1990;**66**:1434–7.
43. Mittleman MA, Mintzer D, Maclure M, et al. Triggering of myocardial infarction by cocaine. *Circulation* 1999;**99**:2737–41.
44. Gitter MJ, Goldsmith SR, Dunbar DN, et al. Cocaine and chest pain: clinical features and outcome of patients hospitalized to rule out myocardial infarction. *Ann Intern Med* 1991; **115**:277–82.
45. Gamouras GA, Monir G, Plunkitt K, et al. Cocaine abuse: repolarization abnormalities and ventricular arrhythmias. *Am J Med Sci* 2000;**320**:9–12.
46. Brickner ME, Willard JE, Eichhorn EJ, et al. Left ventricular hypertrophy associated with chronic cocaine abuse. *Circulation* 1991;**84**:1130–5.
47. Kontos MC, Jesse RL, Tatum JL, et al. Coronary angiographic findings in patients with cocaine-associated chest pain. *J Emerg Med* 2003;**24**(1):9–13.
48. Weber JE, Shofer FS, Larkin GL, et al. Validation of a brief observation period for patients with cocaine-associated chest pain. *N Engl J Med* 2003;**348**:510–7.
49. Hoffman RS, Hollander JE. Thrombolytic therapy and cocaine-induced myocardial infarction. *Am J Emerg Med* 1996;**14**:693–5.
50. Hollander JE. The management of cocaine-associated myocardial ischemia. *N Engl J Med* 1995;**333**:1267–72.
51. Baumann BM, Perrone J, Hornig SE, et al. Randomized, double-blind, placebo-controlled trial of diazepam, nitroglycerin, or both for treatment of patients with potential cocaine-associated acute coronary syndromes. *Acad Emerg Med* 2000;**7**:878–85.
52. Ferdinand KC. Substance abuse and hypertension. *J Clin Hypertens* 2000;**2**(1):37–40.
53. Ghuran A, Nolan J. Recreational drug misuse: issues for the cardiologist. *Heart* 2000; **83**(6):627–33.
54. Lange RA, Cigarroa RG, Yancy Jr CW, et al. Cocaine-induced coronary-artery vasoconstriction. *N Engl J Med* 1989;**321**:1557–62.
55. Guinn MM, Bedford JA, Wilson MC. Antagonism of intravenous cocaine lethality in nonhuman primates. *Clin Toxicol* 1980;**16**:499–508.
56. Lange RA, Cigarroa RG, Flores ED, et al. Potentiation of cocaine-induced coronary vasoconstriction by beta-adrenergic blockade. *Ann Intern Med* 1990;**112**:897–903.
57. Boehrer JD, Moliterno DJ, Willard JE, et al. Influence of labetalol on cocaine-induced coronary vasoconstriction in humans. *Am J Med* 1993;**94**:608–10.

58. Dattilo PB, Hailpern SM, Fearon K, et al. Beta-blockers are associated with reduced risk of myocardial infarction after cocaine use. *Ann Emerg Med* 2008;**51**(2):117–25.
59. Littmann L, Narveson SY, Fesel NM, et al. Beta blocker treatment of heart failure patients with ongoing cocaine use. *Int J Cardiol* 2013;**168**(3):2919–20.
60. Shih RD, Hollander JE, Burstein JL, et al. Clinical safety of lidocaine in patients with cocaine-associated myocardial infarction. *Ann Emerg Med* 1995;**26**:702–6.
61. Chambers HF, Morris DL, Tauber MG, et al. Cocaine use and the risk for endocarditis in intravenous drug users. *Ann Intern Med* 1987;**106**:833–6.
62. Hsue PY, Salinas CL, Bolger AF, et al. Acute aortic dissection related to crack cocaine. *Circulation* 2002;**105**:1592–5.
63. Singh S, Trivedi A, Adhikari T, et al. Cocaine-related acute aortic dissection: patient demographics and clinical outcomes. *Can J Cardiol* 2007;**23**(14):1131–4.
64. Hollander JE, Hoffman RS, Gennis P, et al. Cocaine-associated chest pain: one-year follow-up. *Acad Emerg Med* 1995;**2**:179–84.
65. Drugscope [Internet]. Ecstasy. [accessed 4.29.13]. Available from: <www.drugscope.org.uk/resources/drugsearch/drugsearchpages/ecstasy>.
66. Wong GTC, Irwin MG. Poisoning with illicit substances: toxicology for the anaesthetist. *Anaesthesia* 2013;**68**(1):117–24.
67. Hill SL, Thomas SHL. Clinical toxicology of newer recreational drugs. *Clin Toxicol* 2011;**49**(8): 705–19.
68. Quinton MS, Yamamoto BK. Causes and consequences of methamphetamine and MDMA toxicity. *AAPS J* 2006;**8**(2):E337–47.
69. Hall AP, Henry JA. Acute toxic effects of "Ecstasy" (MDMA) and related compounds: overview of pathophysiology and clinical management. *Br J Anaesth* 2006;**96**(6): 678–85.
70. Furlanello F, Serdoz LV, Cappato R, et al. Illicit drugs and cardiac arrhythmias in athletes. *G Ital Cardiol* 2010;**11**(10 Suppl 1):118S–21S.
71. Kaumann AJ, Levy FO. 5-hydroxytryptamine receptors in the human cardiovascular system. *Pharmacol Ther* 2006;**111**(3):674–706.
72. Cosyns B, Droogmans S, Rosenhek R, et al. Drug-induced valvular heart disease. *Heart* 2013;**99**:7–12.
73. Mizia-Stec K, Gasior Z, Wojnicz R, et al. Severe dilated cardiomyopathy as a consequence of ecstasy intake. *Cardiovasc Pathol* 2008;**17**(4):250–3.
74. Hall AP. Ecstasy and the anaesthetist. *Br J Anaesth* 1997;**79**:697–8.
75. Hysek CM, Vollenweider FX, Liechti ME. Effects of a β-blocker on the cardiovascular response to MDMA (ecstasy). *Emerg Med J* 2010;**27**:586–9.
76. Lipski J, Stimmel B, Donoso E. The effect of heroin and multiple drug abuse on the ECG. *Am Heart J* 1973;**86**:663–8.
77. Routsi C, Kolias S, Kaskarellis I, et al. Acute cardiomyopathy and cardiogenic pulmonary edema after inhaled heroin. *Acta Anaesthesiol Scand* 2007;**51**(2):262–4.
78. Montecucco F, Di Marzo V. At the heart of the matter: the endocannabinoid system in cardiovascular function and dysfunction. *Trends Pharmacol Sci* 2012;**33**(6):331–40.
79. Singla S, Sachdeva R, Mehta J. Cannabinoids and atherosclerotic coronary heart disease. *Clin Cardiol* 2012;**35**(6):329–35.
80. Mittleman MA, Lewis RA, Maclure M, et al. Triggering myocardial infarction by marijuana. *Circulation* 2001;**103**(23):2805–9.
81. Pratap B, Korniyenko A. Toxic effects of marijuana on the cardiovascular system. *Cardiovasc Toxicol* 2012;**12**(2):143–8.

82. Basnet S, Mander G, Nicolas R. Coronary vasospasm in an adolescent resulting from marijuana use. *Pediatr Cardiol* 2009;**30**(4):543–5.
83. Deusch E, Kress HG, Kraft B, et al. The procoagulatory effects of delta-9-tetrahydrocannabinol in human platelets. *Anesth Analg* 2004;**99**:1127–30.
84. Kochar M, Hosko MJ. Electrocardiographic effects of marijuana. *JAMA* 1973;**225**:25–7.
85. Fisher BA, Ghuran A, Vadamalai V, et al. Cardiovascular complications induced by cannabis smoking: a case report and review of the literature. *Emerg Med J* 2005;**22**(9): 679–80.
86. Passie T, Halpern JH, Stichtenoth DO, et al. The pharmacology of lysergic acid diethylamide: a review. *CNS Neurosci Ther* 2008;**14**(4):295–314.
87. Passie T, Seifert J, Schneider U, et al. The pharmacology of psilocybin. *Addict Biol* 2002; **7**(4):357–64.
88. Hasler F, Grimberg U, Benz MA, et al. Acute psychological and physiological effects of psilocybin in healthy humans: a double-blind, placebo-controlled dose–effect study. *Psychopharm* 2004;**172**(2):145–56.
89. Raval MV, Gaba RC, Brown K, et al. Percutaneous transluminal angioplasty in the treatment of extensive LSD-induced lower extremity vasospasm refractory to pharmacologic therapy. *J Vasc Interv Radiol* 2008;**19**(8):1227–30.
90. Lieberman AN, Bloom W, Kishore PS, et al. Carotid artery occlusion following ingestion of LSD. *Stroke* 1974;**5**(2):213–5.
91. Borowiak KS, Ciechanowski K, Waloszczyk P. Psilocybin mushroom (*Psilocybe semilanceata*) intoxication with myocardial infarction. *J Toxicol Clin Toxicol* 1998;**36**(1–2): 47–9.
92. Bickel M, Ditting T, Watz H, et al. Severe rhabdomyolysis, acute renal failure and posterior encephalopathy after "magic mushroom" abuse. *Eur J Emerg Med* 2005;**12**(6): 306–8.
93. Klock JC, Boerner U, Becker CE. Coma, hyperthermia and bleeding associated with massive LSD overdose: a report of eight cases. *West J Med* 1974;**120**:183–8.
94. Williams JF, Storck M. Inhalant abuse. American Academy of Pediatrics Committee on Substance Abuse; American Academy of Pediatrics Committee on Native American Child Health. *Pediatrics* 2007;**119**(5):1009–17.
95. Cunningham SR, Dalzell GWN, McGirr P, et al. Myocardial infarction and primary ventricular fibrillation after glue sniffing. *Br Med J* 1987;**294**:739–40.
96. Bass M. Sudden sniffing death. *JAMA* 1970;**212**:2075–9.
97. Boon NA. Solvent abuse and the heart. *Br Med J* 1987;**294**(6574):722.
98. Cruz SL, Orta-Salazar G, Gauthereau MY, et al. Inhibition of cardiac sodium currents by toluene exposure. *Br J Pharmacol* 2003;**140**(4):653–60.
99. Khedun SM, Maharaj B, Naicker T. Hexane cardiotoxicity: an experimental study. *Isr J Med Sci* 1996;**32**(2):123–8.
100. Alper AT, Akyol A, Hasdemir H, et al. Glue (toluene) abuse: increased QT dispersion and relation with unexplained syncope. *Inhal Toxicol* 2008;**20**(1):37–41.
101. Olgar S, Dindar A, Ertugrul T, et al. Electrocardiographic and echocardiographic findings in street children known to be substance abusers. *Cardiol Young* 2008;**18**(1):58–61.
102. Wiseman MN, Banim S. "Glue sniffer's" heart? *Br Med J* 1987;**294**:739.
103. Carder JR, Fuerst RS. Myocardial infarction after toluene inhalation. *Pediatr Emerg Care* 1997;**13**(2):117–9.
104. Einav S, Amitai Y, Reichman J, et al. Bradycardia in toluene poisoning. *J Toxicol Clin Toxicol* 1997;**35**:295–8.

105. Craven R. Ketamine. *Anaesthesia* 2007;**62**(Suppl 1):48–53.
106. Strayer RJ, Nelson LS. Adverse events associated with ketamine for procedural sedation in adults. *Am J Emerg Med* 2008;**26**(9):985–1028.
107. Chan WM, Liang Y, Wai MS, et al. Cardiotoxicity induced in mice by long term ketamine and ketamine plus alcohol treatment. *Toxicol Lett* 2011;**207**(2):191–6.
108. Apps A, Matloob S, Dahlal MT, et al. Khat: an emerging threat to the heart in the UK. *Postgrad Med J* 2011;**87**:387–8.
109. Al-Motarreb A, Al-Kebsi M, Al-Adhi B, et al. Khat chewing and acute myocardial infarction. *Heart* 2002;**87**(3):279–80.
110. Schifano F, Corkery J, Ghodse AH. Suspected and confirmed fatalities associated with mephedrone (4-methylmethcathinone, "meow meow") in the United Kingdom. *J Clin Psychopharmacol* 2012;**32**(5):710–4.
111. Nicholson PJ, Quinn MJ, Dodd JD. Headshop heartache: acute mephedrone "meow" myocarditis. *Heart* 2010;**96**(24):2051–2.
112. Hill SL, Harbon SC, Coulson J, et al. Methoxetamine toxicity reported to the National Poisons Information Service: clinical characteristics and patterns of enquiries (including the period of the introduction of the UK's first Temporary Class Drug Order). *Emerg Med J* 2014;**31**(1):45–7.
113. Wikström M, Thelander G, Dahlgren M, et al. An accidental fatal intoxication with methoxetamine. *Anal Toxicol* 2013;**37**(1):43–6.
114. Seely KA, Lapoint J, Moran JH, et al. Spice drugs are more than harmless herbal blends: a review of the pharmacology and toxicology of synthetic cannabinoids. *Prog Neuro-psychopharmacol Biol Psychiatry* 2012;**39**(2):234–43.
115. Traub SJ, Hoffman RS, Nelson LS. Body packing—the internal concealment of illicit drugs. *N Engl J Med* 2003;**349**:2519–26.
116. Bulstrode N, Banks F, Shrotria S. The outcome of drug smuggling by "body-packers": the British experience. *Ann R Coll Surg Eng* 2002;**84**:35–8.
117. Coughlin PA, Mavor AI. Arterial consequences of recreational drug use. *Eur J Vasc Endovasc Surg* 2006;**32**(4):389–96.

Chapter 16

Pediatric Cardiovascular Toxicity: Special Considerations

Jennifer A. Lowry
Division of Clinical Pharmacology, Toxicology and Therapeutic Innovations, Children's Mercy Hospital, and University of Missouri, School of Medicine, Kansas City, Missouri

16.1 INTRODUCTION

Pediatric exposures to xenobiotics are more frequent than in any other age group. Because of this and the potential for lifelong harm, pediatricians were instrumental in establishing and promoting the practice of medical toxicology. Additionally, they were the founders of the poison control systems that are used today worldwide. Although the practice of medical toxicology and approaches to the diagnosis and treatment of poisoning are similar in children and adults, understanding the differences in exposures during fetal and child development is imperative in the care of the poisoned child.

16.2 PEDIATRIC POISONINGS

According to the Centers for Disease Control and Prevention (CDC) in the United States, the drug poisoning death rate in 2008 was four times the rate it was in 1999 and the leading cause of injury and deaths in 2008.[1] In addition, unintentional poisoning is the tenth leading cause of nonfatal injury seen in emergency departments (ED). However, it is the second leading cause of injury seen in EDs in 1- to 4-year-olds. This is surprising because the focus of toxicologists has been on preventing unintentional ingestions in young children. However, the emergence and increased utilization of poison control centers, as well as the development of child-resistant closures on pill bottles, has resulted in a rapid decline of morbidity and mortality of young children attributed to poisoning.

In 2011, 57 poison control centers (PCC) submitted data to the American Association of Poison Control Centers (AAPCC). Poison control centers cared for 2,333,004 human exposures with 1,449,186 (62%) occurring in children 19 years of age and under.[2] This is likely an underestimation

The Heart and Toxins. DOI: http://dx.doi.org/10.1016/B978-0-12-416595-3.00016-5

because poison control centers rely on voluntary reporting. By far, the majority of calls and exposures occurred in the home environment. Approximately 18% of calls to PCCs were from health care professionals, as patients may present to health care facilities due to decreased knowledge of the PCC or the severity of the exposure. Ingestions account for the majority (83%) of exposures; dermal and inhalation exposures occur less frequently (7% and 6.1%, respectively).

Similarly, the United Kingdom's National Poisons Information Service (NPIS) is a network of four units that provide poison information to health care workers by phone and supply information through the information database (TOXBASE®) to assist in the management of poisoned patients.[3] In 2012–2013, 607,163 inquiries were made to NPIS (an increase of 4.1%) using TOXBASE and phone consultations. More than a quarter (28.2%) involved children under the age of five. Similar to the AAPCC data, this is an underestimation of poisonings as NPIS also relies on voluntary reporting and only accepts inquiries from health care workers.

Data from the United States show that management largely occurs at the site of exposure (69%) with the primary site occurring at the patient's residence. Approximately 25% of patients were seen in an emergency department; however, these patients may not have required emergency care. Of those patients seen in the ED, the majority were treated and released, with approximately one quarter admitted for hospitalization.

According to the AAPCC, children are more likely to be poisoned by substances that are readily available. The most common substance categories resulting in exposures are cosmetics and/or personal care products (13.95%), analgesics (9.85%), household cleaning substances (9.18%), foreign bodies (6.9%), and topical preparations (6.55%). While these products, with the exception of analgesics, are not likely to result in significant toxicity, significant effects have been seen with pharmaceutical substances that are less likely to be ingested by the younger age group. Toxicity is more associated with the intent of the child, and the reason for exposures changes with age. Pharmaceutical agents are more likely to be the cause for inquiries to NPIS. The discrepancy between the two databases can be attributed to the user, in that the AAPCC is largely contacted by the public whereas NPIS is dedicated to health care professionals' inquiries.

16.2.1 Unintentional Exposures

Children younger than three were involved in 36.2% of exposures, and children younger than six accounted for approximately half of all of the human exposures (48.9%) called in to U.S. poison control centers. Of the pediatric exposures, approximately 80% of them occur in children under the age of six, with the majority as a result of unintentional exposures. Ingestions in young children are secondary to development of exploration behaviors and a tendency for hand-to-mouth activity. This results in pills, toys, paint chips, and other substances being ingested by children. Pediatricians and other health care professionals are encouraged to discuss these risks at the one-year well-child check-up to provide anticipatory guidance for prevention of future poisonings.

While the majority of ingestions in young children result in minor symptoms, they are at higher risk for the adverse effects that result from these exposures. It is commonly believed that small quantities of a drug would not cause harm, but select agents may cause morbidity and mortality.[4] These include substances such as calcium channel blockers, centrally acting alpha agonists, methyl salicylate, sulfonylureas, tricyclic antidepressants, chloroquine, diphenoxylate-atropine, toxic alcohols, and methadone. It is imperative for health care providers to consider poisoning in their differential and to understand the presenting signs and symptoms of these exposures so that appropriate care can be started in a timely manner.

The prescription drug abuse epidemic has resulted in unintentional exposures of young children to these medications. For example, the number of buprenorphine exposures of children younger than six reported to PCCs increased significantly from 2 cases (2002) to 907 cases (2008).[5] During this same period, exposures to methadone in the same age group doubled from 155 to 332 cases.

Even though not specific to children, NPIS receives inquiries in regard to the misuse of prescription and novel psychoactive drugs used in the United Kingdom. For the year 2012–2013, cocaine, methylenedioxymethamphetamine, heroin, cannabis, amphetamines, and mephedrone were the most commonly reported substances. The number of inquiries related to recreational substances increased compared to previous years. Likewise, the United Nations estimates that global drug use has remained stable.[6] However, as increased global trafficking occurs in unstable or developing countries, the number of illicit drug users may increase. Thus, it can be estimated that children, globally, are likely to have unintentional exposures to such drugs.

16.2.2 Intentional Exposures

As children age, the reasons for toxin exposures shift from unintentional to intentional. Adolescents (children over 12) are more likely to intentionally ingest drugs for abuse and suicide. One of the best long-term studies of adolescent substance use in the United States is the Monitoring the Future survey.[7] This survey has been conducted annually by The University of Michigan's Institute for Social Research since 1975 and is supported by grants from the National Institute on Drug Abuse. The 2012 data are the result of surveying approximately 45,400 8th-, 10th-, and 12th-grade students nationwide. While the most recent data suggest decreased drug use among high school teenagers, there is concern for persistently high rates of marijuana and nonmedical prescription drugs. Specifically, the use of nonmedically prescribed opioids occurs at high rates (e.g., hydrocodone use is estimated in 7.5% of all U.S. high school seniors). The use of prescription stimulants is also a cause for concern. Since 2009, the nonmedical use of prescription stimulants, such as Adderall® by 12th graders, has increased from 5.4 to 7.6% in 2012. In addition, approximately 68% of 12th graders admit to obtaining prescription pain relievers for free from friends and family.

16.2.3 Pediatric Poisoning Fatalities

According to the CDC National Vital Statistics 2012 Report, the percent of deaths from poisonings increased 80% between the years 2000 and 2009. In 2010, the CDC reported that 832 unintentional and 121 suicide poisoning deaths occurred in children between 1 and 19 years of age. Notably, in children 12 to 19 years of age, unintentional poison deaths were second only to motor vehicle accidents and were attributed to prescription drug abuse.[8]

In comparison, more than 1200 poisoning deaths were reported to poison control centers in 2011 with 70 occurring in the pediatric age group.[2,9] As previously mentioned, pharmaceutical products were more likely to result in fatalities compared to nonpharmaceutical products. Analgesics accounted for the majority of pediatric deaths (21%) followed by stimulants and street drugs (11.5%) and cold and cough preparations (9.6%). Of the pediatric deaths, 63% occurred in children under six. Notably, several cases were due to the lack of child-resistant packaging on older adult medications. Of the deaths in children over the age of six, 46.4% were the result of suicide with another 14.8% from substance abuse.

In recent years, pediatric deaths following the use of cough and cold preparations prompted the U.S. Food and Drug Administration (FDA) to recommend prohibiting their use by children younger than six. In response to this decision, a review panel of medical toxicologists and clinical pharmacologists assessed 189 cases and concluded that the majority of fatalities from nonprescription cough and cold preparations were uncommon and associated with overdoses rather than therapeutic doses.[10] Most of the fatalities were in children less than two years of age and were associated with intent, including malicious use or to induce sedation. The authors agree with the decision to limit its use in children due to the frequency of misuse of the medications.

16.2.4 Diagnosis and Management

The diagnosis of a poisoned child should be largely based on the presenting history, physical, and basic laboratory results. Although a urine drug screen is commonly employed in a broad variety of clinical contexts, studies confirm that using it for drugs of abuse will rarely offer new information to the practitioner that wasn't already suspected. In most instances, the screening test is a qualitative detection of specific drugs in the urine or blood by immunoassay, a technique fraught with many false-positive and false-negative result errors. These tests have many limitations because most immunoassays are capable of screening for commonly abused drugs in adults, including marijuana, opiates, amphetamines, benzodiazepines, and cocaine. The reference drug in a particular class (e.g., morphine for opiates) may have a different structure than more commonly used drugs (e.g., oxycodone) resulting in a negative drug screen. Alternatively, some pharmaceuticals may cause false-positive results due to the similarity in structures between drugs; for example, phencyclidine (PCP) positive with the therapeutic use of dextromethrophan. Additionally, children are less likely to be exposed to these drugs, which will result in a negative test and the erroneous belief that a poisoning has not occurred. Most pharmaceuticals and chemicals will not result in a positive urine drug screen suggesting that the utility of these tests by pediatricians may not be useful.

Hypoglycemia is one of the most easily detected and treated effects of poisons. All patients with altered mental status or ingestion of an agent with hypoglycemic potential should have a blood glucose concentration obtained. In addition, other chemistry tests can be helpful in the diagnosis of the poisoned patient. Serum electrolytes and a calculated high anion gap can guide the health care provider in determining that a poisoning has occurred and the resulting agent. Poisons that result in a high anion gap include salicylates, ethanol, ethylene glycol, methanol, isoniazid, and iron.

Management of poisonings should always begin with the ABCs of medicine: airway, breathing, and circulation. Decontamination with activated charcoal should be considered for the child who presents with a potentially harmful dose within an hour of the ingestion; has not ingested a hydrocarbon or seizure-causing medication; and is awake, alert, and maintaining the airway. Activated charcoal is not without complications (e.g., chemical pneumonitis) resulting in death. The use of emetics, such as syrup of ipecac, is no longer recommended.[11]

For the majority of poisonings, treatment is largely symptomatic and supportive. The key is recognizing the patient as poisoned and, if unknown, determining the etiology by the history, physical, and basic laboratory analysis. Thus, discussion with the PCC or a medical toxicologist is advised. Antidotes are available for very few poisons, and it is important to recognize that their inappropriate overuse can complicate the clinical situation.

16.3 DRUG DISPOSITION IN INFANTS AND CHILDREN

Pediatrics encompasses a broad range of ages at which certain stages of life profoundly influence drug response and disposition.[12] Dramatic pharmacokinetic, pharmacodynamic, and psychosocial changes occur as preterm infants mature toward term, as infants mature through the first few years of life, and as children reach puberty and adolescence (Figure 16.1). Dr. Abraham Jacobi,

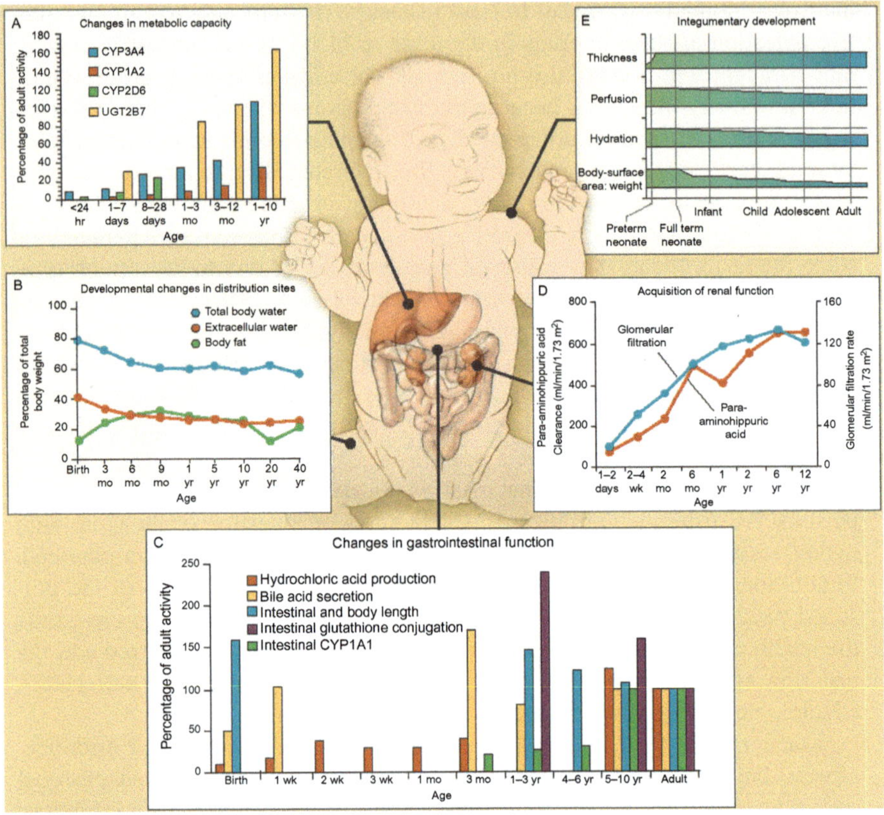

FIGURE 16.1 Developmental changes in physiologic factors that influence drug disposition in infants, children, and adolescents. Physiologic changes in multiple organ systems during development are responsible for age-related differences in drug disposition.

As shown in panel A, the activity of many cytochrome P450 (CYP) isoforms and a single glucuronsyltransferase (UGT) isoform is markedly diminished during the first two months of life. In addition, the acquisition of adult activity over time is enzyme- and isoform-specific.

Panel B shows the effect of postnatal development on the processes of active tubular secretion, represented by the clearance of para-aminohippuric acid and the glomerular filtration rate, both of which approximate adult activity by 6 to 12 months of age.

Panel C shows age-dependent changes in body composition, which influence the apparent volume of distribution of drugs. Infants during the first 6 months of life have markedly expanded total body water and extracellular water, expressed as a percentage of total body weight, as compared with older infants and adults.

Panel D summarizes the age-dependent changes in both the structure and function of the gastrointestinal tract. As with the hepatic drug-metabolizing enzymes (panel A), the activity of cytochrome P450 1A1 (CYP1A1) in the intestine is low during early life.

Panel E shows age dependence in the thickness, extent of perfusion, and extent of hydration of the skin and the relative size of the skin-surface area (reflected by the ratio of body-surface area to body weight). Although skin thickness is similar in infants and adults, the extent of perfusion and hydration diminishes from infancy to adulthood.

Source: *Reprinted with permission from Kearns et al. and the Massachusetts Medical Society.*[12]

the father of American pediatrics, recognized the need for specific dose therapy for children when he wrote: "Pediatrics does not deal with miniature men and women, with reduced doses with the same class of disease in smaller bodies, but ... has its own independent range and horizon." Unfortunately, this was not fully understood until the late twentieth century as drug development realized the need for the addition of children in clinical trials.

One exemplary episode from medical history is the case of the *gray baby syndrome*, a life-threatening condition produced by the administration of the antibiotic chloramphenicol to neonates and young infants.[13] The many neonatal deaths associated with this syndrome were the direct result of using chloramphenical at dosing regimens which, while safe and effective for older children, were not appropriately "adjusted" for the delayed maturation of a metabolic enzyme (glucuronosyltransferase) responsible for catalyzing the biotransformation of the drug.[14] Despite this oversight, therapeutic advances have been made in regard to children due to investigations in the field of pediatric clinical pharmacology with a focus on developmental changes on drug disposition and action. These differences can influence basic pharmacokinetic properties of a given medication and how development can impact drug disposition. Thus, it is imperative to examine the impact of development on pharmacokinetics, especially absorption, distribution, metabolism, and elimination.

16.3.1 Absorption

Largely, absorption occurs via passive diffusion, but active transport or facilitated diffusion may also be necessary for drug entry into cells. Several physiologic factors affect this process, one or more of which may be altered in the face of certain disease states (e.g., inflammatory bowel disease, diarrhea) and, consequently, produce changes in drug bioavailability. The most important factors that influence drug absorption from the gastrointestinal (GI) tract are related to the physiology of the stomach, intestine, and biliary tract (see Figure 16.1, panel C). The rate and extent of absorption of ingested drugs depends primarily on the pH-dependent passive diffusion and motility of the stomach and intestinal tract as both of these factors will influence transit time of the drug.[15] Gastric pH changes significantly throughout development with the highest (alkaline) values occurring during the first hours of life. However, it drops to 2 to 3 within a few hours of birth but increases during the neonatal period due do the immaturity of the parietal cells.[12,16,17]

Over the first few months of life, the gastric acid secretory capacity increases (pH decreases) and reaches adult levels by 3 to 7 years of age. As a result, the peroral bioavailability of acid-labile drugs, such as penicillin or ampicillin, is increased in neonates compared to adults. In contrast, the absorption of weak organic acids (e.g., phenobarbital and phenytoin) is relatively decreased, a condition that may necessitate the administration of larger doses in the very young to achieve therapeutic plasma levels.[12] Gastric emptying time is prolonged throughout infancy and childhood resulting in decreased drug delivery to the intestine where the majority of absorption takes place. Gastric emptying rates reach or exceed adult values by 6 to 8 months of life. Like other factors, the rate of drug absorption depends on the age of the child.[12,18]

Absorption of drugs from tissues and organs (e.g., intramuscular, transdermal, and rectal) can also be affected by development. Intramuscular (IM) blood flow changes with an increase in age, which can result in variable and unpredictable absorption. Reduced muscular blood flow in the first few days of life and the relative inefficiency of muscular contractions may delay the absorption rate and/or extent of drugs given intramuscularly to the neonate.[12,19] Muscular blood flow increases during infancy and, consequently, the bioavailability of drugs given by the IM route is comparable to that seen in children and adolescents.

Transdermal drug absorption in the neonate and very young infant is increased as a result of the thinner and more hydrated stratum corneum (see Figure 16.1, panel E). Additionally, the ratio of body surface area to body weight is greater in infants and children compared to adults.[12,20,21] Consequently, these developmental differences may predispose the infant and toddler to increased exposure and risk for toxicity from topical application of drugs and/or chemicals (e.g., silver sulfadiazine, topical corticosteroids, isopropyl alcohol, lindane, lidocaine, benzocaine, diphenhydramine).

16.3.2 Distribution

Drug distribution is influenced by drug transporters, blood/tissue protein binding, blood and tissue pH, and perfusion. However, age-related changes to drug distribution are primarily related to developmental changes in body composition and the quantity of plasma proteins capable of drug binding. The absolute amounts and distribution of body water and fat depend on a child's age and nutritional status. As well, certain disease states (e.g., ascites, dehydration, burn injuries) can influence body water compartment sizes and further impact the volume of distribution for certain drugs.

The most dynamic changes occur during the first year of life with the exception of total body fat, which has its greatest change between the second and third decade of life. Due to the larger extracellular and total body water compartments in the neonate and young infant, lower plasma concentrations may occur for hydrophilic drugs that diseminate to these compartments. Alternatively, lipophilic drugs are not affected by age-related differences.[12]

Albumin, total proteins, and total globulins are the most important circulating proteins responsible for drug binding in plasma. The concentrations of almost all circulating plasma proteins are reduced in the neonate and young infant ($\sim$80% of adult) and reach adult values by 1 year of age.[12] The extent of drug binding to proteins in the plasma may influence distribution characteristics as only a free, unbound, drug can be distributed from the vascular space to tissues where drug–receptor interaction occurs. Developmentally associated changes in drug binding can occur as a consequence of altered protein concentrations and/or binding affinity. For example, circulating fetal albumin in the neonate has significantly reduced binding affinity for acidic drugs, such as phenytoin, which is extensively (~94–98%) bound to albumin in adults as compared to 80 to 85% in the neonate. The resultant 6- to 8-fold difference in the free fraction can result in central nervous system (CNS) adverse effects in the neonate when total plasma phenytoin concentrations are within the generally accepted "therapeutic range" (10–20 mg/L).[20]

16.3.3 Metabolism

The primary organ responsible for drug metabolism is the liver, although the kidney, intestine, lung, adrenals, circulating phosphatases and esterases, and skin all have detoxication capabilities to a lesser extent for select xenobiotics. Drug metabolism occurs primarily through two general classes of enzymatic processes: phase I, nonsynthetic, and phase II, synthetic, reactions. Phase I reactions include oxidation, reduction, hydrolysis, and hydroxylation reactions, whereas phase II reactions primarily involve conjugation with an endogenous ligand (e.g., glucuronide, glutathione, or sulfate). Many drug metabolizing enzymes demonstrate a developmental profile with generally low activity present at birth, which matures over a period of months to years (see Figure 16.1, panel A).

While many enzymes are capable of drug biotransformation, the quantitatively most important are represented by the cytochrome P450 (CYP450) enzymes. The specific CYP450 isoforms responsible for the majority of human drug metabolism are represented by CYP1A2, CYP2C9, CYP2C19, CYP2D6, CYP2E1, and CYP3A4. These enzymes represent the products of genes that may be polymorphically expressed with variants producing enzymes capable of reduced or no normal activity. At birth, the activity of these oxidizing enzyme systems is reduced, resulting in slow clearance (and prolonged elimination) of many drugs. After birth, the hepatic CYP450s appear to mature at different rates. Within hours after birth, CYP2E1 activity increases rapidly with CYP2D6 being detectable soon thereafter. CYP2C9, CYP2C19, and CYP3A4 are present within the first month of life, a few months before CYP1A2. CYP3A4 activity in young infants may exceed that observed in adults as reflected by the clearance of drugs that are substrates for this enzyme (e.g., cyclosporine, tacrolimus).[12]

The impact of development on the activity of phase II enzymes (i.e., acetylation, glucuronidation, sulfation) is not characterized as well. Phase II enzyme activity is decreased in the newborn and increases into childhood. For example, conjugation of compounds metabolized by isoforms of glucuronosyltransferase

(UGT) (e.g., morphine, bilirubin, and chloramphenicol) is reduced at birth but can exceed adult values by 3 to 4 years of age.[21,22] Newborns and infants primarily metabolize the commonly used analgesic acetaminophen by sulfate conjugation because the UGT isoforms responsible for its glucuronidation have markedly reduced activity. As children age, the glucuronide conjugate becomes predominant in the metabolism of therapeutic doses of acetaminophen.[23]

The activity of certain hydrolytic enzymes, including blood esterases, is also reduced during the neonatal period. Blood esterases are important for the metabolic clearance of cocaine, and the reduced activity of these plasma esterases in the newborn may account for the delayed metabolism (prolonged effect) of local anesthetics in the neonate.[24] In addition, this may account for the prolonged effect that cocaine has on the fetus with prenatal exposures. Adult esterase activity is achieved by 10 to 12 months of age.

16.3.4 Elimination

The kidney is the primary organ responsible for the excretion of drugs and their metabolites and depends exquisitely on gestational age and developmental changes for its function. The development of renal function begins during early fetal development and is complete by early childhood (see Figure 16.1, panel D).[12] Renal clearance is limited in the newborn due to both anatomical and functional immaturity of the kidney. Nephrogenesis is not fully complete until 36 weeks of gestation resulting in a marked decrease in the glomerular filtration rate (GFR) in preterm infants. During the first few days of life, changes in renal and intrarenal blood flow occur resulting in an increase in GFR.[25] In term neonates, GFR increases rapidly over the first few months of life until adult values are reached by 8 to 12 months of age.[26,27] In young children between 2 and 5 years of age, GFR may exceed adult values, especially during periods of increased metabolic demand (e.g., during a fever).

Altered renal drug clearance in newborns and infants is the major determinant of medication dosing in pediatrics. Without this knowledge, the result can be damaging as drug concentrations may become toxic. The aminoglycoside antibiotic gentamicin provides an illustrative example.[28] In adolescents and young adults with normal values for GFR (85–130 ml/min/1.73 m^2), the recommended dosing interval for the drug is 8 hours. In young children who may have a GFR > 130 ml/min/1.73 m^2, a gentamicin dosing interval of every six hours may be necessary in selected patients. However, use of similar dosing intervals in the neonate may result in serum concentrations in the toxic range. Thus, longer intervals (e.g., every 24 hours) may be necessary.

16.4 EMBRYONIC AND FETAL CARDIAC DEVELOPMENT

Understanding normal cardiac development is essential prior to discussing the adverse effects that toxins have on the developing heart. Embryology is

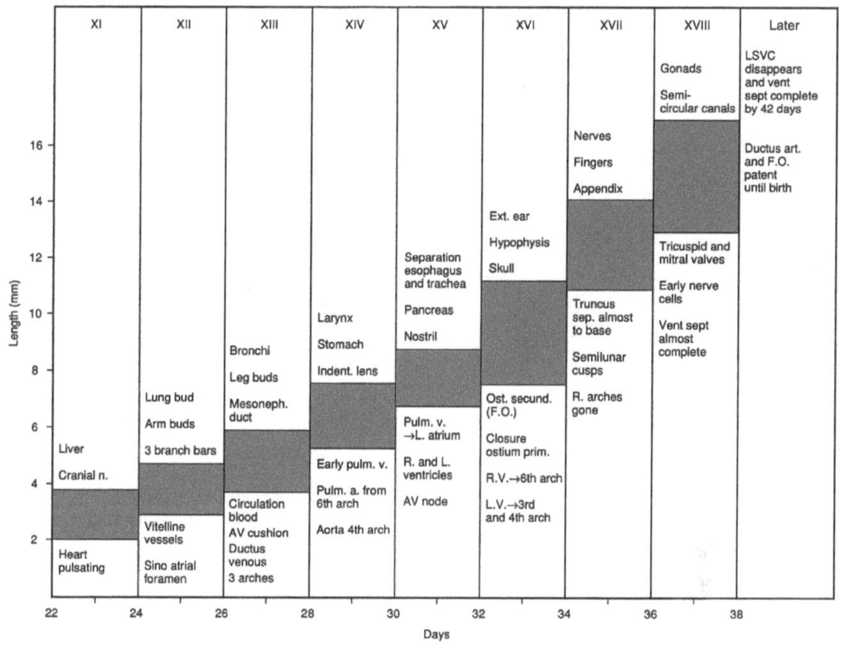

FIGURE 16.2 **Developmental horizons of the human embryo.** Beginning in the third week of life, cells that will eventually become the cardiovascular system are formed. By 20 days after conception, a heartbeat begins. The fourth week after conception includes the development of the morphologically left and right ventricles, circulation commences, and the evolution of the aortic arch begins. Between days 29 and 35, changes occur that can result in common congenital cardiac anomalies. Thus, developmental problems during this stage may lead to malformation such as d-transposition of the great arteries, Taussig–Bing malformation, and tetralogy of Fallot. By the end of the fifth week, cardiovascular separation is nearly complete. Cardiovascular maturation continues until well after birth. **Source:** *Reprinted with permission from* NADAS' Pediatric Cardiology.[29]

important to pediatric cardiology as it helps to clarify the etiology and pathogenesis of cardiac malformations (Figure 16.2).[29]

During the first two weeks after conception, the developing human embryo lacks a heart and vascular system. Beginning in the third week of life, cells that will eventually become the cardiovascular system are formed. The mesoderm develops from the ectoderm to create the cardiogenic crescent, which is the immediate precursor of the heart. By 20 days after conception, a heartbeat begins. The fourth week after conception includes the development of the morphologically left and right ventricles, circulation commences, and the evolution of the aortic arch begins. Neural crest cells give rise to pharyngeal (aortic) arches 3, 4, and 6 that then participate in the formation of the great arteries (Figure 16.3).[30]

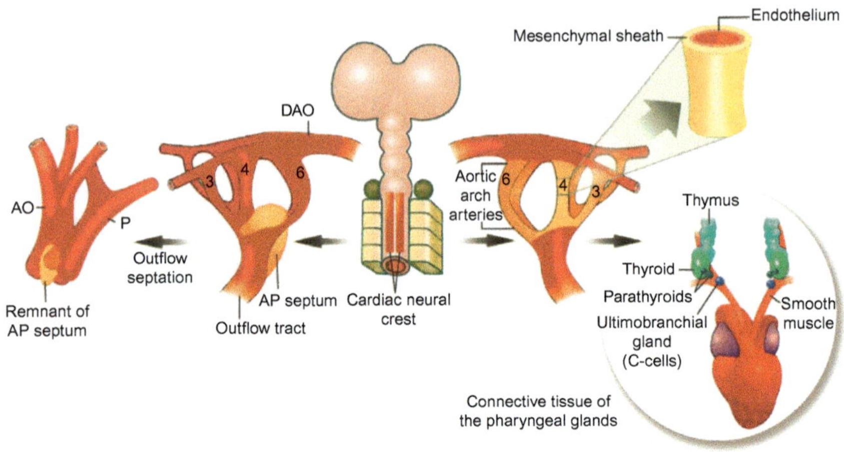

FIGURE 16.3 Neural crest development. Derivatives of the cardiac neural crest cells support development of and patterning of the persisting aortic arch arteries into the great arteries of the thorax and form their smooth muscle tunics. In addition, the cells contribute to the development of the thymus, parathyroids, and thyroid glands and provide stromal cells after gland development. **Source:** *Reprinted with permission from Keyte et al.*[30]

Between 29 and 35 days after conception, changes occur that can result in common congenital cardiac anomalies. The right and left heart begins separation between the ascending aorta and main pulmonary artery and between the mitral and tricuspid valves. The muscular ventricular septum moves from right to left beneath the atriventricular canal. Thus, developmental problems during this stage may lead to malformation such as d-transposition of the great arteries, Taussig–Bing malformation, and tetralogy of Fallot (TOF). By the end of the fifth week, cardiovascular separation is nearly complete with the exception of the presence of the interventricular foramen, which may not close until after birth. Cardiovascular maturation continues until well after birth.

Congenital heart defects (CHD) are thought to be the most common congenital defect worldwide, with 1 to 10% of live births affected. In fact, CHD is the most common cause of infant mortality in Europe and the United States.[31] It has been suggested that up to 80% of congenital anomalies are due to a fetal interaction of genetics and the environment.[32] It is believed that most cases of CHD are the result of disruption of the development of the neural crest cells by interactions with the environment (e.g., medications or other toxins) or other etiologies.[31]

16.4.1 Vital Signs and the Electrocardiogram

Evaluation of a pediatric patient in a resuscitative setting begins with the initial assessment of airway, breathing, and circulation, followed by a primary

TABLE 16.1 Normal Vital Signs According to Age

Age	Heart Rate (beats/min)	Blood Pressure (mmHg)	Respiratory Rate (breaths/min)
Premature	120–170[a]	55–75/35–45[b]	40–70[c]
0–3 mo	100–150[a]	65–85/45–55	35–55
3–6 mo	90–120	70–90/50–65	30–45
6–12 mo	80–120	80–100/55–65	25–40
1–3 yr	70–110	90–105/55–70	20–30
3–6 yr	65–110	95–110/60–75	20–25
6–12 yr	60–95	100–120/60–75	14–22
12+ yr	55–85	110–135/65–85	12–18

[a]*In sleep, infant heart rates may drop significantly lower, but if perfusion is maintained, no intervention is required.*
[b]*A blood pressure cuff should cover approximately two-thirds of the arm; too small a cuff yields spuriously high pressure readings, and too large a cuff yields spuriously low pressure readings.*
[c]*Many premature infants require mechanical ventilatory support, making their spontaneous respiratory rate less relevant.*

Source: Reprinted with permission from *Nelson Textbook of Pediatrics.*[33]

assessment that includes a brief, hands-on assessment of neurologic function and disability. The examination and vital signs can only be interpreted with the understanding of normal values, which are dynamic as children age. Normal respiratory rate, heart rate, blood pressure, and global cognitive function have age-specific normal values and ranges.[33] Failure to remember this can lead to misdiagnosis of abnormality when there may or may not be one (Table 16.1). During the first year, blood pressure increases significantly during the first 6 months and then slows down between 6 to 12 months. Children do not achieve adult blood pressure norms until approximately 12 years of age.[33]

A normal pulse rate in the neonate ranges from 125 to 145 bpm. Normal "adult" heart rate is not achieved until approximately 12.[34] Children also have electrocardiogram (ECG) patterns distinct from adults. Several studies have found discrepancies in the interpretation of pediatric ECGs between adult practitioners and pediatric cardiologists.[35,36] The interpretation of the pediatric ECG should thus be done based on the patient's age. Other than the heart rate changes that occur in children, many of the changes on the pediatric ECG are a reflection of the dominance of the right ventricle at birth. During infancy, the pulmonary pressures fall, resulting in decreasing right ventricular wall stress and thickness.[34] As such, the determination of right ventricular hypertrophy in children cannot be made by an ECG study alone.

Additionally, differences are obvious when comparing the ECG waveforms of adults and children.[34] The PR interval is shorter in children compared to adults and is presumed to be secondary to the decreased muscle mass in children. Infants and children have, on average, a PR interval of 100 msec, which increases during childhood to the normal adult average of 150 msec. Thus, what may appear as normal to the adult practitioner, may, in actuality, be a prolonged PR interval in the child indicating an atrioventricular conduction defect. Comparatively, the QRS duration is shorter in children compared to adults and can be attributed, again, to the decreased muscle mass. The normal QRS ranges from 50 to 80 msec in children. Similarly, the corrected QT interval is age dependent with the longest occurring in the infant (>6 months of age) at 490 msec and decreasing over time to 440 msec. Lastly, the T wave, which is an important indicator for ischemic heart disease in adults, may be affected by electrolyte abnormalities and right ventricular hypertrophy. In the infant and young child, the T wave is typically upright in most leads. During childhood, it is deeply inverted at 7 years of age or later when the upright deflection is established.

16.5 THE PEDIATRIC HEART AND TOXINS

Although advances in genetic etiologies of congenital heart defects have occurred, similar advances to determine other risks for malformations have not. The majority of our knowledge is the result of case reports of children born with CHD and further epidemiological studies identifying an association with exposures and outcomes. Similarly, medications that may cause harm to children and adolescents, unfortunately, cannot be discovered until drugs are approved and have resulted in morbidity and mortality from adverse cardiac events.

16.5.1 Prenatal Exposures

Since David Smith coined the term *dysmorphology* in 1966, the study of genetics has expanded exponentially. While the field currently includes genetic evaluation as an etiology for congenital defects, one cannot discount the effect that other etiologies have in the abnormalities in morphology. Exposures to toxins are known to result in cardiac malformations, and women should be counseled against their use during pregnancies if not during their entire reproductive years given the early formation of the heart during fetal development.

Teratogens are chemical, physical, radiological, or biological agents that cause congenital malformations. Some teratogens are associated with specific syndromes that have recognizable features. As not all exposures to teratogens result in dysmorphology, the etiology of many of these syndromes may be multifactorial with both genetic and environmental causes.

Exposure to medications and other xenobiotics are common during pregnancy. Analgesics, antipyretics, antimicrobials, and vitamins are commonly

TABLE 16.2 U.S. Food and Drug Administration Pharmaceutical Pregnancy Categories

Category A	*Controlled studies show no risk.* Adequate, well-controlled human studies fail to demonstrate a risk to the fetus and, if available, animal studies do not demonstrate a risk. *Example:* multivitamins.
Category B	*No evidence of risk in humans.* Animal studies have failed to demonstrate a risk to the fetus and there are no adequate and well-controlled studies of pregnant women; or animal studies have shown an adverse effect, but adequate and well-controlled studies of pregnant women have failed to demonstrate a risk. *Example:* amoxicillin and acetaminophen.
Category C	*Risk cannot be excluded.* Animal studies have shown an adverse effect on the fetus and there are no adequate and well-controlled studies of humans, but potential benefits may justify the potential or unknown risks. *Example:* rifampicin.
Category D	*Positive evidence of risk.* Investigational or marketing experience show evidence of human fetal risk, but potential benefits may warrant use of the drug for pregnant women despite potential risks. *Example:* phenytoin and tetracycline.
Category X	*Contraindicated during pregnancy.* Investigational or marketing experiences in animals or humans have demonstrated fetal abnormalities, and the risks involved in use of the drug for pregnant women clearly outweigh potential benefits. *Example:* isotretinoin.

used throughout pregnancy as well as other, potentially, dangerous chemicals such as caffeine, nicotine, and ethanol. By law, pharmaceutical companies in the United States are required to label their products by risk to the pregnant woman[37] (Table 16.2). Similar classification systems have been developed in other countries (e.g., Sweden, Germany, and Australia). The original intent of the U.S. FDA was to place a medication in a generalized risk category for pregnancy. However, these categories have been mistaken to infer teratogenicity and, thus, may be conservative because there is discordance between the use-in-pregnancy label and teratogenic risk.

A meta-analysis in 2007 by the International Clearinghouse of Birth Defects Surveillance and Research estimated that 16% of children who have major malformations have been exposed to medications within the first trimester.[38] The difficulty lies with determining association versus causation. For example, use of insulins were associated with septal defects, cardiac outflow tract defects, and left ventricle obstructive defects. However, later studies have confirmed that hyperglycemia is itself a major teratogen.[39] Despite these uncertainties, some exposures have been conclusively linked to exposure to xenobiotics occurring within the first trimester.

Fetal Alcohol Syndrome

One of the most common malformation syndromes and a leading cause of learning disabilities is fetal alcohol syndrome (FAS). With an estimated frequency greater than 1 in 1000 live births, FAS is characterized by prenatal and postnatal growth retardation, microcephaly, specific facial characteristics, developmental delay, and CHD. Its most dire consequences are associated with neurodevelopmental disabilities; however, approximately half of individuals with FAS have congenital cardiac malformations. Animal studies have suggested that alcohol can adversely impact neural crest development and migration resulting in structural abnormalities of the arches and septum.[30] A recent meta-analysis found 29 studies had assessed the prevalence of CHD in children with FAS. The proportion of cases with CHD, including atrial (ASD) and ventricular (VSD) septal defects, range from 33 to 100%.[40] The rate of a comorbid diagnosis of CHD and FAS is 28.6%. Most commonly, a ventricular septal defect is present. Atrial septal defect and tetralogy of Fallot can also occur with increased frequency. The difficulty of making a diagnosis of FAS during the perinatal period leads to a general underestimation of this syndrome.

Antiepileptics

Anticonvulsants have long been known to be teratogens. In 1979 the American Academy of Pediatrics Committee on Drugs released a Policy Statement regarding anticonvulsants during pregnancy, introducing commonly used diagnoses such as *fetal hydantoin syndrome, fetal valproate syndrome*, and *fetal carbamazepine syndrome*; the diagnoses are still relevant.[41] It is understood that seizures in the mother during pregnancy can cause harm to the fetus. However, no medication should be given to a pregnant woman unless it is necessary for the health and well-being of both the mother and the child. The risk of major malformations in the fetus born to mothers on antiepileptics ranges from 4 to 8% compared to 2 to 4% in the general population.[42] While the majority of malformations are due to neural tube defects and developmental delays, use of antiepileptics have been associated with CHD including atrial and ventricular septal defects, tetralogy of Fallot, pulmonary stenosis, patent ductus arteriosis, and coarctation of the aorta.[43,44]

Retinoic Acid Embryopathy

Retinoic acid, an analog of vitamin A, is teratogenic in animals and humans. Several case studies have been reviewed previously[43] and have documented malformations associated with the use of high-dose vitamins and the synthetic isomers isotretinoin and etretinate. Similarly, a retrospective analysis has shown that exposure to isotretinoin is associated with a high risk for major malformations including micrognathia, cleft palate, thymic defects, optic nerve defects, CNS malformations and conotruncal heart defects, and aortic-arch abnormalities.[45,46]

Thalidomide Embryopathy

Thalidomide, once commonly prescribed as an antiemetic for hyperemesis gravidarum, is a teratogen that has been found to be the cause of more severe malformations than any other known drug.[43] The most striking malformations were bilateral and symmetrical limb reduction defects known as phocomelia. Thousands of "thalidomide babies" were born before it was removed from use in most countries during the early 1960s, and this agent remains a potent symbol of iatrogenic birth defects. However, it has recently been approved in the United States (for the first time) as an immunomodulatory agent for the treatment of erythema nodosum leprosum.

The critical period for malformations is during the first 20 to 36 days after conception, a period of significant iatrogenesis. The suspected lesion involves disruption of genes specifically involved in limb development.[47,48] In addition to the limb reduction defects, cardiac malformations (e.g., VSD, ASD, TOF, hypertrophy of the atrium and ventricle, coarctation of the aorta, and systolic murmurs) have also been identified in neonates with prenatal exposures.[43,49]

Fetal Warfarin Syndrome

Pregnancies affected by exposure to coumadin-derivative medications, such as warfarin, can result in fetal malformations depending on when the exposure occurs. As with most teratogens, the critical period of exposure occurs within the first 6 to 9 weeks after conception. *Fetal warfarin syndrome* occurs with exposure during the first trimester but is not seen later in pregnancy.[43,50] This syndrome consists of nasal hypoplasia and may include the following: low birth weight, eye defects (seen in warfarin exposure throughout a pregnancy), hypoplasia of the extremities, developmental delay, seizures, scoliosis, hearing loss, congenital heart disease, and death. Exposure in the second and third trimesters may lead to CNS defects as a result of early fetal hemorrhage. Cardiac defects that have been associated with coumadin use include patent ductus arterosis and peripheral pulmonary stenosis.[44]

Lithium

Lithium is commonly used for the treatment of manic-depressive disorder in children and adults. While the drug has a narrow therapeutic index, its efficacy and low cost make it a widely prescribed pharmaceutical especially in refractory cases. Unfortunately, use of lithium during the first trimester of pregnancy has been associated with an increased incidence of congenital defects, particularly involving the cardiovascular system.

The majority of the data surrounding cardiac malformations associated with prenatal lithium use is a result of the Lithium Baby Register, which was founded in 1968 to follow children born to mothers exposed to lithium during the first trimester of pregnancy.[43] Within the first 10 years of the register's establishment, 20 out of 183 infants (11%) were documented to have a major malformation. These defects included cardiovascular defects ($n = 15$), including Ebstein's anomaly ($n = 5$) (i.e., a downward displacement of the tricuspid valve resulting in a poorly functioning right ventricle), tricuspid valve regurgitation, and obstruction of the right outflow tract. Other defects noted included septal defects; coarctation of the aorta, mitral, and tricuspid valve atresia; patent ductus arteriosus; and others. Since that time, other studies have documented CHD with lithium use early in pregnancy;[51,52] however, more recent data suggest a lower frequency than has been previously reported.[53,54]

Selective Serotonin Reuptake Inhibitors

As many pregnancies are unplanned and approximately 10% of pregnant women are depressed, there is a need to evaluate antidepressant use during pregnancy. Effects have been noted in neonates and children who have been exposed to antidepressants in utero.[55,56] These include increased rates of prematurity, admission to special care nurseries, decreased weight and size, minor anomalies, and poor neonatal adaptation. While primarily focused on the normal development of the brain and the neuropsychological effects after delivery, teratogenicity has been associated with these exposures.

Specifically for congenital heart defects, GlaxoSmithKline released safety findings regarding paroxetine and major congenital malformations. Of the 527 fetuses exposed to paroxetine in the first trimester, 23 had congenital malformations with the majority having ventricular septal defects. As a result, the U.S. Food and Drug Administration requested product label changes from a pregnancy category C (studies in animals have demonstrated adverse fetal effects or there are no controlled data from studies in women) to a category D (studies in pregnant women have demonstrated risk to the fetus).[57] A subsequent study found an increased risk of congenital malformations after exposure to selective serotonin reuptake inhibitors (SSRIs) in early pregnancy compared to those pregnancies without SSRIs (4.9% and 3.4%, respectively).[58] However, the study did not distinguish between the SSRIs. Other epidemiologic studies have found other associations. Compared to healthy neonates, persistent pulmonary hypertension was six times more likely to occur in those exposed to an SSRI during the second half of pregnancy.[59]

A subsequent prospective study of women who were exposed to SSRIs in the first trimester compared to control patients found an increased incidence of nonsyndromic congenital heart defects in the exposed population compared to those who were not (3.4% and 1.6%, respectively).[60] Paroxetine and fluoxetine exposures resulted in the highest prevalence. Other studies have found similar associations, especially as it relates to paroxetine exposure.[61,62]

Exposures from the Environment

Exposures from the surrounding environment may also affect embryogenesis resulting in neural tube defects, cleft lip and palate, and CHD. These may be due to direct toxicity during cell division or from gene–environmental interactions. Environmental toxins that have been studied in regard to CHD include ambient air pollution,[63] chlorinated hydrocarbons,[64,65] lead,[66] and pesticides.[67] While the results are mixed for most environmental toxins as direct causal agents for CHD, trichloroethylene (TCE) has been associated with cardiac malformations. Trichloroacetic acid (TCAA), a metabolite of TCE, is a specific cardiac teratogen when given to pregnant rats.[68] In humans, there may be an association between CHD incidence in children and a combination of increased maternal age and proximity to TCE source sites during the course of the pregnancy.[69]

16.5.2 Special Considerations in Children

Adverse effects of medications and other toxins are not unique to children. However, children are uniquely vulnerable given their developing organ systems and behaviors that increase risk for exposures. Because of developmental changes in pharmacokinetic, pharmacodynamic, and pharmacogenetic response, medications that may be safe for adults can have adverse consequences for children.

One Pill Can Kill

Approximately 60% of poison exposures reported to the U.S. poison control centers occur in children. While the majority of these have minor or no adverse effects, there are toxins that can result in significant morbidity and mortality in children in only small doses.[70,71] There are several medications that may cause significant toxicity with very little ingested (Table 16.3). With regard to cardiac toxicity, common agents include the antiarrhythmics (e.g., flecainide), beta blockers (BBs), calcium channel blockers (CCBs), clonidine and other imidazolines, cyclic antidepressants, diphenoxylate and atropine, opioids (e.g., methadone and buprenorphine), phenothiazines, antimalarials (e.g., quinine and chloroquine), sulfonylureas, and theophylline. Sustained-release preparations pose a concern for delayed effects, requiring at least 24 observation periods in a continuously monitored setting. While the mechanisms leading to fatality in children vary widely, many of these drugs result in fatal dysrhythmias. Others may result in respiratory and CNS depression, ultimately leading to bradycardia and/or asystole. The effects of these drugs are described elsewhere, and are frequently similar to signs observed with adult exposures.

Prolonged QT Syndrome in Children

Prolonged QT interval is a disorder of delayed ventricular repolarization. Long QT syndrome (LQTS) may be congenital or acquired. Congenital LQTS is the result of genetic polymorphisms in ion channels that affect the action potential of the heart.[72,73] Multiple genotypes have been discovered

TABLE 16.3 Medicinal Preparations that Can Be Fatal for a 10-kg Toddler on Ingestion of a One-Dose Unit

Drug	Minimal Potential Fatal Dose (mg/kg)	Maximal Dose Unit Available (mg unless otherwise stated)	No. of Tablets that Can Cause Fatality
Tricyclic antidepressants			
amitriptyline	15	100	1–2
imipramine	15	150	1
desipramine	15	75	1–2
Antipsychotics			
loxapine	30–70	50	1–2
thioridazine	15	200	1
chlorpromazine	25	200	1–2
Antimalarials			
chloroquine	20	500	1
hydroxycholoroquine	20	200	1
quinine	80	650	1–2
Antiarrhythmics			
quinidine	15	324	1
disopyramide	15	150	1
procainamide	70	1000	1
flecainide	25	150	1–2
Calcium channel blockers			
nifedipine	15	90	1–2
verapamil	15	360	1
diltiazem	15	360	1
Camphor	100	1 g/5 mL	1 tsp
Methyl salicylate	200	1.4 g/mL	<1 tsp
Theophylline	8.4	500	1
Narcotics			
codeine	7–14	60	1–2
hydrocodone	1.5	60 mg/5 mL	<1 tsp
methadone	1–2	40	1
morphine		200	1
Oral hypoglycemics			
chlorpropamide	5	25	1
glibenclamide	0.1	2.5	1
glipizide	0.1	5	1
Podophyllin 25%	15–20	1.25 g/5 mL	1 mL

tsp = teaspoon

Source: Reprinted with permission from Bar-Oz et al.[70]

TABLE 16.4 Common Classes for Drugs that May Result in Acquired Long QT Syndrome

Drugs	Class IA, IC, and III Antidysrhythmics
Antidepressants	Tricyclic antidepressants, SSRIs (e.g., citalopram), monoamine oxidase inhibitors
Antifungals	Itraconazole, ketoconazole
Antihistamines	Astemizole, terfenadine
Antimicrobials	Erythromycin, trimethoprim-sulfamethoxazole, chloroquine
Antipsychotics	Atypical antipsychotics, haloperidol, risperidone, thioridazine
Opioids	Methadone

and sporadic cases of QT prolongation have been reported with new gene mutations.

There are many causes of acquired long QT syndrome in adults and children;[74] however, medications are the most common. Additionally, disease states, such as anorexia nervosa and bulimia, can result in acquired LQTS. Patients with eating disorders may also be using medications (e.g., serotonin reuptake inhibitors), which in turn can add to the QT prolongation effects and increase the risk of fatal dysrhythmias. Common classes of drugs that may result in prolonged QT are listed in Table 16.4. A complete list can be found at *http://crediblemeds.org/everyone/composite-list-all-qtdrugs/*.

Congenital LQTS most commonly presents during childhood or adolescence (9–15 years of age) with syncopal (or near-syncopal) episodes. Events such as these are usually precipitated by exercise or extreme emotions (e.g., fear, excitement).[73] These episodes may be misdiagnosed as vasovagal events or seizures. A thorough medication history and 12-lead ECG are required elements in the workup for all syncopal events.

Approximately 10% of children with LTQS will present with sudden cardiac death. These may include cases of sudden infant death syndrome, near-drownings, exertional sudden death, and other situations that arise from adrenal arousal. All children who have LTQS should be referred to a pediatric cardiologist for treatment that may include medications and/or a pacemaker.[72,75,76]

Attention Deficit and Hyperactivity and Dysrhythmias

Attention deficit/hyperactivity disorder (ADHD) is the most prevalent neurobehavioral disorder among school-age children in many countries, affecting between 3% and 7% of school children in the United States. This chronic disorder severely impairs function both at home and at school, and its symptoms persist into adolescence and adulthood in 37 to 85% of children.[77] Stimulants are the most common class of drug prescribed for ADHD. It is estimated that approximately 2.5 million children in the United States currently receive a stimulant for this disorder.[78]

Within the past decade, there has been increasing concern about the cardiovascular risks with use of stimulant drugs in children with ADHD. There are numerous reports of sudden death of children treated with methylphenidate or amphetamines.[79,80] Many of these cases occurred in children with risk factors (e.g., patients with potentially life-threatening arrhythmias such as supraventricular tachycardia, long QT syndrome, and hypertrophic cardiomyopathy) for sudden cardiac death. Since these deaths, the FDA has assigned a "black box" warning that patients with underlying heart disease might be at risk for sudden death when taking stimulant drugs.[81] In 2008, the American Heart Association published a scientific statement suggesting that in addition to a thorough personal history, family history, and physical examination, universal ECG screenings ought to be performed on patients who are prescribed stimulants.[82] This recommendation has been controversial,[83–85] as subsequent studies have found that the screening ECGs are not helpful for identifying higher-risk patients in the context of prescription stimulant use.

Intentional Poisoning and Child Abuse

Intentional poisoning of children, while rare, is a significant form of child abuse and one that is rarely suspected. This may be combined with overt signs of physical or sexual abuse, but often it is a child who presents with a diagnostic dilemma. Diagnosis is more commonly made after a patient dies and an autopsy is performed, a comprehensive drug screen is ordered, or the history does not account for the symptoms and raises suspicions. Whereas pediatric fatalities from unintentional poisonings rarely result in death, intentional poisonings by a caregiver may result in fatality rates as high as 20 to 30%.[86]

Originally termed *Munchausen syndrome by proxy*, factitious illnesses in children can occur from either caregiver-inflicted illnesses (e.g., surreptitious salt, syrup of ipecac, or insulin administration) or fabrication of the symptoms altogether to achieve medical attention. The fabricated illness often leads to multiple medical evaluations; frequent hospitalizations, unnecessary diagnostic studies (including surgeries); unnecessary pharmacologic treatments; and, possibly, the death of the child.[71] Child abuse and neglect should always be on the differential in the poisoned child especially in very young children with a large ingestion, patients with a confusing history, children and siblings with repeated or similar exposures, an exposure to a xenobiotic to which a child would not normally have access, "accidental" ingestion in a school-aged or adolescent child, and a history of an unexplained death in the family. A high clinical index of suspicion is almost always needed to confirm this difficult diagnosis, and poisoning should be in the differential of any pediatric patient with an unusual or inconsistent historical presentation.

16.6 CONCLUSION

Poisonings are more common in children compared to any other age group. Fortunately, fatalities are rare. However, serious toxic exposures do occur and can cause significant harm to the developing fetus or young child. The correct evaluation and treatment of children with cardiotoxic exposures begins with a thorough understanding of normal cardiac development and proceeds with an appreciation of how the heart and detoxification systems can change normally over time and abnormally in response to a variety of toxins. The heart is specifically vulnerable to toxins during the prenatal period as it is completely formed at a time in which many women are unaware of their pregnancy. Additionally, the child continues to develop its systems for absorption, distribution, metabolism, and elimination after birth and into late childhood. Cardiac development follows a similar developmental pattern, with normal changes in heart rate and blood pressure over the first decade of life. Education about poisoning risks to parents and women of child-bearing age continues to be an important public health intervention by poison control centers and other agencies.

REFERENCES

1. Warner M, Chen L, Makuc D, et al. Drug poisoning deaths in the United States, 1980–2008. *NCHS Data Brief* 2011;81.
2. Bronstein A, Spyker D, Cantilena L, et al. 2011 Annual Report of the American Association of Poison Control Centers' National Poison Data System (NPDS). *Clin Toxicol* 2012; **50**(10):911–1164.
3. National Poisons Information Service. Annual Report. London: Public Health England; 2012 [accessed 12/02/13]; available from: <www.npis.org/NPISAnnualReport2012-13.pdf> and <www.hpa.org.uk/webc/HPAwebFile/HPAweb_C/1317139808601>.
4. Henry K, Harris C. Deadly ingestions. *Pediatr Clin N America* 2006;**53**(2):293–315.
5. Boyer E, McCance-Katz E, Marcus S. Methadone and buprenorphine toxicity in children. *Am J Addict* 2009;**9**(1):89–95.
6. United Nations Office on Drugs and Crime. *World Drug Report 2012. Vienna, Austria: UN; 2013*. [accessed 12/02/13]; available from: <www.unodc.org/unodc/secured/wdr/wdr2013/World_Drug_Report_2013.pdf>.
7. Monitoring the future: A continuing study of American youth. Ann Arbor: The University of Michigan; 2012 [accessed 10/01/13]; available from: <www.monitoringthefuture.org>.
8. CDC, Vital Signs: Unintentional injury deaths among persons aged 0–19 years — United States, 2000–2009. *Morb Mortal Wkly Rep* 2012;**61**:270–6.
9. Fine J, Callelo D, Marcus S, et al. 2011 Pediatric fatality review of the National Poison Data System. *Clin Toxicol* 2012;**50**(10):872–4.

10. Dart R, Paul I, Bond G. Pediatric fatalities associated with over the counter (nonprescription) cough and cold medications. *Ann Emerg Med* 2009;**53**(4):411–7.
11. Hojer J, Troutman WG, Hoppu K, et al. Position paper update: ipecac syrup for gastrointestinal decontamination. *Clin Toxicol* 2013;**51**(3):134–9.
12. Kearns G, Abdel-Rahman S, Alander S, et al. Developmental pharmacology: drug disposition, action, and therapy in infants and children. *N Engl J Med* 2003;**349**(12):1157–67.
13. Mulhall A, de Louvois J, Hurley R. Chloramphenicol toxicity in neonates: its incidence and prevention. *Br Med J* 1983;**287**(6403):1424–7.
14. Brown R, Campoli-Richards D. Antimicrobial therapy in neonates, infants and children. *Clin Pharmacokinet* 1989;**17**(Suppl 1):105–15.
15. Ritschel W, Kearns G. Pediatric pharmacokinetics. In: Ritschel W, Kearns G, editors. *Handbook of basic pharmacokinetics... including clinical applications*. Washington, DC: American Pharmacists Association; 2004. p. 227–39.
16. Agunod M, Yamaguchi N, Lopez R, et al. Correlative study of hydrochloric acid, pepsin, and intrinsic factor secretion in newborns and infants. *Am J Dig Dis* 1969;**14**(6):400–14.
17. Rodbro P, Krasilnikoff P, Christiansen P. Parietal cell secretory function in early childhood. *Scand J Gastroenterol* 1967;**2**(3):209–13.
18. Heimann G. Enteral absorption and bioavailability in children in relation to age. *Eur J Clin Pharmacol* 1980;**18**(1):43–50.
19. Greenblatt D, Koch-Weser J. Intramuscular injection of drugs. *N Engl J Med* 1976;**295**(10):542–6.
20. Lowry JA, Vandover J, De Greeff J, et al. Unusual presentation of iatrogenic phenytoin toxicity in a newborn. *J Med Toxicol* 2005;**1**(1):26–9.
21. Hartley R, Green M, Quinn M, et al. Development of morphine glucuronidation in premature neonates. *Biol Neonate* 1994;**66**(1):1–9.
22. Scott C, Riggs K, Ling E, et al. Morphine pharmacokinetics and pain assessment in premature newborns. *J Pediatr* 1999;**135**(4):423–9.
23. Miller R, Roberts R, Fischer L. Acetaminophen elimination kinetics in neonates, children and adults. *Clin Pharmacol Ther* 1976;**19**(3):284–94.
24. Hall R. Anesthesia and analgesia in the NICU. *Clin Perinatol* 2012;**39**(1):239–54.
25. Robillard J, Guillery E, Petershack J. Renal function during fetal life. In: Barratt T, Avner E, Harmon W, editors. *Pediatric nephrology*. Philadelphia: Lippincott Williams & Wilkins; 1999.
26. van den Anker J, Shoemaker R, Hop W, et al. Ceftazidime pharmacokinetics in preterm infants: effects of renal function and gestational age. *Clin Pharmacol Ther* 1995;**58**(6):650–9.
27. Arant B. Developmental patterns of renal functional maturation compared in the human neonate. *J Pediatr* 1978;**92**(5):705–12.
28. Davies M, Cartwright D. Gentamicin dosage intervals in neonates: longer dosage interval-less toxicity. *J Paediatr Child Health* 1998;**34**(6):577–80.
29. Van Praagh R. Embryology. In: Keane J, Lock J, Fyler D, editors. *NADAS' Pediatric Cardiology*. Philadelphia: Saunders Elsevier; 2006; p. 13–25.
30. Keyte A, Hutson M. The neural crest in cardiac congenital anomalies. *Differentiation* 2012;**84**(1):25–40.
31. van Mil N, Oosterbaan A, Steegers-Theunissen R. Teratogenicity and underlying mechanisms of homocysteine in animal models: a review. *Reprod Toxicol* 2010;**30**(4):520–31.
32. Hartman M, Cheifetz I. Pediatric emergencies and resuscitation. In: Kliegman R, Stanton B, Schor N, St. Geme III J, Berhrman R, editors. *Nelson textbook of pediatrics*. 19th ed. Philadelphia: Elsevier Saunders; 2011; p. 279–96.

33. O'Connor M, McDaniel N, Brady W. The pediatric electrocardiogram. Part I: Age-related interpretation. *Am J Emerg Med* 2008;**26**(4):506–12.
34. Horton LA, Mosee S, Brenner J. Use of the electrocardiogram in a pediatric emergency department. *Arch Pediatr Adolesc Med* 1994;**148**(2):184–8.
35. Wathen J, Rewers A, Yetman A, et al. Accuracy of ECG interpretation in the pediatric emergency department. *Ann Emerg Med* 2005;**46**(6):507–11.
36. Food and Drug Administration. *Federal Register* 1980;**44**:37434–67.
37. Lisi A, Botto LD, Robert-Gnansia E, et al. Surveillance of adverse fetal effects of medications (SAFE-Med): findings from the International Clearinghouse of Birth Defects Surveillance and Research. *Reprod Toxicol* 2010;**29**(4):433–42.
38. Allen VM, Armson BA, Wilson RD, et al. Teratogenicity associated with pre-existing and gestational diabetes. *J Obstet Gynaecol Can* 2007;**29**(11):927–44.
39. Rosenquist TH. Folate, homocysteine and the cardiac neural crest. *Dev Dyn* 2013;**242**(3): 201–18.
40. Burd L, Deal E, Rios R, et al. Congenital heart defects and fetal alcohol spectrum disorders. *Congenit Heart Dis* 2007;**2**(4):250–5.
41. Segal S. Drugs Co. Anticonvulsants and pregnancy. *Pediatrics* 1979;**63**(2):331–3.
42. Beghi E, Annegers J. Epilepsy CGftPRi. Pregnancy Registries in Epilepsy. *Epilepsia* 2001; **42**(11):1422–5.
43. Briggs G, Freeman R, Yaffe S, editors. *Drugs in pregnancy and lactation*. 9th ed. Philadelphia: Lippincott Williams & Wilkins; 2011.
44. Lacro R. Dysmorphology and genetics. In: Keane J, Lock J, Fyler D, editors. *NADAS' pedi atric cardiology*. 2nd ed. Philadelphia: Sanders Elsevier; 2006. p. 49–72.
45. Lammer E, Chen D, Hoar R, et al. Retinoic acid embryopathy. *N Engl J Med* 1985;**313** (14):837–41.
46. de la Cruz E, Sun S, Vangvanichyakorn K, et al. Multiple congenital malformations associated with maternal isotretinoin therapy. *Pediatrics* 1984;**74**(3):428–30.
47. Stephens T, Bunde C, Fillmore B. Mechanism of action in thalidomide teratogenesis. *Biochem Pharmacol* 2000;**59**(12):1489–99.
48. Stephens T, Fillmore B. Hypothesis: Thalidomide embryopathy-proposed mechanism of action. *Teratol* 2000;**61**(3):189–95.
49. Newman C. The thalidomide syndrome: risks of exposure and spectrum of malformations. *Clin Perinatol* 1986;**13**(3):555–73.
50. Hall J, Pauli R, Wilson K. Maternal and fetal sequelae of anticoagulation during pregnancy. *Am J Med* 1980;**68**(1):122 40.
51. Rane A, Tomson G, Bjarke B. Effects of maternal lithium therapy in a newborn infant. *J Pediatr* 1978;**93**(2):296–7.
52. Arnon R, Marin-Garcia J, Peeden J. Tricuspid valve regurgitation and lithium carbonate toxicity in a newborn infant. *Am J Dis Child* 1981;**135**(10):941–3.
53. Jacobson S, Jones KL, Johnson KA, et al. Prospective multicentre study of pregnancy outcome after lithium exposure during first trimester. *Lancet* 1992;**339**(8792):350–3.
54. Eikmeier G. Fetal malfromations under lithium treatment. *Eur Psychiatry* 1996;**11**(7): 376–7.
55. Chambers C, Johnson KA, Dick LM, et al. Birth outcomes in pregnant women taking fluoxetine. *N Engl J Med* 1996;**335**(14):1010–5.
56. De las Cuevas C, Sanz EJ. Safety of selective serotonin reuptake inhibitors in pregnancy. *Curr Drug Saf* 2006;**1**(1):17–24.

57. FDA. GlaxoSmithKline (GSK) and FDA notified healthcare professionals of changes to the pregnancy/PRECAUTIONS section of the prescribing information for Paxil and Paxil CR controlled-release tablets, 2005 [accessed 10/01/13]; available from: <www.fda.gov/Safety/MedWatch/SafetyInformation/SafetyAlertsforHumanMedicalProducts/ucm152310.htm>.
58. Wogelius P, Norgaard M, Gislum M, et al. Maternal use of selective serotonin reuptake inhibitors and risk of congenital malformations. *Epidemiol* 2006;**17**(6):701–4.
59. Chambers C, Hernandez-Diaz S, Van Marter L, et al. Selective serotonin reuptake inhibitors and risk of persistent pulmonary hypertension in the newborn. *N Engl J Med* 2006;**354**(6): 579–87.
60. Merlob P, Birk E, Sirota L, et al. Are selective serotonin reuptake inhibitors cardiac teratogens? Echocardiographic screening of newborns with persistent heart murmur. *Birth Defects Res A Clin Mol Teratol* 2009;**85**(10):837–41.
61. Tuccori M, Montagnani S, Testi A, et al. Use of selective serotonin reuptake inhibitors during pregnancy and risk of major and cardiovascular malformations: an update. *Postgrad Med* 2010;**122**(4):49–65.
62. Tuccori M, Testi A, Antonioli L, et al. Safety concerns associated with the use of serotonin reuptake inhibitors and other serotonergic/noradrenergic antidepressants during pregnancy: a review. *Clin Ther* 2009;**31**(1):1426–53.
63. Patel S, Burns T, Botto LD, et al. Analysis of selected maternal exposures and non-syndromic atrioventricular septal defects in the National Birth Defects Prevention Study, 1997–2005. *Am J Med Genet A* 2012;**158A**(10):2447–55.
64. Fraumeni J. Chemicals in human teratogenesis and transplacental carcinogenesis. *Pediatrics* 1974;**53**(5):817–22.
65. Wigle D, Arbuckle T, Turner M, et al. Epidemiologic evidence of relationships between reproductive and child health outcomes and environmental chemical contaminants. *J Toxicol Environ Health B Crit Rev* 2008;**11**(5–6):373–517.
66. Bellinger D. Teratogen update: lead in pregnancy. *Birth Defects Res A Clin Mol Teratol* 2005; **73**(6):409–20.
67. Romero P, Barnett P, Midtling J. Congenital anomalies associated with maternal exposure to oxydemeton-methyl. *Environ Res* 1989;**50**(2):256–61.
68. Johnson PD, Dawson BV, Goldberg SJ. Cardiac teratogenicity of trichloroethylene metabolites. *J Am Coll Cardiol* 1998;**32**(2):540–5.
69. Yauck JS, Malloy ME, Blair K, et al. Proximity of residence to trichloroethylene-emitting sites and increased risk of offspring congenital heart defects among older women. *Birth Defects Res A Clin Mol Teratol* 2004;**70**(10):808–14.
70. Bar-Oz B, Levichek Z, Koren G. Medications that can be fatal for a toddler with one tablet or teaspoonful: A 2004 update. *Pediatr Drugs* 2004;**6**(2):123–6.
71. Fine J. Pediatric principles. In: Nelson LS, Hoffman RS, Lewin N, Goldfrank L, Howland M, Flomenbaum, editors. *Goldfrank's toxicologic emergencies*. 9th ed. New York: McGraw-Hill; 2011; p. 447–60.
72. Collins KK, Van Hare GF. Advances in congenital long QT syndrome. *Curr Opin Pediatr* 2006;**18**(5):497–502.
73. Webster G, Berul CI. Congenital long-QT syndromes: a clinical and genetic update from infancy through adulthood. *Trends Cardiovasc Med* 2008;**18**(6):216–24.
74. Friedman MJ, Mull CC, Sharieff GQ, et al. Prolonged QT syndrome in children: an uncommon but potentially fatal entity. *J Emerg Med* 2003;**24**(2):173–9.
75. Fish FA, Kannankeril PJ. Diagnosis and management of sudden death in children. *Curr Opin Pediatr* 2012;**24**(5):592–602.

76. Kapetanopoulos A, Kluger J, Maron BJ, et al. The congenital long QT syndrome and implications for young athletes. *Med Sci Sports Exerc* 2006;**38**(5):816–25.
77. CDC. Mental Health in the United States: prevalence of diagnosis and medication treatmetn for attention-deficit/hyperactivity disorder. *MMWR Morb Mortal Wkly Rep* 2005;**54**(34):842–7.
78. Nissen S. ADHD drugs and cardiovascular risk. *N Engl J Med* 2006;**354**(14):1445–8.
79. U.S. Food and Drug Administration. Pediatric Advisory Committee: briefing information; 2006 [accessed 05/10/13]; available from: <www.fda.gov/ohrms/dockets/ac/06/slides/2006-4210s-index.htm>.
80. Biederman J, Spencer TJ, Wilens TE, et al. Treatment of ADHD with stimulant medications: response to Nissen perspective in the *New England Journal of Medicine. J Am Acad Child Adolesc Psychiatry* 2006;**45**(10):1147–50.
81. U.S. Food and Drug Administration. Detailed view: safety labeling changes approved by FDA Center for Drug Evaluation and Research (CDER); 2005 [accessed 10/05/13]; available from: <www.fda.gov/medwatch/safety/2005/aug05.htm>.
82. Vetter VL, Elia J, Erickson C, et al. Cardiovascular monitoring of children and adolescents with heart disease receiving stimulant drugs: a scientific statement from the American Heart Association Council on Cardiovascular Disease in the Young Congenital Cardiac Defects Committee and the Council on Cardiovascular Nursing. *Circulation* 2008;**117**(18):2407–23.
83. Thomas PE, Carlo WF, Decker JA, et al. Impact of the American Heart Association scientific statement on screening electrocardiograms and stimulant medications. *Arch Pediatr Adolesc Med* 2011;**165**(2):166–70.
84. Martinez-Raga J, Knecht C, Szerman N, et al. Risk of serious cardiovascular problems with medications for attention-deficit hyperactivity disorder. *CNS Drugs* 2013;**27**(1):15–30.
85. Olfson M, Huang C, Gerhard T, et al. Stimulants and cardiovascular events in youth with attention-deficit/hyperactivity disorder. *J Am Acad Child Adolesc Psychiatry* 2012;**51**(2):147–56.
86. Squires J, Squires R. A review of Munchausen syndrome by proxy. *Ped Annals* 2013;**42**(4):67–71.

Chapter 17

Cardiovascular Toxicity as a Result of Radiological Imaging

Shah Sweni,[1] Ramachandran Meenakshisundaram,[2] and Ponniah Thirumalaikolundusubramanian[1]
[1]*Chennai Medical College Hospital and Research Centre, Trichy, India*
[2]*Epsom and St. Helier University Hospitals, NHS Trust, Epsom, United Kingdom*

17.1 INTRODUCTION

This chapter deals with the health effects of imaging (ionizing and nonionizing) radiation with special relevance to the cardiovascular system and interventional cardiology procedures, so as to increase awareness and implications of health risk with preventive measures that can be undertaken. Medical imaging involves the use of electromagnetic fields, which are classified into two groups: ionizing and nonionizing. Ionizing radiation (IR) involves the use of highly energetic particles or rays to ionize (or remove) one or more electrons from an atom or molecule. The ionizing ability is based on the energy and frequency of the impinging particles or rays. The produced (removed) ions are chemically reactive and, thus, potentially capable of cell damage[1]; for example, IR is an established carcinogen.[2] Nonionizing radiation (NIR) involves the use of acoustic waves for ultrasound or electromagnetic waves in magnetic resonance imaging (MRI). NIR has the capacity to change the position of atoms; however, it cannot alter their structure, composition, and properties. Based on frequency, NIR is divided into static, extremely low frequency, intermediate frequency, and radiofrequency fields. Figure 17.1 depicts the electromagnetic spectrum.[3]

17.2 IONIZING RADIATION

IR has been used as an indispensable tool in medical diagnosis and treatment since its introduction by Roentgen in 1895. Images are created by the use of X-rays (high-speed electrons bombarding a tungsten anode) or gamma rays nuclear decay of radioactive tracers introduced into the body. Very little is known about the risks of diagnostic low-dose radiation (LDR) imaging technologies; some of the radiation doses can reach higher than cumulative

The Heart and Toxins. DOI: http://dx.doi.org/10.1016/B978-0-12-416595-3.00017-7

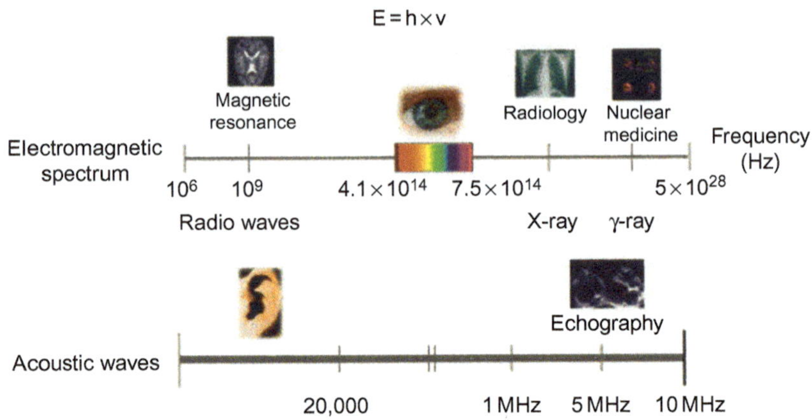

FIGURE 17.1 **The electromagnetic spectrum and acoustic waves. Source:** *Figure reproduced with permission from Andreassi et al., 2007.*[3]

occupational exposures.[4] Advances in imaging technology have resulted in an increase in radiation exposure in cardiovascular patients. The graphical representation of the doses in various imaging procedures and the risk is illustrated in Figure 17.2.[5] It is important to note that most cardiovascular procedures involving IR are at the highest end of the spectrum.

The biological effects of IR are divided into two types: deterministic and stochastic effects. Deterministic effects have a threshold dose below which response does not occur.[6] Thus, interventional procedures, such as percutaneous coronary intervention, radio-frequency ablation with long screening times, and multiple image acquisition, give rise to deterministic effects in both staff and patients.[7]

A stochastic effect is a probabilistic event and there is no known threshold dose. The likelihood of inducing the effect may differ among individuals. Thus, in the following subsections we review:

- The possible effects of imaging-induced radiation on the cardiovascular system
- Occupational exposure in cardiology practice
- Strategies to minimize exposure of patients and staff

17.2.1 Indices Used in Ionizing Radiation

Some of the frequent indices and units that are used in ionizing radiation are described in the following subsections.

Exposure Dose

Exposure dose is the extent of ionizing events taking place when air is irradiated by ionizing photons (X-rays and γ rays). It is expressed in Roentgen (R) or

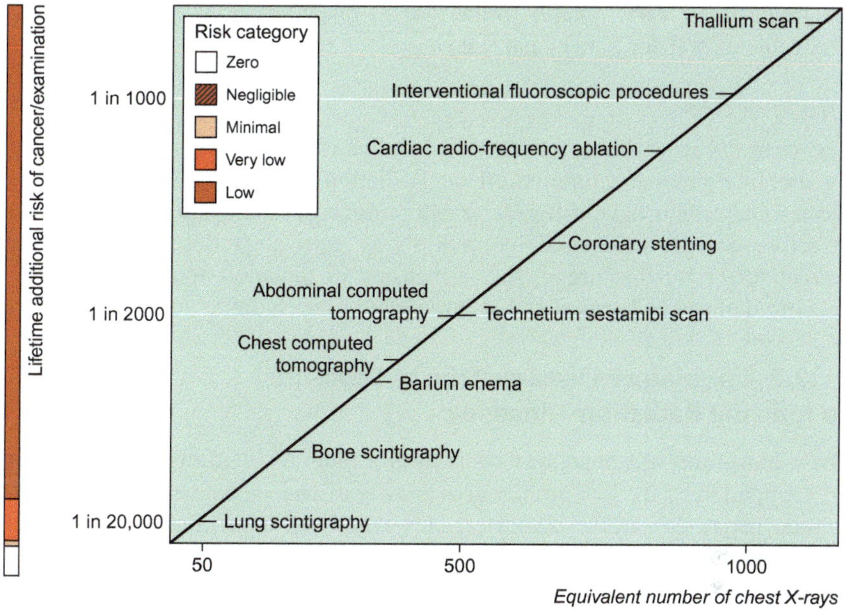

FIGURE 17.2 Presentation of cancer risk and radiation dose for some common radiological and nuclear medicine examinations. Data are from multiples of a dose from a single chest X-ray. **Source:** *Figure reproduced with permission from Picano, 2004.*[5]

coulombs/kg (c/kg). Exposure rate is the rate of charge production per unit mass of air and is expressed in R/min. Dose area product (DAP) is the total radiation energy delivered to the patient, whereas exposure is radiation concentration only at a particular point, which is expressed in gray-cm^2 (Gy-cm^2).

Absorbed Dose

The absorbed dose is the quantity of radiation energy absorbed per unit mass of tissue (i.e., concentration of radiation energy actually absorbed at a specific point in the tissue). It is directly related to the biologic tissues and is expressed in gray (Gy). Computed tomography dose index (CTDI) is the absorbed dose in a phantom model, determined at the center of the slice. It is a good indicator of tissue dose. Dose length product (DLP) is the total radiation to the patient during that CT procedure. DLP is a product of CTDI and length of the body area scanned. It is expressed in Gy-cm.

Equivalent Dose

Equivalent dose is used to express the biological impact of radiation in humans. It relates absorbed dose to effective biological damage of radiation and is expressed in sievert (Sv). The absorbed dose (Gy) multiplied by the radiation weighting factor (WR; unique to the type of radiation) is the equivalent dose. The value of WR for X-rays and γ rays is 1.

Effective Dose

The term *effective dose* expresses the relative risk to humans and is defined by the International Commission on Radiation Protection (ICRP). Various tissues and organs have different sensitivities to the same equivalent dose, so effective dose is calculated by multiplying equivalent dose by the tissue weight factor for that organ. It is a measure of radiation and organ system-specific damage in humans and is measured in the Sv unit.

17.2.2 Circulatory Diseases Due to Exposure to Ionizing Radiation—Imaging

For a long time, the heart was considered a radio-resistant organ. This belief was supported by the low proliferative activity of endothelial cells and the post-mitotic nature of cardiac myocytes. However, both experimental and human studies have demonstrated the effect of high-dose ionizing radiation on the heart and large arteries (i.e., coronaries, carotids, and aorta). These effects include direct damage; marked diffuse fibrotic damage of the pericardium, myocardium, and pericardial adhesions; microvascular damage; and stenosis of the valves.[8] Moreover, excess cardiovascular disease (CVD) morbidity is seen in post-radiotherapy in breast cancer survivors[9–11] (with high-dose radiation to the heart), Hodgkin's lymphoma[12,13] (with mediastinal irradiation), and childhood cancer survivors.[14,15] An increased risk of cardiovascular morbidity and mortality is suggested by studies in moderate (<0.5 Sv/day) and low-dose (<0.01 Sv/day) Japanese atomic bomb survivors[16] and even in occupationally exposed groups such as nuclear workers and uranium miners.[17–19] Cardiovascular diseases are some of the leading causes of death worldwide,[20] so just a small rise in the relative risks are of considerable public health concern.

Low-Dose Radiation for Benign Conditions

In a cohort of patients treated with low-dose ionizing irradiation to the chest for peptic ulcer disease, there was a significant increase in the trend of coronary heart disease.[21] However, a study from the United Kingdom demonstrated that ankylosing spondylitis patients did not show any such increasing trend in a similar situation.[22]

17.3 RADIATION FROM MEDICAL PROCEDURES IN PATHOGENESIS OF ISCHEMIC HEART DISEASE

Mortality from ischemic heart disease (IHD) has been shown to be significantly related to medical radiation even at very low doses, as suggested by Gofman.[23] The study used a premise that average per-capita radiation dose per

100,000 population is directly proportional to the number of practicing physicians per 100,000 population to see whether there was any dose–response relationship between noncancer deaths and IHD. It was conducted under the premise that "physicians order X-rays, even if others perform them."

17.3.1 Theories and Hypothesis about Pathogenesis for Low-Dose Radiation-Induced Cardiovascular Effects

Although knowledge of the cellular and molecular effects of LDR is quite limited, this section summarizes the possible mechanisms. The sequence of events postradiation is very complex and involves a series of biological transformations that evolve over different time scales. IR is a uniquely powerful mutagen capable of inducing every known type of deoxyribonucleic acid (DNA) damage, and there is no threshold dose for induction of this genetic damage that is not repairable by IR. LDR causes cancers not only through mutational DNA damage,[6] other non-DNA targeted effects may also be involved.[24] LDR used in cardiac imaging is energetic enough to induce DNA damage.[25] One of the most important effects of low-radiation doses of ≤ 1 Gy is the induction of persistent genomic instability. Increased levels of genomic instability in atherosclerotic plaques[26] may form a synergistic interaction between radiation-induced effects and pathogenetic events unrelated to radiation.[27]

Thus, ionizing radiation may induce mutations in the coronary arteries, forming dysfunctional clones, thereby contributing to IHD. Proliferation of vascular smooth muscle cells (VSMC) in an atherosclerotic plaque are noted to be far more monoclonal compared to adjacent nonatherosclerotic tissue.[23,28] Clonality suggests that multiple rounds of division between 7 and 13 cumulative doublings occur in the VSMC.[29] Furthermore, plaque VSMCs show enhanced apoptosis[30] and VSMC apoptosis is known to accelerate atherosclerotic lesion development.[31] Based on the "unified model" suggested by Gofman,[23] mutations causing a clone of dysfunctional cells failing to clear the lipids out of dysfunctional tissue may thus lead to formation of the arterial plaque. This explains why plaques occur only in discrete patches, surrounded by normal tissue. Interestingly, radioactivity tends to follow aortic calcification. Calcium deposits have been shown to have α-activity.[32]

Proinflammatory cytokines, such as like interleukin 6 (IL-6), C-reactive protein (CRP), tumor necrosis factor α (TNF-α), and interferon γ (IFN-γ), are increased in Japanese atomic bomb survivors.[33,34] In addition, a dose-related increase in erythrocyte sedimentation rate and immunoglobulins are seen.[33,34] According to a prediction by Little et al.,[35] multiple small radiation doses increases mean chemoattractant (MCP-1) concentration linearly with cumulative dose. The model predicts MCP-1 increases more or less directly with levels of low-density lipoprotein (LDL). Furthermore, there are common mechanisms involved in normal arteriosclerotic processes and those induced by radiation, as reviewed previously and shown in Figure 17.3.[36]

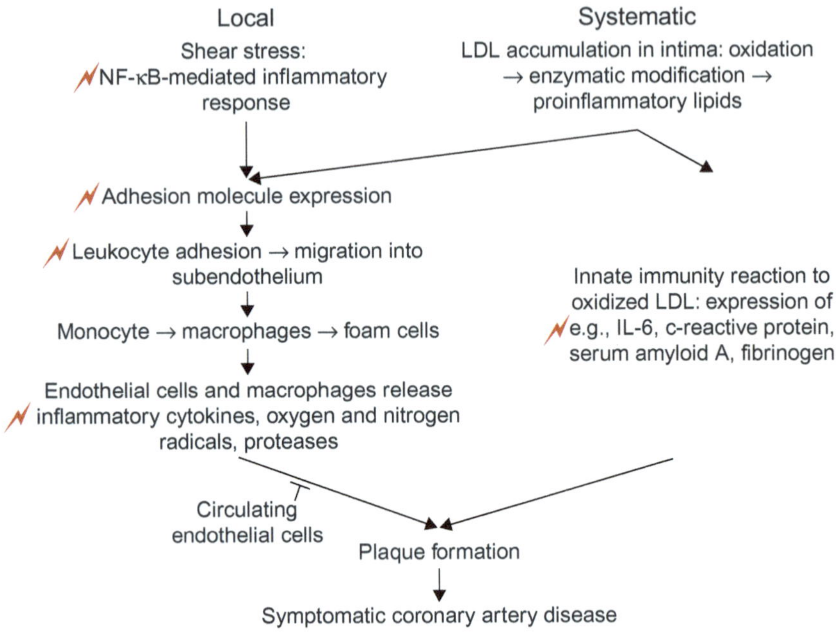

FIGURE 17.3 **Steps in pathogenesis of coronary artery disease.** Shown are possible interactions with radiation effects. Flash symbols indicate events that are also seen after radiation. **Source:** *Figure reproduced with permission from Schultz-Hector and Trott, 2007.*[36]

Post-LDR and E-selectin, the endothelial adhesion molecule mediating leukocyte rolling, are upregulated, both in vitro[37] and in mice.[38] P-selectin, an early proinflammatory factor, is also increased on the luminal vessel surface. Radiation-mediated NF-κB regulated intercellular adhesion molecule (ICAM) is also seen in the endothelium.[39] Expression of adhesion molecules (i.e., CD31, ICAM, VCAM) starts from doses as low as 2 Gy in mice.[40] Apart from adhesion molecules, upregulation of IL-6 and IL-8 cytokines have been seen in time- and dose-related manners[41] and IL-8 induces endothelial cell proliferation.[42] Further, endothelial apoptosis is known to be a potential trigger of arteriosclerotic lesions.[43] These proinflammatory events may be considered as molecular-correlated, radiation-induced ultrastructural changes as seen in myocardial microvasculature.[36] Interaction of radiation-induced changes in endothelial function with the initial events of atherosclerotic lesion results in chronic inflammation, favoring the development of a vulnerable plaque.[44]

In summary, our knowledge regarding cellular response to LDR is very limited. Further research is required to better understand the nature of epidemiological associations; knowing whether they represent a direct causal relationship or some other explanation is necessary.

17.4 OCCUPATIONAL EXPOSURE OF RELEVANCE TO CARDIOLOGY PRACTICE

Increasing workloads and complexity of procedures over the past few years has led to radiation exposure concerns for cardiologists. Fluoroscopic procedures are by far the largest source of occupational exposure in medicine.[45] Interventional cardiology (IC) staff is exposed to two or three times more radiation per year than radiologists.[46] Electrophysiologists rely only on fluoroscopy. Cardiologists and interventionists use fluoroscopy for central catheter and pacemaker placements, whereas nuclear cardiologists are exposed to radioactive materials. Fellowship trainees are exposed to 60% more radiation compared to their seniors.[47] Moreover, nurses and support personnel are also exposed to radiation in the cardiac laboratory and through patients who have been injected with radioactive materials.[48] Fluoroscopic procedures may increase the risk for brain tumors in interventional cardiologists.[49] Furthermore, technical limitations of working with the smaller patient may adversely affect radiation exposure of pediatric cardiologists. Thus, it is important for cardiologists to be aware of the radiation risks of the various procedures to which they subject themselves and patients.

The main differences between cardiac diagnostic and interventional procedures is the increased duration of fluoroscopy due to a longer procedure and the altered ratio of fluoroscopic to cine time during cardiac intervention.[48] The operator dose per procedure in IC may range from 3 to 450 μSv at the neck, from less than 0.1 to 32 μSv at the waist or chest, and from 48 to 1280 μSv at the hands, depending on the type of procedure and the technique used.[50] Radiation doses differ based on the different catheter insertion sites. The radial approach (RA) requires the cardiologist to work closer to the X-ray beam,[51] thereby increasing operator exposure by 100% in the radial versus femoral approach (FA) during diagnostic catheterization and by 51% during coronary interventions.[52] During implantation of pacemakers and similar devices, the subclavian approach results in higher exposure compared to RA and FA due to the operator's proximity to the X-ray beam.[53] Electrophysiologic ablation of atrial fibrillation is a long procedure and the operator is exposed to scattered X-rays.[54] Occupational exposure rates with various procedures used in cardiology have been extensively mentioned in the consensus statement.[48]

An increased incidence of cataracts were reported for IC staff. Thus, ICRP recommends an equivalent annual dose limit for the lens of 20 mSv for five years, with no single annual exposure exceeding 50 mSv. Chronic exposure to LDR increases hydrogen peroxide levels and alters redox balance in IC staff.[55] Cellular responses to radiation effects include antioxidant defense and increased susceptibility to apoptosis induction. LDR is associated with increased micronuclei frequency in IC and this risk correlates with years of professional activity.[56] Furthermore, IC with an average exposure of 4 mSv per year shows an increase in circulating lymphocytes, chromosome aberrations, and micronuclei. These may represent an intermediate risk of carcinogenesis[56–58]

17.4.1 Strategies to Minimize Exposure

Many interventionists are not aware of the potential for injury from procedures, their occurrence, or methods for decreasing their incidence by using dose-control strategies.[59] Moreover, types of X-ray equipment (film based vs. digital), particular settings of the equipment, use of radiation protection facilities, personal dosimeters, and operators' workload affect radiation doses received.[60]

Access to medical physics advisors and the use of dose parameters in quality-assurance programs can significantly limit patient and operator exposures.[61] Rapid exchange coronary angioplasty systems, low-framing speed, pressure tubing, electronic radiation dosimeters, mediated self-surveillance, and rotational coronary angiography are beneficial in limiting radiation.[62] With the use of a movable floor shield for radiation protection, the exposure in RA can be decreased to levels less than from an FA.[63] Furthermore, it is vital to closely supervise cardiology fellows and/or trainees early in their training to minimize radiation doses both to patients and staff.[47] Fluoroscopy times can be minimized by pulsing the X-ray beam and software with road-mapping facilities.[64] Approaches, such as pulse-mode fluoroscopy, progressive television scanning, fiber optic coupling, and filmless cardiac catheterization, represent other potential means of reducing radiation exposure.[48]

The *as low as reasonably achievable* (ALARA) principle should be followed to minimize the risk of radiation exposure. Some precautions that can be used are as follows[48]:

- Increase operator's distance from the patient (radiation source). Working at 80 cm from the isocenter instead of 40 cm can decrease scattered dose to a quarter.
- Use a ceiling-suspended screen and personal shielding tools (e.g., two-piece wraparound aprons, thyroid shields, eye protection).
- Use newer shielding equipment such as a lead cap.
- Two monitoring batches should be worn (one inside the lead apron at waist level, one outside the lead apron at collar level).
- Use sterilizable ring batches for monitoring hand exposure during peripheral angiography.
- Minimize the use of fluoroscopy and use low fluoroscopy modes (e.g., pulsed fluoroscopy) whenever possible.
- A scattered dose during cine is much higher than during fluoroscopy; acquire only the necessary number of frames per series and limit the number of series.
- Because dose levels are more important for large patients, and during very angulated projections, staff should be better protected in these cases.

- Scattered radiation is higher at the side of the X-ray tube and less important at the side of the image intensifier during lateral projections.
- Collimation of the radiation field (and generally all other factors reducing patient dose) decreases the level of scattered dose.
- Being aware of the radiation to the patient will improve operators' occupational protection.

The combination of limiting radiation exposure and following optimal practices (e.g., the use of appropriate equipment, maintaining distance from X-ray sources, and using shielding) will prevent excessive exposure or every operator. If national and international agencies responsible for medical and radiation safety establish exposure limits for all of the interventional cardiology procedures, it will be an added advantage.

17.5 CARDIOVASCULAR EFFECTS OF CONTRAST MEDIA USED WITH IONIZING RADIATION

Contrast media (CM) are among the most widely used pharmaceuticals for intravascular administration.[65] The aim of the CM is to absorb IR and to create a distinction between the organ and its surrounding tissues. Thus, CM are necessary to provide X-ray absorption of the bloodstream. However, all of its other observed effects are adverse in nature.

Contrast media are typically classified according to their osmolality: the ratio of iodine atoms to osmotically active particles.[66] Because much of the toxicity of these agents is related to their osmolality, this parameter is often used.[66] The classes of CM in use are

- Ionic, high-osmolar monomeric contrast media (HOCM)
- Ionic, low-osmolar dimeric CM
- Nonionic low-osmolar monomeric CM (LOCM)
- Nonionic iso-osmolar dimeric CM (IOCM)

Ionic monomers are HOCM with an iodine ratio of 1.5, whereas both the nonionic monomers and the ionic dimer (ioxaglate) are contrast media that have an iodine ratio of 3. HOCM has an osmolality nearly 5 to 8 times that of blood.[67] LOCM are 2 to 3 times more hyperosmolar whereas IOCM (iodixanol) is isotonic with blood (290 mOsm/kg). Viscosity of CM—the resistance to flow of a contrast agent—varies according to molecular structure, concentration in solution, and the temperature of the solution.[68] Low viscosity is preferable because it facilitates rapid and homogeneous mixing of CM with blood.[69] Various CM and their osmolality are shown in Figure 17.4.[70]

The cardiac effects of CM have been extensively studied because of the large numbers of patients who receive CM during invasive angiographic procedures. Contrast media affects cardiac electrophysiology in many ways. It increases stroke volume, cardiac output, and left ventricular contraction, and CM

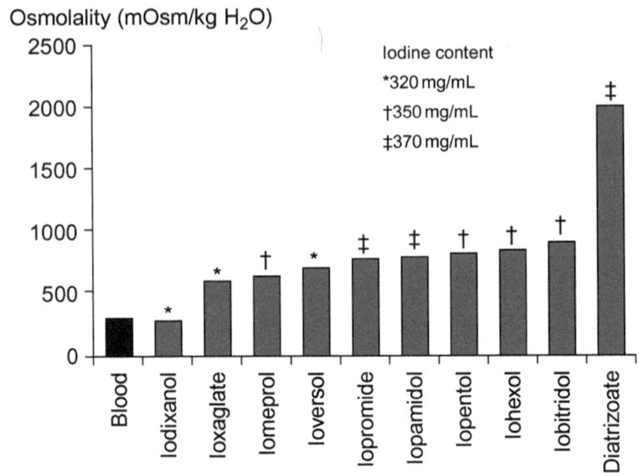

FIGURE 17.4 **Various contrast media and their osmolality. Source:** *Figure reproduced with permission from Davidson et al., 2006.*[70]

decreases peripheral resistance and causes bradycardia. Increased blood pressure in the pulmonary artery and varying changes in the systemic blood pressure are also observed. These changes can cause myocardial ischemia and ventricular fibrillation.[71] Patients suffering from heart failure are at greater risk of CM-related side effects during coronary arteriography. LOCM or IOCM results in less pronounced electrophysiologic and hemodynamic alterations than HOCM.[71] LOCM or ICOM are more efficacious and tolerable in general. Contrast media osmolality, ionicity (nonionic is less toxic), electrolyte composition of the CM, presence of calcium-chelating agents in solution, and the ability of CM to bind calcium have all been implicated in cardiac electrophysiology changes.[72]

Contrast media exert other hemodynamic effects by one of the following mechanisms.

- Hypervolemia due to injected volume and the osmotic load[73]
- Reduction of myocardial contractility due to changes in blood ionicity and myocardial chemotoxicity[74]
- Hematological changes secondary to altered plasma viscosity[75]
- Erythrocyte deformation and aggregation[76]
- Changes in vascular tone as a result of shear stresses and hemodilution-mediated hypoxia[77]

Addition of small amounts of key electrolytes to nonionic CM can minimize the risk of cardiac complications during coronary arteriography.[78] The side effects of CM during coronary arteriography are related to the passive wash-out of cardiac electrolytes. Electrolyte shifts may change the myocardial Na/Ca balance and calcium control.

Iodixanol was associated with less pronounced effects and a better profile on cardiac electrophysiology.[79,80] During cardioangiography and ventriculography in a randomized double-blind trial, patients receiving iodixanol demonstrated less changes in vector cardiography parameters compared to those receiving iohexol.[81] Furthermore, iodixanol was better compared to ioxaglate. Iohexol caused the greatest changes in the ST segment and T wave.[82] Iodixanol had smaller changes in systemic blood pressure compared to other LOCM such as iohexol, ioxaglate, iopromide, and iopamidol.[79] In addition, heart rate increases were smaller with iodixanol than with ioxaglate in all three cardioangiography trials included in a European review.[72,79]

17.5.1 Thrombotic and Hematologic Complications of Contrast Media

Risk of thrombosis precipitation by CM depends on osmolality and ionicity.[83] Iodinated contrast media has an anticoagulant property and this effect is greater with ionic CM. However, the antithrombotic advantages of ionic over nonionic low-osmolar monomeric CM are balanced by a greater frequency of contrast-induced symptoms such as nausea, vomiting, or rashes. Thus, LOCM or IOCM is preferred over HOCM. However, meticulous angiographic technique is the most important factor in reducing thrombotic complications associated with angiography.[83]

17.6 NONIONIZING RADIATION

Nonionizing radiation (NIR) includes ultrasound (i.e., using acoustic waves) or magnetic resonance imaging (MRI) using electromagnetic waves. Various effects of these imaging techniques on the heart and its vessels are reviewed in the following subsections.

17.6.1 Magnetic Resonance Imaging

Magnetic resonance imaging is a powerful diagnostic technique for mapping the structure of various tissues and organs. It does not use ionizing radiation, so it is considered relatively safe with few known adverse effects.[84] Nevertheless, it is an imaging modality through which patients and health care personnel are exposed to three different types of magnetic fields (MF) that are used to create the MR images: a static magnetic field (SMF), a gradient MF, and radio-frequency (RF) electromagnetic waves.

An SMF generates a net magnetization vector in the human body by aligning the proton spins. The static field is located at 0 Hz in the electromagnetic spectrum and does not vary with time. The SMF is expressed in tesla (T), where 1 T = 10,000 Gauss. In clinical use, the commonly used

MRI magnets generate about 1.5 to 3 T in the central field. At the operator's console, magnetic flux density is typically 0.5 mT.[85]

A gradient magnetic field (100–1000 Hz) produces different resonant frequencies for aligned protons based on their spatial positions, thus allowing spatial reconstruction of two-dimensional tissue sections into three-dimensional MRI images. The gradient MF is switched on and off during the image reconstruction process, thus it varies between extremely low frequency (ELF) and intermediate frequency (IF).

A RF electromagnetic wave (i.e., 10–400 MHz) rotates the magnetization vector out of direction of the SMF and the time during which this vector returns to equilibrium is different for each tissue, thereby resulting in two different imaging parameters—T1 and T2—relating to contrast.

The use of MRI in cardiology evaluations is growing rapidly. Various cardiac devices, such as heart valve prostheses, coronary artery stents, aortic stent grafts, pacemakers, and implantable cardioverter-defribrillators, are potential hazards. Interaction of these devices with the MF can cause dislodgment, dysfunction, or damage to the device.[86]

A World Health Organization issued document[87] states that there are no short- or long-term adverse effects of static MF on human health. However, with the increasing use of higher static magnetic field MRI systems, assurance about their safety arises. This subsection reviews the evidence of adverse effects on the cardiovascular system from exposure to MF. We also hypothesize the mechanisms through which these events occur.

In Vitro Studies

Cardiac tissue is composed of myocytes and fibrocytes that are electrically coupled into a functional syncytium by gap junctions. Currents flow through these resistive intracellular, myoplasmic, extracellular pathways as well as across the cell membranes, thereby causing a spread of excitation in the cardiac tissue. MF can affect these structures vital for heart functioning as demonstrated by the following studies.

The various effects of SMF at a cellular level—including calcium signaling, DNA damage, morphology, apoptosis, and micronuclei (MN) formation—have been extensively reviewed earlier.[88] Importantly, SMF generated during a routine cardiac MRI has genotoxic effects on human lymphocytes in vitro. It causes an increased MN frequency, a marker of DNA damage, and genomic instability. However, this damage does seem to be reversible, with MN returning to control values within 24 to 48 hours.[89] Exposure of cultured epithelial cells, fibroblasts, and differentiating neurons to MF of 10 to 17 T have shown deleterious effects on cell organization, viability, and differentiation.[90]

An extremely low-frequency MF is "possibly carcinogenic" to humans, as classified by the International Agency for Research on Cancer (IARC), due to risk of childhood leukemia.[91] In fibroblasts, ELF induces single-strand breaks, MN, and aberrations, even with small exposures in the range of microteslas.[92] ELF is known to affect calcium signaling, induce apoptosis, and affect cytoskeletal structure.[93] The effects of RF related to the MRI procedures have not been studied as much. There was no increase in genotoxicity in mammalian cell cultures exposed to RF at MRI procedure levels.

We hypothesize, thus, that exposure to various types of MF during an MRI can affect calcium signaling, cytoskeletal structure, cell–cell adhesion, and gap junction functioning, influencing current conduction in the myocardial tissue, thereby provoking arrhythmias.

Animal Studies

Exposure of monkeys to an SMF of up to 1.5 T produced a field strength-dependent increase in electrocardiographic (ECG) T-wave signal amplitude.[94] This has been demonstrated previously for other animals[95] such as rats,[96] baboons, and rabbits.[97] Moreover, a very minimal effect on blood pressure is exerted by field strengths up to 1.5 T. This effect is too small for measurement by conventional techniques.[94] Interestingly, a decrease in heart rate has been reported in squirrel monkeys[98] versus an increase in heart rate reported in rabbits[99] and rats.[100] In contrast, anesthetized rats demonstrated no changes in ECG on exposure to pulsed magnetic fields.[101] Exposure of a snail heart to a 1.5 T magnetic field exhibited transitory arrhythmias, a decrease in diastolic action and myocardial hyperpolarization.[96,102] In dogs, a pulsed magnetic field has been shown to induce arrhythmias in diastole, and the type of arrhythmias varied with the coil's position. If placed just above the heart, atrial ectopics were induced, whereas if the coil was placed at the apex, ventricular ectopics were induced.[103] Further, exposure to high-strength MF could exert an effect on vagal stimulation thereby modulating the heart rate. Also, large blood-flow potentials can be induced by an MF in the proximity of the ventricles, thereby affecting the polarization of the heart. This may show distortions in ECG, provoke arrhythmias, and affect the heart rate.[96]

Frog hearts exposed to "nonthermal" action of pulsed radiation (RFR) leads to tachycardia and arrhythmias.[104] Studies using a rat model for effects of heat-inducing levels of RFR exposure showed that, in general, a greater degree of heating leads to a larger increase in heart rate, as reviewed extensively earlier.[105] MRI frequency RF radiation can cause a rise in body temperature.[106] Nevertheless, it is important to realize that RF absorption varies according to body size, anatomical features, and tissue sensitivity; thus, animal studies cannot be directly extrapolated to humans.

Human Studies

In human volunteers, significant increases in systolic blood pressure have been demonstrated, with field strengths up to 8 T.[107] However, this change in blood pressure was in the range of normal physiological variability.

Furthermore, in susceptible workers, ectopics and reversible ventricular arrhythmia can occur. This possibly can lead to ventricular fibrillation.[108] A study of volunteers[109] who were exposed to MR procedures found significant changes in skin blood flow, systolic blood pressure, and heart rate, but all of these changes were within acceptable safety levels. Exposure to very high magnetic fields of 10 T can lead to about a 10% increase in blood pressure[110] and a 10% decrease in blood flow.[111]

The effects of current MRI strength magnetic fields and mechanisms have been reviewed elsewhere;[105] however, we summarize them briefly here. It is important to note that with the advent of higher strengths the following effects would increase and may cause a significant concern.

Electrocardiogram Changes

As mentioned earlier, T-wave changes are seen on electrocardiograms and the reasons are hypothesized as follows.

Aortic blood flow is associated with increased electrical potential due to MF. Repolarization of the ventricle coincides temporally with ejection of blood into the ascending aorta; thus, it is expected that aortic blood flow potential induced by a magnetic field is recorded in the T-wave region of the ECG trace. Combined signals from a naturally occurring T wave and the magnetically induced flow potential appear as an augmented T-wave amplitude.[94–97] The ECG change is due to *magnetohydrodynamic voltage* generated by blood flow in the presence of a MF, and this voltage is unrelated to the pacemaker activity of the heart.[110] Nevertheless, a pulsatile pulmonary artery flow and high field-associated carotid and subclavian artery flow also may contribute to these changes. When blood flows in, an applied magnetic field a transverse EMF is developed. This leads to a small induced current density in the tissues, which in turn leads to a small electric voltage on the body surface that can be detected by the use of metal electrodes on the skin.[84]

Heart Rate Variability

Even if there are minute differences or minimal changes in heart rate under the magnetic field, these may be considered to be within a normal range of variation.[105]

Cardiac Musculature

On exposure to a standard 1.5 T magnetic field, stimulation of a heart cell would not occur.[112] Field-strength thresholds of pulsed magnetic stimuli for cardiac effects are much higher than thresholds for neuromuscular effects. Thus, violent muscle contraction would be expected to occur before any cardiac abnormality.[113]

Myocardial Infarction

Myocardial infarctions based on geomagnetic activity: the Earth's MF is the most prevalent source of magnetic field exposure for humans (ranging from 30–70 μT). The MF of an MRI is much stronger than the geomagnetic field. Geomagnetic storms have been shown to cause increased rates of myocardial infarctions.[114]

Cardiac Magnetic Resonance Imaging

During cardiac MRI, a strong and fast switching electromagnetic gradient is applied. Thus, it uses the highest amount of energy allowed by regulatory authorities. These fast gradient-induced electric fields can stimulate nerves and skeletal and cardiac muscles.[1]

Effects on Blood Flow

The effects on tissues and/or blood vessel exposure on circulation and microvasculature are reviewed in detail by McKay et al.[115] Magnetic fields can trigger either vasoconstriction or vasodilation depending on the initial tone of the vessel. Cellular effects, such as an increase in local nitric oxide and angiogenesis, have been reported.

Miscellaneous

The RF produced during MRI procedures can lead to accidents (e.g., burns) in presence of conducting materials (e.g., leads of physiological parameters such as heart rate, blood pressure, oxygen saturation, and temperature monitoring equipment).[116] This risk is more serious in the case of internal biomedical implants (e.g., aneurism clips, stents), especially ones that are electronically activated (cardiac pacemakers).[117,118]

The strength of the magnetic field declines with distance to the isocenter where the MF is isotropic and homogeneous. However, most studies analyze the effects while participants were not placed in the isocenter, so were not directly exposed to the gradient of the static magnetic field. How patients and health personnel move in this field is also of importance. One must understand that "effects" are not necessarily "hazardous." Some studies previously described may have shown changes or effects that may be within the capacity of an individual to maintain a normal homeostasis. However, if the ability to function properly is compromised, then it becomes "hazardous." Thus, any measurable effect may not actually lead to an adverse outcome. Magnetic fields cause many minor, subtle, adverse effects on the cardiovascular system; how hazardous these are is yet to be thoroughly monitored.

17.6.2 Echocardiography/Diagnostic Ultrasound

Ultrasound imaging, or *sonography*, is a method of obtaining body images through the use of high-frequency sound waves. Ultrasounds are mechanical vibrations with frequencies above the human audibility limit. It is convenient to use owing to its portability and the inherent safety of low-power acoustic interrogation of the tissue. However, even with a diagnostic range of frequency, intensity, and duration, ultrasound can have significant molecular, cellular, and organ effects. Understanding the complex effect of ultrasound on tissues may open avenues toward understanding the possible side effects in patients and the precautionary measures that one needs to take to prevent harm. Health care professionals working in cardiology with color Doppler ultrasonic equipment showed increased genotoxicity (i.e., chromosome aberrations) versus controls.[119] In the following discussion, we highlight such effects on tissues.

The bioeffects of ultrasound depend on the characteristics of the sound wave, the sensitivity of the exposed tissue, and the duration of exposure. The intensity of ultrasound, the frequency of the transducer, and the mode (continuous or pulsed) influence bioeffects on tissues.[120] Ultrasound can produce elevations of temperature and damage the biological tissues in vivo. Apart from thermal mechanisms, nonthermal mechanisms also occur. Both thermal and nonthermal mechanisms have been extensively reviewed earlier.[121,122] The negative effects of contrast-enhanced diagnostic ultrasound have also been described and reviewed thoroughly.[123,124]

Ultrasound contrast agents consist of suspensions of gas bodies (i.e., stabilized microbubbles) that pass through the lungs after intravenous injection and persist in the circulation for some time. They can enhance B-mode and Doppler images, and, thus, special imaging methods can show blood distribution at the capillary level to reveal tissue perfusion.[125] Echocardiography contrast agents are used for left ventricular opacification and border delineation. Contrast echocardiography can also be used for measurement of myocardial perfusion.[126]

The following summarizes some of the effects of relevance to the heart and its vessels. Ultrasonic cavitation is the interaction of the ultrasound with a gas body. Cavitation involving microbubbles is the most likely potential mechanism for the nonthermal bioeffects of a diagnostic ultrasound. Biological tissues that contain gas-filled microbubbles (e.g., lung, intestines, injected microbubbles, and injected ultrasound contrast agents) interact with the acoustic waves of the ultrasound.[121] On exposure to the ultrasound's oscillating acoustic wave, these bubbles undergo alternate expansion and contraction with gradual diffusion of gas at low acoustic outputs. All bubbles produce acoustic radiation forces and microstreaming, while only those significantly affected will exhibit reactions such as free radicals or shock-wave generation (Figure 17.5).[119]

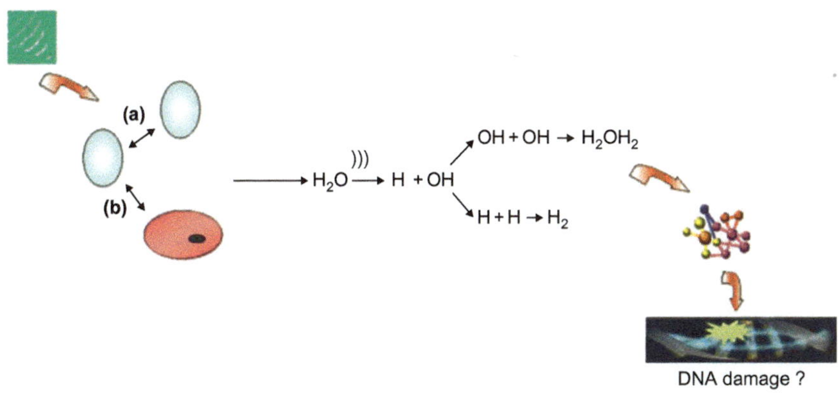

FIGURE 17.5 **The ultrasound's oscillating acoustic wave.** A high acoustic pressure causes rapid bubbles that grow and collapse among them (a) and the cells (b). This produces sufficient energy to disrupt chemical bonds and form reactive free radicals that may interfere with DNA. **Source:** *Figure reproduced with permission from Andreassi, 2004.*[119]

The radiation force can cause a bubble to move at high speed, and the cells near the path of a speeding bubble may be damaged by exposure to high shear stresses as it passes.[127] Oscillating bubbles tend to attract nearby cells, thus collecting them into highly concentrated groups; thereby they can be more easily destroyed. Bubbles oscillating on a solid surface produce a vigorous circulatory motion in the surrounding fluid called microstreaming.[128,129] The velocity of the fluid flowing around the bubble is greatest near the bubble's surface and gradually decreases as the distance from the bubble increases, thus forming a gradient of flow and force around the bubble. This unequal distribution of forces on the exterior of the cell results in shearing forces that distort and tear the cellular membrane, causing hemolysis. Cavitation formed during the process produces sufficient energy to disrupt chemical bonds and produce reactive free radicals.[1] These free radicals may interfere with DNA, thereby leading to chromosomal damage, theoretically. However, this microbubble-induced damage is extremely localized, being confined to the immediate vicinity of the bubble (a blood vessel). Gas bodies suitable for nucleation of ultrasonic cavitation are normally absent from the human body. They are introduced during contrast echocardiography (i.e., ultrasound) procedures.

Experimental Studies of the Effects on the Lung and the Heart

In frog heart, even a single millisecond pulse of ultrasound delivered during systole (moment of contraction of the heart) produces a reduction in the developed aortic pressure and a reduction in the strength of contraction, whereas a pulse delivered during diastole can produce premature ventricular contraction.[130,131] The same results were duplicated in the murine heart.[132]

No damage to cardiac tissue was observed grossly, although significant hemorrhage occurred to adjacent lung tissue. The threshold is a pulse duration of 5 msec. This is longer than the typical diagnostic pulses. Thus, it is unlikely for this effect to occur in the human heart under diagnostic conditions.

The effects of diagnostic contrast ultrasound/echo on the myocardium have been studied both in animals and in humans (Table 17.1). It is known to induce premature ventricular contractions (PVCs), petechiae, troponin elevation, microvascular leakage, histological changes, and so on.[123] Rarely, serious cardiopulmonary complications have occurred, leading to fatalities, during or within 30 min after use of ultrasound contrast agents.[123] Lung hemorrhage can occur during realistic diagnostic echocardiogram exposures, though it is unlikely to occur during the majority of routine examinations. To be completely certain that hemorrhage will not occur, output levels that compromise the quality of the diagnostic information will be required.[121]

The bioeffects of contrast ultrasound include sonoporation, microvascular leakage, capillary rupture microlesions, inflammatory cell infiltration, and

TABLE 17.1 Effects of Various Echocardiographic Contrast Agents in Human and Animal Myocardium

Frequency (MHz)	Contrast Agent	Human/ Animal	Effects on Myocardium	References
1.66	AIP101	Human	PVCs	133
1.7–1.9	Optison	Human	Negative for PVCs, troponin I, CK, CK-MB	134
NS	PB127	Human	Negative for PVCs	135
1.8	PESDA, Sonazoid, Optison, Levovist	Rabbit	Depression of contractile function, increased lactate, capillary rupture, increased coronary perfusion pressures	136
1.3	Optison, Definity	Rat	Troponin T elevation, negative histologic findings, negative LV dysfunction	137
1.7	Optison	Rat	Microvascular leakage <20 min	138
1.7	Optison, Imagent, Definity	Rat	PVCs, petechiae	139
1.5	Optison	Rat	Histologic microlesions	140

**PVCs, premature ventricular contractions; CK, creatine kinase; LV, left ventricle; NS, not significant.*

PVCs during scanning.[123] These effects are primarily in scattered microscopic injuries, which are usually not clinically detectable in the short term. PVCs during contrast echocardiography, which are clinically observable but end on cessation of the ultrasound, are an exception. The longer-term medical significance of these bioeffects is uncertain. The actual risk of permanent harm from diagnostic ultrasound is indeed negligible. In vitro studies (Chinese hamster V-79) demonstrate genetic transformation with ultrasound only at levels above the diagnostic range. In contrast, diagnostic levels of X-rays produce transformation in the same cell line.[141] The theoretical probability of one inertial cavitation under normal conditions, in the absence of external artificial microbubbles for noncontrast ultrasound examination, is 1 per 10,000,000,000 examinations.[142]

Recommendations

The European Medical Imaging guidelines clearly state that a medical ultrasound is safer than other imaging techniques and is preferred over ionizing imaging techniques if it provides similar clinical information. The National Council on Radiation Protection and Measurements advocates further studies of ultrasound safety, improvements in its safety features, and more safety education for operators to ensure that the benefits outweigh any risks.[143] Medical professionals using ultrasound as a diagnostic modality should apply the ALARA principle, especially if exposed tissues (e.g., lung) contain gas bodies. This is of particular relevance in echocardiography. During high MI contrast, cardiac-gated perfusion echocardiography is vital to monitor ECG, especially in cases with myocardial infarction or unstable cardiovascular disease.

Key Learning Points

- IR may induce mutations in the coronary arteries, forming dysfunctional clones thereby contributing to IHD.
- Contrast media, which are in frequent use for cardiovascular investigations, have been implicated in myocardial ischemia and arrhythmias.
- MRI and ultrasound use have been demonstrated to influence cardiovascular parameters.
- The ALARA principle should be routinely employed in radiological investigations to keep the radiation exposure to a minimum level.
- In conclusion, it is vital for physicians to be aware of the possible hazardous effects of radiological investigations, especially with the advent of stronger fields in medical imaging.

Acknowledgment We would like to thank the authors Andreassi et al.,[3,119] Picano,[5] Schultz-Hector et al.,[36] Davidson et al.,[70] and their publishers for giving us permission to include their figures in this chapter.

REFERENCES

1. Landini L, Ripoli A, Santarelli MF, et al. The biological effects of diagnostic cardiac imaging. *Curr Pharm Des* 2009;**15**(10):1123–30.
2. United Nations Scientific Committee on the Effects of Atomic Radiation (UNSCEAR). Sources and effects of ionizing radiation, Volume 1, Annex A: UNSCEAR 2006 Report, Epidemiological Studies of Radiation and Cancer. New York: UN; 2008.
3. Andreassi MG, Venneri L, Picano E. Cardiac imaging: the biological effects of diagnostic cardiac ultrasound. *Prog Biophys Mol Biol* 2007;**93**(1–3):399–410.
4. Berrington de Gonzalez A, Mahesh M, Kim KP, et al. Projected cancer risks from computed tomographic scans performed in the United States in 2007. *Arch Intern Med* 2009; **16922**:2071–7.
5. Picano E. Informed consent and communication of risk from radiological and nuclear medicine examinations: how to escape from a communication inferno. *BMJ* 2004;**329**(7470):849–51.
6. United Nations Scientific Committee on the Effects of Atomic Radiation. *Sources and effects of ionizing radiation. Volume 2: effects*. New York: UN; 2000.
7. International Commission on Radiological Protection. ICRP Publication 59: The biological basis for dose limitation in the skin. *Ann ICRP* 1991;22.
8. Adams MJ, Hardenbergh PH, Constine LS, et al. Radiation-associated cardiovascular disease. *Crit Rev Oncol Hematol* 2003;**45**(1):55–75.
9. Cuzick J, Stewart H, Rutqvist L, et al. Cause-specific mortality in long-term survivors of breast cancer who participated in trials of radiotherapy. *J Clin Oncol* 1994;**12**(3):447–53.
10. Darby SC, McGale P, Taylor CW, et al. Long-term mortality from heart disease and lung cancer after radiotherapy for early breast cancer: Prospective Cohort Study of about 300,000 women in U.S. SEER cancer registries. *Lancet Oncol* 2005;**6**(8):557–65.
11. Darby SC, Ewertz M, McGale P, et al. Risk of ischemic heart disease in women after radiotherapy for breast cancer. *N Engl J Med* 2013;**368**(11):987–98.
12. Castellino SM, Geiger AM, Mertens AC, et al. Morbidity and mortality in long-term survivors of Hodgkin lymphoma: a report from the Childhood Cancer Survivor Study. *Blood* 2011; **117**(6):1806–16.
13. Swerdlow AJ, Higgins CD, Smith P, et al. Myocardial infarction mortality risk after treatment for Hodgkin disease: a collaborative British cohort study. *J Natl Cancer Inst* 2007; **99**(3):206–14.
14. Tukenova M, Guibout C, Oberlin O, et al. Role of cancer treatment in long-term overall and cardiovascular mortality after childhood cancer. *J Clin Oncol* 2010;**28**(8):1308–15.
15. Mulrooney DA, Yeazel MW, Kawashima T, et al. Cardiac outcomes in a cohort of adult survivors of childhood and adolescent cancer: retrospective analysis of the Childhood Cancer Survivor Study cohort. *BMJ* 2009;**339**:b4606.
16. Shimizu Y, Kodama K, Nishi N, et al. Radiation exposure and circulatory disease risk: Hiroshima and Nagasaki atomic bomb survivor data, 1950–2003. *BMJ* 2010;**340**:b5349.
17. Kreuzer M, Kreisheimer M, Kandel M, et al. Mortality from cardiovascular diseases in the German uranium miners cohort study, 1946–1998. *Radiat Environ Biophys* 2006;**45**(3): 159–66.
18. Howe GR, Zablotska LB, Fix JJ, et al. Analysis of the mortality experience amongst U.S. nuclear power industry workers after chronic low-dose exposure to ionizing radiation. *Radiat Res* 2004;**162**(5):517–26.
19. McGeoghegan D, Binks K, Gillies M, et al. The non-cancer mortality experience of male workers at British Nuclear Fuels plc, 1946–2005. *Int J Epidemiol* 2008;**37**(3):506–18.

20. World Health Organization. Fact sheet no. 310. The top 10 causes of death. Geneva, Switzerland: WHO; 2013. Available from: <http://who.int/mediacentre/factsheets/fs310/en/index.html>.
21. Carr ZA, Land CE, Kleinerman RA, et al. Coronary heart disease after radiotherapy for peptic ulcer disease. *Int J Radiat Oncol Biol Phys* 2005;**61**(3):842–50.
22. Darby SC, Doll R, Gill SK, et al. Long-term mortality after a single treatment course with X-rays in patients treated for ankylosing spondylitis. *Br J Cancer* 1987;**55**(2):179–90.
23. Gofman J, editor. *Radiation from medical procedures in the pathogenesis of cancer and ischemic heart disease: dose-response studies with physicians per 100,000 population.* San Francisco: C.N.R. Books; 1999.
24. Morgan WF. Non-targeted and delayed effects of exposure to ionizing radiation: I. Radiation-induced genomic instability and bystander effects in vitro. *Radiat Res* 2003;**159**(5):567–80.
25. Löbrich M, Rief N, Kühne M, et al. In vivo formation and repair of DNA double-strand breaks after computed tomography examinations. *Proc Natl Acad Sci USA* 2005;**102**(25):8984–9.
26. Hatzistamou J, Kiaris H, Ergazaki M, et al. Loss of heterozygosity and microsatellite instability in human atherosclerotic plaques. *Biochem Biophys Res Commun* 1996;**225**(1):186–90.
27. Andreassi MG, Botto N. DNA damage as a new emerging risk factor in atherosclerosis. *Trends Cardiovasc Med* 2003;**13**(7):270–5.
28. Benditt EP, Benditt JM. Evidence for a monoclonal origin of human atherosclerotic plaques. *Proc Natl Acad Sci USA* 1973;**706**:1753–6.
29. Matthews C, Gorenne I, Scott S, et al. Vascular smooth muscle cells undergo telomere-based senescence in human atherosclerosis: effects of telomerase and oxidative stress. *Circ Res* 2006;**99**(2):156–64.
30. Schwartz SM, Murry CE. Proliferation and the monoclonal origins of atherosclerotic lesions. *Annu Rev Med* 1998;**49**:437–60.
31. Clarke MCH, Littlewood TD, Figg N, et al. Chronic apoptosis of vascular smooth muscle cells accelerates atherosclerosis and promotes calcification and medial degeneration. *Circ Res* 2008;**102**(12):1529–38.
32. Elkeles A. Radioactivity in calcified atherosclerosis. *Br J Radiol* 1961;**34**:602–5.
33. Hayashi T, Kusunoki Y, Hakoda M, et al. Radiation dose-dependent increases in inflammatory response markers in A-bomb survivors. *Int J Radiat Biol* 2003;**79**(2):129–36.
34. Hayashi T, Morishita Y, Kubo Y, et al. Long term effects of radiation dose on inflammatory markers in atomic bomb survivors. *Am J Med* 2005;**118**(1):83–6.
35. Little MP, Gola A, Tzoulaki I. A model of cardiovascular disease giving a plausible mechanism for the effect of fractionated low-dose ionizing radiation exposure. *PLoS Comput Biol* 2009;**5**(10):e1000539.
36. Schultz-Hector S, Trott K-R. Radiation-induced cardiovascular diseases: Is the epidemiologic evidence compatible with the radiobiologic data? *Int J Radiat Oncol Biol Phys* 2007;**67**(1):10–8.
37. Hallahan D, Clark ET, Kuchibhotla J, et al. E-selectin gene induction by ionizing radiation is independent of cytokine induction. *Biochem Biophys Res Commun* 1995;**217**(3):784–95.
38. Hallahan DE, Virudachalam S. Ionizing radiation mediates expression of cell adhesion molecules in distinct histological patterns within the lung. *Cancer Res* 1997;**57**(11):2096–9.
39. Baeuml H, Behrends U, Peter RU, et al. Ionizing radiation induces, via generation of reactive oxygen intermediates, intercellular adhesion molecule-1 ICAM-1 gene transcription

and NF-κB-like binding activity in the ICAM-1 transcriptional regulatory region. *Free Radic Res* 1997;**27**(2):127–42.

40. Patties I, Habelt B, Rosin B, et al. Late effects of local irradiation on the expression of inflammatory markers in the *Arteria saphena* of C57BL/6 wild-type and ApoE-knockout mice. *Radiat Environ Biophys* 2013;**27**:1–8.
41. van der Meeren A, Squiban C, Gourmelon P, et al. Differential regulation by IL-4 and IL-10 of radiation-induced IL-6 and IL-8 production and ICAM-1 expression by human endothelial cells. *Cytokine* 1999;**11**(11):831–8.
42. Li A, Dubey S, Varney ML, et al. IL-8 directly enhanced endothelial cell survival, proliferation, and matrix metalloproteinases production and regulated angiogenesis. *J Immunol* 2003;**170**(6):3369–76.
43. Reidy MA. A reassessment of endothelial injury and arterial lesion formation. *Lab Invest* 1985;**53**(5):513–20.
44. Stewart FA, Heeneman S, te Poele J, et al. Ionizing radiation accelerates the development of atherosclerotic lesions in ApoE$^{-/-}$ mice and predisposes to an inflammatory plaque phenotype prone to hemorrhage. *Am J Pathol* 2006;**168**(2):649–58.
45. International Commission on Radiological Protection. *ICRP publication 85: Avoidance of radiation injuries from medical interventional procedures*. London: Elsevier Science Ltd; 2000.
46. Picano E, Santoro G, Vañó E. Sustainability in the cardiac cath lab. *Int J Cardiovasc Imaging* 2007;**23**(2):143–7.
47. Watson LE, Riggs MW, Bourland PD. Radiation exposure during cardiology fellowship training. *Health Phys* 1997;**73**(4):690–3.
48. Limacher MC, Douglas PS, Germano G, et al. ACC expert consensus document. Radiation safety in the practice of cardiology. *J Am Coll Cardiol* 1998;**31**(4):892–913.
49. Finkelstein MM. Is brain cancer an occupational disease of cardiologists? *Can J Cardiol* 1998;**14**(11):1385–8.
50. Miller D, Vañó E, Bartal G, et al. Occupational radiation protection in interventional radiology: a joint guideline of the Cardiovascular and Interventional Radiology Society of Europe and the Society of Interventional Radiology. *Cardiovasc Intervent Radiol* 2010;**33**(2): 230–9.
51. Whitby M, Martin CJ. A study of the distribution of dose across the hands of interventional radiologists and cardiologists. *Br J Radiol* 2005;**78**(927):219–29.
52. Lange HW, von Boetticher H. Randomized comparison of operator radiation exposure during coronary angiography and intervention by radial or femoral approach. *Catheter Cardiovasc Interv* 2006;**67**(1):12–6.
53. Kesavachandran C, Haamann F, Nienhaus A. Radiation exposure and adverse health effects of interventional cardiology staff. In: Whitacre DM, editor. *Reviews of environmental contamination and toxicology*. New York: Springer; 2013. p. 73–91.
54. Padovani R, Vañó E, Trianni A, et al. Reference levels at European level for cardiac interventional procedures. *Radiat Prot Dosimetry* 2008;**129**(1-3):104–7.
55. Russo GL, Tedesco I, Russo M, et al. Cellular adaptive response to chronic radiation exposure in interventional cardiologists. *Eur Heart J* 2012;**33**(3):408–14.
56. Andreassi MG, Cioppa A, Botto N, et al. Somatic DNA damage in interventional cardiologists: a case-control study. *FASEB J* 2005;**19**(8):998–9.
57. Zakeri F, Assaei RG. Cytogenetic monitoring of personnel working in angiocardiography laboratories in Iran hospitals. *Mutat Res* 2004;**562**(1-2):1–9.

58. Maffei F, Angelini S, Forti GC, et al. Spectrum of chromosomal aberrations in peripheral lymphocytes of hospital workers occupationally exposed to low doses of ionizing radiation. *Mutat Res* 2004;**547**(1-2):91–9.
59. Valentin J. Abstract: Avoidance of radiation injuries from medical interventional procedures, ICRP Publication 85. *Ann ICRP* 2000;**30**(2):7.
60. Vañó E, González L, Guibelalde E, et al. Radiation exposure to medical staff in interventional and cardiac radiology. *Br J Radiol* 1998;**71**(849):954–60.
61. Kuon E, Birkel J, Schmitt M, et al. Radiation exposure benefit of a lead cap in invasive cardiology. *Heart* 2003;**89**(10):1205–10.
62. Vañó E, González L, Fernandez JM, et al. Occupational radiation doses in interventional cardiology: a 15-year follow-up. *Br J Radiol* 2006;**79**(941):383–8.
63. Mann 3rd JT, Cubeddu G, Arrowood M. Operator radiation exposure in PTCA: comparison of radial and femoral approaches. *J Invasive Cardiol* 1996;**8**(Suppl D):22D–25DD.
64. Martin CJ, Sutton DG, Sharp PF. Balancing patient dose and image quality. *Appl Radiat Isot* 1999;**50**(1):1–19.
65. Christiansen C. X-ray contrast media: an overview. *Toxicology* 2005;**209**(2):185–7.
66. Maeder M, Klein M, Fehr T, et al. Contrast nephropathy: review focusing on prevention. *J Am Coll Cardiol* 2004;**44**(9):1763–71.
67. Brinker J. What every cardiologist should know about intravascular contrast. *Rev Cardiovasc Med* 2003;**4**(Suppl 5):S19–27.
68. Dawson PCD, Grainger R. *Textbook of Contrast Media.* Oxford: Isis Medical Media Ltd; 1999.
69. Stacul F. Current iodinated contrast media. *Eur Radiol* 2001;**11**(4):690–7.
70. Davidson C, Stacul F, McCullough PA, et al. Contrast medium use. *Am J Cardiol* 2006;**98** (6A):42K–58K.
71. Morcos SK, Dawson P, Pearson JD, et al. The haemodynamic effects of iodinated water soluble radiographic contrast media: a review. *Eur J Radiol* 1998;**29**(1):31–46.
72. McClennan BL, Stolberg HO. Intravascular contrast media. Ionic versus nonionic: current status. *Radiol Clin North Am* 1991;**29**(3):437–54.
73. Morissette M, Gagnon RM, Lamoureux J, Lemire J. Effects of angiographic contrast media on colloid oncotic pressure. *Am Heart J* 1980;**100**(3):319–22.
74. Newell JD, Higgins CB, Kelley MJ, et al. The influence of hyperosmolality on left ventricular contractile state: disparate effects of nonionic and ionic solutions. *Invest Radiol* 1980; **15**(5):363–70.
75. Reinhart WH, Pleisch B, Harris LG, Lutolf M. Influence of contrast media (iopromide, ioxaglate, gadolinium-DOTA) on blood viscosity, erythrocyte morphology and platelet function. *Clin Hemorheol Microcirc* 2005;**32**(3):227–39.
76. Dawson P, Harrison MJ, Weisblatt E. Effect of contrast media on red cell filtrability and morphology. *Brit J Radiol* 1983;**56**(670):707–10.
77. Morris TW, Kern MA, Katzberg RW. The effects of media viscosity on hemodynamics in selective arteriography. *Invest Radiol* 1982;**17**(1):70–6.
78. Pedersen HK. Electrolyte addition to nonionic contrast media. Cardiac effects during experimental coronary arteriography. *Acta Radiol Suppl* 1996;**405**:1–31.
79. Grynne BH, Nossen JO, Bolstad B, et al. Main results of the first comparative clinical studies on Visipaque. *Acta Radiol Suppl* 1995;**399**:265–70.
80. Fransson SG, Stenport G, Andersson M. Immediate and late adverse reactions in coronary angiography. *Acta Radiol* 1996;**37**(2):218–22.

81. Flinck A, Selin K, Björneld L, et al. Iodixanol and iohexol in cardioangiography: a comparative vectorcardiographic study. *Acta Radiol* 2000;**41**(4):384–9.
82. Flinck A, Selin K. Vectorcardiographic changes during cardioangiography with iodixanol and ioxaglate. *Int J Cardiol* 2000;**76**(2–3):173–80.
83. Aspelin P, Stacul F, Thomsen H, et al. Effects of iodinated contrast media on blood and endothelium. *Eur Radiol* 2006;**16**(5):1041–9.
84. Schenck JF. Safety of strong, static magnetic fields. *J Magn Reson Imaging* 2000;**12**(1):2–19.
85. World Health Organization. Environmental Health Criteria: Static Fields, Monograph No. 232-2006. Geneva, Switzerland: WHO. Available from: <www.who.int/peh-emf/publications/reports/ehcstatic/en/index.html>.
86. Shellock FG. Magnetic resonance safety update 2002: implants and devices. *J Magn Reson Imaging* 2002;**16**(5):485–96.
87. World Health Organization. Environmental Health Criteria 232. Static fields. Geneva, Switzerland: WHO; 2006.
88. Miyakoshi J. Effects of static magnetic fields at the cellular level. *Prog Biophys Mol Biol* 2005;**87**(2–3):213–23.
89. Simi S, Ballardin M, Casella M, et al. Is the genotoxic effect of magnetic resonance negligible? Low persistence of micronucleus frequency in lymphocytes of individuals after cardiac scan. *Mutat Res/Fundam Mol Mech Mutagenesis* 2008;**645**(1–2):39–43.
90. Valiron O, Peris L, Rikken G, et al. Cellular disorders induced by high magnetic fields. *J Magn Reson Imaging* 2005;**22**(3):334–40.
91. International Agency for Research on Cancer. Nonionizing radiation, Part 1: Static and extremely low-frequency (ELF) electric and magnetic fields, Vol. 80. IARC monographs on the evaluation of carcinogenic risks to humans. Lyon, France: IARC;2002, 429 p.
92. Crumpton MJ, Collins AR. Are environmental electromagnetic fields genotoxic? *DNA Repair* 2004;**3**(10):1385–7.
93. Dini L, Abbro L. Bioeffects of moderate-intensity static magnetic fields on cell cultures. *Micron* 2005;**36**(3):195–217.
94. Tenforde TS, Gaffey CT, Moyer BR, et al. Cardiovascular alterations in Macaca monkeys exposed to stationary magnetic fields: experimental observations and theoretical analysis. *Bioelectromag* 1983;**4**(1):1–9.
95. Togawa T, Okai O, Oshima M. Observation of blood flow E.M.F. in externally applied strong magnetic field by surface electrodes. *Med Biol Eng* 1967;**5**(2):169–70.
96. Gaffey CT, Tenforde TS. Alterations in the rat electrocardiogram induced by stationary magnetic fields. *Bioelectromag* 1981;**2**(4):357–70.
97. Beischer D. Vectorcardiogram and aortic blood flow of squirrel monkeys *Saimiri sciureus* in a strong superconductive electromagnet. In: Barnothy M, editor. *Biological effects of magnetic fields*. New York: Springer; 1995. p. 241–59.
98. Beischer DE, Knepton Jr. JC. Influence of strong magnetic fields on the electrocardiogram of squirrel monkeys *Saimiri sciureus*. *Aerosp Med* 1964;**35**:939–44.
99. Nakagawa M. Changes of the cardiovascular system of rabbits subjected to static magnetic field of 600 Oe (author's transl.). *Sangyo Igaku* 1978;**20**(2):11–23.
100. Lazetic B, Pekaric-Nadj N. Heart rate in rats exposed to constant magnetic fields. *Electromag Biol Med* 1993;**12**(2):117–23.
101. McRobbie D, Foster MA. Cardiac response to pulsed magnetic fields with regard to safety in NMR imaging. *Phys Med Biol* 1985;**30**(7):695–702.
102. Chalazonitis N, Arvanitaki A. *Effets du champ magnetique constant sur l'autoactivite des fibres myocardiques d'Helix pomatia*. *CR Soc Biol Paris* 1964;**158**:190–228.

103. Yamaguchi M, Andoh T, Goto T, et al. Effects of strong pulsed magnetic fields on the cardiac activity of an open chest dog. *IEEE Trans Biomed Eng* 1994;**41**(12):1188–91.
104. Frey AH, Seifert E. Pulse modulated UHF energy illumination of the heart associated with change in heart rate. *Life Sci* 1968;**7**(10):505–12.
105. Jauchem JR. Exposure to extremely-low-frequency electromagnetic fields and radiofrequency radiation: cardiovascular effects in humans. *Int Arch Occup Environ Health* 1997; **70**(1):9–21.
106. Shellock FG. Radiofrequency energy-induced heating during MR procedures: a review. *J Magn Reson Imaging* 2000;**12**(1):30–6.
107. Chakeres DW, Kangarlu A, Boudoulas H, et al. Effect of static magnetic field exposure of up to 8 tesla on sequential human vital sign measurements. *J Magn Reson Imaging* 2003;**18**(3):346–52.
108. Franco GPR, Murolo A. Health effects of occupational exposure to static magnetic fields used in magnetic resonance imaging: a review. *Med Lav* 2008;**99**(1):16–28.
109. Shellock FG, Schaefer DJ, Kanal E. Physiologic responses to an MR imaging procedure performed at a specific absorption rate of 6.0 W/kg. *Radiol* 1994;**192**(3):865–8.
110. Miller G. Exposure guidelines for magnetic fields. *Am Ind Hyg Assoc J* 1987;**48**(12): 957–68.
111. Kinouchi Y, Yamaguchi H, Tenforde TS. Theoretical analysis of magnetic field interactions with aortic blood flow. *Bioelectromag* 1996;**17**(1):21–32.
112. Simunic DWP, Renhart W, Stollberger R, et al., editors. Determination of stimulation hazard for heart cells during MRI using FEM simulation. Abstracts of the 17th Annual Meeting of the Bioelectromagnetics Society. Boston; 1995.
113. Reilly JP. Magnetic field excitation of peripheral nerves and the heart: a comparison of thresholds. *Med Biol Eng Comput* 1991;**29**(6):571–9.
114. Tiasto MI, Ptitsyna NG, Kopytenko IUA, et al. *Vliianie elektromagnitnykh polei qestestvennogo i antropogennogo proiskhozhdeniia na chastotu proiavleniia razlichnykh patologii v Sankt-Peterburge*. [Effect of electromagnetic fields of natural and anthropogenic origin on the frequency of pathology occurrence in St. Petersburg.] *Biofizika* 1995;**40**(4): 839–47.
115. McKay JC, Prato FS, Thomas AW. A literature review: the effects of magnetic field exposure on blood flow and blood vessels in the microvasculature. *Bioelectromag* 2007; **28**(2):81–98.
116. Formica D, Silvestri S. Biological effects of exposure to magnetic resonance imaging: an overview. *Biomed Eng Online* 2004;**3**(1):11.
117. Shellock FG. Magnetic resonance safety update 2002: implants and devices. *J Magn Reson Imaging* 2002;**165**:485–96.
118. Shellock FG, Slimp GL. Severe burn of the finger caused by using a pulse oximeter during MR imaging. *Am J Roentgenol* 1989;**153**(5):1105.
119. Andreassi M. The biological effects of diagnostic cardiac imaging on chronically exposed physicians: the importance of being non-ionizing. *Cardiovasc Ultrasound* 2004;**2**(1):25.
120. Dalecki D. Mechanical bioeffects of ultrasound. *Annu Rev Biomed Eng* 2004;**6**(1):229–48.
121. Church CC, Carstensen EL, Nyborg WL, et al. The risk of exposure to diagnostic ultrasound in postnatal subjects: nonthermal mechanisms. *J Ultrasound Med* 2008;**27**(4):565–92.
122. O'Brien Jr. WD, Deng CX, Harris GR, et al. The risk of exposure to diagnostic ultrasound in postnatal subjects: thermal effects. *J Ultrasound Med* 2008;**274**:517–35.
123. Miller DL, Averkiou MA, Brayman AA, et al. Bioeffects considerations for diagnostic ultrasound contrast agents. *J Ultrasound Med* 2008;**27**(4):611–32.

124. Kobayashi N, Yasu T, Yamada S, et al. Influence of contrast ultrasonography with perflutren lipid microspheres on microvessel injury. *Circ J* 2003;**67**(7):630–6.
125. Dawson D, Vincent MA, Barrett EJ, et al. Vascular recruitment in skeletal muscle during exercise and hyperinsulinemia assessed by contrast ultrasound. *Am J Physiol Endocrinol Metab* 2002;**282**(3):E714–20.
126. Lindner JR. Contrast echocardiography. *Curr Probl Cardiol* 2002;**27**(11):454–519.
127. Miller DL, Thomas RM, Williams AR. Mechanisms for hemolysis by ultrasonic cavitation in the rotating exposure system. *Ultrasound Med Biol* 1991;**17**(2):171–8.
128. Rooney JA. Shear as a mechanism for sonically induced biological effects. *J Acoust Soc Am* 1972;**52**(6B):1718–24.
129. Rooney JA. Hemolysis near an ultrasonically pulsating gas bubble. *Science* 1970;**169**(3948):869–71.
130. Dalecki D, Keller BB, Carstensen EL, et al. Thresholds for premature ventricular contractions in frog hearts exposed to lithotripter fields. *Ultrasound Med Biol* 1991;**17**(4):341–6.
131. Dalecki D, Keller BB, Raeman CH, et al. Effects of pulsed ultrasound on the frog heart: I. Thresholds for changes in cardiac rhythm and aortic pressure. *Ultrasound Med Biol* 1993;**19**(5):385–90.
132. MacRobbie AG, Raeman CH, Child SZ, et al. Thresholds for premature contractions in murine hearts exposed to pulsed ultrasound. *Ultrasound Med Biol* 1997;**23**(5):761–5.
133. van der Wouw PA, Brauns AC, Bailey SE, et al. Premature ventricular contractions during triggered imaging with ultrasound contrast. *J Am Soc Echocardiogr* 2004;**13**(4):288–94.
134. Borges AC, Walde T, Reibis RK, et al. Does contrast echocardiography with optison induce myocardial necrosis in humans? *J Am Soc Echocardiogr* 2002;**15**(10):1080–6.
135. Raisinghani A, Wei KS, Crouse L, et al. Myocardial contrast echocardiography MCE with triggered ultrasound does not cause premature ventricular complexes: evidence from PB127 MCE studies. *J Am Soc Echocardiogr* 2003;**16**(10):1037–42.
136. Ay T, Havaux X, van Camp G, et al. Destruction of contrast microbubbles by ultrasound: effects on myocardial function, coronary perfusion pressure, and microvascular integrity. *Circulation* 2001;**104**(4):461–6.
137. Chen S, Kroll MH, Shohet RV, et al. Bioeffects of myocardial contrast microbubble destruction by echocardiography. *Echocardiogr* 2002;**19**(6):495–500.
138. Miller D, Li P, Armstrong WF. The effect of time and of vasoactive drugs on capillary leakage induced during myocardial contrast echocardiography. *Echocardiogr* 2004;**21**(2):125–32.
139. Li P, Cao L-Q, Dou C-Y, et al. Impact of myocardial contrast echocardiography on vascular permeability: an in vivo dose-response study of delivery mode, pressure amplitude and contrast dose. *Ultrasound Med Biol* 2003;**29**(9):1341–9.
140. Miller DL, Li P, Gordon D, et al. Histological characterization of microlesions induced by myocardial contrast echocardiography. *Echocardiogr* 2005;**22**(1):25–34.
141. Doida Y, Miller MW, Cox C, et al. Confirmation of an ultrasound-induced mutation in two in-vitro mammalian cell lines. *Ultrasound Med Biol* 1990;**16**(7):699–705.
142. Church CC. Spontaneous homogeneous nucleation, inertial cavitation and the safety of diagnostic ultrasound. *Ultrasound Med Biol* 2002;**28**(10):1349–64.
143. National Council on Radiation Protection & Measurements (NCRP). Exposure criteria for medical diagnostic ultrasound: II. Criteria based on all known mechanisms. Report No. 140. Bethesda, MD: NCRP; 2002.

Chapter 18

Nanomaterials and Cardiovascular Toxicity

Arghya Paul,[1] Hamood Al Kindi,[2] Ragini Medhi,[3] Vijayan Manoharan,[4] Satya Prakash,[5] and Dominique Sum-Tim[2]

[1]*Harvard-MIT Division of Health Sciences and Technology, Harvard University, Cambridge, Massachusetts; Department of Chemical and Petroleum Engineering, University of Kansas, Lawrence, Kansas*

[2]*The Royal Victoria Hospital, McGill University Health Centre, Montreal, Quebec*

[3]*Department of Genetic Engineering, School of Bioengineering, SRM University, Chennai, India*

[4]*Center for Nanotechnology and Advanced Biomaterials, School of Chemical and Biotechnology, SASTRA University, Tanjore, India*

[5]*Biomedical Technology and Cell Therapy Research Laboratory, Department of Biomedical Engineering, Faculty of Medicine, McGill University, Montreal, Quebec*

18.1 CARDIOVASCULAR NANOMEDICINE AND NANOTOXICITY

The rapid advances and growing applications of nanomedicine have resulted in a vast array of nanomaterials that vary in shape, size, charge, chemistry, surface topography, and solubility.[1,2] This evolving field deals with materials with structures that exhibit distinct physical, chemical, and biological properties due to their nanoscale. They have unique optical, magnetic, and thermoresponsive properties and a large surface area. The features of nanomaterials make them extremely valuable in biomedical applications as biomarkers, contrast agents for biological imaging, diagnositic tools, and drug delivery applications.[3,4]

Nanomaterials are considered the primary vehicles for delivery of drug molecules because they can pass biological barriers, get internalized, and be distributed within cells by energy-dependent pathways. Polymeric nanoparticles (NPs), composed of biodegradable materials, have been the most suitable and preferred system for the sustained delivery of therapeutic biomolecules and drugs. Recent studies have demonstrated that these materials can help with creating bioresponsive smart drugs that can specifically target the diseased organ and deliver its cargo when prompted by a specific biological stimulus.[5,6] Such targeting becomes very effective when NPs, of 20 to 200 nm

The Heart and Toxins. DOI: http://dx.doi.org/10.1016/B978-0-12-416595-3.00018-9

in diameter, are administered systemically.[7] This avoids their renal filtration and leads to prolonged residence times within the bloodstream.

NPs have demonstrated immense potential in the field of cardiovascular nanomedicine. They can be prepared from different polymeric biocompatible and biodegradable materials, such as polycaprolactone (PCL), poly lactic-co-glycolic acid (PLGA), hyperbranched poly (amido amine) dendrimers, micelles, and so on, for diverse biomedical applications.[8–10] Brenneman et al. reported that quantum dot (Q-dot) nanocrystal-labeled bone marrow-derived mononuclear stem cells can be efficiently used for the treatment of rats after acute ischemic stroke.[11] Because of their high surface area, efficient drug-loading capacity, and ability to target specific organs, biofunctionalized NPs offer immense potential for molecular imaging of atherosclerosis,[12] as well as targeted delivery of drugs and genes to the diseased cardiovascular sites with reduced side effects.[1,13,14]

Designing strategies for noninvasive evaluation of atherosclerotic plaque are ideal for clinical applications and have been a subject of active research. Ligand conjugated NPs and Q-dots have been shown to effectively recognize and visualize stage-specific atherosclerotic markers.[12,15] Paclitaxel drug-eluting therapeutic "nanoburrs" are also considered as a potential replacement for drug-eluting stents to treat injured vasculature.[16] Polyelectrolyte gold nanorods have shown potential in antifibrotic therapies by regulating cardiac fibroblast matrix remodeling.[17] However, to achieve significant therapeutic efficacy, these NPs are directly administered to achieve a high concentration of the targeted nanoparticles in the target cardiovascular sites.

Exposure to such a high concentration of nanoparticles can induce adverse effects in the target organs, as has been seen from accumulated evidence.[18] The NPs are normally trapped in the reticuloendothelial (RES) system, primarily in lungs, liver, and spleen. However, RES uptake can be evaded to some extent by surface coating the NPs with polyethylene glycol (PEG) molecules. Although pegylation may also attenuate the cardiotoxicity of NPs and augment their targeting, pegylating nanoparticles may lose their full potential of therapeutic drug delivery, particularly if a heavy molecule (e.g., an antibody) is labeled, which may lose its normal binding characteristics.

With increasing applications of NPs in cardiovascular medicine and for diagnostic purposes,[19] patients will be highly exposed to a wide range of synthetic nanomaterials with diverse properties. However, little is known about their side effects and clinical risks. Recent studies show that inhalation of NPs enhances the risk for cardiorespiratory diseases.[20,21] These NPs, entering the body through other routes such as by inhalation, can also nonspecifically target the cardiovascular system.[22] Apart from direct cardiac exposure to them, the cardiovascular system also is affected by the secondary effects of NPs on other organs, especially through pulmonary NP exposure leading to increased risk of cardiopulmonary toxicity and systemic cardiovascular toxicity, as well as atherosclerosis.[23,24] Translocation of a deposited high concentration of NPs from the lung to systemic circulation plays an important role in such cases. This includes NPs from polluted air emitted from transport vehicles and industrial plumes, which eventually lead to the progression of cardiovascular toxicity and diseases such as atherosclerosis.[25,26] Finally, the systemic distribution of local oxidative stresses and proinflammatory conditions, caused by the NPs, result in alteration of cardiac autonomic function.[27] Thus, there is an urgent need for the establishment of principles and test procedures to confirm optimal use of these nanomaterials for health care applications.

18.1.1 Role of the Reactive Oxygen Species in Nanoparticle-Based Toxicity

In general, the reactive oxygen species (ROS) plays an important role in nanomaterial-based cytotoxicity. Nanoparticle overexposure can lead to asymmetry between ROS production and antioxidant defense systems. Key factors in such a response include change in functionality and particle–cell interactions. Size reduction is frequently compensated by structural defects with the generation of a number of reactive donor or acceptor sites.[28] On reaction with molecular oxygen, a large number of ROSs are generated. Those that are newly generated due to NP add up to the already available endogenous ROS population and contribute to overproduction. Ideally, to detoxify the reactive oxygen species, antioxidant enzyme systems are triggered and activated. Signaling molecules (e.g., MAPK, NF-κB) are induced, which potentiate high and active transcription leading to increasing downstream expression of antioxidants.[29] However, overproduction of ROS causes disarray in cellular signaling resulting in transcriptional error, increased proinflammatory response, DNA damage, and cytotoxicity.

NPs penetrate cells through already-existing, energy-dependent pathways. Entry of nanomaterials, such as carbon nanospheres (CSP), into cells happens via an adenosine triphosphate (ATP)-dependent endocytotic process, as evidenced in mammalian cells. The entry is time-dependent. As large as 100 to 500 nm CSP can permeate through the cell membrane. This process has been identified to be clathrin mediated with the involvement of glucose receptors.[30] Consequently, cell trespass is followed by significant production of ROS owing to numerous surface-reactive sites. Mitochondrion, known for oxidative phosphorylation and apoptotic defenses, becomes an easy target.

In addition to the endogenous reactive oxygen species, ROS due to NP takes a toll on antioxidant defenses and the mitochondria transmembrane potential, $\Delta\psi$, collapses resulting in considerable swelling. ROS decays the mitochondrial Ca^{2+} buffer capacity, so sensitivity to cytoplasmic Ca^{2+} levels increases.[31] However, the extent to which these nanomaterials can cause health hazards by generating ROS can be easily detected by examining their ability to cause oxidative injury within cells using presently available methods.[18]

18.1.2 Molecular Mechanism of Nanoparticle-Based Cardiotoxicity

The ROS plays a crucial role in modulating the cardiomyocyte microenvironment. The primary molecular mechanism of cardiovascular toxicity due to NPs is a result of free radical overproduction, which compromises the mitochondrial bioenergetics; the overproduction of ROS and the reactive nitrogen species (RNS); and the synthesis of peroxynitrite ions ($ONOO^-$) by OH and NO radicals in the presence of iron (Fe^{3+}) by a Fenton reaction.[32] $ONOO^-$ induce not only oxidative but also nitrative stress, leading to further augmentation of ischemia-reperfusion injury and myocardial damage as observed in myocardial infarction, congestive heart failure, and dilated cardiomyopathy. However, NP introduction into the oxygen-rich myocardial environment manifests rapid exhaustive reactions with the highly reactive sites on the particle surface. This results in rapid mitochondrial matrix calcium overload, triggering overproduction of electron transport chain (ETC) ROS.[33] Finally, the tricarboxylic acid cycle and ETC enzymes together undergo allosteric activation leading to rigorous oxidative phosphorylation and high ROS generation as a byproduct.[34] Such cellular deregulations trigger acute cardiovascular apoptotic stimulus and cytotoxicity.

Another important factor for cardiotoxicity is formation of the Charnoly body (CB). CB is a multilameller (usually penta or heptalamellar) electron-dense, pleomorphic structure that is formed by condensation of degenerated mitochondrial membranes. The CB is formed during nutritional insult, toxic NP exposure, drugs, or poisons. These deleterious agents induce oxidative and nitrative stress, induce free radical overproduction, and are inhibited by antioxidants such as superoxide dismutase (SOD), glutathione, and catalase. When levels of SOD, glutathione, and catalase are depleted, metallothioneins (MTs) are induced to prevent CB formation by acting as potent free radical scavengers. MTs are at least 30 times more potent as glutathione. Thus, MTs can serve as excellent cardioprotectants during NP toxicity. Recently, three types of NPs have been proposed:

1. Toxic NPs, which augment CB production
2. Inert NPs, which have no effect on CB formation
3. Protective NPs, which inhibit CB formation by augmenting MTs

18.2 EXPLORING NANOPARTICLE–CELL INTERACTIONS

Understanding the way NPs, with different geometries, interact and get internalized in cells can help us predict their cytocompatibility toward a particular cell type. This requires detailed study and analysis of the molecular events involved in NP–membrane receptor binding, internalization, and subsequent cell-signaling activation.[35] Studies, so far, have demonstrated that cellular behavior and uptake of NPs mainly depends on the size, shape, charge, dispersivity, and sedimentation effects of dense particles, as well as the surface biochemical properties of the NPs in question.[36] In particular, NP size can have a significant influence the mechanism of cellular internalization, which can be phagocytosis, macropinocytosis, caveolar-mediated endocytosis, or clathrin-mediated endocytosis.[37] Detailed knowledge of the modes of cell entry of different NPs can facilitate the design of engineered NPs tailored toward particular applications.

Cells control their biofunctionality through numerous cascades of intracellular signals triggered by binding of ligands to the cell surface receptors. The extent of receptor–ligand binding markedly affects cell behavior.[36] To exploit these properties, NPs can be coated with binding moities that can potentially function as ligands to bind to the surface receptors. By changing the size, shape, and material properties of these engineered NPs, the degree of receptor binding at the nano–biological interface and subsequent cell responses can be controlled. Figure 18.1 illustrates that the NPs can be designed with different ligands for specific applications. However, special attention has to be given while engineering NPs, as it can cause damage to DNA and chromosomes across an intact cellular barrier.[38] In fact, the cell cycle phase is equally important when considering how NPs are to interact with the cells.[39]

When engineering nanoparticles for in vivo applications, it is important to assess the host's immune response to the nanoscale formulations. Hamad et al. reported that the exposure of polymeric NPs can drastically alter the complement activation pathways of the host immune system.[40] Modifying NP surfaces with different densities of copolymers did not prevent activation of the complement system. To address this problem and evade NP recognition by the host complement system, nanoparticle surfaces can be coated with complement-regulating protein such as factor H.[41] Thus, proper understanding of the cell–NP interactions are needed to successfully design biologically safe NPs for clinical applications and reduce potential nanotoxic effects.

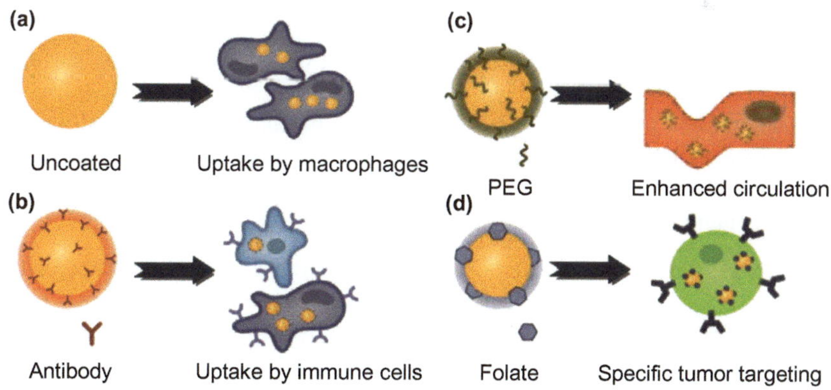

FIGURE 18.1 Nanoparticle–cell interactions. Engineering an NP surface for different applications such as (a) imaging by MRI, (b) tracking immune cells, and drug delivery with (c) enhanced circulation and (d) tumor-specific NP internalizations. (PEG: polyethylene glycol)

18.3 DIFFERENT ROUTES OF NANOPARTICLE ENTRY AND POTENTIAL TOXICITY

Exposure to NPs can occur unintentionally from natural phenomena or by occupational exposure at the workplace in research facilities and pharmaceutical companies. It can also occur intentionally because of products used in biomedical therapy and medical imaging.[42,43] Human exposure to NPs can occur via different routes: inhalation, intravenous, dermal, and oral. After entering the human body, they can be either metabolized or partially metabolized in the liver, accumulate in various tissue and organs, or be excreted in the urine or feces.[44] Several studies have suggested that intravenous injection of certain NPs may cause them to accumulate within the lungs and result in pulmonary toxicity, while others showed that some adverse systemic effects occur after NP inhalation. This demonstrates the ability for NPs to enter and exit the vascular system and be distributed throughout the circulation.[45] This section discusses various routes by which NPs access the human body and eventually circulate.

18.3.1 Inhalation Route

This is the most common route by which nanoparticles enter the human body. The location of inhaled NP deposition in the respiratory tract varies according to size. Two size ranges can lead to maximum deposition in the lungs: 50 to 100 nm, corresponding to the alveoli, and 1 to 5 μm for the lower airways. Small NPs within 1 nm in diameter are deposited mainly in the nasopharyngeal compartment.[46] Three deposition mechanisms are dominant for inhaled NPs:[47]

- *Diffusional deposition*: This occurs due to displacement when NPs collide with air molecules. This is the main mechanism of NP deposition in small airways and alveoli.
- *Inertial forces*: Deposition after the collision of particles with the surface of the airways.
- *Gravitational sedimentation*: An important mechanism for deposition in the smaller bronchi, bronchioles, and alveolar spaces, where the airways are small and air velocity is low. Sedimentation is more effective for aerodynamic diameters of >0.5 μm.

Electrostatic precipitation occurs when nanoparticles carry significant electric charges.

Several parameters influence nanoparticle deposition retention and efficiency, including air velocity, respiratory time and flow rate, tidal volume, and airway structure.[46] These different deposition efficiencies may have consequences on the potential effects of NPs as well as for their disposition to extrapulmonary organs. The fate of NPs after deposition in the lungs depends mainly on their landing zone and their physicochemical properties. Particles deposited in extrathoracic airways or ciliated thoracic airways will have a faster clearance phase than in the nonciliated airway spaces. Soluble particles will be readily translocated into blood circulation as early as 6 hours after exposure, in contrast to insoluble particles that will have a local effect or will be translocated outside of the respiratory tract.[48] Several studies have shown that both the size and material of the NPs determines the degree of translocation and deposition in extrapulmonary organs. The small diameter of these particles allows them to escape from the defense and clearance mechanisms found in the respiratory tract. In addition, the electrical charge of NPs may influence their translocation and toxicity. Inhaled NPs with a negative charge have negligible systemic or pulmonary toxic effects compared to positively charged NPs of the same material. The clearance of deposited particles in the respiratory tract is basically due to two processes:

- Physical translocation of particles
- Chemical clearance processes

Several mechanisms are involved in the clearance of deposited NPs. One involves transcytosis across the epithelia of the respiratory tract into the interstitium and access to the blood circulation directly or via lymphatic system. The other mechanism involves their uptake by sensory nerve endings embedded in airway epithelia, followed by axonal translocation to other central nervous system (CNS) organs.

18.3.2 Dermal Route

Nanomaterials used in cosmetics and dermatological therapy can gain access to the human body through the skin. The skin is composed of two main layers: epidermis and dermis. The epidermis has three layers: stratum corneum, stratum spinosum, and stratum basale. These layers form a very tight protective layer for the underlying dermis. The dermis has a rich supply of blood and tissue macrophages, lymph vessels, dendritic cells, and different types of sensory nerve endings. In general, there are four pathways for particles to penetrate the skin: intercellular, transcellular, through hair follicles, or through sweat glands.[49] Multiple factors can influence dermal uptake: skin integrity, anatomical site, and presence of skin disease.[50]

Dermal penetration and toxicity depend on nanoparticle size, shape, coating, presence of catalysts, and multiple exposure variables.[43] These are important considerations when studying the interaction of new nanocompounds and human skin.[50] As NPs can move freely compared to larger particles, they are more prone to creating toxic effects than bulky materials. In fact, taking advantage of their easy penetration ability into skin, different types of NPs have been widely used to study NP–skin interactions for drug penetration applications.[51]

18.3.3 Oral Route

NPs enter the gastrointestinal (GI) tract via direct ingestion if they are contained in food or water. Alternatively, they can be cleared from the respiratory tract via the mucociliary escalator and subsequently ingested into the GI tract.[48] The mucinous layer and the multiple layers of GI epithelium prevent various toxins and foreign material from entering the body. It was thought that NPs are unable to penetrate the mucus layer until recent studies demonstrated that specific viruses and nanostructures diffuse in human mucus as rapidly as they do in water. The surface of these small particles was densely coated with positive and negative charges, which prevented adhesion and aggregation with the mucus layer.[52] Moreover, the pore size in the mucus layer is about 100 nm, which suggests that small NPs might also be capable of diffusing through the mucus layer. Furthermore, ingested NPs can penetrate the mucus layer by inducing PH and viscosity and structural changes.[53]

The GI epithelium generally represents the highest resistance barrier against the passage of chemical compounds and nanomaterials. Epithelial cells have an apical surface facing the lumen and a basal site facing the underlying tissue. Epithelia may vary in height, type, and depth according to the site in the GI tract. Penetration through a monostratified squamous epithelium is easier than through the simple columnar epithelium in the stomach and intestine. The basal membrane separates the epithelium from the underlying connective tissue that contains capillaries, lymphatics, and neural plexuses. NPs have to cross the basal membrane and connective tissue before they can reach the systemic circulation.[44,48,53]

The assessment of nanoparticles' penetration through the gastrointestinal tract is a complex problem because of interindividual and interspecies variations. Regardless of this difficulty, nanotoxicity via the GI route is an important consideration because NPs acquire unpredicted new physicochemical properties that are different from their bulky materials. This phenomenon is called the *Trojan horse effect*. For instance, silver NPs can be internalized by GI epithelial cells, increasing their intracellular silver concentration, in contrast to silver cations that are restricted from accessing the plasma membrane.[54]

18.3.4 Intravenous Route

The intravenous (IV) route is a common way for drug delivery using nanotechnology. Delivering drugs via an IV allows quick translocation in the circulatory system compared to intradermal or intramuscular injection where the uptake is through the lymphatic system or neuronal system, respectively.[44] Several NP drug-delivery systems have been approved by the U.S. Food and Drug Administration (FDA). Most of the drugs are used in the field of oncology; for example, PEG-stabilized liposomal doxorubicin (e.g., Doxil, Evacet), protein-bound paclitaxel (e.g., Abraxane), and pemetrexed (e.g., Alimta).[55] The advantage of NPs in drug delivery is that they can help to increase circulation time due to their small particle size. It has been shown that particles under 200 nm had a longer circulation time regardless of any modification on the surface.[56]

The targeting strategy for drug delivery using NPs relies on the presence of endothelial fenestrations in tumor or inflamed tissue that will allow passive and selective diffusion to target tissue. This is known as the *enhanced retention and permeability* effect. Clearance of NPs from the circulation is usually by opsonization and further destruction by the reticuloendothelial system (RES). It has been shown also that coating NPs with hydrophilic polymers, such as PEG, can increase circulation time and decrease their clearance by RES. After injecting NPs, their intravascular journey can be divided into three parts:[57]

1. *Margination*: A dynamic process that results in lateral drift of NPs from the blood vessel core to the walls.
2. *Adhesion*: When suitable conditions occur in terms of shear-stress hydrodynamics, NPs will interact specifically or nonspecifically with endothelial cells and adhere firmly to them.
3. *Internalization*: After firm adhesion, NPs should be internalized to release drugs and therapeutic agents or resist internalization if the target cell is used as a docking site.

Precise knowledge about endothelial diversity in different pathological conditions and accurate control of the geometry and physiochemical properties of NPs are essential for optimal design of nanodrug delivery systems.

18.4 EFFECTS OF NANOMATERIALS ON THE PROGRESSION OF CARDIOVASCULAR TOXICITY

This section focuses on exploring carbon, airborne, and metal-based nanomaterials, which have wide-ranging applications in biomedicine; however, they have significant risks of cardiovascular toxicity.

18.4.1 Carbon-Based Nanomaterials in Cardiovascular Diseases

Carbon-based NPs, mainly carbon nanotubes (CNTs) and graphene oxide (GO), have provided a feasible platform for drug delivery because of their high aspect ratio and modifiable surface chemistry. However, they can also induce cytotoxic effects.[58] For some time, CNTs, fullerenes, and GO have been used in many in vitro and in vivo therapeutic applications.[1,2] With its hexagonal and pentagonal arrangement of 60 carbons, fullerene is nonreactive but its surface group modification adds functionality. CNTs commonly used for biomedical purposes include single-walled nanotubes (SWNT), composed of a single wall of cylinder, and multiwalled nanotubes (MWNT), composed of multiple walls of SWNTs. From a structural standpoint, CNTs have always been compared to asbestos due to the similarity in shape and insolubility in water.[59] This leads to speculation regarding toxicity similar to that of asbestos. Both CNTs and GO have been reported to have high bacterial toxicity.[60,61] Table 18.1 summarizes the different carbon-based cardiovascular nanotoxic materials.

Carbon Nanotubes

The main mechanism of toxicity of carbon-based nanomaterials varies, but the most popular belief is that it is through oxidative stress. CNTs are known to induce the release of the reactive oxygen species due to their physicochemical properties.[71] The accumulating ROS eventually induces cellular apoptosis. Toxicity also arises because of the process of synthesis of CNT nanomaterials. Functionalizing the surface chemistry may alter the sensitivity of the CNTs toward the cells. Adding hydroxyl groups to CNTs and fullerenes makes them more soluble and increases reactivity. Single-walled carbon nanotubes (SWCNT) and multiwalled carbon nanotubes (MWCNT) are mostly based on reactive metal species. Nonremoval of the metallic species increases the cytotoxicity by many folds. Catalytic removal, however, reduces the toxicity to a greater extent.[72] Replacement of the net surface charge has been reported to alter reactivity and toxicity. However, cytotoxicity of CNTs defers on many levels based on the organ of accumulation and method of transfer (i.e., intravenous, intratracheal) and the functionality of the particles.

Present methods of detection are not adequate enough to determine the toxic effects of NPs. Recent reports have shown the deleterious effects of such particles on health, especially in lung and cardiovascular diseases. This necessitates drawing more attention to the toxic nature of the CNTs and development of a sound screening system to determine their biological activity. Lacerda et al. showed that MWCNT also stimulates platelet aggregation by upregulating the influx of calcium into the cells.[73] CNTs and fullerenes were compared against their platelet-aggregating property with thrombin receptor-activating peptide as the positive control. Analysis of the electron-microscopic images showed that samples treated with fullerenes showed almost no aggregation of platelets. The suggested mechanism of aggregation was the influx of calcium into the cells by store-operated Ca^{2+} entry (SOCE). The activation of SOCE is controlled by the dense tubular system (DTS) of the platelets, which serves as a major reservoir of calcium.

Depletion of calcium from the DTS triggers SOCE. Thus, it results in accumulation of platelets. The macrophages are recruited to the site of damage. They grow and rupture to cause more aggregation. This in turn triggers another chain of immune responses resulting in more aggregation. While performing cardiotoxicity studies with CNTs, it is important to consider that because of the manufacturing process they may contain high levels of several

TABLE 18.1 Summary of Different Carbon-Based Nanomaterials with Cardiovascular Nanotoxicity

Carbon-based Nanomaterial	Characteristics	Applications	Toxicity	References
Fullerenes	3D covalent attachment of C_{60} rings, distortion (due to covalent bond) increases strain energy	Antiviral activity, as photosensitizers, antioxidant activity	Inflammatory responses, oxidative damage, high cytotoxicity	62,63
SWCNT	Graphene sheets folded to form a single-layered cylinder	pH-based drug delivery systems, antibacterial activity	Phagolysosome of macrophages (due to nonbiodegradability unlike the carboxylated counterpart)	64,65
MWCNT	Extension of SWCNT but with multiple layers	Biosensors, water treatment	Platelet aggregation	64,66
Graphene	Honeycomb lattice (sublattices with σ bond), topological vacancies, and absorbed impurities	Sensors, biological applications as a result of high biocompatibility in catalysis	High intracellular generation of ROS, mitochondrial injury	67
Graphene oxide (GO)	Highly oxidized modified form of graphene, polar hydroxyl, and epoxide functional groups	Drug delivery, tissue engineering, anticancer effects, other biomedical applications	High oxidative damage, low biodegradability, antibacterial property	68,69
Reduced graphene (rGO)	Chemical treatment of GO with reducing agents (hydrazine), low surface charge, large number of vacancies, high hydrophobicity	Biosensors, cellular and bioimaging, nanocarriers in drug delivery, biosensors	Higher antibacterial property (than GO)	68,69
Carbon black	High surface-to-volume ratio, SPM	Electronics, additives	Apoptotic cell death induction, lysosomal destabilization, high intracellular ROS	25,70

impurities, including toxic metals (e.g., Ni, Co, Fe, and so on) that can influence experimental results.

Graphene and Graphene Oxide Nanosheets

During the last few years, graphene and GO, a one-atom-thick two-dimensional layer of sp_2-bonded carbon, have been the subject of a lot of research as a novel carbon-based nanomaterial for diverse biomedical applications.[74,75] However, Singh et al. reported that graphene oxide induces thrombus formation leading to a narrowing of the arterial walls and atherosclerosis.[76] It triggers platelet accumulation with the activation of Src kinases and release of calcium. Aggregation was mainly mediated by $\alpha_{IIb}\beta_3$ integrin relayed by the adhesion of cells to immobilized fibrinogen. To make the graphene oxide biocompatible, the group further chemically modified the GO and demonstrated that modifying it with amine groups can reduce in vivo thrombogenic effects.[77] Replacement of the existing functional group with amine renders a net positive surface charge and lowers cytotoxicity. In another study, the effects of graphene sheets and graphene oxide on human erythrocytes and skin fibroblasts were investigated.[78] Results showed that the high oxygen contents of GO markedly increases its hemolytic activity compared to graphene sheets. Further studies with thoroughly characterized GO and its derivatives may facilitate a better understanding of the cardiotoxic potential of GO nanosheets.

Pollution-Borne Nanoparticles

Air pollution, due to a high concentration of proatherogenic NPs, constitutes a significant cardiovascular risk to human health.[79] Exposure to such ultrafine NPs increases ischemic cardiovascular events and promotes atherosclerosis. Moreover, there is increasing evidence that the smallest pollutant particles pose the greatest danger because of their high content of organic chemicals and prooxidative potential. Campen et al. conducted in vivo animal studies on mice to evaluate the effects of such NPs in air pollutants.[80] The results demonstrated a significant increase in heart rate in hypersensitive rats, associated with myocytic degeneration and inflammation. Thus, it is extremely important to understand the effect of different airborne NPs on cardiotoxicity.

Toward this goal, an in vitro model was developed to facilitate a better understanding of the effect of different NPs on the heart.[81] This in vitro model, known as the Langendorff heart, is an isolated animal heart with a myocardial rhythm similar to humans. The heart rate of the in vitro model was monitored on introduction of particles; it was found that the heart rate increased on the administration of them. This suggested a direct interaction of the particles with the organ, which resulted in an increase in heart rate due to catecholamine release from adrenergic cardiac nerves and increased endothelin from stressed endothelial cells (Figure 18.2).

In a separate study, the cytotoxic effects of carbon black (CB), a component of diesel exhaust particles, on human endothelial cells was

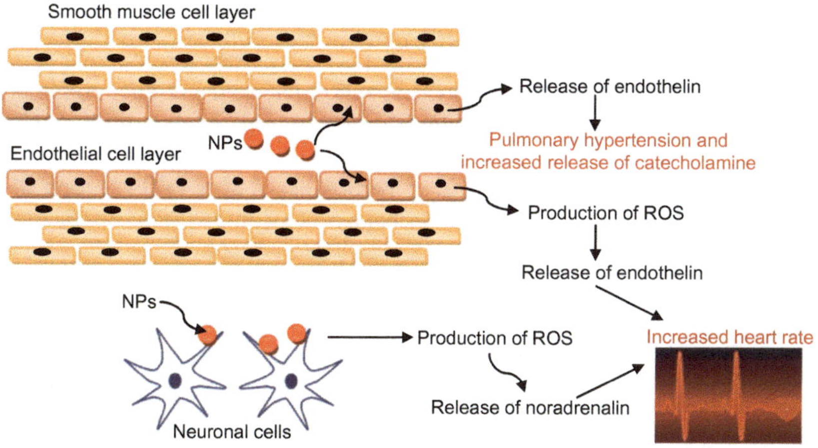

FIGURE 18.2 Effect of NP exposure on the cardiovascular system. NPs' mediated release of endothelin from the vascular endothelial cells. Endothelins are potent vasoconstrictors that increase pulmonary hypertension. They act on the chromaffin cells of the sympathetic nerves, stimulating them to release high amounts of intracellular calcium. This in turn triggers the release of catecholamine. However, NPs induce generation of the ROS causing mitochondrial and DNA damage. They target the neurons and the endothelial cells causing them to release noradrenaline and endothelin, respectively. Consequently, the heart rate rises.

investigated.[25] The study attempted to analyze the harmful health effects of suspended NPs in polluted air. CB imparted morphological changes in the vascular endothelial cells. The cell density was reduced in a dose-dependent manner. Higher concentrations of lactate dehydrogenase (LDH) were observed on increasing the concentration of CB. LDH is a marker of cellular death and damage of the plasma membrane. It is interesting to note that the level of expression of endothelial NO synthase, which happens to be antithrombogenic and antiatherogenic, decreased considerably. The investigation also revealed a decrease in the expression of endothelial gap junctions and inhibition of the expression of connexin 37. Such a cellular environment produces a platform favorable for initiation of atherosclerosis.[82] NPs can themselves aggregate and accumulate inside the cells and initiate cell death.

The biocompatibility and the possible interactions of NPs with biological entities are vital because their biomedical applications are, by and large, increasing. Byproducts released by the nanomaterials in lieu of their decomposition may produce an antagonistic effect. Moreover, because of their small dimension they are unnoticed by the immune system and are absorbed easily. This increases their chance of entering the systemic circulatory system which in turn makes organ targets readily accessible. Thus, future studies definitely need to include development of a better screening system for evaluating the cytotoxicity of nanomaterials.

18.4.2 Metallic and Metallic Oxide Nanoparticles

Metallic elements, especially their oxides, play a significant role in nanotoxicology. Due to their popular use in therapeutics and their presence in the air, pollutant toxicity assessment of metal NPs has become extremely important. Transition metal oxides, such as TiO_2 and Fe_2O_3, have been used as NPs and their properties are similar to those of carbon-based NPs. This section discusses the various metallic particles in nanodimension and their application. The main focus is on their role in cardiovascular diseases.

Silica-Based Nanoparticles

The oxides of silica, particularly silicon dioxide (SiO_2), have been used in many applications such as in the manufacture of paper, rubber, drugs, and cosmetics. The low-density SiO_2 NPs can easily mix in the air leading to air pollution. There are many reports of their involvement in a number of health problems. Chen et al. showed age-dependent toxicity toward SiO_2.[83] Old, adult, and young rats were taken as experimental animals for study. After exposure to the NPs, serum histamine, fibrinogen, polymononuclear neutrophils, and lymphocytes increased significantly.

Histamine is generally an indication of inflammation; high levels denote significant inflammation. The level of histamine was higher in old rats than their adult and young counterparts. In the case of tissue damage, fibrinogen levels increased in old rats leading to increased blood viscosity, blood-cell adhesion, and coagulation. This raises the possibility of myocardial infarction and thrombotic stroke. It increases the levels of polymononuclear neutrophils and lymphocytes. It also induces the release of increasing amounts of chemokines and cytokines, causing more inflammation and tissue damage. This in turn increases the synthesis of fibrinogen in the hepatocytes, which is indicative of atherosclerosis and may lead to long-term cardiovascular diseases.[84]

Transition Metal Oxide Nanoparticles

The oxides of the d-block of the periodic table have gained popularity in their numerous applications in nanotechnology. TiO_2, CuO, and ZnO are a few examples, and their respective applications[85] are listed in Table 18.2. Their use in different products increases the chances of their uptake. A number of factors contribute to NP uptake, mainly their size, shape, structure, solubility, surface area, surface charge, and coating. Among the toxicity factors, the major ones are: element-specific surface catalytic activity (e.g., metallic, semi-conducting), the amount of NP uptake, and dissolution. Their major route of entry is through inhalation.

Wagner et al. reported elevation in the levels of chemokine and inflammatory markers, intercellular adhesion molecule 1 (ICAM-1), interleukin 8 (IL-8), and monocyte chemotactic protein 1 (MCP-1) on exposure to transition metal

TABLE 18.2 Summary of Different Metallic Nanomaterials with Cardiovascular Nanotoxicity

Nanoparticle	Characteristics	Applications	Toxicity	References
TiO_2	Crystalline phase structure, spherical to rod-like structure, highly hydrophobic	Osteointegration of medical implants and bone UV absorber in cosmetics and skin care products	Celluar death (through apoptosis), lipid peroxidation	70
ZnO	Wurtzite structure, photoresponse in UV region, photocatalytic activity	Biomimic membrane to immobilize and modify biomolecules Sunscreens and cosmetics Catalysis reactions	Cytotoxic, oxidative DNA damage and resulting chromosome damage	86
CuO	Crystallite size and lattice parameters increases with temperature	Disposal of hazardous materials Pigment in ceramics Abrasive to polish optical equipments	Microbial toxicity, nucleic acid strand cross-linking, protein denaturation	87
Au	Inert, low reactivity, high surface-to-volume ratio	Sensors Therapeutic drug delivery Diagnostics	Cellular damage (cytotoxic), oxidative stress, organelle damage	88

NPs.[89] ICAM-1 and IL-8 are involved in inflammatory responses.[90] ICAM-1 is a glycoprotein expressed on the surface of cells of the immune system and on endothelial cells. It has been known to stabilize cell–cell interactions. Its high expression is indicative of inflammation. IL-8 is a chemokine released by macrophages. MCP-1 is a chemokine released by the macrophages in cases of tissue damage. High expression of ICAM-1, IL-8, and MCP-1 mRNA relates to inflammation and tissue damage. For the study, the group used human aortic endothelial cells (HAEC) as an in vitro model. HAEC was exposed to Fe_2O_3, Y_2O_3, and ZnO NPs. Fe_2O_3 particles did not stimulate any change in ICAM-1, IL-8, or MCP-1 mRNA levels. However, Y_2O_3 increased the levels of ICAM-1, IL-8, and MCP-1 mRNA levels. The increase was dose-dependent; however, ZnO caused high cell death and cellular inflammation.

Iron-Based Nanoparticles

Magnetic NPs, especially their iron derivatives, are popularly used in biomedical and therapeutic applications. In their study, Kahn et al. reported in detail how iron NPs indirectly affect cardiac cells.[91] The group used mouse cardiac cells, HL1-NB, for this research. Iron nanoparticles are redox reactive and interact with the cardiomyocytes. A biomolecule, 7-ketocholesterol is involved in a wide variety of cardiovascular diseases. Such iron NPs enhance the inflammatory and cytotoxic effects of 7-ketocholesterol. Because lactate dehydrogenase is a marker of cellular damage, the changes in levels were noted. Interestingly, iron NPs did not directly induce the release of ROS or RNS, contrary to the normal belief. They did not induce an increase in the levels of IL-8 or MCP-1. However, NPs synergistically enhanced the cytotoxicity and cellular damage caused by 7-ketocholesterol with a slight increase in the levels of LDH. In general, however, the main mechanism of action for metal oxide NP-based toxic effects is via oxidative stress, which results from excessive generation of the reactive oxygen species. Knaapen et al. reported oxidative stress caused by both ROS and RNS.[92] The release of such reactive molecules causes transcription error and mitochondrial damage, resulting in high inflammatory reactions. Some NPs are also sensitive to changes, such as fluctuation in the magnetic field and laser and ultrasound signals, in their environment.[93,94]

Gold Nanoparticles

Gold NPs have tremendous applications in the field of drug delivery and therapeutics. This is mainly because of their inertness, less reactivity, and high surface-to-volume ratio. Applications include usage as diagnostics for the detection of biomarkers in various diseases; as drug delivery vehicles; as probes to detect biomolecules; and, very recently, as cardioprotective nanomaterials.[95] They have been more successful as drug delivery agents because of their high surface charge-to-volume ratio. As a result, a large number of molecules can bind to them. However, cytotoxic effects of gold NPs on rat cardiac cells have also been critically investigated.[88] The group conducted an in vivo study by intraperitoneally administering rats with varying concentrations of gold NPs. The heart, in comparison to the controls, was congested with dilated blood vessels due to induction of oxidative stress. The gold NPs interacted with the hepatic tissue and interfered with the available antioxidant mechanisms. Consequently, the levels of ROS shot up, leading to cellular damage and cytotoxic effects.

Silver Nanoparticles

Recently, silver NPs have been used extensively due to their high antimicrobial characteristics. In turn, their applications in shampoos, soaps, detergents, and for many hygienic purposes increased. Thus, exposure to such NPs is inevitable. Hussain et al. emphasized that the cytotoxic effects of silver NPs cause considerable mitochondrial shrinkage and apoptosis in vitro.[96] In another study, the toxicity by silver NPs after exposure to various organs was explored. Results showed significant alteration in expression of various genes that cause inflammatory responses and thrombosis.[97]

Cerium Oxide

Very recently, cerium oxide (CeO_2), a rare-earth oxide found in the lanthanide series of the periodic table, has been increasingly used as a nanotherapeutic material.[98] The ability of CeO_2, also known as nanoceria, to switch between mixed oxidation states is similar to that of biological antioxidants. This imparts nanoceria with the crucial biological property of radical scavenging. The reversibility of the oxidation state of nanoceria makes it a potent antioxidant, thereby eliminating the need for frequent repeated doses. Nanoceria has been efficiently used as a cardioprotective[99] drug in the murine model with cardiomyopathy. It also reduced myocardial oxidative stress and inflammatory response processes by its autoregenerative antioxidant properties. However, in another group of studies, Srinivas et al. reported that acute exposure of nanoceria particles through inhalation in experimental rats can incur cytotoxicity effects through oxidative stress.[100] This may eventually lead to chronic inflammatory responses. Similar results were noticed in another study where repeated chronic nanoceria inhalation led to accumulation of the NPs in the body, leading to systemic toxicological effects.[101]

18.5 RECENT ADVANCES IN NANOMEDICINE AND NANOTOXICOLOGY

18.5.1 Virus-Like Nanoparticles

Virus-like NPs have been shown to have immense importance in biomedical applications.[102] These virus-like particles are hollow protein cages derived from viruses but do not contain viral genomes. They are ideal nanoscale biomaterial as a result of their biocompatibility, regular structure, and homogenous size. Virus-like particles are noninfectious, but by nature they inherit the essential features of the viruses from which they are derived, including a cell-penetrating ability, the capacity to pack nucleic acids or structurally similar molecules, and a specific tissue tropism. These features make virus-like particles excellent nanocarriers for delivery of therapeutics. They can be employed as an alternative to currently used carbon or metallic oxide-based NPs for potential cardiovascular therapy applications.

18.5.2 Insect Cell-Originated Baculovirus Nanobiohybrids

The baculovirus nanobiohybrid complex has been shown to have advanced gene delivery features and demonstrated improved potential to treat ischemic heart disease with a high therapeutic index for direct gene delivery, as well as in combination with stem cells.[14,103] Baculovirus is an insect cell-originated viral vector that is considered nonpathogenic to humans as it cannot replicate in mammalian cells. The viral genome remains in episomal form within the nucleus and is degraded within mammalian cells over time, eliminating the chance of potential mutagenesis. It can efficiently transduce a wide range of mammalian cells with minimal cytotoxicity, including stem cells, and reports confirm the safety of transplanting baculovirus-engineered stem cells into immunocompetent animals for cell-based gene therapy. These studies demonstrate baculovirus nanohybrids as a potential alternative to widely used CNTs and GO for cardiovascular therapeutic applications.

18.5.3 Bioresponsive Nanodelivery Systems

A new concept in the field of nanomaterials is the inclusion of biological and/or material interactions into nanomaterials, whereby biomolecules or cells trigger the behavior of the nanomaterials.[6] These biointeractive nanomaterials contain receptors for biomolecules that, when stimulated, cause localized or bulk changes in the material properties. The nanomaterials automatically detect disease markers and respond to them to repair the diseased area. An example of such a nanomaterial is mesoporous silica NPs, which can release anticancer drugs in response to proteases present at a tumor site.[104] Application of such smart bioresponsive nanomaterials in cardiovascular therapy is expected to achieve better therapeutic efficacy with a minimum of side effects.

18.5.4 Strategy to Evade Immune Attack on Nanoparticles

Recent studies have shown that altering the configuration of polymer coatings of NPs, with the aim of circumventing the body's immune system, actually triggers the immune system through a different complement pathway. Surface coating the NPs with complement-regulating proteins, such as factor H, can be a potential way of circumventing the host immune response for the intravascular route of NP administration to the target site.[41]

18.6 CONCLUSION

Nanoparticles are being used increasingly in medicinal applications. They have a high retention time in the systemic circulation, which leads to increased exposure of these NPs to other nonspecific organs including the heart. This frequently leads to unwanted cardiotoxic effects. Thus, apart from determining the therapeutic efficacies of NPs, understanding their mode of interaction with different organs at the cellular level and examining their biosafety profile has now become imperative. In addition, their genotoxic potential for both direct and indirect effects in different organs should be evaluated. Future research work needs to focus on developing strategies and experimental techniques that will enable researchers and clinicians to properly investigate the short- and long-term nanotoxic effects of NPs before going into clinical trials.

Key Learning Points

- FDA-approved synthetic materials (e.g., PLGA, PCL) and natural materials (e.g., collagen, alginate, or chitosan) can be safely used as NPs in cardiovascular applications.
- Although there is no standard protocol to properly assess cardiovascular nanotoxicity, the key elements of a toxicity-screening strategy should include physicochemical characterization of NPs, in vitro cellular assays, and in vivo safety and mechanistic studies.
- NP size can significantly influence the mechanism of cellular internalization and potential cytotoxicity, which can be by phagocytosis, macropinocytosis, caveolar-mediated endocytosis, or clathrin-mediated endocytosis.
- The main mechanism to induce cardiovascular nanotoxicity by different nanomaterials is by developing the ROS within the cells causing oxidative stress.
- Apart from direct cardiac exposure to toxic carbon or metallic NPs, the cardiovascular system also is affected by the secondary effects of these NPs on other organs, especially through pulmonary nanoparticle exposure, leading to increased risk of cardiopulmonary toxicity and systemic cardiovascular toxicity, as well as atherosclerosis.

Acknowledgments Dominique Sum-Tim acknowledges support from the Collaborative Health Research Project, Canada; and Satya Prakash acknowledges support from the Canadian Institutes of Health Research (CIHR) (MOP 93641). Arghya Paul acknowledges a postdoctoral award from Fonds de Recherche du Québec–Santé (FRQS), Canada.

REFERENCES

1. Binsalamah ZM, Paul A, Prakash S, et al. Nanomedicine in cardiovascular therapy: recent advancements. *Expert Rev Cardiovasc Ther* 2012;**10**(6):805–15.
2. Dvir T, Timko BP, Kohane DS, et al. Nanotechnological strategies for engineering complex tissues. *Nat Nanotechnol* 2011;**6**(1):13–22.
3. Xia Y. Nanomaterials at work in biomedical research. *Nat Mater* 2008;**7**(10):758–60.
4. Winter PM, Morawski AM, Caruthers SD, et al. Molecular imaging of angiogenesis in early-stage atherosclerosis with alpha(v)beta3-integrin-targeted nanoparticles. *Circulation* 2003;**108**(18):2270–4.
5. Kamaly N, Fredman G, Subramanian M, et al. Development and in vivo efficacy of targeted polymeric inflammation-resolving nanoparticles. *Proc Natl Acad Sci USA* 2013;**110**(16): 6506–11.
6. Stuart MA, Huck WT, Genzer J, et al. Emerging applications of stimuli-responsive polymer materials. *Nat Mater* 2010;**9**(2):101–13.
7. Park JH, Gu L, von Maltzahn G, et al. Biodegradable luminescent porous silicon nanoparticles for in vivo applications. *Nat Mater* 2009;**8**(4):331–6.

8. Paul A, Elias CB, Shum-Tim D, et al. Bioactive baculovirus nanohybrids for stent based rapid vascular re-endothelialization. *Sci Rep* 2013;**3**:2366.
9. Paul A, Shao W, Abbasi S, et al. PAMAM dendrimer-baculovirus nanocomplex for microencapsulated adipose stem cell-gene therapy: in vitro and in vivo functional assessment. *Mol Pharm* 2012;**9**(9):2479–88.
10. Alexis F, Pridgen E, Molnar LK, et al. Factors affecting the clearance and biodistribution of polymeric nanoparticles. *Mol Pharm* 2008;**5**(4):505–15.
11. Brenneman M, Sharma S, Harting M, et al. Autologous bone marrow mononuclear cells enhance recovery after acute ischemic stroke in young and middle-aged rats. *J Cereb Blood Flow Metab* 2010;**30**(1):140–9.
12. Douma K, Prinzen L, Slaaf DW, et al. Nanoparticles for optical molecular imaging of atherosclerosis. *Small* 2009;**5**(5):544–57.
13. Paul A, Shao W, Shum-Tim D, et al. The attenuation of restenosis following arterial gene transfer using carbon nanotube coated stent incorporating TAT/DNA(Ang1 + Vegf) nanoparticles. *Biomaterials* 2012;**33**(30):7655–64.
14. Paul A, Binsalamah ZM, Khan AA, et al. A nanobiohybrid complex of recombinant baculovirus and Tat/DNA nanoparticles for delivery of Ang-1 transgene in myocardial infarction therapy. *Biomaterials* 2011;**32**(32):8304–18.
15. Rzigalinski BA, Strobl JS. Cadmium-containing nanoparticles: perspectives on pharmacology and toxicology of quantum dots. *Toxicol Appl Pharmacol* 2009;**238**(3):280–8.
16. Chan JM, Zhang L, Tong R, et al. Spatiotemporal controlled delivery of nanoparticles to injured vasculature. *Proc Natl Acad Sci USA* 2010;**107**(5):2213–8.
17. Sisco PN, Wilson CG, Mironova E, et al. The effect of gold nanorods on cell-mediated collagen remodeling. *Nano Lett* 2008;**8**(10):3409–12.
18. Stone V, Donaldson K. Nanotoxicology: signs of stress. *Nat Nanotechnol* 2006;**1**(1): 23–4.
19. Patel DN, Bailey SR. Nanotechnology in cardiovascular medicine. *Catheter Cardiovasc Interv* 2007;**69**(5):643–54.
20. Stone V, Johnston H, Clift MJ. Air pollution, ultrafine and nanoparticle toxicology: cellular and molecular interactions. *IEEE Trans Nanobiosci* 2007;**6**(4):331–40.
21. Seaton A, Donaldson K. Nanoscience, nanotoxicology, and the need to think small. *Lancet* 2005;**365**(9463):923–4.
22. Mills NL, Donaldson K, Hadoke PW, et al. Adverse cardiovascular effects of air pollution. *Nat Clin Pract Cardiovasc Med* 2009;**6**(1):36–44.
23. Pope CA 3rd, Burnett RT, Thurston GD, et al. Cardiovascular mortality and long-term exposure to particulate air pollution: epidemiological evidence of general pathophysiological pathways of disease. *Circulation* 2004;**109**(1):71–7.
24. Hoek G, Krishnan RM, Beelen R, et al. Long-term air pollution exposure and cardio- respiratory mortality: a review. *Environ Health* 2013;**12**(1):43.
25. Yamawaki H, Iwai N. Mechanisms underlying nano-sized air-pollution-mediated progression of atherosclerosis: carbon black causes cytotoxic injury/inflammation and inhibits cell growth in vascular endothelial cells. *Circ J* 2006;**70**(1):129–40.
26. Brook RD. Cardiovascular effects of air pollution. *Clin Sci* 2008;**115**(6):175–87.
27. Nurkiewicz TR, Porter DW, Barger M, et al. Systemic microvascular dysfunction and inflammation after pulmonary particulate matter exposure. *Environ Health Perspect* 2006;**114**(3): 412–9.
28. Manke A, Wang L, Rojanasakul Y. Mechanisms of nanoparticle-induced oxidative stress and toxicity. *BioMed Res Int* 2013;**2013**:15.

29. Huang CC, Aronstam RS, Chen DR, et al. Oxidative stress, calcium homeostasis, and altered gene expression in human lung epithelial cells exposed to ZnO nanoparticles. *Toxicol In Vitro* 2010;**24**(1):45–55.
30. Selvi RB, Chatterjee S, Jagadeesan D, et al. ATP driven clathrin dependent entry of carbon nanospheres prefer cells with glucose receptors. *J Nanobiotechnol* 2012;**10**:35.
31. Fernandez-Gomez FJ, Galindo MF, Gomez-Lazaro M, et al. Malonate induces cell death via mitochondrial potential collapse and delayed swelling through an ROS-dependent pathway. *Br J Pharmacol* 2005;**144**(4):528–37.
32. Sharma S, Rais A, Sandhu R, et al. Clinical significance of metallothioneins in cell therapy and nanomedicine. *Int J Nanomed* 2013;**8**:1477–88.
33. Sack MN. Mitochondrial depolarization and the role of uncoupling proteins in ischemia tolerance. *Cardiovasc Res* 2006;**72**(2):210–9.
34. Duchen MR. Mitochondria and calcium: from cell signalling to cell death. *J Physiol* 2000; **529**(1):57–68.
35. Xu M, Li J, Iwai H, et al. Formation of nano-bio-complex as nanomaterials dispersed in a biological solution for understanding nanobiological interactions. *Sci Rep* 2012;**2**:406.
36. Jiang W, Kim BY, Rutka JT, et al. Nanoparticle-mediated cellular response is size-dependent. *Nat Nanotechnol* 2008;**3**(3):145–50.
37. Petros RA, DeSimone JM. Strategies in the design of nanoparticles for therapeutic applications. *Nat Rev Drug Discov* 2010;**9**(8):615–27.
38. Bhabra G, Sood A, Fisher B, et al. Nanoparticles can cause DNA damage across a cellular barrier. *Nat Nanotechnol* 2009;**4**(12):876–83.
39. Kim JA, Aberg C, Salvati A, et al. Role of cell cycle on the cellular uptake and dilution of nanoparticles in a cell population. *Nat Nanotechnol* 2012;**7**(1):62–8.
40. Hamad I, Al-Hanbali O, Hunter AC, et al. Distinct polymer architecture mediates switching of complement activation pathways at the nanosphere-serum interface: implications for stealth nanoparticle engineering. *ACS Nano* 2010;**4**(11):6629–38.
41. Sim RB, Wallis R. Surface properties: Immune attack on nanoparticles. *Nat Nanotechnol* 2011;**6**(2):80–1.
42. Klien K, Godnic-Cvar J. Genotoxicity of metal nanoparticles: focus on in vivo studies. *Arh Hig Rada Toksikol* 2012;**63**(2):133–45.
43. Wani MY, Hashim MA, Nabi F, et al. Nanotoxicity: dimensional and morphological concerns. *Adv Phys Chem* 2011;**2011**:(450912).
44. Sahu SC, Casciano DA. *Nanotoxicity: from in vivo and in vitro models to health risks.* New York: Wiley; 2009.
45. Mann EE, Thompson LC, Shannahan JH, et al. Changes in cardiopulmonary function induced by nanoparticles. *Wiley Interdiscip Rev Nanomed Nanobiotechnol* 2012;**4**(6): 691–702.
46. Rogueda PG, Traini D. The nanoscale in pulmonary delivery. Part 1: deposition, fate, toxicology and effects. *Expert Opin Drug Deliv* 2007;**4**(6):595–606.
47. Lippmann M, Yeates DB, Albert RE. Deposition, retention, and clearance of inhaled particles. *Br J Ind Med* 1980;**37**(4):337–62.
48. Oberdorster G, Oberdorster E, Oberdorster J. Nanotoxicology: an emerging discipline evolving from studies of ultrafine particles. *Environ Health Perspect* 2005;**113**(7):823–39.
49. Scheuplein RJ. Mechanism of percutaneous absorption. II. Transient diffusion and the relative importance of various routes of skin penetration. *J Invest Dermatol* 1967;**48**(1):79–88.
50. Crosera M, Bovenzi M, Maina G, et al. Nanoparticle dermal absorption and toxicity: a review of the literature. *Int Arch Occup Environ Health* 2009;**82**(9):1043–55.

51. Schneider M, Stracke F, Hansen S, et al. Nanoparticles and their interactions with the dermal barrier. *Dermato-Endocrinol* 2009;**1**(4):197–206.
52. Olmsted SS, Padgett JL, Yudin AI, et al. Diffusion of macromolecules and virus-like particles in human cervical mucus. *Biophys J* 2001;**81**(4):1930–7.
53. Frohlich E, Roblegg E. Models for oral uptake of nanoparticles in consumer products. *Toxicol* 2012;**291**(1–3):10–7.
54. Navarro E, Piccapietra F, Wagner B, et al. Toxicity of silver nanoparticles to *Chlamydomonas reinhardtii*. *Environ Sci Technol* 2008;**42**(23):8959–64.
55. Faraji AH, Wipf P. Nanoparticles in cellular drug delivery. *Bioorg Med Chem* 2009; **17**(8):2950–62.
56. Galvin P, Thompson D, Ryan KB, et al. Nanoparticle-based drug delivery: case studies for cancer and cardiovascular applications. *Cell Mol Life Sci* 2012;**69**(3):389–404.
57. Decuzzi P, Pasqualini R, Arap W, et al. Intravascular delivery of particulate systems: does geometry really matter? *Pharm Res* 2009;**26**(1):235–43.
58. Magrez A, Kasas S, Salicio V, et al. Cellular toxicity of carbon-based nanomaterials. *Nano Lett* 2006;**6**(6):1121–5.
59. Uo M, Akasaka T, Watari F, et al. Toxicity evaluations of various carbon nanomaterials. *Dent Mater J* 2011;**30**(3):245–63.
60. Hu W, Peng C, Luo W, et al. Graphene-based antibacterial paper. *ACS Nano* 2010; **4**(7):4317–23.
61. Akhavan O, Ghaderi E. Toxicity of graphene and graphene oxide nanowalls against bacteria. *ACS Nano* 2010;**4**(10):5731–6.
62. Bakry R, Vallant RM, Najam-ul-Haq M, et al. Medicinal applications of fullerenes. *Int J Nanomedicine* 2007;**2**(4):639–49.
63. Sayes CM, Fortner JD, Guo W, et al. The differential cytotoxicity of water-soluble fullerenes. *Nano Lett* 2004;**4**(10):1881–7.
64. Zhang B-T, Zheng X, Li H-F, et al. Application of carbon-based nanomaterials in sample preparation: A review. *Anal Chim Acta* 2013;**784**(0):1–17.
65. Wang L, Shi J, Jia X, et al. NIR-/pH-responsive drug delivery of functionalized single-walled carbon nanotubes for potential application in cancer chemo-photothermal therapy. *Pharm Res* 2013;**30**:2757–71.
66. Zhao B, Yan J, Wang D, et al. Carbon nanotubes multifunctionalized by rolling circle amplification and their application for highly sensitive detection of cancer markers. *Small* 2013;**9**:2595–601.
67. Zhang Y, Ali SF, Dervishi E, et al. Cytotoxicity effects of graphene and single-wall carbon nanotubes in neural phaeochromocytoma-derived PC12 cells. *ACS Nano* 2010;**4**(6):3181–6.
68. Sanchez VC, Jachak A, Hurt RH, et al. Biological interactions of graphene-family nanomaterials: an interdisciplinary review. *Chem Res Toxicol* 2011;**25**(1):15–34.
69. Shen H, Zhang L, Liu M, et al. Biomedical applications of graphene. *Theranostics* 2012; **2**(3):283–94.
70. Hussain S, Thomassen LC, Ferecatu I, et al. Carbon black and titanium dioxide nanoparticles elicit distinct apoptotic pathways in bronchial epithelial cells. *Part Fibre Toxicol* 2010;**7**:10.
71. Shvedova AA, Pietroiusti A, Fadeel B, et al. Mechanisms of carbon nanotube-induced toxicity: focus on oxidative stress. *Toxicol Appl Pharmacol* 2012;**261**(2):121–33.
72. Pulskamp K, Diabate S, Krug HF. Carbon nanotubes show no sign of acute toxicity but induce intracellular reactive oxygen species in dependence on contaminants. *Toxicol Lett* 2007;**168**(1):58–74.

73. Lacerda SH, Semberova J, Holada K, et al. Carbon nanotubes activate store-operated calcium entry in human blood platelets. *ACS Nano* 2011;**5**(7):5808–13.
74. Feng L, Wu L, Qu X. New horizons for diagnostics and therapeutic applications of graphene and graphene oxide. *Adv Mater* 2013;**25**(2):168–86.
75. Li N, Zhang Q, Gao S, et al. Three-dimensional graphene foam as a biocompatible and conductive scaffold for neural stem cells. *Sci Rep* 2013;**3**:1604.
76. Singh SK, Singh MK, Nayak MK, et al. Thrombus inducing property of atomically thin graphene oxide sheets. *ACS Nano* 2011;**5**(6):4987–96.
77. Singh SK, Singh MK, Kulkarni PP, et al. Amine-modified graphene: thrombo-protective safer alternative to graphene oxide for biomedical applications. *ACS Nano* 2012;**6**(3): 2731–40.
78. Liao KH, Lin YS, Macosko CW, et al. Cytotoxicity of graphene oxide and graphene in human erythrocytes and skin fibroblasts. *ACS Appl Mater Interfaces* 2011;**3**(7):2607–15.
79. Araujo JA, Barajas B, Kleinman M, et al. Ambient particulate pollutants in the ultrafine range promote early atherosclerosis and systemic oxidative stress. *Circ Res* 2008;**102**(5): 589–96.
80. Campen MJ, McDonald JD, Gigliotti AP, et al. Cardiovascular effects of inhaled diesel exhaust in spontaneously hypertensive rats. *Cardiovasc Toxicol* 2003;**3**(4):353–61.
81. Stampfl A, Maier M, Radykewicz R, et al. Langendorff heart: a model system to study cardiovascular effects of engineered nanoparticles. *ACS Nano* 2011;**5**(7):5345–53.
82. Boerma M, Forsberg L, Van Zeijl L, et al. A genetic polymorphism in connexin 37 as a prognostic marker for atherosclerotic plaque development. *J Intern Med* 1999;**246**(2): 211–8.
83. Chen Z, Meng H, Xing G, et al. Age-related differences in pulmonary and cardiovascular responses to SiO_2 nanoparticle inhalation: nanotoxicity has susceptible population. *Environ Sci Technol* 2008;**42**(23):8985–92.
84. Pearson TA, LaCava J, Weil HF. Epidemiology of thrombotic-hemostatic factors and their associations with cardiovascular disease. *Am J Clin Nutr* 1997;**65**(5 Suppl): 1674S–1682SS.
85. Huang Y-W, Wu C-h, Aronstam RS. Toxicity of transition metal oxide nanoparticles: recent insights from in vitro studies. *Materials* 2010;**3**(10):4842–59.
86. Mishra SK, Srivastava RK, Prakash SG. ZnO nanoparticles: Structural, optical and photo-conductivity characteristics. *J Alloys Comp* 2012;**539**(0):1–6.
87. Azam A, Ahmed AS, Oves M, et al. Size-dependent antimicrobial properties of CuO nano-particles against Gram-positive and -negative bacterial strains. *Int J Nanomed* 2012; **7**:3527–35.
88. Abdelhalim MA. Exposure to gold nanoparticles produces cardiac tissue damage that depends on the size and duration of exposure. *Lipids Health Dis* 2011;**10**:205.
89. Wagner M, Klein CL, van Kooten TG, et al. Mechanisms of cell activation by heavy metal ions. *J Biomed Mater Res* 1998;**42**(3):443–52.
90. Gojova A, Guo B, Kota RS, et al. Induction of inflammation in vascular endothelial cells by metal oxide nanoparticles: effect of particle composition. *Environ Health Perspect* 2007;**115**(3):403–9.
91. Kahn E, Baarine M, Pelloux S, et al. Iron nanoparticles increase 7-ketocholesterol-induced cell death, inflammation, and oxidation on murine cardiac HL1-NB cells. *Int J Nanomedi* 2010;**5**:185–95.
92. Knaapen AM, Borm PJ, Albrecht C, et al. Inhaled particles and lung cancer. Part A: Mechanisms. *Int J Cancer* 2004;**109**(6):799–809.

93. Arora S, Rajwade JM, Paknikar KM. Nanotoxicology and in vitro studies: the need of the hour. *Toxicol Appl Pharmacol* 2012;**258**(2):151–65.
94. Ai J, Biazar E, Jafarpour M, et al. Nanotoxicology and nanoparticle safety in biomedical designs. *Int J Nanomed* 2011;**6**:1117–27.
95. Spivak MY, Bubnov RV, Yemets IM, et al. Development and testing of gold nanoparticles for drug delivery and treatment of heart failure: a theranostic potential for PPP cardiology. *EPMA J* 2013;**4**(1):20.
96. Hussain SM, Hess KL, Gearhart JM, et al. In vitro toxicity of nanoparticles in BRL 3A rat liver cells. *Toxicol In Vitro* 2005;**19**(7):975–83.
97. Kim E, Chu Y, Han J, et al. Proteomic analysis of silver nanoparticle toxicity in rat. *Toxicol Environ Health Sci* 2010;**2**(4):251–62.
98. Giri S, Karakoti A, Graham RP, et al. Nanoceria: a rare-earth nanoparticle as a novel anti-angiogenic therapeutic agent in ovarian cancer. *PloS One* 2013;**8**(1):e54578.
99. Niu J, Azfer A, Rogers LM, et al. Cardioprotective effects of cerium oxide nanoparticles in a transgenic murine model of cardiomyopathy. *Cardiovasc Res* 2007;**73**(3):549–59.
100. Srinivas A, Rao PJ, Selvam G, et al. Acute inhalation toxicity of cerium oxide nanoparticles in rats. *Toxicol Lett* 2011;**205**(2):105–15.
101. Geraets L, Oomen AG, Schroeter JD, et al. Tissue distribution of inhaled micro- and nano-sized cerium oxide particles in rats: results from a 28-day exposure study. *Toxicol Sci* 2012;**127**(2):463–73.
102. Shao W, Paul A, Abbasi S, et al. A novel polyethyleneimine-coated adeno-associated virus-like particle formulation for efficient siRNA delivery in breast cancer therapy: preparation and in vitro analysis. *Int J Nanomed* 2012;**7**:1575–86.
103. Paul A, Nayan M, Khan AA, et al. Angiopoietin-1-expressing adipose stem cells genetically modified with baculovirus nanocomplex: investigation in rat heart with acute infarction. *Int J Nanomed* 2012;**7**:663–82.
104. Singh N, Karambelkar A, Gu L, et al. Bioresponsive mesoporous silica nanoparticles for triggered drug release. *J Am Chem Soc* 2011;**133**(49):19582–5.

Chapter 19

Forensic Pathology Related to Cardiovascular Toxicity

Magdy A. Kharoshah,[1] Mohammed Saleh Madadin,[2] Sherien Salah Ghaleb,[3] Hanan Hosny Hassan,[4] Kholoud Samy Alsowayigh,[5] and Ritesh George Menezes[2]

[1]*Consultant Forensic Medical Examiner, Forensic Medicine Authority, Ministry of Justice, Cairo, Egypt*
[2]*Forensic Medicine Division, Department of Pathology, College of Medicine, University of Dammam, Dammam, Saudi Arabia*
[3]*Forensic Medicine and Clinical Toxicology, Cairo University and BeniSuef University, Cairo, Egypt*
[4]*Forensic Pathology Consultant and Forensic Medicine Authority, Ministry of Justice, Cairo, Egypt*
[5]*Forensic Medicine Specialist, Forensic Medicine Center, Jeddah, Saudi Arabia*

19.1 INTRODUCTION

Toxicology is an area that combines various aspects of chemistry, biology, and medicine. It focuses on the adverse effects of chemicals, drugs, and other substances on the human body. Toxicology is further subdivided into several branches, including clinical toxicology, a branch mainly associated with diagnosis and therapeutic approaches toward patients. Another division of toxicology is associated with testing and laboratory work and is known as analytical toxicology.[1] However, the branch that is the main focus of this chapter is *forensic toxicology*, which involves the implementation and utilization of toxicology in the legal system for the provision of justice either in living cases (e.g., drunk driving or illicit drug use) or postmortem medicolegal cases.

In the process of opining this diagnosis, forensic pathologists are required to assess and confirm that there were no other potential causes of death.[2] Pathologists may make the diagnosis of intoxication based on autopsy findings in combination with the results of a histopathological examination, which sometimes includes immunohistochemistry, nueropharmacological analysis, and toxicological analysis, together with the circumstantial and background information from the police report and/or medical records.[2]

The Heart and Toxins. DOI: http://dx.doi.org/10.1016/B978-0-12-416595-3.00019-0

The exposure of humans to toxins is common and can be confirmed by the 55 U.S. Poison Centers that disclosed approximately 2.3 million incidences of previous exposures to various toxins and pharmaceuticals. Many of the cases reviewed by forensic pathologists involve suspected toxicological death.[3]

The majority of the cases reviewed are often associated with crime investigation. The diagnostic process involves several stages and a wide variety of events associated with the patient. As mentioned before, these involve the history of the patient, clinical presentation and information about the circumstances surrounding death. All of these are important for the pathologist to know to make the appropriate diagnosis.[3]

19.2 POSTMORTEM TOXICOLOGY

During the postmortem diagnostic process, the deceased can present with signs that can be either external or internal. Such presentations also may be macroscopical or microscopical and offer assistance in detecting the toxins that are present in the individual's body. The forensic pathologist is required to assess whether there are other underlying causes of death, or not, and must perform a complete autopsy and carefully assess the findings. In addition, different body tissues and fluids must be examined, preserved, and analyzed. In addition to providing further insight on the disease or injury, the autopsy will also provide a clear presentation of the extent of damage to the body that was caused by the toxins.[1,2]

Other examples of inconsistencies involve situations when there are no indications of death from drug toxicity during an autopsy; however, certain signs and indications, such as froth at the mouth, can indicate drug-related deaths as well as drowning. Other signs include needle track marks that indicate intravenous (IV) use of a drug, packs of drugs in the intestines and pills in the stomach in the case of oral use, pulmonary edema, and congestion of the lungs. Liver necrosis is also common in subjects who have died from a drug overdose and, in particular, this is a typical finding in acetaminophen intoxication. The smell of bitter almonds and hemorrhage in the gastric mucosa are signs of cyanide poisoning. Bluish discoloration of the gastric mucosa may indicate pesticide ingestion, and black discoloration of the gastric mucosa may be found in deaths from alkaline fluids.

Histopathological findings in forensic autopsies reveal both nonspecific and specific characteristics of diseases depending on the toxin and the organ affected. Most toxins are transported through the circulatory system and the heart is one of the major organs affected by toxic substances. Cardiovascular toxins are classified into a wide array of pharmacological actions. This chapter, however, focuses on the toxins that are commonly found in forensic practice. Morphological changes occur in cardiovascular tissues that are similar to those from other pathological processes that are not drug-induced, and this makes the diagnostic procedure in forensic pathology more difficult because pathologists are required to distinguish between the different changes.[4–6]

Postmortem redistribution potentially affects drug concentration as a result of diffusion after death and cellular membrane disruption.[7] The specimens of choice in postmortem cases are blood and urine. However, other specimens (e.g., liver and vitreous humor) can be used in routine cases, while other specimens are used in specific cases, such as collecting muscle tissue, gastric contents, and hair, particularly in putrefied bodies. The vitreous humor has a great value in postmortem toxicology, especially for the estimation of alcohol as an alternative to blood and urine (and due to delay in delivery more usually reflects the urinary concentration), as well as of drugs of abuse and endogenous compounds.

Gastric contents are particularly important to analyze if the drug in question is very potent and difficult to detect in blood, or when it is important to determine whether a poisoning was due to oral intake, transdermal exposure, smoking, or injection. Sometimes the maggots in a decomposed body can be analyzed to find evidence of drug presence, and this field is specifically dubbed entomotoxicology. Sampling, preservation, storage, and analysis of postmortem specimens must be done carefully. Most forensic departments will add 1 to 2% sodium (or potassium) fluoride to fluid samples to reduce postmortem generation of alcohols and to limit breakdown of certain drugs (e.g., zopiclone).[8] Special knowledge is necessary because stability of the drugs postmortem differs and depends on many factors such as the structure of the drug and the storing temperature, as well as duration of storing.[1,2,9]

This chapter focuses on the forensic pathology of the heart and relevant clinical discussions about drugs, toxins, and recreational drugs that induce cardiac damage. Clinical manifestations of their toxicities, however, are explained in this book's other chapters.

19.3 PSYCHOSTIMULANTS

Psychostimulants are substances that increase psychomotor activity. Cocaine and amphetamines are commonly abused psychostimulants.

19.3.1 Cocaine

Cocaine is the second most common drug that is abused among the population masses today and is responsible for a large number of drug-induced deaths. Cocaine has a higher level of popularity among individuals ranging from 18 to 25 years of age.[10,11] There are several ways by which cocaine can be administrated: nasal inhalation ("snorting"), smoking, and injection. When the drug is in the system, it follows various pharmacokinetics that result in an increase in the peak levels of its concentration in the blood. These concentrations may remain at high levels for as long as 90 min.

The abusive use of cocaine is associated with complications that affect different systems in the body and can be of either an acute or chronic nature.

The most important system affected by the abuse of cocaine is the cardiovascular system.[11–14] Other diseases or predisposition to certain cardiovascular conditions are not part of the pathogenesis of cocaine-related complications. The rate of cocaine abuse has made the study of associated complications an area of major concern.[14] Morbidity related to cocaine abuse and misuse of the substance as a whole is important in the differential diagnostic process. Individuals who are abusers of cocaine often experience a wide array of symptoms that result in the loss of productivity and, in a substantial number of cases, even death. Thus, it is necessary that these individuals receive the appropriate diagnosis before abuse becomes too extreme.

Cocaine abuse is associated with a wide variety of cardiovascular effects, including

- *Cardiac symptoms*: coronary artery spasms, myocardial infarction, atherosclerosis, cardiomyopathy, myocarditis, hypertension, arrhythmia, and endocarditis
- *Vascular symptoms*: vasculitis, aortic rupture, and dissection[15]

Cocaine also can have some neurologic effects and acts as a sympathomimetic agent, in particular by the inhibition of the dopamine transporter protein (DAT) and by inhibition of norepinephrine, causing a significant increase in the synaptic levels of dopamine and norepinephrine.[11] It also increases the release of catecholamines from peripheral and central stores.

Further, cocaine can act as a local anesthetic by inhibiting the transmission and initiation of electrical impulses by blocking membrane permeability to Na^+ during depolarization.[16] Most cocaine toxicity deaths are thought to be the result of a sympathetic overstimulation that causes cardiac insufficiency, but some severe intoxications may be due to excited delirium, affecting subjects with a genetic or acquired deficiency in upregulating DAT during chronic abuse.[17]

Cocaine causes tachycardia and hypertension and the rate of increase is dependent on the dose of cocaine present in the blood. When minimal amounts of cocaine are introduced into the body, the increase of these parameters is usually within the physiological range. At the cellular level, the sympathomimetic actions of cocaine are mediated by stimulation of alpha and beta adrenergic receptors. In addition, cocaine interacts with the muscarinic receptors that block the uptake of serotonin and dopamine at the nerve endings.[18–20]

Endocarditis is one of the most common diseases associated with IV drug abuse including cocaine, and it represents one of the most severe consequences. In addition, the endocarditis that is developed as a result of cocaine abuse is different from the one that is observed in abuse of other drugs; it is mainly associated with left-sided cardiac valves (Figure 19.1).

The pathogenesis of cocaine-related disorders include tachycardia and hypertension, which result in the injury of vessels and valves with IV use of

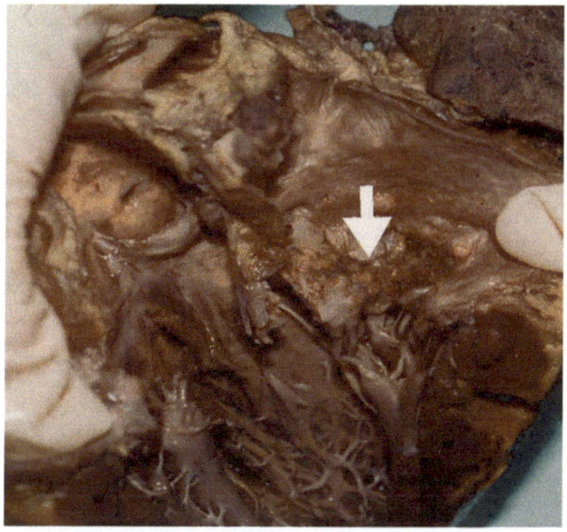

FIGURE 19.1 **Vegetations on the upper aspect of the mitral valve (endocarditis).**

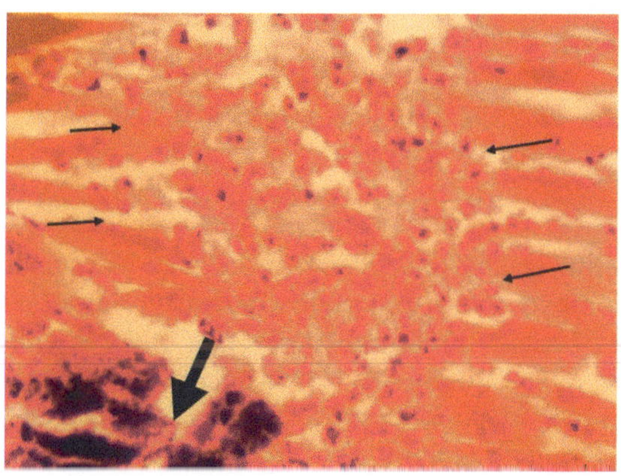

FIGURE 19.2 **Septic myocarditis, bacterial colonies (*thick arrow*) with neutrophiic infiltrate (*thin arrows*).**

the drug and predisposes an individual to bacterial infections and septic myocarditis (Figure 19.2). Cocaine also has immunosuppressive effects that have been associated with a higher incidence of infection among some individuals.[21–23] The abuse of cocaine also may lead to other disorders such as aortic dissection (Figure 19.3). This is mainly a result of elevated arterial pressure, which is one of the major reasons for chest pain among cocaine addicts.

(a)
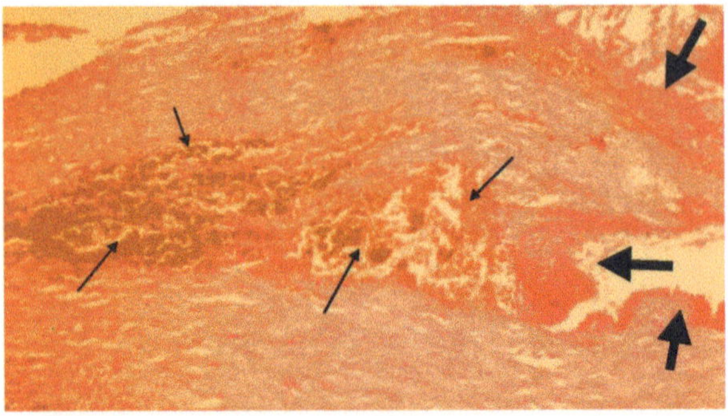

(b)
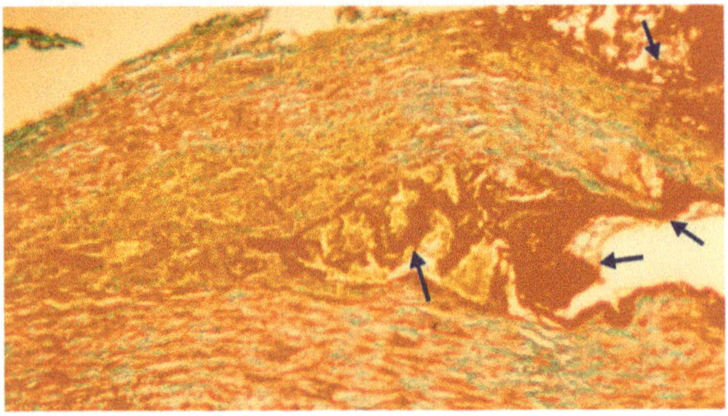

FIGURE 19.3 **Aortic dissection.** (a) Blood dissecting into the aortic media (*thin arrows*) associated with fibrin deposition (*thick arrows*), H&E stain; and (b) fibrin threads stained red with MSB stain (*thin arrows*).

Dilated cardiomyopathy is common for cocaine users and is assumed to be a result of the toxic effect of cocaine. Intoxication from cocaine leads to heart failure as a result of interstitial fibrosis, myofibril destruction, and myocardial dilation. Cocaine is also responsible for the hyperadrenergic state that is present in individuals and contributes to cardiomyopathy that has similar presentations to those seen in pheochromocytoma.[24]

The pathogenesis is assumed to be secondary to hypersensitivity reactions, which lead to myocarditis and vasculitis.[24–26] Cocaine also leads to cardiac toxicity, which is induced by catecholamines. However, myocardial infarction caused by cocaine abuse is reversible with abstinence.[27,28] Cardiomegaly and heart failure in young individuals are warning signs for drug abuse.[29] The presence of these finding without other signs or clinical history may indicate chronic cocaine abuse by them.

A majority of patients present with a wide array of conditions; however, these are often resolved after the cocaine has been metabolized. Sinus bradycardia and tachycardia, bundle branch block, supraventricular arrhythmia, ventricular fibrillation, and ventricular tachycardia have all been recorded.[30–32] There is also autonomic imbalance and deviated cardiac automaticity, which is a result of the elevated adrenergic state of the individual. This status is also one of the trigger factors for arrhythmia.[25,33] Reentrance tachycardia is also a result of electrical inhomogeneity, which is a result of cocaine-induced ischemia.[34] In addition, cocaine abuse leads to the prolongation of QRS and QT duration.

Numerous studies have shown that there is a correlation between cases of cocaine abuse and myocardial infarction.[35] Cocaine has also been discovered to trigger acute myocardial infarction in patients who are experiencing underlying conditions such as coronary atherosclerosis. Cocaine-related myocardial infarction usually presents its symptoms a few hours after use of the substance.[35]

There are three pharmacological mechanisms involved in the pathogenesis of cocaine-related myocardial infarction. The first process involves adrenergic stimulation, which in turn results in the elevation of heart rate, left ventricle contractility, and arterial blood pressure. The second mechanism involves dual action on the coronary vasculature. This includes early vasodilatation, which is followed by sustained vasoconstriction. These effects are mainly dependent on the dosage used and presumably stem from different pharmacological traits. In essence, the vasodilatation is initiated by the anesthetic trait of cocaine, whereas the vasoconstriction is a result of the potentiating of the alpha-adrenergic response on the coronary arteries. The third mechanism involves elevation in the number of thrombolytic aggregation potentials, which in turn increases the rate of intravascular thrombus formation in the pulmonary, coronary, peripheral venous circulation, renal vasculature, and the skin.[35–38]

The relationship between cocaine abuse and atherosclerosis has not been corroborated by substantial evidence.[39] However, it has been reported that cocaine abuse has a significant impact on the reduction of nitrous oxide production, which leads to an increase in the migration and expression of leukocytes. Moreover, the endothelial barrier is compromised by cocaine, which leads to the increase in permeability for lipoproteins into the subintimal space. In addition, there is an elevated amount of adventitial mast cells in the coronary arteries of deceased individuals who have been reported to be cocaine addicts. This indicates that cocaine may be used in the generation of mast cells and the intermediate products from these cells may be the trigger factor for atherosclerosis and thrombus formation.[40]

Microscopical patterns in the myocardium can be used to identify the toxic effects of cocaine use. After an acute exposure to a high dose of cocaine, scattered contraction band necrosis affecting single or groups of cardiomyocytes may be seen, distributed globally across the ventricular walls. In chronic cocaine addicts, the myofibers affected will be replaced by collagen, and such minute scarring in all samples from the myocardium is typical of extensive cocaine (or other psychostimulant) abuse. Thus, this pattern is different from the picture seen in an acute myocardial infarction, where a large area around the nonreperfused area is composed of contraction-band necrosis.[41]

A postmortem diagnosis depends on circumstances and suspicion of the addiction, and cardiac findings, as described, may be found. Toxicological analysis from blood and urine usually are diagnostic and we must be aware of cocaine's metabolism; that is, it is rapidly hydrolyzed by cholinesterases to its major metabolite ecgoninemethylester and to benzoylecgonine by non-enzymatic hydrolysis.

19.3.2 Amphetamine

Amphetamine-related pathological findings are usually found in the liver then the heart. Methamphetamine shares similarities with amphetamine's influence on the cardiovascular system. Chronic abusers of the drug develop cardiac hypertrophy and cardiovascular diseases. Current findings show that the individuals who abuse this substance have a higher risk of developing heart disease, and chronic users have an even greater risk. If individuals have preexisting cardiac conditions, their risk of cardiovascular disease (CVD) development is elevated by the abuse of this substance.[42] Vasculitis, specifically necrotizing arteritis in cerebral arteries, is reported in amphetamine abusers.[43] Microscopic findings in the heart include hypertrophy, interstitial fibrosis, and microvascular disease.[41]

19.3.3 Ecstasy

3,4 methylendioxymethylamphetamine (MDMA, or "ecstacy") is a very popular party drug and has continued to be abused for recreational purposes due to the misconception that a majority of its abusers believe it is a "safe drug." There are numerous adverse effects that have been reported with the abuse of ecstasy, including rhabdomyolysis, psychiatric complications, stroke, hepatic and renal failure, and disseminated intravascular coagulation (DIC) syndrome. Cardiac hypertrophy is associated with other illicit drug abuse and the individuals who abuse ecstasy have also been found to possess a risk of developing this condition. MDMA also has adverse effects on the sympathetic system (Table 19.1).

TABLE 19.1 Toxic and Lethal Concentrations of Drugs of Abuse

	Toxic			Lethal/Postmortem		
Drug (mg/L)	Whole Blood	Serum/ Plasma	Urine	Whole Blood	Serum/ Plasma	Urine
Cocaine	–	0.2	25	1.6	1.8	50
Morphine/heroin	–	0.1	0.5	0.2	0.1	5
Alcohol	1000	–	–	4000	–	5000
Amphetamine	–	0.2	25	0.5	0.5	25–700

In experiments with animals, MDMA has been found to cause myocarditis, fibrosis, and contraction-band necrosis. Endomyocardial biopsy has shown that individuals possess focal myocyte injuries with exacerbation of human leukocyte antigen (HLA) in classes I and II. Other conditions that have also been reported include cardiac arrhythmia, aortic dissection, myocardial infarction, and myocardial damage caused by catecholamines. Morphologically, the findings are similar to those found with cocaine abusers; however, the pathogenesis differs between the two. Acute coronary heart disease is less frequent in ecstasy abusers.[44–46]

19.3.4 Opiates

Opiates comprise natural, synthetic, and semisynthetic opioids. Natural opiates are extractions from opium poppy plants, heroin being produced by acetylation of morphine. Opiates are mainly used as analgesics and anesthetics. Among these drugs, the most commonly abused include codeine, morphine, oxycodone, and heroin.

There are four types of opioid receptors: kappa, delta, mu, and nociceptine. Activation of mu receptors results in the reduction of the cardiomyocytes response to calcium ions. This is also accompanied by inhibition of adenylate cyclase and the elevation of calcium ion concentration in the cells. The drugs also have an effect on the nervous system, including activation of the parasympathetic system while inducing the opposite effect on the sympathetic system.[47] The summation of these effects results in hypotension, bradycardia, respiratory arrest, and peripheral vasodilatation. Certain narcotics act directly on the conductive system and the sinoatrial node.

Opioids may also initiate the release of histamine from mast cells, resulting in anaphylactic responses. About 5% of heroin toxicity deaths have been considered to be due to anaphylactic shock.[47,48] Subjects will have a high amount of blood in the large veins and no blood congestion in the inner organs, and no froth in the airways, but still significant amounts of morphine (or related substances) in the blood.

The effects of narcotics will be further elaborated in the following sections. These effects are mainly presented about individuals who have chronically abused analgesics; they are often drug addicts. They present with clinical manifestations that include apnea and cyanosis. Acute toxicity and death are mainly due to respiratory depression.

19.4 THE EFFECT OF NARCOTICS ON THE HEART

Macroscopically, the right ventricle may be dilated, there will be some thickening and opacity of the endocardium and epicardium, hypertrophy of the left ventricle, and the upper part of the papillary muscles may show focal necrosis as a result of prolonged hypoxia following opiate intoxication. There may also

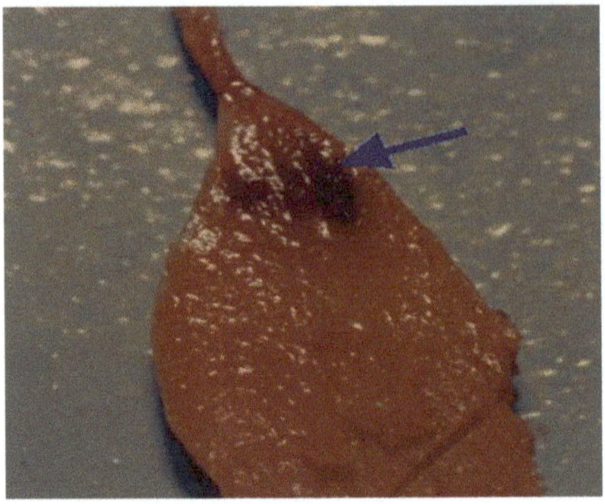

FIGURE 19.4 **Papillary muscle with recent infarction.**

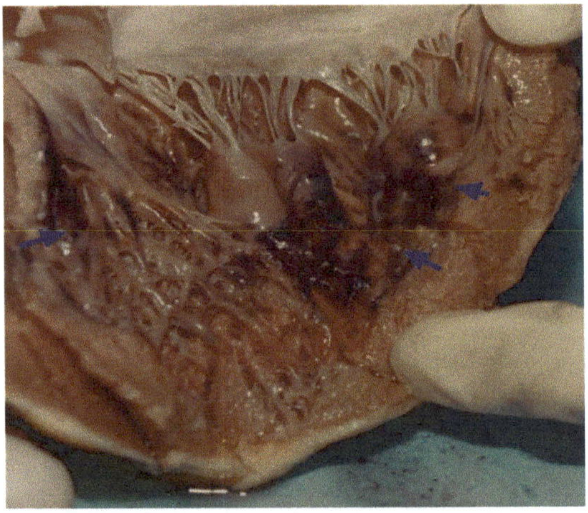

FIGURE 19.5 **Subendocardial hemorrhages in the left ventricle.**

be mononuclear inflammatory cells and necrotic myocardial fibers and congestion (Figure 19.4). The left bundle branches may show diffuse fibrosis and there can be hemorrhages in the subendocardium (Figure 19.5).[49]

Aside from the macroscopical features, there are also several microscopical traits. Opioids' effects are characterized by the infiltration of fibroblasts and histiocytes and collagen fiber formation (Figure 19.6). Therefore, it is essential for pathologists to combine both the macroscopic features and the microscopic pictures to distinguish the etiology of an individual's death. Interstitial edema is also visible in patients who have died from narcotic intoxication. These histological features are seen in other parts of the heart (e.g., the atrioventricular bundle).[48] The abuse of opioids is also associated with the destruction of microvasculature, which may lead to hypoxia.

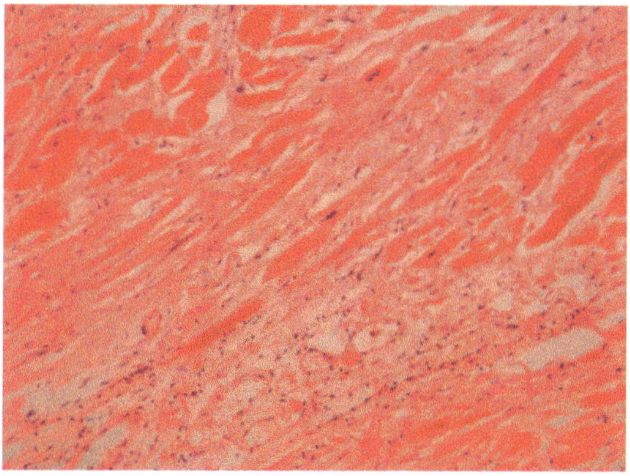

FIGURE 19.6 **Myocardial ischemic changes in the form of granulation tissue formation with fibroblasts and collaged deposition infiltrated by histiocytes.**

19.4.1 Cannabis

Cannabis (marijuana) is one of the most widely abused drugs in the world and research has found that it is the third most popular substance after tobacco and alcohol. It is commonly referred to as marijuana and is a psychoactive substance derived from the *Cannabis sativa* plant. Marijuana is often used in two forms: the herbal form, which is obtained from the leaves of the female plant, and the resinous form, which is derived from the trichomes of the plant. The former is often referred to as weed and the latter is often referred to as hashish. *Cannabis sativa* contains more than 60 cannabinoids; however, the most common include cannabichomene, cannabidiol, and cannabigerol. Another psychoactive substance present in marijuana is tetrahydrocannabinol. In the past, medical professionals assumed that cannabis was not addictive and had no adverse effects on an individual's health. However, this was later proved to be incorrect.

Marijuana was also discovered to be a gateway drug to other more harmful substances such as heroine. Cannabis has an adverse impact on the nervous system and reacts with cannabinoid receptors. Under intoxication, individuals often display impaired cognitive activity and a loss in normal mental function. Other systems are also affected by intoxication and individuals usually have an increased heart rate exhibit mood swings and behavior changes that are dependent on the dose of the drug taken by the person. Although these manifestations have been observed in individuals who are under the influence of marijuana, more research is required to establish withdrawal symptoms in detail.[50]

As mentioned, marijuana has been established as an addictive drug, and it is listed as such in the American Psychological Association's *DSM IV*. Cannabis is introduced into the system through several routes, which include

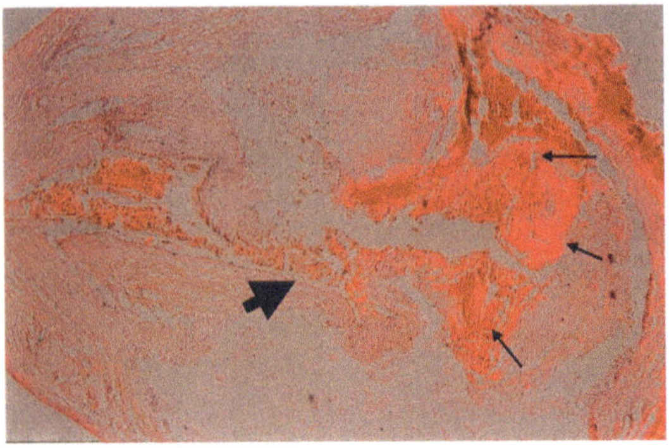

FIGURE 19.7 **Ruptured atheroma (*thick arrow*) with massive hemorrhage dissecting the wall (*thin arrows*).**

sublingual, inhaled, oral, and transdermal. In humans, the effects on the system and therefore the outcomes are dose-dependent; however, it is often accompanied by an increase in arterial pressure. Several researchers have carried out studies and it has been established that individuals develop complete tolerance to all the symptoms of marijuana with prolonged use. In addition, with time, the effects on the cardiac function are reduced. Marijuana abuse leads to the development of acute coronary syndromes and plaques cause minor obstructions in the arteries (Figure 19.7). The plaques lead to the development of lipid-rich lesions that can be viewed in the postmortem assessment of the individuals. Smoking marijuana, cigarettes, and high lipoprotein concentration in the blood all lead to a disruption in the hemodynamic balances and result in the development of infarction.[51–54]

The adverse effects of marijuana abuse are poorly understood and there is a need for further research. In addition, there has been debate over the effects of marijuana compared to nicotine and other substances in cigarettes. However, both have been found to predispose an individual to cardiovascular diseases.

Even though marijuana increases the risk of individuals developing cardiovascular diseases, the incidence of myocardial infarction among these individuals is low. In addition, individuals rapidly develop a tolerance to CVD-related symptoms; thus, this should be considered when determining the cause of death. Cannabinoids have been found to be the major cause of cardiovascular symptoms as they interact with centers of the circulatory system in the central nervous system (CNS). They also interact with receptors that directly influence vascular muscle tone.

Deaths have been reported among marijuana smokers with preexisting CVDs; however, it is difficult for pathologists to base their conclusions on these results as they lack depth in research. Other effects that have been associated with the CNS include the inhibition of the sympathetic system, which leads to hypotension and bradycardia. In summation, the autopsy findings from marijuana abuse are unspecific and several other factors have to be considered. Atheroma and myocardial infarction are the most common findings.[55]

19.4.2 Synthetic Cannabinoid

Synthetic cannabinoids, because of their psychoactive properties, are considered recreational drugs of abuse. Recently, many new cannabis-like compounds have been synthesized, such as those dubbed "Spice" and K2, as alternatives to marijuana. These synthetic cannabinoid compounds are introducing a new challenge for forensic scientists and toxicologists because many of them will undergo detection at routine screenings for drugs.[56]

Most synthetic cannabinoid compounds are receptor agonists and mainly act as potent agonists on cannabinoid receptor type 1 (CB_1) and hence produce cannabis-like effects even after low doses. In addition to cannabis-like toxicity, these compounds are often associated with stimulant effects such as tachycardia, hypertension, agitation, convulsions, and hypokalemia.[56]

Although pharmacokinetic and the pharmacodynamic properties of most synthetic cannabinoids still remain to be clarified in detail, most of them have longer half-lives leading to prolonged toxic effects. Unlike Δ9-tetrahydrocannabinol (THC) metabolites, synthetic cannabinoid metabolites are often pharmacologically active and may act as agonists, neutral antagonists, or inverse agonists at the cannabinoid 1 receptor.[57,58]

The clinical presentation of patients of synthetic cannabinoid use is variable. Usually they present psychoactive and physical effects. Psychoactive effects include euphoria, anxiety, psychosis, cognitive disturbances, seizures, and CNS depression.[59–62] Physical effects vary in severity from sympathomimetic-like symptoms (e.g., psychomotor agitation, diaphoresis, and palpitations) to emesis, appetite changes, tolerance, withdrawal, and drug dependence.[60,63–66]

Cardiovascular Effects

Absence of confirmed synthetic cannabinoid exposure limits the studies and conclusions regarding synthetic cannabinoid use. There are few reports about its effect on the heart. Clinically, chest pain, tachycardia, and bradycardia have been described.[67]

Hypertension and tachyarrhythmia are also reported to be common in patients exposed to synthetic cannabinoid compounds.[68–72] Recently, cases of acute myocardial infarction (MI) have been reported after use of Spice.[68] Postmortem signs are not specific but due to increased relative risk of MI,[73] infarction can be seen at autopsy.

19.4.3 Alcohol

In the United States, cardiac disorders associated with alcohol are the leading cause of toxic cardiomyopathy. Alcoholic cardiomyopathy (AC) is a type of dilated cardiomyopathy (Figure 19.8). AC has been widely researched and a connection between excessive alcohol consumption and cardiomyopathy has been established.[74–76] In the chronic form of alcohol abuse, the extent of pathological

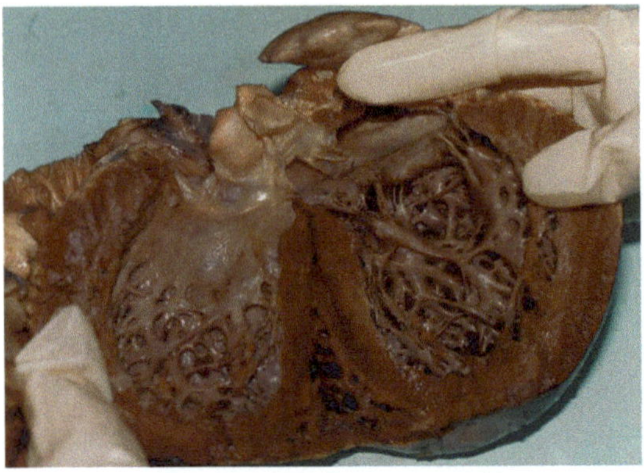

FIGURE 19.8 Dilated cardiomyopathy that shows a markedly dilated left ventricle.

damage has been found to be dose-dependent. AC tends to present contractile abnormalities in the myocardium, cardiomegaly, and alterations in the ventricular structure and hemodynamic properties. Microscopical observations in AC include chronic inflammation and diffuse interstitial fibrosis. There are also atrophic changes in the myocardium, particularly degenerated hypertrophy of myocytes. The pathogenesis of the disease has not yet been established; however, genetics have been found to play a role in the development of the disease.[76–79]

Alcohol also causes peripheral vasodilatation, which in turn leads to ischemia and cardiac myopathy. The microscopic presentations observed in AC are similar to those in anoxia/ischemia.[74,80] Death due to alcohol can occur due to acute toxicity or to a related pathology of chronic abuse. Acute toxicity death is typically considered to be due to depressive effects on the CNS or indirectly due to aspiration of vomit when the cough reflex is inhibited. Postmortem findings may include external signs (e.g., malnutrition, aspiration, acute inflammation of gastric mucosa), liver pathology (e.g., fatty changes, cirrhosis), and described heart pathology.[1,81–86]

19.4.4 Anabolic Androgenic Steroids

Anabolic androgenic steroids (AAS) are derivatives of testosterone and are the most widely abused doping agents. Long-term use of steroids leads to several negative effects on different systems of the body. The use of steroids stimulates protein synthesis which leads to growth in organs. Some of these are also used as therapeutic agents (e.g., to compensate low endogenous production in cases of hypogonadism). In addition, they are used for enhancement of nitrogen balance and to increase erythropoiesis during treatment of bone marrow disorders. The most widely known dark side of these compounds is their use by athletes to enhance performance.[87,88]

In the majority of sporting activities, AAS are illegal. One reason is that sports authorities are concerned about their possible performance-enhancing properties, but also due to the fact that their abuse may lead to illness and even death. Anabolic androgenic steroids have been associated with cardiac failure, arrhythmia, myocardial hypertrophy, and myocardial infarction. The effects increase with duration of usage, and more risk was found in older users. There has been a wide range of histological findings reported among subjects with a record of steroid abuse. Histologically, some subjects may develop left ventricular hypertrophy and postmortems show microscopical changes in the myocardium such as interstitial fibrosis, cardiomyocyte broadening, and nuclear enlargement. Some of these subjects may therefore present with signs of ischemia and myocardial infarction.[89–92]

The hypertrophic changes in the cardiac muscles of individuals who are involved in AAS abuse are attributed to adaptation to individuals' mass and other factors related to steroids. In addition, the hypertrophic changes are attributed to myocytolysis and fibrosis, and this in turn leads to changes that may increase the risk for arrhythmias. These cardiac changes have been discovered in approximately 35% of the patients. Prolonged treatment with AAS also results in an elevated resistance of peripheral vessels; however, the extent of damage is dose-dependent. AAS may further lead to reduced contraction capacity of the heart, resulting to an increase in lactase dehydrogenase release, which in turn will increase cellular injury.[93–95]

The causes of deaths that have been reported among individuals abusing AAS include congestive heart failure, sudden cardiac death, myocardial hypertrophy, areas of cardiomyocyte death, and eosinophilic myocarditis (Figure 19.9). The pathogenesis of these conditions involves a reduction of proteins that bind to insulin-like growth factor (IGF), leading to an increase in growth hormone and other factors. The steroids also bind to corticosteroid receptors and decrease the globulins that bind to sex hormones; this leads to elevated testosterone levels.

In AAS abusers, there is an increase in cholesterol, LDL, TG, and a reduction in HDL. There is also a disturbance in the clotting factors and vascular nitric oxide system that leads to vasospasm. The injury to the myocardium can include fibrosis, edema, and hypertrophy, which result in reduction in contractility. Indirectly, AAS increase the need for O_2, peripheral resistance, and red cell mass. They also increase the risk for developing atherosclerosis. Other changes that have been recorded in AAS users during examination include testicular atrophy and atrophy of the thyroid gland. In addition, skeletal muscle hypertrophy has been discovered along with gynecomastia.[95–97]

19.5 THERAPEUTIC DRUGS

Virtually all therapeutic drugs can produce toxic effects when used above the recommended dose level. Several therapeutic drugs may confer cardiac toxicity; such drugs may affect cardiac preload, afterload, or cardiac contractility. The effects of these drugs depend on preexisting heart conditions as well as on the dose administered.

(a)
(b)
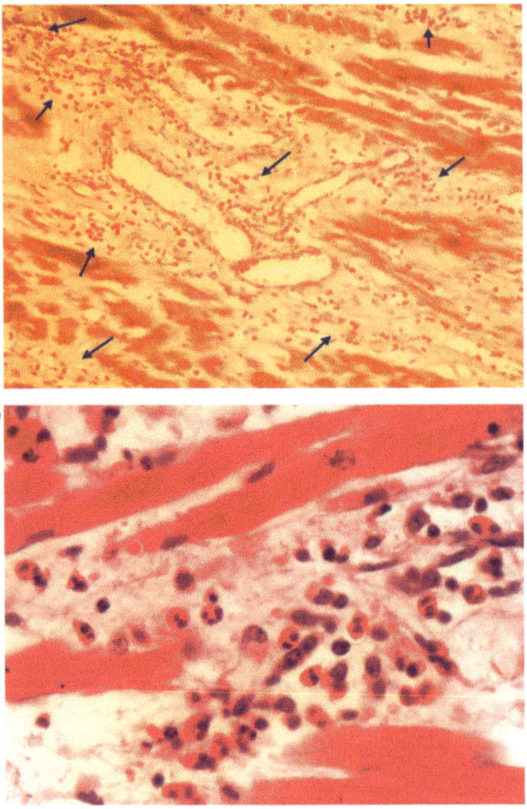

FIGURE 19.9 **Eosinophilic myocarditis with abundant eosinophilic infiltrate.** (a) low power and (b) high power.

19.5.1 Antiarrhythmics

The effects of antiarrhythmic drugs on the heart are cardiodepressant, mainly due to the negative inotropic properties of this type of drug. Antiarrhythmics can induce or aggravate heart failure, particularly in patients with preexisting left ventricular failure.

The degree of negative inotropic effect, which is mediated by intracellular calcium alteration, varies from one drug to the next. It should be noted that class III antiarrhythmics lack these negative inotropic properties. Antiarrhythmics often affect peripheral circulation, which can lead to adverse effects on myocardial contractility. Preexisting cardiac condition is an important factor that may aggravate the effects of antiarrhythmics; the risk of both ventricular and supraventricular arrhythmias increase with decreased left ventricular function and sudden death may result.[98–99]

Death can occur as a result of antiarrhythmic drugs, but it is rare in forensic practice since most patients are under clinical supervision. Postmortem diagnosis depends on circumstances, as well as toxicological analysis, because postmortem findings are often nonspecific.

19.5.2 Beta-Adrenoceptor Antagonists

Beta-adrenoceptor antagonists (beta blockers), such as metoprolol, atenolol, carvedilol, and propranolol, have negative chronotropic and inotropic properties that can easily induce or aggravate heart failure. Systemic effects can occur even after topical use of a beta blocker, although this drug class has a limited effect on the heart when used locally.[100–105]

The clinical manifestations of beta-blocker toxicity may vary depending on the drug's solubility and metabolism. This toxicity can affect the heart and can lead to QRS interval widening.[106] Other clinical features associated with beta-blocker toxicity include bronchospasm, hypoglycemia, hyperkalemia, mental status alteration, and seizures.[107] Again, death can and does occur frequently after intake of a too large dose of beta blockers.

19.5.3 Calcium Channel Blocking Agents

Cardiovascular agents, such as calcium channel blockers, are among the most common cardiovascular drugs prescribed to adults.[108] Various classes of calcium channel blockers are recommended for clinical use: phenylalkylamines (e.g., verapamil), dihydropyridines (e.g., nifedipine), and benzodiazepines (e.g., diltiazem). Negative inotropic and peripheral vasodilatory effects differ from one beta blocker to another; however, in overdose conditions, this selectivity is lost and all will induce severe myocardial depression.[109]

Several pathophysiologic mechanisms that contribute to the potential effects of calcium channel blocking on left ventricular function have been described, such as a negative inotropic effect on the heart as well as the activation of endogenous neurohormonal systems (i.e., renin–angiotensin system) and the sympathetic nervous system.[109–110]

19.5.4 Antidepressant Drugs

Antidepressants are the third leading cause of overdose-related deaths. The most common antidepressant drug classes, in terms of toxicity level, are tricyclic antidepressants (classical antidepressants), selective serotonin reuptake inhibitors (SSRIs), and monoamine oxidase (MAO) inhibitors.[111–113] The main antidepressant toxicity categories include anticholinergic effects, cardiovascular effects, and CNS effects.

The cardiovascular effects of tricyclic antidepressants (TCAs) are well known. Sinus tachycardia occurs in most TCA users; to a lesser extent, patients may complain of postural hypotension. These side effects occur because of a combination of multiple effects characteristic of TCAs. This drug class features a peripheral antiadrenergic action, a myocardial depressant effect, and an alpha-adrenergic blocking effect on the central nervous system.[114]

Another effect of tricyclic antidepressants on the heart is atrioventricular blocking, which can lead to asystole and sudden death. TCAs usually concentrate at high levels in myocardial tissues affecting cardiac rate and rhythm. TCAs' effects on myocardial function are mediated via several mechanisms: anticholinergic action, quinidine-like action, and direct myocardial depression, among others. Fatalities from TCA overdose are primarily caused by cardiac toxicity and tend to occur within 24 hours of ingestion. The cardiac toxicity is often characterized by sinus tachycardia, accompanied by QRS and QTc prolongation associated with ventricular arrhythmias and seizures.[115,116]

From a medicolegal point of view, the postmortem findings in poisonings with these pharmaceutical drugs are nonspecific and to assess whether a person has died from such an intoxication, the pathologist is highly dependent on the toxicological results. These drugs usually can easily be detected and quantified by gas chromatography–mass spectrometry (GC-MS) and liquid chromatography–mass spectrometry (LC-MS) analyses, and a fairly large body of reference information about blood drug concentrations is available. Due to the large propensity for postmortem distribution,[7] it is, however, important to analyze peripheral blood when available.

19.6 CONCLUSION

Forensic pathology related to cardiovascular toxicity deals with the observation and interpretation of macroscopic and microscopic findings in deaths due to drugs causing cardiovascular toxicity, quantitative drug analysis, and substance identification, in addition to an understanding of the mechanisms of death in such cases. Various substances of abuse (e.g., cocaine, 3,4 methylendioxymethylamphetamine, cannabis) and therapeutic agents (e.g., antidepressant drugs, antiarrythmics) were discussed in this chapter. In recent times, forensic pathologists are increasingly confronted with investigating deaths due to substances of abuse and therapeutic agents that pose cardiovascular toxicity. Advances in analytical methodologies have certainly eased the toxicological analyses of postmortem samples, aiding the diagnosis of poisoning. However, postmortem redistribution of substances of abuse and therapeutic agents should be cautiously considered before interpreting toxicological analyses findings in cases of death due to cardiovascular toxicity.

A cautious approach should be considered in order to exclude poisoning or drug overdose as the cause of death whenever there is a case of unexpected sudden death. Thus, it is necessary to obtain a detailed history in all such cases. At autopsy, the nonspecific gross morphological and histopathological findings in the field of forensic pathology related to cardiovascular toxicity should be carefully interpreted before opining on the cause of death.

Key Learning Points

- Forensic pathologists are increasingly confronted with investigating deaths due to substances of abuse and therapeutic drugs, including those with cardiovascular toxicity.
- Postmortem analytical toxicology deals with analyses of samples obtained at autopsy and certainly can prove invaluable in cases of death due to cardiovascular toxicity.
- It is important to have a cautious approach to exclude poisoning or drug overdose as the cause of death whenever there is a case of unexpected sudden death.
- It is important to obtain a detailed history in all cases of death due to cardiovascular toxicity.
- Nonspecific gross morphological autopsy findings are not uncommon in the field of forensic pathology related to cardiovascular toxicity and such findings should be carefully interpreted.
- Postmortem redistribution of substances of abuse and therapeutic drugs should be cautiously considered before interpreting the toxicological analyses findings in cases of death due to cardiovascular toxicity.

REFERENCES

1. Saukko P, Knight B. *Knights forensic pathology*. 3rd ed. London: Hodder Arnold; 2004.
2. Dolinak D, Matshes E, Lew E. *Forensic pathology principle and practice*. Boston: Elsevier; 2005.
3. Hendrickson RG, Osterhoudt KC. Reflections on the 2011 report of the US National Poison Data System. *Clin Toxicol (Phila)* 2012;**50**(10):869–71.
4. Fineschi V, Baroldi G, Silver MD. *Pathology of the heart and sudden death in forensic medicine*. Boca Raton, FL: Taylor & Francis Group; 2006.
5. Dettmeyer RB. *Forensic histopathology, fundamentals and perspectives*. New York: Springer; 2011.
6. Silver LW, Gotlieb AI, Schoen FJ, editors. *Effects of drugs on the cardiovascular system. Cardiovascular pathology*. 3rd ed. New York: Churchill Livingstone; 2011.
7. Pounder DJ, Jones GR. Postmortem redistribution, toxicological nightmare. *Forensic Sci Int* 1990;**45**(3):253–63.
8. Nilsson GH, Kugelberg FC, Kronstrand R, et al. Stability test of zopiclone in whole blood. *Forensic Sci Int* 2010;**200**(1–3):130–5.
9. Leikin JB, Paloucek FP. *Poisoning and toxicology compendium*. 1st ed. Hudson, OH: Lexi-Comp, Inc; 1997.
10. Substance Abuse and Mental, Health Services Administration. *National household survey on drug abuse: main findings*. Rockville, MD: Department of Health and Human Services; 2000.
11. Lange RA, Hillis LD. Cardiovascular complications of cocaine use. *N Engl J Med* 2001;**345**: 351–8.
12. Substance Abuse and Mental Health Services Administration. *National household survey on drug abuse: population estimates 1998*. Rockville, MD: Department of Health and Human Services; 1999.

13. Brody SL, Slovis CM, Wrenn KD. Cocaine-related medical problems: consecutive series of 233 patients. *Am J Med* 1990;**88**:325–31.
14. Goldfrank LR, Hoffman RS. The cardiovascular effects of cocaine. *Ann Emerg Med* 1991;**20**:165–75.
15. Krendel DA, Ditter SM, Frankel MR, et al. Biopsy-proven cerebral vasculitis associated with cocaine abuse. *Neurol* 1990;**40**:1092–4.
16. Niazi M, Kondru A, Levy J, et al. Spectrum of ischemic colitis in cocaine users. *Dig Dis Sci* 1997;**42**:1537–41.
17. Mash DC, Duque L, Pablo J, et al. Brain biomarkers for identifying excited delirium as a cause of sudden death. *Forensic Sci Int* 2009;**190**:e13–9.
18. Mouhaffel AH, Madu EC, Satmary WA, et al. Cardiovascular complications of cocaine. *Chest* 1995;**107**:1426–34.
19. Orr D, Jones I. Anesthesia for laryngoscopy: a comparison of the cardiovascular effects of cocaine and lignocaine. *Anesthesia* 1968;**23**:194–202.
20. Xiao YF, Morgan JP. Cocaine blockade of the acetylcholine-activated muscarinic K^+ channel in ferret cardiac myocytes. *J Pharmacol Exp Ther* 1998;**284**:10–8.
21. Chambers HF, Morris DL, Tauber MG, et al. Cocaine use and the risk for endocarditis in intravenous drug users. *Ann Intern Med* 1987;**106**:833–6.
22. Mao JT, Zhu LX, Sharma S, et al. Cocaine inhibits human endothelial cell IL-8 production: the role of transforming growth factor-beta. *Cell Immunol* 1997;**181**:38–43.
23. Jeong TC, Jordan SD, Matulka RA, et al. Immunosuppression induced by acute exposure to cocaine is dependent on metabolism by cytochrome P-450. *J Pharmacol Exp Ther* 1996; **276**(3):1257–65.
24. Virmani R, Robinowitz M, Smialek JE, et al. Cardiovascular effects of cocaine: an autopsy study of 40 patients. *Am J Med* 1986;**81**:699–701.
25. Kloner RA, Hale S, Alker K, et al. The effects of acute and chronic cocaine use of the heart. *Circulation* 1992;**85**:407–19.
26. Peng SK, French WJ, Pelikan PC. Direct cocaine cardiotoxicity demonstrated by endomyocardial biopsy. *Arch Pathol Lab Med* 1989;**113**:842–5.
27. Isner JM, Chokshi SK. Cardiovascular complications of cocaine. *Curr Probl Cardiol* 1991;**16**:89–123.
28. Chokshi SK, Moore R, Pandian NG, et al. Reversible cardiomyopathy associated with cocaine intoxication. *Ann Intern Med* 1989;**111**:1039–40.
29. Willens HJ, Chakko SC, Kessler KM. Cardiovascular manifestations of cocaine abuse. A case of recurrent dilated cardiomyopathy. *Chest* 1994;**106**:594–600.
30. Isner JM, Estes NA, Thompson PD, et al. Acute cardiac events temporally related to cocaine abuse. *N Engl J Med* 1986;**315**:1438–43.
31. Bauman JL, Grawe JJ, Winecoff AP, et al. Cocaine-related sudden cardiac death: a hypothesis correlating basic science and clinical observations. *J Clin Pharmacol* 1994;**34**:902–11.
32. Nanji AA, Filipenko JD. Asystole and ventricular fibrillation associated with cocaine intoxication. *Chest* 1984;**85**:132–3.
33. Lange RA, Willard JE. The cardiovascular effects of cocaine. *Heart Dis Stroke* 1993; **2**:136–41.
34. Billman GE, Lappi MD. Effects of cocaine on cardiac vagal tone before and during coronary artery occlusion: cocaine exacerbates the autonomic response to myocardial ischemia. *J Cardiovasc Pharmacol* 1993;**22**:869–76.
35. Gradman AH. Cardiac effects of cocaine: a review. *Yale J Biol Med* 1988;**61**:137–47.

36. Tonga G, Tempesta E, Tonga AR, et al. Platelet responsiveness and biosynthesis of thromboxane and prostacyclin in response to in vitro cocaine treatment. *Haemostasis* 1985; **15**:100–7.
37. Kugelmass AD, Shannon RP, Yeo EL, et al. Intravenous cocaine induces platelet activation in the conscious dog. *Circulation* 1995;**91**:1336–40.
38. Heesch CM, Withelm CR, Ristich J, et al. Cocaine activates platelet and increases the formation of circulating platelet containing microaggregates in humans. *Heart* 2000;**83**: 688–95.
39. Erwin MB, Hoyle JR, Smith CH, et al. Cocaine and accelerated atherosclerosis: insights from intravascular ultrasound. *Int J Cardiol* 2004;**93**:301–3.
40. Kolodgie FD, Virmani R, Cornhill JF. Increase in atherosclerosis and adventitial mast cells in cocaine abusers: an alternative mechanism of cocaine-associated coronary vasospasm and thrombosis. *J Am Coll Cardiol* 1991;**17**:1553.
41. Karch SB. *Karch's pathology of drug abuse*. 4th ed. New York: CRC Press; 2008.
42. Rajs J, Falconer B. Cardiac lesions in intravenous drug addicts. *Forensic Sci Int* 1979; **13**(3):193–209.
43. Citron BP, Halpern M, Mccarron M, et al. Necrotizing angiitis associated with drug abuse. *N Engl J Med* 1970;**283**:1003–11.
44. Patel MM, Belson MG, Wright D, et al. Methylenedioxymethamphetamine (ecstasy)-related myocardial hypertrophy: an autopsy study. *Resuscitation* 2005;**66**(2):197–202.
45. Mizia-Stec K, Gasior Z, Wojnicz R, et al. Severe dilated cardiomyopathy as a consequence of Ecstasy intake. *Cardiovasc Pathol* 2008;**17**(4):250–3.
46. Milroy CM, Clark JC, Forrest AR. Pathology of deaths associated with "ecstasy" and "eve" misuse. *J Clin Pathol* 1996;**49**(2):149–53.
47. Afonso L, Mohamad T, Badheka A. *Comprehensive toxicology: drugs of abuse and cardiotoxicity*. Oxford: Elsevier Ltd; 2010.
48. Darke S, Kaye S, Duflou J. Comparative cardiac pathology among deaths due to cocaine toxicity, opioid toxicity and non-drug-related causes. *Addiction* 2006;**101**(12):1771–7.
49. Dettmeyer R, Friedrich K, Schmidt P, et al. Heroin associated damages—conventional and immunohistochemical investigations. *Forensic Sci Int* 2009;**187**(1–3):42–6.
50. Benowitz NL, Jones RT. Cardiovascular effects of prolonged delta-9-tetrahydrocannabinol ingestion. *Clin Pharmacol Ther* 1975;**18**(3):287–97.
51. Servoss SJ, Januzzi JL, Muller JE. Triggers of acute coronary syndromes. *Prog Cardiovasc Dis* 2002;**44**(5):369–80.
52. Khiabani HZ, Mørland J, Bramness JG. Frequency and irregularity of heart rate in drivers suspected of driving under the influence of cannabis. *Eur J Intern Med* 2008;**19**(8): 608–12.
53. Montisci M, Thiene G, Ferrara SD, et al. Cannabis and cocaine: a lethal cocktail triggering coronary sudden death. *Cardiovasc Pathol* 2008;**17**(5):344–6.
54. Verster JC, Brady K, Galanter M, et al. *Drug abuse and addiction in medical illness*. New York: Springer; 2012.
55. Karch SB. Cannabis and Cardiotoxicity. *Forensic Sci Med Pathol* 2006;**2**(1):213–8.
56. Spaderna M, Addy PH, D'Souza DC. Spicing things up: synthetic cannabinoids. *Psychopharmacol* 2013;**228**:525–40.
57. Brents LK, Reichard EE, Zimmerman SM, et al. Phase I hydroxylated metabolites of the K2 synthetic cannabinoid JWH-018 retain in vitro and in vivo cannabinoid-1 receptor affinity and activity. *PLoS One* 2011;**6**:e21917.

58. Brents LK, Gallus-Zawada A, Radominska-Pandya A, et al. Monohydroxylated metabolites of the K2 synthetic cannabinoid JWH-073 retain intermediate to high cannabinoid-1 receptor (CB1R) affinity and exhibit neutral antagonist to partial agonist activity. *Biochem Pharmacol* 2012;**83**:952–61.
59. Auwärter V, Dresen S, Weinmann W, et al. "Spice" and other herbal blends: harmless incense or cannabinoid designer drugs. *J Mass Spectrom* 2009;**44**:832–7.
60. Castellanos D, Singh S, Thornton G, et al. Synthetic cannabinoid use: a case series of adolescents. *J Adolesc Health* 2011;**49**:347–9.
61. Müller H, Sperling W, Köhrmann M, et al. The synthetic cannabinoid Spice as a trigger for an acute exacerbation of cannabis induced recurrent psychotic episodes. *Schizophr Res* 2010;**118**:309–10.
62. Vardakou I, Pistos C, Spiliopoulou CH. Spice drugs as a new trend: mode of action, identification and legislation. *Toxicol Lett* 2010;**197**:157–62.
63. Every-Palmer S. Warning: legal synthetic cannabinoid-receptor agonists such as JWH018 may precipitate psychosis in vulnerable individuals. *Addiction* 2010;**105**:1859–60.
64. Every-Palmer S. Synthetic cannabinoid JWH-018 and psychosis: an explorative study. *Drug Alcohol Depend* 2011;**117**:152–7.
65. Schneir AB, Cullen J, Ly BT. "Spice" girls: synthetic cannabinoid intoxication. *J Emerg Med* 2011;**40**:296–9.
66. Zimmermann US, Winkelmann PR, Pilhatsch M, et al. Withdrawal phenomena and dependence syndrome after the consumption of "SpiceGold." *Dtsch Artzebl Int* 2009;**106**:464–7.
67. Young AC, Schwarz E, Medina G, et al. Cardiotoxicity associated with the synthetic cannabinoid, K9, with laboratory confirmation. *Am J Emerg Med* 2012;**30**(7):1320.e5-7
68. Forrester MB, Kleinschmidt K, Schwarz E, et al. Synthetic cannabinoid exposures reported to Texas poison centers. *J Addict Dis* 2011;**30**:351–8.
69. Simmons J, Cookman L, Kang C, et al. Three cases of "spice" exposure. *ClinToxicol* 2011; **49**:431–3.
70. Simmons JR, Skinner CG, Williams J, et al. Intoxication from smoking "Spice." *Ann Emerg Med* 2011;**57**:187–8.
71. Lapoint J, James LP, Moran CL, et al. Severe toxicity following synthetic cannabinoid ingestion. *Clin Toxicol* 2011;**49**:760–4.
72. Hoyte CO, Jacob J, Monte AA, et al. A characterization of synthetic cannabinoid exposures reported to the National Poison Data System in 2010. *Ann Emerg Med* 2012;**60**:435–8.
73. Mir A, Obafemi A, Young A, et al. Myocardial infarction associated with use of the synthetic cannabinoid K2. *Pediatrics* 2011;**128**:1622–7.
74. Capasso JM, Li P, Guideri G, et al. Left ventricular dysfunction induced by chronic alcoholic ingestion in rats. *Am J Physiol* 1991;**261**:H212–9.
75. Fernandez-Sola J, Fatjo F, Sacanella E, et al. Evidence of apoptosis in alcoholic cardiomyopathy. *Hum Pathol* 2006;**37**:1100–10.
76. Fernandez-Sola J, Nicolas JM, Oriola J, et al. Angiotensin converting enzyme gene polymorphism is associated with vulnerability to alcoholic cardiomyopathy. *Ann Intern Med* 2002;**137**:321–6.
77. Frishman WH. Alcohol as a cardio-protective agent. *Heart Dis* 2002;**4**:273–5.
78. Guarnieri T, Lakatta EG. Mechanism of myocardial contractile depression by clinical concentration of ethanol, a study in ferret papillary muscles. *J Clin Invest* 1990;**85**:1462–7.
79. Lazarevic AM, Nakatani S, Neskovic AN, et al. Early changes of left ventricular function in chronic asymptomatic alcoholics: relation to the duration of heavy drinking. *Am Coll Cardiol* 2000;**35**:1599–606.

80. Moriya N, Nihira M, Sato S, et al. Ultra structural changes of liver, heart, lung and kidney of mice in a large dose of ethanol injection. *Arukoru Kenkyuto Yakubutsu Ison* 1992;**27**: 189–200.
81. Piano MR, Schwertz DW. Alcoholic heart disease a review. *Heart Lung* 1994;**23**:3–20.
82. Regan TJ, Ettinger PO, Haider B, et al. The role of ethanol in cardiac disease. *Annu Rev Med* 1977;**28**:393–409.
83. Rimm EB, Klatsky A, Grobbee D, et al. Review of moderate alcohol consumption and reduced risk of coronary heart disease is the effect due to beer, wine, or spirits. *BMJ* 1996; **312**:731–6.
84. Rubin E, Urbano-Marquez A. Alcoholic cardiomyopathy. *Alcohol Clin Exp Res* 1994; **18**:111–4.
85. Urbano-Marquez A, Estruch R, Fernandez-Sola J, et al. The great risk of alcohol cardiomyopathy and myopathy in women compared with men. *JAMA* 1995;**274**:149–54.
86. Urbano-Marquez A, Estruch R, Navarro-Lopez F, et al. The effects of alcoholism on skeletal and cardiac muscles. *N Engl J Med* 1989;**320**:409–15.
87. Ramo P. Anabolic steroids alter the hemodynamic response of the canine left ventricle. *Acta Physiol Scand* 1987;**130**:209–17.
88. Karhunen MK, Ramo MP, Kettunen R. Anabolic steroids and the hemodynamic effects of endurance training and deconditioning in rats. *Acta Physiol Scand* 1988;**133**:297.
89. Melchert RB, Herron TJ, Welder AA. The effect of abolicandrogenic steroids on primary myocardial cell cultures. *Med Sci Sports Exerc* 1992;**24**:206–12.
90. Baggish AL, Weiner RB, Pope HG, et al. Long term anabolic-androgenic steroid use is associated with left ventricular dysfunction. *Circ Heart Fail* 2010;**3**(4):472–6.
91. Thiblin I, Lindquist O, Rajs J. Cause and manner of death among users of anabolic androgenic steroids. *J Forensic Sci* 2000;**45**:16–23.
92. Rothman RD, Weiner RB, Pope H, et al. Anabolic androgenic steroid induced myocardial toxicity: an evolving problem in an ageing population. *BMJ Case Rep* 2011;2011:ii.
93. Far HRM, Ågren G, Thiblin I. Cardiac hypertrophy in deceased users of anabolic androgenic steroids: an investigation of autopsy findings. *Cardiovasc Pathol* 2012;**21**(4): 312–6.
94. Fanton L, Belhani D, Vaillant F, et al. Heart lesions associated with anabolic steroid abuse: comparison of post-mortem findings in athletes and norethandrolone-induced lesions in rabbits. *Exp Toxicol Pathol* 2009;**61**(4):317–23.
95. Nieminen MS, Rämö MP, Viitasalo M, et al. Serious cardiovascular side effects of large doses of anabolic steroids in weight lifters. *Eur Heart J* 1996;**17**(10):1576–83.
96. Brennan BP, Kanayama G, Hudson JI, et al. Human growth hormone abuse in male weightlifters. *Am J Addict* 2010;**20**(1):9–13.
97. Fineschi V, Baroldi G, Monciotti F, et al. Anabolic steroid abuse and cardiac sudden death: a pathologic study. *Arch Pathol Lab Med* 2001;**125**(2):253–5.
98. Schlepper M. Cardiodepressive effects of antiarrhythmic drugs. *Eur Heart J* 1989; **10**(Suppl. E):73–80.
99. Hammermeister KE. Adverse hemodynamic effects of antiarrhythmic drugs in congestive heart failure. *Circulation* 1990;**81**:1151–3.
100. Ball S. Congestive heart failure from betaxolol: Case report. *Arch Ophthalmol* 1987; **105**:320.
101. Britman NA. Cardiac effects of topical timolol (letter). *N Engl J Med* 1979;**300**:566.
102. Atkins JM, Pugh Jr BR, Timewell RM. Cardiovascular effects of topical beta-blockers during exercise. *Am J Ophthalmol* 1985;**99**:173–5.

103. McMahon CD, Shaffer RN, Hoskins Jr HD, et al. Adverse effects experienced by patients taking timolol. *Am J Ophthalmol* 1979;**88**:736–8.
104. Monane M, Bohn RL, Gurwitz JH, et al. Topical glaucoma medications and cardiovascular risk in the elderly. *Clin Pharmacol Ther* 1994;**55**:76–83.
105. Lam SW, Engebretsen KM, Bauer SR. Toxicology today: what you need to know now. *J Pharm Pract* 2011;**24**:174.
106. Love BJN, Howell JM, Litovitz TL, et al. Acute beta blocker overdose: factors associated with the development of cardiovascular morbidity. *J Toxicol Clin Toxicol* 2000;**38**(3): 275–81.
107. Reith DM, Dawson AH, Epid D, et al. Relative toxicity of beta blockers in overdose. *J Toxicol Clin Toxicol* 1996;**34**(3):273–8.
108. Ramoska EA, Spiller HA, Myers A. Calcium channel blocker toxicity. *Ann Emerg Med* 1990;**19**(6):649–53.
109. Packer M. Pathophysiological mechanisms underlying the adverse effects of calcium channel-blocking drugs in patients with chronic heart failure. *Circulation* 1989;**80**(Suppl. 6): IV59–67.
110. Bronstein AC, Spyker DA, Cantilena Jr LR, et al. Annual Report of the American Association of Poison Control Centers' National Poison Data System (NPDS): 26th Annual Report. *Clin Toxicol* 2009;**47**(10):911–1084.
111. Sheline YI, Freedland KE, Carney RM. How safe are serotonin reuptake inhibitors for depression in patients with coronary heart disease? *Am J Med* 1997;**102**:54–9.
112. Glassman AH, Bigger Jr JT, Giardina EV, et al. Clinical characteristics of imipramine-induced orthostatic hypotension. *Lancet* 1979;**1**:468–72.
113. Roos JC. Cardiac effects of antidepressant drugs. A comparison of the tricyclic antidepressants and fluvoxamine. *Br J Clin Pharmacol* 1983;**15**(Suppl. 3):439S–445SS.
114. Dukes MNG. *Meyler's side effects of drugs*. 14th ed. Amsterdam: Elsevier Science; 1996.
115. Callaham M, Kassel D. Epidemiology of fatal tricyclic antidepressant ingestion: implications for management. *Ann Emerg Med* 1985;**14**(1):1–9.
116. Boehnert MT, Lovejoy Jr FH. Value of the QRS duration versus the serum drug level in predicting seizures and ventricular arrhythmias after an acute overdose of tricyclic antidepressants. *N Engl J Med* 1985;**313**(8):474–9.

Chapter 20

Ethics, Legality, and Education in the Practice of Cardiology

Ponniah Thirumalaikolundusubramanian,[1] Ramachandran Meenakshisundaram,[2] and Subramanian Senthilkumaran[3]

[1]*Chennai Medical College Hospital and Research Centre, Irungalur, Trichy, India*
[2]*Epsom and St. Helier University Hospitals NHS Trust, Epsom, United Kingdom*
[3]*Sri Gokulam Hospitals and Research Centre, Salem, India*

20.1 INTRODUCTION

Cardiology is a branch of medicine characterized by state-of-the-art biomedical technology. Evidence-based medicine is incorporated in the practice of cardiology. Ethical dilemmas known to the specialty arise from the decisions to be made between what can be done and what should be done for varieties of cardiovascular pathologies. Some of the ethical deliberations are resurrection of a patient with acute coronary syndrome or malignant arrhythmias.[1] The practice of cardiology is increasingly constrained by guidelines, regulations, and legal considerations.[2] All doctors, including cardiologists, have a primary duty of care to individual patients, in addition to their responsibilities to society, institutions, and colleagues. Among cardiologists, the propensity to test and treat is closely associated with fear of malpractice suits.[3]

Cardiology practice has gained importance as a result of the introduction of advanced diagnostics, drugs, and devices. Early recognition and appropriate intervention minimizes the consequences of cardiac illnesses, which in turn reduces the cardiovascular mortality and morbidity. At the same time, awareness about cardiac care among community members has increased their expectations. When the expectations are not achieved, legal issues creep in. In an analysis of 1899 judicial sentences related to cardiology in Spain, 31% of the decisions were made for plaintiffs and the defects were misdiagnosis including gaps in communication and informed consent.[4,5]

Misuse and unethical practices related to state health insurance in India was reported by a weekly magazine,[6] and it urged medical associations to curtail such violations. Factors linked to malpractice claims are patient dissatisfaction,

The Heart and Toxins. DOI: http://dx.doi.org/10.1016/B978-0-12-416595-3.00023-2

lack of physician communication, and poor interpersonal skills. The trial attorneys believe that the threat of litigation makes doctors discharge their services more safely.[7] A deep-seated tension exists between the malpractice system and the goals and initiatives of patient safety.[7] Interestingly, improvements in patient safety have been accompanied by an increase in claims of professional medical liability.[4] The number of medical negligence cases against doctors, hospitals, and nursing homes is increasing day by day, globally.

Ethical analysis in general is accused of being abstract and abstruse, as practitioners face difficulties in making decisions while handling cardiac emergencies. At the same time, analysis of the moral concepts of clinical problems, issues, challenges, and arguments among professionals based on scientific evidence help to move the field forward and/or provide clarity. Doctors face ethical complexities among the cases handled by them, and while most of the issues are very similar, they receive less attention or go unnoticed, are not recorded, and are rarely documented. Because ethics pervades all areas of practice in one or other way, it has to be considered, realized, studied, and practiced.

Medicine, law, and social values are dynamic. New additions are constantly incorporated into health laws because of new issues and technological advances. The American College of Physicians *Ethics Manual* covers emerging issues in medical ethics[8] and reiterates ethical principles. Moral principles form the basis for ethics and are incorporated into the ethical rules of different professions. Historically, the practice of medicine was subject to legal restrictions in ancient Egypt. Also the right to practice was restricted to certain members. Doctors had to learn and follow certain percepts laid down by their predecessors. All these were practiced probably with an intension to protect the public from quackery.

Although many professions deal with people who are in vulnerable states and have access to the inner lives of other human beings through service or trust, the impact of medical service on individuals is one reason why medicine and physicians exist in the first place.[9] The main goals of medical ethics are to improve the standard of care by identifying, analyzing, and attempting to resolve ethical problems that arise in medical practice.[10]

Organizations that lack adequate role-modeling have resulted in a lack of ethical concern for patients, inadequate monitoring of institutional ethics, and a lack of improvement in patient care resulting in deteriorating outcomes of health care.[10] This chapter discusses ethical and legal aspects based on generally accepted principles, including cardiology-related emerging issues for developing appropriate patient care. In addition, the educational aspects related to law and ethics are briefly highlighted.

20.2 HEALTH CARE DELIVERY

Health care delivery has to be tailored with regard to recipients' status such as attitude, knowledge, beliefs, understanding, social system position, resources, culture, views, religious sentiments, and so on. Thus, the influencing factors for delivery of services are status and the channel adopted by the health care team. The quality of care and the quality of caring are integrated with each other in many ways.

Medicine has to be practiced by knowledgeable, competent, and compassionate physicians who have high-quality therapeutic and healing relationships with patients and their families or caregivers in the setting of a safe and efficient health care system.[11] With good documentation, legal issues are less likely to occur. Analyzing the existing status of health care delivery, the Supreme Court of India has stated that doctors, hospitals, nursing homes, and other connected health care establishments are to be dealt with strictly if found to be negligent. Patients, irrespective of their social, cultural, and economic background, are entitled to be handled with dignity.

20.3 PHYSICIAN BEHAVIOR

Seven ideal physician behaviors were identified by Bendapudi et al.:[12] physicians should be confident, empathetic, humane, personal, forthright, respectful, and thorough. These seven ideal behaviors reflect deep-seated character traits that the physician should have, as well as more superficial behaviors or skills.[11] Physicians should remember that the opposite of desired physician behavior—for example, poor service, arrogance in dismissing patients' input, disinterest in the patient, impatience in answering a patient's questions, and callousness in discussing the patient's prognosis—are considered negatively by patients.[12]

Pellegrino's virtues of a good physician are benevolence, compassion, courage, fidelity to trust, intellectual honesty, prudence, and truthfulness.[13] All of these behaviors are essential for delivering appropriate, effective, efficient, equitable, patient-centered, safe, and easily accessible health care.[14] Ayurveda, one of the ancient systems of medicine in India, insists on medical ethics; doctor–patient relationships; and the qualities of teacher, physician, and student. The cornerstones of medical ethics in Ayurveda medicine are compassion, integrity, respect, honesty, courage, and conscientiousness.[15] Ayurveda's paternal beneficence is different from modern medical ethics.

Individual factors that impair the decision-making process and their failure to maintain technical competency, such as health status, stress, alcohol, and drugs, may be seen in some physicians. A number of people in India believe that doctors are negligent, not courteous enough, not easily accessible, and prescribe unnecessary investigations and drugs;[16] they attribute medical negligence to lack of knowledge on the part of doctors. Despite this, many cases go unnoticed and very few come to consumer courts. Doctors have long been portrayed as wise and dedicated healers; however, this positive image has been in decline since 1970. Recent movies have portrayed doctors as masters, commanders, greedy, egotistical, uncaring, and unethical.[17,18] Complaint letters about physicians often relate to their inappropriate attitude, lack of satisfaction, noncompliance with patients' demands, incorrect diagnosis, complication(s) related to procedures, medical malpractice, and so forth. Even though many physicians recognize impairment and incompetency among coprofessionals, they generally do not bring them to the attention of the respective authorities or report them.[19]

20.4 MEDICAL ETHICS

Medical ethics is a dynamic field that plays an important role in cardiovascular medicine. Codes of ethics, advanced directives, informed consent, privacy rules, and disclosure of conflicts of interest have changed the practice of cardiology.[20] Ethical issues are concerned with the ideas of right or wrong, duties or obligations, and rights or responsibilities. There are always gray areas in ethics. Ethical norms derived from various laws and federal and state constitutions (fundamental laws of nations) are related to citizens' welfare, safety, and security; professional councils and statutory organizations that deal with practitioners, policies of professional organizations, professional standards of care, fiduciary obligations, institutional policies, and judgments delivered; and public health regulations and other laws related to patient care and the hospital environment.

20.4.1 Principles of Health Care Ethics

The fundamental principles and the expectations for ethical conduct are:

- Beneficence
- Fidelity
- Nonmaleficence
- Autonomy
- Justice
- Social interest

Codes of ethics provide a wealth of information on a variety of aspects such as virtue-based ethical decision making, professional responsibilities, counseling relationships, consulting, private practice, evaluation and assessment, research and publications, counselor education, training, and supervision. Keith-Spiegel and Tabachnick offer eight general ethical principles: respect for autonomy of others, doing no harm, benefit to others, fairness and equity, fidelity and honesty, dignity, caring, and doing one's best.[21]

Trust

Patients and the public trust doctors and health care systems, despite limitations. Trust has various dimensions and determinants.[22] Developments in sociophysical scenarios, market-based diagnostics, drugs and devices, economic crises, and the needs of providers have had an affect on systems and services. In addition to blind trust, Gopichandran described four more types of trust:[23] calculated trust, trust but with verification, skeptical trust, and impersonal trust (with some overlapping features among them).

Communication

Information given to patients and/or caregivers should have certainty, clarity, and eliminate divergent meaning; whereas Komesaroff suggested that clinical communication often requires the deliberate preservation of uncertainty.[24]

Doctor–Patient Relationship

The doctor–patient relationship is the cornerstone in health care. This helps with diagnosis, therapy, prevention, and compliance, as well as improvement of psychosocial quality of life and patient satisfaction.[25] Medicolegal duties are essential components to maintain a physician–patient relationship.

20.4.2 Ethics Related to Medical Tourism

Web-based medical tourism facilitators are the connectors between foreign patients and host countries that identify concerns regarding the information displayed.[26] Patients or caregivers are attracted by price comparisons, perceived quality of care, additional support services, other social websites, and patient blogs. These sites do not mention patients' rights.

E-health is a multidisciplinary field. The facilitators of web-based services have to be educated about the ethical aspects of patient care, be prevented from exploiting patients or clients, and be required to respect privacy and confidentiality.[26] International organizations, such as the World Health Organization (WHO) and the United Nations Educational, Scientific and Cultural Organization (UNESCO), have formulated standard ethical guidelines for the quality of health information available on the Internet.[27]

The various organizations need to focus on four basic principles of e-health information: self-regulation, evaluation of information, regulation, and awareness of users.[28] Facilitators of medical tourism have to ensure confidentiality; disclose health regulations, ethical guidelines, and laws related to compensation; provide information on potential risks of travel, facility stays, and procedures; and divulge the correct details of hospitals and providers.

20.4.3 Ethics of Ubiquitous Health Care

Modern medicine is supported by advanced technologies and devices used for the purposes of diagnosis, prevention, monitoring, alleviation of symptoms, and therapy. Each hospital has a plethora of machines. Ubiquitous health care is an emerging field of technology. It uses a variety of environmental and patient sensors and actuators to monitor and improve patients' physical and mental conditions.[29] Such machine-dependent health care poses challenging ethical questions related to trust, efficacy, and societal issues; longevity gaps related to economic status;[29] and clinical problems related to failure of medical devices have led to mounting concerns in the regulatory processes.[30–32] Medical, technical, and computer professionals have to collaborate to develop and deploy ubiquitous health care systems.

20.4.4 Consultation Ethics

Geppert and Shelton have described various aspects of consultation. It is a service and the professionals who provide consultation should follow certain principles and norms.[33]

20.4.5 Wrongdoing in Medicine

DuBois et al. identified 10 environmental factors that enable wrongdoing in medical practice and research.[34] To avoid wrongdoing, education and policy of ethics have to be strengthened.

20.4.6 Ethical Issues of Managing End-Stage Heart Failure

Health care providers face ethical, moral, psychological, and medicolegal challenges while providing care for end-stage heart failure. Although cardiac transplantation may have limitations, left ventricular assist devices are becoming a more prominent therapeutic option for some. While dealing with such cases, physicians find themselves in a peculiar situation of having a professional obligation to perform whatever is the best for the patient, but doing it in the context of shared decision making.[35] At the same time, health care providers must refrain from doing anything that might conflict with the law or raise legal concerns.

20.4.7 Ethics Related to Futile Medical Treatment

In some clinical cases, discussion of medical futility reflects an apparent conflict between respect for patient autonomy and the preservation of ethical integrity of the medical profession.[36] The danger of futility-based exception to informed consent includes the emergence of paternalism and physicians' avoidance of difficult discussions with patients about dying and death. Hariharan et al. have described indifference toward ethical and legal issues among doctors and nurses, and the need for appropriate training.[37]

20.4.8 Integration of Complementary Therapy

Integrating the use of complementary therapies in medical practice increases the physicians' risk for license suspension or revocation.[38,39] It is suggested that physicians should continue to provide conventional medical care.

20.4.9 Medical Errors

Medical errors are bound to happen. Even though they may be known, errors often do not come to light or go undocumented because of the inherent threatening attitude of the existing health care system and the possibility of self-harm. Moreover, medical errors are not discussed as a part of medical education. In most situations, nurses and paramedical professionals do not have the liberty to write medical errors in their reports and are not trained to document such incidents because they are considered suboptimal. As a result, prevalence cannot be estimated, occurrence cannot be prevented, and root caused cannot be assessed. The complexity of the current health care system contributes to various types of medication errors, leading to legal liability. Reducing such errors is an ongoing process of quality improvement, which is an essential component of clinical audits. In some cases, root cause analysis and redesigning faulty systems have reduced medication errors.[40]

20.4.10 Ethical Practice in Laboratory Medicine

Medical laboratories have responsibilities to three main groups: patients, professional colleagues, and society. They are required to maintain high standards in sample collection, good laboratory practices, reporting, monitoring, maintenance of records, and confidentiality.

20.4.11 Ethics Related to Treatment

While making decisions about resuscitation, health care providers should be guided by science; individual patient or surrogate preferences; local policy and legal requirements with reference to respect for autonomy, advanced directives, living wills, patient self-determination, do not attempt resuscitation (DNAR) orders, withholding and withdrawing cardiopulmonary resuscitation (CPR), and terminating resuscitation efforts. All these should include providing emotional support to family, limitations of care and withdrawal of life-sustaining therapies, ethics of organ donation, and privacy.

20.4.12 Ethical Issues in Cardiac Surgery

Deliberations about ethical issues among cardiothoracic surgeons have gained prominence in recent years and is starting to become part of medical schools' curriculum. A recent review described various ethical issues related to cardiac surgery in a detailed manner.[41]

20.4.13 Medically Acceptable Wait Time

Wait time for access to a cardiovascular specialist's evaluation and therapy has an impact on a vulnerable population in many ways including lawsuits. Shortage of cardiologists and demands from the community have been discussed by professional societies, universities, and health ministries.[42]

20.4.14 Ethical Issues in Fetal Management

If a fetus is found to have a cardiac anomaly, the ethical principle of autonomy creates a responsibility for the physician to help the pregnant woman make an informed decision based on her values and aspirations. A decision to terminate is partly a medical matter and partly a personal decision bound by legal, cultural, and religious constraints.[43]

20.4.15 Do No (Financial) Harm

"Do no harm" is an established mantra of medical ethics and the profession. In the modern era, medical bills are escalating and are the leading cause of financial harm.[44] Physicians, while providing patient-centered care, should also consider patients' financial well-being.[45] They should understand financial ramifications, provide appropriate care, and optimize care plans for individual patients considering their socioeconomic status.

20.4.16 Shift from Individual Autonomy

In order to achieve patient-centered care, family members and/or caregivers need unprecedented education and assistance. This will entail a fundamental shift from individual autonomy to family- and caregiver-centered care.[46]

20.4.17 Ward Rounds

Ward rounds provide a benefit to patients that is safe, effective, and efficient. In addition, rounds can lead to staff and patient satisfaction.[47] When various levels of health care providers see and talk to patients, they gain confidence and are more satisfied. In addition, ward rounds involve teamwork, where senior and junior doctors, medical students, nurses, and paramedical professionals share patient information and gain knowledge.

20.4.18 Ethics Related to Health and the Media

Standard ethical guidelines need to be practiced while disseminating information about health, disease, institutions, and physicians to the media. The media (i.e., print, television, Internet) has become a powerful source for sharing information and knowledge.[48] Issues related to ethics and the media, including showing patients in news stories or taking pictures of them, have to be evaluated, and students of health sciences and practitioners/physicians should be aware of these issues. The importance of the media was revealed during the epidemics of severe acute respiratory syndrome (SARS) and the H1N1 influenza infection. It has been suggested that the media become a partner in health care delivery to carry messages quickly to the population, to enforce prevention or therapeutic strategies,[49] and to protect communities from hazards.[50]

20.4.19 Ethics for Medical Education

The medical profession is considered a noble profession. Ethics is important for the correct conduct of the professionals and practitioners. There are two levels in a teacher's ethics: legal and/or administrative and personal. Medical educators have various roles including teacher, manager, administrator, researcher, and physician, and all roles are governed by and require a code of ethics.[51]

20.4.20 Involving Patients in Medical Education

In clinical teaching, patients are brought to a consultation room or students are taken to the patient for bedside instruction. In developing countries, many times ethical principles are violated or not followed when patients are involved in medical education.[52] Principles of ethics have to be discussed at every stage of clinical teaching.

20.4.21 Research and Publication Ethics

Like the fact that human beings need oxygen to breathe, science needs research to live and progress. Thus, research is an essential component in the science. Publications help exchange and share knowledge for the betterment and advancement of medicine. Therefore, authors are expected to follow standard ethical guidelines when submitting research for publication.[53]

20.4.22 Ethics Related to Patients with Heart Disease

Hall et al.[54] have noted that cardiac patients should be assessed by cardiologists as a matter of course. The authors suggested that health care providers listen to patients' various needs; provide access to information technology (IT) for audit, clinical governance, and continuous professional development (CPD); and collect valuable outcome data. The authors, while discussing human rights, stated that cardiac cases should not be delayed when starting treatment or denied ethics for want of resources because of an unpredictable response from the illness.[54]

20.4.23 Defensive Medicine and Consumer-Driven Health Care

Defensive medicine is defined as the ordering of tests, procedures, and visits and/or referrals, or avoidance of high-risk patients or procedures, primarily to reduce malpractice liability. A study from Israel revealed that defensive medicine is prevalent, especially among surgical specialties, and more so among those exposed to lawsuits.[55] Studies have revealed that defensive medicine is a well-known practice in industrialized countries.

Emergency medicine, general surgery, neurosurgery, obstetrics, gynecology, orthopedic surgery, and radiology specialists were common in defensive medicine and were affected more by liability cases.[56] The influencing factors for defensive medical practice are increasing demands of the patients, lack of resources, work and patient loads, deviation from standard guidelines, and increasing lawsuits and compensation amounts. Measures to reduce lawsuits and complaints are improving the quality of care, clinical audits, anonymous disclosure of medical mistakes, resources, practicing standard guidelines, lawsuit immunity, and no-fault compensation.

Current public reports compare health care providers in terms of quality or cost to help consumers, who in turn decide where and from whom to seek care.[57] In view of the changing trends of consumers, health care providers need to become accustomed to patient empowerment and meet their expectations. The public expects standard guidelines for treatment; however, they too have limitations,[58] and conflict of interest (CoI) is prevalent in the ones related to cardiology.[59]

20.5 LEGAL ISSUES

Sophistication of medical practice, increasing costs of health care, involvement of insurance systems, increased awareness and high expectations from the community, and increased participation of media have enhanced society's attitude toward claiming compensation and acquisition of health care delivery. This section outlines the various ways by which hospitals and health care providers can minimize their exposure to legal liability or consequences.

20.5.1 Sources of Legal Obligations

There are numerous sources for obligation in emergencies and health care providers should identify legal issues and adopt a risk management approach,[60] including characterizing the hazards, establishing a community profile, determining vulnerability, analyzing risk, and identifying and evaluating the management plan. An alternative way to avoid legal liability is to use the PPRR (prevention, preparedness, response, and recovery) framework.[60]

20.5.2 Law and the Legal System

Laws include federal and state constitutions, statutes, and administrative and judicial decisions, which may vary from country to country and/or state to state; thus, a legal cause of action may be created by a constitutional right, by a statute, or by a common law. A cause of action consists of several distinct elements; however, the following must exist before there is legal liability:

- Plaintiff must have an interest that is protected by law
- There must be a legal duty
- A breach of duty by the defendant must be provable
- An injury must be shown as damage to the protected party
- It must be proved that the breach of a law caused injury

20.5.3 Negligence

Negligence is the most common allegation (cause of action) in medical practice cases. It is legally defined as the omission of something that a reasonable person, guided by ordinary considerations of someone who regulates human affairs, would do; or the doing of something that a reasonable and prudent person would not do. Therefore, negligence is a violation of the duty to use care. It arises when injury results from the failure of the wrongdoer ("tortfeasor") to exercise due care. The four elements required to establish a prima facie case for negligence are duty, breach, causation, and compensable injury.

20.5.4 Malpractice

Malpractice is defined as a tort or civil wrong committed by a professional acting in his or her professional capacity. The term *malpractice* refers to any misconduct that encompasses an unreasonable lack of skill or unfaithfulness in carrying out professional or fiduciary duties.[61] Malpractice law is a part of tort or personal injury law. The three goals of malpractice litigation are to deter unsafe practices, to compensate persons injured through negligence, and to exact corrective justice.

Plaintiffs' legal theories or causes of action against a physician are ethical negligence, wrongful death, loss of a chance of recovery or survival, battery and assault, lack of informed consent, abandonment, breach of privacy and confidentiality, product liability because of drugs and medical devices, vicarious responsibility for the acts of others, negligent referral, false imprisonment, defamation, failure to warn or control, negligent infliction of emotional distress, outrage, failure to report, fraud or misrepresentation (deceit), and loss of consortium.[61]

20.5.5 Competency

The issue of determining patient competence for medical decision making is often difficult. No absolute definition exists.[62] Competency is essentially the ability of a person to make a decision, and it is where understanding, appreciation, nationality, and religion begin.[62]

20.5.6 Duties of Prescribers

The prescriber exercises clinical skill in diagnosing a condition and determines which drug is indicated and in what dose. Also, the prescriber has to provide sufficient information to the patient so as to make an informed decision whether or not to take the drug in the light of any possible adverse effects or risks that may be associated with the choice.[63] Illegible prescriptions have been a source of legal liability for both pharmacists and prescribers. Therefore, prescribers should respond to queries from pharmacists to avoid the possiblity of injury to the patient and expensive litigation for all concerned. Patients may be motivated to follow the directions on product information (PI). A consumers' movement view is that an off-label prescription indicates it as experimental; under such circumstances, the prescriber has to provide complete information to obtain the patients valid informed consent.[64]

Prescribers should focus on evidence and drug quality, as opposed to commercial influences. Both prescribers and pharmaceutical companies need ethical regulation.[65] The practice of gifts from pharmaceutical companies to doctors alters the physicians' decision in prescribing the drug, which may be expensive and in turn could cause poor patient compliance.[65] Such practice should not be encouraged. Pharmaceutical companies should disclose payments made to doctors above a certain value in their accounts, as is already practiced in some countries. In the United States, according to federal law, drug companies must disclose payments made over a certain value to physicians for research and other miscellaneous expenses.[65]

20.5.7 Legal Action against Health Care Providers

A health care provider is likely to be charged for civil actions against injuries resulting from lack of informed consent and standard care. Medical treatment and malpractice laws are specific to each state and may vary with different countries.

Informed Consent

Ethical and legal standards require a health care provider to obtain informed consent (IC) before delivering care. If the patient cannot provide it, a legally authorized surrogate decision maker may do so. In an emergency, when the patient is not legally competent to give IC and no surrogate decision maker is readily available, the law implies consent on behalf of the patient.[66] Information should be conveyed in a language and at a level that the patient can understand, and consent should be obtained for diagnostic tests, treatment's nature and character, anticipated results, and risks and benefits of treatment (or no treatment), along with other alternative options if available. The informed consent must be voluntary.

The physician is responsible for obtaining IC and delineating the potential risks, benefits, and alternatives. Physicians must be transparent for all potential ethical or financial conflicts concerning therapies and various other matters related to them or the devices employed in patient care.[67] Patients' reaction to confidentiality, liability, and the financial aspects of IC in cardiology research have been described by Fortune-Greeley et al.[68] In a 2012 study, informed consent forms for cardiology practice were found to be suboptimal and lacking in clarity.[69]

The number of patients coming to the emergency department (ED) for diversities of toxicological emergencies is on the rise. Each patient requires immediate care and yet is unable to give consent due to impaired consciousness—that is, prevents the patient from making informed decisions. Some of the challenges faced by emergency physicians are acute organic impairment manifested by confusion, irrational thoughts, or dangerous behavior at times. They should familiarize themselves with relevant laws, which vary from state to state and country to country, and carry out immediate management to avoid liability for negligence and abandonment.

There has been much discussion about IC and its types. Medical ethicists claim that informed consent is valuable because it supports individual autonomy. In addition, it provides an assurance that patients and others are neither deceived nor coerced.[70] In real clinical scenarios, there are variations among patients. Since there are many distinct conceptions of individual autonomy, ethical importance varies. Consent forms should be designed to give patients and others control over the amount of information they receive and the opportunity to rescind consent given already.[70] Limitations of informed consent are getting informed consent from the young, very ill, mentally impaired, or unconscious patients; or patients brought to the ED. Other limitations are nonapplicability for public health policies, access to a patient's personal information by third parties, and duress or constraints experienced by patients.

Failure to Follow the Standard of Care

A patient has to prove the provider's failure to follow the accepted standard of care. There are various sources for standard of care. Many states apply different standards to specialists and practitioners with accommodations for practice limitations such as availability of medical facilities, services, devices, and the like. Common ethical issues are related to medical decision making and good patient–physician communication.

Laws and regulations on surrogate decision making are slowly evolving due to societal changes. It is highly recommended that readers become familiar with the various laws and regulations related to decision making according to their country or state. Other areas of ethics with regard to treatment are advanced directives, futile resuscitation, withdrawal of treatment, missed diagnosis, medical issues, incorrect procedures or drugs, prescription-related matters, adverse effects to drugs, failure to screen family members, withholding diagnosis and information, organ transplantation or harvesting, and breach of confidentiality.

20.5.8 Vicarious Liability

Hospital authorities are vicariously liable for the unlawful act of physicians or nurses, and/or others employed by them when such acts occur during the course and scope of individuals' work.

20.5.9 Patient's Bill of Rights

Doctors are often alarmed by the patient's bill of rights but should remember it is for everyone, including doctors. Familiarizing oneself with the patient's bill of rights is especially important in this era of medical tourism. Care providers should disclose patients' rights to their clients before entering a contract. Benefits of a patient's bill of rights are:

- Improved awareness within the health care system
- Increased expectations of the patients
- Empowerment of patients
- Changes and improvement in the attitude of health care practitioners, professionals, and providers
- Partnership between consumers and providers
- Improvement in the quality of health care

A bill of rights should be revised or reconsidered periodically so as to benefit the patients and address the concerns of the health care system.

20.5.10 Abuse of Power

Physicians have the privilege of making decisions to investigate, treat, or refer cases. Decisions are influenced by clinical, psychological, economic, and other issues. During clinical practice, however, doctors may be at risk of abusing this privilege.[71] Similarly, in recommending leave on medical grounds, some doctors are likely to overuse their privilege for want of definitive guidelines. Granting medical leave has an impact on productivity; thus, there is a need to develop standard guidelines for recommending a medical leave.

Doctor and Patient Realization of Risk

Both doctors and patients should realize that sickness, pain, and death are part of life and medicine, that diagnostics and therapy have limitations, and that predictions of clinical course and outcome may not always be certain. All these may become less or averted once personalized medicine becomes a reality. Doctors and patients have to accept the risk of developing a drug reaction (i.e., drug reactions can neither be predicted nor prevented on all occasions) and treatment complications. Good communication and understanding of all these aspects is a part of ethics and, if well understood, litigations are less likely to arise.

20.6 EDUCATION AND TRAINING

On most occasions, health care professionals are unaware of the importance of ethical practice. It needs to be introduced in education curricula, and training on clinical ethics in health care settings is also beneficial.[72,73] Medical schools organize clinico-pathological conferences (CPC) where students and faculty interact with each other, but discussions on judgments related to health care or liability cases and ethics are not often presented. Interactive programs in medical schools and clinical practices about such issues will help students and physicians understand not only the intricacies of laws related to health care and the shortcomings of health professionals and health systems, but also strengthen their understanding of ethics and laws, improve their practice and standard of care, and minimize litigation issues.

20.6.1 Teaching Ethics and Laws in Medical Schools

Medical students are trained and exposed to CPCs regularly during their studies. They are confident about substantiating their statements with evidence. However, most students do not have the confidence to do a clinical audit, to find deficiencies in medical records, to correlate the laws related to patient care, and/or to provide legal support to patients. This is because knowledge about the legal issues is suboptimal among medical teachers, and legal and ethical aspects are not discussed during regular ward rounds or grandrounds, nor are they incorporated into lecture classes. Anything that is not learned in medical school is less likely to be remembered and practiced or inculcated. The University of Iowa College of Medicine is among a relative few to offer separate courses in ethics and law as part of their curriculum.[74]

20.6.2 Continuing Medical Education

The European Society of Cardiology (ESC) has made recommendations about how to minimize bias in all scientific communications and continuing medical education (CME), and how to ensure proper ethical standards and transparency between the medical profession and industry.[75,76] Medical education and training do not focus on the intricacies and importance of consultation aspects and, thus, these have to be incorporated into the curriculum. Suitable modules have to be prepared for training doctors on the principles of consultation. Current medical curricula in India lacks ethics[77] and medical humanities; training programs to acquire such skills were proposed by the Medical Council of India,[78] but implementing them awaits permission from the Indian government.

Law and ethics support health care professionals and services rendered.[79] The United Kingdom's National Health Service (NHS) has introduced programs to teach and train health care professionals about law, ethics, education, and development (LEED) based on surveys. A central thrust of the program was the initiation of an "Introduction to Health Care Law and Ethics" study day.[79] In this survey, 74 to 100% of participants rated the educational value as excellent. Although the program has now been stopped, there is a need for all to have access to law and ethics. By and large, there is no dispute for dissemination of knowledge on ethics and law to health professionals.

20.6.3 A Model Curriculum for Legal Medicine and Medical Ethics

Some of the subjects that are included in teaching and training for legal medicine and medical ethics are medical humanities, public health policies, the business aspects of medicine, forensic sciences, end-of-life issues, how to care of special patients, the liability of specialties, and medical liability. Also such a curriculum should cover hospitals' reputation in medical practice, nursing facilities, managed care organizations, laws, courts and judicial processes, consumer protection acts, human rights, state medical boards, hospital ethics committees, and medical examiners.

Cardiovascular training should recognize and address ethical issues, the rational use of resources, and conflicts related to end-of-life care. A cardiovascular advance directive should address the deactivation of a pacemaker, use of an implantable cardioverter defibrillator (ICD), or a left ventricular assist device (LVAD). In addition, cardiovascular clinicians should understand ethical dynamics and equip trainees with the tools of ethical reasoning because the complexity of the specialty continues to increase.[80–86] Medical students and physicians-in-training should be trained in the various components of medical ethics—for example, definition, approaches, perspectives, ethics of individuals, ethics of human life, ethics related to family and society, ethics related to death and dying, research ethics, ethical work-up of cases, doctor—patient relationships, doctor—doctor relationships, pharmaceutical ethics, transplant ethics, resource allocation, end-of-life ethics, truth and confidentiality, decision making, teamwork, social responsibility, and so on.[74,87]

20.6.4 Early Exposure to Ethics

Medical students are exposed to the intricacies and interconnections between physiological systems and their influence on living organisms. There are also opportunities available to introduce and discuss ethics and ethical issues and reinforce ethical principles.[88] Introduction of ethics in physiology[89] has led to a greater awareness of the importance of ethical issues and are likely to be more widely practiced in clinical practices too.

20.6.5 Medicolegal Guidelines

Even though doctors make appropriate efforts to treat patients, they have to be trained and retrained about proper documentation from the point of view of litigation; encouraged to record dying declarations and obtain valid consents; and trained on the preservation and confidentiality of medical records, on the avoidance of hiding facts, and on adherence to standard guidelines for therapy. The latter includes selection of the therapeutic agents, methods, and/or procedures and substantiating written statements if there are any deviations. In addition, guidance is needed for the reporting of illnesses to public health authorities as per the regulations of the county or state and hospitals.

20.6.6 Clinical Audit

A clinical audit is an essential component of clinical governance. It considers procedures adopted to diagnose, methods used to deliver health care and treatment, utilization of resources, and the effect of outcome along with quality of life of the patient. Principles for the best practice in clinical audit, published by the National Institute for Health and Clinical Excellence (NICE),[90] defines it as a quality improvement process that seeks to improve patient care and outcomes through systemic care delivered against explicit criteria and the implementation of change.

Even though the terms *medical audit* and *clinical audit* are used interchangeably, clinical audit has multiple components related to patients, education and training, health care delivery, resources for health, working relationships, and so on. In addition, a clinical audit helps when listening to patients; understanding patients' expectations; deals with evidence-based practice; assists in the development of local guidelines or protocols; minimizes error or harm to patients; and reduces incidents, complaints, and claims. Because a clinical audit involves patients and patient-related matters, confidentiality of information has to be adhered to strictly. For the review of case notes during a clinical audit, there is no need for informed consent from patients as long as confidentiality and anonymity are preserved.[91]

20.6.7 Teaching and Assessing Medical Ethics

Medical ethics has been incorporated into medical education and training. However, the shortfall in ethics core competencies does not preclude graduation.[92] There is a need for greater integration of ethics in theory and practice. Major concerns are how to make a medical faculty impart medical ethics to students of health care sciences. The "Cardiology Ethics Curriculum" at the University of Toronto is a useful resource.[93] Ethical components are rarely discussed in during ward rounds.[94] Teaching ethics should be incorporated into all clinically oriented teaching and training activities. Students can also learn ethics through role models and interaction with patients, instead of formal lectures.[95] Because medicine and ethics are inseparable, teaching ethics at the bedside becomes essential but is seldom done.[94] There is a need to introduce bioethics as part of the curriculum at all health science education institutions.[96]

The complexity of the technology involved in making a diagnosis and therapy requires a careful search for the right answer under some very difficult situations. Training on descriptive bioethics[96] using various modern medical education technologies and discussion with fellow doctors, hospital administrators, religious leaders, and social scientists in difficult situations, can help to make policy decisions amicable to the patient, family, and society. Achievements in medical practice and health care delivery depend on the status of the medical education system, the nature of medical manpower, economic resources, the health care system, and, importantly, medicolegal law. The quality of medical manpower and medical services are determined by policy, findings, regulatory mechanisms, professional organizations, direction from judiciary, and expectations and participation of the community.

A common difficulty in the current medical education system is that it fails to inculcate appropriate skills and competencies to recognize and tackle ethical issues among the learners. Although developing an ethics-based medical education system is a challenging task, we have to take initiatives to deal with a variety of issues.

Prevention

A. Knowledge in medicine and timely application in a given situation to assess the case at the bedside meticulously with empathy; analysis of clinical issues and challenges, and differentiating one from another with or without the help of laboratory data; application of appropriate specific and nonspecific measures to alleviate the symptoms or to treat cases with currently available guidelines; acceptance of limitations of knowledge and skills, resources, and constraints; arrangement for referral or consultation; provision of follow-up care, rehabilitation services, or welfare program depending on the case; proper documentation, adequate informed consent, continuous monitoring to strengthen physician–patient relationship, patient satisfaction, and quality of life, thereby avoiding any conflicts or liability issues.

B. The attitude of health care professionals toward the patients shall be:
 1. Receive them with care and concern
 2. Respond with empathy
 3. Realize the underlying clinical problem(s) and manage with appropriate measures
 4. Review the clinical course, therapeutic strategies, and outcome
 5. Revise the plan and management with shared decision
 6. Seek further help, if needed

C. Cardiologists should be informed to establish the kind of actions, communication skills, and empathy that are required to build a stronger patient–health care professional relationship, which improves prognosis, treatment efficacy, and therapeutic adhesion.[97]

D. Cardiac report cards should be developed that incorporate an ethical framework, which can identify points of ethical concern for practitioners, patients, policymakers, and researchers[98]

E. The integrity and behavior of an individual physician plays a major role in the prevention of liability cases. Professional associations play a major role in the regular monitoring process and for initiating self-regulation of professionals.

F. Doctors involved in handling insured patients are encouraged to consult Booth, 2008,[99] which has extensive information on prevention. Also, physicians are encouraged to familiarize themselves with local guidelines because policies and practices of health care insurers are extremely varied.

G. The role of community participation and utilization of existing laws and regulations in India toward clinical trials have alerted the highest health care bodies in the country, such as the Ministry of Health and Family Welfare (MOHFW), Central Drugs Standards Control Organization (CDSCO), and policymakers, to introduce new regulations on compensation for injury and death occurring during drug trials.[100]

H. The Clinical Establishment Act[101] of India, if implemented, will improve the dignity of patients and ensure safe services.

I. To overcome the legal issues, one has to follow ethics and maintain professionalism and team spirit at all times. Respect, dedication, sincerity, compassion, good communication skills, prudence, meticulous documentation, and acknowledging the limitations of one's abilities help avoid legal liability.

J. In India, the Supreme Court included the health profession under the 1986 Consumer Protection Act (CPA) of India [Supreme Court judgment AIR 1995 SC 550, dated 13.11.1995. New Delhi, India]. As a result, doctors can be charged for medical negligence liability under CPA.[102]

K. *Responsibilities of the health care institutions*: Health care institutions have to provide a written notice of their privacy policy to all those who seek medical care for the first time. Patients must be informed of how the institution may use and disclose information. The written notice must describe patient's rights, including the right to access their medical information and their right to provide feedback or complaint, if they believe their rights were violated. The notice shall be in plain language and presented in an understandable manner. An acknowledgment of receipt of the information notice must be obtained from the patient.[103] The institution shall have a close working relationship among medical personnel, legal, and risk management team. The team, through interaction on case by case basis, can acquire a wealth of information, learn, and meet the needs of clinical dilemma, which are evolving constantly. Such interactions enhance health care and ensure patient safety, and thereby establish a high standard of care. Also, the institution shall ensure quality assurance, organize risk management programs, and conduct routine audits or monitoring as suggested by a specialist's team. Laws generally require that therapeutic interventions be done in a reasonable and medically approved manner, preferably based on clinical evidence or standard guidelines. Medical staff dealing with cases should remember that the issues raised by any one case are complex and application in a real situation is difficult. Hence, staff should review the hospital counsel and the local laws and regulations pertaining to these issues.[103]

L. *Medical records*: Medical personnel should be trained and motivated for proper documentation, adherence to risk management principles, and entry of every aspect related to the case in the respective medical record. The physician is required to write a medical record that will support the basis for the appropriate medical judgment made. The medical record should reveal supporting clinical data or facts, the patient's history available at that time, hemodynamic status, physical examination, investigation reports, and treatment plans administered. One must remember that hospital records are prima facie evidence.[104] Therefore physicians who do not record supporting clinical data and history deprive themselves of a strong "medical judgment defense."[103] Moreover, inappropriate entries or markings on the medical record can weaken the defense in a liability case.[104]

If any corrections are required in the medical record, the preferable method is to draw a single line over the word or value to be changed; and at the place where such correction is made, initials or signature should be affixed. However, total obliteration of a number or word may indicate that obliteration was done intentionally. Quality-assurance reviews of records of EDs often reveal inadequate charting by physicians and nurses.[103] A lapse of documentation of patient's clinical condition for 4 hours or more after initial physician and nursing assessment will be utilized by the plaintiffs attorney in a lawsuit to develop the theory that no care was given. The uncooperative acts of the patient and the difficulties encountered in providing care to the patient because of his or her actions should be recorded. Poorly written physicians' notes may become an issue in a medical malpractice action and considered a less-than-caring attitude by the doctor. Descriptions of a patient's behavior and lifestyles should depict compassion and a professional manner. Health care institutions and cardiovascular physicians should realize consumer rights and responsibilities such as information disclosure, choice of providers and plans, access to emergency services, participation in a treatment discussion, respect and nondiscrimination, confidentiality of health information, and complaints and appeals.[83]

M. *Incentives and ethics:* Incentives, in different forms, have been introduced for various services and are not accessible in health care delivery.[105] Incentive mechanisms are based on four types of motives: traditional, self-interest, affective, and shared purposes. Financial incentives have helped to achieve the quality or efficiency of related targets but have potential adverse consequences including conflicts of interest that threaten patient–physician trust. The ethical allegations, implications for effectiveness, and mechanisms of different motives were highlighted. Incentive schemes based on a robust sense of shared purposes protect and promote physicians' sense of moral responsibilities and ethical standards. These also enable physicians to take ownership of it rather than make them feel it is imposed on them.[105]

N. *Patient-centered specialty practice*: As a part of a new phase of health care delivery systems, the National Committee for Quality Assurance (NCQA) launched patient-centered specialty practice (PCSP) in 2013. The objectives of the committee are to reinforce care coordination, improve access to specialty care, reduce the use of unnecessary or duplicate tests, enhance communication, and assess and improve performance.[106]

O. Provision of safe prescriptions is already included in the curriculum of health sciences. However, prescription errors remain a significant cause of patient harm. Lum et al.[107] identified 12 core competencies for safe prescribing and listed them under 4 stages: information-gathering, decision making, communication and decisions, and monitoring and review.

- Overall, medical students and physicians should be trained and motivated to maintain:
  - Physician–patient relationship on all aspects of patient care
  - Holistic approach
  - Empathy
  - Respect for patient autonomy
  - Involvement of patients or, where appropriate, family members or caregivers as partners in therapeutic and management decisions
  - Respect for patient culture, values, views, and beliefs or religious sentiments
  - Consideration of patient's views on the use of alternative treatment options
  - Nonjudgmental attitude in all aspects of work
  - Engagement in reflective practice
  - Clinical audit

Delay in diagnostic investigations, the nonavailability of interventional procedures, recurrent and repeated system failures, and so on, are real aspects of all our daily lives, yet rarely are considered in mitigation. Careful documentation of these failures is paramount, if we need to defend ourselves against a system that seeks criminal prosecution or compensation. A legal decision for violation of ethics may lead to loss of status, career, and livelihood.

20.7 FUTURE DIRECTIONS

It is worth studying the patterns and prevalence of defensive medicine and the status of lawsuits in different countries to find any economical impacts and adverse effects. Patterns of malpractice insurance systems adopted in various countries and the compensation paid to patients should be analyzed and the information shared among health care professionals, community members, and payers, so as to design modalities to reduce expenses.

Periodic analyses of physician behaviors and attitudes, along with health care utilization and lawsuits, are likely to help professional organizations and health science educators introduce teaching and training modules for reduction in malpractice lawsuits. The consequences of wait time for health care services in terms of quality of life and health care economics should be studied for the effective introduction of remedial measures.

The attitude of physicians and their beliefs, preparedness, and experiences related to impaired and incompetent colleagues should be explored, as well as suggestions to overcome them. Issues related to devices, drugs, and diagnoses with reference to ethical and legal aspects and alternatives for better treatment of the patient should be analyzed. Periodical analyses of judgments related to health care issues, health care professionals, and health services at national and international levels, as well as identification of the causes, to find methods to avert them can be undertaken at educational and industrial levels.

The capacity of medical students and practitioners to get informed consent should be evaluated and suitable measures introduced to teach and train them. Analyses of curriculum and methods of teaching and training should be undertaken, and medical students should be assessed on their knowledge of ethics and laws. A common curriculum should be introduced so that medical students can be trained uniformly to deliver health care effectively and as expected. Moreover, such a curriculum will help patients and practitioners, as patients start moving between countries for medical treatment (medical tourism).

The current status of consultation and its impact on the part of referring doctors and their patients have to be studied to find lacunae. Suitable modules should be prepared for training doctors on consultation. There are inter- and intracountry variations in one or more components of bioethics, thus the boundaries are disputable. Disagreements about the relationship between ethics and law should be addressed.

Future directions in therapeutics should focus on the role of pluripotent stem cells in cardiotoxicity,[108] the feasibility implications of personalizing health care,[109] and the assessment of myocardial neuronal function.[110] Health science research explores basics, epidemiology, risk, prevention, diagnostics and management, financial burdens of disease, and behavioral aspects. Each component of health science should focus on ethical aspects before asking for funding. Cardiovascular research should produce guidelines on management of end-of-life care, palliative care, and support for patients' decisions to forgo treatment.

Cardiovascular physicians should prepare and produce modules to facilitate the patient and faculty decision-making process about cardiovascular health.

Managed care organizations (MCO) and cardiovascular physicians should plan and develop guidelines to handle ethical issues or problems related to MCO and standard of ethics for hospital administrators and all others involved directly or indirectly with patient care. Protocols and algorithms need to be developed to curtail inappropriate prescribing, unnecessary testing, overuse of technologies, diversion of scarce resources, and nonrecognition of complications.

20.8 CONCLUSION

The public has realized that "doctors are not always to be trusted, as they have fallen off their pedestal."[2] Doctors, therefore, have to remain up to date and demonstrate their continuing competency. Currently the duties and responsibilities of doctors have been defined by medical councils belonging to the individual country where they practice. As medical tourism has developed and patients are moving from one country to another for treatment, there is a need to develop an international norm to overcome inter- and intracountry variations.

Cardiovascular care is increasingly complicated, device-oriented, and expensive, and it requires striking a balance for quality of life and longevity. The modern cardiovascular specialist has to explain these spectra, considering individual patients' wishes and best interests, the maleficence of side effects, and the impact of therapeutic interventions in the larger social context. To achieve these things, physicians should acquire skills in the art of medicine and interact with patients and caregivers with empathy, astute observations, acccurate interpretation of clinical and laboratory data, and application of up-to-date knowledge with evidence and ethics. This can strengthen insight into disease and suffering and provide a means to diagnose, treat, and prevent disease; if followed meticulously and documented sincerely, legal liability is less likely to develop.

Treatment decisions, in general, should be based on a profound respect for the patient's autonomy, balanced by an appreciation of what may be in the patient's best interest. Decision making involves a wide array of personal, professional, institutional, spiritual, and social values. In routine practice, clinical management decisions should be coupled with outcome analyses and guide physicians' approach to best address the needs and expectations of patients. To improve the quality of the physician–patient encounter, we need to develop formal and informal curricula at medical institutions on communication skills and ethics.

Key Learning Points

- Illness remains universal; however, human expectations and experience may vary personally and geographically.

- Principles of ethics and legal aspects have to be incorporated into health science education and discussed at every stage of clinical teaching.
- Ethics and legal aspects related to emerging areas of health care, such as medical tourism, ubiquitous health care, consumer-driven health care, integration of complementary therapy, and so on, need to be addressed.
- Curriculum about medical ethics should be included in all aspects of medical education and training, including continuing education. Such measures should reduce the possibilities of medical liability cases.

REFERENCES

1. Calinas-Correia J. Cardiopulmonary resuscitation ethics: a response to Michael Ardagh. *J Med Ethics* 2001;**27**(1):64–5.
2. Petch MC. Heart disease, guidelines, regulations, and the law. *Heart* 2002;**87**(5):472–9.
3. Lucas FL, Sirovich BE, Gallagher PM, et al. Variation in cardiologists' propensity to test and treat: is it associated with regional variation in utilization? *Circ Cardiovasc Qual Outcomes* 2010;**3**(3):253–60.
4. Arimany-Manso J. Professional liability in cardiology. *Rev Esp Cardiol (Engl Ed)* 2012;**65**(9):788–90.
5. Santiago-Saez A, Perea-Perez B, Albarran-Juan ME, et al. Analysis of judgments in the practice of cardiology resolved on appeal in Spain between 1992 and 2007. *Rev Esp Cardiol (Engl Ed)* 2012;**65**(9):801–6.
6. Similima.com. Health insurance schemes abused to profit private providers. India: Simila.com; 2012 Vol. XLVII, No. 38 Available from: <www.similima.com/health-insurance-schemes-abused-to-profit-private-providers> [accessed 03.23.14].
7. Studdert DM, Mello MM, Brennan TA. Medical malpractice. *N Engl J Med* 2004;**350**(3):283–92.
8. Snyder L, American College of Physicians Ethics, Professionalism, Human Rights Committee. American College of Physicians Ethics Manual: sixth edition. *Ann Intern Med* 2012;**156**(1 Pt 2):73–104.
9. Pellegrino ED. The internal morality of clinical medicine: a paradigm for the ethics of the helping and healing professions. *J Med Phil* 2001;**26**(6):559–79.
10. Benatar SR, Bhutta ZA, Daar AS, et al. Clinical ethics revisited: responses. *BMC Med Ethics* 2001;**2**:E2.
11. Li JT. The quality of caring. *Mayo Clin Proc* 2006;**81**(3):294–6.
12. Bendapudi NM, Berry LL, Frey KA, et al. Patients' perspectives on ideal physician behaviors. *Mayo Clin Proc* 2006;**81**(3):338–44.
13. Pellegrino ED. Professionalism, profession and the virtues of the good physician. *Mt Sinai J Med* 2002;**69**(6):378–84.
14. Committee on Quality of Health Care in America, Institute of Medicine. *Crossing the quality chasm: A new health system for the 21st century*. Washington, DC: National Academies Press; 2001.
15. Jayasundar R. Healthcare the Ayurvedic way. *Indian J Med Ethics* 2012;**9**(3):177–9.
16. Jindal RP. The truth about medical negligence. *Indian J Med Ethics* 2014;**11**(1):42–6.
17. Dong E. The physician at the movies: Seabiscuit. The Pharos of Alpha Omega Alpha-Honor Medical Society. *Alpha Omega Alpha* 2005;**68**(1):51.

18. Golden G. The physician at the movies: Master and Commander. *The Pharos of Alpha Omega Alpha-Honor Medical Society Alpha Omega Alpha* 2005;**68**(1):51.
19. DesRoches CM, Rao SR, Fromson JA, et al. Physicians' perceptions, preparedness for reporting, and experiences related to impaired and incompetent colleagues. *JAMA* 2010;**304**(2):187–93.
20. Kirkpatrick JN, Fields AV, Ferrari VA. Medical ethics and the art of cardiovascular medicine. *Lancet* 2010 14;**376**(9740):508–9.
21. Keith-Spiegel P, Tabachnick B. What scientists want from their research ethics committee. *Journal of Empirical Research on Human Research Ethics: JERHRE* 2006;**1**(1):67–82.
22. Pearson SD, Raeke LH. Patients' trust in physicians: many theories, few measures, and little data. *J Gen Intern Med* 2000;**15**(7):509–13.
23. Gopichandran V. Trust in healthcare: an evolving concept. *Indian J Med Ethics* 2013;**10**(2): 79–82.
24. Komesaroff PA. Uses and misuses of ambiguity: uses of ambiguity. *Intern Med J* 2005;**35** (10):632–3.
25. Dorr Goold S, Lipkin Jr. M. The doctor–patient relationship: challenges, opportunities, and strategies. *J Gen Intern Med* 1999;**14**(Suppl. 1):S26–33.
26. Wagle S. Web-based medical facilitators in medical tourism: the third party in decision-making. *Indian J Med Ethics* 2013;**10**(1):28–33.
27. Dyer KA. Ethical challenges of medicine and health on the internet: a review. *J Med Internet Res* 2001;**3**(2):E23.
28. Eysenbach G. Towards ethical guidelines for dealing with unsolicited patient emails and giving teleadvice in the absence of a pre-existing patient-physician relationship systematic review and expert survey. *J Med Internet Res* 2000;**2**(1):E1.
29. Brown I, Adams AA. The ethical challenges of ubiquitous healthcare. *Int Rev Inf Ethics* 2007; **8**(12):53–60.
30. Fraser AG, Daubert JC, Van de Werf F, et al. Clinical evaluation of cardiovascular devices: principles, problems, and proposals for European regulatory reform. Report of a policy conference of the European Society of Cardiology. *Eur Heart J* 2011;**32**(13):1673–86.
31. Kramer DB, Mitchell SL, Brock DW. Deactivation of pacemakers and implantable cardioverter-defibrillators. *Prog Cardiovasc Dis* 2012;**55**(3):290–9.
32. Swetz KM, Mansel JK. Ethical issues and palliative care in the cardiovascular intensive care unit. *Cardiol Clin* 2013;**31**(4):657–68.
33. Geppert CM, Shelton WN. A comparison of general medical and clinical ethics consultations: what can we learn from each other? *Mayo Clin Proc* 2012;**87**(4):381–9.
34. Dubois JM, Carroll K, Gibb T, et al. Environmental factors contributing to wrongdoing in medicine: a criterion-based review of studies and cases. *Ethics Behav* 2012;**22**(3):163–88.
35. Schwarz ER, Philip KJ, Simsir SA, et al. Maximal care considerations when treating patients with end-stage heart failure: ethical and procedural quandaries in management of the very sick. *J Relig Health* 2011;**50**(4):872–9.
36. Smith ML. Futile medical treatment and patient consent. *Cleve Clin J Med* 1993;**60**(2): 151–4.
37. Hariharan S, Jonnalagadda R, Walrond E, et al. Knowledge, attitudes and practice of healthcare ethics and law among doctors and nurses in Barbados. *BMC Med Ethics* 2006;**7**:E7.
38. Barrette J. *Complementary and alternative medicine*. In: Sanbar SS, Fiscina S, LeBlang TR, et al., (editors). Legal Medicine. Philadelphia: Mosby-Elsevier; 2007. p. 67–73.

39. Schouten R, Cohen MH. Legal issues in integration of complementary therapies into cardiology practice. In: Frishman W, Weintraub MI, Micozzi MS, (editors). *Complementary and integrative therapies for cardiovascular disease*. Elsevier; 2005. p. 29–55.
40. Benjamin DM, O'Donnell JT. Medication errors. In: Sanbar SS, Fiscina S, LeBlang TR, et al., (editors). *Legal medicine*. Philadelphia: Mosby-Elsevier; 2007. p. 271–84.
41. Kavarana MN, Sade RM. Ethical issues in cardiac surgery. *Future Cardiol* 2012;**8**(3): 451–65.
42. Knudtson ML, Beanlands R, Brophy JM, et al. Treating the right patient at the right time: access to specialist consultation and non-invasive testing. *Can J Cardiol* 2006;**22**(10):819–24.
43. Malhotra A, Menahem S, Gillam L. Ethical issues in fetal management: a cardiac perspective. *Int J Pediatr* 2010;**2010**:857460.
44. Schoen C, Collins SR, Kriss JL, et al. How many are underinsured? Trends among U.S. adults, 2003 and 2007. *Health Aff* 2008;**27**(4):298–309.
45. Moriates C, Shah NT, Arora VM. First, do no (financial) harm. *JAMA* 2013;**310** (6):577–8.
46. Gillick MR. The critical role of caregivers in achieving patient-centered care. *JAMA* 2013; **310**(6):575–6.
47. Launer J. What's wrong with ward rounds? *Postgrad Med J* 2013;**89**(1058):733–4.
48. Lahariya C. Health and media: should there be an internal check mechanism in and by media? *Natl Med J India* 2012;**25**(4):250–1.
49. Thirumalaikolundusubramanian P, Srinivasan M, Vinodhkumaradithyaa A, Uma A. Chikungunya epidemic: Analysis of reports of a lay press. *Ann Trop Med Public Health* 2008;**1**(1):25–8.
50. Thirumalaikolundusubramanian P, Meenakshisundaram R. Measles Vaccination and deaths: analysis of lay press reports. *Internet J World Health Societal Polit* 2007;**5**:2.
51. Singh A. Ethics for medical educators: an overview and fallacies. *Ind J Psychol Med* 2010;**32**(2):83–6.
52. Bashour H, Sayed-Hassan R, Koudsi A. Involving patients in medical education: ethical issues experienced by Syrian patients. *Educ Health* 2012;**25**(2):87–91.
53. Masic I. Ethical aspects and dilemmas of preparing, writing and publishing of the scientific papers in the biomedical journals. *J Soc Med Informatics Bosnia & Herzegovina* 2012;**20** (3):141–8.
54. Hall R, More R, Camm J, et al. Fifth report on the provision of services for patients with heart disease. *Heart* 2002;**88**(Suppl. 3):iii, 1–56.
55. Asher E, Greenberg-Dotan S, Halevy J, et al. Defensive medicine in Israel-a nationwide survey. *PloS One* 2012;**7**(8):e42613.
56. Studdert DM, Mello MM, Sage WM, et al. Defensive medicine among high-risk specialist physicians in a volatile malpractice environment. *JAMA* 2005;**293**(21):2609–17.
57. Huckman RS, Kelley MA. Public reporting, consumerism, and patient empowerment. *N Engl J Med* 2013;**369**(20):1875–7.
58. de Boer MJ, van der Wall EE. Choosing wisely or beyond the guidelines. *Netherlands Heart* 2013;**21**(1):1–2.
59. Mendelson TB, Meltzer M, Campbell EG, et al. Conflicts of interest in cardiovascular clinical practice guidelines. *Arch Intern Med* 2011;**171**(6):577–84.
60. Dunlop C. Legal issues in Emergency Management: Lessons from the last decade. *Aust J Emerg Manag* 2004;**19**(1):26–33.
61. Sanbar SS. Medical malpractice. In: Sanbar SS, Fiscina S, LeBlang TR, et al., (editors). *Legal medicine*. Philadelphia: Mosby-Elsevier; 2007. p. 253–64.

62. Chell B. Competency: What it is, what it is not, and why it matters. In: Morrison EE, editor. *Health care ethics: critical issues for the 21st century*. 2nd ed. Sudbury, MA: Jones and Bartlett; 2009. p. 110–22.
63. Reynolds C. Common law duties of prescribers. *Aust Pract* 1996;**19**(1):18–20.
64. Graham JD. What consumers want to know about medicines. *Aust Pract* 1995;**18**:10–1.
65. Mukherjee R. Can India stop drug companies giving gifts to doctors? *BMJ* 2013;**346**: f2635.
66. Brook LV. Clinical ethics and law. University of Washington School of Medicine: University of Washington School of Medicine; 2013 Available from: <https://depts.washington.edu/bioethx/topics/law.html> [accessed 07.03.14].
67. Bashore TM, Balter S, Barac A, et al. Interventions expert consensus document on cardiac catheterization laboratory standards update: A report of the American College of Cardiology Foundation Task Force on Expert Consensus documents developed in collaboration with the Society of Thoracic Surgeons and Society for Vascular Medicine. *J Am Coll Cardiol* 2012;**59**(24):2221–305.
68. Fortune-Greeley AK, Hardy NC, Lin L, et al. Patient reactions to confidentiality, liability, and financial aspects of informed consent in cardiology research. *Circ Cardiovasc Qual Outcomes* 2010;**3**(2):151–8.
69. Terranova G, Ferro M, Carpeggiani C, et al. Low quality and lack of clarity of current informed consent forms in cardiology: how to improve them. *JACC Cardiovasc Imaging* 2012;**5**(6):649–55.
70. O'Neill O. Some limits of informed consent. *J Med Ethics* 2003;**29**(1):4–7.
71. Launer J. Doctors as victims. *Postgrad Med J* 2013;**89**(1049):182.
72. Subramanian T, Mathai AK, Kumar N. Knowledge and practice of clinical ethics among healthcare providers in a government hospital, Chennai. *Indian J Med Ethics* 2013;**10**(2):96–100.
73. Qidwai W, Tabassum R, Khan FH, et al. Informed consent, privacy and confidentiality practised by doctors of a tertiary care hospital in a developing country. *Indian J Med Ethics* 2013;**10**(1):36–40.
74. Olick RS. It's ethical, but is it legal? Teaching ethics and law in the medical school curriculum. *Anat Rec* 2001;**265**(1):5–9.
75. Board ESC. Relations between professional medical associations and the health-care industry, concerning scientific communication and continuing medical education: a policy statement from the European Society of Cardiology. *Eur Heart J* 2012;**33**(5):666–74.
76. Luscher TF. Good publishing practice. *Eur Heart J* 2012;**33**(5):557–61.
77. Ravi M. How to teach ethics to those who need to learn. *Indian J Med Ethics* 2011;**8**(2):117–9.
78. Singh M, Talwar KK. Putting the humanities back into medicine: some suggestions. *Indian J Med Ethics* 2013;**10**(1):54–5.
79. Hagger LE, Woods S. Law and ethics support for health professionals: an alternative model. *J Med Ethics* 2005;**31**(2):111.
80. Popp RJ, Smith Jr. SC, Adams RJ, et al. ACCF/AHA Consensus Conference Report on Professionalism and Ethics. *Circulation* 2004;**110**(16):2506–49.
81. Jacobs AK, Smith Jr. SC, Popp RL, et al. Ethics and professionalism: can there be consensus? An editorial comment on the report of the ACCF/AHA Consensus Conference on Professionalism and Ethics. *Circulation* 2004;**110**(16):2278–9.
82. NCBI Abstract. 21st Bethesda Conference on Ethics in Cardiovascular Medicine. *J Am Coll Cardiol* 1990;**16**(1):1–36.

83. Parmley WW, Passamani ER, Lo B. Abstract from 29th Bethesda Conference on Ethics in Cardiovascular Medicine, 1997. *J Am Coll Cardiol* 1998;**31**(5):917–49.
84. Kirkpatrick JN, Kim AY. Ethical issues in heart failure: overview of an emerging need. *Perspect Biol Med* 2006;**49**(1):1–9.
85. Kirkpatrick JN, Fedson SE, Verdino R. Ethical dilemmas in device treatment for advanced heart failure. *Curr Opin Support Palliat Care* 2007;**1**(4):267–73.
86. Goldstein NE, Lampert R, Bradley E, et al. Management of implantable cardioverter defibrillators in end-of-life care. *Ann Intern Med* 2004;**141**(11):835–8.
87. Rameshkumar K. Ethics in medical curriculum; Ethics by the teachers for students and society. *Ind J Urol* 2009;**25**(3):337–9.
88. Weibel ER. The physiologist's ethical dilemmas. *News Physiol Sci* 2002;**17**:43–6.
89. Goswami N, Batzel JJ, Hinghofer-Szalkay H. Assessing formal teaching of ethics in physiology: an empirical survey, patterns, and recommendations. *Adv Physiol Educ* 2012;**36** (3):188–91.
90. NHS. *Principles for best practice in clinical audit.* London NHS; 2002. Available from: www.nice.org.uk/media/796/23/BestPracticeClinicalAudit.pdf.
91. Morerell C, Harvey G. *Improving the quality of health care.* 3rd ed. Beijing: Bailliere Tindall; 1999. p. 1–216.
92. Mattick K, Bligh J. Teaching and assessing medical ethics: where are we now? *J Med Ethics* 2006;**32**(3):181–5.
93. Ross H. *Cardiology ethics curriculum toronto.* University of Toronto; 2002 [accessed 03.04.14]. Available from: www.utoronto.ca/pgme/documents/curricula/cardio_curr.pdf.
94. Parker L, Watts L, Scicluna H. Clinical ethics ward rounds: building on the core curriculum. *J Med Ethics* 2012;**38**(8):501–5.
95. Silverman HJ. Description of an ethics curriculum for a medicine residency program. *West J Med* 1999;**170**(4):228–31.
96. Selvakumar D, Joseph LB. The importance of including bio-medical ethics in the curriculum of health education institutes. *Educ Health* 2004;**17**(1):93–6.
97. de Hoyos A, Nava-Diosdado R, Mendez J, et al. Cardiovascular medicine at face value: a qualitative pilot study on clinical axiology. *PEHM* 2013;**8**:3.
98. Richard SA, Rawal S, Martin DK. An ethical framework for cardiac report cards: a qualitative study. *BMC Med Ethics* 2005;**6**:E3.
99. Booth KA. *Medical assisting: Administrative and clinical procedures including anatomy and physiology.* New York: McGraw-Hill; 2008.
100. Jesani A. New regulations on compensation for injury and death in drug trials. *Indian J Med Ethics* 2013;**10**(2):76–9.
101. Ministry of Law and Justice. *The clinical establishments (registration and regulations) Act.* New Delhi, India: Ministry of Law and Justice; 2010 [accessed 03.03.14]. Available from: http://clinicalestablishments.nic.in/WriteReadData/969.pdf.
102. Joga Rao SV. Medical negligence liability under the consumer protection act: a review of judicial perspective. *Indian J Urol* 2009;**25**(3):361–71.
103. Kirrane BK, Drukteinis DA. Risk management and legal principles. In: Nelson LS, Lewin NA, Howland MA, et al., (editors). *Goldfrank's Toxicologic Emergencies.* 9th ed. New York: McGraw-Hill Medical; 2011. p. 1831–7.
104. Quinn C. *The Medical Record as a Forensic Resource.* Sudbury, MA: Jones and Bartlett; 2005.
105. Biller-Andorno N, Lee TH. Ethical physician incentives: from carrots and sticks to shared purpose. *N Engl J Med* 2013;**368**(11):980–2.

106. Huang X, Rosenthal MB. Transforming specialty practice--the patient-centered medical neighborhood. *N Engl J Med* 2014;**370**(15):1376–9.
107. Lum E, Mitchell C, Coombes I. The competent prescriber: 12 core competencies for safe prescribing. *Aust Prescr* 2013;**36**(1):13–6.
108. Wobus AM, Loser P. Present state and future perspectives of using pluripotent stem cells in toxicology research. *Arch Toxicol* 2011;**85**(2):79–117.
109. Godman B, Finlayson AE, Cheema PK, et al. Personalizing health care: feasibility and future implications. *BMC Med* 2013;**11**:179.
110. Carrio I. Cardiac neurotransmission imaging. *J Nucl Med* 2001;**42**(7):1062–76.

Index

A

Note: Page numbers followed by "*f*" and "*t*" refer to figures and tables, respectively.

B

E

I

J

K

L

M

N

O

Q

R

T

U

V

FSC
www.fsc.org
MIX
Papier | Fördert gute Waldnutzung
FSC® C083411

Kolibri 360 GmbH
Ferdinand-Jühlke-Straße 7
99095 Erfurt, Deutschland
produktsicherheit@kolibri360.de